AF778261

Advances in PGPR Research

Advances in PGPR Research

Edited by

Harikesh B. Singh

Department of Mycology and Plant Pathology, Institute of Agricultural Sciences, Banaras Hindu University, Varanasi India

Birinchi K. Sarma

Department of Mycology and Plant Pathology, Institute of Agricultural Sciences, Banaras Hindu University, Varanasi India

and

Chetan Keswani

Department of Biochemistry, Institute of Science, Banaras Hindu University, Varanasi India

CABI

CABI is a trading name of CAB International

CABI
Nosworthy Way
Wallingford
Oxfordshire OX10 8DE
UK

Tel: +44 (0)1491 832111
Fax: +44 (0)1491 833508
E-mail: info@cabi.org
Website: www.cabi.org

CABI
745 Atlantic Avenue
8th Floor
Boston, MA 02111
USA

Tel: +1 (617)682-9015
E-mail: cabi-nao@cabi.org

© CAB International 2017. All rights reserved. No part of this publication may be reproduced in any form or by any means, electronically, mechanically, by photocopying, recording or otherwise, without the prior permission of the copyright owners.

A catalogue record for this book is available from the British Library, London, UK.

Library of Congress Cataloging-in-Publication Data

Names: Singh, H. B., Dr., editor.
Title: Advances in PGPR research / edited by Harikesh Bahadur Singh, Banaras Hindu University, India, Birinchi Kumar Sarma, Banaras Hindu University, India, Chetan Keswani, Banaras Hindu University, India.
Description: Boston, MA : CABI, 2017. | Includes bibliographical references and index.
Identifiers: LCCN 2017016222 (print) | LCCN 2017034348 (ebook) | ISBN 9781786390332 (ePDF) | ISBN 9781786390349 (ePub) | ISBN 9781786390325 (hbk : alk. paper)
Subjects: LCSH: Plant growth-promoting rhizobacteria. | Plant-microbe relationships.
Classification: LCC QR351 (ebook) | LCC QR351 .A384 2017 (print) | DDC 579/.178--dc23
LC record available at https://lccn.loc.gov/2017016222

ISBN-13: 978 1 78639 032 5

Commissioning editor: Rachael Russell
Editorial assistant: Emma McCann
Production editor: Shankari Wilford

Typeset by SPi, Pondicherry, India
Printed and bound in the UK by CPI Group (UK) Ltd, Croydon, CR0 4YY, UK

Contents

Contributors

P.C. Abhilash, Institute of Environment and Sustainable Development, Banaras Hindu University, Varanasi 221005, India. E-mail: pcabhilash@hotmail.com

Betina Cecilia Agaras, Laboratorio de Bioquímica, Microbiología e Interacciones Biológicas en el Suelo, Departamento de Ciencia y Tecnología, Universidad Nacional de Quilmes, Roque Sáenz Peña 352 - Bernal B1876BXD - Buenos Aires, Argentina. E-mail: betina_agaras@yahoo.com.ar

Laith K. Tawfeeq Al-Ani, School of Biological Sciences, Universiti Sains Malaysia, 11800 Pinang, Malaysia and Department of Plant Protection, College of Agriculture-Baghdad University, 10071 Baghdad, Iraq. E-mail: cmv_virus2002@yahoo.com and laith.kt77@gmail.com

Gautam Anand, Department of Biochemical Engineering and Biotechnology, Indian Institute of Technology Delhi, Hauz Khas, New Delhi 110016, India. E-mail: bez138024@dbeb.iitd.ac.in

Fabiola Padilla Arizmendi, Bio-Protection Research Centre, PO Box 85084, Lincoln University, Lincoln 7647, Canterbury, New Zealand. E-mail: fabiola.padillaarizmendi@lincoln.ac.nz

Mansi Bakshi, Institute of Environment and Sustainable Development, Banaras Hindu University, Varanasi 221005, India. E-mail: mansibakshi51@gmail.com

E.J. Bedmar, Department of Microbiology, Faculty of Pharmacy, Campus of Cartuja, University of Granada, 18071-Granada, Spain. E-mail: eulogio.bedmar@eez.csic.es

Adrien Biessy, Université de Moncton, Département de biologie, 18 Antonine-Maillet Ave, Moncton, NB E1A 3E9, Canada. E-mail: eab1826@umoncton.ca

Virendra S. Bisaria, Department of Biochemical Engineering and Biotechnology, Indian Institute of Technology Delhi, Hauz Khas, New Delhi 110016, India. E-mail: vbisaria@dbeb.iitd.ac.in

Kartikay Bisen, Department of Mycology and Plant Pathology, Institute of Agricultural Sciences, Banaras Hindu University, Varanasi-221005, India. E-mail: kartikaybisen@rediffmail.com

Marieta Marin Bruzos, Belowground Ecosystem Group, Department of Forest and Conservation Sciences, University of British Columbia, Vancouver, BC, Canada. E-mail: mmarinb@mail.ubc.ca

Antonio Castellano-Hinojosa, Department of Soil Microbiology and Symbiotic Systems, Estación Experimental del Zaidín-CSIC. E-419, 18080-Granada, Spain and Department of Microbiology, Faculty of Pharmacy, Campus of Cartuja, University of Granada, 18071-Granada, Spain. E-mail: ach@ugr.es

Niladri Chaudhry, Department of Pathology, Institute of Medical Sciences, Banaras Hindu University, Varanasi, Uttar Pradesh 221005, India. E-mail: niladri0225@gmail.com

Vasvi Chaudhry, Bacterial Genomics and Evolution Laboratory, CSIR-Institute of Microbial Technology, Chandigarh-160036, India. E-mail: dr.vasvi.india@gmail.com

M.K. Chitara, Department of Mycology and Plant Pathology, Institute of Agricultural Sciences, Banaras Hindu University, Varanasi 221005, India. E-mail: manojchitara01@gmail.com

Pooja Choudhary, Department of Biotechnology, ITS Paramedical College, Ghaziabad, U.P. – 201 206, India. E-mail: choudharypuja55@gmail.com

Natalia Cripps-Guazzone, Bio-Protection Research Centre, PO Box 85084, Lincoln University, Lincoln 7647, Canterbury, New Zealand. E-mail: natalia.cripps-guazzone@lincoln.ac.nz

Manish Kumar Dubey, Department of Botany, Institute of Science, Banaras Hindu University, Varanasi-221005, India. E-mail: mkmkdubey@gmail.com

Pradeep Kumar Dubey, Institute of Environment and Sustainable Development, Banaras Hindu University, Varanasi 221005, India. E-mail: pdubeybhu@gmail.com

Rama Kant Dubey, Institute of Environment and Sustainable Development, Banaras Hindu University, Varanasi 221005, India. E-mail: ramakant.sls@gmail.com

Sheikh Adil Edirisi, Institute of Environment and Sustainable Development, Banaras Hindu University, Varanasi 221005, India. E-mail: adi13acme@gmail.com

Martin Filion, Université de Moncton, Départemement de Biologie, 18 Antonine-Maillet Ave, Moncton, NB E1A 3E9, Canada. E-mail: martin.filion@umoncton.ca

Elisabetta Franchi, Eni S.p.A, Renewable Energy and Environmental R & D, S. Donato Milanese (MI), Italy. E-mail: elisabetta.franchi@eni.com

Deepika Goyal, Department of Biotechnology, School of Life Sciences, Central University of Rajasthan, Bandarsindri, NH-8, Kishangarh, Ajmer- 305817 Rajasthan, India. E-mail: deepika_rs@curaj.ac.in

Robert Hill, Bio-Protection Research Centre, PO Box 85084, Lincoln University, Lincoln 7647, Canterbury, New Zealand. E-mail: Robert.hill@lincoln.ac.nz

Angela Cristina Ikeda, Federal University of Parana, Department of Forest Science, Curitiba, Brazil. E-mail: angela.ikeda@gmail.com

Madhu Kamle, Department of Forestry, North Eastern Regional Institute of Science and Technology (Deemed University), Nirjuli (Itanagar)-791109, Arunachal Pradesh, India. E-mail: madhu.kamle18@gmail.com

Diwakar Kandula, Bio-Protection Research Centre, PO Box 85084, Lincoln University, Lincoln 7647, Canterbury, New Zealand. E-mail: Diwakar.Kandula@lincoln.ac.nz

Chetan Keswani, Department of Biochemistry, Institute of Science, Banaras Hindu University, Varanasi-221005, India. E-mail: chetankeswani@rediffmail.com

Gagan Kumar, Department of Mycology and Plant Pathology, Institute of Agricultural Sciences, Banaras Hindu University, Varanasi 221005, India. E-mail: chauhangagan.chauhan@gmail.com

Pradeep Kumar, Department of Forestry, North Eastern Regional Institute of Science and Technology (Deemed University), Nirjuli (Itanagar)-791109, Arunachal Pradesh, India. E-mail: pkbiotech@gmail.com

Punam Kumari, Department of Botany, Institute of Science, Banaras Hindu University, Varanasi-221005, India. E-mail: poonam1928@gmail.com

Richard Daniel Lally, EnviroCORE, The Dargan Centre, Department of Science and Health, Institute of Technology Carlow, Co. Carlow, Ireland and Alltech, Summerhill Road, Sarney, Dunboyne, Co. Meath, Ireland. E-mail: lallyr43@yahoo.com

Robert Lawry, Bio-Protection Research Centre, PO Box 85084, Lincoln University, Lincoln 7647, Canterbury, New Zealand. E-mail: vorlor@gmail.com

Jesús Salvador López-Bucio, Instituto de Biotecnología, Universidad Nacional Autónoma de México, Apartado Postal 510-3, 62250 Cuernavaca, Morelos, México. E-mail: lopbucio@ibt.unam.mx

José López-Bucio, Instituto de Investigaciones Químico-Biológicas, Universidad Michoacana de San Nicolás de Hidalgo, Edifico B3, Ciudad Universitaria, C.P. 58030 Morelia, Michoacán, México. E-mail: jbucio@umich.mx

Anupam Maharshi, Department of Mycology and Plant Pathology, Institute of Agricultural Sciences, Banaras Hindu University, Varanasi-221005, India. E-mail: anupammaharshi@gmail.com

Shrikant S. Mantri, Computational Biology Laboratory, National Agri-Food Biotechnology Institute (NABI), Mohali, Punjab- 160071, India. E-mail: shrikant@nabi.res.in

Jorge Armando Mauricio-Castillo, Unidad Académica de Agronomía, Universidad Autónoma de Zacatecas, Km. 15.5 Carr. Zacatecas-Guadalajara 98170, Cieneguillas, Zacatecas, México. E-mail: jamaca1209@hotmail.com

Dmitri V. Mavrodi, Department of Biological Sciences, the University of Southern Mississippi, Hattiesburg, MS, 39406, USA. E-mail: Dmitri.Mavrodi@usm.edu

Olga V. Mavrodi, Department of Biological Sciences, the University of Southern Mississippi, Hattiesburg, MS, 39406, USA. E-mail: Olga.mavrodi@usm.edu

Mukesh Meena, Department of Botany, Institute of Science, Banaras Hindu University, Varanasi-221005, India. E-mail: mukeshmeenabhu@gmail.com

Artemio Mendoza-Mendoza, Bio-Protection Research Centre, PO Box 85084, Lincoln University, Lincoln 7647, Canterbury, New Zealand. E-mail: artemio.mendoza@lincoln.ac.nz

Pallavi Mittal, Department of Biotechnology, ITS Paramedical College, Ghaziabad, U.P. – 201206, India. E-mail: mittal_pallavi@yahoo.com

Átila Francisco Mógor, Federal University of Parana, Department of Plant Science and Crop Protection, Curitiba, Brazil. E-mail: afmogor@gmail.com

Rafael Jorge León Morcillo, Department of Plant Growth Promotion Rhizobacteria. Plant Stress Centre for Biology (PSC), Chinese Academy of Sciences (CAS), Shanghai, China. E-mail: rafael@sibs.ac.cn

Arpan Mukherjee, Department of Botany, Institute of Science, Banaras Hindu University, Varanasi-221005, India. E-mail: arpan.mukherjee55@gmail.com

María Fernanda Nieto-Jacobo, Bio-Protection Research Centre, PO Box 85084, Lincoln University, Lincoln 7647, Canterbury, New Zealand. E-mail: mafeni@hotmail.com

Guillermo Nogueira-López, Bio-Protection Research Centre, PO Box 85084, Lincoln University, Lincoln 7647, Canterbury, New Zealand. E-mail: guillermo.nogueiralopez@lincoln.uni.ac.nz

Randy Ortiz-Castro, Instituto de Ecología, Xalapa, México. E-mail: randyortizcastro@yahoo.com.mx

Sumita Pal, Department of Mycology and Plant Pathology, Institute of Agricultural Sciences, Banaras Hindu University, Varanasi-221005, India. E-mail: sumita.kgp.06@gmail.com

Janmejay Pandey, Department of Biotechnology, School of Life Sciences, Central University of Rajasthan, Bandarsindri, NH-8, Kishangarh, Ajmer-305817, Rajasthan, India. E-mail: janmejay@curaj.ac.in

Jai Singh Patel, Department of Botany, Institute of Science, Banaras Hindu University, Varanasi-221005, India. E-mail: jayp94@gmail.com

Gianniantonio Petruzzelli, Institute of Ecosystem Study, National Council of Research, Pisa, Italy. E-mail: gianniantonio.petruzzelli@ise.cnr.it

Om Prakash, Microbial Culture Collection, National Centre for Cell Sciences, Pune, Maharashtra 411007, India. E-mail: prakas1974@gmail.com

Rahul Singh Rajput, Department of Mycology and Plant Pathology, Institute of Agricultural Sciences, Banaras Hindu University, Varanasi-221005, India. E-mail: rahulsinghr829@gmail.com

Devendra Pratap Rao, Department of Chemistry, D. A. V. College, Uttar Pradesh, Kanpur 208001, India. E-mail: devendrapratapra o@yahoo.com

Amitava Rakshit, Department of Soil Science and Agricultural Chemistry, Institute of Agricultural Sciences, Banaras Hindu University, Varanasi 221005, India. E-mail: amitavabhu@gmail.com

Shatrupa Ray, Department of Botany, Institute of Science, Banaras Hindu University, Varanasi-221005, India. E-mail: shatrupa.ray@gmail.com

Jorge Sáenz-Mata, Laboratorio de Ecología Microbiana, Facultad de Ciencias Biológicas, Universidad Juárez del Estado de Durango Av. Universidad S/N, Fracc. Filadelfia, Durango, México. E-mail: jsaenz_mata@ujed.mx

Abhineet Sain, Department of Biochemical Engineering and Biotechnology, Indian Institute of Technology Delhi, Hauz Khas, New Delhi 110016, India. E-mail: bb5120005@dbeb.iitd.ac.in

Silvia Salas-Muñoz, CONACYT-Campo Experimental Zacatecas, Instituto de Investigaciones Forestales, Agrícolas y Pecuarias, Km 24.5 Carr. Zacatecas-Fresnillo 98500, Calera de V. R. Zacatecas, México. E-mail: silviasm0207@gmail.com

Fatima Berenice Salazar-Badillo, Unidad Académica de Agronomía, Universidad Autónoma de Zacatecas, Zacatecas, Mexico. E-mail: fatimiuxb681@gmail.com

Deepranjan Sarkar, Department of Soil Science and Agricultural Chemistry, Institute of Agricultural Sciences, Banaras Hindu University, Varanasi 221005, India. E-mail: deep.gogreen@gmail.com

B.K. Sarma, Department of Mycology and Plant Pathology, Institute of Agricultural Sciences, Banaras Hindu University, Varanasi 221005, India. E-mail: birinchi_ks@yahoo.com

Shilpi Sharma, Department of Biochemical Engineering and Biotechnology, Indian Institute of Technology Delhi, Hauz Khas, New Delhi 110016, India. E-mail: shilpi@dbeb.iitd.ac.in

Shubhangini Sharma, Department of Biotechnology, ITS Paramedical College, Ghaziabad, U.P. – 201 206, India. E-mail: sharmashubh4@gmail.com

Irina Sidorova, Lomonosov Moscow State University, Moscow, Russia. E-mail: irsidor2008@yandex.ru

Ajeet Singh, Institute of Environment and Sustainable Development, Banaras Hindu University, Varanasi 221005, India. E-mail: ajeetbhu97943@gmail.com

Akanksha Singh, Microbial Technology and Nematology Division, CSIR-Central Institute of Medicinal and Aromatic Plants, Lucknow, India. E-mail: bhuaks29@gmail.com

D.P. Singh, ICAR-National Bureau of Agriculturally Important Microorganisms, Kushmaur, Mau Nath Bhanjan, Mau, India. E-mail: dpsfarm@rediffmail.com

H.B. Singh, Department of Mycology and Plant Pathology, Institute of Agricultural Sciences, Banaras Hindu University, Varanasi 221005, India. E-mail: hbs1@rediffmail.com

Jyoti Singh, Department of Botany, Institute of Science, Banaras Hindu University, Varanasi-221005, India. E-mail: singh.jyoti146@gmail.com

S.P. Singh, Department of Biochemistry, Institute of Science, Banaras Hindu University, Varanasi-221005, India. E-mail: suryasingh@hotmail.com

Shalini Singh, Department of Botany, Institute of Science, Banaras Hindu University, Varanasi-221005, India. E-mail: shalini222003@gmail.com

Surendra Singh, Department of Botany, Institute of Science, Banaras Hindu University, Varanasi-221005, India. E-mail: surendrasingh.bhu@gmail.com

Upma Singh, Department of Biochemical Engineering and Biotechnology, Indian Institute of Technology Delhi, Hauz Khas, New Delhi 110016, India. E-mail: bez158004@dbeb.iitd.ac.in

Vivek Singh, Department of Botany, Institute of Science, Banaras Hindu University, Varanasi-221005, India. E-mail: vvksigh@gmail.com

Alison Stewart, SCION, 49 Sala Street, Rotorua 3010, New Zealand. E-mail: alison.stewart@scionresearch.com

Johanna Steyaert, Bio-Protection Research Centre, PO Box 85084, Lincoln University, Lincoln 7647, Canterbury, New Zealand. E-mail: johanna.steyaert@lincolngritech.co.nz

Prashant Swapnil, Department of Botany, Institute of Science, Banaras Hindu University, Varanasi-221005, India. E-mail: mbhuprashant@gmail.com

Vishal Tripathi, Institute of Environment and Sustainable Development, Banaras Hindu University, Varanasi 221005, India. E-mail: vishalbiotechbhu@gmail.com

R.S. Upadhyay, Department of Botany, Institute of Science, Banaras Hindu University, Varanasi-221005, India. E-mail: upadhyay_bhu@yahoo.co.uk

Claudio Valverde, Laboratorio de Bioquímica, Microbiología e Interacciones Biológicas en el Suelo, Departamento de Ciencia y Tecnología, Universidad Nacional de Quilmes, Roque Sáenz Peña 352 - Bernal B1876BXD - Buenos Aires, Argentina. E-mail: valverdecl @hotmail.com

Jay Prakash Verma, Institute of Environment and Sustainable Development, Banaras Hindu University, Varanasi 221005, India. E-mail: jayprakashbhu@gmail.com

Juan Ignacio Vílchez, Department of Plant Growth Promotion Rhizobacteria, Plant Stress Centre for Biology (PSC), Chinese Academy of Sciences (CAS), Shanghai, China. E-mail: juan@sibs.ac.cn

Elena Voronina, Lomonosov Moscow State University, Moscow, Russia. E-mail: mvsadnik@ list.ru

Luis Gabriel Wall, Laboratorio de Bioquímica, Microbiología e Interacciones Biológicas en el Suelo, Departamento de Ciencia y Tecnología, Universidad Nacional de Quilmes, Roque Sáenz Peña 352 - Bernal B1876BXD - Buenos Aires, Argentina. E-mail: lgwall@ hotmail.com

Shanshan Wen, Department of Agronomy, Northwest A & F University, Yangling, Shaanxi, 712100, China. E-mail: sswen812@yahoo.com

Ranjeet Singh Yadav, Department of Soil Science and Agricultural Chemistry, Institute of Agricultural Sciences, Banaras Hindu University, Varanasi 221005, India. E-mail: singhyadav. ranjeet@gmail.com

Mingming Yang, Department of Agronomy, Northwest A & F University, Yangling, Shaanxi, 712100, China. E-mail: myang@nwsuaf.edu.cn

Antoine Zboralski, Université de Moncton, Département de biologie, 18 Antonine-Maillet Ave, Moncton, NB E1A 3E9, Canada. E-mail: eaz6186@umoncton.ca

Vivian Jaskiw Szilagyi Zecchin, Federal University of Parana, Department of Plant Science and Crop Protection, Curitiba, Brazil. E-mail: vivian.szilagyi@gmail.com

Andleeb Zehra, Department of Botany, Institute of Science, Banaras Hindu University, Varanasi-221005, India. E-mail: zandleeb143@gmail.com

Foreword

Today, many economically important agricultural, horticultural and ornamental crop plants are attacked by various soilborne and foliar diseases, resulting in billions of dollars in crop losses. Currently, the most widely used disease management strategy is the use of chemical fungicides. However, the use of these fungicides has encountered problems, such as development by pathogens of resistance to fungicides, and rapid degradation of the chemicals. Other factors leading to increased interest in alternatives include the increasing cost of soil fumigation, lack of suitable replacements for methyl bromide and public concerns over exposure to fungicides. Both the agriculture and agri-food sector are now expected to move toward environmentally sustainable development, while maintaining productivity. These concerns and expectations have led to renewed interest in the use of "biologically based pest management strategies". One approach to such biologically based strategies is the use of naturally occurring and environmentally safe products such as PGPR.

It has long been known that many microorganisms in the soil–root ecosystem are attracted by nutrients exuded by plant roots. This soil–root ecozone is called the rhizosphere. Many bacteria from the rhizosphere can influence plant growth and plant health positively, and we refer to them as PGPR – Plant Growth-Promoting Rhizobacteria, defined as root-colonizing bacteria (biofertilizers and biofungicides) that exert beneficial traits on plant growth and development. Root colonization comprises the ability of PGPR to establish on or in the root or rhizosphere to multiply, survive and colonize along the growing root in the presence of the indigenous microflora. PGPR are considered as efficient microbial competitors in the soil–root zone. In addition to plant growth promotion, PGPR are also used for controlling several plant pathogens, enhancement of nutrient up-take and in rhizomediation. PGPR colonize plant roots and exert beneficial effects on plant growth and development by a wide variety of mechanisms. To be an effective PGPR, bacteria must be able to colonize roots, because bacteria need to establish in the rhizosphere at population densities sufficient to produce the beneficial effects.

The exact mechanism by which PGPR stimulate plant growth is not clearly established, although several hypotheses such as production of phytohormones, suppression of deleterious organisms, activation of phosphate solubilization, and promotion of the mineral nutrient uptake are usually believed to be involved.

In the context of increasing international concern for food and environmental quality, the use of PGPR for reducing chemical inputs in agriculture is a potentially important issue. PGPR have gained worldwide importance and acceptance for sustainable agricultural benefits. PGPR are the potential tools for the future of sustainable agriculture. Currently, there is an active and growing group of researchers working on fundamental and applied aspects of PGPR. The application and commercialization of PGPR for sustainable agriculture is a growing and demanding market around the world.

Worldwide, PGPR technology is being considered as the latest pursuit for expertise in knowledge-intensive sectors. Currently, the global agriculture biotech industry is valued at an estimated US$ 45 billion and is expected to grow at 25% annually.

The green revolution of agriculture brought an enormous increase in food production. It not only made the world self-sufficient in food but also gave the world's scientists and farmers an immense amount of self-respect. Though the green revolution did increase food production, the productivity levels have remained low and the increased production was achieved at a cost of intensive use of water, fertilizer and other inputs which have caused problems of soil salinity, groundwater pollution, nutrient imbalances, emergence of new pests and diseases, and environmental degradation. To feed the ever increasing population globally and particularly in Asia more and more food now has to be produced from less and less land, water and other natural resources. It is therefore apparent that we have to do things differently and doing more of what we did yesterday will not take us forward. With the advent of PGPR technology and its use on crops, we can achieve higher productivity, better quality, improved nutrition, improved storage properties, increased resistance to pests and disease, and achieve higher prices for farmers in the global marketplace. PGPR technology has immense potential for eradicating rural poverty and fuelling global GDP growth.

In this context, *Advances in PGPR Research* includes contributions from vastly experienced, global experts in PGPR research in a comprehensive and influential manner, with the most recent facts and extended case studies. My heartfelt congratulations to the editors for synchronizing with global authorities on the subject to underline the upcoming challenges and present most viable options for translating commercially viable ideas into easily affordable products and technologies.

Prof. M.S. Reddy
Chairman, Asian PGPR Society
Department of Entomology and Plant Pathology
Auburn University
Auburn, Alabama
USA

Preface

Rhizosphere biology is approaching a century of investigations wherein growth-promoting rhizomicroorganisms (PGPR) have attracted special attention for their beneficial skills. Considering the priorities of food security and enhancing productivity, profitability and sustainable rural livelihoods at farm level, developing a new order of farm inputs has become imperative. In this perspective, bio-inputs – either directly in the form of microbes or their by-products – are gaining tremendous momentum. The global market for biopesticides was valued at $1,796.56 million in 2013 and is expected to reach $4,369.88 million by 2019, growing at a compound annual growth rate of 16.0% from 2013 to 2019. The PGPR industry is just coming out of its infancy. Its potential is being tested, realized and used. Public awareness and acceptance of PGPR will accelerate the process. Currently these are being supplemented by individual private entrepreneurs for developing PGPR products for local needs as well as for the export market. Harnessing the potential of agriculturally important microorganisms could help in providing low-cost and environmentally safe technologies to the farmers, especially those who cannot afford expensive technologies. Considering recent developments in biopesticide research and their implications in sustainable productivity we have included a list of 25 chapters which address the current global issues in biopesticide research.

Harikesh B. Singh
Birinchi K. Sarma
Chetan Keswani
Editors

1 Mechanisms of Growth Promotion by Members of the Rhizosphere Fungal Genus *Trichoderma*

Artemio Mendoza-Mendoza,[1,*] Guillermo Nogueira-López,[1] Fabiola Padilla Arizmendi,[1] Natalia Cripps-Guazzone,[1] María Fernanda Nieto-Jacobo,[1] Robert Lawry,[1] Diwakar Kandula,[1] Fatima Berenice Salazar-Badillo,[2] Silvia Salas-Muñoz,[3] Jorge Armando Mauricio-Castillo,[2] Robert Hill,[1] Alison Stewart[4] and Johanna Steyaert[1]

[1]Bio-Protection Research Centre, Lincoln University, Canterbury, New Zealand; [2]Unidad Académica de Agronomía, Universidad Autónoma de Zacatecas, Zacatecas, México; [3]CONACYT- Campo Experimental Zacatecas, Instituto de Investigaciones Forestales, Agrícolas y Pecuarias, Calera de V. R. Zacatecas, México; [4]SCION, Rotorua, New Zealand

1.1 Introduction

Trichoderma species are cosmopolitan filamentous fungi found in agricultural, native prairie, forest, salt marsh, and desert soils of all biomes (rainforests, savannas, deserts, grasslands, temperate deciduous forest, temperate, conifer forest, Mediterranean scrub, taiga and tundra), as well as in lake water, dead plant material, living roots of virtually any plant species, seeds and air (Atanasova *et al.*, 2013; Mukherjee *et al.*, 2013; Waghund *et al.*, 2016). The ability of *Trichoderma* spp. to thrive in such a wide range of habitats is linked to their capability to produce a number of bioactive molecules, such as lytic enzymes, antibiotics and multiple other secondary metabolites.

Rhizosphere competency is widespread among the *Trichoderma* and many strains are considered opportunistic plant endophytes frequently found in symbiotic relationships with diverse crops (including maize, tomato, cucumber, cotton, cocoa, etc.), ornamental flowers, grasses, palms, ferns, trees, etc. (Harman, 2000; Harman *et al.*, 2004; Sobowale *et al.*, 2007; Hohmann *et al.*, 2011; Keswani *et al.*, 2013; Cripps-Guazzone, 2014; Singh *et al.*, 2016). Successful rhizosphere competence and endophytism are subject to host specificity and changes in abiotic environmental factors (Cripps-Guazzone, 2014).

Trichoderma spp. induce plant growth by direct and indirect mechanisms. Direct mechanisms include the facilitation or

*E-mail: artemio.mendoza@lincoln.ac.nz

© CAB International 2017. *Advances in PGPR Research* (eds H.B. Singh, B.K. Sarma and C. Keswani)

increment of nutrients uptake such as phosphate solubilization, iron sequestration and production of secondary metabolites, including phytohormones and volatile or nonvolatile compounds (Vinale *et al.*, 2012; Qi and Zhao, 2013; Saravanakumar *et al.*, 2013; Zhao *et al.*, 2014; Borges Chagas *et al.*, 2015; Lee *et al.*, 2015; Li *et al.*, 2015; Bisen *et al.*, 2016; Garnica-Vergara *et al.*, 2016). Indirect mechanisms include biocontrol activity against plant pathogens (bacteria, fungi and nematodes) and the ability to impart abiotic stress tolerance within the plant (Bruce *et al.*, 1984; Bae *et al.*, 2009; Shukla *et al.*, 2012; Yang *et al.*, 2012; Qi and Zhao, 2013; Vinale *et al.*, 2013; Contreras-Cornejo *et al.*, 2014; Stewart and Hill, 2014; Zhao *et al.*, 2014; Kottb *et al.*, 2015; Shukla *et al.*, 2015; Pandey *et al.*, 2016; Zachow *et al.*, 2016) (Fig. 1.1).

1.2 *Trichoderma* Plant Growth Promotion: Direct Mechanisms

1.2.1 Nutrient acquisition

Phosphorus and iron are the key elements for plant growth and, although these nutrients are abundant in nature, they are poorly accessible to plants (de Santiago *et al.*, 2013).

Fig. 1.1. Plant growth promotion induced by *Trichoderma spp.* (A-B) Direct mechanisms of induction (A). Plant growth promotion induced by *Trichoderma* (+) in *Acacia mangium* in comparison to control plants (–). In this experiment plants were free of any chemical pesticide or fertilizer. (B) Root growth promotion induced by *Trichoderma* in canola. Roots of plants untreated (left) and roots of plants inoculated with a mixture of *T. atroviride* (right). (C-E) Indirect mechanisms of induction (C), effect of leaf spot disease on untreated oil palm seedlings (D) and treated with endophytic *Trichoderma* spp. (E). Silvergrass roots grown in soil infested with *Rhizoctonia solani* (left) and treated plants with commercial product based on a mixture of *Trichoderma* isolates (right).

Microbial communities modify nutrient cycling in the rhizosphere, affecting nutrient availability to plants. *Trichoderma* spp. secrete diverse molecules, including siderophores, organic acid compounds and proteins that contribute to the solubility of inorganic phosphate and iron (Kapri and Tewari, 2010; Khan *et al.*, 2010; de Santiago *et al.*, 2013; Saravanakumar *et al.*, 2013; Borges Chagas *et al.*, 2015).

Phosphate solubilization

Phosphorus (P) is the second most limiting nutrient to plant growth behind nitrogen (N) (Condron, 2004) and, as a result, pasture and crops require the input of organic P through fertilizers (Koning *et al.*, 2008; Simpson *et al.*, 2014; Desmidt *et al.*, 2015) to reach the growth levels needed for sustainable farming. Phosphate fertilizers contain phosphate rock which is mined from natural deposits and is therefore non-renewable (Desmidt *et al.*, 2015). The uptake of organic P by plants is low, somewhere between 5% and 30% depending on the soil alkalinity (Condron, 2004). The rest forms insoluble inorganic compounds with aluminium (Al), iron (Fe) and calcium (Ca), unavailable for plant uptake (Ward *et al.*, 1996; Heffer and Prud'homme, 2008). Phosphate-solubilizing micro-organisms, such as fungi and bacteria, play a major role in the transformation of insoluble soil P into soluble available forms (Rodriguez and Fraga, 1999; Khan, 2009), and therefore are potential bio-fertilizers. *Trichoderma* spp. have known P solubilizing activity (Kapri and Tewari, 2010; Mukherjee *et al.*, 2013; Borges Chagas *et al.*, 2015). The potential mechanism for phosphate solubilization might be acidification either by proton extrusion or association with ammonium assimilation.

Siderophores

Iron (Fe) is one of the most abundant elements on earth, however it is present as ferric ions in the soil which are not very soluble and are consequently inaccessible to plants (Lehner *et al.*, 2013). Siderophores are molecules that solubilize Fe and as a consequence alter nutrient availability in soil environments for microorganisms and plants (Vinale *et al.*, 2013). Lehner *et al.* (2013) analysed eight different strains of *Trichoderma*, including *T. atroviride* IMI206040, *T. asperellum*, *T. gamsii*, *T. hamatum*, *T. virens* Gv29.8, *T. harzianum*, *T. polysporum* and *T. reesei* QM6a and observed that on average *Trichoderma* spp. produced 12 to 14 siderophores, with six common to all species. In *Trichoderma* spp., intracellular siderophores are synthesized by three non-ribosomal protein synthases (NRPs), which are present as a cluster in the genome (Mukherjee *et al.*, 2012a; Zeilinger *et al.*, 2016). The role of the NRP6 from *T. virens* has been related to the biosynthesis of 10 of 12 extracellular secreted siderophores. Harzianic acid is a secreted siderophore molecule synthesized by *T. harzianum* and this molecule has plant growth-promoting and antifungal activity (Vinale *et al.*, 2013). The role of siderophores in aiding competition with other microbes in the rhizosphere or in providing Fe to the plants has not been completely explored and there is still much work to be done to understand the role of these molecules in the plant–microbe–rhizosphere interaction and its relation to plant growth promotion.

Synthesis of secondary metabolites

Secondary metabolites produced by plant-associated microbes change the chemical and physical properties of soil, increasing iron, nitrogen or phosphorus availability (Bitas *et al.*, 2013). Moreover, beneficial microorganisms are able to manipulate hormone signalling pathways in the host plant and as a consequence enhance plant growth (Kunkel and Brooks, 2002; Sofo *et al.*, 2011; Spaepen and Vanderleyden, 2011). The chemical composition of secondary metabolites produced by *Trichoderma* is diverse (Vinale *et al.*, 2012; Keswani *et al.*, 2014; Bansal and Mukherjee, 2016; Zeilinger *et al.*, 2016). *Trichoderma* metabolites directly influence plant physiology by modulating hormone activity in the plant, affecting nutrient solubility or by combating plant pathogens (Keswani *et al.*, 2014).

HORMONES. *Trichoderma* synthesizes 3-indoleacetic acid (IAA), the major auxin in plants (Yue *et al.*, 2014; Enders and Strader, 2015), acting as a plant growth promoter (Contreras-Cornejo *et al.*, 2009). *T. virens* synthesizes indolic compounds, viz. IAA, indole-3-acetaldehyde (IAAld), indole-3-ethanol (tryptophol) and indole-3-carboxaldehyde (ICAld) (Contreras-Cornejo *et al.*, 2009; 2011). IAA, IAAld and ICAld synthesized by *T. virens* have auxin activity in *A. thaliana*; however tryptophol did not show significant auxin activity in this model plant (Contreras-Cornejo *et al.*, 2009; 2011). The exact mechanism and enzymes involved in the synthesis of IAA are unknown but in *T. virens* it has been suggested that indole-3-ethanol and indole-3-acetaldehyde are key components for the IAA biosynthetic pathway (Contreras-Cornejo *et al.*, 2009). Nevertheless, not all plant growth-promoting *Trichoderma* strains synthesize auxins, at least under laboratory conditions, suggesting that additional mechanisms are involved (Hoyos-Carvajal *et al.*, 2009). Inhibition of ethylene represents the best studied mechanism of plant growth promotion induced by microorganisms (Nascimento *et al.*, 2014). *T. asperellum* produces 1-aminocyclopropane-1-carboxylate (ACC) deaminase, which cleaves ACC, the immediate precursor of the plant hormone ethylene, to produce α-ketobutyrate and ammonia (Todorovic and Glick, 2008). Inoculation of microorganisms that synthesize ACC deaminase, such as *Trichoderma* spp., induces plant growth promotion by the reduction of ethylene (Viterbo *et al.*, 2010).

VOLATILE ORGANIC COMPOUNDS (VOCS). Plant-associated fungi produce a great variety of VOCs which comprise mainly pyrones, terpenes, alcohols, ketones, alkanes and alkenes (Korpi *et al.*, 2009). The biosynthesis of fungal VOCs often coincides with certain developmental stages like spore formation and some of these compounds are produced in association with mycotoxins (Wilkins *et al.*, 2003). VOCs synthesis depends on nutrient availability, pH, temperature and light, and is species/strain-specific (Zeilinger *et al.*, 2016). Overall, microbial VOCs promote plant growth, increase crop yield and protect host plants against pathogenic organisms.

Trichoderma volatiles are able to induce beneficial effects on *A. thaliana* seedlings. VOCs emitted by *T. viride*, *T. atroviride* and *T. virens* cultures in a shared atmosphere with *A. thaliana,* without direct contact, resulted in larger plants, earlier flowering, and enhancement of lateral root development (Hung *et al.*, 2013; Contreras-Cornejo *et al.*, 2014; Salazar-Badillo *et al.*, 2015). However, Kottb *et al.* (2015) reported that after the interaction of *A. thaliana* with VOCs emitted by *T. asperellum* IsmT5, there was an accumulation of anthocyanin pigments, a rise by 47% of the trichome density, an increased level of H_2O_2 as a sign of the activation of plant defence responses, 97% increase in camalexin accumulation, a higher respiration activity (40% more than the control group) and greater concentrations of salicylic acid and abscisic acid. Overall *A. thaliana* plants exposed to *Trichoderma* volatiles showed improved survival strategies and defence responses in these two different experiments.

The emission of biologically active VOCs by *Trichoderma* has been reported frequently since the 1950s. For several years, the plant growth promotion effects and fungal plant-pathogen inhibition were attributed to carbon dioxide, ethanol, acetaldehyde and acetone (Tamimi and Hutchinson, 1975), but improvements in analytical techniques revealed the diversity in volatile profiles of several species of *Trichoderma.* Analysis of VOCs produced by *T. atroviride*, demonstrated the presence of 25 different fungal metabolites including alcohols, ketones, alkenes, furanes, pyrenes, monoterpenes and sesquiterpenes (Stoppacher *et al.*, 2010).

Within the range of metabolites produced by genus *Trichoderma*, 6-pentyl-2H-pyron-2-one (6-PP) is often reported as the major volatile produced by this fungus which promotes plant growth and influences root architecture. *A. thaliana* root response to 6-PP involves components of auxin transport and signalling and the ethylene response modulator EIN2 (Garnica-Vergara *et al.*, 2016).

1.3 *Trichoderma* Plant Growth Promotion: Indirect Mechanisms

1.3.1 Biocontrol of plant disease

The potential of *Trichoderma* species as biocontrol agents of plant diseases was first recognized in the early 1930s (Weindling, 1932) and since then *Trichoderma* species have been reported to control many plant diseases of fruit and vegetable crops. For example, Matei and Matei (2008) reported that *T. harzianum* P8 was able to control *Botrytis cinerea* on strawberry cultivars by hyperparasitism. Perello *et al.* (2009) used *T. harzianum* and *T. koningii* to protect wheat from leaf blotching caused by *Septoria tritici* in Argentina. The same authors also used these species to control tan spot, caused by *Pyrenophora tritici-repentis* on wheat (Perello *et al.*, 2008). Harman (2000) reported that *T. harzianum* T22 controlled diseases caused by *Fusarium* species on tomatoes. *Trichoderma* has also been used in wood preservation, with Vanneste *et al.* (2002) reporting that *T. harzianum* could provide better control of sapstain than the standard fungicide on *Pinus radiata*. These studies and others have culminated in the development of commercial products/prototype formulations of several *Trichoderma* species for the protection of a number of crops in the USA, India, Israel, New Zealand and Sweden (Howell, 2003; Kandula *et al.*, 2015).

Trichoderma achieves successful biocontrol through a multitude of mechanisms including induced systemic resistance, mycoparasitism, antibiosis, microbial competition and direct growth promotion. Direct growth promotion results in stronger and healthier plants, which, in turn, are better able to cope with disease and abiotic stresses (Bisen *et al.*, 2015; Mishra *et al.*, 2015). Biocontrol itself in turn promotes plant growth indirectly by protecting the plant from pathogens and subsequent yield/growth loss.

Induced systemic resistance

There are three recognized pathways of induced resistance in plants (Harman *et al.*, 2004): the salicylic acid pathway, the jasmonic acid pathway and the non-pathogenic root-associated bacteria induced pathway. In the salicylic and jasmonic acid pathways the production of pathogenesis-related proteins (PR) (antifungal chitinases, glucanases, thaumatins, and oxidative enzymes) are triggered by the attack of pathogenic microorganisms and the wounding or necrosis-inducing plant pathogens (herbivory by insects). In the non-pathogenic root-associated bacteria induced pathway, the PR proteins are not induced by root colonization in the absence of attack by plant-pathogenic microorganisms.

De Meyer *et al.* (1998) were the first to demonstrate that *Trichoderma* spp. could induce resistance in plants. They reported that bean plants grown in soil treated with *T. harzianum* T39 showed fewer disease symptoms after *B. cinerea* inoculation to the leaves compared with untreated control plants, even though T39 was only present on the roots and not on the foliage. Yedidia *et al.* (2003) presented conclusive evidence for the induction of a systemic response against angular leaf spot of cucumber caused by *Pseudomonas syringae* pv. *lachrymans* following application of *T. asperellum* to the root system. Disease symptoms were reduced by as much as 80%, corresponding to a reduction of two orders of magnitude in numbers of bacterial cells in leaves of plants pre-treated with *T. asperellum*. Similar studies have now been carried out with a wide range of plants, including both monocotyledons and dicotyledons and with multiple *Trichoderma* species and strains (Harman *et al.*, 2004), and these have demonstrated that induced resistance can be mediated by *Trichoderma* spp.

Mycoparasitism

A key characteristic of members of the genus *Trichoderma* is their ability to parasitize other fungi, some of which are plant pathogens, and many instances of successful biocontrol with *Trichoderma* species have been ascribed to this mechanism. Mycoparasitism occurs in several steps, first, *Trichoderma* spp. detect other fungi and grow tropically towards them (Chet *et al.*, 1981). Uncharacterised diffusible factors act as elicitors of proteases (*prb1*) which are directly associated with the mycoparasitic activity of *T. atroviride* (Geremia *et al.*, 1993; Cortes *et al.*, 1998; Olmedo-Monfil *et al.*, 2002; Steyaert *et al.*, 2004). Brunner *et al.* (2003) suggested that diffusion of low levels of an extracellular exochitinase catalyses the release of cell-wall oligomers from target fungi, and this in turn induces the expression of fungitoxic endochitinases which also diffuse and begin the attack on the target fungus before contact is actually made (Viterbo *et al.*, 2002). Once in direct contact, *Trichoderma* produces several fungitoxic cell wall degrading enzymes (CWDEs) (Chet *et al.*, 1998; Steyaert *et al.*, 2003; Lorito *et al.*, 2010). These enzymes function by breaking down the polysaccharides, chitin, and β-glucans which form fungal walls, thereby destroying the cell wall integrity of pathogenic fungi (Howell, 2003). Signalling genes/pathways involved in mycoparasitism include the kinase Tvk1/TmkA from *T. virens* and Tmk1 from *T. atroviride*, which are negative regulators of hydrolytic enzymes and antibiotics. The corresponding gene deletion mutants were more effective in controlling plant disease caused by *R. solani* than the commercial chemical fungicides in beans (Reithner *et al.*, 2007; Mukherjee *et al.*, 2013).

Howell (1982) observed *T. virens* (formerly *Gliocladium virens*) parasitizing *R. solani* by coiling around and penetrating the hyphae. In *T. atroviride* IMI206040, Tga1 and Tga3, two G-protein α subunits from the cAMP signalling pathway, regulate coiling (Rocha-Ramirez *et al.*, 2002). In addition, Tga1 regulates the production of lytic enzymes and biosynthesis of antifungal metabolites that impact mycoparasitism (Reithner *et al.*, 2005) while Tga3 regulates secretion of CWDEs but not their biosynthesis (Zeilinger *et al.*, 2016). Recent comparative analysis of the genome, secretome and transcriptome of the three species: *T. atroviride* IMI206040, *T. virens* Gv29.8, and *T. reesei* QM6a, indicated mycoparasitism as the ancestral lifestyle of *Trichoderma* (Kubicek *et al.*, 2011; Atanasova *et al.*, 2013).

Antibiosis

Antibiosis is the process of secretion of anti-microbial compounds by antagonistic microbes to suppress and/or kill pathogenic microbes in the vicinity of their growing area (Schirmbock *et al.*, 1994). *Trichoderma* produces many secondary metabolites with antibiotic activities and their production is species/strain dependent (Mukherjee *et al.*, 2012b, Zeilinger *et al.*, 2016). Ghisalberti and Sivasithamparam (1991) classified the secondary metabolites into three categories: (i) volatile antibiotics, eg. 6-pentyl-α-pyrone (6-PP) and most of the isocyanide derivates; (ii) water-soluble compounds, i.e. heptelidic acid or koningic acid; and (iii) peptaibols, which are linear oligopeptides of 12–22 amino acids rich in α-aminoisobutyric, N-acetylated at the N-terminus and containing an amino alcohol at the C-terminus. Lorito *et al.* (1996) investigated the activity of peptaibols and cell wall hydrolytic enzymes produced by *T. harzianum* in the antagonism of *B. cinerea*. Peptaibols trichorzianin TA and TB inhibited β-glucan synthase activity in the host fungus. The inhibition was synergistic with *T. harzianum* β-1, 3-glucanase and prevented the reconstruction of the pathogen cell wall, which facilitated the action of the glucanase and enhanced the fungicidal activity. Antibiotics probably act synergistically with lytic enzymes.

Competition

Competition for carbon, nitrogen and other growth factors, together with competition for space or specific infection sites, is an indirect mechanism by which *Trichoderma* controls plant pathogens (Vinale *et al.*, 2008). Gullino (1992) reported that *T. harzianum* was able to control *B. cinerea* on grapes by colonizing blossom tissue and excluding the pathogen from infection sites. Competition for nutrients is the major mechanism used by *T. harzianum* to control *Fusarium oxysporum* f. sp. *melonis* (Sivan and Chet, 1989). Benitez *et al.* (2004) showed that *Trichoderma* has a strong capacity to mobilize and take up soil nutrients which make it more efficient and competitive than other soil microbes.

1.3.2 Abiotic Stress Tolerance

There is increasing evidence to show that *Trichoderma* can protect plants from the adverse effects of abiotic stress. Stress tolerance in turn results in promotion of growth. Drought tolerance induced by *Trichoderma* have been observed in multiple host plants, including rice, maize, cocoa, wheat and *A. thaliana* (Bae *et al.*, 2009; Shukla *et al.*, 2012; Zaidi *et al.*, 2014; Contreras-Cornejo *et al.*, 2015; Shukla *et al.*, 2015; Chandra and Gaur (2016); Pandey *et al.*, 2016; Rawat *et al.*, 2016). Drought tolerance by *Trichoderma* appears to be strain-specific (Shukla *et al.*, 2012; Rawat *et al.*, 2016). *Trichoderma virens* and *T. atroviride* synthesize abscisic acid (ABA) that modulate stomatal aperture closure and consequently protection against loss of water (Contreras-Cornejo *et al.*, 2015). In wheat, maize and rice, H_2O_2 content significantly increases in response to drought, however inoculation with *Trichoderma* spp. can significantly reduce the H_2O_2 content as compared with the control plants (Shoresh and Harman, 2008; Rawat *et al.*, 2012; Rawat *et al.*, 2016). During the plant–*Trichoderma* interaction, *Trichoderma* induces an increased synthesis of antioxidative enzymes in the host plants, these include superoxide dismutases (SOD), peroxidases, glutathione-reductases and glutathione-S-transferases (GST), as well as other detoxifying enzymes in leaves (Shoresh and Harman, 2008).

Another strategy used by *Trichoderma* to provide stress tolerance to its host plant is via the ethylene pathway, where *Trichoderma* mutants unable to synthesise ACC deaminase are less effective in providing tolerance to salt stress, suggesting that *Trichoderma*, similarly to ACC deaminase-producing bacteria, can ameliorate plant growth under conditions of abiotic stress, by lowering detrimental increases in ethylene levels (Brotman *et al.*, 2013).

1.4 The 'Omics' of *Trichoderma*

Despite the importance of *Trichoderma* only seven species of *Trichoderma* corresponding to 10 strains have been fully sequenced and are publicly available (http://genome.jgi-psf.org/) (Table 1.1).

Some members of the genus *Trichoderma* such as *T. virens, T. harzianum, T. atroviride, T. hamatum, T. asperellum* and *T. ovalisporum* have the capacity to colonise roots and develop a close interaction with their host plant (Bailey *et al.*, 2006; Alfano *et al.*, 2007; Shoresh and Harman, 2008; Moran-Diez *et al.*, 2015); however, the exact mechanisms that regulate these symbiotic interactions are not fully characterised. More recently, high-dimensional biology, transcriptomics and proteomics have been used to unravel the regulatory mechanisms of *Trichoderma* spp. as plant symbionts (Bailey *et al.*, 2006; Marra *et al.*, 2006; Alfano *et al.*, 2007; Chacon *et al.*, 2007; Segarra *et al.*, 2007; Shoresh and Harman, 2008; Samolski *et al.*, 2009; Mehrabi-Koushki *et al.*, 2012; Lamdan *et al.*, 2015; Moran-Diez *et al.*, 2015; Schmoll *et al.*, 2016). However, much is still unknown and further high-throughput omics technologies are essential to understanding the complexity of biological processes that drive *Trichoderma*–plant interactions and plant growth promotion.

1.4.1 *Trichoderma*–plant interaction transcriptomics

In general, it has been observed that root colonization by *Trichoderma* causes transcriptional changes in genes involved in metabolism and stress resistance in both the plant and the fungus (Bailey *et al.*, 2006; Samolski *et al.*, 2009; Shoresh *et al.*, 2010; Hermosa *et al.*, 2012; Brotman *et al.*, 2013), which in turn promotes growth of the host plant. For example, a macroarray study showed that cacao gene expression profiles in response to endophytic association with four different growth-promoting strains of *Trichoderma* were highly similar (Bailey *et al.*, 2006). The majority of up-regulated plant genes were related to environmental stress response. In contrast, the *Trichoderma* expressed genes were mainly involved in nutrient acquisition and cell functionality (Bailey *et al.*, 2006). Using the plant model *A. thaliana*, Brotman *et al.* (2013)

Table 1.1. Publicly available *Trichoderma genomes*.

	Genome Assembly size (Mbp)	No. gene models[a]	Read coverage depth	Genome source
Trichoderma asperellum CBS 433.97	40.87	13932	100X	http://genome.jgi.doe.gov/Trias1/Trias1.home.html
Trichoderma asperellum TR356	35.39	12320	120X	http://genome.jgi.doe.gov/Triasp1/Triasp1.home.html
Trichoderma atroviride IMI206040	36.1	11863	~8.26X	http://genome.jgi.doe.gov/Triat2/Triat2.home.html
Trichoderma citrinoviride TUCIM 6016	33.22	9737	63.1X	http://genome.jgi.doe.gov/Trici4/Trici4.home.html
Trichoderma harzianum CBS 226.95	40.98	14095	120X	http://genome.jgi.doe.gov/Triha1/Triha1.home.html
Trichoderma harzianum TR274	40.87	13932	100X	http://genome.jgi.doe.gov/Trihar1/Trihar1.home.html
Trichoderma longibrachiatum ATCC 18648	40.87	13932	100X	http://genome.jgi.doe.gov/Trilo3/Trilo3.home.html
Trichoderma reesei RUT C-30	32.69	9852	47.6X	http://genome.jgi.doe.gov/TrireRUTC30_1/TrireRUTC30_1.home.html
Trichoderma reesei QM6a	34.1	9129		http://genome.jgi.doe.gov/Trire2/Trire2.home.html
Trichoderma virens Gv29-8	39	12427	~8.05X	http://genome.jgi.doe.gov/TriviGv29_8_2/TriviGv29_8_2.home.html

[a]Predicted and annotated using the JGI annotation pipeline

observed that *Trichoderma* stimulated plant growth and resistance to saline stress and significantly improved seed germination. Analysis of the up-regulated plant genes show that they were related mainly to osmo-protection and general stress response. Other authors observed the regulation of the WRKY transcription factors and the ADC genes are related with several important biological functions (Sáenz-Mata *et al.*, 2014; Salazar-Badillo *et al.*, 2015). In addition, it has been suggested that similarly to ACC deaminase-producing bacteria, *Trichoderma* can enhance plant growth under conditions of abiotic stress, by lowering the levels of ethylene as well as promoting an increment in antioxidative activity and by the modulation of polyamine content (Brotman *et al.*, 2013; Salazar-Badillo *et al.*, 2015).

1.4.2 Proteomics

To understand the changes occurring in the plant in response to interacting with *Trichoderma*, several studies have been carried out to identify proteome and secretome profiles using gel-based 2-DE gel analysis coupled with LC-MS/MS or MALDI-TOF MS and gel free-based proteomics coupled with LC-MS/MS (Marra *et al.*, 2006; Segarra *et al.*, 2007; Shoresh and Harman, 2008; Lamdan *et al.*, 2015). Shoresh and Harman (2008) demonstrated that colonization of maize roots by *T. harzianum* altered the shoot proteome in terms of carbohydrate metabolism, photosynthesis and stress, and this up-regulation may correspond to the enhanced growth promotion response and induce systemic resistance.

Induced systemic resistance is an indirect plant growth promotion mechanism that *Trichoderma* confer to plants. The presence of *T. asperellum* in cucumber roots triggers the salicylic acid and jasmonate pathways in the plant, and increase peroxidases activity, hence conferring protection to cucumber plants against foliar pathogens (Segarra *et al.*, 2007). Remarkably, *T. harzianum* did not alter plant secondary metabolism and protein biosynthesis compared to *T. asperellum*, suggesting that *Trichoderma* spp. may have different strategies to induce plant immune changes. Currently diverse reports indicate that *Trichoderma* induce systemic resistance by releasing not only proteins, but also secondary metabolites (Reithner *et al.*, 2007; Luo *et al.*, 2010; Mukherjee *et al.*, 2012b; Cai *et al.*, 2013; Martinez-Medina *et al.*, 2013; Harel *et al.*, 2014; Sáenz-Mata *et al.*, 2014; Lamdan *et al.*, 2015; Salas-Marina *et al.*, 2015; Salazar-Badillo *et al.*, 2015; Keswani *et al.*, 2016). Overall, the current proteomic studies from *Trichoderma*–plant interaction give us insight of how *Trichoderma* induces changes in plant metabolism that leads to enhanced growth and immunity to plant pathogens.

1.5 Conclusion

Trichoderma spp. are best known for their biocontrol capabilities against a range of phytopathogenic microorganisms and increased plant drought tolerance. However, all the attributes of *Trichoderma* are also related to their ability to induce plant growth promotion by direct or indirect mechanisms. The activation of these mechanisms might be dependent on the ability of *Trichoderma* to respond to the environmental conditions and host plant.

Acknowledgments

Our research work on *Trichoderma* has been supported by the Tertiary Education Commission, New Zealand through The Bio-Protection Research Centre, Marsden Fund, and Lincoln University Research Fund. GNL received a scholarship from Conacyt-Mexico and Meadow, New Zealand for his PhD studies.

References

Alfano, G., Ivey, M.L., Cakir, C., Bos, J.I., Miller, S.A. *et al.* (2007) Systemic modulation of gene expression in tomato by *Trichoderma hamatum* 382. *Phytopathology* 97, 429–437.

Atanasova, L., Druzhinina, I. and Jaklitsch, W. (2013) Two hundred *Trichoderma* species recognized on the basis of molecular phylogeny. In: Mukherjee, P.K., Horwitz, B.A., Singh, U.S., Mukherjee, M. and S.M (eds) *Trichoderma: Biology and Applications*. CAB International, Wallingford, UK, pp. 10–42.

Bae, H., Sicher, R.C., Kim, M.S., Kim, S.-H., Strem, M.D. *et al.* (2009) The beneficial endophyte *Trichoderma hamatum* isolate DIS 219b promotes growth and delays the onset of the drought response in *Theobroma cacao*. *Journal of Experimental Botany* 60, 3279–3295.

Bailey, B.A., Bae, H., Strem, M.D., Roberts, D.P., Thomas, S.E. *et al.* (2006) Fungal and plant gene expression during the colonization of cacao seedlings by endophytic isolates of four *Trichoderma* species. *Planta* 224, 1449–1464.

Bansal, R. and Mukherjee, P.K. (2016) Identification of novel gene clusters for secondary metabolism in *Trichoderma* genomes. *Microbiology* 85, 185–190.

Benitez, T., Rincon, A.M., Limon, M.C. and Codon, A.C. (2004) Biocontrol mechanisms of *Trichoderma* strains. *International Microbiology* 7, 249–260.

Bisen, K., Keswani, C., Mishra, S., Saxena, A., Rakshit, A. and Singh, H.B. (2015) Unrealized potential of seed biopriming for versatile agriculture. In: Rakshit, A., Singh, H.B. and Sen, A. (eds) *Nutrient Use Efficiency: From Basics to Advances*. Springer, New Delhi, pp. 193–206.

Bisen, K., Keswani, C., Patel, J.S., Sarma, B.K. and Singh, H.B. (2016) *Trichoderma* spp.: efficient inducers of systemic resistance in plants. In: Chaudhary, D.K. and Verma, A. (eds) *Microbial-Mediated Induced Systemic Resistance in Plants*. Springer, Singapore, pp. 185–195.

Bitas, V., Kim, H.S., Bennett, J.W. and Kang, S. (2013) Sniffing on microbes: diverse roles of microbial volatile organic compounds in plant health. *Molecular Plant–Microbe Interactions* 26, 835–843.

Borges Chagas, L.F., Chagas Junior, A.F., Rodrigues de Carvalho, M., de Oliveira Miller, L. and Orozco Colonia, B.S. (2015) Evaluation of the phosphate solubilization potential of *Trichoderma* strains (Trichoplus JCO) and effects on rice biomass. *Journal of Soil Science and Plant Nutrition* 15, 794–804.

Brotman, Y., Landau, U., Cuadros-Inostroza, Á., Tohge, T., Fernie, A.R. *et al.* (2013) *Trichoderma*-plant root colonization: escaping early plant defense responses and activation of the antioxidant machinery for saline stress tolerance. *PLoS Pathogen* 9, e1003221.

Bruce, A., Austin, W.J. and King, B. (1984) Control of growth of *Lentinus lepideus* by volatiles from *Trichoderma*. *Transactions of the British Mycological Society* 82, 423–428.

Brunner, K., Peterbauer, C.K., Mach, R.L., Lorito, M., Zeilinger, S. and Kubicek, C.P. (2003) The Nag1 *N*-acetylglucosaminidase of *Trichoderma atroviride* is essential for chitinase induction by chitin and of major relevance to biocontrol. *Current Genetics* 43, 289–295.

Cai, F., Yu, G., Wang, P., Wei, Z., Fu, L. *et al.* (2013) Harzianolide, a novel plant growth regulator and systemic resistance elicitor from *Trichoderma harzianum*. *Plant Physiology and Biochemistry* 73, 106–113.

Chacon, M.R., Rodriguez-Galan, O., Benitez, T., Sousa, S., Rey, M. *et al.* (2007) Microscopic and transcriptome analyses of early colonization of tomato roots by *Trichoderma harzianum*. *International Microbiology* 10, 19–27.

Chandra, S. and Gaur, A.K. (2016) Dose-dependent response of *Trichoderma harzianum* in improving drought tolerance in rice genotypes. *Planta* 243, 1251–1264.

Chet, I., Harman, G.E. and Baker, R. (1981) *Trichoderma hamatum*: its hyphal interactions with *Rhizoctonia solani* and *Pythium* spp. *Microbial Ecology* 7, 29–38.

Chet, I., Benhamou, N. and Haran, S. (1998) Mycoparasitism and lytic enzymes. In: Harman, G.E. and Kubicek, C.P. (eds) *Trichoderma and Gliocladium Vol. 2*. Taylor and Francis Ltd, London, pp. 153–169.

Condron, L.M. (2004) Phosphorus – surplus and deficiency. In: Schjønning, P., Elmholt, B. and Christensen, S. (eds) *Managing Soil Quality: Challenges in Modern Agriculture*. CAB International, Wallingford, UK, pp. 69–84.

Contreras-Cornejo, H.A., Macias-Rodriguez, L., Cortes-Penagos, C. and Lopez-Bucio, J. (2009) *Trichoderma virens*, a plant beneficial fungus, enhances biomass production and promotes lateral root growth through an auxin-dependent mechanism in *Arabidopsis*. *Plant Physiology* 149, 1579–1592.

Contreras-Cornejo, H.A., Macias-Rodriguez, L., Beltran-Pena, E., Herrera-Estrella, A. and Lopez-Bucio, J. (2011) *Trichoderma*-induced plant immunity likely involves both hormonal- and camalexin-dependent mechanisms in *Arabidopsis thaliana* and confers resistance against necrotrophic fungi *Botrytis cinerea*. *Plant Signalling Behavior* 6, 1554–1563.

Contreras-Cornejo, H.A., Macías-Rodríguez, L., Herrera-Estrella, A. and López-Bucio, J. (2014) The 4-phosphopantetheinyl transferase of *Trichoderma virens* plays a role in plant protection against *Botrytis cinerea* through volatile organic compound emission. *Plant and Soil* 379, 261–274.

Contreras-Cornejo, H.A., Macías-Rodríguez, L., Vergara, A.G. and López-Bucio, J. (2015) *Trichoderma* modulates stomatal aperture and leaf transpiration through an abscisic acid-dependent mechanism in arabidopsis. *Journal of Plant Growth Regulation* 34, 425–432.

Cortes, C., Gutierrez, A., Olmedo, V., Inbar, J., Chet, I. and Herrera-Estrella, A. (1998) The expression of genes involved in parasitism by *Trichoderma harzianum* is triggered by a diffusible factor. *Molecular and General Genetics* 260, 218–225.

Cripps-Guazzone, N. (2014) Rhizosphere competence of selected *Trichoderma* species. PhD thesis, Lincoln University, Canterbury, New Zealand.

De Meyer, G., Bigirimana, J., Elad, Y. and Höfte, M. (1998). Induced systemic resistance in *Trichoderma harzianum* T39 biocontrol of *Botrytis cinerea*. *European Journal of Plant Pathology* 104, 279–286.

de Santiago, A., García-López, A.M., Quintero, J.M., Avilés, M. and Delgado, A. (2013) Effect of *Trichoderma asperellum* strain T34 and glucose addition on iron nutrition in cucumber grown on calcareous soils. *Soil Biology and Biochemistry* 57, 598–605.

Desmidt, E., Ghyselbrecht, K., Zhang, Y., Pinoy, L., Van der Bruggen, B. *et al.* (2015) Global phosphorus scarcity and full-scale P-recovery techniques: a review. *Critical Reviews in Environmental Science and Technology* 45, 336–384.

Enders, T.A. and Strader, L.C. (2015) Auxin activity: past, present, and future. *American Journal of Botany* 102, 180–196.

Garnica-Vergara, A., Barrera-Ortiz, S., Munoz-Parra, E., Raya-Gonzalez, J., Mendez-Bravo, A. *et al.* (2016) The volatile 6-pentyl-2H-pyran-2-one from *Trichoderma atroviride* regulates *Arabidopsis thaliana* root morphogenesis via auxin signaling and *ETHYLENE INSENSITIVE 2* functioning. *New Phytologist* 209, 1496–1512.

Geremia, R.A., Goldman, G.H., Jacobs, D., Ardiles, W., Vila, S.B. *et al.* (1993) Molecular characterization of the proteinase-encoding gene, *prb1*, related to mycoparasitism by *Trichoderma harzianum*. *Molecular Microbiology* 8, 603–613.

Ghisalberti, E.L. and Sivasithamparam, K. (1991) Antifungal antibiotics produced by *Trichoderma* spp. *Soil Biology and Biochemistry* 23, 1011–1020.

Gullino, M.L. (1992) Control of *Botrytis* rot of grapes and vegetables with *Trichoderma* spp. In: Tjamos, E.C., Papavizas, G.C. and Cook, R.J. (eds) *Biological Control of Plant Diseases, Progress and Challenges for the Future*. Plenum Press, New York, pp. 125–132.

Harel, Y.M., Mehari, Z.H., Rav-David, D. and Elad, Y. (2014) Systemic resistance to gray mold induced in tomato by benzothiadiazole and *Trichoderma harzianum* T39. *Phytopathology* 104(2), 150–157.

Harman, G.E. (2000) Myths and dogmas of biocontrol: changes in perceptions derived from research on *Trichoderma harzianum* T-22. *Plant Disease* 84, 377–393.

Harman, G.E., Howell, C.R., Viterbo, A., Chet, I. and Lorito, M. (2004) *Trichoderma* species – opportunistic, avirulent plant symbionts. *Natural Review of Microbiology* 2, 43–56.

Heffer, P. and Prud'homme, M. (2008) Outlook for world fertilizer demand, supply, and supply/demand balance. *Turkish Journal of Agriculture and Forestry* 32, 159–164.

Hermosa, R., Viterbo, A., Chet, I. and Monte, E. (2012) Plant-beneficial effects of *Trichoderma* and of its genes. *Microbiology* 158, 17–25.

Hohmann, P., Jones, E.E., Hill, R.A. and Stewart, A. (2011) Understanding *Trichoderma* in the root system of *Pinus radiata*: associations between rhizosphere colonisation and growth promotion for commercially grown seedlings. *Fungal Biology* 115, 759–767.

Howell, C.R. (1982) Effect of *Gliocladium virens* on *Pythium ultimum*, *Rhizoctonia solani* and damping-off of cotton seedlings. *Phytopathology* 72, 496–498.

Howell, C.R. (2003) Mechanisms employed by *Trichoderma* species in the biological control of plant diseases: the history and evolution of current concepts. *Plant Disease* 87, 4–10.

Hoyos-Carvajal, L., Orduz, S. and Bissett, J. (2009) Growth stimulation in bean (*Phaseolus vulgaris* L.) by *Trichoderma*. *Biological Control* 51, 409–416.

Hung, R., Lee, S. and Bennett, J.W. (2013) *Arabidopsis thaliana* as a model system for testing the effect of *Trichoderma* volatile organic compounds. *Fungal Ecology* 6, 19–26.

Kandula, D., Jones, E., Stewart, A., McLean, K. and Hampton, J. (2015) *Trichoderma* species for biocontrol of soil-borne plant pathogens of pasture species. *Biocontrol Science and Technology* 25, 1052–1069.

Kapri, A. and Tewari, L. (2010) Phosphate solubilization potential and phosphatase activity of rhizospheric *Trichoderma* spp. *Brazilian Journal of Microbiology* 41, 787–795.

Keswani, C., Singh, S.P. and Singh, H.B. (2013) A superstar in biocontrol enterprise: *Trichoderma* spp. *Biotech Today* 3, 27–30.

Keswani, C., Mishra, S., Sarma, B.K., Singh, S.P. and Singh, H.B. (2014) Unraveling the efficient applications of secondary metabolites of various *Trichoderma* spp. *Applied Microbiology Biotechnology* 98, 533–544.

Keswani, C., Bisen, K., Singh, S.P., Sarma, B.K. and Singh, H.B. (2016) A proteomic approach to understand the tripartite interactions between plant–*Trichoderma*–pathogen: investigating the potential for efficient biological control. In: Hakeem, K.R. and Akhtar, M.S. (eds) *Plant, Soil and Microbes Vol. 2. Mechanisms and Molecular Interactions*. Springer, Switzerland, pp. 79–93.

Khan, M.S., Zaidi, A., Wani, P.A., Ahemad, M., Oves, M. (2009) Functional diversity among plant growth-promoting rhizobacteria. In Khan, M.S., Zaidi, A. and Musarrat, J. (eds) *Microbial Strategies for Crop Improvement*. Springer, Berlin, pp. 105–132.

Khan, M.S., Zaidi, A., Ahemad, M., Oves, M. and Wani, P.A. (2010) Plant growth promotion by phosphate solubilizing fungi–current perspective. *Archives of Agronomy and Soil Science* 56, 73–98.

Koning, N.B.J., Van Ittersum, M.K., Becx, G.A., Van Boekel, M., Brandenburg, W.A. *et al.* (2008) Long-term global availability of food: continued abundance or new scarcity? *Njas-Wageningen Journal of Life Sciences* 55, 229–292.

Korpi, A., Jarnberg, J. and Pasanen, A.L. (2009) Microbial volatile organic compounds. *Critical Review of Toxicology* 39, 139–193.

Kottb, M., Gigolashvili, T., Grosskinsky, D.K. and Piechulla, B. (2015) *Trichoderma* volatiles effecting *Arabidopsis*: from inhibition to protection against phytopathogenic fungi. *Frontiers in Microbiology* 6, 995.

Kubicek, C.P., Herrera-Estrella, A., Seidl-Seiboth, V., Martinez, D.A., Druzhinina, I.S. *et al.* (2011) Comparative genome sequence analysis underscores mycoparasitism as the ancestral life style of *Trichoderma*. *Genome Biology* 12, R40.

Kunkel, B.N. and Brooks, D.M. (2002) Cross talk between signaling pathways in pathogen defense. *Current Opinion in Plant Biology* 5, 325–331.

Lamdan, N.L., Shalaby, S., Ziv, T., Kenerley, C.M. and Horwitz, B.A. (2015) Secretome of the biocontrol fungus *Trichoderma virens* co-cultured with maize roots: role in induced systemic resistance. *Molecular Cell Proteomics* 14, 1054–1063.

Lee, S., Hung, R., Yap, M. and Bennett, J.W. (2015) Age Matters: the effects of volatile organic compounds emitted by *Trichoderma atroviride* on plant growth. *Archives of Microbiology* 197, 723–727 doi:10.1007/s002003-015-1104-5.

Lehner, S.M., Atanasova, L., Neumann, N.K., Krska, R., Lemmens, M. *et al.* (2013) Isotope-assisted screening for iron-containing metabolites reveals a high degree of diversity among known and unknown siderophores produced by *Trichoderma* spp. *Applied Environmental Microbiology* 79, 18–31.

Li, R.X., Cai, F., Pang, G., Shen, Q.R., Li, R. and Chen, W. (2015) Solubilisation of phosphate and micronutrients by *Trichoderma harzianum* and its relationship with the promotion of tomato plant growth. *PLoS One* 10, e0130081.

Lorito, M., Farkas, V., Rebuffat, S., Bodo, B. and Kubicek, C.P. (1996) Cell wall synthesis is a major target of mycoparasitic antagonism by *Trichoderma harzianum*. *Journal of Bacteriology* 178, 6382–6385.

Lorito, M., Woo, S.L., Harman, G.E. and Monte, E. (2010) Translational research on *Trichoderma*: from 'omics to the field. *Annual Review Phytopathology* 48, 395–417.

Luo, Y., Zhang, D.D., Dong, X.W., Zhao, P.B., Chen, L.L. *et al.* (2010) Antimicrobial peptaibols induce defense responses and systemic resistance in tobacco against tobacco mosaic virus. *FEMS Microbiology Letters* 313, 120–126.

Marra, R., Ambrosino, P., Carbone, V., Vinale, F., Woo, S.L. *et al.* (2006) Study of the three-way interaction between *Trichoderma atroviride*, plant and fungal pathogens by using a proteomic approach. *Current Genetics* 50, 307–321.

Martinez-Medina, A., Fernandez, I., Sanchez-Guzman, M.J., Jung, S.C., Pascual, J.A. and Pozo, M.J. (2013) Deciphering the hormonal signalling network behind the systemic resistance induced by *Trichoderma harzianum* in tomato. *Frontiers in Plant Science* 4, 206.

Matei, G.M. and Matei, S. (2008) Research on isolation, characterization and testing the interaction between *Trichoderma harzianum* and *Botrytis cinerea* for biological control of gray mold in strawberry. *Lucrari Stiintifice – Universitatea de Stiinte Agronomice si Medicina Veterinara Bucuresti. Seria B, Horticultura* 51, 653–657.

Mehrabi-Koushki, M., Rouhani, H. and Mahdikhani-Moghaddam, E. (2012) Differential display of abundantly expressed genes of *Trichoderma harzianum* during colonization of tomato-germinating seeds and roots. *Current Microbiology* 65, 524–533.

Mishra, S., Singh, A., Keswani, C., Saxena, A., Sarma, B.K. and Singh, H.B. (2015) Harnessing plant–microbe interactions for enhanced protection against phytopathogens. In: Arora, N.K. (ed.) *Plant Microbes Symbiosis: Applied Facets*. Springer, New Delhi, pp. 111–125.

Moran-Diez, M.E., Trushina, N., Lamdan, N.L., Rosenfelder, L., Mukherjee, P.K. *et al.* (2015) Host-specific transcriptomic pattern of *Trichoderma virens* during interaction with maize or tomato roots. *BMC Genomics* 16, 8.

Mukherjee, P.K., Buensanteai, N., Moran-Diez, M.E., Druzhinina, I.S. and Kenerley, C.M. (2012a) Functional analysis of non-ribosomal peptide synthetases (NRPSs) in *Trichoderma virens* reveals a polyketide synthase (PKS)/NRPS hybrid enzyme involved in the induced systemic resistance response in maize. *Microbiology* 158, 155–165.

Mukherjee, P.K., Horwitz, B.A. and Kenerley, C.M. (2012b) Secondary metabolism in *Trichoderma* – a genomic perspective. *Microbiology* 158, 35–45.

Mukherjee, P.K., Horwitz, B.A., Herrera-Estrella, A., Schmoll, M. and Kenerley, C.M. (2013) *Trichoderma* research in the genome era. *Annual Review of Phytopathology* 51, 105–129.

Nascimento, F.X., Rossi, M.J., Soares, C.R., McConkey, B.J. and Glick, B.R. (2014) New insights into 1-aminocyclopropane-1-carboxylate (ACC) deaminase phylogeny, evolution and ecological significance. *PLoS One* 9, e99168.

Olmedo-Monfil, V., Mendoza-Mendoza, A., Gomez, I., Cortes, C. and Herrera-Estrella, A. (2002) Multiple environmental signals determine the transcriptional activation of the mycoparasitism related gene *prb1* in *Trichoderma atroviride*. *Molecular Genetics and Genomics* 267, 703–712.

Pandey, V., Ansari, M.W., Tula, S., Yadav, S., Sahoo, R.K. *et al.* (2016) Dose-dependent response of *Trichoderma harzianum* in improving drought tolerance in rice genotypes. *Planta*, 243, 1251. doi:10.1007/s00425-016-2482-x

Perello, A., Moreno, V., Monaco, C. and Simon, M.R. (2008) Effect of *Trichoderma* spp. isolates for biological control of tan spot of wheat caused by *Pyrenophora tritici-repentis* under field conditions in Argentina. *BioControl* 53, 895–904.

Perello, A.E., Moreno, M.V., Monaco, C., Simon, M.R. and Cordo, C. (2009) Biological control of *Septoria tritici* blotch on wheat by *Trichoderma* spp. under field conditions in Argentina. *BioControl* 54, 113–122.

Qi, W. and Zhao, L. (2013) Study of the siderophore-producing *Trichoderma asperellum* Q1 on cucumber growth promotion under salt stress. *Journal of Basic Microbiology* 53, 355–364.

Rawat, L., Singh, Y., Shukla, N. and Kumar, J. (2012) Seed biopriming with salinity tolerant isolates of *Trichoderma harzianum* alleviates salt stress in rice: growth, physiological and biochemical characteristics. *Journal of Plant Pathology* 94, 353–365.

Rawat, L., Singh Bisht, T., Kukreti, A. and Prasad, M. (2016) Bioprospecting drought tolerant *Trichoderma harzianum* isolates promote growth and delay the onset of drought responses in wheat (*Triticum aestivum* L.). *Molecular Soil Biology* 7, 1–15.

Reithner, B., Brunner, K., Schuhmacher, R., Peissl, I., Seidl, V. *et al.* (2005) The G protein alpha subunit *Tga1* of *Trichoderma atroviride* is involved in chitinase formation and differential production of antifungal metabolites. *Fungal Genetics and Biology* 42, 749–760.

Reithner, B., Schuhmacher, R., Stoppacher, N., Pucher, M., Brunner, K. and Zeilinger, S. (2007) Signaling via the *Trichoderma atroviride* mitogen-activated protein kinase *Tmk1* differentially affects mycoparasitism and plant protection. *Fungal Genetics and Biology* 44, 1123–1133.

Rocha-Ramírez, V., Omero, C., Chet, I., Horowitz, B.A. and Herrera-Estrella, A. (2002) *Trichoderma atroviride* G-protein α-subunit gene tga1 is involved in mycoparasitic coiling and conidiation. *Eukaryot Cell* 1, 594–605.

Rodriguez, H. and Fraga, R. (1999) Phosphate solubilizing bacteria and their role in plant growth promotion. *Biotechnology Advances* 17, 319–339.

Sáenz-Mata, J., Salazar-Badillo, F.B. and Jiménez-Bremont, J.F. (2014) Transcriptional regulation of *Arabidopsis thaliana* WRKY genes under interaction with beneficial fungus *Trichoderma atroviride*. *Acta Physiologiae Plantarum* 36, 1085–1093.

Salas-Marina, M.A., Isordia-Jasso, M.I., Islas-Osuna, M.A., Delgado-Sanchez, P., Jimenez-Bremont, J.F. *et al.* (2015) The Epl1 and Sm1 proteins from *Trichoderma atroviride* and *Trichoderma virens* differentially modulate systemic disease resistance against different life style pathogens in Solanum lycopersicum. *Frontiers in Plant Science* 6, 77.

Salazar-Badillo, F.B., Sanchez-Rangel, D., Becerra-Flora, A., Lopez-Gomez, M., Nieto-Jacobo, F. *et al.* (2015) *Arabidopsis thaliana* polyamine content is modified by the interaction with different *Trichoderma* species. *Plant Physiology Biochemistry* 95, 49–56.

Samolski, I., de Luis, A., Vizcaino, J.A., Monte, E. and Suarez, M.B. (2009) Gene expression analysis of the biocontrol fungus *Trichoderma harzianum* in the presence of tomato plants, chitin, or glucose using a high-density oligonucleotide microarray. *BMC Microbiology* 9, 217.

Saravanakumar, K., Shanmuga Arasu, V. and Kathiresan, K. (2013) Effect of *Trichoderma* on soil phosphate solubilization and growth improvement of *Avicennia marina*. *Aquatic Botany* 104, 101–105.

Schirmbock, M., Lorito, M., Wang, Y.L., Hayes, C.K., Arisanatac, I. *et al.* (1994) Parallel formation and synergism of hydrolytic enzymes and peptaibol antibiotics, molecular mechanisms involved in the antagonistic action of *Trichoderma harzianum* against phytopathogenic fungi. *Applied Environmental Microbiology* 60, 4364–4370.

Schmoll, M., Dattenböck, C., Carreras-Villaseñor, N., Mendoza-Mendoza, A., Tisch, D. *et al.* (2016) The genomes of three uneven siblings: footprints of the lifestyles of three *Trichoderma* species. *Microbiology and Molecular Biology Review* 80, 205–327.

Segarra, G., Casanova, E., Bellido, D., Odena, M.A., Oliveira, E. and Trillas, I. (2007) Proteome, salicylic acid, and jasmonic acid changes in cucumber plants inoculated with *Trichoderma asperellum* strain T34. *Proteomics* 7, 3943–3952.

Shoresh, M. and Harman, G.E. (2008) The molecular basis of shoot responses of maize seedlings to *Trichoderma harzianum* T22 inoculation of the root: a proteomic approach. *Plant Physiology* 147, 2147–2163.

Shoresh, M., Harman, G.E. and Mastouri, F. (2010) Induced systemic resistance and plant responses to fungal biocontrol agents. *Annual Review Phytopathology* 48, 21–43.

Shukla, N., Awasthi, R.P., Rawat, L. and Kumar, J. (2012) Biochemical and physiological responses of rice (*Oryza sativa* L.) as influenced by *Trichoderma harzianum* under drought stress. *Plant Physiology and Biochemistry* 54, 78–88.

Shukla, N., Awasthi, R.P., Rawat, L. and Kumar, J. (2015) Seed biopriming with drought tolerant isolates of *Trichoderma harzianum* promote growth and drought tolerance in *Triticum aestivum*. *Annals of Applied Biology* 166, 171–182.

Simpson, R.J., Richardson, A.E., Nichols, S.N. and Crush, J.R. (2014) Pasture plants and soil fertility management to improve the efficiency of phosphorus fertiliser use in temperate grassland systems. *Crop and Pasture Science* 65, 556–575.

Singh, V., Upadhyay, R.S., Sarma, B.K. and Singh, H.B. (2016) *Trichoderma asperellum* spore dose depended modulation of plant growth in vegetable crops. *Microbiological Research* 193, 74–86.

Sivan, A. and Chet, I. (1989) The possible role of competition between *Trichoderma harzianum* and *Fusarium oxysporum* on rhizosphere colonization. *Phytopathology* 79, 198–203.

Sobowale, A., Cardwell, K., Odebode, A., Bandyopadhyay, R. and Jonathan, S. (2007) Persistence of *Trichoderma* species within maize stem against *Fusarium verticillioides*. *Archives of Phytopathology and Plant Protection* 40, 215–231.

Sofo, A., Scopa, A., Manfra, M., De Nisco, M., Tenore, G. *et al.* (2011) *Trichoderma harzianum* strain T-22 induces changes in phytohormone levels in cherry rootstocks (*Prunus cerasus* × *P. canescens*). *Plant Growth Regulation* 65, 421–425.

Spaepen, S. and Vanderleyden, J. (2011) Auxin and plant–microbe interactions. *Cold Spring Harbor Perspectives in Biology* 3:a001438.

Stewart, A. and Hill, R. (2014) Applications of *Trichoderma* in plant growth promotion. *Biotechnology and Biology of Trichoderma*. Elsevier, Amsterdam, pp. 415–428.

Steyaert, J., Ridgway, H., Elad, Y. and Stewart, A. (2003) Genetic basis of mycoparasitism: a mechanism of biological control by species of *Trichoderma*. *New Zealand Journal of Crop and Horticultural Science* 31, 281–291.

Steyaert, J.M., Stewart, A., Jaspers, M.V., Carpenter, M. and Ridgway, H.J. (2004) Co-expression of two genes, a chitinase (*chit42*) and proteinase (*prb1*), implicated in mycoparasitism by *Trichoderma hamatum*. *Mycologia* 96, 1245–1252.

Stoppacher, N., Kluger, B., Zeilinger, S., Krska, R. and Schuhmacher, R. (2010) Identification and profiling of volatile metabolites of the biocontrol fungus *Trichoderma atroviride* by HS-SPME-GC-MS. *Journal of Microbiology Methods* 81, 187–193.

Tamimi, K.M., and Hutchinson, S.A. (1975) Differences between the biological effects of culture gases from several species of *Trichoderma*. *Transactions of the British Mycological Society* 64, 455–463.

Todorovic, B. and Glick, B.R. (2008) The interconversion of ACC deaminase and D-cysteine desulfhydrase by directed mutagenesis. *Planta* 229, 193–205.

Vanneste, J.L., Hill, R.A., Kay, S.J., Farrell, R.L. and Holland, P.T. (2002) Biological control of sapstain fungi with natural products and biological control agents: a review of the work carried out in New Zealand. *Mycological Research* 106, 228–232.

Vinale, F., Sivasithamparam, K., Ghisalberti, E.L., Marra, R., Woo, S.L. and Lorito, M. (2008) Trichoderma–plant–pathogen interactions. *Soil Biology and Biochemistry* 40, 1–10.

Vinale, F., Sivasithamparam, K., Ghisalberti, E.L., Ruocco, M., Wood, S. and Lorito, M. (2012) *Trichoderma* secondary metabolites that affect plant metabolism. *Natural Production Communication* 7, 1545–1550.

Vinale, F., Nigro, M., Sivasithamparam, K., Flematti, G., Ghisalberti, E.L. *et al.* (2013) Harzianic acid: a novel siderophore from *Trichoderma harzianum*. *FEMS Microbiology Letter* 347, 123–129.

Viterbo, A., Montero, M., Ramot, O., Friesem, D., Monte, E. *et al.* (2002) Expression regulation of the endochitinase *chit36* from *Trichoderma asperellum* (*T. harzianum* T-203). *Current Genetics* 42, 114–122.

Viterbo, A., Landau, U., Kim, S., Chernin, L. and Chet, I. (2010) Characterization of ACC deaminase from the biocontrol and plant growth-promoting agent *Trichoderma asperellum* T203. *FEMS Microbiology Letter* 305, 42–48.

Waghund, R.R., Shelake, R.M. and Sabalpara, A.N. (2016) *Trichoderma*: a significant fungus for agriculture and environment. *African Journal of Agricultural Research* 11, 1952–1965.

Ward, C.R., Corcoran, J.F., Saxby, J.D. and Read, H.W. (1996) Occurrence of phosphorus minerals in Australian coal seams. *International Journal of Coal Geology* 30, 185–210.

Weindling, R. (1932) *Trichoderma lignorum* as a parasite of other soil fungi. *Phytopathology* 22, 837–845.

Wilkins, K., Larsen, K. and Simkus, M. (2003) Volatile metabolites from indoor molds grown on media containing wood constituents. *Environmental Science and Pollution Research* 10, 206–208.

Yang, Z., Yu, Z., Lei, L., Xia, Z., Shao, L. *et al.* (2012) Nematicidal effect of volatiles produced by *Trichoderma* sp. *Journal of Asia-Pacific Entomology* 15, 647–650.

Yedidia, I., Shoresh, M., Kerem, Z., Benhamou, N., Kapulnik, Y., and Chet, I. (2003) Concomitant induction of systemic resistance to *Pseudomonas syringae* pv. *lachrymans* in cucumber by *Trichoderma asperellum* (T-203) and accumulation of phytoalexins. *Applied Environmental Microbiology* 69, 7343–7353.

Yue, J., Hu, X. and Huang, J. (2014) Origin of plant auxin biosynthesis. *Trends in Plant Sciences* 19, 764–770.

Zachow, C., Berg, C., Muller, H., Monk, J. and Berg, G. (2016) Endemic plants harbour specific *Trichoderma* communities with an exceptional potential for biocontrol of phytopathogens. *Journal of Biotechnology* 235, 162–170.

Zaidi, N.W., Dar, M.H., Singh, S. and Singh, U.S. (2014) Chapter 38 – *Trichoderma* species as abiotic stress relievers in plants. In: Gupta, Schmoll,Herrera-Estrella, Upadhyay, Druzhinina, Tuohy (eds) *Biotechnology and Biology of Trichoderma*. Elsevier, Amsterdam, pp. 515–525.

Zeilinger, S., Gruber, S., Bansal, R. and Mukherjee, P.K. (2016) Secondary metabolism in *Trichoderma* – chemistry meets genomics. *Fungal Biology Reviews* 30, 74–90.

Zhao, L., Wang, F., Zhang, Y. and Zhang, J. (2014) Involvement of *Trichoderma asperellum* strain T6 in regulating iron acquisition in plants. *Journal of Basic Microbiology* 54 Suppl 1, S115–124.

2 Physiological and Molecular Mechanisms of Bacterial Phytostimulation

Randy Ortiz-Castro,[1] Jesús Salvador López-Bucio[2] and José López-Bucio[3]*

[1]*Red de estudios moleculares avanzados, Instituto de Ecología A.C., Carretera Antigua a Coatepec 351, Veracruz, México;* [2]*Instituto de Biotecnología, Universidad Nacional Autónoma de México, Morelos, México;* [3]*Instituto de Investigaciones Químico-Biológicas, Universidad Michoacana de San Nicolás de Hidalgo, Ciudad Universitaria, Michoacán, México*

2.1 Introduction

Plants and bacteria have coexisted for millions of years. As a result, sophisticated signalling mechanisms allow cross-kingdom communication, which benefits plant health, growth and productivity (Singh *et al.*, 2014). Rhizobacteria sense roots via chemotaxis systems and chemoreceptors, which have been identified in the genomes of several plant-associated species (Scharf *et al.*, 2016). Chemotaxis provides a competitive advantage to motile flagellated bacteria in colonization of root epidermis, as it enables cells to sense and respond to gradients of chemical compounds released by plants (Scharf *et al.*, 2016).

Research from the last two decades, mainly using the model plant *Arabidopsis thaliana*, and development of plant–bacteria co-cultivation systems under axenic conditions increased the understanding of the physiological and developmental aspects of the plant–PGPR relationship. The beneficial effects of PGPR, including Gram-positive and Gram-negative species, are ubiquitous for crops such as wheat, soybean, lettuce, bean, maize and barley (Kloepper *et al.*, 1989; Barazani and Friedman, 1999).

Despite different bacterial species acting as PGPR, the most bioactive strains on plant functioning include different *Pseudomonas* species such as *P. fluorescens*, *P. putida*, *P. aureofaciens* and *P. chlororaphis*, as well as *Bacillus*, *Rhizobium* and Actinobacteria; these bacteria can act directly as PGPR with a predominant biostimulant action, via the production of compounds that increase plant growth, while others may antagonize pathogens or activate plant immunity (Calvo *et al.*, 2014; Pieterse *et al.*, 2014).

Roots release sugars, amino acids, organic acids and other essential nutrients which are used as nutrients or signal to attract bacteria (Rudrappa *et al.*, 2008; Badri *et al.*, 2009; Moe, 2013). This process allows bacteria to actively swim towards roots and is critical for competitive colonization. On the other hand, bacteria release phytohormones, diffusible bioactive molecules and volatile compounds, which locally regulate root branching and increase the absorptive potential to take up water and minerals (Ortiz-Castro *et al.*, 2009).

*E-mail: jbucio@umich.mx

© CAB International 2017. *Advances in PGPR Research* (eds H.B. Singh, B.K. Sarma and C. Keswani)

PGPR typically harbour more than one plant-beneficial property and it is possible that the selection of genes contributing directly to growth promotion relies on fine recognition events. Interestingly, *N*-acyl-L-homoserine lactones (AHLs), cyclodipeptides (CDPs), volatile organic compounds (VOCs), and virulence factors, which comprise a large family of natural compounds biosynthesized by bacteria, modify root developmental programs (Zhang *et al.*, 2007; Ortiz-Castro *et al.*, 2011).

Plants use protein receptors and downstream signalling effectors such as kinases and transcription factors to recognize and interact with their bacterial partners. The canonical auxin receptor *Transport Inhibitor Response 1* (TIR1), and the jasmonic acid receptor *Coronatine Insensitive 1* (COI1) bind bacterial molecules that mimic the endogenous plant regulators (Yan *et al.*, 2009; Ortiz-Castro *et al.*, 2011). Indeed, transcriptomic and metabolomic approaches are starting to reveal the cellular responses during several stages of the interaction in model and crop plants.

Given the comprehensive reviews already published summarizing the suitability of bacteria as bioinoculants in horticultural crops, for enhanced crop resistance to abiotic, or biotic stresses (Dimpka *et al.*, 2009; Pieterse *et al.*, 2014; Bisen *et al.*, 2015, 2016; Mishra *et al.*, 2015; Ruzzi and Aroca, 2015), this review will focus on what is known about the plant–bacteria recognition mechanisms, the recently discovered molecules from bacteria influencing root growth and plant development, and the genes and proteins whose expression changes in plants and bacteria during the interaction.

2.2 Chemical Recognition between Plants and Bacteria

Plants recognize structural components of bacterial cells, the microbe-associated molecular patterns (MAMPs), and also other small organic molecules including secondary metabolites and quorum-sensing signals that can induce positive or negative effects during plant development. MAMPs consist of ubiquitous protein motifs such as glycans and glycol-conjugates, for example, the flagellins, which are protein subunits from flagella of motile bacteria, as well as lipopolysaccharides (LPSs), which are constituents of the bacterial envelope (Boller and Felix, 2009). Small organic molecules include peptides, AHLs, CDPs, VOCs, aminolipids, and virulence factors (Blom *et al.*, 2011; Venturi and Keel, 2016). Additionally, different kinds of phytohormones, mainly auxins, cytokinins and gibberellins are produced by PGPR, which reprograms growth and developmental patterns (Dodd *et al.*, 2010; Sukumar *et al.*, 2012; Kurepin *et al.*, 2014; Ludwig-Müller, 2015).

PGPR form biofilms on roots, which are assemblages of cells embedded in a matrix composed of exopolysaccharides, proteins, and sometimes DNA (Zhang *et al.*, 2014). Plants release exudates that influence root colonization and biofilm formation (de Weert *et al.*, 2002; Chen *et al.*, 2012; Dutta *et al.*, 2013). Amino acids, organic acids and aromatic compounds in root exudates from different plant species recruit *P. putida* and *B. amyloliquefaciens*, and thus, represent chemotactic substances (Matilla *et al.*, 2007; Rudrappa *et al.*, 2008; Ling *et al.*, 2011; Chen *et al.*, 2012; Yuan *et al.*, 2015).

In response to *B. cereus*, roots alter their exudate-chemodiversity changing the proportions of carbohydrates, organic acids, alkanes, and polyols (Dutta *et al.*, 2013). Arabinogalactan proteins, pectin and xylan isolated from *A. thaliana* trigger the formation of robust biofilms in both *B. subtilis* GB03 and *B. amyloliquefaciens* FZB42, which suggest that the ability to recognize plant polysaccharides is conserved in PGPR (Beauregard *et al.*, 2013). Interestingly, during biofilm formation, plant polysaccharides can be metabolized and used as a carbohydrate source to build the exopolysaccharide component of the *B. subtilis* matrix. Thus, plant polysaccharides act as signals that stimulate biofilm formation while acting as a substrate that is processed and incorporated into the biofilm matrix.

2.2.1 Plant developmental and genetic responses to PGPR

The use of co-cultivation systems of bacteria with *Arabidopsis* developed in the last decade, enabled dissection of major developmental and physiological plant responses to PGPR both *in vitro* and in soil. Most studies revealed the modulation of phytohormone signalling mainly auxin, cytokinin and gibberellins by bacterial effectors (Fig. 2.1).

Pseudomonas spp. promote plant growth and drive developmental plasticity in the roots of *Arabidopsis* by inhibiting primary root growth and promoting lateral root and root hair formation. By studying cell type-specific developmental markers and employing genetic and pharmacological approaches, Zamioudis *et al.* (2013) demonstrated the crucial role of auxin signalling and transport in PGPR-stimulated changes of root architecture. An ongoing report showed that plant growth stimulation by *Pseudomonas simiae* WCS417r is partly caused by volatile organic compounds produced by the bacterium and that several *Arabidopsis* ecotypes are able to respond. However, there was a large variation between accessions in the increase in shoot fresh weight, primary root growth and lateral root formation (Wintermans *et al.*, 2016). These results indicate that plants possess natural genetic variation for the capacity to profit from the PGPR action.

Plant growth enhancement by *Bacillus megaterium* was reported by López-Bucio *et al.* (2007), who showed that growth promotion is independent of auxin and ethylene signalling, but required cytokinin signalling as single, double and triple *Arabidopsis* mutants defective on the cytokinin receptors CRE1, AHK2 and AHK3 showed reduced bacterial biostimulation (Ortiz-Castro *et al.*, 2009). *B. megaterium* BP17 (BmBP17), an isolate that endophytically colonized *Arabidopsis* plants increased both root and shoot growth (Vibhuti *et al.*, 2017).

Fig. 2.1. Chemical signalling in the rhizosphere influences root developmental patterning. Beneficial bacteria (PGPR) are attracted to roots via chemoreceptors, which perceive nutrients and bioactive molecules from root exudates. On the other hand, bacterial molecules including auxins, quorum-sensing inducers, volatiles and virulence factors are sensed by roots to coordinate the transitions from indeterminate growth to slow growth or determinate growth. In the latter case, root branching and root hair development are promoted increasing water and nutrient uptake potential. Root–shoot long distance signalling may impact on photosynthesis and shoot patterns such as phyllotaxis, stem branching and flowering, which may be dependent on the multiple feedback loops established by the root microbiome.

The PGPR *Bacillus phytofirmans* PsJN affected the whole life cycle of *Arabidopsis*, accelerated flowering time and shortened its vegetative growth period; these modifications correlated with the early up-regulation of flowering control genes and genes implicated in auxin and gibberellin pathways (Poupin *et al.*, 2013). When *Arabidopsis* seedlings were inoculated with *Gluconacetobacter diazotrophicus*, a root endophyte, growth promotion was consistently observed for up to 50 days, which correlated with higher canopy photosynthesis, lower plant transpiration, and increased water-use efficiency (Rangel de Souza *et al.*, 2016). Thus, single inoculations with a PGPR could affect the whole life cycle of a plant, accelerating its growth rate, improving photosynthesis and water use efficiency.

2.2.2 Plant molecular responses to PGPR

The molecular responses to PGPR are emerging mainly due to transcriptomic and metabolomics approaches. Vibhuti *et al.* (2017) performed microarray-based gene expression profiling during the *Arabidopsis–B. megaterium* BmBP17 interaction, which revealed the up-regulation of nutrient uptake-associated genes and down-regulation of genes coding for transcription factors of ethylene-responsive genes. A total of 150 *Arabidopsis* genes were differentially expressed, which represented 80 up-regulated and 70 down-regulated genes. Key up-regulated genes were NIR1, AMT1-5, TIP2-3 and SULTR1-2 that are likely involved in the transport of nutrients through membranes; SHV3, MMP, RLP44, PROPEP4, AGL42, SCPL30, ANAC010 and KNAT7 participate in cell organization, biogenesis and transcription. On the other hand, ethylene-responsive genes such as ERF5, ERF71, ERF104, ERF105, TEM1 and RAP2.6, and salicylic acid and jasmonic acid-responsive genes such as BAP1, SIB1, BT4, MKK9 and PLA2A were down-regulated. This study shows the coordination of growth and defence through hormonal signalling pathways in response to a PGPR.

The *Arabidopsis* leaf transcriptome reveals distinct but also overlapping responses to *Sphingomonas melonis* Fr1 and *Methylobacterium extorquens* PA1. *M. extorquens* only marginally affected the expression of 10 plant genes, whereas *S. melonis* colonization changed the expression of almost 400 genes (Vogel *et al.*, 2016), suggesting that plants are able to respond differently to members of its natural microbiome. This conclusion is strengthened by comparison of gene expression changes in *Arabidopsis* roots after inoculation with *A. brasilense*, which increases the number of lateral roots and root hairs and also increases auxin concentration in plant tissues, whereas an auxin biosynthesis mutant did not elicit these transcriptional changes (Spaepen *et al.*, 2014).

Changes in proteome correlates well with already reported transcriptomes performed in plants co-cultivated with PGPR. *Paenibacillus polymyxa* E681 PGPR increased *Arabidopsis* shoot and root dry weights that correlated with primary root growth inhibition. A proteomic study via a 2D approach in conjunction with MALDI-TOF/TOF analysis, revealed a total of 41 proteins that were differentially regulated in plants. Of these, 36 proteins related to amino acid metabolism, antioxidant systems, stress response, photosynthesis and hormonal response were up-regulated, which correlated with highly increased plant levels of tryptophan, indole-3-acetonitrile (IAN), indole-3-acetic acid (IAA) and camalexin, with potential roles in plant–bacteria interaction (Kwon *et al.*, 2016).

2.3 Bacterial Signals Regulate Root Morphogenesis

Bacteria may produce either growth-repressing or -promoting molecules for the major traits that determine root system architecture, namely primary root growth, lateral root formation, and root hair development. Moreover, recent reports point to bacterial quorum-sensing signals playing a key role in plant signal transduction.

2.3.1 N-acyl-L-homoserine lactones

Quorum-sensing (QS) regulates several bacterial processes, such as biofilm formation, virulence, production of antimicrobial compounds and also modulation of symbiosis traits between *Rhizobium* and its legume hosts (Mommer *et al.*, 2016). Gram-negative bacteria produce different AHLs as quorum-sensing signals; these compounds contain a conserved homoserine lactone (HL) ring and an amide (N)-linked acyl side chain. The acyl groups of naturally occurring AHLs range from 4 to 18 carbons in length and drive effects on plants that vary in intensity with the length of the acyl group (Camilli and Bassler, 2006; Ortiz-Castro *et al.*, 2008; Schikora *et al.*, 2016).

Application of micromolar concentrations of AHLs to *Arabidopsis* seedlings inhibited primary root growth and stimulated lateral root formation in a dose-dependent manner by modulating cell division and differentiation programmes. Root growth reprogramming and leaf development are regulated by AHLs via hydrogen peroxide and nitric oxide signalling. Several *Arabidopsis* mutants have been identified that define novel mechanisms for perception of these molecules (Morquecho-Contreras *et al.*, 2010; Bai *et al.*, 2012). Noteworthy, *N*-3-oxo-hexanoyl-homoserine lactone (3-oxo-C6-HSL) promotes primary root growth (von Rad *et al.*, 2008; Zhao *et al.*, 2016). Such stimulatory effect was abolished in AtMY*B*44 *Arabidopsis* mutant. In contrast, an enhanced promoting-effect of 3-oxo-C6-HSL was observed in AtMYB44 over-expressing seedlings via regulating the expression of cytokinin- and auxin-related genes (Zhao *et al.*, 2016). These results indicate the critical role of AtMYB44 in connecting phytohormone-related gene expression to perception of QS molecules.

2.3.2 Cyclodipeptides

An increasing number of natural, organic molecules modulate QS and cross-kingdom reactions in evolutionarily distant organisms, notably the cyclodipeptides (CDPs) and 2, 5-diketopiperazines (DKPs). CDPs are cyclized molecules comprising two amino acids linked by peptide bonds, which are produced by a wide range of bacterial species from different environments (Zhang *et al.*, 2007; Ortiz-Castro *et al.*, 2011; Seguin *et al.*, 2011; Abbamondi *et al.*, 2014). CDPs belong to the non-ribosomal peptides that are synthesized by peptide synthases, which use free amino acids, or CDP synthases, which utilize amino-acylated transfer RNAs (aa-tRNAs) as substrates (Bonnefond *et al.*, 2011). The CDPs' role on QS signalling has been demonstrated, since the CDPs' cyclo(D-Ala-L-Val) and cyclo(L-Pro-L-Tyr) inhibit the activity of regulatory LuxR-type proteins important for AHL-dependent QS regulation (Degrassi *et al.*, 2002; Campbell *et al.*, 2009; Galloway *et al.*, 2011).

A major breakthrough in understanding the relevance of chemical signalling in PGPR was the finding that AHLs from *P. aeruginosa* repress the biosynthesis of CDPs. When AHL signalling is down-regulated by mutation of AHL synthases, cyclodipeptide biosynthesis is activated attenuating *P. aeruginosa* virulence. Increasing CDP abundance in the *Arabidopsis* rhizosphere promotes lateral root formation through direct binding to the auxin receptor TIR1. Computational molecular docking analysis revealed CDP affinity to TIR1, via its interaction with amino acids located at the same pocket where natural and synthetic auxins bind (Ortiz-Castro *et al.*, 2011; Ortiz-Castro *et al.*, 2014). Plant growth promotion mediated by CDPs seems to be ubiquitous to horticultural plant species, since the *P. aeruginosa LasI* mutant, which overproduces CDPs, dramatically promotes root branching and root, shoot biomass in tomato (Fig. 2.2). Application of purified CDPs is the next challenge towards improving crop productivity.

2.3.3 Volatile compounds

Bacterial volatiles and aminolipids are emitted by PGPR and are perceived by plants through complex molecular mechanisms (Mathesius *et al.*, 2003; Ortiz-Castro

Fig. 2.2. *Pseudomonas aeruginosa LasI* mutant promotes growth and induces root branching in tomato. Bacterial co-cultivation with tomato seedlings *in vitro* induces root branching. Seedling growth on 0.2x MS medium (A), and co-cultivated with *P. aeruginosa LasI* mutant (B). (C-F) After transfer to pots and grown under greenhouse conditions, plants bacterized with *LasI* mutant show an increased leaf area and improved root development. Images show representative axenic control (left) and inoculated (right) plants.

et al., 2011; Hartmann and Schikora, 2012). In particular, the volatiles 2, 3-butanediol and acetoin were released from *B. subtilis* GB03 and *B. amyloliquefaciens* IN937a. Application of 2,3-butanediol increased plant biomass, whereas bacterial mutants blocked in 2,3-butanediol and acetoin synthesis were devoid in their growth-promotion capacity (Ryu *et al.*, 2003).

The evidence that PGPR strains release different volatile blends and that plant growth is stimulated by differences in volatiles' composition establishes an important function for volatile organic compounds such as signalling molecules mediating plant–microbe interactions. *N,N*-dimethyl amines (DMAs) are amino-containing lipids, from which *N,N*-dimethyl-hexadecylamine (C16-DMA) was identified from volatile blends of the plant-growth-promoting actinobacterium *Arthrobacter agilis* UMCV2 (Velázquez-Becerra *et al.*, 2011). C16-DMA has been found in the VOCs blend produced by different rhizobacteria, including *B. subtilis* G8, *Sinorhizobium meliloti* 1021 and *P. fluorescens* UM270, indicating that C16-DMA emission might be ubiquitous in several microorganisms (Liu *et al.*, 2008; Orozco-Mosqueda *et al.*, 2013; Hernández-León *et al.*, 2015). C16-DMA affects growth and development in evolutionarily distant plant species, such as *Medicago sativa*, *Sorghum bicolor* and *Pinus devoniana*, regulating shoot biomass, stem length, chlorophyll production and root system architecture (RSA) (Velázquez-Becerra *et al.*, 2011; Castulo-Rubio *et al.*, 2015).

A recent report by Raya-González *et al.* (2017), examined the bioactivity of C16-DMA and other related molecules with varied length. C16-DMA inhibited primary root growth, promoted lateral root formation and induced the expression of the jasmonic acid (JA)-responsive gene marker *pLOX2:uidA* in *Arabidopsis* wild-type seedlings. In contrast, JA-related *jar1*, *coi1-1* and *myc2* mutants defective on JA biosynthesis and perception, respectively, are compromised in C16-DMA responses. Comparison of root architectural responses in wild types (WT) and auxin-related mutants *aux1-7, tir1/afb2/afb3*, and *arf7-1/arf19-1* to C16-DMA showed that the effects on root morphogenesis did not involve auxin signalling, but occurs predominantly via jasmonic acid. The current hypothesis is that AHLs, CDPs and volatile compounds could participate in cross-kingdom signalling while they can be directly perceived by plants to adjust functional and adaptive traits.

2.3.4 Virulence factors

Several bacterial species produce secondary metabolites, some of which act as virulence factors to help host colonization. Coronatine is a phytotoxin produced by some plant pathogenic strains of *P. syringae* which exerts its virulence by activating plant JA signalling, also known to repress root growth (Feys *et al.*, 1994; Bender *et al.*, 1999; Raya-González *et al.*, 2012, 2017). The insensitivity of *coi1* mutants of *Arabidopsis* and tomato to the toxin by direct binding assays demonstrated that the JA receptor COI1 is required for the action of the toxin. Intriguingly, coronatine is about a thousand times more active in binding COI1 than the endogenous ligand JA-Ile (Katsir *et al.*, 2008; Yan *et al.*, 2009). These observations, together with the structural similarity of coronatine to JA-Ile, support the notion that this virulence factor acts as a strong molecular mimic of JA-Ile.

P. aeruginosa is most recognized for its importance as a human and plant pathogen. Many studies have revealed extensive conservation in its virulence mechanisms to infect evolutionary divergent hosts. One of these conserved virulence factors is pyocyanin, which participates in the fast killing of *Caenorhabditis elegans*, *Drosophila melanogaster* and *Mus musculus* (Lau *et al.*, 2004). Pyocyanin itself functions as a QS signal, because it accumulates in a cell-density-dependent manner, diffuses freely through membranes, recognized by neighbour cells and triggers a specific transcriptional response (Dietrich *et al.*, 2006).

In contrast to other eukaryotes, treatment of *Arabidopsis* seedlings with pyocyanin did not cause toxic symptoms, but instead repressed primary root growth without affecting meristem viability or causing cell death. These effects correlated with altered accumulation of hydrogen peroxide and superoxide in root tips. Mutant analyses showed that pyocyanin modulation of growth was likely independent of auxin, cytokinin, and abscisic acid, but required ethylene signalling because the *Arabidopsis etr1-1*, *ein2-1*, and *ein3-1* ethylene-related mutants were less sensitive to pyocyanin-induced root growth inhibition and reactive oxygen species (ROS) accumulation (Ortiz-Castro *et al.*, 2014). These findings suggest that pyocyanin is an important factor modulating the interplay between ROS production and root system architecture by an ethylene-dependent mechanism.

2.4 Molecular Responses of Bacteria to Root Exudates

2.4.1 Exudate-induced changes in PGPR gene expression

PGPR associate with roots as a response to the accumulation of root exudates, which changes the transcriptional and phenotypical cell population behaviour. A total of 176 genes showed significantly altered expression in *B. subtilis* okb105 after 2 h co-cultivation with rice seedlings. Among these, 52 were up-regulated, the majority of which are involved in metabolism and transport of nutrients and stress responses, including *araA*, *ywkA*, *yfls*, *mtlA*, and *ydgG*. The 124 genes that were down-regulated included *cheV*, *fliL*, *spmA* and *tua*, possibly involved in chemotaxis, motility, sporulation and teichuronic acid biosynthesis, respectively (Xie *et al.*, 2015).

The gene expression profiles of the plant biostimulant strains *B. amyloliquefaciens* FZB42 and SQR9 and *B. atrophaeus* UCMB-5137 were studied in response to maize root exudates (Fan *et al.*, 2012; Zhang *et al.*, 2015; Mwita *et al.*, 2016). Commonalities and differences arose during the cellular response of these bacteria to the exudates, in which glucose, citric acid, and fumaric acid stimulated biofilm formation via extracellular matrix production and promoted *B. amyloliquefaciens* SQR9 growth increasing the expression of metabolism-associated genes, which are considered critical for root colonization and rhizosphere competence.

In *B. amyloliquefaciens* FZB42 a total of 302 genes representing 8.2 % of the bacterial transcriptome showed significantly altered expression levels in the presence of root exudates, of which 261 genes were up-regulated and 41 genes were down-regulated. Several groups of the induced genes were involved in metabolic pathways relating to nutrient utilization, bacterial chemotaxis, motility and non-ribosomal synthesis of antimicrobial peptides and polyketides (Fan *et al.*, 2012). In addition, numerous groups of genes were involved in rhizosphere adaptation and in interactions with plants such as polysaccharide utilization and plant growth promotion by maize root exudates (Zhang *et al.*, 2015). The gene regulation in *B. atrophaeus* UCMB-5137 in response to the root exudate stimuli differed from *B. amyloliquefaciens* and was more sensitive to the chemical composition of the exudates (Mwita *et al.*, 2016).

2.4.2 Exudates modulate the protein profile

To correlate gene expression with the extracellular proteome maps of *B. amyloliquefaciens* FZB42, Kierul *et al.* (2015) analyzed the changes in bacterial secreted proteins during the late exponential and stationary growth phases by 2D gel electrophoresis. Out of the 121 proteins identified by MALDI-TOF MS, 34 proteins were differentially secreted in response to maize root exudates. These were mainly involved in nutrient utilization and transport. The protein with the highest fold change in the presence of maize root exudates during the late exponential growth phase was acetolactate synthase (AlsS), an enzyme involved in the synthesis of the volatile acetoin, known as an inducer of systemic resistance against plant pathogens and as a trigger of plant growth.

2.5 Conclusion

The bacterial microbiome is integral to plant functioning and root exudates attract PGPR to roots via chemical sensing. The study of *P. aeruginosa* and related QS mutants reveals that many molecules, including AHLs, CDPs, volatiles and virulence factors are important for growth promotion in *Arabidopsis* and crop plants. Sustainable crop productivity depends not only on the availability and application of fertilizers, which are costly and dangerous to the environment, but also on the PGPR associated with roots. Recent experimentation shows the highly promising potential of selected bacterial strains towards improving fertilizer use efficiency (Fig. 2.3). Thus, rhizosphere signalling research will provide new tools to increase crop productivity and reduce the application of agrochemicals.

The potential of single bacterial strains to interfere with plant hormone levels remains one of the major challenges toward better understanding, predicting and possibly controlling plant hormone responses in complex plant-associated bacterial communities. Many PGPRs produce auxins as part of their metabolism. The finding that *Azospirillum* affects the auxin content of roots reveals a critical facet of auxin in mediating plant–microbe interactions. Thus, sensing of rhizobacteria coordinates shoot patterns such as phyllotaxis, branching and stem initiation, which may be dependent on the multiple feedback loops in the auxin machinery (Leyser, 2010).

The rhizosphere provides the conditions to strengthen mutual benefits through complex networks of molecular interactions. Ongoing research exploiting the metabolic methods currently available will be instrumental in unravelling the signalling interactions with plants and bacteria and their response to the environment.

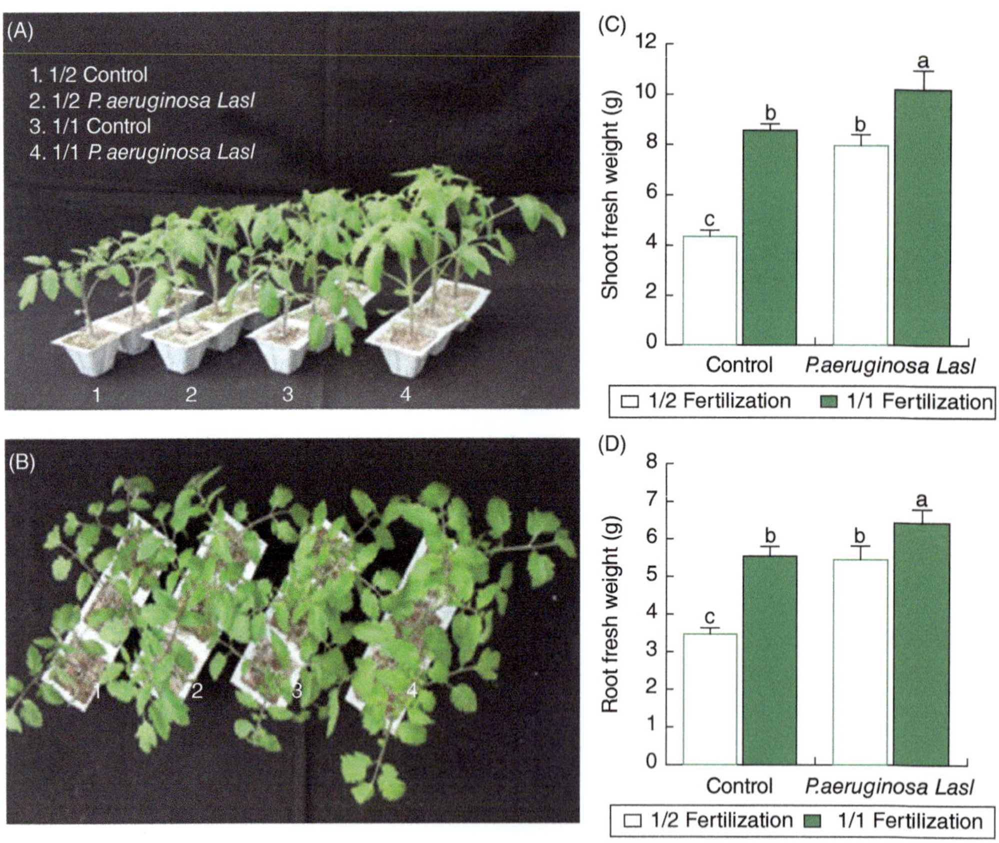

Fig. 2.3. *Pseudomonas aeruginosa LasI* improves fertilizer use efficiency. (A-B) Representative photographs of tomato plants irrigated with half or complete fertilization dosage. Leaf area in plants irrigated with the complete fertilization treatment is similar to that of plants irrigated with half fertilization when bacterized with *LasI* mutants. (C-D) Shoot and root biomass production in tomato plants co-cultivated with *LasI* mutant increases under optimal fertilizer supplementation or at medium fertilizer dosage. Data points indicate the mean ± standard deviation, $n = 30$. Different letters indicated statistical differences ($P < 0.05$).

References

Abbamondi, G.R., de Rosa, S., Iodice, C. and Tommonaro, G. (2014) Cyclic dipeptides produced by marine sponge-associated bacteria as quorum sensing signals. *Natural Products Communications* 9, 229–232.

Bai, X., Todd, C.D., Desikan, R., Yang, Y. and Hu, X. (2012) *N*-3-oxo-decanoyl-L-homoserine-lactone activates auxin-induced adventitious root formation via hydrogen peroxide- and nitric oxide-dependent cyclic GMP signaling in Mung Bean. *Plant Physiology* 158, 725–736.

Badri, D.V., Weir, T.L., van der Lelie, D., and Vivanco, J.M. (2009) Rhizosphere chemical dialogues: plant–microbe interactions. *Current Opinion in Biotechnology* 20, 642–650.

Barazani, O. and Friedman, J. (1999) Is IAA the major growth factor secreted from plant growth mediating bacteria? *Journal of Chemical Ecology* 25, 2397–2406.

Beauregard, P.B., Chai, Y., Vlamakis, H., Losick, R. and Kolter, R. (2013) *Bacillus subtilis* biofilm induction by plant polysaccharides. *Proceedings of the National Academy of Sciences of the United States of America* 110, E1621–E1630.

Bender, C.L., Alarcón-Chaidez, F. and Gross, D.C. (1999) *Pseudomonas syringae* phytotoxins: mode of action, regulation and biosynthesis by peptide and polyketide synthetases. *Microbiology and Molecular Biology Reviews* 63, 266–292.

Bisen, K., Keswani, C., Mishra, S., Saxena, A., Rakshit, A. and Singh, H.B. (2015) Unrealized potential of seed biopriming for versatile agriculture. In: Rakshit, A., Singh, H.B. and Sen, A. (eds) *Nutrient Use Efficiency: from Basics to Advances*. Springer, New Delhi, India, pp. 193–206.

Bisen, K., Keswani, C., Patel, J.S., Sarma, B.K. and Singh, H.B. (2016) *Trichoderma* spp.: efficient inducers of systemic resistance in plants. In: Chaudhary, D.K. and Verma, A. (eds) *Microbial-Mediated Induced Systemic Resistance in Plants*. Springer, Singapore, pp. 185–195.

Blom, D., Fabbri, C., Connor, E.C., Schiestl, F.P., Klauser, D.R. *et al.* (2011) Production of plant growth modulating volatiles is widespread among rhizosphere bacteria and strongly depends on culture conditions. *Environmental Microbiology* 13, 3047–3058.

Boller, T. and Felix, G. (2009) A renaissance of elicitors: perception of microbe-associated molecular patterns and danger signals by pattern-recognition receptors. *Annual Review of Plant Biology* 60, 379–406.

Bonnefond, L., Arai, T., Sakaguchi, Y., Suzuki, T., Ishitani, R. and Nureki, O. (2011) Structural basis for nonribosomal peptide synthesis by an aminoacyl-tRNA synthetase paralog. *Proceedings of the National Academy of Sciences of the United States of America* 108, 3912–3917.

Calvo, P., Nelson, L. and Kloepper, J.W. (2014) Agricultural uses of plant biostimulants. *Plant and Soil* 383, 3–41.

Camilli, A. and Bassler, B.L. (2006) Bacterial small-molecule signaling pathways. *Science* 311, 1113–1116.

Campbell, J., Lin, Q., Geske, G.D. and Blackwell, H.E. (2009) New and unexpected insights into the modulation of LuxR-type quorum sensing by cyclic dipeptides. *ACS Chemical Biology* 4, 1051–1059.

Castulo-Rubio, D.Y., Alejandre-Ramírez, N.A., Orozco-Mosqueda, M.C., Santoyo, G., Macías-Rodríguez, L.I. and Valencia-Cantero, E. (2015) Volatile organic compounds produced by the rhizobacterium *Arthrobacter agilis* UMCV2 modulate *Sorghum bicolor* (strategy II plant) morphogenesis and *SbFRO1* transcription in vitro. *Journal of Plant Growth Regulation* 34, 611–623.

Chen, Y., Cao, S., Chai, Y., Clardy, J., Kolter, R. *et al.* (2012) A *Bacillus subtilis* sensor kinase involved in triggering biofilm formation on the roots of tomato plants. *Molecular Microbiology* 85, 418–430.

Degrassi, G., Aguilar, C., Bosco, M., Zahariev, S., Pongor, S. and Venturi, V. (2002) Plant growth-promoting *Pseudomonas putida* WCS358 produces and secretes four cyclic dipeptides: cross-talk with quorum sensing bacterial sensors. *Current Microbiology* 45, 250–254.

de Weert, S., Ermeiren, H., Mulders, I.H., Kuiper, I., Hendrickx, N. *et al.* (2002) Flagella-driven chemotaxis towards exudate components is an important trait for tomato root colonization by *Pseudomonas fluorescens*. *Molecular Plant Microbe Interactions* 15, 1173–1180.

Dietrich, L.E.P., Price-Whelan, A., Petersen, A., Whiteley, M. and Newman, D.K. (2006) The phenazine pyocyanin is a terminal signaling factor in the quorum sensing network of *Pseudomonas aeruginosa*. *Molecular Microbiology* 61, 1308–1321.

Dimpka, C., Weinand, T. and Asch, F. (2009) Plant–rhizobacteria interactions alleviate abiotic stress conditions. *Plant Cell and Environment* 32, 1682–1694.

Dodd, I.C., Zinovkina, N.Y., Safronova, V.I. and Belimov, A.A. (2010) Rhizobacterial mediation of plant hormone status. *Annals of Applied Biology* 157, 361–379.

Dutta, S., Rani, T.S. and Podile, A.R. (2013) Root exudate-induced alterations in *Bacillus cereus* cell wall contribute to root colonization and plant growth promotion. *PLOS ONE* 8, e78369.

Fan, B., Carvalhais, L.C., Becker, A., Fedoseyenko, D., von Wirén, N. and Borriss R. (2012) Transcriptomic profiling of *Bacillus amyloliquefaciens* FZB42 in response to maize root exudates. *BMC Microbiology* 12, 116.

Feys, B., Benedetti, C.E., Penfold, C.N. and Turner, J.G. (1994) *Arabidopsis* mutants selected for resistance to the phytotoxin coronatine are male sterile, insensitive to methyl jasmonate, and resistant to a bacterial pathogen. *Plant Cell* 6,751–759.

Galloway, W., Hodgkinson, J.T., Bowden, S.D., Welch, M. and Spring, D.R. (2011) Quorum-sensing in Gram-negative bacteria: small-molecule modulation of AHL and AI-2 quorum sensing pathways. *Chemical Reviews* 111, 28–67.

Hartmann, A. and Schikora, A. (2012) Quorum sensing of bacteria and trans-kingdom interactions of *N*-acyl homoserine lactones with eucaryotes. *Journal of Chemical Ecology* 38, 704–713.

Hernández-León, R., Rojas-Solís, D., Contreras-Pérez, M., Orozco-Mosqueda, M.C., Macías-Rodríguez, L.I. *et al.* (2015) Characterization of the antifungal and plant growth-promoting effects of diffusible and volatile organic compounds produced by *Pseudomonas fluorescens* strains. *Biological Control* 81, 83–92.

Katsir, L., Schilmiller, A., Staswick, P.E., Yang-He, S. and Howe, G.A. (2008) COI1 is a critical component of a receptor for jasmonate and the bacterial virulence factor coronatine. *Proceedings of the National Academy of Sciences of the United States of America* 105, 7100–7105.

Kierul, K., Voigt, B., Albrecht, D., Chen, X.H., Carvalhais, L.C. and Borriss, R. (2015) Influence of root exudates on the extracellular proteome of the plant growth-promoting bacterium *Bacillus amyloliquefaciens* FZB42. *Microbiology* 161, 131–147.

Kloepper, J.W., Lifshitz, R. and Zablotowicz, R.M. (1989) Free-living bacterial inocula for enhancing crop productivity. *Trends in Biotechnology* 7, 39–44.

Kurepin, L.V., Zaman, M. and Pharis, R. (2014) Phytohormonal basis for the plant growth promoting action of naturally occurring biostimulators. *Journal of the Science of Food and Agriculture* 94, 1715–1722.

Kwon, Y.S., Lee, D.Y., Rakwal, D., Baek, S., Lee, J.H. *et al.* (2016) Proteomic analyses of the interaction between the plant-growth promoting rhizobacterium *Paenibacillus polymyxa* E681 and *Arabidopsis Thaliana*. *Proteomics* 16,122–135.

Lau, G.W., Hassett, D.J., Ran, H. and Kung, F. (2004) The role of pyocyanin in *Pseudomonas aeruginosa* infection. *Trends in Molecular Medicine* 10, 599–606.

Leyser, O. (2010) The power of auxin in plants. *Plant Physiology* 154, 501–505. DOI: 10.1104/pp.110.161323.

Ling, N., Raza, W., Ma, J.H., Huang, Q.W. and Shen, Q.R. (2011) Identification and role of organic acids in watermelon root exudates for recruiting *Paenibacillus polymyxa* SQR-21 in the rhizosphere. *European Journal of Soil Biology* 47, 374–379.

Liu, W., Wei, M., Bingyu, Z. and Feng, L. (2008) Antifungal activities and components of VOCs produced by *Bacillus subtilis* G8. *Current Research in Bacteriology* 1, 28–34. DOI: 10.3923/crb.2008.28.34.

López-Bucio, J., Campos-Cuevas, J.C., Hernández-Calderón, E., Velásquez-Becerra, C., Farías-Rodríguez, R. *et al.* (2007) *Bacillus megaterium* rhizobacteria promote growth and alter root-system architecture through an auxin- and ethylene-independent signaling mechanism in *Arabidopsis thaliana*. *Molecular Plant-Microbe Interactions* 20, 207–217.

Ludwig-Müller, J. (2015) Bacteria and fungi controlling plant growth by manipulating auxin: balance between development and defense. *Journal of Plant Physiology*, 172, 4–12.

Mathesius, U., Mulders, S., Gao, M., Teplitski, M., Caetano-Anollés, G. *et al.* (2003) Extensive and specific responses of a eukaryote to bacterial quorum-sensing signals. *Proceedings of the National Academy of Sciences of the United States of America* 100, 1444–1449.

Matilla, M.A., Espinosa-Urgel, M., Rodríguez-Herva, J.J., Ramos, J.L. and Ramos-González, M.I. (2007) Genomic analysis reveals the major driving forces of bacterial life in the rhizosphere. *Genome Biology* 8, R179.

Mishra, S., Singh, A., Keswani, C., Saxena, A., Sarma, B.K. and Singh, H.B. (2015) Harnessing plant–microbe interactions for enhanced protection against phytopathogens. In: Arora, N.K. (ed.) *Plant Microbes Symbiosis: Applied Facets*. Springer, New Delhi, India, pp. 111–125.

Moe, L.A. (2013) Amino acids in the rhizosphere: from plants to microbes. *American Journal of Botany* 100, 1692–1705.

Mommer, L., Hinsinger, P., Prigent-Combaret, C. and Visser, E.J.W. (2016) Advances in the rhizosphere: stretching the interface of life. *Plant and Soil* 407, 1–8.

Morquecho-Contreras, A., Méndez-Bravo, A., Pelagio-Flores, R., Raya-González, J., Ortiz-Castro, R. and López-Bucio, J. (2010) Characterization of *drr1*, an alkamide resistant mutant of *Arabidopsis* reveals an important role for small lipid amides in lateral root development and plant senescence. *Plant Physiology* 152, 1659–1673.

Mwita, L., Chan, W.Y., Pretorius, T., Lyantagaye, S.L., Lapa, S.V. *et al.* (2016) Gene expression regulation in the plant growth promoting *Bacillus atrophaeus* UCMB-5137 stimulated by maize root exudates. *Gene* 590, 18–28.

Orozco-Mosqueda, M.C., Macías-Rodríguez, L.I., Santoyo, G., Flores-Cortez, I., Farías-Rodríguez, R. and Valencia-Cantero, E. (2013) *Medicago truncatula* increases its iron-uptake mechanisms in response to volatile organic compounds produced by *Sinorhizobium meliloti*. *Folia Microbiologica* 58, 579–585.

Ortiz-Castro, R., Martínez-Trujillo, M. and López-Bucio, J. (2008) *N*-acyl-L-homoserine lactones: a class of bacterial quorum-sensing signals alter post-embryonic root development in *Arabidopsis thaliana*. *Plant Cell and Environment* 31, 1497–1509.

Ortiz-Castro, R., Contreras-Cornejo, H.A., Macías-Rodríguez, L.I. and López-Bucio, J. (2009) The role of microbial signals in plant growth and development. *Plant Signaling and Behavior* 4, 701–712.

Ortiz-Castro, R., Díaz-Pérez, C., Martínez-Trujillo, M., del Río, R.E., Campos-García, J. and López-Bucio, J. (2011) Transkingdom signaling based on bacterial cyclodipeptides with auxin activity in plants. *Proceedings of the National Academy of Sciences of the United States of America* 108, 7253–7258.

Ortiz-Castro, R., Pelagio-Flores, R., Méndez-Bravo, A., Ruiz-Herrera, L.F., Campos-García, J. and López-Bucio, J. (2014) Pyocyanin, a virulence factor produced by *Pseudomonas aeruginosa*, alters root development through reactive oxygen species and ethylene signaling in *Arabidopsis*. *Molecular Plant–Microbe Interactions* 27, 364–378.

Pieterse, C.M.J., Zamioudis, C., Berendsen, R.L., Weller, D.M., Van Wees, S.C.M. and Bakker, P.A.H.M. (2014) Induced systemic resistance by beneficial microbes. *Annual Review of Phytopathology* 52, 347–375.

Poupin, M.J., Timmerman, T., Vega, A., Zuñiga, A. and González, B. (2013) Effects of the plant growth-promoting bacterium *Burkholderia phytofirmans* PsJN throughout the life cycle of *Arabidopsis thaliana*. *PLOS ONE* 8, e69435.

Rangel de Souza, A.L.S., De Souza, S.A., De Oliveira, M.V.V., Ferraz, T.M., Figueiredo, F.A.M.M.A. *et al.* (2016) Endophytic colonization of *Arabidopsis thaliana* by *Gluconacetobacter diazotrophicus* and its effect on plant growth promotion, plant physiology, and activation of plant defense. *Plant and Soil* 399, 257–270.

Raya-González, J., Pelagio-Flores, R. and López-Bucio, J. (2012) The jasmonate receptor COI1 plays a role in jasmonate-induced lateral root formation and lateral root positioning in *Arabidopsis thaliana*. *Journal of Plant Physiology* 169, 1348–1358.

Raya-González, J., Velázquez-Becerra, C., Barrera-Ortiz, S., López-Bucio, J. and Valencia-Cantero, E. (2017) *N,N*-dimethyl hexadecylamine and related amines regulate root morphogenesis via jasmonic acid signaling in *Arabidopsis thaliana*. *Protoplasma*, 254, 1399. DOI: 10.1007/s00709-016-1031-6.

Rudrappa, T., Czymmek, K.J., Pare, P.W. and Bais, H.P. (2008) Root-secreted malic acid recruits beneficial soil bacteria. *Plant Physiology* 148, 1547–1556.

Ruzzi, M. and Aroca, R. (2015) Plant growth-promoting rhizobacteria act as biostimulants in horticulture. *Scientia Horticulturae* 196, 124–134.

Ryu, C.M., Farag, M.A., Hu, C.H., Reddy, M.S., Wei, H.X. *et al.* (2003) Bacterial volatiles promote growth in *Arabidopsis*. *Proceedings of the National Academy of Sciences of the United States of America* 100, 4927–4932.

Scharf, B.E., Hynes, M.F. and Alexandre, G.M. (2016) Chemotaxis signaling systems in model beneficial plant–bacteria associations. *Plant Molecular Biology* 90, 549–559.

Schikora, A., Schenk, S.T. and Hartmann, A. (2016) Beneficial effects of bacteria–plant communication based on quorum sensing molecules of the *N*-acyl homoserine lactone group. *Plant Molecular Biology* 6, 605–612.

Seguin, J., Moutiez, M., Li, Y., Belin, P., Lecoq, A. *et al.* (2011) Nonribosomal peptide synthesis in animals: the cyclodipeptide synthase of *Nematostella*. *Chemistry and Biology* 18, 1362–1368.

Singh, H.B., Jain, A., Saxena, A., Singh, A., Keswani, C. *et al.* (2014). Deciphering the Pathogenic Behaviour of Phyto-Pathogens Using Molecular Tools. In: Sharma, N. (ed.) *Biological Controls for Preventing Food Deterioration: Strategies for Pre- and Postharvest Management*. John Wiley & Sons, Chichester, UK, pp. 377–408.

Spaepen, S., Bossuyt, S., Engelen, K., Marchal, K. and Vanderleyden, J. (2014) Phenotypical and molecular responses of *Arabidopsis thaliana* roots as a result of inoculation with the auxin-producing bacterium *Azospirillum brasilense*. *New Phytologist* 201, 850–861.

Sukumar, P., Legué, V., Vayssieres, A., Martin, F., Tuskan, G.A. and Kalluri, U.C. (2012) Involving of auxin pathways in modulating root architecture during beneficial plant–microorganism interactions. *Plant, Cell and Environment* 36, 909–919.

Velázquez-Becerra, C., Macías-Rodríguez, L.I., López-Bucio, J., Altamirano-Hernández, J., Flores-Cortez, I. and Valencia-Cantero, E. (2011) A volatile organic compound analysis from *Arthrobacter agilis* identifies dimethylhexadecylamine, an amino-containing lipid modulating bacterial growth and *Medicago sativa* morphogenesis in vitro. *Plant and Soil* 339, 329–340.

Venturi, V. and Keel, C. (2016) Signalling in the rhizosphere. *Trends in Plant Science* 21, 187–198.

Vibhuti, M., Kumar, A., Sheoran, N., Valiya, A., Nadakkakath, S. and Eapen, S.J. (2017) Molecular basis of endophytic *Bacillus megaterium*-induced growth promotion in *Arabidopsis thaliana*: revelation by microarray-based gene expression analysis. *Journal of Plant Growth Regulation* 36, 118. DOI: 10.1007/s00344-016-9624-z.

Vogel, C., Bodenhausen, N., Gruissem, W. and Vorholt, J.A. (2016) The *Arabidopsis* leaf transcriptome reveals distinct but also overlapping responses to colonization by phyllosphere commensals and pathogen infection with impact on plant health. *New Phytologist* 212, 192–207.

von Rad, U., Klein, I., Dobrev, P.I., Kottova, J., Zazimalova, E. *et al.* (2008) Response of *Arabidopsis thaliana* to *N*-hexanoyl-DL-homoserine-lactone, a bacterial quorum sensing molecule produced in the rhizosphere. *Planta* 229, 73–85.

Wintermans, P.C.A., Bakker, P.A.H.M. and Pieterse, C.M.J. (2016) Natural genetic variation in *Arabidopsis* for responsiveness to plant growth-promoting rhizobacteria. *Plant Molecular Biology* 90, 623–634.

Xie, S., Wu, H., Chen, L., Zang, H., Xie, Y. and Gao X. (2015) Transcriptome profiling of *Bacillus subtilis* OKB105 in response to rice seedlings, *BMC Microbiology* 15, 21.

Yan, J., Zhang, C., Gu, M., Bai, Z., Zhang, W. *et al.* (2009). The *Arabidopsis* CORONATINE INSENSITIVE1 protein is a jasmonate receptor. *Plant Cell* 21, 2220–2236.

Yuan, J., Zhang, N., Huang, Q., Raza, W., Li, R. *et al.* (2015) Organic acids from root exudates of banana help root colonization of PGPR strain *Bacillus amyloliquefaciens* NJN-6. *Scientific Reports* 5, 13438.

Zamioudis, C., Mastranesti, P., Dhonukshe, P., Blilou, I., and Pieterse, C.M.J. (2013) Unraveling root developmental programs initiated by beneficial *Pseudomonas* spp. bacteria. *Plant Physiology* 162, 304–318.

Zhang, H., Kim, M., Krishnamachari, V., Payton, P., Sun, Y. *et al.* (2007) Rhizobacterial volatile emissions regulate auxin homeostasis and cell expansion in *Arabidopsis*. *Planta* 226, 839–851.

Zhang, N., Wang, D., Liu, Y., Li, S., Shen, Q. and Zhang, R. (2014) Effects of different plant root exudates and their organic acid components on chemotaxis, biofilm formation and colonization by beneficial rhizosphere-associated bacterial strains. *Plant and Soil* 374, 689–700.

Zhang, N., Yang, D., Wang, D., Miao, Y., Shao, J. *et al.* (2015) Whole transcriptomic analysis of the plant-beneficial rhizobacterium *Bacillus amyloliquefaciens* SQR9 during enhanced biofilm formation regulated by maize root exudates. *BMC Genomics* 16, 685.

Zhao, Q., Li, M., Jia, Z., Liu, F., Ma, H. *et al.* (2016) AtMYB44 positively regulates the enhanced elongation of primary roots induced by *N*-3-oxo-hexanoyl-homoserine lactone in *Arabidopsis thaliana*. *Molecular Plant–Microbe Interactions* DOI: 10.1094/MPMI-03-16-0063-R.

3 Real-time PCR as a Tool towards Understanding Microbial Community Dynamics in Rhizosphere

Gautam Anand, Upma Singh, Abhineet Sain, Virendra S. Bisaria and Shilpi Sharma*
Department of Biochemical Engineering and Biotechnology, Indian Institute of Technology Delhi, New Delhi, India

3.1 Introduction

Soil is a complex amalgam of minerals, organic phase, porous phase and diverse life forms. Soil processes, such as nutrient cycling, are of prime importance for the maintenance of our ecosystem (Keswani *et al.*, 2013; 2016; Bisen *et al.*, 2015; Mishra *et al.*, 2015). Microorganisms play a pivotal role in these soil processes. Changes in the soil microbial community have been linked with varying soil functional capabilities (Torsvik and Øvreås, 2002; Nannipieri *et al.*, 2003; Singh *et al.*, 2014) that are still poorly understood. Despite the meticulous efforts of scientists to unravel the vast expanse of the microbial community in soil, till now only 1–2% of the total microorganisms present in the soil have been cultured in the laboratory (Amann *et al.*, 1995). With recent improvements in media preparation and optimization the limits have only marginally increased (Davis *et al.*, 2005). Therefore, direct extraction and analysis of the microbial community presents an excellent alternative to bypass the method of culturing microorganisms in order to assess their diversity. Directly lysing the cells present in the soil releases molecular markers such as PLFAs (phospholipid fatty acids), PLELs (phospholipid etherlipids), ergosterol and nucleic acid (DNA and RNA) for evaluation of the soil microbial community. Techniques using lipids as biochemical markers have been commonly used (Zelles *et al.*, 1992; Gattinger *et al.*, 2002) but suffer the limitation of not being as specific as nucleic acid markers (Hirsch *et al.*, 2010). With the advent of DNA sequencing nucleic acid databases have provided a solid background for assessment of soil microbial community structure and function. Over the decades DNA-based molecular biology tools have helped a lot in understanding the role of microorganisms in biogeochemical processes, and how the relationship changes with various biotic and abiotic factors. Microbial gene expression data of taxonomic genes, such as 16S rRNA, 23S rRNA, etc., and functional genes responsible for biogeochemical cycles can bridge the gap in our understanding of the underlying mechanisms of various soil processes. Techniques such as Denaturing Gradient Gel Electrophoresis (DGGE), Temperature Gradient Gel Electrophoresis (TGGE), Terminal Restriction Fragment Length Polymorphism (T-RFLP), real-time PCR and Reverse Transcription Quantitative

*E-mail: shilpi@dbeb.iitd.ac.in

© CAB International 2017. *Advances in PGPR Research* (eds H.B. Singh, B.K. Sarma and C. Keswani)

PCR (qRT-PCR), used for such studies rely on PCR amplification for analysis of the microbial community.

Real-time PCR presents an efficient tool to analyse the soil microbial structure and function. Real-time PCR saw its first appearance in the field of microbial ecology in 2000 (Becker *et al.*, 2000; Suzuki *et al.*, 2000; Takai and Horikoshi, 2000). It can assess microbial community structure with the help of genes such as 16S rRNA, 23S rRNA and 16S rRNA-23S rRNA Intergenic Spacer Region (ISR), and microbial community function using functional genes pertaining to important genes in the biogeochemical cycles. This throws light on the potential of the system under study. In qRT-PCR cDNA serves as a template, which provides an image of the actively transcribing population in the microbial community. In comparison with other molecular techniques, real-time PCR has numerous advantages such as high sensitivity, low detection limit, high throughput and rapid data analysis (Saleh-Lakha *et al.*, 2005; Sharma *et al.*, 2007; Saleh-Lakha *et al.*, 2011). Many studies have considered real-time PCR as "the" standard technique to analyse soil microbial community structure using 16S rRNA gene as marker, and to enumerate the abundance of functional genes involved in biogeochemical cycles. Together with the numerous advantages of the technique, it suffers from several challenges, viz. the extraction of nucleic acid from the soil presents an uphill task owing to numerous reasons such as differential treatment for cell lysis for different microorganisms, inhibitors binding to nucleic acids, etc. Apart from the extraction of nucleic acid, setting up of a real-time PCR reaction, assessing its sensitivity and efficiency, and data analysis require moderate levels of skill. The main scope of this chapter is to provide a comprehensive knowledge of real-time PCR as a technique for assessment of microbial community dynamics, including the sequential methodological issues surrounding it. Various case studies that are related to rhizospheric bacterial community will be enumerated to evaluate its potential in studying soil microbial community structure and function.

3.2 Extraction of Metagenomic Nucleic Acid from Environment

A cultivation-independent approach offers a more comprehensive look into the diversity of microorganisms. Molecular biology techniques that require the extraction of nucleic acids from soil, without having the need to culture, have proven to be a comprehensive tool. The extraction of total nucleic acid from soil is a cumbersome process requiring constant standardization and optimization, owing to the vast degree of heterogeneity present in soil. Organic components of soil such as humic acid, fulvic acid, etc. reduce the purity of nucleic acid and also inhibit enzymes in subsequent processing such as PCR and other downstream enzymatic analysis (Tebbe and Vahjen, 1993). Therefore no universal method has been documented that works for all soil types. In this chapter we briefly describe different methods that have been used depending upon the soil types and numerous strategies employed for removing the contaminants that can otherwise reduce the efficiency of real-time PCR.

3.2.1 Cell lysis

There are two approaches for cell lysis: cell extraction and subsequent lysis, and direct lysis. Cell extraction is the extraction of cells from the sample before lysis. The cell extraction method was firstly introduced by Fægri *et al.* (1977) and Torsvik and Goksøyr (1978). Dispersion of soil is usually done by both mechanical and chemical methods: mechanically by using Waring blenders (Fægri *et al.*, 1977), sonication (Ramsay, 1984), etc.; chemical dispersion has been carried out using cation exchange resin (Chelex 100) (Jacobsen and Rasmussen, 1992), detergents such as sodium cholate (McDonald, 1986), and SDS-PEG with PVP (Steffan *et al.*, 1988). Centrifugation methods have also been used for soil dispersion by employing differential centrifugation (Steffan *et al.*, 1988), and a modified sucrose gradient centrifugation (Jacobsen and Rasmussen, 1992). The cell extraction method reported lower recovery of cells, as the cells tightly adhering to the soil surfaces are protected from both mechanical and chemical treatments and thus exhibited bias for bacterial cells (Steffan *et al.*, 1988).

The direct lysis method has comparatively greater yields and involves *in-situ* lysis of the cells with the soil matrix. Direct lysis was first introduced by Ogram *et al.* (1987). Modes of disruption in this method can be mechanical, chemical and enzymatic. Physical disruption methods include freeze thawing, freeze boiling and bead beating methods. Bürgmann *et al.* (2001) showed that the bead beating method gave the highest yield of DNA compared with other methods such as mortar mill grinding, grinding with liquid nitrogen, microwave thermal heating with smaller fragments (Smalla *et al.*, 1993) not more than 20 kb (Robe *et al.*, 2003). Chemical methods include usage of detergents such as SDS for lysis in combination with EDTA and sodium phosphate buffers (Robe *et al.*, 2003). CTAB and PVP are also commonly used as they help in removal of humic acid. However, Zhou *et al.* (1996) showed that PVP resulted in DNA loss. In co-extraction methods PVP has been shown to have higher ability to absorb humic acid without loss of RNA (Mettel *et al.*, 2010; Sharma *et al.*, 2012). Lysozyme and proteinase K have also been used as a part of lysis buffers. It has been reported that increasing the concentration of chelating agents and detergents for stronger lysis resulted in lower DNA purity and shearing (Robe *et al.*, 2003). Thus the choice of lysis mainly depends on the soil type, the soil matrix, the techniques for which the nucleic acid fraction is required, and the probable soil microbial community being targeted.

Studies have shown that extraction efficiency differs depending on the type of dominant microbes (Zhou *et al.*, 1996; Kuske *et al.*, 1998), and also with growth stages (Frostegård *et al.*, 1999). In downstream techniques like metagenomic library construction, the indirect cell extraction method is preferable as the fragment size of extracted nucleic acid is longer, though it shows bias for certain components of the community. For quantitative PCR purposes the direct lysis method is more widely used, as nucleic acid from the majority of the microbial community can be extracted. However, with the direct lysis method there are higher chances of co-extraction of humic acid; this method should be paired with appropriate nucleic acid purification steps before any downstream application can be done (Sharma *et al.*, 2012).

3.2.2 Purification of nucleic acid

Purification of nucleic acid is influenced by the soil organic content (Roose-Amsaleg *et al.*, 2001; Sharma *et al.*, 2007; Saleh-Lakha *et al.*, 2011). For the purification of nucleic acid the two main contaminants are proteins and humic acid. For protein aggregation solvent extraction (Ogram *et al.*, 1987; Smalla *et al.*, 1993) and salting out methods (Selenska and Klingmüller, 1991) have been used. Humic acid, being polyphenolic in nature, is inhibitory as the phenols bind to proteins via hydrogen bonds resulting in altered conformation of the enzymes used in downstream analysis (Kreader, 1996; Saleh-Lakha *et al.*, 2011) and physiochemical properties similar to those of nucleic acid (Roose-Amsaleg *et al.*, 2001). The presence of humic acid can be marked with the brownish colouration of the extracted DNA (Roose-Amsaleg *et al.*, 2001; Robe *et al.*, 2003). Tebbe and Vahjen (1993) showed that the minimum inhibitory concentration (MIC) of humic acid varies with its composition, source and the enzyme used in downstream applications. For further nucleic acid purification caesium chloride density gradient ultracentrifugation (Ogram *et al.*, 1987; Steffan *et al.*, 1988; Tebbe and Vahjen, 1993), chromatography using Sephadex columns (Jackson *et al.*, 1997), electrophoresis using low-melting-point agarose (Harry *et al.*, 1999), and dialysis (Porteous *et al.*, 1997) have been employed. Numerous studies have used multiple combined strategies also (Ogram *et al.*, 1987; Steffan *et al.*, 1988; Smalla *et al.*, 1993). For the precipitation of nucleic acid, polyethylene glycol (PEG) has been shown to reduce the co-extraction of humic acid (Porteous *et al.*, 1997; Cullen and Hirsch, 1998; Roose-Amsaleg *et al.*, 2001; Robe *et al.*, 2003). Thus the purification of nucleic acid from different soil types requires diverse purification strategies or a combination of strategies.

3.2.3 Extraction of RNA from soil

Extraction of RNA, both mRNA and rRNA, has been a wearisome process. Lower RNA stability and the ubiquitous presence of RNases in soil complicate the extraction and purification of RNA in sufficient yields and purity (Ogram *et al.*, 1995). Extraction of RNA from soil has been less documented than the strategies for isolating DNA. Different strategies for isolation of RNA (Moran *et al.*, 1993; Felske *et al.*, 1996; Miskin *et al.*, 1999) and co-extraction of RNA and DNA (Duarte *et al.*, 1998; Griffiths *et al.*, 2000; Hurt *et al.*, 2001; Costa *et al.*, 2004; Peršoh *et al.*, 2008; Mcllroy *et al.*, 2009; Towe *et al.*, 2011; Sharma *et al.*, 2012) have been published in the last two decades. The co-extraction of DNA and RNA removes the bias developed due to the different extraction procedures if gene abundances are linked to transcript rates (Towe *et al.*, 2011). The various studies enlisted have used diverse methods for extraction of RNA along with different strategies for removal of humic acid including centrifugation with PVP and BSA (Felske *et al.*, 1996), and employing G75 columns (Moran *et al.*, 1993). In co-extraction studies reported by Griffiths *et al.* (2000) a direct lysis bead beating method of extraction in CTAB, NaCl and potassium phosphate buffer using PEG for precipitation of nucleic acid was adopted. Peršoh *et al.* (2008) and Fang *et al.* (2014) utilized aluminium sulphate ($Al_2(SO_4)_3$) for flocculation of humic acid prior to lysis. Sharma *et al.* (2012) improved the Griffith's protocol by the addition of PVP along with CTAB-NaCl for lysis, and doubling the duration of bead beating, followed by precipitation of nucleic acid using PEG on melting ice. With increasing numbers of researchers interested not only in gene abundance but also in the transcript analysis using qRT-PCR we would expect to see the advent of newer protocols for RNA extraction, not withstanding that the present protocol has substantial success in isolation of RNA (mRNA and rRNA).

3.3 Real-time PCR

The advent of real-time PCR using DNA and RNA as molecular markers has given researchers an efficient tool to decipher not only the structure of the microbial community but also the functioning of an ecosystem. Real-time PCR and real-time RT-PCR are techniques that provide quantification of abundance of genes and transcripts, respectively. For real-time PCR the different fluorescent chemistries that are most widely used include SYBR green and TaqMan. SYBR dye binds non-specifically to any double-stranded DNA. A prerequisite of employing TaqMan is designing a probe that can be hydrolysed by the 5′ nuclease ability of the DNA polymerase during the extension step which puts a brake on FRET (Fluorescence Resonance Energy Transfer) and thus upon DNA synthesis emits a fluorescence. The probe is labelled with a reporter dye at the 5' end and a quencher at the 3' end and binds to the target DNA in between target sites (Heid *et al.*, 1996). Data acquisition for analysis is done where amplification first detected is higher than the background fluorescence known as the cycle threshold (Wittwer *et al.*, 1997).

3.3.1 q-PCR: Setting up the reaction

The first step when working with real-time PCR is deciding upon the fluorescent chemistry that is to be used. SYBR dye is the most basic dye used for most of the real-time PCR studies. The SYBR dye method is said to be non-specific as it binds to dsDNA. The method is economically efficient as only designing of primers is a prerequisite. The TaqMan method is specific and requires designing of an additional specific probe, which can be expensive. Other fluorescent dyes using advanced probes are also available that can be utilized (Murray *et al.*, 2014). As most studies have utilized SYBR and/or TaqMan for real-time PCR we shall limit our discussion to these two methods in this chapter.

3.3.2 Primer designing

After deciding the type of fluorescent chemistry, primer designing is the next important step. For SYBR-based chemistry 70-200 bp is an ideal amplicon length. The primers must be so designed that they do not have any secondary structures, the ideal T_m (melting temperature) condition lies between 50 and 60°C, and GC content in the range 50–60 % (Bustin *et al.*, 2009). Also, in case of primers 3' end complementarity should be checked to avoid any primer dimers. The specificity of a primer pair and all the conditions mentioned above can be tested using any of the available online tools. Designing real-time PCR primers for assessment of microbial diversity using a conserved gene such as 16S rRNA can be relatively easier because of the huge repertoire of taxonomic data available; this will be dealt with later in the chapter. Designing primers for group-specific study, and functional gene analysis in a microbial community can be a challenging task owing to the vast diversity of microorganisms. Degenerate primer sets, as mostly designed, have a lot of non-specificity, therefore designing multiple primer pairs may be warranted for one successful primer pair. Sometimes for certain groups or for particular genes, the variation in the genome can be overwhelming enough to make designing a primer pair next to impossible. For TaqMan-based chemistry an additional task is to design a specific probe. Aspects that need to be kept in mind while designing a probe are: (1) the probe's T_m should be 5–10°C higher than that of the primers, (2) its length should not be more than 30 nucleotides, (3) there should be no G at the 5' end of the probe as it would quench the fluorescent signal, (4) GC content of the target should be between 30–80 %, and (5) the choice of reporter and quencher. FAM-labelled probes are the most commonly used.

3.3.3 Optimizing real-time PCR conditions

Optimizing real-time PCR for soil samples takes a lot of effort. Several precautions have to be kept in mind to ensure that the real-time PCR gives accurate results. The major problem is contamination with organic matter such as humic acid, as it binds to the enzymes and destabilizes them. To overcome the problem of inhibition post nucleic acid extraction it is advisable to dilute the template (Sharma *et al.*, 2007). Polymerase stabilizing substances such as BSA and T4 protein can also be added, however in quantitative studies such additions interfere with the results as they themselves bind to the DNA or cDNA (Poussier *et al.*, 2002; Jiang *et al.*, 2005). C_q is the cycle number at which the SYBR Green or TaqMan probe-bound fluorescence of amplicons can be detected, and is a means to quantify the original template copy number. Miniscule amounts of inhibitor leads to erroneously low estimates of template copy number as it delays the C_q of each sample. The next step is to optimize the annealing conditions, which is done using melt curve analysis in real-time PCR. In an optimized reaction a single peak should show up in melt curve analysis which signifies that a single specific product has been amplified. The melt curve analysis can also give us the differences in the GC content of a specific gene which might result in multiple or blunted peaks (Sharma *et al.*, 2007).

3.3.4 Standards for quantification, calibration curve generation and normalization

For preparations of standards, the (normal) PCR amplified product can be cloned into a plasmid. The quality and quantity of a plasmid can be determined spectrophotometrically. Its copy number can be calculated by adding plasmid length and the amplicon length, and the quantity of the observed plasmid. Once the theoretical copy number is calculated, a calibration curve is generated by serial dilutions of the plasmid (10^1–10^9). The experimental copy numbers are calculated using a linear regression curve of the acquired C_q values. The C_q values are inversely proportional to the amount of target nucleic acid in the sample, i.e. the lower the C_q level the greater the amount of target nucleic acid in the sample. The copy number is determined by the real-time PCR experimentally, and the theoretical copy number yields the PCR amplification efficiency. Real-time PCR efficiency can be calculated from the calibration curve slope as follows:

$$E = 10^{(-1/\text{slope})}-1$$

For an efficiency of 100% the slope is –3.32. The coefficient of correlation (R^2) obtained from the standard curve should be >0.99. Soil spiking is done when there are chances of erroneous results because of inhibitors. In such a case spiking a foreign DNA or a reference plasmid that is not usually present in that particular environment is employed. Standard curves are generated with and without extracted nucleic acid. Sequentially the copy number in the presence and the absence of the foreign DNA, in presence and absence of extracted metagenomic DNA/RNA gives us the percentage inhibition. Data normalization can be done by absolute methods by comparison with a standard curve of diluted template. Relative methods involve the use of an internal standard or housekeeping gene. However there is no such natural internal standard that can be used in case of the rhizosphere microbial community. Many studies have used samples spiked and non-spiked with the target sequence to normalize the real-time PCR and qRT-PCR data (Daniell *et al.*, 2012). The relative quantification is normalized against a unit mass and normally expressed as gene copy number per gram dry soil.

3.4 Microbial Gene Abundance and Expression Studies in Rhizosphere Biology

Microbial community studies using real-time PCR have gained momentum in the last decade, following the work of Hermansson and Lindgren (2001) using 16S rRNA specific primers for quantification of ammonium-oxidizing bacteria. Since then there have been numerous studies quantifying gene abundances to assess the diversity of soil microbial communities in particular environments and soil types. 16S rRNA emerged at the forefront, being used as a universal marker to assess the structure of a microbial community. Owing to its conserved nature, 16S rRNA gene has now become the most comprehensive tool for bacterial identification, supported by a vast sequencing database. Various studies have harnessed the power of real-time PCR and have been successful in quantifying specific genera or groups and also specific genes. The sections that follow are an attempt to collate primer pairs used for quantification of different markers employed in the rhizosphere, together with enumerating studies addressing an array of ecological questions by using the technique of real-time PCR and RT-PCR.

3.4.1 16S rRNA as a molecular chronometer for total bacteria

The first marker to be employed for quantitative bacterial community analysis was the gene for 16S rRNA. Several questions, including attaining a better understanding of the ecology of the bacterial community in a particular region or in the rhizosphere of economically important crops, and comparing the rhizospheres of plants in different geographic location and different climates, etc., have been addressed with this marker. One of the shortcomings of working with 16S rRNA gene is its redundancy in its operon. The number of the 16S rRNA gene varies from 1 to 15 in many organisms (Klappenbach *et al.*, 2001). Researchers have indicated that some bacterial species cannot be identified and characterized solely on the basis of 16S rRNA gene; in such cases the 23S rRNA gene and the 16S rRNA-23S rRNA Intergenic Spacer Region (ISR) can also be used for confirmation (Liu *et al.*, 2012). There are numerous universal primer pairs that have been used by various studies (Table 3.1). The most widely used primer pair for 16S rRNA gene as a universal marker for total bacterial community quantification is 338F/518R

Table 3.1. Primer pairs used for total bacterial, and taxon-specific quantification.

Name of Primer	Primer Sequence	Target region in 16S rRNA	Amplicon size (bp)	PCR Chemistry / Probe	Reference
BACT 1369F- PROK 1492R	Forward: CGGTGAATACGTTCYCGG Reverse: GGWTACCTTGTTACGACTT	V8-V9	123	Taqman Probe: 1369F CTTGTACACACCGCCCGTC	Suzuki *et al.* (2000)
338F-518R	Forward: CCT ACG GGA GGC AGC AG Reverse: ATT ACC GCG GCT GCT GG	V3	180	SYBR	Muyzer *et al.* (1993)
968 F- 1401 R	Forward: CA CGG GGG GAA CGC GAA GAA CCT TAC Reverse: CCG TGT GTA CAA GAC CC	V6-V8	423	SYBR	Nubel *et al.* (1996)
1055-1070 F 1392-1406 R	Forward: ATGGCTGTCGTCAGCT Reverse: ACGGGCGGTGTGTAC	V6-V8	342	SYBR	Ferris *et al.* (1996)
341 F 534 R	Forward: CCT ACG GGA GGC AGC AG Reverse: ATT CCG CGG CTG GCA	V3	174	SYBR	López-Gutiérrez *et al.* (2004)
Bac349F Bac806R	Forward: AGGCAGCAGTDRGGAAT Reverse: GGACTACYVGGGTATCTAAT	V3-V4	406	Taqman Probe: Bac516F TGCCAGCAGCCGCGGTA ATACRDAG	Takai and Horikoshi (2000)
Taxon/Group Specific Primer					
α-Proteobacteria Eub338 (F) Alf685 (R)	Forward ACTCCTACGGGAGGCAGCAG Reverse TCTACGRATTTCACCYCTAC	V3–V4	365	SYBR	Fierer and Jackson (2005)
β-Proteobacteria Eub338 (F) Bet680 (R)	Forward ACTCCTACGGGAGGCAGCAG Reverse TCACTGCTACACGYG	V3–V4	360	SYBR	Fierer and Jackson (2005)
Actinobacteria Actino235 (F) Eub518 (R)	Forward CGCGGCCTATCAGCTTGTTG Reverse ATTACCGCGGCTGCTGG	V2–V4	300	SYBR	Fierer and Jackson (2005)
Firmicutes Lgc353 (F)	Forward GCAGTAGGGAATCTTCCG Reverse ATTACCGCGGCTGCTGG	V3	180	SYBR	Fierer and Jackson (2005)
Eub518 (R) Firm350f Firm814r	Forward GCCAGCAGTRGGGAATCTTC Reverse ACACYTAGYACTCATCGTTT	V3–V4	464	SYBR	Muhling *et al.* (2008)
Bacteroidetes Cfb319 (F)	Forward GTACTGAGACACGGACCA Reverse ATTACCGCGGCTGCTGG	V3	220	SYBR	Fierer and Jackson (2005)
Eub518 (R) CFB555f (F) CFB968r (R)	Forward CCGGAWTYATTGGGTTTAAAGGG Reverse GGTAAGGTTCCTCGCGTA	V4–V5	413	SYBR	Muhling *et al.* (2008)

Continued

Table 3.1. Continued.

Name of Primer	Primer Sequence	Target region in 16S rRNA	Amplicon size (bp)	PCR Chemistry / Probe	Reference
Acidobacteria Acid31 (F) Eub518 (R)	Forward GATCCTGGCTCAGAATC Reverse ATTACCGCGGCTGCTGG	V1–V3	500	SYBR	Fierer and Jackson (2005)
Holophagae (Acidobacteria) Acg8f (F) Acg8r (R)	Forward TGGGATGTTGATGGTGAAAC Reverse AGTCTCGGATGCAGTTCCTG		470	SYBR	Da Rocha *et al.* (2010)
Pseudomonas PsF (F)	Forward: GGTCTGAGAGGATGATCAGT- Reverse: TTAGCTCCACCTCGCGGC		~1000	SYBR	Drigo *et al.* (2009)
PsR (R) Pse435F (F) Pse686R (R)	Forward ACTTTAAGTTGGGAGGAAGGG Reverse ACACAGGAAATTCCACCACCC	V3–V4	251	Taqman Pse449 Fam-ACAGAATAAG CACCGGCTAACBHQ	Bergmark *et al.* (2012)
Burkholderia Burk3 (F) BurkR (R)	Forward: CTGCGAAAGCCGGAT Reverse: TGCCATACTCTAGCYYGC3		460	SYBR	Drigo *et al.* (2009)
Bacillus BacF (F) 1378 (R)	Forward: GGGAAACCGGGGCTAA TACCGGAT Reverse: CGGTGTGTACAAGGCCC GGGAACG		1300	SYBR	Drigo *et al.* (2009)
Luteolibacter VS1Af (F) VS1Ar (R)	Forward CAGCTCGTGTCGTGAGATGT Reverse TCTCGGTTCTCATTGTGCTG		199	SYBR	Da Rocha *et al.* (2010)
renarchaeota 771F (F) 957R (R)	Forward ACG GTGAGGGATGAAAGCT Reverse CGGCGTTGACTCCAATTG	V5	220	SYBR	Ochsenreiter *et al.* (2003)
Verrucomicrobia Verr349 Eub518	Forward GYGGCASCAGKCGMGAAW Reverse ATTACCGCGGCTGCTGG	V3	~ 169	SYBR	Philippot *et al.* (2009)

(Muyzer *et al.*, 1993) targeting the V3–V4 region of the 16S rRNA gene.

Recently there has been much focus on the burning problem of bioremediation of contaminated soils. Real-time PCR has served as an efficient way to monitor the soil microbial community in such systems. The essentiality of the rhizospheric soil community in the attenuation of organic matter contamination was shown by Kaplan *et al.* (2016). The detrimental effect of heavy metals in the soil rhizospheric community has been reported by employing real-time PCR quantification using 16S rRNA gene (bacteria) and 18S rRNA gene (fungi and other eukaryotes) (Deng *et al.*, 2015). The phytoremediation of mercury-contaminated soil by rooted macrophyte (*Elodea nuttallii*) altered the soil microbial community and created a microenvironment that enhanced mercury methylation. This was noted by monitoring the rhizospheric microbial community by quantification of 16S rRNA gene and *dsrA* gene (Regier *et al.*, 2012). Exploring the microbial community of re-vegetated mine tailing dumps by quantification of 16S rRNA gene and partial segment of 16S rRNA gene specific for *Pseudomonas* led to the conclusion that environmental filtering occurs by activity of trees' roots rather than soil characteristics (Zappelini *et al.*, 2015).

Agricultural practices impact the soil microbial community, which in turn leads to beneficial or detrimental effect on the crop turnover. Vega-Avila *et al.* (2015) compared the rhizospheric community of *Vitis vinifera* L. cultivated under distinct agricultural practices using 16S rRNA and *nifH* gene quantification, and complementing it with group-specific and whole community DGGE along with high-throughput sequencing. Effects of intercropping and rhizobial inoculation on the ammonia oxidizing microorganisms in rhizospheres of maize and faba bean plants were assessed by using TaqMan probes for 16S rRNA (total Bacteria), total archaea and *amoA* gene (Zhang *et al.*, 2014). Land-use changes and agricultural management of soybean in Amazon forest soils was studied by Navarrete *et al.* (2013) wherein abundance and composition of the acidobacterial community and total bacteria was assessed using 16S rRNA primers specific for acidobacterial community and universal bacterial primers, respectively, together with pyrosequencing. Seasonal variations constrain the soil bacterial community as shown by Taketani *et al.* (2016) who reported lower phylogenetic diversity over a period of dryness, and how the constraint is removed with the onset of rain. Consecutive monoculture is a negative agricultural practice that has a detrimental effect on the soil microbial community as shown by Zhou *et al.* (2015). They reported that in the case of a continuous cycle of monocropped cucumber (*Cucumis sativus L.*) system the soil microbial community is altered with fewer beneficial microorganisms and more pathogenic microorganisms. Similar findings were advocated by other groups (Zhou *et al.*, 2014). A study elucidating the rhizosphere microbial community underlying *Rhizoctonia* suppressive soil (Avon, South Australia) aimed to investigate how this community may develop agricultural soils conducive to disease (Donn *et al.*, 2014). Inoculations of bioinoculants for improved grain yield and soil fertility has been widely practised. Gupta *et al.* (2015) compared the effects of microbial consortium (*Bacillus megaterium, Pseudomonas fluorescens* and *Trichoderma harzianum*) inoculation in comparison to the chemical fertilizers in the rhizosphere of *Cajanus cajan* (pigeon pea). The consortium was reported to perform better than the chemical fertilizer with no adverse effects on soil microbial community. Real-time PCR was used for quantification of total bacteria and genes involved in nitrogen cycle using both DNA and RNA as markers to elucidate non-target effects in this study.

Numerous studies have also correlated the soil microbial community with different plant growth stages. Zhang *et al.* (2016) studied the dynamics of eubacterial, fungal and actinomycetes populations in the rhizosphere of the *Bt* cotton at different growth stages under field conditions. The effect on the bacterial community structure was assessed in the tuber rhizosphere of field-grown sweet potato plants with different plant ages and genotypes (Marques *et al.*, 2014). Dynamics of microbial community structure and function associated with rhizosphere over periods of rice growth were illustrated by Hussain *et al.* (2012).

3.4.2 Quantification of specific microbial taxa

The resolution of the technique of real-time PCR is further improved with information regarding abundance of specific bacterial taxa/genera. This information can be synthesized by quantification at the taxon level using appropriate primers targeting the group. A literature survey of qPCR studies revealed a database of several primers used for group-specific quantification (Table 3.1). The use of taxon-specific qPCR has found importance in various studies that have evaluated the effect of environmental factors on rhizosphere microbial community, e.g. assessment of the effect of certain chemicals/bio-inoculants, the changes brought about by bioremediation, or the effect of certain climatic conditions.

A series of studies by Da Rocha *et al.* (2010; 2013) had designed and tested novel group-specific primers for *Holophagae* (Acidobacteria), *Luteolibacter/Prosthecobacter* and unclassified Verrucomicrobiaceae subdivision 1, and later employed them to determine the distribution of these genera in different regions of soil and rhizosphere of leek (*Allium porrum*) plants. The numbers of *Holophagae* were maximum for the outer rhizosphere, followed by bulk soil, and were minimum for the inner rhizosphere. In contrast, *Luteolibacter/Prosthecobacter* decreased as distance from plant root increased (highest for inner rhizosphere, lowest for bulk soil). For the case of unclassified Verrucomicrobiaceae subdivision 1, the numbers in the rhizosphere were greater than those in the bulk soil. The latter study went a step further and designed (and tested) primers specific for different classes within Acidobacteria and Verrucomicrobia subdivision 1 for analysing shifts in numbers, caused by different plants (grass, potato and leek) and seasons, within these genera.

Gupta *et al.* (2014) studied the non-target effects of a consortium of bio-inoculants on the rhizospheric microbial community of *Cajanus cajan*. Taxon-specific real-time PCR assay was conducted to target Actinomycetes and β-Proteobacteria for evaluation of the population shifts occurring in the rhizosphere community. In the case of triple inoculation with *B. megaterium*, *Pseudomonas fluorescens* and *T. harzianum*, β-proteobacteria reduced in number to the levels in unplanted soil at maturity stage. The abundance of phosphate-solubilizing *Bacillus* sp., *Pseudomonas* sp. and fungal population was found to increase at the maturity stage of the plant in case of triple inoculation.

Studies on soil have focused not just on widely used pesticides and bio-inoculants but also on certain specific compounds that are applied in soil management strategies. One such compound is chitin, which has been known to increase the suppression of pathogens in soil. Its significance for the group Actinobacteria has been established in a series of studies by Cretoiu *et al.* (2013) and Kielak *et al.* (2013), wherein the effectiveness of chitin amendment in suppressing soil pathogens was reported. While the abundance of Actinobacteria did not vary significantly in control samples, over a period of 3 years their numbers were positively correlated with the quantity of chitin for chitin-amended soil (increase in abundance with chitin treatment).

Moving on to studies that have analysed the effect of other environmental factors, an extensive study by Drigo *et al.* (2009) examined (using existing and newly designed primer pairs) the effect of elevated CO_2 levels on the rhizosphere community of different plant systems. An abundance of *Pseudomonas*, *Burkholderia* and *Bacillus* were found to be resistant to elevated CO_2 levels in the rhizosphere of *Carex arenaria* plants. However, *Burkholderia* was affected when the plant system was *Festuca rubra*.

3.4.3 Functional genes as markers

Till now we have dealt with assessment of the structure of the microbial community in rhizospheres employing the technique of real-time PCR. To address questions related to the functionality of the system it is important to gain insight into specific processes by targeting functional markers. Table 3.2 lists different representative primers pairs that

Table 3.2. Primer pairs used for real-time PCR quantification of different functional genes.

Gene	Function	Name of Primer	Primer Sequence	Amplicon Size (bp)	PCR Chemistry / Probe	Reference
nifH	Nitrogenase reductase part of the nitrogenase complex	FPGH19 – PolR	Forward: TACGGCAA(GA)GGTGG(TCGA) AT(TCA)G Reverse: ATSGCCATCATYTCRC CGGA	~400 430 342 321	SYBR	Simonet *et al.* (1991) Rösch *et al.* (2002) Poly *et al.* (2001) Poly *et al.* (2001)
		nifH-F- nifH-R	Forward: AAAGGYGGWATCG GYAARTCCACCAC Reverse: TTGTTSGCSGCR TACATSGCCATCAT			
		PolF-PolR	Forward: TGCGAYCCSAARG CBGACTC Reverse: ATSGCCATCATYTCR CCGGA			
		PolF -AQER	Forward: TGCGAYCCSAARGC BGACTC Reverse: GACGATGTAGATYTCCTG			
narG	Membrane-bound nitrate reductase	narG 1960m2f- narG 2050m2r	Forward: TA(CT)GT(GC)GGGCAGGA(AG) AAACTG Reverse: CGTAGAAGAAGCTG-GTGCTGTT	90 173	SYBR	López *et al.* (2004) Bru *et al.* (2007)
		narGF- narGR	Forward: TCGCCSATYCCGGC SATGTC Reverse: GAGTTGTACCAGTCR GCSGAYTCSG			
nirK	Cu-containing nitrite reductase	nirK876-nirK1040	Forward: ATYGGCGGVAYGGCGA Reverse: GCCTCGATCAGRT TRTGGTT	164	SYBR	Henry *et al.* (2004)
nifA	Nitrogen-sensitive protein NifA	AznifAF12-AznifAF19	Forward: CGCAGCAACTGATAT GCAAAA Reverse: GCGTGCTTCCGTGAC AAGT	439	SYBR	Faleiro *et al.* (2013)

Continued

Table 3.2. Continued.

Gene	Function	Name of Primer	Primer Sequence	Amplicon Size (bp)	PCR Chemistry / Probe	Reference
napA	Periplasmic nitrate reductase	V17m- narA4R	Forward: TGGACVATGGGYTTYAAYC Reverse: ACYTCRCGHGCVG TRCCRCA	152	SYBR	Bru *et al.* (2007)
nirS	Cytochrome cd1 containing nitrite reductase	nirS4QF- nirS6QR	Forward: AACGYSAAGGARACSGG Reverse: GASTTCGGRT GSGTCTTSAYGAA		SYBR	Kandeler *et al.* (2006)
nosZ	Nitrous oxide reductase	nosZ1840F- nosZ2090R	Forward: CGCRACGGCAA SAAGGTSMSSGT Reverse: CAKRTGCAKSGCRT GGCAGAA	250	SYBR	Henry *et al.* (2004)
Nitrobacter *nxrA*	Nitrite Oxidoreductase	F1norA –R1norA	Forward: CAGACCGACGTG TGCGAAAG Reverse: TCYACAAGGAACGG AAGGTC	322	SYBR	Poly *et al.* (2008)
Anammox bacteria (*AMX*)	Anaerobic ammonia oxidation	Pla46f-AMX820r	Forward: GACTTGCATGCCTAATCC Reverse: AAAACCCCTCTA CTTAGTGCCC	774	SYBR	Zhang *et al.* (2007)
Archaea (*amoA*) Bacterial *amoA*	Ammonia-oxidizing gene	Arch-amoAF-AR amoA19F-	Forward: TAATGGTCTGGCT TAGACG Reverse: GCGGCCATCCATCTG TATGT	635 624 340	SYBR	Francis *et al.* (2005) Leininger *et al.* (2006) Holmes *et al.* (1995) Rotthauwe *et al.* (1997)
		CrenamoA616r48x	Forward: ATGGTCTGGCTWAGACG Reverse: GCCATCCABCKRTANG TCCA			
		A189- amoA-2R′	Forward: GAASGCNGAGAAGAASGC Reverse: CCCCTCKGSAAAGCCT TCTTC			
		amoA-1F - amoA-2R	Forward: GGGGTTTCTACTGGTGGT Reverse: CCCCTCKGSAAAGCC TTCTTC			

phoD	Alkaline phosphatase gene	ALPS-F730 and ALPS-1101	Forward: CAGTGGGACGACCACGAGGT Reverse: GAGGCCGATCGGCATGTCG	371	SYBR	Sakurai *et al.* (2008)
b-propeller phytase (*BPP*)	Myo-inositol hexakisphosphate-phosphohydrolases and acidphosphatase	MQHf - MQHr	Forward: TTCCTATCCTACCGGGAAGC Reverse: TGCTTTGTAATGTGCCGTTT	158	SYBR	Jorquera *et al.* (2013)
Organophosphorus hydrolase (*opd*)	Opd gene encoding OPH in indigenous plasmids of bacteria	OPD-forward - OPD-reverse	Forward: CACACTGACTCACGAG Reverse: GGCCAATAAACTGACG		SYBR	Kwak *et al.* (2013)
dsrA	Sulfite reductase gene	DSR3FRExt-DSR1Fext	Forward: GGAACGGCTGCTACGCAGTCATTCGGGCAG Reverse: CGCTATTCAGACCTGCCGGAAGAATTTCCT	440	SYBR	Tang *et al.* (2004)
soxA and *rdsrAB*	S-oxidation pathway genes		Various primer sets	Different sizes	SYBR	Thomas *et al.* (2014)
apr *npr*	Alkaline metallo-peptidase gene Neutral-metallopeptidases gene	Faprl-Raprll Fnprl-Rnprll	Forward: TAYGGBTTCAAYTCCAAYAC Reverse: VGCGATSGAMACRTTRCC Forward: GTDGAYGCHCAYTAYTAYGC Reverse:ACMGCATGBGTYADYTCATG	194 233	SYBR	Bach *et al.* (2001)
pmoA	Subunit of particulate methane monooxygenase	A189 -mb661	Forward: GGNGACTGGGACTTCTGG Reverse: GGTAARGACGTTGCNCCGG	432	SYBR	Kolb *et al.* (2003)
chiA	Chitinase enzyme	GA1F- GA1R	Forward: GTCGACATCGACTGGGARTDBCC Reverse: ACGCCGGTCCAGCCNCKNCCRTA		SYBR	Williamson *et al.* (2000)

have been employed to characterize functional aspects of microbial communities in the rhizosphere.

One of the most widely used functional markers has been *nifH* encoding for nitrogenase reductase. Various real-time PCR studies have utilized different primer pairs for quantifying abundance of the *nifH* gene in the microbial communities in the rhizosphere. Abundance of *nifH* gene was studied using real-time PCR (Primer pair: FPGH19-PolR) in the rhizosphere of vines cultivated in San Juan under different agricultural practices (Vega-Avila *et al.*, 2015). Nitrogen is the regulating factor for the productivity of arid terrestrial ecosystems. Ecological restoration processes recovered the damaged ecosystem, and an increase in *nifH* gene copies in qRT-PCR (Primer pair: FPGH19-PolR) was observed along with increments in diazotroph diversity (Lopez-Lozano *et al.*, 2016). Similar work conducted in mine tailings (Nelson *et al.*, 2015) demonstrated that *amoA* and *nifH* genes can be employed as *in situ* indicators of biological soil responses to phytoremediation using real-time PCR. Soil management has also been shown to specifically affect abundance of *nifH* gene. Shu *et al.* (2012) reported organically managed soil to have higher nitrogen-fixing bacterial diversity, microbial activity and biomass compared with conventionally managed soil. Application of a microbial consortium comprising three bio-inoculants was shown to exert positive non-target impact on rhizospheric N-cycling microbial communities of *Cajanus cajan* (Gupta *et al.*, 2012). This was observed at both DNA and mRNA levels. The study also included quantification of ammonia oxidation gene (*amoA*) and genes involved in denitrification (*narG, napA, nirK, nirS, nosZ*) with respective gene-specific primers. Adverse non-target impacts of chemical and biological pesticides on the rhizosphere of *Cajanus cajan* were evaluated by Singh *et al.* (2015), employing cultivation-dependent methodologies as well as real-time PCR to quantify the gene abundance and transcript copy number for nitrogen cycling genes (*nifH, amoA, nirK, nirS, narG*). Mårtensson *et al.* (2009) studied the diurnal variation in the diazotrophic community structure along with *nifH* gene expression and nitrogenase activity using qRT-PCR, where no significant diurnal variation was observed in total *nifH* expression; however the *nifH* expressing diazotrophic community showed high diurnal variations. In nitrogen starving condition, NifA protein gets activated and then activates the rest of the *nif* genes. Because of the importance of the *nifA* gene, real-time PCR using AznifAF12-AznifAF19 primer set was performed by Faleiro *et al.* (2013) to quantify *nifA* in *Azospirillum brasilense* in maize seedlings.

Dynamics of microbial communities involved in denitrification and nitrification has also been the focus of various studies, as they are considered to be sensitive markers for environmental fluctuations. Hussain *et al.* (2011) demonstrated that the prevalence of AOB (ammonia-oxidizing bacteria) and denitrifying bacterial communities in the rhizosphere of rice plants was highly dynamic compared to AOA (ammonia-oxidizing archaea). Intercropping of legumes and inoculation of *Rhizobium leguminosarum* caused various dynamic shifts in the rhizobial community which aided in decreasing nitrification and also improved the plant nitrate uptake efficiency. Amendment like sulfadiazine-contaminated pig manure was shown to impair abundance of ammonia-oxidizing (*amoA* gene of AOA, AOB) and *Nitrobacter*-like nitrite oxidizing (*nxrA* gene) microorganisms in the rhizosphere of pasture plants using qRT-PCR (Ollivier *et al.*, 2013).

In soil phosphorus is available as mineral phosphorus or organic phosphorus, but is insoluble and inaccessible to plants. It is often one of the critical elements deciding the growth of plants. Markers employed to target the P cycle are: alkaline phosphatase (*phoD*), phytase (*phy*), organophosphorus hydrolase (*opd*), exopolyphosphatase (*ppx*), pyrroloquinoline quinone biosynthesis gene (*pqqC*) and gene for C-P lyase. Anthropogenic changes in soil management influence the nutrient cycle and also affect the physical, chemical and biological properties. Fraser *et al.* (2015) reported this influence on bacterial diversity using RT-PCR of *phoD* gene (ALPS-F730/ALPS-1101 primer pair), which was further validated with phosphatase activity assay. Intercropping has been reported to enhance abundance and diversity of phosphobacteria in the rhizosphere of phosphorus-rich soil as demonstrated by a study done in China using real-time PCR to quantify *phoD* gene (Wu *et al.*, 2016). Terrestrial extreme environments have a wide range of salinity, temperatures and pH and low nutrient availabilities. Acuña *et al.* (2016) shed light upon the negative correlation of total bacterial diversity, APase (alkaline phosphomonoesterase) harbouring bacterial diversity and APase activity with phosphorus availability in the rhizosphere of plants grown in Chilean extreme environments using16S rRNA gene (total bacteria), *phoD* and *phoX* gene (APase abundance) quantifications by real-time PCR. Phytate is a major component of organic P forms in soil and microorganisms make it accessible for plants. Plant growth promoting bacteria possess the phytase gene, *phyA*, for this function (Behera *et al.*, 2014). Phytate-mineralizing rhizobacteria also perform an important role in this process. Jorquera *et al.* (2013) used *Bacillus* ß-propeller phytase gene (*BPP*) as a molecular marker to evaluate the role of phytate mineralization in the rhizosphere, and employed real-time PCR for quantification of the induced changes in abundance and expression of the *BPP* gene after addition of phytate in rhizospheric soil.

Other than characterization of microbial communities involved in the N and P cycle, functional markers have been applied for other functions in the rhizosphere. Transfer of sulphur from organic form (like protein amino acids, etc.) to inorganic form (sulphur, sulphates, sulphite, thiosulphate, etc.) in soil is metabolized with the help of soil biota and makes it accessible to plants. Thomas *et al.* (2014) employed the markers *soxB* and *rdsrAB* genes (genes of the S-oxidation pathway) to demonstrate that small-scale heterogeneities of the rhizosphere alter the abundance and potential activity of the S-oxidizers in salt marsh sediments colonized in the rhizosphere of plant *Spartina alterniflora*. This was done using multiple primer pair sets. Upon targeting the chiA gene (GA1F/ GA1R primer) in real-time PCR quantification, the abundance of chitin-degrading microbial communities was assessed in terrestrial and aquatic habitats (Cretoiu *et al.*, 2012). Ma *et al.* (2013) quantified the *pmoA* gene (A189/ mb661 primers) and revealed the abundance of methanotrophs in rice fields.

3.5 Conclusion

Real-time PCR has emerged as an efficient tool in recent decades, and has consolidated its importance in the field of microbial ecology. Recent studies have unanimously used real-time PCR for studying microbial structure and function. Real-time PCR studies have mainly focused on DNA as a molecular marker, but steadily there are studies coming up that employ transcriptomic data using qRT-PCR. Complementing real-time data with fingerprinting and high-throughput sequencing techniques has deepened the understanding of microbial ecology. Real-time PCR has now become a benchmark in the field of rhizosphere biology and with increasing number of taxonomic studies and bioinformatics tools it will aid convergence on improvements in accuracy and efficiency.

Acknowledgement

The authors wish to thank the Science and Engineering Research Board for funding the study (Grant No.YSS/2015/001437). US wishes to acknowledge the scholarship awarded by UGC, India to support her doctoral work.

References

Acuña, J.J., Duran, P., Lagos, M.L., Ogram, A., Luz-Mora, M. and Jorquera, M.A. (2016) Bacterial alkaline phosphatase in the rhizosphere of plants grown in Chilean extreme environments. *Biology and Fertility of Soils* 52, 763–773.

Amann, R.I., Ludwig, W. and Schleifer, K. (1995) Phylogenetic identification and *in situ* detection of individual microbial cells without cultivation. *Microbiological Reviews* 59, 143–169.

Bach, H.J., Hartmann, A., Schloter, M. and Munch, J.C. (2001) PCR primers and functional probes for amplification and detection of bacterial genes for extracellular peptidases in single strains and in soil. *Journal of Microbiological Methods* 44, 173–182.

Becker, S., Böger, P., Oehlmann, R. and Bo, P. (2000) PCR bias in ecological analysis: a case study for quantitative Taq nuclease assays in analyses of microbial communities. *Applied and Environmental Microbiology* 66, 4945–4953.

Behera, B.C., Singdevsachan, S.K., Mishra, R.R., Dutta, S.K. and Thatoi, H.N. (2014) Diversity, mechanism and biotechnology of phosphate solubilising microorganism in mangrove – a review. *Biocatalysis and Agricultural Biotechnology* 3, 97–110.

Bergmark, L., Poulsen, P.H.B., Al-Soud, W.A., Norman, A., Hansen, L.H. and Sørensen, S.J. (2012) Assessment of the specificity of *Burkholderia* and *Pseudomonas* qPCR assays for detection of these genera in soil using 454 pyrosequencing. *FEMS Microbiology Letters* 333, 77–84.

Bisen, K., Keswani, C., Mishra, S., Saxena, A., Rakshit, A. and Singh, H.B. (2015) Unrealized potential of seed biopriming for versatile agriculture. In: Rakshit, A., Singh, H.B., and Sen, A. (eds) *Nutrient Use Efficiency: From Basics to Advances*. Springer, New Delhi, India, pp. 193–206.

Bru, D., Sarr, A. and Philippot, L. (2007) Relative abundances of proteobacterial membrane-bound and periplasmic nitrate reductases in selected environments. *Applied and Environmental Microbiology* 73, 5971–5974.

Bürgmann, H., Pesaro, M., Widmer, F. and Zeyer, J. (2001) A strategy for optimizing quality and quantity of DNA extracted from soil. *Journal of Microbiological Methods* 45, 7–20.

Bustin, S.A., Benes, V., Garson, J.A., Hellemans, J., Huggett, J. *et al.* (2009) The MIQE guidelines: minimum information for publication of quantitative real-time PCR experiments. *Clinical Chemistry* 55, 611–622.

Costa, R., Gomes, N.C.M., Milling, A. and Smalla, K. (2004) An optimized protocol for simultaneous extraction of DNA and RNA from soils. *Brazilian Journal of Microbiology* 35, 230–234.

Cretoiu, M.S., Kielak, A.M., Al-Soud, W.A., Sørensen, S.J., van Elsas, J.D. (2012) Mining of unexplored habitats for novel chitinases—chiA as a helper gene proxy in metagenomics. *Applied Microbiology and Biotechnology* 94, 1347–1358.

Cretoiu, M.S., Korthals, G.W., Visser, J.H.M. and Van Elsas, J.D. (2013) Chitin amendment increases soil suppressiveness toward plant pathogens and modulates the actinobacterial and oxalobacteraceal communities in an experimental agricultural field. *Applied and Environmental Microbiology* 79, 5291–5301.

Cullen, D.W. and Hirsch, P.R. (1998) Simple and rapid method for direct extraction of microbial DNA from soil for PCR. *Soil Biology and Biochemistry* 30, 983–993.

Daniell, T.J., Davidson, J., Alexander, C.J., Caul, S. and Roberts, D.M. (2012) Improved real-time PCR estimation of gene copy number in soil extracts using an artificial reference. *Journal of Microbiological Methods* 91, 38–44.

Da Rocha, U.N., Plugge, C.M., George, I., Van Elsas, J.D. and Van Overbeek, L.S. (2010) Real-time PCR detection of *Holophagae* (*Acidobacteria*) and *Verrucomicrobia* subdivision 1 groups in bulk and leek (*Allium porrum*) rhizosphere soils. *Journal of Microbiological Methods* 83, 141–148.

Da Rocha, U.N., van Elsas, J.D. and van Overbeek, L.S. (2013) The rhizosphere selects for particular groups of Acidobacteria and Verrucomicrobia. *PLoS ONE* 8, 16–20.

Davis, K.E.R., Joseph, S.J., Peter, H. and Janssen, P.H. (2005) Effects of growth medium, inoculum size, and incubation time on culturability and isolation of soil bacteria. *Applied and Environmental Microbiology* 71, 826–834.

Deng, L., Zeng, G., Fan, C., Lu, L., Chen, X. *et al.* (2015) Response of rhizosphere microbial community structure and diversity to heavy metal co-pollution in arable soil. *Applied Microbiology and Biotechnology* 99, 8259–8269.

Donn, S., Almario, J., Muller, D., Moënne-Loccoz, Y., Gupta, V.V.S.R. *et al.* (2014) Rhizosphere microbial communities associated with *Rhizoctonia* damage at the field and disease patch scale. *Applied Soil Ecology* 78, 37–47.

Drigo, B., van Veen, J.A. and Kowalchuk, G.A. (2009) Specific rhizosphere bacterial and fungal groups respond differently to elevated atmospheric CO_2. *The ISME Journal* 3, 1204–1217.

Duarte, G.F., Rosado, A.S., Seldin, L., Keijzer-Wolters, A.C. and Van Elsas, J.D. (1998) Extraction of ribosomal RNA and genomic DNA from soil for studying the diversity of the indigenous bacterial community. *Journal of Microbiological Methods* 32, 21–29.

Fægri, A., Torsvik, V.L. and Goksøyr, J. (1977) Bacterial and fungal activities in soil: separation of bacteria and fungi by a rapid fractionated centrifugation technique. *Soil Biology and Biochemistry* 9, 105–112.

Faleiro, A.C., Pereira, T.P., Espindula, E., Brod, F.C.A. and Arisi, A.C.M. (2013) Real time PCR detection targeting nifA gene of plant growth promoting bacteria *Azospirillum brasilense* strain FP2 in maize roots. *Symbiosis* 61, 125–133.

Fang, C., Xu, T., Ye, C., Huang, L., Wang, Q. and Lin, W. (2014) Method for RNA extraction and cDNA library construction from microbes in crop rhizosphere soil. *World Journal of Microbiology and Biotechnology* 30, 783–789.

Felske, A., Engelen, B., Nübel, U. and Backhaus, H. (1996) Direct ribosome isolation from soil to extract bacterial rRNA for community analysis. *Applied and Environmental Microbiology* 62, 4162–4167.

Ferris, M.J., Muyzer, G. and Ward, D.M. (1996) Denaturing gradient gel electrophoresis profiles of 16S rRNA-defined populations inhabiting a hot spring microbial mat community. *Applied and Environmental Microbiology* 62, 340–346.

Fierer, N. and Jackson, J. (2005) Assessment of soil microbial community structure by use of taxon-specific quantitative PCR assays. *Applied and Environmental Microbiology* 71, 4117–4120.

Francis, C.A., Roberts, K.J., Beman, J.M., Santoro, A.E. and Oakley, B.B. (2005) Ubiquity and diversity of ammonia-oxidizing archaea in water columns and sediments of the ocean. *Proceedings of the National Academy of Sciences of the United States of America (PNAS)* 102, 14683–14688.

Fraser, T., Lynch, D. and Entz, M. (2015) Linking alkaline phosphatase activity with bacterial phoD gene abundance in soil from a long-term management trial. *Geoderma* 257–258, 115–122.

Frostegård, Å.S.A., Courtois, S., Ramisse, V., Clerc, S., Bernillon, D. *et al.* (1999) Quantification of bias related to the extraction of DNA directly from soils. *Applied and Environmental Microbiology* 65, 5409–5420.

Gattinger, A., Ruser, R., Schloter, M. and Munch, J.C. (2002) Microbial community structure varies in different soil zones in a potato field. *Journal of Plant Nutrition and Soil Science* 165, 421–428.

Griffiths, R.I., Whiteley, A.S., Anthony, G., Donnell, O., Bailey, M.J. and Donnell, A.G.O. (2000) Rapid method for coextraction of DNA and RNA from natural environments for analysis of ribosomal DNA- and rRNA-based microbial community composition. *Applied and Environmental Microbiology* 66, 5488–5491.

Gupta, R., Bru, D., Bisaria, V.S., Philippot, L. and Sharma, S. (2012) Responses of *Cajanus cajan* and rhizospheric N-cycling communities to bioinoculants. *Plant Soil* 358, 143–154.

Gupta, R., Mathimaran, N., Wiemken, A., Boller, T., Bisaria, V.S. and Sharma, S. (2014) Non-target effects of bioinoculants on rhizospheric microbial communities of *Cajanus cajan*. *Applied Soil Ecology* 76, 26–33.

Gupta, R., Bisaria, V.S. and Sharma, S. (2015) Effect of agricultural amendments on *Cajanus cajan* (pigeon pea) and its rhizospheric microbial communities - a comparison between chemical fertilizers and bioinoculants. *PLoS ONE* 10, 1–17.

Harry, M., Gambier, B., Bourezgui, Y. and Garnier-Sillam, E. (1999) Evaluation of purification procedures for DNA extracted from rich organic samples: interference with humic substances. *Analusis* 27, 439–441.

Heid, C.A., Stevens, J., Livak, K.J. and Williams, P.M. (1996) Real time quantitative PCR. *Genome Research* 6, 986–994.

Henry, S., Baudoin, E., López-Gutiérrez, J.C., Martin-Laurent, F., Brauman, A. and Philippot, L. (2004) Quantification of denitrifying bacteria in soils by nirK gene targeted real-time PCR. *Journal of Microbiological Methods* 59, 327–335.

Hermansson, A. and Lindgren, P.E. (2001) Quantification of ammonia-oxidizing bacteria in arable soil by real-time PCR. *Applied and Environmental Microbiology* 67, 972–976.

Hirsch, P.R., Mauchline, T.H. and Clark, I.M. (2010) Culture-independent molecular techniques for soil microbial ecology. *Soil Biology and Biochemistry* 42, 878–887.

Holmes, A.J., Costello, A., Lidstrom, M.E. and Murrell, J.C. (1995) Evidence that particulate methane monooxygenase and ammonia monooxygenase may be evolutionarily related. *FEMS Microbiology Letters* 132, 203–208.

Hurt, R.A., Qiu, X., Wu, L., Roh, Y., Palumbo, A.V. *et al.* (2001) Simultaneous recovery of RNA and DNA from soils and sediments. *Applied and Environmental Microbiology* 67, 4495–4503.

Hussain, Q., Liu, Y., Jin, Z., Zhang, A., Pan, G. *et al.* (2011) Temporal dynamics of ammonia oxidizer (*amoA*) and denitrifier (*nirK*) communities in the rhizosphere of a rice ecosystem from Tai Lake region, China. *Applied Soil Ecology* 48, 210–218.

Hussain, Q., Pan, G.X., Liu, Y.Z., Zhang, A., Li, L.Q. *et al.* (2012) Microbial community dynamics and function associated with rhizosphere over periods of rice growth. *Plant Soil and Environment* 58, 55–61.

Jackson, C.R., Harper, J.P., Willoughby, D., Roden, E.E. and Churchill, P.F. (1997) A simple, efficient method for the separation of humic substances and DNA from environmental samples. *Applied and Environmental Microbiology* 63, 4993–4995.

Jacobsen, C.S. and Rasmussen, O.F. (1992) Development and application of a new method to extract bacterial DNA from soil based on separation of bacteria from soil with cation-exchange resin. *Applied and Environmental Microbiology* 58, 2458–2462.

Jiang, J., Alderisio, K.A., Singh, A. and Xiao, L. (2005) Development of procedures for direct extraction of *Cryptosporidium* DNA from water concentrates and for relief of PCR inhibitors. *Applied and Environmental Microbiology* 71, 1135–1141.

Jorquera, M.A., Saavedra, N., Maruyama, F., Richardson, A.E., Crowley, D.E. *et al.* (2013) Phytate addition to soil induces changes in the abundance and expression of *Bacillus* propeller phytase genes in the rhizosphere. *FEMS Microbiology Ecology* 83, 352–360.

Kandeler, E., Deiglmayr, K., Tscherko, D., Bru, D. and Philippot, L. (2006) Abundance of narG, nirS, nirK, and nosZ genes of denitrifying bacteria during primary successions of a glacier foreland. *Applied and Environmental Microbiology* 72, 5957–5962.

Kaplan, D.I., Xu, C., Huang, S., Lin, Y., Tolić, N. *et al.* (2016) Unique organic matter and microbial properties in the rhizosphere of a wetland soil. *Environmental Science and Technology* 50, 4169–4177.

Keswani, C., Singh, S.P. and Singh, H.B. (2013) A superstar in biocontrol enterprise: *Trichoderma* spp. *Biotech Today* 3, 27–30.

Keswani, C., Bisen, K., Singh, V., Sarma, B.K. and Singh, H.B. (2016) Formulation technology of biocontrol agents: present status and future prospects. In: Arora, N.K., Mehnaz, S. and Balestrini, R. (eds) *Bioformulations: For Sustainable Agriculture*. Springer, New Delhi, India, pp. 35–52.

Kielak, A.M., Cretoiu, M.S., Semenov, A.V., Sørensen, S.J. and Van Elsas, J.D. (2013) Bacterial chitinolytic communities respond to chitin and pH alteration in soil. *Applied and Environmental Microbiology* 79, 263–272.

Klappenbach, J.A., Saxman, P.R., Cole, J.R. and Schmidt, T.M. (2001) rrndb: the ribosomal RNA operon copy number database. *Nucleic Acids Research* 29, 181–184.

Kolb, S., Knief, C., Stubner, S. and Conrad, R. (2003) Quantitative detection of methanotrophs in soil by novel pmoA-targeted real-time PCR assays. *Applied and Environmental Microbiology* 69, 2423–2429.

Kreader, C.A. (1996) Relief of amplification inhibition in PCR with bovine serum albumin or T4 gene 32 protein. *Applied and Environmental Microbiology* 62, 1102–1106.

Kuske, C., Banton, K., Adorada, D., Stark, P., Hill, K. and Jackson, P. (1998) Small-scale DNA sample preparation method for field PCR detection of microbial cells and spores in soil. *Applied and Environmental Microbiology* 64, 2463–2472.

Kwak, Y., Rhee, I.K. and Shin, J.H. (2013) Expression pattern of recombinant organophosphorus hydrolase from *Flavobacterium* sp. ATCC 27551 in *Escherichia coli*. *Applied Microbiology and Biotechnology* 97, 8097–8105.

Leininger, S., Urich, T., Schloter, M., Schwark, L., Qi, J. *et al.* (2006) Archaea predominate among ammonia-oxidizing prokaryotes in soils. *Nature* 442, 806–809.

Liu, W., Li, L., Khan, M.A. and Zhu, F. (2012) Popular molecular markers in bacteria. *Molecular Genetics, Microbiology and Virology* 27, 14–17.

López-Gutiérrez, J.C., Henry, S., Hallet, S., Martin-Laurent, F., Catroux, G. and Philippot, L. (2004) Quantification of a novel group of nitrate-reducing bacteria in the environment by real-time PCR. *Journal of Microbiological Methods* 57, 399–407.

Lopez-Lozano, N.E., Carcaño-Montiel, M.G. and Bashan, Y. (2016) Using native trees and cacti to improve soil potential nitrogen fixation during long-term restoration of arid lands. *Plant Soil* 403, 317–329.

Ma, K., Conrad, R. and Lu, Y. (2013) Dry/wet cycles change the activity and population dynamics of methanotrophs in rice field soil. *Applied Environmental Microbiology* 79, 4932–4939.

Marques, J.M., da Silva, T.F., Vollu, R.E., Blank, A.F., Ding, G.C. *et al.* (2014) Plant age and genotype affect the bacterial community composition in the tuber rhizosphere of field-grown sweet potato plants. *FEMS Microbiology Ecology* 88, 424–435.

Mårtensson, L., Díez, B., Wartiainen, I., Zheng, W., El-Shehawy, R. and Rasmussen, U. (2009) Diazotrophic diversity, nifH gene expression and nitrogenase activity in a rice paddy field in Fujian, China. *Plant Soil* 325, 207–218.

McDonald, R.M. (1986) Sampling soil microfloras: dispersion of soil by ion exchange and extraction of specific microorganism from suspension by elutriation. *Soil Biology and Biochemistry* 18, 399–406.

McIlroy, S.J., Porter, K., Seviour, R.J. and Tillett, D. (2009) Extracting nucleic acids from activated sludge which reflect community population diversity. *Antonie van Leeuwenhoek, International Journal of General and Molecular Microbiology* 96, 593–605.

Mettel, C., Kim, Y., Shrestha, P.M. and Liesack, W. (2010) Extraction of mRNA from soil. *Applied and Environmental Microbiology* 76, 5995–6000.

Mishra, S., Singh, A., Keswani, C., Saxena, A., Sarma, B.K. and Singh, H.B. (2015) Harnessing plant-microbe interactions for enhanced protection against phytopathogens. In: Arora, N.K. (ed.) *Plant Microbes Symbiosis: Applied Facets*. Springer, New Delhi, India, pp. 111–125.

Miskin, I.P., Farrimond, P. and Head, I.M. (1999) Identification of novel bacterial lineages as active members of microbial populations in a freshwater sediment using a rapid RNA extraction procedure and RT-PCR. *Microbiology* 145, 1977–1987.

Moran, M.A., Torsvik, V.L., Torsvik, T. and Hodson, R.E. (1993) Direct extraction and purification of rRNA for ecological studies. *Applied and Environmental Microbiology* 59, 915–918.

Muhling, M., Woolven-Allen, J., Murrell, J.C. and Joint, I. (2008) Improved group-specific PCR primers for denaturing gradient gel electrophoresis analysis of the genetic diversity of complex microbial communities. *The ISME Journal* 2, 379–392.

Murray, J.L., Hu, P. and Shafer, D.A. (2014) Seven novel probe systems for real-time PCR provide absolute single-base discrimination, higher signaling and generic components. *The Journal of Molecular Diagnostics* 16, 627–638.

Muyzer, G., Waal, E.C.D.E. and Uitierlinden, A.G. (1993) Profiling of complex microbial populations by denaturing gradient gel electrophoresis analysis of polymerase chain reaction-amplified genes coding for 16S rRNA. *Applied and Environmental Microbiology* 59, 695–700.

Nannipieri, P., Ascher, J., Ceccherini, M.T., Landi, L., Pietramellar, G. and Renell, G. (2003) Microbial diversity and soil functions. *European Journal of Soil Science* 54, 655.

Navarrete, A.A., Kuramae, E.E., de Hollander, M., Pijl, A.S., van Veen, J.A. and Tsai, S.M. (2013) Acidobacterial community responses to agricultural management of soybean in Amazon forest soils. *FEMS Microbiology Ecology* 83, 607–621.

Nelson, K.N., Neilson, J.W., Root, R.A., Chorover, J. and Maier, R.M. (2015) Abundance and activity of 16S rRNA, amoA and nifH bacterial genes during assisted phytostabilization of mine tailings. *International Journal of Phytoremediation* 17, 493–502.

Nubel, U., Engelen, B., Felske, A., Jiri, S., Weishuber, A. *et al.* (1996) Sequence heterogeneities of genes encoding 16S rRNAs in *Paenibacillus polymyxa* detected by temperature gradient gel electrophoresis. *Journal of Bacteriology* 178, 5636–5643.

Ochsenreiter, T., Selezi, D., Quaiser, A., Bonch-Osmolovskaya, L. and Schleper, C. (2003) Diversity and abundance of Crenarchaeota in terrestrial habitats studied by 16S RNA surveys and real time PCR. *Environmental Microbiology* 5, 787–797.

Ogram, A., Sayler, G.S. and Barkay, T. (1987) The extraction and purification of microbial DNA from sediments. *Journal of Microbiological Methods* 7, 57–66.

Ogram, A., Sun, W. and Brockman, F.J. (1995) Isolation and characterization of RNA from low biomass deep subsurface sediments. *Applied and Environmental Microbiology* 61, 763–768.

Ollivier, J., Schacht, D., Kindler, R., Groeneweg, J., Engel, M. *et al.* (2013) Effects of repeated application of sulfadiazine-contaminated pig manure on the abundance and diversity of ammonia and nitrite oxidizers in the root–rhizosphere complex of pasture plants under field conditions. *Frontiers in Microbiology* 4, 1–14.

Peršoh, D., Theuerl, S., Buscot, F. and Rambold, G. (2008) Towards a universally adaptable method for quantitative extraction of high-purity nucleic acids from soil. *Journal of Microbiological Methods* 75, 19–24.

Philippot, L., Bru, D., Saby, N.P.A., Čuhel, J., Arrouays, D. *et al.* (2009) Spatial patterns of bacterial taxa in nature reflect ecological traits of deep branches of the 16S rRNA bacterial tree. *Environmental Microbiology* 11, 3096–3104.

Poly, F., Monrozier, L.J. and Bally, R. (2001) Improvement in the RFLP procedure for studying the diversity of *nifH* genes in communities of nitrogen fixers in soil. *Research in Microbiology* 152, 95–103.

Poly, F., Wertz, S., Brothier, E. and Degrange, V. (2008) First exploration of *Nitrobacter* diversity in soils by a PCR cloning-sequencing approach targeting functional gene nxrA. *FEMS Microbiology Ecology* 63, 132–140.

Porteous, L.A., Seidler, R.J. and Watrud, L.S. (1997) An improved method for purifying DNA from soil for polymerase chain reaction amplification and molecular ecology applications. *Molecular Ecology* 6, 787–791.

Poussier, S., Cheron, J.J., Couteau, A. and Luisetti, J. (2002) Evaluation of procedures for reliable PCR dectection of *Ralstonia solanacearum* in common natural substrates. *Journal of Microbiological Methods* 51, 349–359.

Ramsay, A.J. (1984) Extraction of bacteria from soil: efficiency of shaking or ultrasonication as indicated by direct counts and autoradiography. *Soil Biology and Biochemistry* 16, 457–481.

Regier, N., Frey, B., Converse, B., Roden, E., Grosse-Honebrink, A. *et al.* (2012) Effect of *Elodea nuttallii* roots on bacterial communities and MMHg proportion in a Hg polluted sediment. *PLoS ONE* 7, 1–8.

Robe, P., Nalin, R., Capellano, C., Vogel, T.M. and Simonet, P. (2003) Extraction of DNA from soil. *European Journal of Soil Biology* 39, 183–190.

Roose-Amsaleg, C.L., Garnier-Sillam, E. and Harry, M. (2001) Extraction and purification of microbial DNA from soil and sediment samples. *Applied Soil Ecology* 18, 47–60.

Rösch, C., Mergel, A. and Bothe, H. (2002) Biodiversity of denitrifying and dinitrogen-fixing bacteria in an acid forest soil. *Applied Environmental Microbiology* 68, 3818–3829.

Rösch, C. and Bothe, H. (2009) Diversity of total, nitrogen-fixing and denitrifying bacteria in an acid forest soil. *European Journal of Soil Science* 60, 883–894.

Rotthauwe, J.H., Witzel, K.P. and Liesack, W. (1997) The ammonia monooxygenase structural gene *amoA* as a functional marker: molecular fine-scale analysis of natural ammonia-oxidizing populations. *Applied and Environmental Microbiology* 63, 4704–4712.

Sakurai, M., Wasaki, J., Tomizawa, Y., Shinano, T. and Osaki, M. (2008) Analysis of bacterial communities on alkaline phosphatase genes in soil supplied with organic matter. *Soil Science and Plant Nutrition* 54, 62–71.

Saleh-Lakha, S., Miller, M., Campbell, R.G., Schneider, K., Elahimanesh, P. *et al.* (2005) Microbial gene expression in soil: methods, applications and challenges. *Journal of Microbiological Methods* 63, 1–19.

Saleh-Lakha, S., Shannon, K.E., Goyer, C. and Trevors, J.T. (2011) Challenges in quantifying microbial gene expression in soil using quantitative reverse transcription real-time PCR. *Journal of Microbiological Methods* 85, 239–243.

Selenska, S. and Klingmüller, W. (1991) Direct detection of nif-gene sequences of *Enterobacter agglomerans* in soil. *FEMS Microbiology Letters* 80, 243–246.

Sharma, S., Radl, V., Hai, B., Kloos, K., Mrkonjic Fuka, M. *et al.* (2007) Quantification of functional genes from procaryotes in soil by PCR. *Journal of Microbiological Methods* 68, 445–452.

Sharma, S., Mehta, R., Gupta, R. and Schloter, M. (2012) Improved protocol for the extraction of bacterial mRNA from soils. *Journal of Microbiological Methods* 91, 62–64.

Simonet, P., Grosjean, M., Misra, A.K., Nazaret, S., Cournoyer, B. and Normand, P. (1991) *Frankia* genus-specific characterization by polymerase chain reaction. *Applied and Environmental Microbiology* 57, 3278–3286.

Shu, W., Pablo, G.P., Jun, Y. and Dangfeng, H. (2012) Abundance and diversity of nitrogen-fixing bacteria in rhizosphere and bulk paddy soil under different duration of organic management. *World Journal of Microbiological Biotechnology* 28, 493–503.

Singh, H.B., Jain, A., Saxena, A., Singh, A., Keswani, C. *et al.* (2014) Deciphering the pathogenic behaviour of phyto-pathogens using molecular tools. In: Sharma, N. (ed.) *Biological Controls for Preventing Food Deterioration: Strategies for Pre-and Postharvest Management.* John Wiley & Sons, Chichester, UK, pp. 377–408.

Singh, S., Gupta, R., Kumari, M. and Sharma, S. (2015) Non target effects of chemical pesticides and biological pesticide on the rhizospheric microbial community structure and function in *Vigna radiata*. *Environmental Science and Pollution Research* 22, 11290–11300.

Smalla, K., Cresswell, N., Mendonca Hagler, L.C., Wolters, A. and Van Elsas, J.D. (1993) Rapid DNA extraction protocol from soil for polymerase chain reaction mediated amplification. *Journal of Applied Bacteriology* 74, 78–85.

Steffan, R.J., Goksøyr, J., Bej, A.K. and Atlas, R.M. (1988) Recovery of DNA from soils and sediments. *Applied and Environmental Microbiology* 54, 2908–2915.

Suzuki, M.T., Taylor, L.T. and DeLong, E.F. (2000) Quantitative analysis of small-subunit rRNA genes in mixed microbial populations via 5′-nuclease assays. *Applied and Environmental Microbiology* 66, 4605–4614.

Takai, K. and Horikoshi, K. (2000) Rapid detection and quantification of members of the archaeal community by quantitative PCR using fluorogenic probes. *Applied and Environmental Microbiology* 66, 5066–5072.

Taketani, R.G., Lanconi, M.D., Kavamura, V.N., Durrer, A., Andreote, F.D. and Melo, I.S. (2016) Dry season constrains bacterial phylogenetic diversity in a semi-arid rhizosphere system. *Microbial Ecology* 72, 1–9.

Tang, Y., Shigematsu, T., Ikbal, Morimura, S. and Kida, K. (2004) The effects of micro-aeration on the phylogenetic diversity of microorganisms in a thermophilic anaerobic municipal solid-waste digester. *Water Research* 38, 2537–2550.

Tebbe, C.C. and Vahjen, W. (1993) Interference of humic acids and DNA extracted directly from soil in detection and transformation of recombinant DNA from bacteria and a yeast. *Applied and Environmental Microbiology* 59, 2657–2665.

Thomas, F., Giblin, A.E., Cardon, Z.G. and Sievert, S.M. (2014) Rhizosphere heterogeneity shapes abundance and activity of sulfur-oxidizing bacteria in vegetated salt marsh sediments. *Frontier Microbiology* 5, 1–14.

Torsvik, V. and Øvreås, L. (2002) Microbial diversity and function in soil: From genes to ecosystems. *Current Opinion in Microbiology* 5, 240–245.

Torsvik, V.L. and Goksøyr, J. (1978) Determination of bacterial DNA in soil. *Soil Biology and Biochemistry* 10, 7–12.

Towe, S., Wallisch, S., Bannert, A., Fischer, D., Hai, B. *et al.* (2011) Improved protocol for the simultaneous extraction and column-based separation of DNA and RNA from different soils. *Journal of Microbiological Methods* 84, 406–412.

Vega-Avila, A.D., Gumiere, T., Andrade, P.A.M., Lima-Perim, J.E., Durrer, A. *et al.* (2015) Bacterial communities in the rhizosphere of *Vitis vinifera* L. cultivated under distinct agricultural practices in Argentina. *Antonie van Leeuwenhoek* 107, 575–588.

Williamson, N., Brian, P. and Wellington, E.M.H. (2000) Molecular detection of bacterial and streptomycete chitinases in the environment. *Antonie van Leeuwenhoek, International Journal of General and Molecular Microbiology* 78, 315–321.

Wittwer, C.T., Hermann, M.G., Moss, A.A. and Rasmussen, R.P. (1997) Continuous fluorescence monitoring of rapid cycle DNA amplification. *BioTechniques* 22, 130–138.

Wu, X., Wu, F., Zhou, X., Fu, X., Tao, Y. and Xu, W. (2016) Effects of intercropping with potato onion on the growth of tomato and rhizosphere alkaline phosphatase genes diversity. *Frontiers in Plant Science* 7, 1–13.

Zappelini, C., Karimi B., Foulon, J., Lacercat-Didier, L., Maillard, F. *et al.* (2015) Diversity and complexity of microbial communities from a chlor-alkali tailings dump. *Soil Biology and Biochemistry* 90, 101–110.

Zelles, L., Bai, Q., Beck, T. and Beese, F. (1992) Signature fatty acids in phospholipids and lipopolysaccharides as indicators of microbial biomass and community structure. *Soil Biology and Biochemistry* 24, 317–323.

Zhang, N.N., Sun, Y.M., Wang, E.T., Yang, J.S., Yuan, H.L. and Scow, K.M. (2014) Effects of intercropping and rhizobial inoculation on the ammonia-oxidizing microorganisms in rhizospheres of maize and faba bean plants. *Applied Soil Ecology* 85, 76–85.

Zhang, Y., Ruan, X.H., Op Den Camp, H.J.M., Smits, T.J.M., Jetten, M.S.M. and Schmid, M.C. (2007) Diversity and abundance of aerobic and anaerobic ammonium-oxidizing bacteria in freshwater sediments of the Xinyi River (China). *Environmental Microbiology* 9, 2375–2382.

Zhang, Y.J., Xie, M., Peng, D.L., Zhao, J.J. and Zhang, Z.R. (2016) Dynamics of microbial population size in rhizosphere soil of Monsanto's *Cry1*Ac cotton. *Plant, Soil and Environment* 62, 92–97.

Zhou, J., Bruns, M.A. and Tiedje, J.M. (1996) DNA recovery from soils of diverse composition. *Applied and Environmental Microbiology* 62, 316–322.

Zhou, X., Gao, D., Liu, J., Qiao, P., Zhou, X. *et al.* (2014) Changes in rhizosphere soil microbial communities in a continuously monocropped cucumber (*Cucumis sativus* L.) system. *European Journal of Soil Biology* 60, 1–8.

Zhou, X., Zhang, J., Li, Y., Liu, B., Chu, J. *et al.* (2015) Distribution characteristics of ammonia oxidizing microorganisms in rhizosphere sediments of cattail. *Ecological Engineering* 88, 99–111.

4 Biosafety Evaluation: A Necessary Process Ensuring the Equitable Beneficial Effects of PGPR

Juan Ignacio Vílchez,[1]* Richard Daniel Lally[2] and Rafael Jorge León Morcillo[1]
[1]*Department of Plant Growth Promotion Rhizobacteria, Plant Stress Centre for Biology (PSC), Chinese Academy of Sciences (CAS), Shanghai, China;* [2]*EnviroCORE, The Dargan Centre, Department of Science and Health, Institute of Technology Carlow, County Carlow, Ireland and Alltech, Dunboyne, County Meath, Ireland*

4.1 Biosafety of PGPR in Soil

Today bio-inoculants capable of stimulating plant growth and providing plant protection against environmental stresses are sought with the aim to isolate efficient commercial products for field effective application (Niranjan Raj *et al.*, 2006; Turan *et al.*, 2010; Keswani *et al.*, 2014; Singh *et al.*, 2016). Plant growth-promoting rhizobacteria (PGPR) applied as biofertilizers and biocontrol agents have been used broadly both in natural and agricultural soils. To date, PGPR products have only been perceived to contribute positive effects as a result of their use in plant growth promotion (Niranjan Raj *et al.*, 2006; Gupta *et al.*, 2015; Bisen *et al.*, 2016; Keswani *et al.*, 2016a). In some situations, the products must satisfy quality criteria, such as the minimal number of viable cells that provide product storage stability over time (Turan *et al.*, 2010; Malusá *et al.*, 2012; Malusá and Vassilev, 2014; Saranraj, 2014; Keswani *et al.*, 2016b). However, the biosafety and environmental considerations following the mass application of these products in the environment are rarely known and are almost never scientifically validated. Safety testing and risk assessment now need to be standard practice, to ensure security for both exposed people and the environment, in order to help manage hazardous secondary effects (Berg *et al.*, 2005; Berg, 2009). Although there are directly positive effects of bio-inoculants, it must be acknowledged that the large-scale application of PGPR in the environment can lead to a series of conditions or side effects on humans and the environment that are currently uncontrolled and only assumed innocuous.

4.1.1 Risk groups and biosafety levels

In the interest of human safety, the use of plant growth-promoting bacteria or biocontrol strains are restricted to the pathogenicity risk groups. These are based on criteria set for the protection of human health such as those described by the World Health Organization (Taylor *et al.*, 2001; WHO, 2015). Thus, isolates belonging to the risk 2 group

*E-mail: juan@sibs.ac.cn

© CAB International 2017. *Advances in PGPR Research* (eds H.B. Singh, B.K. Sarma and C. Keswani)

or higher, such as pathogenic and opportunistic strains *Burkholderia cepacia*, *Bacillus weihenstephanensis*, *Klebsiella pneumonia* (Sachdev *et al.*, 2009) or *Pseudomonas aeruginosa* (Ganesan, 2008; Braud *et al.*, 2009), are not accepted for use as bio-inoculants due to human health concerns, despite being capable of generating positive plant effects (Berg, 2009).

After the exception of human safety, the application of other risk group strains in the environment and its consequences are not subject to any particular safety criterion. Thus, applications are only constrained by possible negative effects on human health (Berg *et al.*, 2005; Berg, 2009). In spite of this, some strains are still used in the laboratory as indicators of plant growth promotion, but not commercially owing to risk of pathogenic outbreaks. *Burkholderia* is an example of such an inoculum. Its genus contains effective PGPR, but is also studied as model organisms for their pathogenicity interactions (Govan *et al.*, 1996; LiPuma *et al.*, 1999). *Burkholderia cepacia* complex (BCC) strains define a group of opportunistic pathogenic strains that can serve as an example of a strain that should be considered limited in use. This particular bacterial species group is one of the few that has been examined for its influence on the soil microbiota (Nacamulli *et al.*, 1997).

Recent studies have also suggested a widespread mechanism of plant growth promotion by volatile organic compounds, including beneficial, neutral and known pathogenic bacterial strains (Blom *et al.*, 2011; Sánchez-López *et al.*, 2016). While this conservative mechanism is still not entirely understood there is a controversial discussion about the true effect of these volatile compounds in plant growth promotion. Some evidences indicate that phytopathogenic microorganisms like *Aternaria alternata* may be utilising this proposed plant growth-promoting (PGP) mechanism (Sánchez-López *et al.*, 2016).

Controversy surrounds the use of certain strains in biocontrol such as selective antagonists or selective pathogens, since many of them reside in high-risk groups. Metabolite products of strains from *Bacillus* or *Pseudomonas* genera are used in the control of plant pathogenic microorganisms, insects or even other competing plants (Beneduzi *et al.*, 2012); some of these are even currently marketed for use. However, the risk assessment of these merely considers safety standards for human health. In many cases, bacterial virulence towards humans is allocated to particular genera or species and identified by genetic characterization. In many cases, laboratory tests focus on known effectiveness of a biocontrol strain and the target species, regardless of any other organisms that may be affected. These applications could be potentially harmful to resident bacteria that are also beneficial to plants (Raaijmakers *et al.*, 2009). Current isolation methods focus on the effect of specialised PGP mechanisms. Unknown to the user these favoured traits may be having detrimental side effects on beneficial species.

4.1.2 Ecological interactions

PGPR-based bio-inoculants are commonly applied to ensure effective contact between plant root systems and the soil. Other formulation addition methods include foliar applications and spray inoculation. In either case, the ecosystem associated with each plant compartment may be affected by application of product. Resident organisms associated with the soil, rhizosphere, aerial plant parts and local plant surroundings can be disrupted more than is usually assumed. In order to protect against this, it is necessary to accept that plant-associated organism populations linked to the soil ecosystem form a complex and dynamic equilibrium (Kokalis-Burelle *et al.*, 2006; Hayat *et al.*, 2010). The number of interconnected ecological niches possibly impacted by the implementation of PGPR strains can cause a broad spectrum of issues which should be accounted for when assessing precisely the possible effects caused by such inoculum for large-scale application (Coenye and Vandamme, 2003).

Soil indigenous populations

The soil is a living system irregularly distributed. The diversification of life in the soil largely depends on its composition (texture, water content, pH and organic matter). One of the factors influencing the zoning of life in the soil is the presence of plant species (Drążkiewicz, 1994; Agnelli *et al.*, 2004; Nicol *et al.*, 2008). The concentration of living organisms in the soil increases with respect to the proximity area of the roots; this is referred to as the rhizosphere (Barea *et al.*, 2005; Hawkes *et al.*, 2007). The rhizosphere is the volume of soil directly influenced by the roots of a plant or population thereof. It can be a few millimetres to several centimetres around the roots and sometimes extends to extremely large bulk volumes of soil (Raaijmakers *et al.*, 2009).

It can be characterized through stable conditions of pH, moisture and chemical composition and is physio-chemically distinct from the overall soil chemical characteristics located outside the rhizosphere's proximity. Moreover, plant roots generate a series of exudate compounds favouring nutritional aspects of soil (Walker *et al.*, 2003). The rhizosphere provides substrates and optimal conditions that support large populations of microorganisms and other fauna. Beneficial bacterial strains are attracted by specific plant exudates which provide the basis of plant–microbe symbiotic interactions. In this sense, soil is the home to one of the most complex and interconnected symbiotic systems of the naturally existing environment. Plants establish beneficial contacts with one or more different species of microorganisms simultaneously which delivers positive interactions (Schippers *et al.*, 1987; Somers *et al.*, 2004; Raaijmakers *et al.*, 2009; Vacheron *et al.*, 2013).

However, not all microorganisms existing in a rhizospheric population are beneficial or mutualistic; the nutrient-rich niche also provides a habitable zone for pathogenic microorganisms and other non-beneficial microorganisms (Schippers *et al.*, 1987; Raaijmakers *et al.*, 2009). In this area, populations of insects, nematodes, protozoa, fungi and soil bacteria establish both positive and negative relationships with plants (Ingham *et al.*, 1985; Rosenheim, 1998; Schmelz *et al.*, 2003; Griffiths *et al.*, 2007). In some cases it could involve herbivores such as insect larvae (beetles, moths, etc.), phytoparasitism, as with protozoa and fungi or plant infection from fungi and certain classes of nematodes. In these examples the organisms benefit from the plant without returning any benefit to the host or in the case of mutualistic interactions do not cause any significant impact on the relationship.

Soil essential populations: the beneficial organisms

Beneficial relationships between PGPR and plants established in the soil are the result of thousands of years of co-evolution of mutualistic and symbiotic processes (Raven, 2002; Provorov and Vorobyov, 2010; Provorov and Vorob'ev, 2012; Ma *et al.*, 2016). To date several different mechanisms of beneficial relationships have been distinguished through the experimentation in the area influenced by the interaction. Examination of these interactions between the host and the microorganism can help in understanding the evolution of these specialized interactions (Bryan *et al.*, 1996; Santi *et al.*, 2013). However, not every relationship will give rise to better growth promotion in each environmental condition. Each variable considered will result in different interactions for that soil condition, this includes different soil microbial populations and balances that maintain the availability of water and nutrients, or competition with them and the subsequent interaction (Hayat *et al.*, 2010; Vacheron *et al.*, 2013).

With this considered, there exists a number of bacterial genera that directly influence growth promotion and positive interactions for beneficial plant development (Egamberdieva *et al.*, 2008; Ahemad and Kibret, 2014). To determine this it is necessary to classify the relationship based on the location where the interaction takes place. Thus, when assessing the implication of ecosystem biosafety using PGPR, two interaction sites are identified: interaction without plant contact and interaction with plant contact. The first group includes strains that live in the soil almost ubiquitously without ever coming in contact with plant roots, but that still enhance plant development. Some of them are soil strains involved in biogeochemical cycles that favour the presence of chemical forms available to plants. This can be considered indirectly beneficial to plants. Furthermore there exist rhizospheric strains which do not come in contact with plant roots but that are capable of producing beneficial plant compounds. Examples of such bacteria are nitrogen fixers, trace metal chelators or volatile organic compound (VOC) synthesizers (Marschner *et al.*, 2004). The metabolites produced lead to factors such as increased plant-available nutrients, hormones and growth-promoting substances. In turn this improves and supports plant development (van Loon *et al.*, 1998; Marschner *et al.*, 2004; Berg, 2009). In general, these strains are involved in what can be described as general soil health and equilibrium formation. This has not only been linked to a positive function with plants, but also to various ecological systems within the pedological cycles and the ecosystem in general (Zaidi *et al.*, 2004; Berg and Smalla, 2009). Thus, they carry out cross functions along different population balances depending on seasonality. The condition of the community can affect many levels of the ecosystem.

However, most of the PGPR strains maintain a closer relationship with the roots of plants maintaining direct contact with the rich extruding secretions. This usually means greater adaptation requirements of bacteria in ecological networks to enhance their colonization of plants. Thus, the more specialized the rhizobacteria, the higher the affinity between the plant and microbe. At a biosafety level epiphytic PGPR refers to strains with stabilized contact on the outer surface of the roots, and endophytic bacteria refers to PGPR with stabilized contact within the root tissue (Rosenblueth and Martínez-Romero, 2006; Ryan *et al.*, 2008; Compant *et al.*, 2010; Jha *et al.*, 2011; Reinhold-Hurek and Hurek, 2011). Epiphytic strains are able to settle on the surface of the roots, both in preferential areas and in structural root holes or on rootlets, or in areas adapted for interaction such as trichomes or twisted rootlets that favour the establishment of colonization. These strains obtain access to root exudates more efficiently than rhizosphere free-living bacteria. They are also equally capable of providing nutrients and plant hormones in a controlled and efficient way (Gaiero *et al.*, 2013; Nongkhlaw and Joshi, 2014). This lifestyle is generalized but is an advantage to the plant roots to which they are attached. Endophytic strains may or may not be free-living bacteria, and have adapted to interact with a number of plant species, some endophytes colonize a broad range of plant genera and species. Endophytes establish colonization in plants' internal tissues and can colonize specialized regions for symbiotic interactions such as thickening or nodule production in roots (Saikkonen *et al.*, 2004; Compant *et al.*, 2010; Reinhold-Hurek and Hurek, 2011; Gaiero *et al.*, 2013; Hardoim *et al.*, 2015). Nodule formation is considered a very specialized adaptation because the strains directly signal plant roots through the production of bacteroids and as a result lose the ability to live independently (Larrainzar *et al.*, 2007; Bianco and Defez, 2009). Generally rhizosphere bacteria are part of the bulk microbial community but are not necessarily predominant. However, their adaptations enable the interaction of a greater host plant range; this increases chances of productivity and enhances its presence in the ecosystem through natural selection. Therefore, the presence of certain plants can consequently influence the residential microbial community. Plant distribution and habitat formation will inevitably influence these symbiotic interactions. The more specialized this interaction, the greater the co-dependency will be between the plant–microbe interactions, thus resulting in a much more vulnerable ecological balance in which they are integrated, as will be the ecosystem where this interaction is present (Provorov and Vorobjev, 2008; Provorov and Vorob'ev, 2012).

4.1.3 Hidden dangers in the use of PGPR

The mass application of PGPR strains in the environment leads to a series of possible consequences that must be accounted for. Most PGPR inoculants have at least

10^6–10^9 CFU ml^{-1} and during the application this could significantly affect the environment in which it is being applied (Çakmakçi *et al.*, 2006; Kidoglu *et al.*, 2008; Almaghrabi *et al.*, 2013). In general, PGPR strains are used to promote plant growth, improve crop productivity as well as alleviate plant stress responses to environmentally challenging conditions (Timmusk *et al.*, 2013; Islam *et al.*, 2014; Sun *et al.*, 2015). However, there are only a number of studies that have defined the impact of mass and repetitive inoculum application in the environment by a product strain that's deemed beneficial. Even when assuming PGPR strains are safe, their introduction to soil for PGP, biocontrol and stress protection will inevitably cause some change in the proportion and balance of the residential soil microbial community considering the inoculum is capable of staying alive. Even if the strain is unable to compete, the applied formulation (including the microbially produced metabolites in the supernatant) will in some way alter the balance of the soil community (De Leij *et al.*, 1995; Bergsma-Vlami *et al.*, 2005; Kokalis-Burelle *et al.*, 2006; Marulanda *et al.*, 2009).

In most cases, the inoculated strain will become the most abundant bacteria proportionally, considering inocula are developed in order to promote growth and ensure effective colonization with target plants. However, it will also interact with non-target plants, leading in some scenarios to change in composition, prevalence or even a loss of plant biodiversity in nearby environments (Bever, 2002; Kardol *et al.*, 2007; Sanon *et al.*, 2009). In crop agriculture this may seem an indirect positive effect, but it must be taken into account that the less diverse the environment (where agricultural activity is carried out) the poorer is the soil due to nutrient deficiency, chemical depletion and it could be more susceptible to plagues and infestations. If plant diversity is lost, so is the influence of that diversity on the soil. This includes root exudates, organic compounds in the soil, texture and associated beneficial microbiota such as nitrogen fixers linked to specific plant biological control in the ecosystem. This imbalance may affect the same relationship between indigenous microorganisms, causing resident species to compete for limited ecological niches in the system.

The lack of specificity between introduced PGPR strains and the environment they are being used in could explain why they are less effective, considering there are probably large differences in the environment from which they were originally derived. If this is the case, these PGPR strains are considered poor competitors *in situ*. This results in a lack of functionality of the strains in the purpose for which they are applied and have been recorded in various laboratory tests. However, this is not a generalized result. Many PGPR strains are poor competitors in some soils and very effective in others. Production of antibiotic or growth-regulatory substances can dramatically affect the structure of the local microbial activity, even if the colonization time is not typically long. *Pseudomonas*, *Bacillus*, *Azospirillum* and *Serratia*, for example, are capable of producing antibiotics which could affect both harmful and beneficial bacteria of the plants retained in the ecosystem in which it is used (Kloepper and Schroth, 1981; Zhang *et al.*, 2000; Fernando *et al.*, 2006). When antibiotic-producing PGPR change the microbial local community assemblages, bacteria resistant to antibiotic substances will become more prevalent. These strains can be harmful to plants or cause detrimental effects as a result of metabolism such as acid production and the rapid consumption of soil organic matter. In extreme cases, this could affect the microbiota depending on the cycle of nutrients such as nitrogen or phosphorus, and when the nutrient balance is disrupted it may have the opposite effect to that intended. There have also been documented cases of increasing soil number of phytopathogenic nematodes as a result of feeding on the cells of the inoculant compound (Ingham *et al.*, 1985). Effects on organisms that are feeding or that are in contact with the inoculant strain can cause obvious changes in the ecosystem. Following application there could be a temporary loss to stability which may be recovered or a perpetuating long-term negative impact that if unaddressed may be irreversible.

Recently, many microbial strains that were assumed to be beneficial were identified to be in fact negatively impacting their surroundings. One of these is a widely commercialised mutualistic fungal symbiont, arbuscular mycorrhizal (AM) fungus which was identified as being deleterious to its host plant due to competition. The growth promoter characterized from the soil community context produced changes in the growth promotion of some plants and caused a reduction in growth for other related species (Bever, 2002). Regardless of the unexamined theoretical beneficial effects of PGPR application, the same effects may have a broader negative impact. A study carried out by Sharma and Nowak (1998) identified that strains of genus *Pseudomonas* were able to trigger inhibition and enhancement of mechanisms plants utilise to establish growth as well as regulate plant disease. These considerations show that abiotic environmental conditions as well as biotic composition and relative abundance are factors that can cause changes in one way or another (Sharma and Nowak, 1998). Further to this, it is not only native microbial flora that is altered in a deleterious way through the application of hypothetically beneficial PGPR, but the soil can be affected through structure change, texture shifts and overall chemical composition. Irresponsible application of the bioformulations may indefinitely impact the natural state of the entire soil ecosystem (Kohler *et al.*, 2010; Gupta *et al.*, 2015).

Contrary to this, several strains characterized as opportunistic pathogens, such as enterobacteria, may have great potential if their biosafety is ensured. Risk groups need to be approved as safe to soil and live communities in order to represent good PGPR candidates (Farooq *et al.*, 2014). Some of these strains are innocuous to human and wildlife health and are potentially easier and cheaper to culture. If these strains are assessed and identified as biologically and environmentally safe it will open up the choice of strains, in particular to those not usually considered for plant growth promotion. For this reason, specifically designed analysis of the inoculant should assess the effects on the entire soil species to ensure subsequent inter-specific interactions. Actually, complete knowledge about full community relationship is the only way to determine the potential consequences of PGPR applications (Gaiero *et al.*, 2013; Kristin and Miranda, 2013; Vacheron *et al.*, 2013).

Independent of this, complex analysis of community interactions can reveal the role and interactions within the soil for each potential PGPR strain. This ensures the responsible application of each inoculant in a safe way and can also help in deciding if a product should be used or disregarded. Finally, PGPR strains that may cause harm don't necessarily have to be banned by strictly one-way comparison standard. It is envisaged that each strain will be contextualized within the soil ecosystem. Each case should be considered to decide if collateral damage as a result of application is going to produce even worse conditions (Schmitt *et al.*, 2005; Salles *et al.*, 2006; Raaijmakers *et al.*, 2009; Gaiero *et al.*, 2013; Coats and Rumpho, 2014).

4.1.4 Economic impact of inattentive application

The annual market for PGPR is expanding globally. Today it is estimated that nearly 80% of the global market revenues are recorded in Europe and Latin America, in particular Argentina. Increased use of biofertilizers comes as a response to the reduction in use of harmful chemical fertilizers and pesticides (Allied Market Research, 2016; Brisk Insights, 2016). According to Markets and Markets (2015), the estimated market for biofertilizers has grown at a rate of 14% and is predicted to generate US$ 1.88 billion worldwide by 2020 (Markets and Markets, 2013, 2015). Another report by Brisk Insights (2016) agrees with this trend, predicting the value of this industry at US$ 1.95 billion by 2022 (Allied Market Research, 2016). This growth is expected to be predominantly in North America, Europe and Asia–Pacific (Brisk Insights, 2016). Countries like China and India will respond with greater impetus to these technologies. The global economic impact of biofertilizer sales is increasing significantly as they offer an ecofriendly alternative, consistent with the requirements of sustainable practices, alleviating pressure on producers (Markets and Markets, 2015; Allied Market Research, 2016). Similarly, effective use significantly reduces the burden of pesticides and fertilizer, as a general rule lowering the cost of production (Allied Market Research, 2016). However, the beneficial short-term effects have not always been as desired. In most cases, biofertilizer application is an incorrect formula, costs too much to produce or performs poorly during application. The ability to apply the product, strain survival and strain environmental factors are major concerns to most commercial product producers (Stephens and Rask, 2000; Nelson, 2004; Malusá *et al.*, 2012; Bashan *et al.*, 2014).

However, there are other potentially hazardous effects. In certain cases, the strain has no problem adapting or colonizing, and may become invasive, affecting not only local microbial communities, but also other organisms knitting together the localized ecosystem such as nematodes, worms and insects (Castro-Sowinski *et al.*, 2007). Certain strains may utilise high levels of nutrient, monopolising a normally balanced ecological niche and could limit access to other plant-associated organisms in the environment. They may also be pathogenic to non-target plants and animals, causing losses in biodiversity (Enebak *et al.*, 1998). This kind of interaction could result in leakage phenomena of surrounding organisms through pressure shifts, changes in environmental conditions and other non-pathogenic factors affecting the growth and development of local organisms (Castro-Sowinski *et al.*, 2007). In such cases, these beneficial soil organisms may be driven away or reduced in numbers. Evidence suggests that an organism sensitive to habitat change, like earthworms, are less likely to stay or will seek alternative habitats more suitable in composition or microbial condition. This is also known as deworming. The loss or decrease in the number of beneficial organisms causes increases in pest population, nutritional depletion and loss of quality such as aeration and texture. In addition, other organisms within the food chain and ecosystem could be disrupted. Moreover, not only inoculated environments are affected. Following extended periods of application, community dilution and the migration of substances produced by the strains can occur. Ultimately, these conditions can extend to sensitive ecosystems such as freshwater resources (rivers, ponds and lakes) (Vílchez *et al.*, 2016).

Pathogenic strains cannot be used in agriculture due to human health concerns, but many recent isolates not currently identified in a distinct risk category are used in the laboratory or in the greenhouse. Recent articles provide evidence that soil and plants are in fact good reservoirs for pathogenic enterobacteria and other opportunistic pathogens (Berg, 2009; Berg and Smalla, 2009; Islam *et al.*, 2014). In addition, some of these strains are being described as PGPR; repetitive exposure and safety-in-use knowledge is urgent in such cases (Berg, 2009; Berg and Smalla, 2009; Farooq *et al.*, 2014). Similarly, recent studies have shown that neutral or phytopathogenic microorganisms such as *Alternaria alternata* can have a positive effect on plant growth through the emission of volatile organic compounds (Sánchez-López *et al.*, 2016). There is a need for strict control of their use and disposal to prevent leakage or contamination as a result of misuse; this can be achieved by following standard biosafety measures. Moreover, many strains that are considered low risk or no risk could be mass produced for wide-scale use. Prolonged contact with them is not expected to cause long-term chronic health effects in workers or exposed animals and the wider human population (Horrigan *et al.*, 2002; Berg *et al.*, 2005).

Not considering these serious effects when planning large-scale applications of PGPR can lead to a number of negative effects of economic, agricultural and environmental concern. Through the use of PGPR the advantage is the ability to alleviate the usual negative effects associated with the use of pesticides and chemical fertilizers since the green revolution (Ju *et al.*, 2007; Vitousek *et al.*, 2009). It is therefore necessary to make a study of potential environmental vulnerabilities to avoid costly ramifications, complex issues and economic losses. Anything else would be contradictory to the spirit of the beneficial and sustainable advantages that biofertilizers are capable of providing and are normally marketed towards (Nelson, 2004).

The first economic knee-jerk reaction is to initially invest in products which are prepared for their particular PGP properties and are not selected based on biosafety. This will inevitably incur undesirable side effects on crops, soils, the environment and animals. Long term this will hurt the economy as the costs of repairing the incurred damage will outweigh the potential profitable benefits. Assessing such risk factors alongside the PGP benefits of PGPR use should be taken into account to help avoid potential devastating consequences like the aforementioned chemical fertilisers. As explored above, an understudied application can change the nutritional balance in the soil; this could increase the need for spending on chemical fertilizers to remedy the situation, negating the benefit of its initial application.

4.2 Mechanisms Involved

Each soil bacterium interacts with the environment and other organisms that inhabit it, modifying through many mechanisms its conditions and can influence nutrient availability, pH, concentration of gases and organic matter content, etc. These changes may be large or small, but will always condition the community living in the soil. With this considered, changes in pH have been described as having one of the greatest influences on the distribution and diversity of soil microbial communities, thus influencing other organisms. Besides changes from metabolites, processes of competition have a large role to play affecting resources and habitat availability (Castro-Sowinski *et al.*, 2007). Finally, negative interactions between soil organisms through predation, parasitism, commensalism and pathogenesis, will select for the populations that impose on the others and undesirable community structures will prevail.

4.2.1 Antigenic substances

Soil organisms change habitat conditions through basal metabolism to improve their environment. Organisms and soil microorganisms change various characteristics to facilitate the survival process (Zhang *et al.*, 2000; Schmitt *et al.*, 2005; Albareda *et al.*, 2006; Berg, 2009). In this regard, increased pressure on the soil is driven by the concentration of nutrients in their various chemical forms. Many ecological niches in soil often overlap for different groups of microorganisms. In this regard, certain groups of microorganisms are able to generate compounds that favour the presence in the medium conditioning the presence of potential opponents (Azad *et al.*, 1985). These antigenic compounds or substances can be specific or non-specific and often serve more than one function. The most non-specific antigenic substances are those that affect the pH of the environment where they are excreted. The most common in this area are acids from basal metabolism of soil microorganisms. Most products come from sugar metabolism and the electron transport chain of aerobic microorganisms. These acids act locally on nearby populations by selecting tolerance to acidification of the environment and can poorly solubilize accessible nutrients at a higher pH as well as certain phosphates or some metallic trace elements. Thus, microorganisms capable of living at a lower pH are favoured and access nutrients whose habitual chemical form becomes limiting to the growth factors. Similarly, some microorganisms are able to basify soil, stabilizing carbonated structures or mineralizing organic substances in the environment (Bulgarelli *et al.*, 2013; Turner *et al.*, 2013). This situation favours microorganisms capable of forming biofilms on an inorganic support medium.

Moreover, some microorganisms are able to indirectly affect the pH, mediating a modification to plant ability to release protons to the soil. Plant growth promotion is usually associated with cell proliferation and elongation through a plasma membrane proton gradient produced by plasma membrane proton pumps. Alteration in the activity of these proton pumps can affect not only cell elongation but also the gradient of protons and acidification of the medium. In addition, proton pumps can also play an important role in other processes such as nutrient acquisition, which can also affect the environment and microorganism population around the roots; pathogen perception, with evident modifications in the interaction between plant and microorganisms including stomatal regulation or root gravitropism (Palmgren, 2001; Liu *et al.*, 2009; Haruta and Sussman, 2012; Lanteigne *et al.*, 2012).

On the other hand, PGPR strains produce certain enzymes such as chitinases, dehydrogenases, β-glucanases, lipases, phosphatases, proteases and other kinds of hydrolases (Lanteigne *et al.*, 2012). Most of these enzymes are effective against parasites or pathogens since they are active against cell walls. These mechanisms can be important in defence against biotic stresses and plant protection from pathogenic fungi or nematodes (Ingham *et al.*, 1985; Saxena and Stotzky, 2001). Their function is not necessarily specific or using specialized mechanisms, and can have a widespread effect on other genera of non-pathogenic bacteria, fungi or nematodes (Tan *et al.*, 1999a; Hawlena *et al.*, 2010). On the other hand, production of specific substances such as bactericides, bacteriostatics and antibiotics could also have an impact over native bacterial populations since their targets could include several species in the same genus (Kloepper and Schroth, 1981; Compant *et al.*, 2010). This exerts greater control over competition for habitat and soil nutrients by conditioning the development or even existence of determinate groups of bacteria in the soil environment. In this regard, antibiotic production is one of the most common mechanisms to evaluate as biocontrol strains. These antibiotic-producing plant growth-promoting rhizobacteria employed as biocontrol agents are usually selected because of their potential to produce more than one antibiotic substance in order to avoid resistances that could be developed by some phytopathogens (Kloepper and Schroth, 1981; Compant *et al.*, 2010). Despite getting consistent results in biological control and subsequent plant growth promotion the collateral damage, ecological and health impacts of their use have not yet been studied (de Souza *et al.*, 2003). The effects of these substances are able to seriously affect bacterial populations, so much so that they can regulate the nutritional capacity of the soil and therefore the containing ecosystem.

4.2.2 Biological control agents

The performance of PGPR in biocontrol has been shown to protect plants from pathogenic bacteria, fungi, entomopathogenic nematodes and herbivores (Blackman and Eastop, 1994; van Loon *et al.*, 1998; Cross *et al.*, 1999; Kloepper *et al.*, 2004; Herman *et al.*, 2008; Pechy-Tarr *et al.*, 2008; Berg, 2009). These are usually mechanisms associated with the production of metabolites capable of causing a disruption to the normal health of a competing organism. These agents are antimicrobials, lytic enzymes, compounds that restrict nutrient availability and that disrupt antagonistic effects. Antimicrobials have been identified in a range of PGPR. These are important for bacterial competition within the plant environment. The front line of defence for plants often depends on mechanisms of antibiosis, as it is useful in antagonism towards invading plant pathogens. They are usually coded for by non-ribosomal means as secondary metabolites and can depend on the nutrient availability in the soil and so the condition of the environment is important in synthesis. *Pseudomonas* strains alone have been found to produce amphisin, 2, 4-di-acetylphloroglucinol (DAPG), hydrogen cyanide, oomycin A, phenazine, pyoluteorin, pyrrolnitrin, tensin, tropolone and cyclic lipopeptides; other bacteria such as *Bacillus, Streptomyces* and *Stenotrophomonas* sp. have been found to produce oligomycin A, kanosamine, zwittermicin A and xanthobaccin (Raaijmakers *et al.*, 2002; King *et al.*, 2006; Compant *et al.*, 2010; Beneduzi *et al.*, 2012; Glick, 2012).

Hydrolases can reduce the impact of pathogens when produced by bacterial strains, as they have the ability to degrade cell wall structures. These hydrolases include chitanases, glucanases and proteases (Hamid *et al.*, 2013; Radif and Hassan, 2014). Chitanase can degrade fungal cell walls and has been shown to disrupt pathogenic fungal growth *in vitro* and has been seen to inhibit spore germination and germ tube formation in *Botrytis cinerea* (Jijakli and Lepoivre, 1998) and may have a role in the control of insects (Kramer and Muthukrishnan, 1997). Proteases can contribute to the control of plant parasitic nematodes. Then, 1,3-glucanase, a bacterial-produced glucanase, was observed to limit the robustness of *R. solani*, *S. rolfsii* and *Pythium ultimum* cell walls (Compant *et al.*, 2010).

Bacteria can produce compounds that bind to ferric iron helping them transport iron across cell membranes. Iron is an essential nutrient in all life forms and so is important to competing community dynamics, but its bioavailability is limited. Siderophores are the extracellularly secreted molecules that bacteria use to bind iron (Crosa and Walsh, 2002). This is usually produced under iron-limited conditions. This mechanism restricts iron availability in organisms unable to transport siderophore-bound iron in the plant environment. Siderophore production has been associated with biocontrol for this reason (Kloepper *et al.*, 1980; Loper, 1988; Beneduzi *et al.*, 2012). Bacteria producing siderophores sequester the iron available to other strains and can deprive pathogenic organisms of iron as the pathogens produce siderophores with low iron affinities or don't produce them at all. Some siderphore producers obtain iron by removing it from existing ferric siderophores; this is achieved by siderophores with a higher iron affinity than the competing compound. Evidence suggests that siderophore-producing strains can contribute to the availability of iron to plants and it is not only involved in biocontrol but plant stress regulation and nutrition (Bar-Ness *et al.*, 1992; Trapet *et al.*, 2016).

The most characterised microbial insecticide is the toxin produced from *Bacillus thuringiensis* (also called Bt toxin). This bacteria produces endotoxin within its endospores as a crystal. When it is ingested by the pest host it is activated in alkaline conditions damaging the cells lining the animal's gut (Weinzierl *et al.*, 1995). *Xenorhabdus* sp. and *Photorhabdus* sp. also produce toxin complexes (Tc) (also called "makes caterpillars floppy" or MCF toxin (Ffrench-Constant *et al.*, 2007)) and may have future use in agriculture in pest control applications considering there is currently concern arising over resistance to the *B. thuringiensis* produced Bt (Chattopadhyay *et al.*, 2004). Other mechanisms of insecticidal activity have also been reported and these include mechanisms produced by plant-associated bacteria. Some *Pseudomonas fluorescens* bacteria can produce FIT (Fluorescens Insect Toxin) which is similar to MCF. *P. fluorescens* strains that produce this toxin have been found to induce lethal effects in *Drosophila melanogaster* and *Manduca sexta* (Pechy-Tarr *et al.*, 2008; Olcott *et al.*, 2010). Vodovar *et al.* (2006), analysed the genome *Pseudomonas entomophila*, a bacterium that, when ingested, is fatal to *Drosophila melanogaster* as well as insects from different orders (Vodovar *et al.*, 2006). They reported several features of the genome that could be contributing to the strain's entomopathogenic properties including a number of potential virulence factors such as toxins, proteases, putative hemolysins, hydrogen cyanide and novel secondary metabolites. *Pseudomonas fluorescens* F113 has also recently had its genome sequenced and annotated (Shanahan *et al.*, 1992; Redondo-Nieto *et al.*, 2013). Similarly, its sequence revealed a range of plant-protective traits including insecticidal and anti-microbial metabolites which have been presented in a range of publications. These include hemolysin, hemagglutinins, adhesion agglutination proteins, RTX toxins, Rhs-family proteins and YD-repeat-containing proteins. This research indicates the potential role of microorganisms in the future of agricultural pest management.

4.2.3 Competence

For successful plant colonization bacterial cells need to have the capabilities of competitiveness. The plant root surface and immediate environments are sites of extreme competition, as nutrients are limited. Thus PGPB depend on mechanisms to invade these niches and outcompete the competing bacteria. Mechanisms that have been identified are motility and chemotaxis. These first allow the bacteria to detect chemical signals from the plant and effective motility aids the movement of the strains to the rhizosphere and root area. Once in the vicinity of the root the bacteria can utilise biofilm-forming mechanisms to help establish themselves with the community (de Weert and Bloemberg, 2006).

4.2.4 Virulence

Attachment and colonization is vital for PGPR to compete within the rhizospheric environment. The mechanisms by which the organisms attach to surfaces have often been referred to as virulence factors in humans, other animals and plants. The mechanisms can be a result of extracellular secretions, secretion system structures or other molecules that enable the bacteria to optimise their environment. Extracellular polysaccharides (EPS) are one of these factors. Bacterial cells secrete them to aid community assemblages, biofilm formations and surface adhesion. Such characterized EPS include alginate, cellulose, polysaccharide synthesis locus (PSL) and pellicle (PEL) formation. These structures have been identified in organisms associated with human disease such as *P. aeruginosa* but also in many PGPR. The secretion systems enable bacteria to engage directly with their extracellular environment offering the cell passage mechanisms for enzymes and proteins. These systems have been associated with various virulence factors as they often dispense disease-causing effectors and molecules.

4.2.5 Alteration of plant-associated mechanisms

Naturally, soil beneficial microorganisms interact with plants through a wide variety of mechanisms that can produce changes in the environment of the rhizosphere or affect directly the plant. The proximity to the root is usually determining the type of relationship that those microorganisms will establish with plants. Certain neutral or beneficial microorganisms can alter the soil around the root by releasing compounds able to modify different characteristics of the field, such as pH or nutrient disposition. However, other microorganisms need to be physically in contact with plants to generate the benefit, either growth promotion or tolerance to biotic or abiotic stresses. An exception is those microorganisms, normally bacteria, able to produce volatile organic compounds that can affect plants directly without being physically in contact with them. Those volatiles can be perceived for the plant and activate different metabolism or signalling pathways related with hormones, as auxins or cytokinins that are involved in cell elongation and proliferation; nutrient uptake, such as iron or sulphur; or increases in photosynthetic efficiency (Ryu *et al.*, 2003; Zhang *et al.*, 2008; Blom *et al.*, 2011).

Either those microorganisms that need to be in contact with the plant or those that can release volatile organic compounds, can produce drastic changes in the plant status that could take effect at physiological or molecular level. Even when these changes are destined to improve plant growth or stress tolerance, in specific conditions, they could become a dangerous factor that may produce undesirable effects in plants.

Several PGPRs may promote plant growth at the expense of some plant biological processes, which can be observed under certain stress conditions, preventing the promotion of plant growth, for example, under deficient nutritional conditions. Plants need a good balance between photosynthetic product generation and nutrient assimilation to ensure healthy growth. In natural soil, where the production is not an important factor, this equilibrium tends to remain unaltered and the risk of producing an imbalance between nutrient acquisition/assimilation and photosynthates is low. However, in agrarian fields, where the yield is the aim, this balance could be affected producing undesirable effects in plant growth. Some PGPRs can promote plant growth through augmentation of photosynthesis capacity/efficiency or optimization of iron homeostasis which will demand, mainly, phosphate and nitrate to convert photosynthetic products to sugar and amino acids (Zhang *et al.*, 2008; Shi *et al.*, 2010; Niu *et al.*, 2013). Actually, phosphate is a critical nutrient directly involved in photosynthesis, because ATP is necessary in the late stage of the process, and sugar metabolism, where some of the intermediate products need to be phosphorylated to ensure sucrose synthesis (Calvin, 1956). Augmenting photosynthesis by PGPR will require enough phosphate in the soil to guarantee plant growth promotion and prevent nutritional stress. In this sense, it is necessary to understand the mechanisms involved in plant growth promotion by different PGPRs or other beneficial microorganisms, and consider the possibility of using more than one PGPR that, in another hand, can interact synergistically with the plant.

In addition, PGPRs may also promote plant growth through modulation of different nutrient transporters of the plant. Nevertheless, this regulation could also affect other minority elements, including heavy metals that can accumulate in plants and, thus, enter into the food chain. Normally, those transporters are designed to uptake specific micro-nutrients, such as iron, manganese or copper. However, the similar equivalence of certain elements under high up-regulation of different transporters could produce an over-accumulation of other micro-elements toxic for plants, animals and humans, such as cadmium or chromium (Clemens, 2006; Mendoza-Cozatl *et al.*, 2014; Clemens and Ma, 2016). In this sense, it is known that some PGPRs may regulate plant acquisition of iron via a deficiency-inducible mechanism, augmenting chlorophyll content and photosynthesis (Zhang *et al.*, 2009). Some components of this pathway have broad specificity for divalent heavy metals, mediating the transport of zinc, manganese, cobalt and cadmium under iron-deficient conditions, and may produce recurring accumulation of toxic elements.

The importance of understanding these mechanisms behind plant growth promotion need to become a necessary task to ensure a proper utilization of these promising beneficial microorganisms, as well as the study of the conditions of the environment and soil where those microorganism will be used.

4.3 Determining the Biosafety of PGPR

As commented above, nowadays disposal of several tools or mechanisms specifically dedicated to PGPR or biocontrollers (bioinoculants) are quite limited. In spite of this some protocols have been adapted and performed in order to offer a significant way to gauge the impact in order to consider a strain's effect on the environment, biological control or human health. On the other hand, new formulations are now focused on being ecofriendly, sustainable and biodegradable (Gupta *et al.*, 2015). Finally, new protocols have to be improved in order to get more information, real-time monitoring and fast easy ways to facilitate biosafety identification and to take correction measures as soon as possible to avoid or control collateral or secondary damages.

4.3.1 *In vitro* bioassays

To date, few reports have addressed these issues and have not been developed in order to begin with biosafety considerations in PGPR and bio-inoculants with a view to use in the natural environment. Most of them are mainly focused on direct or mediated human pathogenicity. Recently some work has focused on the use of coliforms or enterobacteria as PGPR due to their good results under laboratory controlled conditions (Mayak *et al.*, 2001; Holden *et al.*, 2009). To assess their safety they proposed to test antibiotic sensibility of potential PGPR strains as well as antigenic assays to ensure that *E. coli* strains were not O157 pathogenic strains. These trials were carried out to ensure that human potential exposure or manipulation to inoculated seed or plants are safe (Farooq *et al.*, 2014). However, some tests have been employed to assess the safety of using *Bacillus thuringiensis* as a biocontrol strain. For this objective, several target species of ecosystem that could be influenced by their crystal proteins in transgenic plants were used in bioassays. These assays could be considered part of a whole-ecosystem assessment ensuring safety characteristics at the time of the bacterial application or their derivatives in soil (Saxena and Stotzky, 2001).

4.3.2 Environmental and human safety (EHSI) index as a new biosafety tool

As we have described, some PGPR strains may represent a potential threat to human, animal or plant health at different levels; however, their use should be approved if they have been recommended as plant growth enhancers (Berg *et al.*, 2005; Berg, 2009). In general, most of the current regulatory frameworks result in fragmented or contradictory evaluation systems. Nowadays ecofriendly criteria need to establish harmonized protocols for the safe use of PGPR for human and animal health and in the environment (Nelson, 2004). European and American regulations are currently more and more interested in updating biosafety policies and providing alternative methods to replace the use of vertebrate animals (Malusá *et al.*, 2012; Malusá and Vassilev, 2014). In this process a set of new tools has been devised as a panel of tests, while an evaluation system to reliably determine the biosafety of bacterial strains used as PGPB has been recently proposed in order to complete a new European regulation for use and distribution of bio-inoculants and biofertilizers, the environmental and human safety index or EHSI (Vílchez *et al.*, 2016).

This new system is based on a statistical scoring system using a number of bioassays and tests of PGPR based on previous tests assessing harmful chemicals or other potentially dangerous agents or pathogens. It assesses the potential impacts of the products released by the microorganisms' metabolism. It employs Microtox® testing in *Vibrio fischeri* (Onorati and Mecozzi, 2004), microbial viability using the *Escherichia coli* MC4100 sensitivity test as indicator of local soil microfauna (Small *et al.*, 1994; Vassilev *et al.*, 2006), the survival and viability of soil nematodes in *Caenorhabditis elegans* bioassay (Ruiz-Diez *et al.*, 2003; Navas *et al.*, 2007) and a bioassay on the earthworm *Eisenia foetida* (OECD, 2004). The additional aim of these tests was to assess potential harm to the organisms at the second trophic level of the soil cycle (primary consumers) as well as undertaking assays to monitor the effect on organisms from the third trophic level (secondary consumers), including the arthropods *Adalia bipunctata* (neuropteran) and *Chrysoperla carnea* (colleoptera) (Medina *et al.*, 2004; Alvarez-Alfageme *et al.*, 2011). Other assays trial other delicate aspects of the ecosystem that may be indirectly affected due to inoculum leaching in wet environments. For this the EHSI index uses *Daphnia magna* as an indicator of such ecosystems (OECD, 2008, 2012). Although the aim is to eventually develop an alternative assay method that does not require experimentation with mammals or other vertebrates, this index does implement a parallel comparison bioassay using laboratory mice *Mus musculus* (Brenner, 1974; Stelma *et al.*, 1987; Tan *et al.*, 1999a,b; Zachow *et al.*, 2009). This test uses *C. elegans* as an indicator and to compare effects in animal models to judge safety and the risk of chronic and severe health damage to humans. This compilation of tests aims to assess species with habitats both within the soil and in the upper soil ecosystem. The results obtained from the test indicate certain problematic conditions and components, and so ensures the detection of possible environmental impacting factors and human health effectors as well. The results of each test are statistically assessed to generate an indication score. This offers relative weighting for each test.

Based on those scores the Environmental and Human Safety Index (EHSI) is scored from 0 to 100, where a higher value indicates the likelihood that the bacterial strain under investigation would be a safe PGPB (see Table 4.1). Definitive parameters for EHSI are divided into separate indicators: mortality (M, around 50%) which is the main factor; reproduction (R, around 30%),

Table 4.1. Scores for Environmental and human safety index (EHSI). The values shown are EHSI score related to each parameter and test under correspondent weighting correction for a set of commercial PGPR strains or strains recognized as PGPR.

Environmental and safety human index		Score for test strains. Modeling for EHSI Categories							
Bioassay	Parameter	*S. marcescens* 615	*S. proteamaculans* 28115	*S. entomphila* A1	*P. aeruginosa* P14	*P. fluorescens* IABPF05	*A. vinelandii* IABAV02	*R. legominosarum* IABRL05	*B. subtilis* IABBS05
Sensitivity test with *E. coli* MC4100	CFUs/mL	5	5	5	2.5	10	10	10	10
Microtox® Test (*V. fischeri*)	EC_{50}	1.25	1.25	1.25	1.25	2.5	2.5	2.5	2.5
Bioassay with *C. elegans*	No. Adults	1.5	1.5	1.5	1.5	3	3	3	3
	No. Juveniles	2.625	2.625	2.625	2.625	5.25	5.25	7.875	5.25
	No. Eggs	1.5	1.5	1.5	1.5	3	3	4.5	3
	No. Deaths	0	0	0	0	6.25	6.25	12.25	6.25
Bioassay with *C. carnea*	Length	0.75	1	1	1	1	0.75	1	1
	Weight	1.5	1.5	1	1	2	1.5	2	2
	No. Deaths	0	0	0	0	2.8125	2.8125	3.75	2.8125
Bioassay with *A. bipunctata*	Length	0.5	0.5	0.5	0.5	0.75	0.75	0.75	0.75
	Weight	0.25	0.25	0.25	0.25	1	1.5	2	1.5
	No. Deaths	0	0	0	0	1.875	2.8125	3.75	2.8125
Bioassay/Ecotoxicity test with *E. foetida*	Length	2	2	2	2	2	2	2	2
	Weight	3	3	3	3	3	3	3	3
	No. Juveniles	3.375	3.375	2.25	2.25	2.25	3.375	4.5	3.375
	No. Ootheca	2.25	2.25	2.25	1.125	3	2.25	2.25	2.25
DaphtoxKit® Test (*D. magna*)	EC_{50}	1.875	1.875	1.875	1.875	3.75	3.75	3.75	3.75
Test of bacterial effects on plants (based on pepper, *Capsicum annuum*)	Shoot length	1	1	1	1	1	1	1	1
	Dry weight	2	2	2	2	2	2	2	2
	RWC	1	1	1	1	1	1	1	1
Final score		31.375	31.625	30	26.375	57.4375	58.5	72.875	59.25

Target area covered by each test. The table shows the results used to obtain the total EHSI score for the tested strains.

relative to the future dynamics of populations; and development (D, around 20%) of target organisms, related to fulfilling their specific roles in the environment.

The system is based on quartiles that determine the effect compared to a neutral base-level impact over the organism tested. Thus, a PGPR candidate that does not alter any of the specified values would obtain the maximum score of 100 and is therefore considered safe under these assay conditions. Intermediate scores vary depending on the magnitudes of effects in the various assays, and so it is necessary to monitor these affected variables to ensure safe use of this strain following application. Finally, lower scores indicate strong environmental impacts, and those strains should not be considered for large-scale use as it is highly probable that non-reparable damage or collateral effects will occur (Fig. 4.1).

Employing this system reduces the focus on animal models and instead surveys different re-weighting parameters, focusing on other vulnerable points in different aspects of the environment. Inclusion of the EHSI index in new regulations could improve certification processes for PGPR strains. It is a low-cost modular system that provides informed management and safety monitoring data in a very short time. These types of models need to increase with the growth in the PGPB market to ensure better scientifically advised and more easily accessed information regarding the safety concerns of PGPR strains in different ecosystems. This will ensure that each isolate or potential product is used in a responsible manner and with an ecofriendly attitude (Sundh *et al.*, 2011; Li *et al.*, 2013; Selvakumar *et al.*, 2014; OECD, 2015; Vílchez *et al.*, 2016).

It may also be possible to correlate high-risk or low-risk groups of microorganisms using a phylogenetic approach. This may lead to a rapid screening process that could quickly identify strains of high-risk groups providing an intermediate screening step to help discard high-risk strains prior to safety testing and

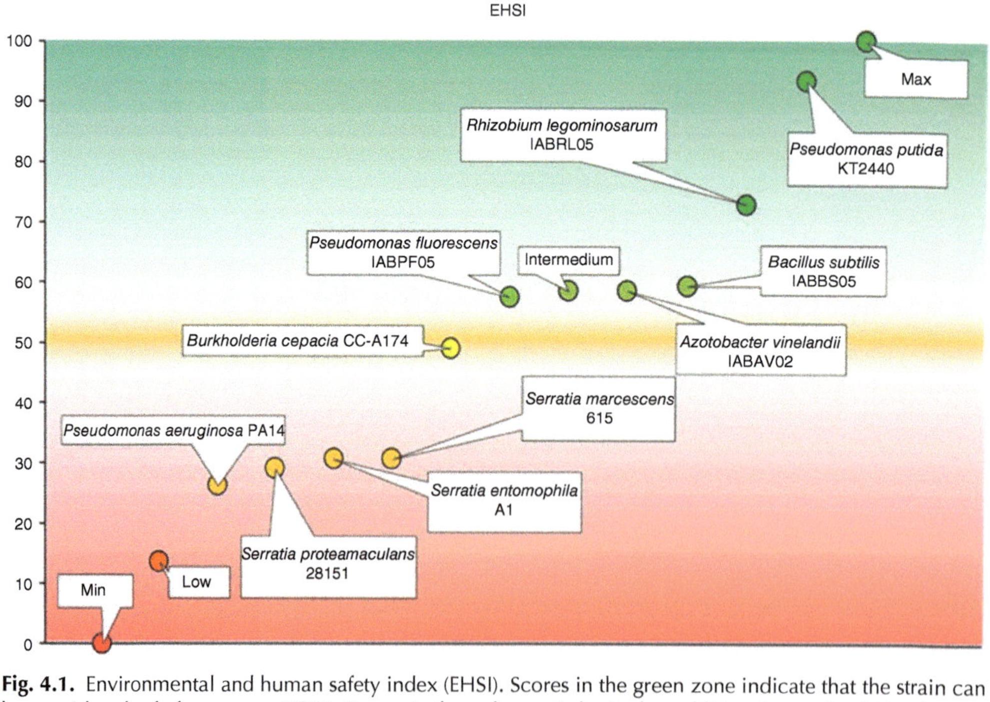

Fig. 4.1. Environmental and human safety index (EHSI). Scores in the green zone indicate that the strain can be considered safe for use as a PGPB. Scores in the red zone indicate that additional tests should be done before the strain can be considered safe for use as a PGPB. The yellow area is considered a transition region of uncertainty. The values shown are the mean and standard deviation of three measurements.

other investment. This approach may also identify bacterial sub-species clades associated with useful metabolite production, this could prove to be significant for bioprocessing purposes. Locating molecular risk group indicators could minimise the time and cost to the researchers or business, helping protect human health and the environment.

4.4 Conclusions and Future Prospects of Biosafety Screening

Advances in research techniques over the past three decades have provided applications that generate large data sets at relatively low costs. Such technologies include DNA sequencing, transcriptomic methods and proteome profiling that can rapidly uncover the cellular capabilities of any organism of interest. The emergence of techniques like these alongside developments in internet technology has generated multiple freely accessible public databases and software. This offers the analyst an opportunity to easily explore genome active traits and pathways important in the central functions of biology (Collins *et al.*, 2003). The development of microbiome research has also been transformed by sequencing technology. The ability to sequence amplified 16S DNA has allowed the modern biologist to analyse the microbial community in any experimental sample, quickly and cheaply. This has provided insights into metabolic and symbiotic relationships between hosts and their commensal flora (Caporaso *et al.*, 2010). Whole genome research has offered new insights into the functionality and adaption mechanisms of a bacterium to its niche environment and has revealed unique differences within the phylogenies of single bacterial species or between species (Garrido-Sanz *et al.*, 2016). This has helped determine differences between groups of bacteria such as endophytes, rhizosphere bacteria, phytopathogens and soil microorganisms (Hardoim *et al.*, 2015). Microbiome analytics has advanced as a tool to determine the effect of environmental conditions, on bacterial communities of plants including those related to plant pathogens (Schlaeppi and Bulgarelli, 2014). These tools will inevitably have a role to play in the future development of crop management technologies. Work surrounding the identification of various sub-clades of *P. fluorescens* has revealed up to 9 separate groups within the species (Garrido-Sanz *et al.*, 2016). Interestingly each of the groups has PGP mechanisms associated with their clustered phylogenies. For example, some groups are associated with higher gene presence for biocontrol traits than others, where others are associated more with plant benefiting coding regions. This raises two questions: can phylogenies be used to indicate various index scores? Could this potentially indicate hazardous strains?

P. fluorescens are considered highly diverse strains. Their use in crop production has shown potential for use in biocontrol and general plant growth promotion. Both rhizobacteria and endophytic isolates have been associated with this species. Initial genomic characterization conducted by Loper *et al.* (2012) provided evidence that *P. fluorescens* strains shared conserved traits, important for plant commensal lifestyle, across phylogenetic groups (Loper *et al.*, 2012). The results divided ten strains into three clusters. One of the identified groups lacked genes responsible for production of antibiotics, plant stress regulation enzyme aminocyclopropan-1-carboxilic acid (ACC) deaminase and polyketide synthase. Unsurprisingly, this indicated that strains belonging to an evolutionary lineage could share similar phenotypes and traits. Potentially this information could help identify hazardous strains using a molecularly informed approach. Evidence further defining subgroups of the *P. fluorescens* clade was provided by Redondo-Nieto *et al.* (2013) and more recently by Garrido-Sanz *et al.* (2016). Both studies reported the conservation of phenotype controlling traits among various clades of the *P. fluorescens* complex. A detailed analysis and genomic survey displayed that eight groups differed greatly across the complex. The analysis examined genes and clusters of CDSs important in biocontrol, siderophore synthesis, toxin production, denitrification, bioremediation and plant interaction. Two of the subgroups, defined as *Pseudomonas chlororaphis* and *Pseudomonas protegens,* contained the most abundant strains for the presence of clusters responsible for synthesis of various antibiotics and siderophores, with both groups also containing a shared presence of the FIT toxin. These traits are considered positive traits in plant protection and biocontrol. However, they could also be considered virulence factors that may jeopardise beneficial organisms during mass applications. Identifying a unique PGPB isolate within either of these subgroups could indicate a necessity for EHSI testing. In contrast, strains which cluster within the *Pseudomonas jessenii* and *P. fluorescens* groups do not contain as many genetic factors associated with virulence, and so they may be considered low-risk groups. Further diagrammatic representation of this concept is presented in Fig. 4.2.

Further methods could also be coupled with these genomic techniques to highlight strains in niche risk groups. The advancements of transcriptomics and proteomics has enabled comprehensive analysis of the full functional genomic characteristics of many organisms including bacterial strains. Following the initial characterization of virulence genes the downstream application of this data could be applied to new strains to determine the rates at which the PGPR are transcribing virulence traits. Those data could be further correlated with genomic and phylogenetic data to help determine isolates in high-risk categories. For example, when assessing epiphytic populations Delmotte *et al.* (2009) assessed the proteomic profile of plant surface populations. The study identified groups of expressed proteins most commonly associated with *Methylobacterium*, *Sphingomonas* and *Pseudomonas* in response to their environment. The authors reported various functional proteins for use in the epiphytic lifestyle. Likewise this technique could be applied to determine the most abundantly expressed proteins that make strains suitable for safe applications as biofertilizers.

Pseudomonas fluorescens complex	Number of traits identified per group and average percent presence of those traits across genomes						Potential hazard risk
	Biocontrol	Siderophores	Toxins	Denitrification	Bioremediation	Plant interactions	
P. mandelii	1 (59%)	2 (5.9%)	2 (23.5%)	4 (63.5%)	5 (19.9%)	2 (47.5%)	Moderate risk
P. jessenii	0 (0%)	1 (25%)	1 (25%)	0 (0%)	5 (100%)	2 (69%)	Low risk
P. koreensis	1 (91%)	0 (0%)	2 (22.5%)	0 (0%)	4 (20.1%)	3 (45.3%)	Moderate risk
P. corrugata	2 (87.5%)	2 (17%)	2 (75%)	6 (75%)	1 (17%)	1 (25%)	Moderate risk
P. fluorescens	0 (0%)	3 (24%)	3 (60%)	4 (22%)	4 (7%)	0 (0%)	Moderate risk
P. gessardii	0 (0%)	3 (43%)	3 (52%)	4 (29%)	1 (29%)	1 (43%)	Moderate risk
P. chlororaphis	4 (93%)	3 (76.3%)	1 (86%)	2 (71%)	3 (14%)	2 (100%)	High risk
P. protegens	5 (60%)	3 (89%)	2 (33.5%)	1 (17%)	2 (17%)	1 (67%)	High risk

Fig. 4.2. Diagrammatic representation of the potential risk groups associated with phylogenetic clusters within *Pseudomonas fluorescens* species. A risk category has been assigned to each group depending on the presence of potentially hazardous traits. Risk categories are as follows: Low risk: will probably not be harmful to organisms of environmental benefit. Moderate risk: could potentially be harmful on application to normal populations of organisms in the environment. High Risk: Could potentially be devastating upon application in the environment. Data for traits and genomic presence is based on the summary figure from Garrido-Sanz *et al.* (2016).

Characterising a range of bacterial species that represent different EHSI scores (for example 10, 50 and 100), could have their proteomic or transcriptomic profiles assessed to determine the most abundant traits appearing as part of their normal metabolism. This could potentially identify key genes and proteins that characterise bacterial strains or species in high- or low-risk groups. However, the concept presented here only represents one species of bacteria. Further data will be needed to correlate EHSI scores to each group and across other bacterial genomic data to determine if this method is truly viable.

References

Agnelli, A., Ascher, J., Corti, G., Ceccherini, T.T., Nannipieri, P. and Pietramellara, G. (2004) Distribution of microbial communities in a forest soil profile investigated by microbial biomass, soil respiration and DGGE of total and extracellular DNA. *Soil Biology and Biochemistry* 36, 859–868.

Ahemad, M. and Kibret, M. (2014) Mechanisms and applications of plant growth promoting rhizobacteria: current perspective. *Journal of King Saud University – Science* 26, 1–20.

Albareda, M., Dardanelli, M.S., Sousa, C., Megias, M., Temprano, F. and Rodriguez-Navarro, D.N. (2006) Factors affecting the attachment of rhizospheric bacteria to bean and soybean roots. *FEMS Microbiological Letters* 259, 67–73.

Allied Market Research (2016) *World Bio-fertilizers Market – Opportunities and Forecasts, 2014–2022.* Allied Market Research, Portland, OR.

Almaghrabi, O.A., Massoud, S.I. and Abdelmoneim, T.S. (2013) Influence of inoculation with plant growth promoting rhizobacteria (PGPR) on tomato plant growth and nematode reproduction under greenhouse conditions. *Saudi Journal of Biological Sciences* 20, 57–61.

Alvarez-Alfageme, F., Bigler, F. and Romeis, J. (2011) Laboratory toxicity studies demonstrate no adverse effects of Cry1Ab and Cry3Bb1 to larvae of *Adalia bipunctata* (Coleoptera: Coccinellidae): the importance of study design. *Transgenic Research* 20, 467–479.

Azad, H.R., Davis, J.R., Schnathorst, W.C. and Kado, C.I. (1985) Relationships between rhizoplane and rhizosphere bacteria and verticillium wilt resistance in potato. *Archives of Microbiology* 140, 347–351.

Barea, J.-M., Pozo, M.J., Azcón, R. and Azcón-Aguilar, C. (2005) Microbial co-operation in the rhizosphere. *Journal of Experimental Botany* 56, 1761–1778.

Bar-Ness, E., Hadar, Y., Chen, Y., Shanzer, A.A. and Libman, J. (1992) Iron uptake by plants from microbial siderophores. *Plant Physiology* 99, 1329–1335.

Bashan, Y., De-Bashan, L., Prabhu, S.R. and Hernandez, J.P. (2014) Advances in plant growth-promoting bacterial inoculant technology: formulations and practical perspectives (1998–2013). *Plant and Soil* 378, 1–33.

Beneduzi, A., Ambrosini, A. and Passaglia, L.M.P. (2012) Plant growth-promoting rhizobacteria (PGPR): their potential as antagonists and biocontrol agents. *Genetics and Molecular Biology* 35, 1044–1051.

Berg, G. (2009) Plant-microbe interactions promoting plant growth and health: perspectives for controlled use of microorganisms in agriculture. *Applied Microbiology and Biotechnology* 84, 11–18.

Berg, G. and Smalla, K. (2009) Plant species and soil type cooperatively shape the structure and function of microbial communities in the rhizosphere. *FEMS Microbiology Ecology* 68, 1–13.

Berg, G., Eberl, L. and Hartmann, A. (2005) The rhizosphere as a reservoir for opportunistic human pathogenic bacteria. *Environmental Microbiology* 7, 1673–1685.

Bergsma-Vlami, M., Prins, M.E. and Raaijmakers, J.M. (2005) Influence of plant species on population dynamics, genotypic diversity and antibiotic production in the rhizosphere by indigenous *Pseudomonas spp. FEMS Microbiology Ecology* 52, 59–69.

Bever, J.D. (2002) Negative feedback within a mutualism: host–specific growth of mycorrhizal fungi reduces plant benefit. *Proceedings of the Royal Society of London B: Biological Sciences* 269, 2595–2601.

Bianco, C. and Defez, R. (2009) *Medicago truncatula* improves salt tolerance when nodulated by an indole-3-acetic acid-overproducing *Sinorhizobium meliloti* strain. *Journal of Experimental Botany* 60, 3097–3107.

Bisen, K., Keswani, C., Patel, J.S., Sarma, B.K. and Singh, H.B. (2016) *Trichoderma* spp.: efficient inducers of systemic resistance in plants. In: Chaudhary, D.K. and Verma, A. (eds) *Microbial-Mediated Induced Systemic Resistance in Plants*. Springer, Singapore, pp. 185–195.

Blackman, R.L. and Eastop, V.F. (1994) *Aphids on the World's Trees, an Identification and Information Guide.* CAB International, Wallingford, UK.

Blom, D., Fabbri, C., Connor, E.C., Schiestl, F.P., Klauser, D.R. *et al.* (2011) Production of plant growth modulating volatiles is widespread among rhizosphere bacteria and strongly depends on culture conditions. *Environmental Microbiology* 13, 3047–3058.

Braud, A., Hoegy, F., Jezequel, K., Lebeau, T. and Schalk, I.J. (2009) New insights into the metal specificity of the *Pseudomonas aeruginosa* pyoverdine-iron uptake pathway. *Environmental Microbiology* 11, 1079–1091.

Brenner, S. (1974) The genetics of *Caenorhabditis elegans*. *Genetics* 77, 71–94.

Brisk Insights (2016) *Biofertilizer Market by Type (Nitrogen Fixing, Phosphate Solubilizing, Potash-mobilizing), by Application (Seed and Soil Treatment), by Crop Type (Cereals and Grains, Pulses and Oil Seeds, Fruits and Vegetables, Plantations), by Microorganisms (*Azotobacter, Rhizobium, Azospirrilum *and* Cyanobacteria*), Industry Size, Growth, Share and Forecast to 2022*. Brisk Insights, Nottingham, UK.

Bryan, J.A., Berlyn, G.P. and Gordon, J.C. (1996) Toward a new concept of the evolution of symbiotic nitrogen fixation in the Leguminosae. *Plant and Soil* 186, 151–159.

Bulgarelli, D., Schlaeppi, K., Spaepen, S., Van Themaat, E.V.L. and Schulze-Lefert, P. (2013) Structure and functions of the bacterial microbiota of plants. *Annual Review of Plant Biology* 64, 807–838.

Çakmakçi, R., Dönmez, F., Aydın, A. and Şahin, F. (2006) Growth promotion of plants by plant growth-promoting rhizobacteria under greenhouse and two different field soil conditions. *Soil Biology and Biochemistry* 38, 1482–1487.

Calvin, M. (1956) [The photosynthetic cycle]. *Bulletin de la Societé de Chimie Biologique (Paris)* 38, 1233–1244. (In French)

Caporaso, J.G., Kuczynski, J., Stombaugh, J., Bittinger, K., Bushman, F.D. *et al.* (2010) QIIME allows analysis of high-throughput community sequencing data. *Nature Methods* 7, 335–336.

Castro-Sowinski, S., Herschkovitz, Y., Okon, Y. and Jurkevitch, E. (2007) Effects of inoculation with plant growth-promoting rhizobacteria on resident rhizosphere microorganisms. *FEMS Microbiology Letters* 276, 1–11.

Chattopadhyay, A., Bhatnagar, N.B. and Bhatnagar, R. (2004) Bacterial insecticidal toxins. *Critical Reviews in Microbiology* 30, 33–54.

Clemens, S. (2006) Toxic metal accumulation, responses to exposure and mechanisms of tolerance in plants. *Biochimie* 88, 1707–1719.

Clemens, S. and Ma, J.F. (2016) Toxic heavy metal and metalloid accumulation in crop plants and foods. *Annual Review of Plant Biology* 67, 489–512.

Coats, V.C. and Rumpho, M.E. (2014) The rhizosphere microbiota of plant invaders: an overview of recent advances in the microbiomics of invasive plants. *Frontiers in Microbiology* 5, 368.

Coenye, T. and Vandamme, P. (2003) Diversity and significance of *Burkholderia* species occupying diverse ecological niches. *Environmental Microbiology* 5, 719–729.

Collins, F.S., Green, E.D., Guttmacher, A.E. and Guyer, M.S. (2003) A vision for the future of genomics research. *Nature* 422, 835–847.

Compant, S., Clement, C. and Sessitsch, A. (2010) Plant growth-promoting bacteria in the rhizo- and endosphere of plants: their role, colonization, mechanisms involved and prospects for utilization. *Soil Biology and Biochemistry* 42, 669–678.

Crosa, J.H. and Walsh, C.T. (2002) Genetics and assembly line enzymology of siderophore biosynthesis in bacteria. *Microbiology and Molecular Biology Reviews* 66, 223–249.

Cross, J.V., Solomon, M.G., Chandler, D., Jarrett, P., Richardson, P.N. *et al.* (1999) Biocontrol of pests of apples and pears in northern and central Europe: 1. Microbial agents and nematodes. *Biocontrol Science and Technology* 9, 125–149.

De Leij, F., Sutton, E.J., Whipps, J.M., Fenlon, J.S. and Lynch, J.M. (1995) Impact of field release of genetically modified *Pseudomonas fluorescens* on indigenous microbial populations of wheat. *Applied and Environmental Microbiology* 61, 3443–3453.

Delmotte, N., Knief, C., Chaffron, S., Innerebner, G., Roschitzki, B. *et al.* (2009) Community proteogenomics reveals insights into the physiology of phyllosphere bacteria. *Proceedings of the National Academy of Sciences of the United States of America* 106, 16428–16433.

De Souza, J.T., Weller, D.M. and Raaijmakers, J.M. (2003) Frequency, diversity, and activity of 2,4-diacetylphloroglucinol-producing fluorescent *Pseudomonas* spp. in Dutch take-all decline soils. *Phytopathology* 93, 54–63.

De Weert, S. and Bloemberg, G.V. (2006) Rhizosphere competence and the role of root colonization in biocontrol. In: Gnanamanickam, S.S. (ed.) *Plant-Associated Bacteria*. Springer, Dordrecht, Netherlands.

Drążkiewicz, M. (1994) Distribution of microorganisms in soil aggregates: effect of aggregate size. *Folia Microbiologica* 39, 276–282.

Egamberdieva, D., Kamilova, F., Validov, S., Gafurova, L., Kucharova, Z. and Lugtenberg, B. (2008) High incidence of plant growth-stimulating bacteria associated with the rhizosphere of wheat grown on salinated soil in Uzbekistan. *Environmental Microbiology* 10, 1–9.

Enebak, S.A., Wei, G., Kloepper, J.W. and Anandham, R. (1998) Effects of plant growth-promoting rhizobacteria on loblolly and slash pine seedlings. *Forest Science* 44, 139–144.

Farooq, N., Raheem, A. and Ali, B. (2014) Waterborne *Escherichia coli*: biosafety and screening as plant growth promoting rhizobacteria. *Journal of Pure and Applied Microbiology* 8, 3963–3971.

Fernando, W.G.D., Nakkeeran, S. and Zhang, Y. (2006) Biosynthesis of antibiotics by PGPR and its relation in biocontrol of plant diseases. In: Siddiqui, Z. (ed.) *PGPR: Biocontrol and Biofertilization*. Springer, Dordrecht, Netherlands.

Ffrench-Constant, R.H., Dowling, A. and Waterfield, N.R. (2007) Insecticidal toxins from *Photorhabdus* bacteria and their potential use in agriculture. *Toxicon* 49, 436–451.

Gaiero, J.R., Mccall, C.A., Thompson, K.A., Day, N.J., Best, A.S. and Dunfield, K.E. (2013) Inside the root microbiome: bacterial root endophytes and plant growth promotion. *American Journal of Botany* 100, 1738–1750.

Ganesan, V. (2008) Rhizoremediation of cadmium soil using a cadmium-resistant plant growth-promoting rhizopseudomonad. *Current Microbiology* 56, 403–407.

Garrido-Sanz, D., Meier-Kolthoff, J.P., Goker, M., Martin, M., Rivilla, R. and Redondo-Nieto, M. (2016) Genomic and genetic diversity within the *Pseudomonas fluorescens* complex. *PLoS One* 11, e0150183.

Glick, B.R. (2012) Plant growth-promoting bacteria: mechanisms and applications. *Scientifica (Cairo)*, 963401.

Govan, J.R., Hughes, J.E. and Vandamme, P. (1996) *Burkholderia cepacia*: medical, taxonomic and ecological issues. *Journal of Medical Microbiology* 45, 395–407.

Griffiths, B.S., Christensen, S. and Bonkowski, M. (2007) Microfaunal interactions in the rhizosphere, how nematodes and protozoa link above- and belowground processes. In: Whitbeck, J.L. (ed.) *The Rhizosphere – An Ecological Perspective*. Academic Press, Burlington, Massachusetts.

Gupta, G., Parihar, S.S., Ahirwar, N.K., Snehi, S.K. and Singh, V. (2015) Plant growth promoting rhizobacteria (PGPR): current and future prospects for development of sustainable agriculture. *Journal of Microbial Biochemical Technology* 7, 96–102.

Hamid, R., Khan, M.A., Ahmad, M., Ahmad, M.M., Abdin, M.Z. *et al.* (2013) Chitinases: an update. *Journal of Pharmacy and Bioallied Sciences* 5, 21–29.

Hardoim, P.R., Van Overbeek, L.S., Berg, G., Pirttilä, A.M., Compant, S. *et al.* (2015) The hidden world within plants: ecological and evolutionary considerations for defining functioning of microbial endophytes. *Microbiology and Molecular Biology Reviews* 79, 293–320.

Haruta, M. and Sussman, M.R. (2012) The effect of a genetically reduced plasma membrane protonmotive force on vegetative growth of Arabidopsis. *Plant Physiology* 158, 1158–1171.

Hawkes, C.V., Deangelis, K.M. and Firestone, M.K. (2007) Root interactions with soil microbial communities and processes. In: Whitbeck, J.L. (ed.) *The Rhizosphere – An Ecological Perspective*. Academic Press, Burlington, Massachusetts.

Hawlena, H., Bashey, F. and Lively, C.M. (2010) The evolution of spite: population structure and bacteriocin-mediated antagonism in two natural populations of *Xenorhabdus* bacteria. *Evolution* 64, 3198–3204.

Hayat, R., Ali, S., Amara, U., Khalid, R. and Ahmed, I. (2010) Soil beneficial bacteria and their role in plant growth promotion: a review. *Annals of Microbiology* 60, 579–598.

Herman, M.A.B., Nault, B.A. and Smart, C.D. (2008) Effects of plant growth-promoting rhizobacteria on bell pepper production and green peach aphid infestations in New York. *Crop Protection* 27, 996–1002.

Holden, N., Pritchard, L. and Toth, I. (2009) Colonization outwith the colon: plants as an alternative environmental reservoir for human pathogenic enterobacteria. *FEMS Microbiology Reviews* 33, 689–703.

Horrigan, L., Lawrence, R.S. and Walker, P. (2002) How sustainable agriculture can address the environmental and human health harms of industrial agriculture. *Environmental Health Perspectives* 110, 445–456.

Ingham, R.E., Trofymow, J.A., Ingham, E.R. and Coleman, D.C. (1985) Interactions of bacteria, fungi, and their nematode grazers: effects on nutrient cycling and plant growth. *Ecological Monographs* 55, 119–140.

Islam, F., Yasmeen, T., Ali, Q., Ali, S., Arif, M.S. *et al.* (2014) Influence of *Pseudomonas aeruginosa* as PGPR on oxidative stress tolerance in wheat under Zn stress. *Ecotoxicology Environmental Safety* 104, 285–293.

Jha, Y., Subramanian, R.B. and Patel, S. (2011) Combination of endophytic and rhizospheric plant growth promoting rhizobacteria in *Oryza sativa* shows higher accumulation of osmoprotectant against saline stress. *Acta Physiologiae Plantarum* 33, 797–802.

Jijakli, M.H. and Lepoivre, P. (1998) Characterization of an exo-β-1,3-glucanase produced by *Pichia anomala* strain K, antagonist of *Botrytis cinerea* on apples. *Phytopathology* 88, 335–343.

Ju, X.T., Kou, C.L., Christie, P., Dou, Z.X. and Zhang, F.S. (2007) Changes in the soil environment from excessive application of fertilizers and manures to two contrasting intensive cropping systems on the North China Plain. *Environmental Pollution* 145, 497–506.

Kardol, P., Cornips, N.J., Van Kempen, M.M.L., Bakx-Schotman, J.M.T. and Van Der Putten, W.H. (2007) Microbe-mediated plant–soil feedback causes historical contingency effects in plant community assembly. *Ecological Monographs* 77, 147–162.

Keswani, C., Mishra, S., Sarma, B.K., Singh, S.P. and Singh, H.B. (2014) Unraveling the efficient application of secondary metabolites of various *Trichoderma*. *Applied Microbiology and Biotechnology* 98, 533–544.

Keswani, C., Bisen, K., Singh, S.P., Sarma, B.K. and Singh, H.B. (2016a) A proteomic approach to understand the tripartite interactions between plant–*Trichoderma*–pathogen: investigating the potential for efficient biological control. In: Hakeem, K.R. and Akhtar, M.S. (eds) *Plant, Soil and Microbes Vol. 2. Mechanisms and Molecular Interactions*. Springer, Cham, Switzerland, pp. 79–93.

Keswani, C., Bisen, K., Singh, V., Sarma, B.K. and Singh, H.B. (2016b) Formulation technology of biocontrol agents: present status and future prospects. In: Arora, N.K., Mehnaz, S. and Balestrini, R. (eds) *Bioformulations: For Sustainable Agriculture*. Springer, New Delhi, India, pp. 35–52.

Kidoglu, F., Gül, A., Ozaktan, H. and Tüzel, Y. (2008) *Effect of Rhizobacteria on Plant Growth of Different Vegetables*. International Society for Horticultural Science (ISHS), Leuven, Belgium, pp. 1471–1478.

King, T., Seeto, S. and Ferenci, T. (2006) Genotype-by-environment interactions influencing the emergence of rpoS mutations in *Escherichia coli* populations. *Genetics* 172, 2071–2079.

Kloepper, J.W. and Schroth, M.N. (1978) Plant growth promoting rhizobacteria on radishes. *Proceedings of the Fourth International Conference on Plant Pathogen Bacteria, Angers, France,* INRA Angers, France, 2, 879–882.

Kloepper, J.W. and Schroth, M.N. (1981) Relationship of *in vitro* antibiosis of plant growth-promoting rhizobacteria to plant growth and the displacement of root microflora. *Phytopathology* 71, 1020–1024.

Kloepper, J.W., Leong, J., Teintze, M. and Schroth, M.N. (1980) Enhanced plant growth by siderophores produced by plant growth-promoting rhizobacteria. *Nature* 286, 885–886.

Kloepper, J.W., Ryu, C.M. and Zhang, S. (2004) Induced systemic resistance and promotion of plant growth by *Bacillus* spp. *Phytopathology* 94, 1259–1266.

Kohler, J., Caravaca, F. and Roldán, A. (2010) An AM fungus and a PGPR intensify the adverse effects of salinity on the stability of rhizosphere soil aggregates of *Lactuca sativa*. *Soil Biology and Biochemistry* 42, 429–434.

Kokalis-Burelle, N., Kloepper, J.W. and Reddy, M.S. (2006) Plant growth-promoting rhizobacteria as transplant amendments and their effects on indigenous rhizosphere microorganisms. *Applied Soil Ecology* 31, 91–100.

Kramer, K.J. and Muthukrishnan, S. (1997) Insect chitinases: molecular biology and potential use as biopesticides. *Insect Biochemistry and Molecular Biology* 27, 887–900.

Kristin, A. and Miranda, H. (2013) The root microbiota—a fingerprint in the soil? *Plant and Soil* 370, 671–686.

Lanteigne, C., Gadkar, V.J., Wallon, T., Novinscak, A. and Filion, M. (2012) Production of DAPG and HCN by *Pseudomonas* sp. LBUM300 contributes to the biological control of bacterial canker of tomato. *Phytopathology* 102, 967–973.

Larrainzar, E., Wienkoop, S., Weckwerth, W., Ladrera, R., Arrese-Igor, C. and González, E.M. (2007) *Medicago truncatula* root nodule proteome analysis reveals differential plant and bacteroid responses to drought stress. *Plant Physiology* 144, 1495–1507.

Li, G.-X., Wu, X.-Q. and Ye, J.-R. (2013) Biosafety and colonization of *Burkholderia multivorans* WS-FJ9 and its growth-promoting effects on poplars. *Applied Microbiology and Biotechnology* 97, 10489–10498.

Lipuma, J.J., Dulaney, B.J., Mcmenamin, J.D., Whitby, P.W., Stull, T.L. *et al.* (1999) Development of rRNA-based PCR assays for identification of *Burkholderia cepacia* complex isolates recovered from cystic fibrosis patients. *Journal of Clinical Microbiology* 37, 3167–3170.

Liu, B., Yao, J., Wang, Y., Li, H. and Qin, F. (2009) Proton inhibition of unitary currents of vanilloid receptors. *The Journal of General Physiology* 134, 243–258.

Loper, J.E. (1988) Role of fluorescent siderophore production in biological control of *Pythium ultimum* by a *Pseudomonas fluorescens* strain. *Phytopathology* 78, 166–172.

Loper, J.E., Hassan, K.A., Mavrodi, D.V., Davis, E.W., II, Lim, C.K., *et al.* (2012) Comparative genomics of plant-associated *Pseudomonas* spp.: insights into diversity and inheritance of traits involved in multitrophic interactions. *PLoS Genetics* 8, e1002784.

Ma, Y., Rajkumar, M., Zhang, C. and Freitas, H. (2016) Beneficial role of bacterial endophytes in heavy metal phytoremediation. *Journal of Environmental Management* 174, 14–25.

Malusá, E. and Vassilev, N. (2014) A contribution to set a legal framework for biofertilizers. *Applied Microbiology and Biotechnology* 98, 6599–6607.

Malusá, E., Sas-Paszt, L. and Ciesielska, J. (2012) Technologies for beneficial microorganisms inocula used as biofertilizers. *The Scientific World Journal* 2012, 491206.

Markets and Markets (2013) *Global Biofertilizers Market by Types, Applications and Geography—Trends and Forecasts to 2017.* Markets and Markets, Sallas, TX.

Markets and Markets (2015) *Biofertilizers Market by Type (Nitrogen-fixing, Phosphate-solubilizing and Potash-mobilizing), Microorganisms* (Rhizobium, Azotobacter, Azospirillum, Cyanobacteria *and Phosphate-Solubilizing Bacteria), Application, Crop Type and by Region – Global Forecast to 2020.* Markets and Markets, Dallas, TX.

Marschner, P., Crowley, D. and Yang, C.H. (2004) Development of specific rhizosphere bacterial communities in relation to plant species, nutrition and soil type. *Plant and Soil* 261, 199–208.

Marulanda, A., Barea, J.-M. and Azcón, R. (2009) Stimulation of plant growth and drought tolerance by native microorganisms (AM fungi and bacteria) from dry environments: mechanisms related to bacterial effectiveness. *Journal of Plant Growth Regulation* 28, 115–124.

Mayak, S., Tirosh, T. and Glick, B.R. (2001) Stimulation of the growth of tomato, pepper and mung bean plants by the plant growth-promoting bacterium *Enterobacter cloacae* CAL3. *Biological Agriculture and Horticulture* 19, 261–274.

Medina, P., Budia, F., Del Estal, P. and Vinuela, E. (2004) Influence of azadirachtin, a botanical insecticide, on *Chrysoperla carnea* (Stephens) reproduction: toxicity and ultrastructural approach. *Journal of Economic Entomology* 97, 43–50.

Mendoza-Cozatl, D.G., Xie, Q., Akmakjian, G.Z., Jobe, T.O., Patel, A. *et al.* (2014) OPT3 is a component of the iron-signaling network between leaves and roots and misregulation of OPT3 leads to an over-accumulation of cadmium in seeds. *Molecular Plant* 7, 1455–1469.

Nacamulli, C., Bevivino, A., Dalmastri, C., Tabacchioni, S. and Chiarini, L. (1997) Perturbation of maize rhizosphere microflora following seed bacterization with *Burkholderia cepacia* MCI 7. *FEMS Microbiology Ecology* 23, 183–193.

Navas, A., Cobas, G., Talavera, M., Ayala, J.A., Lopez, J.A. and Martinez, J.L. (2007) Experimental validation of Haldane's hypothesis on the role of infection as an evolutionary force for Metazoans. *Proceedings of the National Academy of Sciences of the United States of America* 104, 13728–13731.

Nelson, L.M. (2004) Plant growth promoting rhizobacteria (PGPR): prospects for new inoculants. *Crop Management*, 3.

Nicol, G.W., Leininger, S., Schleper, C. and Prosser, J.I. (2008) The influence of soil pH on the diversity, abundance and transcriptional activity of ammonia oxidizing archaea and bacteria. *Environmental Microbiology* 10, 2966–2978.

Niranjan Raj, S., Shetty, H.S. and Reddy, M.S. (2006) Plant growth promoting rhizobacteria: potential green alternative for plant productivity. In: Siddiqui, Z.A. (ed.) *PGPR: Biocontrol and Biofertilization.* Springer, Dordrecht, Netherlands.

Niu, Y., Chai, R., Dong, H., Wang, H., Tang, C. and Zhang, Y. (2013) Effect of elevated CO_2 on phosphorus nutrition of phosphate-deficient *Arabidopsis thaliana* (L.) Heynh under different nitrogen forms. *Journal of Experimental Botany* 64, 355–367.

Nongkhlaw, F.M. and Joshi, S.R. (2014) Epiphytic and endophytic bacteria that promote growth of ethnomedicinal plants in the subtropical forests of Meghalaya, India. *Revista de Biología Tropical* 62, 1295–1308.

OECD (2004) *Test No. 222: Earthworm Reproduction Test (Eisenia foetida/Eisenia andrei).* OECD Publishing, Paris, France.

OECD (2008) *Test No. 211: Daphnia Magna Reproduction Test.* OECD Publishing, Paris, France.

OECD (2012) *Test No. 211: Daphnia Magna Reproduction Test.* OECD Publishing, Paris, France.

OECD (2015) *Biosafety and the Environmental Uses of Micro-Organisms,* OECD Publishing, Paris, France

Olcott, M.H., Henkels, M.D., Rosen, K.L., Walker, F.L., Sneh, B. *et al.* (2010) Lethality and developmental delay in *Drosophila melanogaster* larvae after ingestion of selected *Pseudomonas fluorescens* strains. *PLoS One* 5, e12504.

Onorati, F. and Mecozzi, M. (2004) Effects of two diluents in the Microtox toxicity bioassay with marine sediments. *Chemosphere* 54, 679–687.

Palmgren, M.G. (2001) Plant plasma membrane H+-ATPases: powerhouses for nutrient uptake. *Annual Review of Plant Physiology and Plant Molecular Biology* 52, 817–845.

Pechy-Tarr, M., Bruck, D.J., Maurhofer, M., Fischer, E., Vogne, C. *et al.* (2008) Molecular analysis of a novel gene cluster encoding an insect toxin in plant-associated strains of *Pseudomonas fluorescens*. *Environmental Microbiology* 10, 2368–2386.

Provorov, N.A. and Vorobyov, N.I. (2008) Simulation of plant-bacteria co-evolution in the mutually beneficial symbiosis. *Ecological Genetics* 6, 14.

Provorov, N.A. and Vorobyov, N.I. (2010) Simulation of evolution implemented in the mutualistic symbioses towards enhancing their ecological efficiency, functional integrity and genotypic specificity. *Theoretical Population Biology* 78, 259–269.

Provorov, N.A. and Vorob'ev, N.I. (2012) Co-evolution of partners and the integrity of symbiotic systems. *Zhurnal Obshchei Biologii* 73, 21–36.

Raaijmakers, J.M., Vlami, M. and DE Souza, J.T. (2002) Antibiotic production by bacterial biocontrol agents. *Antonie van Leeuwenhoek* 81, 537–547.

Raaijmakers, J.M., Paulitz, T.C., Steinberg, C., Alabouvette, C. and Moënne-Loccoz, Y. (2009) The rhizosphere: a playground and battlefield for soilborne pathogens and beneficial microorganisms. *Plant and Soil* 321, 341–361.

Radif, H. and Hassan, S. (2014) Detection of hydrolytic enzymes produced by *Azospirillum brasiliense*. *World Journal of Experimental Biosciences* 2, 36–40.

Raven, J.A. (2002) Evolution of cyanobacterial symbioses. In: Rai, A.N., Bergman, B. and Rasmussen, U. (eds) *Cyanobacteria in Symbiosis*. Springer, Dordrecht, Netherlands.

Redondo-Nieto, M., Barret, M., Morrissey, J., Germaine, K., Martínez-Granero, F. *et al.* (2013) Genome sequence reveals that *Pseudomonas fluorescens* F113 possesses a large and diverse array of systems for rhizosphere function and host interaction. *BMC Genomics* 14, 1–17.

Reinhold-Hurek, B. and Hurek, T. (2011) Living inside plants: bacterial endophytes. *Current Opinion in Plant Biology* 14, 435–443.

Rosenblueth, M. and Martínez-Romero, E. (2006) Bacterial endophytes and their interactions with hosts. *Molecular Plant–Microbe Interactions* 19, 827–837.

Rosenheim, J.A. (1998) Higher-order predators and the regulation of insect herbivore populations. *Annual Review of Entomology* 43, 421–447.

Ruiz-Diez, B., Sanchez, P., Baquero, F., Martinez, J.L. and Navas, A. (2003) Differential interactions within the *Caenorhabditis elegans–Pseudomonas aeruginosa* pathogenesis model. *Journal of Theoretical Biology* 225, 469–476.

Ryan, R.P., Germaine, K., Franks, A., Ryan, D.J. and Dowling, D.N. (2008) Bacterial endophytes: recent developments and applications. *FEMS Microbiology Letters* 278, 1–9.

Ryu, C.M., Farag, M.A., Hu, C.H., Reddy, M.S., Wei, H.X. *et al.* (2003) Bacterial volatiles promote growth in *Arabidopsis*. *Proceedings of the National Academy of Sciences of the United States of America* 100, 4927–4932.

Sachdev, D.P., Chaudhari, H.G., Kasture, V.M., Dhavale, D.D. and Chopade, B.A. (2009) Isolation and characterization of indole acetic acid (IAA) producing *Klebsiella pneumoniae* strains from rhizosphere of wheat (*Triticum aestivum*) and their effect on plant growth. *Indian Journal of Experimental Biology* 47, 993–1000.

Saikkonen, K., Wäli, P., Helander, M. and Faeth, S.H. (2004) Evolution of endophyte–plant symbioses. *Trends in Plant Science* 9, 275–280.

Salles, J.F., Van Elsas, J.D. and Van Veen, J.A. (2006) Effect of agricultural management regime on *Burkholderia* community structure in soil. *Microbial Ecology* 52, 267–279.

Sánchez-López, Á.M., Baslam, M., DE Diego, N., Muñoz, F.J., Bahaji, A. *et al.* (2016) Volatile compounds emitted by diverse phytopathogenic microorganisms promote plant growth and flowering through cytokinin action. *Plant, Cell and Environment* 39(12), 2592–2608.

Sanon, A., Andrianjaka, Z.N., Prin, Y., Bally, R., Thioulouse, J. *et al.* (2009) Rhizosphere microbiota interfers with plant-plant interactions. *Plant and Soil* 321, 259–278.

Santi, C., Bogusz, D. and Franche, C. (2013) Biological nitrogen fixation in non-legume plants. *Annals of Botany* 111(5), 743–767.

Saranraj, P. (2014) Biocontrol potentiality of plant growth promoting rhizobacteria (PGPR) – *Pseudomonas fluorescens* and *Bacillus subtilis*: a review. *African Journal of Microbiology Research* 9, 1265–1277.

Saxena, D. and Stotzky, G. (2001) *Bacillus thuringiensis* (Bt) toxin released from root exudates and biomass of Bt corn has no apparent effect on earthworms, nematodes, protozoa, bacteria, and fungi in soil. *Soil Biology and Biochemistry* 33, 1225–1230.

Schippers, B., Bakker, A.W. and Bakker, P.A.H.M. (1987) Interactions of deleterious and beneficial rhizosphere microorganisms and the effect of cropping practices. *Annual Review of Phytopathology* 25, 339–358.

Schlaeppi, K. and Bulgarelli, D. (2014) The plant microbiome at work. *Molecular Plant–Microbe Interactions* 28, 212–217.

Schmelz, E.A., Engelberth, J., Alborn, H.T., O'donnell, P., Sammons, M. *et al.* (2003) Simultaneous analysis of phytohormones, phytotoxins, and volatile organic compounds in plants. *Proceedings of the National Academy of Sciences of the United States of America* 100, 10552–10557.

Schmitt, H., Haapakangas, H. and Van Beelen, P. (2005) Effects of antibiotics on soil microorganisms: Time and nutrients influence pollution-induced community tolerance. *Soil Biology and Biochemistry* 37, 1882–1892.

Selvakumar, G., Panneerselvam, P. and Ganesamurthy, A.N. (2014) Biosafety of novel bioinoculants. *Journal of Biofertilizers and Biopesticides* 5, 145.

Shanahan, P.J., O'sullivan, D., Simpson, P., Glennon, J.D. and Fergal, O.G. (1992) Isolation of 2,4-diacetylphloroglucinol from a fluorescent Pseudomonad and investigation of physiological parameters influencing its production. *Applied and Environmental Microbiology* 58, 353–358.

Sharma, V.K. and Nowak, J. (1998) Enhancement of verticillium wilt resistance in tomato transplants by *in vitro* co-culture of seedlings with a plant growth promoting rhizobacterium (*Pseudomonas* sp. strain PsJN). *Canadian Journal of Microbiology* 44, 528–536.

Shi, Y., Lou, K. and Li, C. (2010) Growth and photosynthetic efficiency promotion of sugar beet (*Beta vulgaris*, L.) by endophytic bacteria. *Photosynthesis Research* 105, 5–13.

Singh, V., Upadhyay, R.S., Sarma, B.K. and Singh, H.B. (2016) Seed bio-priming with *Trichoderma asperellum* effectively modulate plant growth promotion in pea. *International Journal of Agriculture, Environment and Biotechnology* 9, 361.

Small, P., Blankenhorn, D., Welty, D., Zinser, E. and Slonczewski, J.L. (1994) Acid and base resistance in *Escherichia coli* and *Shigella flexneri*: role of rpoS and growth pH. *Journal of Bacteriology* 176, 1729–1737.

Somers, E., Vanderleyden, J. and Srnivasan, M. (2004) Rhizosphere bacterial signalling: a love parade beneath our feet. *Critical Review in Microbiology* 30, 205–240.

Stelma, G.N., JR., Reyes, A.L., Peeler, J.T., Francis, D.W., Hunt, J.M. *et al.* (1987) Pathogenicity test for *Listeria monocytogenes* using immunocompromised mice. *Journal of Clinical Microbiology* 25, 2085–2089.

Stephens, J.H.G. and Rask, H.M. (2000) Inoculant production and formulation. *Field Crops Research* 65, 249–258.

Sun, C.X., Li, M.Q., Gao, X.X., Liu, L.N., Wu, X.F. and Zhou, J.H. (2015) Metabolic response of maize plants to multi-factorial abiotic stresses. *Plant Biology* Suppl 1, 120–129.

Sundh, I., Hökeberg, M., Levenfors, J.J. and Nilsson, A.I. (2011) Safety assessment of biocontrol and plant growth-promoting pseudomonads useful in crop production. *Annals of Applied Biology* 159, 291–301.

Tan, M.-W., Mahajan-Miklos, S. and Ausubel, F.M. (1999a) Killing of *Caenorhabditis elegans* by *Pseudomonas aeruginosa* used to model mammalian bacterial pathogenesis. *Proceedings of the National Academy of Sciences of the United States of America* 96, 715–720.

Tan, M.W., Rahme, L.G., Sternberg, J.A., Tompkins, R.G. and Ausubel, F.M. (1999b) *Pseudomonas aeruginosa* killing of *Caenorhabditis elegans* used to identify *P. aeruginosa* virulence factors. *Proceedings of National Academy of Sciences of the United States of America* 96, 2408–2413.

Taylor, L.H., Latham, S.M. and Woolhouse, M.E. (2001) Risk factors for human disease emergence. *Philosophical Transactions of the Royal Society London B Biology Science* 356, 983–989.

Timmusk, S., Timmusk, K. and Behers, L. (2013) Rhizobacterial plant drought stress tolerance enhancement: towards sustainable water resource management and food security. *Journal of Food Security* 1, 6–9.

Trapet, P., Avoscan, L., Klinguer, A., Pateyron, S., Citerne, S. *et al.* (2016) The *Pseudomonas fluorescens* siderophore pyoverdine weakens *Arabidopsis thaliana* defense in favor of growth in iron-deficient conditions. *Plant Physiology* 171, 675–693.

Turan, M., Gulluce, M., Cakmakci, R., Oztas, T., Sahin, F. and Gilkes, R. (2010) The effect of PGPR strain on wheat yield and quality parameters. *Proceedings of the 19th World Congress of Soil Science: Soil Solutions for a Changing World, Brisbane, Australia, 1–6 August 2010. Symposium 2.3. 1 The Soil-root Interface.* International Union of Soil Sciences (IUSS), c/o Institut für Bodenforschung, Universität für Bodenkultur, Vienna, Austria, 140–143.

Turner, T.R., James, E.K. and Poole, P.S. (2013) The plant microbiome. *Genome Biology* 14, 209–209.

Vacheron, J., Desbrosses, G., Bouffaud, M.-L., Touraine, B., Moënne-Loccoz, Y. *et al.* (2013) Plant growth-promoting rhizobacteria and root system functioning. *Frontiers in Plant Science* 4, 356.

van Loon, L.C., Bakker, P. and Pieterse, C.M.J. (1998) Systemic resistance induced by rhizosphere bacteria. *Annual Review of Phytopathology* 36, 453–483.

Vassilev, N., Vassileva, M. and Nicolaeva, I. (2006) Simultaneous P-solubilizing and biocontrol activity of microorganisms: Potentials and future trends. *Applied Microbiology Biotechnology* 71, 137–144.

Vílchez, J.I., Navas, A., González-López, J., Arcos, S.C. and Manzanera, M. (2016) Biosafety test for plant growth-promoting bacteria: proposed environmental and human safety index (EHSI) protocol. *Frontiers in Microbiology* 6, 1514.

Vitousek, P.M., Naylor, R., Crews, T., David, M.B., Drinkwater, L.E. *et al.* (2009) Nutrient imbalances in agricultural development. *Science* 324, 1519–1520.

Vodovar, N., Vallenet, D., Cruveiller, S., Rouy, Z., Barbe, V. *et al.* (2006) Complete genome sequence of the entomopathogenic and metabolically versatile soil bacterium *Pseudomonas entomophila*. *Nature Biotechnology* 24, 673–679.

Walker, T.S., Bais, H.P., Grotewold, E. and Vivanco, J.M. (2003) Root exudation and rhizosphere biology. *Plant Physiology* 132, 44–51.

Weinzierl, R., Henn, T., Koehler, P.G. and Tucker, C.L. (1995) Microbial insecticides. *The Institute of Food and Agricultural Sciences* ENY-275 (IN081), 1–13.

WHO (2015) *World Health Report – 2015*. World Health Organization, Geneva, Switzerland.

Zachow, C., Pirker, H., Westendorf, C., Tilcher, R. and Berg, G. (2009) The *Caenorhabditis elegans* assay: a tool to evaluate the pathogenic potential of bacterial biocontrol agents. *European Journal of Plant Pathology* 125, 367–376.

Zaidi, A., Khan, M.S. and Aamil, M. (2004) Bioassociative effect of rhizospheric microorganisms on growth, yield and nutrient uptake of greengram. *Journal of Plant Nutrition* 27, 599–610.

Zhang, B., Zhang, P. and Chen, X. (2000) Factors affecting colonization of introduced microorganisms on plant roots. *Ying Yong Sheng Tai Xue Bao* 11, 951–953.

Zhang, H., Xie, X., Kim, M.S., Kornyeyev, D.A., Holaday, S. and Pare, P.W. (2008) Soil bacteria augment *Arabidopsis* photosynthesis by decreasing glucose sensing and abscisic acid levels in planta. *The Plant Journal* 56, 264–273.

Zhang, H., Sun, Y., Xie, X., Kim, M.-S., Dowd, S.E. and Paré, P.W. (2009) A soil bacterium regulates plant acquisition of iron via deficiency-inducible mechanisms. *The Plant Journal* 58, 568–577.

5 Role of Plant Growth-Promoting Microorganisms in Sustainable Agriculture and Environmental Remediation

Rama Kant Dubey,[1] Vishal Tripathi,[1] Sheikh Adil Edrisi,[1] Mansi Bakshi,[1] Pradeep Kumar Dubey,[1] Ajeet Singh,[1] Jay Prakash Verma,[1] Akanksha Singh,[2] B.K. Sarma,[3] Amitava Rakshit,[4] D.P. Singh,[5] H.B. Singh[3] and P.C. Abhilash[1]*

[1]Institute of Environment and Sustainable Development, Banaras Hindu University, Varanasi, India; [2]Microbial Technology and Nematology Division, CSIR-Central Institute of Medicinal and Aromatic Plants, Lucknow, India; [3]Department. of Mycology and Plant Pathology, Institute of Agricultural Sciences, Banaras Hindu University, Varanasi, India; [4]Department of Soil Science and Agricultural Chemistry, Institute of Agricultural Sciences, Banaras Hindu University, Varanasi, India; [5]ICAR-National Bureau of Agriculturally Important Microorganisms, Kushmaur, Mau Nath Bhanjan, Mau, India

5.1 Introduction: Plant Growth-Promoting Rhizobacteria (PGPR)

The world population is projected to be 9.5 billion by the year 2050 (Godfray *et al.*, 2010) which will demand at least 50% increase in the food production (Abhilash *et al.*, 2016a). This rapid increase in global population will also demand more arable land to meet the challenge of producing food, fodder, fibre and biomass for biofuel. Coupled with the increasing population, changing climate may also induce various biotic and abiotic stresses in the near future. This may lead to an increase in the use of chemical fertilizers and pesticides to protect the crops and increase the agricultural production. The extensive use of agrochemicals, rapid urbanization and industrialization have already polluted and dearly cost the environment and poses even more serious threat to the environment by polluting air, water and soil (Abhilash *et al.*, 2013a; 2016a, b). About 30% of the global land area is already degraded or contaminated due to various anthropogenic activities (Abhilash *et al.*, 2013b). There is an immediate need to take preventive measures and save our soils from further degradation. Besides, we also need to sustainably increase the agricultural production to meet our goal of future demand. Thus, it is imperative to search, develop, implement and adopt novel agricultural tools to meet the requirements of the sustainable food production for future generations. These studies involve observation of the works already performed, management of soil and application of new technologies.

All over the world several agricultural crops are being produced with various distinct nutritional qualities. Although collective improvements through scientific and

*E-mail: pcabhilash@hotmail.com

© CAB International 2017. *Advances in PGPR Research* (eds H.B. Singh, B.K. Sarma and C. Keswani)

technological interventions increased the crop yield, sustainable and equitable feed for overwhelming population is still a global issue (Wu *et al.*, 2014). Food crises, malnutrition, especially that of micronutrients, and current practices heavily reliant on synthetic agrochemicals remain a major challenging health and environmental hazard. To combat all these issues, there is a prime need to increase the crop yield and nutritional make-up of the food crop by adopting more sustainable agricultural approaches (Dubey *et al.*, 2015, 2016; Abhilash *et al.*, 2016a, b).

The rhizospheric domain of the plant supports various forms of life by providing them with nutrients as root exudates. Approximately 40% of the total plant photosynthate secreted as root exudate in the rhizosphere by plants is a rich source of sugars, phenolics, organic acids, amino acids and proteins (Bais *et al.*, 2006). This is the reason why microbial diversity is very high in the rhizosphere as microorganisms feast on these carbon-rich compounds (Philippot *et al.*, 2013). On the other hand, microorganisms promote the plant growth supporting plant life by providing vital nutrients to them and rescue the plant from various biotic and abiotic stresses (Bais *et al.*, 2006; Philippot *et al.*, 2013; Ray *et al.*, 2016a, b). The rhizosphere also harbours microorganisms with ability to degrade and metabolize various xenobiotics. Some of these microorganisms even have the multi-pronged potential of simultaneously supporting plant growth and removing pollutants from soil. Thus, the rhizosphere microorganisms could be harnessed as a tool for sustainably increasing crop productivity and restoring degraded land without harming the ecosystem (Singh *et al.*, 2004; Abhilash *et al.*, 2009; Weyens *et al.*, 2009; Abhilash *et al.*, 2012; Singh *et al.*, 2016a, b).

Microorganisms within the rhizosphere thus hold promising functional attributes that are intrinsic to their pattern of life. They are the most natural inhabitants keeping the whole rhizospheric ecosystem alive in terms of buffering the microenvironment with ionic and metabolic exchanges, enriching roots from nutrient solubilization and mobilization. They also mantain the microbial community dynamics through myriad signalling molecules and provide tolerance to biotic and abiotic stresses systemically (Singh *et al.*, 2014; Saxena *et al.*, 2015). This is why such microorganisms offer great potential for environmentally friendly sustainable crop productivity.

Although extensive findings and applications corresponding to plant growth-promoting microorganisms are available in the scientific literature, comprehensive knowledge of these microbes and their potential for plant growth promotion, enhancing food and nutritional security, plant disease management, sustainable agriculture, carbon sequestration, phytobioremediation of soils mildly, moderately and heavily contaminated by pesticide, organic and heavy metals, and for biomass and biofuel production are not properly discussed by the scientific community. This evaluation aims to present and discuss the growth potential of the plant to promote all these microorganisms mentioned above. The response of the plant growth promoted by microbes under climate interference, agronomic and rhizospheric engineering and molecular approaches can certainly improve the activity of these microorganisms.

5.2 Role of Plant Growth-Promoting Bacteria (PGPR) and Fungi (PGPF) in Sustainable Agriculture

Implications and holistic usage of these plant growth-promoting rhizobacteria (PGPRs) for increasing agricultural produce, regulation of biogeochemical cycles and maintaining homeostasis within the root ecosystem have generated keen attention from the scientific community towards these unseen organisms (Dubey *et al.*, 2015, 2016). PGPRs are a distinct group of microbes that improve plant performance by involving numerous independent or linked mechanisms with their multitrophic participation between the plants and microbial communities. These plant–rhizospheric chemical interactions support the plants during various developmental stages and environmental stress conditions. All these chemical interactions happen in the soil ecological environment known as the rhizosphere and control the plant health as well as soil fertility simultaneously. This parallel between plant and PGPR interaction positively affects the plant growth and could be an excellent option for future needs such as sustainable or ecological intensification and sustainable agriculture (Lugtenberg and Kamilova, 2009). Phyllospheric microbial communities of the plant significantly affect the plant health and growth (Vorholt, 2012; Bulgarelli *et al.*, 2013; Bisen *et al.*, 2016). Rhizospheric PGPR can enhance plant tolerance by promoting plant growth, even in poor growth conditions and increase agricultural produce of different crops under stressful environments (Singh *et al.*, 2011a; Nadeem *et al.*, 2014; Bisen *et al.*, 2015). Apart from the above-mentioned facts, recent reports suggest that application of PGPRs also improves nutritional quality and antioxidant status of the crops (Jain *et al.*, 2014; Singh *et al.*, 2014). Harnessing the above-mentioned plant–microbe interactions can also help in reclamation of degraded lands, reduction in usage of chemical fertilizers and agrochemicals (Mishra *et al.*, 2015).

A prominent agricultural symbiotic association exists between the rhizospheric bacteria and roots of the legumes by the formation of root nodules. Previous studies showed that plant–fungal associations are much older than the rhizobia–legume interaction. In various plant–fungal interactions fungi help in phosphate acquisition and make it available to plants (Marx, 2004). Some reports also indicate that the DMI2 protein is required for the initiation of the plant-arbuscular mycorrhizae interaction, which helps in phosphate solubilization. Although the underlying mechanism of the PGPR and PGPF interactions with the plants are quite different, some studies showed a similarity between them (Fig. 5.1). In *Medicago truncatula,* a type of Arbuscular Mycorrhizal Fungus (AMF), the interaction releases some small diffusible factors to activate the similar genes by the rhizobacterial Nod factor (Marx, 2004). This confirms the analogy of the PGPR and PGPF to some extent and needs further clarification from cutting-edge research on the topic. Mutualistic association (co-inoculation) of PGPR and PGPF (arbuscular mycorrhizae) increases the growth, nutrient uptake potential, and yield of the plants (Rathi *et al.*, 2014). PGPF can directly enhance the nutrient uptake (P, Zn) and water use efficiency of the inhabiting plant by increasing the root surface with hyphal network. With increased water use efficiency AMF also controls the N_2O emission, which is a potent greenhouse gas emitted from agricultural fields (Lazcano *et al.*, 2014). AMF alone or in combination with certain PGPR enhances plant growth indirectly by inhibiting growth of root pathogens and optimizing soil structures (Smith and Read, 2008). Apart from this PGPF can also regulate soil health and fertility by improving the overall soil nutrient dynamics (Berta *et al.*, 2014). The negative effect of climate change is also mitigated by AMF through maintenance of proper soil aggregation and thereby providing another major advantage to agricultural crop production. More and more studies show that the mycorrhizae can play an essential role in plant growth by enhancing plant vigour in poorly performing soils, and through their ability to store large amounts of carbon, which in turn may improve some of the effects of climate change.

In conclusion we can say that application of PGPR and PGPF in combination or alone can negate the hazardous effect of chemical fertilizers, improve soil health, reduce environmental stresses and promote sustainable agriculture (Abd-Alla *et al.*, 2014; Keswani, 2015). The interaction of AMF and rhizobacteria thus can promote plant growth by improving soil structural properties as well as the enhanced availability of nutrients and reduce disease progression in a sustainable manner (Fig. 5.1).

5.2.1 Fixation, solubilization and mineralization of nutrients

In plant–microbe interactions, the rhizospheric region surrounding the rhizoplane in subsoil is the most crucial and active zone of the plant (Singh *et al.*, 2004; Bisen *et al.*, 2015, 2016). This region is nutrient rich (rhizodeposits and root exudates) and

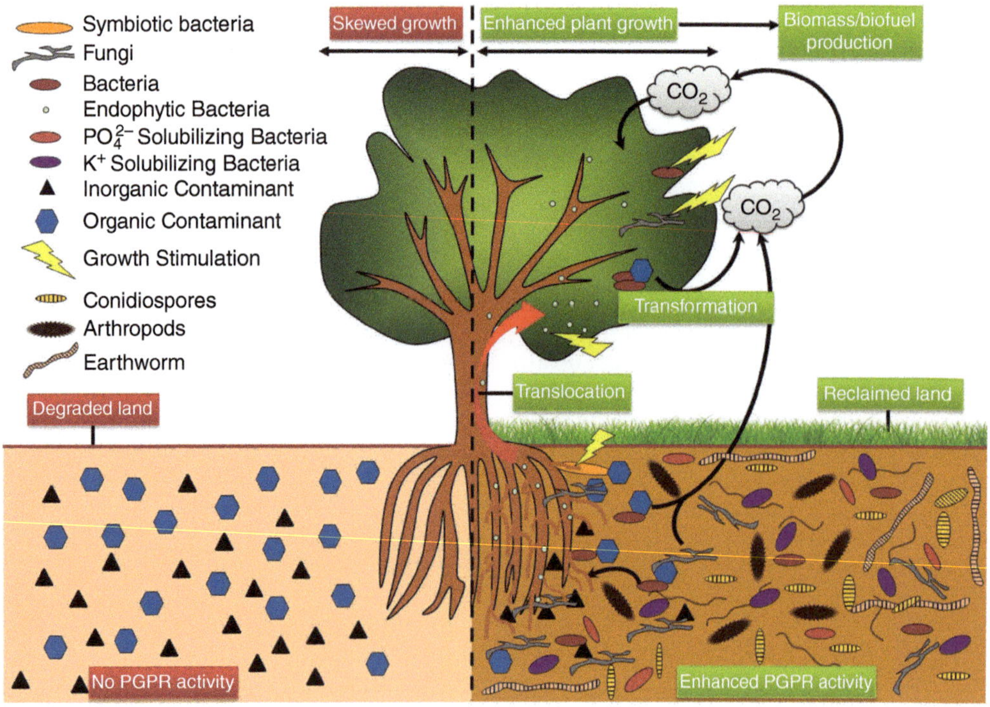

Fig. 5.1. Schematic representation of the role of plant growth-promoting microorganism in remediation and management of contaminated and degraded lands and also for biofuel production.

harbours a variety of microbial life (Bais *et al.*, 2006; Keswani *et al.*, 2016) including PGPR, PGPF and other root-associated microfauna. During plant–microbe interactions, a multitude of the complex reactions are governed by bioactive molecules like phytohormones, secondary metabolites/flavonoids (Lugtenberg and Kamilova, 2009; Hassan and Mathesius, 2012), signalling molecules (Buee *et al.*, 2000), plant/microbial enzymes (Chisholm *et al.*, 2006), etc., that are present in this crucial pocket. The rhizospheric microbial community also actively participates in biogeochemical cycling (Frey-Klett *et al.*, 2011) with the help of flavonoids (Cesco *et al.*, 2012) via solubilization, decomposition and mineralization of nutrients. Such processed nutrients act as energy source and are uptaken by the plant root. PGPR and PGPF are known to facilitate plant growth via certain mechanisms. The key mechanisms include conversion of unavailable forms of nutrients like nitrogen (N), phosphorus (P), potassium (K) and microelements to the available form and thereby making it available to plants (Glick, 2012). Soil fertilization is generally required for crop production; however, its excess application reduces the nutrient use efficiency and hence contaminates surface and ground waters as well as atmospheric systems (Tilman *et al.*, 2001). Smil (2000) concluded that 75% of the phosphate fertilizers applied to soil is reported to be rapidly lost and become unavailable to plants. However, this problem can be overcome by application of phosphate-solubilizing bacteria and fungi. The different populations of the soil microbes (PGPR and PGPF) are affected by a wide range of factors, biotic or abiotic. To unravel the hidden mystery of aboveground–belowground interactions and nutrient turnover in climate change scenarios, biotechnological interventions and integrated modelling should be emphasized (Abhilash and Dubey, 2014).

Biological nitrogen fixation

Nitrogen (N) is an essential nutrient for plant growth and development. In the current scenario nitrogen is becoming a limiting factor for agricultural crop growth. The major nitrogen losses are through runoff from agricultural lands, mineral leaching, lower nutrient use efficiency (Tilman *et al.*, 2002; Bhattacharyya and Jha, 2012), and more N_2O emission from agriculture fields (Reay *et al.*, 2012). Although the use of chemical N fertilizers in agriculture boost the crop yield, only 30–50% of the applied N fertilizer is utilized by crop plants (Smil, 1999). Therefore excessive use of synthetic nitrogenous fertilizer has raised many questions over agricultural sustainability. Another factor is that activity may result in soil property alterations and raising of the microbial biomass, changing the bacterial community structure over time; it can lead to the decrease of specific bacteria relevant to soil activity. A further example is the increase of denitrification against the use of soil N fertilization (Webb *et al.*, 2004). In this context promotion of biological N fixation could be a green technology in replacing/minimizing chemical fertilizers. Biological N fixation converts the atmospheric N into ammonia, which is a highly transcriptionally regulated process (Dixon and Kahn, 2004) and achieved through symbiotic and free-living PGPR.

In the process of symbiotic nitrogen fixation the molecular crosstalk involves the nitrogen fixing (*nif*) (Halbleib and Ludden, 2000) and nodulation (*nod*) genes (Abd-Alla, 2011). The *nif* genes activate the iron–molybdenum cofactor biosynthesis, electron donation and regulatory genes, required for functioning of the nitrogenase enzyme complex (Halbleib and Ludden, 2000). The *nod* genes of the rhizobia get activated by the root-released flavonoids of the plant and help in the downstream process of the symbiosis (Abd-Alla, 2011). The bacteria also shift physiology from glycogen synthesis to oxidative phosphorylation to enhance the ATP

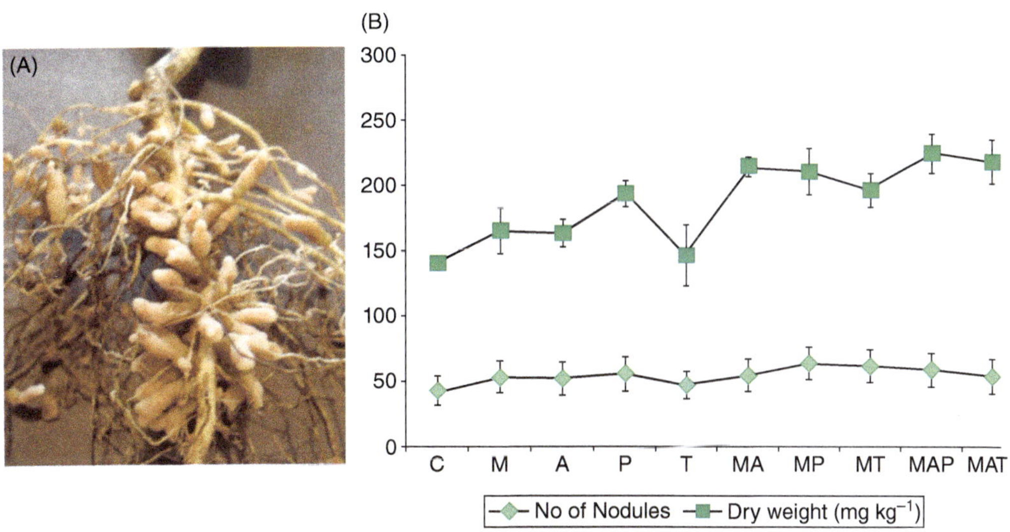

Fig. 5.2. (A) Enhanced root nodulation in chickpea (*Cicer arietinum* L.) observed after the tri-inoculation of microbial consortia MAP (*Mesorhizobium* sp. + *Azotobacter chrococcum* MTCC-446 + *Trichoderma harzianum*. (B) Number of nodules and nodules dry weight in chick pea inoculated with single, double and triple combinations of PGPR (Verma *et al.*, 2014). Figure legends: C (Control (un-inoculated)); M (*Mesorhizobium* sp.); A (*Azotobacter chrococcum* MTCC-446); P (*Pseudomonas aeruginosa* BHU PSB01); T (*Trichoderma harzianum*); MA (*Mesorhizobium* sp. + *Azotobacter chrococcum* MTCC-446); MP (*Mesorhizobium* sp. + *P. aeruginosa* BHU PSB01); MT (*Mesorhizobium* sp. + *T. harzianum*); MAP (*Mesorhizobium* sp. + *A. chrococcum* MTCC-446 +*P. aeruginosa* BHU PSB01); MAT (*Mesorhizobium* sp. + *A. chrococcum* MTCC-446 + *T. harzianum*).

production required for the endergonic reactions of N_2 fixation (Glick, 2012). It was observed that deletion of glycogen synthase in *Rhizobium tropici* enhanced nodulation in bean plants (Marroquí *et al.*, 2001) (Fig. 5.2). Alternatively the increased expression of the H uptake hydrogenase can make the nitrogen fixation machinery more anoxygenic. It utilizes the liberated H^+ quickly to produce more ATP that favours the nitrogen fixation process. In a recent study the change in the expression of hoxB, a small subunit for hydrogenase uptake in *Azoarcus* sp. Strain BH72, helped in more nitrogen accumulation in Kallar grass (Sarkar and Reinhold-Hurek, 2014). Also the insertion of bacterial haemoglobin gene into PGPR strains or transferring nitrogen fixing genes into non-legume plants (Beatty and Good, 2011; Geurts *et al.*, 2012) improved nitrogen use efficiency of the test plants. PGPRs also synthesize rhizobitoxine (Yuhashi *et al.*, 2000) which upregulates the synthesis of ACC deaminase leading to down regulation in synthesis of ethylene hormone, helping the process of N_2 fixation (Glick, 2014). Nitrogen-fixing PGPR enhance the carbon and nitrogen metabolism of the test plants increasing the plant growth and productivity. Application of nitrogen-fixing PGPR *Burkholderia caribensis* XV on *Amaranthus hypochondriacus* enhanced the carbon and nitrogen metabolism by over-expression of the genes like *ahnadh gogat* (NADH dependent glutamate synthase), *AhNRT.1.1* (a nitrate transporter type 1.1), *AhAlaAT* (alanine aminotransferase), *DOF1* (transcription factor AhDOF1) and *AhGS1* (cytosolic glutamine synthase 1) (Parra-Cota *et al.*, 2014).

Apart from the various above-mentioned mechanisms PGPR and PGPF are actively involved in nitrogen fixation. *Rhizobium, Azorhizobium* in legume plants, and *Frankia* in nonlegume can fix atmospheric nitrogen. Dinitrogen fixation, by the legume in a cereal–legume mixed cropping system can concurrently transfer about 20–30% of the fixed N_2 to the nonlegume (Patra *et al.*, 1986). Apart from this, non-symbiotic *Pseudomonas* (Mirza *et al.*, 2006), *Azoarcus, Acetobacter diazotrophicus, Azotobactor, Azospirillum* (de-Bashan *et al.*, 2010), and cyanobacteria also have the ability to fix atmospheric nitrogen and make it available for plants. *Rhizobium daejeonense, P. monteilii* and *Enterobacter cloacae* and *Bacillus pumilus* isolated from rice, wheat and maize, respectively, also possess good nitrogen fixation activity (Habibi *et al.*, 2014). Previous studies also showed that AMF belonging to the order Glomalescan are important contributors to enhanced N acquisition under some conditions (Hodge *et al.*, 2001). Karasu *et al.* (2009) observed that inoculation of chickpea (*Cicer arietinum*) seeds with *Rhizobium ciceri* isolate had a significant effect on seed yield, plant height, first pod height, number of pods per plant, number of seeds per plant, harvest index and 1000 seed weight under different nitrogen doses. Inoculation of *Bradyrhizobium* and *Pseudomonas striata* in *Glycine max* enhanced the rate of nitrogen fixation (Dubey, 1996). Thus, composite inoculation of nitrogen-fixing PGPR is also an option for sustainable agriculture. However, exact estimates of microbial effects on plant productivity in a natural system are often difficult to understand: it is necessary to explore the involved mechanism(s) and molecular dialogue(s) of the nitrogen fixation pathways.

Phosphate solubilization

Phosphorus (P) is a nonrenewable resource and essential macronutrient for plant growth and development. However only 45% of phosphate fertilization is utilized by the plant (Smil, 2000). In the period 1960–1995 the use of phosphorus fertilizer increased 3.5-fold and is expected to be 3-fold more by 2050 with the same fertilizer efficiency (Tilman *et al.*, 2001). Phosphate fertilizer after a complex exchange remains in the soil. This nutrient concentration of the soil gets reduced along with crop harvesting and demands for more N, P chemical fertilizer globally. Rock phosphate is a precious resource for phosphorus but it is not readily available to plants. Only a minor proportion of this nutrient resource is released via biological or chemical processes that are unable to sustain the P requirements of the crop plants. Other forms of insoluble soil phosphorus are inorganic mineral (P_i) forms such as apatite, strengite, variscite; organic forms (P_o) such as soil phytate, phosphonates, phosphomonoesters, triesters (Khan *et al.*, 2007; Shen *et al.*, 2011) and applied chemical fertilizer. These forms are highly stable, unavailable to the plants and limit their growth (O'Rourke *et al.*, 2013). In this critical situation it is necessary to find alternatives to chemical fertilizers and develop some suitable methods for solubilization of the stable phosphorus. In this aspect the PGPR and PGPF could be a green replacement for the chemical fertilizers and solubilize the above-mentioned insoluble phosphate, increasing its bioavailability to the plants (Marschner *et al.*, 2011).

Diverse phosphate-solubilizing bacteria (PSB) and PGPF solubilize the inorganic phosphate (P_i) by producing H^+, OH^- or CO_2, and citric, oxalic and gluconic (de Oliveira Mendes *et al.*, 2014) acetic, succinic, malic, oxaloacetic, pyruvic and α-ketoglutaric acids (Mardad *et al.*, 2013). The organic phosphate (P_o) proportion in agricultural fields is about 50%. Its mineralization utilizes microbial origin phytase (Menezes-Blackburn *et al.*, 2013), phosphatase (Spohn *et al.*, 2013), and phosphoric ester hydrolysis.

In a recent study, the fungi *Aspergillus niger* FS1, *Penicillium canescens* FS23, *Eupenicillium ludwigii FS27* and *Penicillium islandicum* FS30 were analysed for their phosphorus-solubilizing potential. Out of four species *Aspergillus niger* FS1 showed excellent potential to solubilize a variety of insoluble phosphorus, such as $AlPO_4$, $FePO_4$, $Ca_3(PO_4)_2$ and Araxá rock phosphate (de Oliveira Mendes *et al.*, 2014). Other phosphate-solubilizing bacteria such as *Enterobacter hormaechei* sub sp. *Steigerwaltii* strain NM23-1, *Enterobacter* sp. strain TSSAS2-48 and Bacterium DR172, solubilize the inorganic phosphorus by synthesizing gluconic acid (Mardad *et al.*, 2013). Singh and Kapoor (1999) reported that PGPF, *Glomus* sp. 88, *B. circulans* and *Cladosporium herbarum*, (single and consortia inoculation) increased the population of P-solubilizing microorganisms in wheat rhizosphere. Grain yield increased in the treatments that were inoculated with the AMF and Mussoorie rock phosphate. PGPR like *Pantoea cypripedii* and *Pseudomonas plecoglossicida* when inoculated in maize and wheat crops increased crop yield, phosphorus uptake, enzyme activities, P-solubilizing bacterial population and soil organic carbon (SOC) under different agroclimatic regions. The effects of *Pantoea cypripedii* and *Pseudomonas plecoglossicida* become more pronounced when it is amended with rock phosphate (Dharni *et al.*, 2014a; Kaur and Reddy, 2014). Similarly, phosphate-solubilizing diazotrophic bacteria when inoculated with tricalcium phosphate (TCP) promoted phosphate solubilization and nutrient uptake potential of rice plants as well as aromatic crops (Sahay and Patra, 2014). Inoculum of *Herbaspirillum* strains (H18, ZA15) and a *Burkholderia vietaminensis* strain (AR114) when used with TCP improved the rice grain yield by 33–47% (Estrada *et al.*, 2013). *Fusarium verticillioides* RK01 is an endophytic fungus which when inoculated in soybean plants significantly promoted the shoot length by 7.3% over the control plants. It also showed increased phosphate-solubilizing activity with 1.0 U^{-1} g fwt acid and 2.1 U^{-1} g fwt alkaline phosphatase activities (Radhakrishnan *et al.*, 2015). Enhancing nonmycorhhizal microbes by root exudation might be a strategy for increasing the organic phosphate (P_o) mineralization (Spohn *et al.*, 2013).

In some cases microorganisms may also negatively regulate P availability to the plants. The process involves sequestration of phosphate by microbes (Baggie *et al.*, 2005), which may be broken down by phosphate mobilizing molecules secreted by roots, and thereby inhibit root growth. Although a wide variety of microbial strains have been deciphered in response to phosphate solubilization, it is however, time to assess their implications at field scale. It is also necessary to develop some genetically engineered strains of PGPRs and PGPFs with the P_i solubilization and P_o mineralization activity. These developed inoculants can improve yield and nutrient quality of the agricultural produce sustainably.

Potassium solubilization

Potassium (K) is an essential macronutrient among NPK, necessary for plant growth and development. Although the total soil K is generally enough, its forms (exchangeable or soluble, nonexchangeable or insoluble or mineral) are very dynamic, thus often not freely available to plants. Also the potassium pool of the soil is continually depleting due to long natural processes of weathering, leaching, runoff, improper awareness, and removal by crop residues from agricultural fields (Sheng *et al.*, 2002). In low potassium conditions the plants become more susceptible to pests (Troufflard *et al.*, 2010) and diseases (Armengaud *et al.*, 2010). Potassium solubilizing bacteria (KSB) have potential to solubilize rock K (an insoluble form of K) such as micas, feldspars, illite and orthoclases, by producing a group of organic acids (Ullman *et al.*, 1996) like citric, acetic, tartaric, oxalic, lactic, and malic (Hu *et al.*, 2006). K-solubilizing bacteria in the rhizospheric region solubilize insoluble minerals and release potassium, aluminium and silicon and make it available to the plant (Hu *et al.*, 2006). K uptake mechanism is totally governed by different types of K transporters. In *Escherichia coli* K-12 three (Trk, Kdp, and Kup) types of K transporter have been identified. The first two are the major uptake and the third is the minor K uptake system. The *Ktr* gene found in *Bacillus subtilis* regulates K uptake. This gene is homologous to the bacterial KUP (TrkD) potassium transporter. In fungi a P-type ATPase transporter is there for K transport that resembles animal sodium potassium ATPase. In *Ustilago maydis* there are three genes named as *Umacu1, Umacu2* and *PsACU1* encoding for the P-type ATPase.

Application of *Bacillus mucilaginosus* solubilized potassium and increased its availability in the rhizospheric soil leading to increased potassium content in the plant tissues (Sheng *et al.*, 2002). Also *Bacillus mucilaginosus* used as a potassic fertilizer in *Sorghum vulgare* Pers. var. Sudanens showed significant enhancement in biomass, K uptake and yield over the control levels (Basak and Biswas, 2009). Some silicate-dissolving bacteria can also release Si, Fe and K from feldspar (Badr, 2006). Certain PGPR such as *Azospirillum brasilense* sp245 have the attribute to increase the uptake of K, S, Na, Mn, Cd, and Ni elements of the host crop simultaneously with plant growth (Güneş *et al.*, 2014). *Penicillium pinophilum* (NFCCI 2498) when associated with *Punica granatum* solubilizes the insoluble potassium and results in improved potassium and phosphorus uptake potential by 47.47% and 63.44%, respectively. Its inoculation also increased the leaf area index and rate of photosynthesis with improved growth of the plants. There were increased dehydrogenase, alkaline and acid phosphatase activities recorded in inoculated plants rather than the control (Maity *et al.*, 2014). In sustainable agricultural practices dual inoculation of such microbes can enhance the overall plant health. Hence, co-inoculation of PSB (*Bacillus megaterium* var. *phosphaticum*) and KSB (*Bacillus mucilaginosus*) was done in pepper and cucumber to check the potential of this biofertilizer consortium. When this co-inoculation was supplemented with rock mineral it increased uptake of NPK and plant growth (Han and Lee, 2006). Irrespective of various KSM identified, further research is needed for its successful application to the field level and its implication to solubilize more mineral phosphate.

Fe sequestration

In the micronutrient profile iron is an essential element for plant growth. In aerated soils most of the iron is present in Fe^{3+} forms which are less soluble and less available to the plants. Plant and microbes have developed certain mechanisms to combat the issue of iron deficit. Dicots and non-Poaceae monocots synthesize protons, phenolics and organic acid anions which interact with Fe^{3+} and make it soluble and available to plants (Vert *et al.*, 2002). In contrast, members of Poaceae produce phytosiderophores to chelate iron (sometimes Cu and Zn) and make them available to plants. Since the bacteria and fungi also require a certain amount of iron to maintain their physiology they also synthesize a low molecular weight biosynthetic siderophore to chelate iron. There is high affinity (K_a = 10^{30} to 10^{53}) (Matzanke, 1991) between siderophores and siderophore receptors that bind to the Fe–siderophore complex for iron sequestration by microbes (Hider and Kong, 2010). Microbes produce a variety of siderophores (Matzanke, 1991) that may be carboxylates (i.e. rhizobactin), catecholates (i.e. enterobactin) or hydroxamates (i.e. ferrioxamine B) in nature. Dharni *et al.* (2014b) identified *Pseudomonas monteilii* from tannery sludge, which has significant Fe sequestration properties. In bacterial systems some mixed functional groups of siderophores are also found, called pyoverdine (Cornelis, 2010). However, fungal siderophores are mostly hydroxamates belonging to the ferrichrome family (i.e. ferrichrome) (Winkelmann, 2007). Microbes can also utilize siderophores produced by the other species (Raaijmakers *et al.*, 1995) as a bacterial siderophore like pyoverdine has more affinity for iron than the phytosiderophores helping in extraction of iron from Fe^{3+}-phytosiderophore complex.

Siderophore-mediated transport of iron differs among fungi and bacteria and even between Gram-negative and Gram-positive bacteria. Gram-negative bacteria have TonB-dependent outer membrane surface receptors that recognize the Fe^{3+} siderophore complexes (Krewulak and Vogel, 2008). A total of 9 genes and ABC transporter family help in the further translocation of the Fe^{3+}–siderophore complex, with 8 genes helping in regulation of the transport across the membrane in reduction of Fe^{3+} within the cell and the ABC transporter family helping in the cytoplasmic transport of the complex (Crowley *et al.*, 1991). In Gram-positive bacteria periplasmic siderophore binding protein binds directly to the Fe(III)-siderophore complex and the rest of the transport mechanisms are the same as in Gram negatives (Braun and Hantke, 2011). The reduced Fe^{3+} becomes free from the complex and available for bacterial cells. However, in fungi there are four distinct mechanisms involved in siderophore-driven iron sequestration (Van der Helm and Winkelmann, 1994). Once transported inside the cell the Fe^{3+} becomes free from the siderophore complex and available for utilization by the cell.

Various studies have supported the Fe sequestration capacity of PGPR and PGPFs. *Glomus intraradices* when inoculated in *Zea mays* increased the total Fe content of the shoot and less supplementation of the Fe was required. The uptake potential varied with difference in concentrations of the micronutrients and phosphorus (Liu *et al.*, 2000). A *Klebsiella* sp. strain PS19 isolated from mustard rhizosphere having siderophore activity enhanced plant growth even under higher dose of the herbicides (Ahemad and Khan, 2011b). Another bacterium *Pseudomonas fluorescens* enhanced root nodulation and yield of groundnut by producing siderophores and ACC deaminase simultaneously (Dey *et al.*, 2004). Also the siderophore produced by the PGPRs can control the growth of various phytopathogens by depleting the iron from the rhizosphere (Persello-Cartieaux *et al.*, 2003). Iron oxidizing bacteria such as *Paenibacillus cookie* JGR8, (MTCC12002), *Pseudomonas jaduguda* JGR2, (LMG25820) and *Bacillus megaterium* JGR9 (MTCC12001) when inoculated in *Typha angustifolia* grown in iron depleted and excess conditions (all isolates with siderophore activity), affected the iron accumulation in the plant root. Increased shoot iron content was also recorded with *P. pseudoalcaligenes* JGR2 inoculum. All these set of inoculated experiments showed better plant growth with higher iron content than the control plants (Ghosh *et al.*, 2014). *Pseudomonas fluorescens* synthesizes pyoverdine and increases the iron content and growth of *Arabidopsis thaliana* plants (Vansuyt *et al.*, 2007). Certain PGPR like *Burkholderia cepacia* OSU7 can sequester more iron and can be utilized for increasing crop production under sustainable agriculture (Güneş *et al.*, 2014). Application of PGPRs, viz. *Pseudomonas putida*, *Pseudomonas fluorescens* and *Azospirillum lipoferum* and other rhizobacterial isolates from the rhizospheric soils applied to field grown rice enhanced the iron content in plants and grains (Sharma *et al.*, 2013). Translocation efficiency of iron from roots to shoots and grains is also enhanced upon PGPR treatment. This attribute offers the opportunity to produce iron-biofortified crops. Also siderophores can serve as biocontrol and chelating agents, biosensors and address weathering of minerals as well as help in bioremediation (Ahmed and Holmstrom, 2014). Along with plant growth promotion this could be an additional avenue for enhancing sustainability in agriculture.

5.2.2 Phytostimulation by production of hormones

Phytohormones, namely auxins, cytokinins and gibberellins, that are produced by plant-associated microbes frequently stimulate growth and indeed have been considered key players for altered plant growth and development (Tanimoto, 2005; Patel and Patra, 2014). The release of auxin indole-3-acetic acid (IAA) by plant-associated bacteria enhances plant growth and development by improving the root architecture. It increases the root growth as well as root length, along with proliferation and elongation of root hairs (Tanimoto, 2005). Since root tips and root surfaces are the prime locations of nutrient uptake, it is most likely that one key mechanism by which PGPR lead to increase the nutrient uptake is via stimulation of root development. Increase in number of branches and pods per plant, as well as grain yield has been clearly observed in *Brassica juncea* by PGPRs producing auxin (Asghar *et al.*, 2002). Under high levels of salt, IAA was found to stimulate lengthening of the root and shoot of wheat seedlings, thereby increasing and maintaining productivity (Egamberdieva, 2009). IAA is synthesized by three different routes (the indole-3-pyruvic acid, indole-3-acetamide, and indole-3-acetonitrile pathways) and has a dual role in plant–microbe interactions: it may be beneficial or deleterious to the plant. IAA controls phytostimulation, phytopathogenesis as well as the bacterial physiology for IAA degradation (Duca *et al.*, 2014). Bacteria producing ACC (1-aminocyclopropane-1-carboxylate) deaminase downregulate the plant ethylene level, often a result of environmental stresses, and can thereby promote plant growth and productivity (Glick *et al.*, 2007). These bacteria are also key players in protection of plants against heavy metals, organic pollutants, flooding, drought, salt and both bacterial and fungal pathogens (Glick, 2014).

5.2.3 Enhanced resistance against abiotic stresses

PGPR as an elicitor of induced systemic resistance (ISR) has been much talked about. However, fewer reports have been published underlying the mechanisms adopted by PGPR and PGPF for abiotic stresses, such as drought, salt and nutrient deficiency or excess, for maintaining crop productivity. Drought stress is a limiting factor for growth and productivity of crops, particularly in arid and semi-arid areas. Inoculation with the PGPR *Paenibacillus polymyxa* enhanced drought tolerance in *Arabidopsis thaliana* (Timmusk and Wagner, 1999). The role of ACC deaminase in decreasing ethylene levels by the enzymatic hydrolysis of ACC into ketobutyrate and ammonia has been documented. Another PGPR strain, *Achromobacter piechaudii* ARV8 which produces ACC deaminase, conferred tolerance to drought stress in pepper (*Capsicum annuum* L.) and tomato (*Solanum lycopersicum* L.) plants thereby rescuing normal plant growth (Mayak *et al.*, 2004). Similarly a recent study shows that *Phyllobacterium brassicacearum* STM196 strain induces drought tolerance in *Arabidopsis thaliana* by altering the abscissic acid content, transpiration, photosynthesis, increasing the biomass and water use efficiency with the delayed flowering (Bresson *et al.*, 2013).

5.2.4 Role of PGPR and PGPF in disease control

Plant growth that is promoted by bacteria and fungi has key characteristics to regulate plant diseases by various mechanisms (Fig. 5.3) (Table 5.1). Some of the key mechanisms are antagonistic activity, PGPR-mediated cell–cell communication inhibition, induced systemic resistance, and lipopeptides mediated ISR, as described below.

Antagonism against phytopathogenic microbes

Plants are exposed to a vast array of pathogenic microorganisms during their lifetime

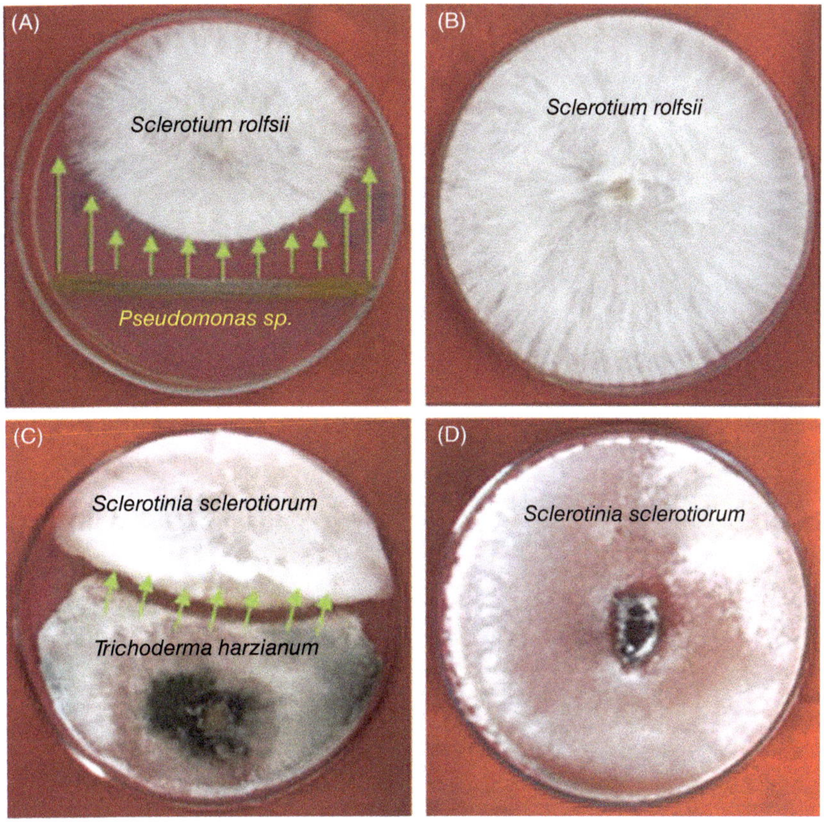

Fig. 5.3. Role of plant growth-promoting microorganisms in disease control. Antagonistic property of *Pseudomonas* sp. and *Trichoderma harzianum* against phytopathogens *Sclerotinia* sp. (A) *Pseudomonas* sp. inhibiting the growth of *Sclerotium rolfsii*. (B) Growth of *Sclerotium rolfsii* without the presence of *Pseudomonas* sp. (C) *T. harzianum* inhibiting the growth of *Sclerotinia sclerotiorum*. (D) *S. sclerotiorum* without the presence of *T. harzianum*. (Arrows indicate the zone of inhibition.)

and the inoculation of PGPR and PGPF potentially benefit plant growth via an indirect route: by fighting "trench warfare" with pathogens for space and nutrients, thus suppressing their growth and activity (Mendes *et al.*, 2011). Dharni *et al.* (2014a) isolated and identified the 2,4 di-tert-butylphenol from a novel strain of *Pseudomonas monteilii*, which was found to be effective against an agriculturally important fungus, viz. *Fusarium oxysporum*, in inhibiting spore germination and hyphal growth. It can be a potent inhibitor of β-tubulin. (Dharni *et al.*, 2014b). According to Shoresh *et al.* (2010) 500 mg of concentrated formulation per hectare was enough for conferring significant advantages to both monocots and dicots by increased plant growth, especially under stress. Priming with *Pseudomonas fluorescens* WCS374 was found to not only suppress *Fusarium* wilt disease but also increase yield in radish (Leeman *et al.*, 1995). In order to overcome the field-level difficulties of PGPF, cell free metabolites have also recently been considered for management of plant pathogens (Keswani *et al.*, 2014). (Table 5.1.).

PGPR-mediated breakdown of pathogen communication

Regulation of virulence in pectinolytic bacteria is mediated through communication of the pathogen via the chemical-signal-based cell–cell communication system known as

Table 5.1. PGPR and its action in plant disease suppression.

S.No.	PGPR/PGPF species	Effect on host	Mode of action	References
	Rhodococcus erythropolis	Suppress the soft rot of potato causing bacterial pathogen *Pectobacterium atrosepticum*	Breakdown of pathogen chemical communication signals	Crépin *et al.* (2012)
	Pseudomonas putida KT2440	Induce plant systemic resistance against *Pseudomonas syringae* pv. *tomato*	Elicit higher root exudation of distinct patterns	Matilla *et al.* (2010)
	Azospirillum species	Stimulate maize defense response and release of defense compounds at higher amount through root exudates	Increase synthesis of defense molecules benzoxazinoids	Walker *et al.* (2011)
	Pseudomonas mosselii	Suppress common scab of potato caused by Streptomyces scabies	Enhanced activation of phenylpropanoid pathway and antioxidant activities	Singhai *et al.* (2011)
	Bacillus subtilis Sb4-23	Excellent biocontrol agent against fugus like *Verticillium dahliae, Fusarium culmorum, Rhizoctonia solani* and nematode *Meloidogyne incognita* simultaneously	Induced systemic resistance (ISR)	Adam *et al.* (2014); Koberl *et al.* (2013)
	Pseudomonas fluorescens HC1-07	Biocontrol agent against *Rhizoctonia solani* AG-8 and *Gaeumannomyces graminis* var. triticicauses Wheat root rot and wheat take-all disease.	Cyclic lipopeptides and visconsin-like protein production	Yang *et al.* (2014)
	Bacillus thuringiensis	Silence the bacterial diseases	Breakdown the cell-cell communication quorum sensing are N-acyl homoserine lactone lactonases	Zhou *et al.* (2008)
	Paenibacillus kribbensis PS04	Have inhibitory effect on the *Rhizoctonia solani* to control sheath blight disease in Rice	Eliciting induced resistance	Guo and Liao (2014)
	Saccharothrix algeriensis NRRL B-24137	Suppress the growth of *Botrytis cinerea* and some other phytopathogens	Activation of Jasmonic acid /Ethylene (JA/ET) dependent ISR but may need the Salicylic acid (SA), NADPH oxidase, UPS1underinducer after pathogen and stress1	Muzammil *et al.* (2014)
	Ampelomyces sp. and *Cladosporium* sp.	Suppress the growth of *Pseudomonas syringae* pv. *tomato* DC3000 (Pst) and reduces the disease risk in *Arabidopsis thaliana*	Synthesizes the Volatile compounds like m-cresol and methyl benzoate respectively. It elicits ISR by involving the SA and JA/ET signalling pathway	Naznin *et al.* (2014)

quorum sensing (QS) (Crépin *et al.*, 2012). The potato soft rot pathogen *Pectobacterium atroseptica* is a Gram-negative bacterium and like many other Gram-negative bacteria it also synthesizes small diffusible signalling molecules that serve in QS. The most common signals deployed by bacteria for QS are through synthesis and perception of *N*-acyl-homoserine lactones (NAHSL) (Waters and Bassler, 2005). The soft rot bacteria coordinate the synthesis of numerous factors involved in pathogenesis through NAHSL (Barnard and Salmond, 2007). Microorganisms including fungi and bacteria are able to perturb QS signalling of plant pathogens through cleavage of NAHSL signals (Uroz *et al.*, 2009). Quorum quenching bacteria of the genus *Rhodococcus* isolated from the potato rhizosphere were able to limit pathogenesis of the soft rot pathogen. The activity and population density of the rhizospheric NAHSL-degrading bacteria may be boosted by introducing NAHSL structural analogs that can be used as nutrients by such bacteria (Cirou *et al.*, 2007).

PGPR-mediated ISR and change in root exudation

The potential biocontrol *P. putida* KT2440 is able to protect *A. thaliana* from infection by *P. syringae* pv. *tomato* DC3000 through modifications in root exudates. The role of extracellular haemperoxidase (PP2561) of *P. putida* was found to be significant for competitive colonization and essential for the induction of plant systemic resistance. Root exudates of plants elicited by the bacterial strain KT2440 exhibited distinct patterns of metabolites compared with those of non-elicited plants. The levels of some of these compounds were dramatically reduced in axenic plants or plants colonized by a mutant defective in PP2561, which has increased sensitivity to oxidative stress with respect to the wild type (Matilla *et al.*, 2010). Similarly, Walker *et al.* (2011) showed that inoculation of maize with strains of PGPR of the genus *Azospirillum* resulted in significant changes in secondary metabolic profiles of root and shoots especially the host defense molecules such as the benzoxazinoids and phenolics. Similarly, fluorescent Pseudomonads elicit metabolic variations in chickpea (Sarma *et al.*, 2002; Singh *et al.*, 2003) with likely impact on root exudation for biological control of the soilborne pathogen *Sclerotium rolfsii*.

Non-pathogen: production of lipopeptides as ISR agents

Lipopeptides (LPs) synthesized by non-pathogenic *Pseudomonas* and *Bacillus* strains are shown to have beneficial effects on plants. In natural habitats, LPs from such bacterial species confer a competitive advantage in interactions with other microorganisms through antagonistic activities (Raaijmakers *et al.*, 2010). Similarly, the LPs are also linked to protection of such bacterial strains from predators like the protozoa (Matz and Kjelleberg, 2005) as well as in facilitating their movement (De Souza *et al.*, 2003). Similarly, the polysaccharides and proteins secreted from bacterial cells form a hydrated gel-like slime that helps in biofilm formation (Stewart and Franklin, 2008). Besides these useful roles of the LPs produced by non-pathogens, a recent study has revealed that they also trigger defense responses in plants against invading fungal and oomycetes pathogens through stimulation of the plant immune system. Activities of two key enzymes of the oxylipin pathway were stimulated in tomato upon treatment with LP-overproducing *Bacillus* isolates (Ongena *et al.*, 2007). LPs from *Bacillus* species were also found to stimulate phytoalexin synthesis in the treated plants (Adam, 2008), and cause modifications in the pattern of phenolics biosynthesis, and activation of defence-related events such as phosphorylation, Ca^{2+}-dependent extracellular alkalinization and oxidative burst without any phytotoxic effect (Jourdan *et al.*, 2009). These studies clearly demonstrated that some LPs from non-pathogenic bacterial species behave as microbial-associated molecular patterns (MAMPs) that can be perceived by plant cells leading to activation of defence responses. Moreover, the LPs produced by such bacterial strains are also reported to chelate metal ions better and aid degradation of xenobiotics (Mulligan, 2005).

Plant-driven recruitment of PGPR for defence

The rhizosphere of plants can turn into a "biased rhizosphere" through modification of the root exudates profile under pathogen-challenged conditions in order to facilitate colonization of specific rhizosphere microbes that can help plants to withstand biotic stresses (Hartmann *et al.*, 2009). The synthesis of specific carbon compounds and their release through roots as exudation favours recruitment of specific microbes which are able to respond with chemotaxis and grow very fast, resulting in replacement of the whole soil microbial diversity by a small community finally colonizing the roots successfully. Hartmann *et al.* (2009) demonstrated that a host-recruited beneficial microbial community can suppress the losses caused by *Verticillium dahlia* in strawberry, potato and oilseed rape. In another study, Rudrappa *et al.* (2008) demonstrated that root secretions of the tricarboxylic acid cycle intermediate L-malic acid (L-MA) are increased by the infection of the foliar pathogen *P. syringae* pv. *tomato* DC3000 in *A. thaliana*. This increased secretion of L-MA from roots of Arabidopsis selectively signals and recruits the beneficial rhizobacteria *Bacillus subtilis* FB17 in a dose-dependent manner. Similarly, Neal *et al.* (2012) demonstrated that defence molecules of the benzoxazinoids group such as 2,4-dihydroxy-7-methoxy-2H-1,4-benzoxazin-3(4H)-one (DIMBOA) released through maize root exudation can also recruit *P. putida* KT2440, a competitive colonizer of the maize rhizosphere with biocontrol traits, through chemotaxis. Beyond all the above-mentioned disease suppression mechanisms, various PGPR and PGPF have the capacity to produce bioactive volatile organic compounds to enhance plant health (Kanchiswamy *et al.*, 2015). These compounds actively help in plant growth promotion and disease suppression (Table 5.1).

5.2.5 Enhancing the nutritional quality and yield of agricultural produce

With the rapid growth of population, global food demands and per capita consumption will increase proportionally which can be met by sustainable intensification (Tilman *et al.*, 2001). To meet this challenge the scientists must think about improving soil health, nutrient use efficiency, and better agronomic practices for cropping systems rather than simply increasing productivity. Along with increased agricultural productivity, attention is also needed for increasing nutritional quality of the crops through a sustainable approach. Improving the agronomic inputs and differentiated crop outputs through biotechnological tools can strengthen the option of naturally biofortified foods, which is currently a high priority as agricultural products are the key primary source of all nutrients for human beings. Malnutrition due to deficiency of micronutrients and vitamins is another major concern. About 805 million people were estimated to be facing undernourishment in 2012–14 (FAO, 2014), with very deleterious effects on humankind, and especially on children's health. In this situation it is necessary to enhance the yield and nutrient quality of food crops. Issues of nutritional deficiency in agricultural produce can be overcome by adopting biotechnological, breeding (Welch and Graham, 2004), and microbial (Güneş *et al.*, 2014) technologies. We could enhance the existing process of biofortification with application of biological inoculum such as PGPR and PGPF (Fig. 5.4).

Inoculation of the *Providencia* sp. PW5 + $N_{60}P_{60}K_{60}$ in wheat results in significant 105.3, 36.7, and 150% increases in Fe, Mn, and Cu accumulation, respectively, compared with the control ($N_{60}P_{60}K_{60}$) (Rana *et al.*, 2012). In a similar study single (*Providencia* sp. PW5+ $N_{60}P_{60}K_{60}$) and double (PW5+ *Anabaena sp.* + $N_{60}P_{60}K_{60}$) inoculation were also applied. These inoculations showed increase in grain yield and protein content by 11–18% over the control ($N_{60}P_{60}K_{60}$). Similarly, in a recent study *Bacillus megaterium* M3 inoculum was applied

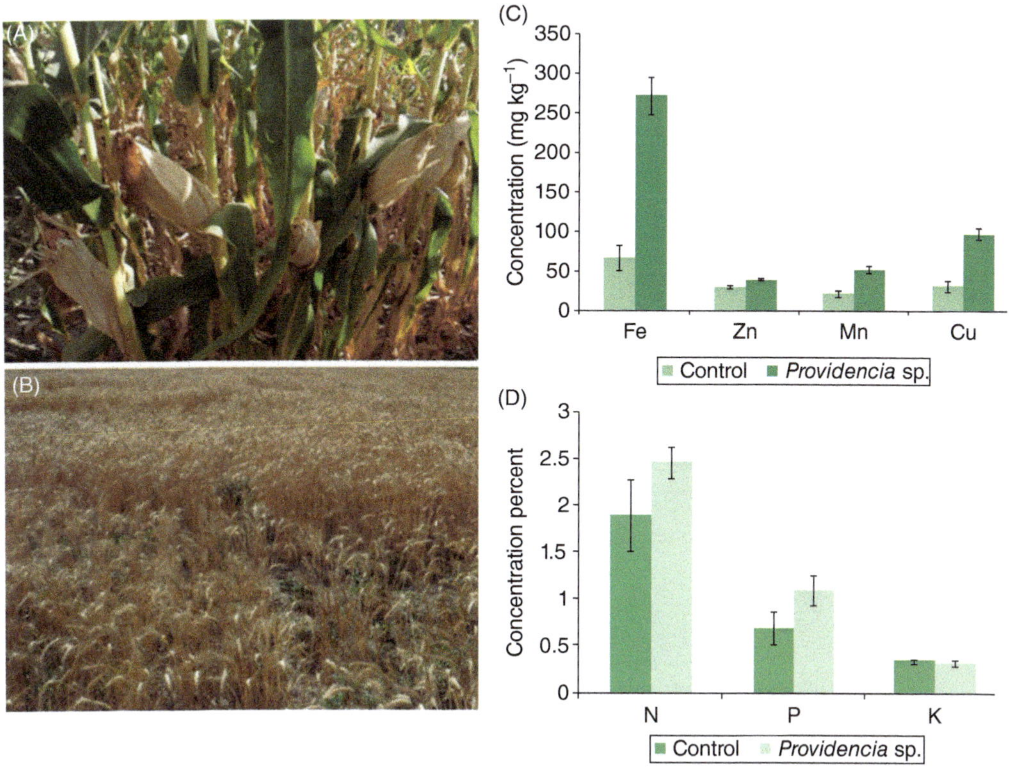

Fig. 5.4. Biofortification of staple food crops such as (A) maize and (B) wheat through the inoculation of PGPB is an innovative strategy for improving the nutritional quality of agricultural produce. The effects of PGPR inoculation (*Providencia* sp.) on (C) micronutrients and (D) NPK content in wheat grains (Rana *et al.*, 2012). The bacterial inoculum contained 10^{11} cells ml^{-1} and applied at a rate of 500 ml culture ha^{-1}.

and higher Ca and P concentrations were observed than in the control plants. The findings also explain the reason for high concentrations of K, S, Na, Mn, Cd, and Ni obtained when *Azospirillum brasilense* Sp 245 was applied (Güneş *et al.*, 2014). Microbial consortia are very effective in promoting plant growth, yield and nutritional makeup in plants. In a study Lavakush *et al.* (2014) showed that consortia of *Pseudomonas aeruginosa* BHUJY16 + *P. aeruginosa* BHUJY20 + *Pseudomonasputida* BHUJY13 + *P. putida* BHUJY23 + *Pseudomonas fluorescens* BHUJY29 + *Azotobacter chroococcum* + *Azospirillum brasilense* + 30 kg ha^{-1} P_2O_5 when formulated and inoculated in rice, gave rise to a significant (30.28 g pot^{-1}) increase in grain yield in comparison to the control (21.17 g pot^{-1}). Nutrient content (N, P, and K) in grain and shoots of the treated plants were also enhanced significantly. In another study a consortium comprising commercially available *Bacillus* sp. and *Glomus intraradices* (PGPR and PGPF) enhanced grain yield and nutrient use efficiency in field corn. Nitrogen content g^{-1} of the grains was significantly enhanced and higher removal of NPK from the inoculated pot was observed than for the control (Adesemoye *et al.*, 2008). Nutrient dynamics and yield of the plants are highly controlled by the association of the PGPR. *Bacillus* sp., *Providencia* sp. and *Brevundimonas* sp. when applied to wheat in combination with ($N_{90}P_{60}K_{60}$) fertilizer dose induces yield and nutrient content of the crop. Inoculation of all bacterial strains alone or in combination enhances the micronutrient concentration 143.6, 193, 63.7 and 45.5% of Fe, Cu, Mn and Zn, respectively, over the full dose of fertilizer in control. Significant increase in NPK was also found (66.7, 100 and 7.5%, respectively) (Rana *et al.*, 2012). In a previous study application of *Pseudomonas syringae* (PUR46) inoculum with 25% vermicompost in *Cicer arietinum* resulted in increased uptake of P, Fe, and Mn along with plant growth promotion (Sahni *et al.*, 2008). Recently, Singh *et al.* (2014) also showed that co-inoculation of compatible beneficial microbes (fluorescent *Pseudomonas*, *Trichoderma harzianum* and *Mesorhizobium* sp.) enhanced antioxidants in chickpea edible parts through synergistic interactions.

Thus application of PGPR and PGPF can improve the nutritional quantity and quality of agricultural produce. It can also lead to savings of 50% nitrogen fertilizer and its environmental hazards. Further extensive research work is needed to develop more sustainable microbial consortia to enhance nutrient use efficiency and health of the crop plants. This in turn can provide an equitable and healthy food for the 8.5 billion people projected for 2025 sustainably.

5.2.6 Beneficial microbiome management and recruitment in the rhizosphere

Plant health is signified by the ultimate productivity of the plants. In plants, diverse microorganisms are associated in spermospheric, rhizospheric, phyllospheric, endospheric, and mycorhizospheric regions (Nelson, 2004; Compant *et al.*, 2010; Vorholt, 2012). The plant–microbe interrelationship is an essential component of plant life. It is mediated by crucial phytohormones like IAA (Duca *et al.*, 2014) which controls plant health and productivity. An increasing body of evidence signifies the importance of the root microbiome, consisting of a complex web of rhizosphere-associated microbes, their genetic elements and their interactions, in determining plant health and productivity. In this context, some recent advances in plant–microbe interactions research pointed out the role of plants' genotype and soil type in shaping their rhizosphere microbiome for their own benefit (Berg and Smalla, 2009; Weinert *et al.*, 2011). However, most of the studies have focused on the recruitment of beneficial microbes on the type of root exudates released and change in community dynamics on pathogen attack (Rudrappa *et al.*, 2008). An attack of a phytopathogen releases some microbe stimulatory compounds from noninfected roots and subsequently recruitment and activation of beneficial microbes takes place. The newly recruited beneficial rhizospheric microbes synthesize pathogen inhibitory compounds and combat the pathogen by induced resistance (Weller *et al.*, 2012). Pepper plants, if feeding *Myzus persicae*, increase the root-associated population of PGPR like *Bacillus subtilis* GB03, with reduced population of the phytopathogen *Ralstonia solanacearum* (Lee *et al.*, 2012). As well as plants, soil physicochemical properties, nutrient availability, environmental conditions (Berg and Smalla, 2009), agricultural and agronomic practices are having a great impact on microbiome management (Jechalke *et al.*, 2014). Microbial community structure analysis has shown that beneficial microbes, when they are compatible with each other, can enhance plant defense responses towards invading pathogens. A recent report shows that the microbial consortium alleviated pathogen stress in chickpea through enhanced antioxidant and phenylpropanoid activities (Singh *et al.*, 2013). Beneficial arbuscular mycorhhizal association with the plants can significantly reduce the N and P loss from the soil thus increasing the nutrients' availability to the crop plants (Cavagnaro *et al.*, 2015). Our current understanding of the complex plant–microbe interactions taking place in the "mysterious underground world" is still in its infancy. Unravelling the mechanisms through which plants control their microbiome and in turn the microbiome controls plant health is still not well explored. This in future will open new avenues for sustainable agriculture to increase crop quality and productivity.

5.3 Importance of PGPR and PGPF in Phyto/Bioremediation

Increasing anthropogenic activities including unsustainable agricultural practices during the last few decades have resulted in the widespread pollution of the planet earth (Tripathi *et al.*, 2014a, b; Tripathi *et al.*, 2015a, b). Among the pollutants, heavy metals and organic pollutants are the major contaminants of the soil. There is an urgent need to focus on the development of clean-up remedies for restoration of such contaminated soil (Banwart, 2011; Tripathi *et al.*, 2015a, b; Abhilash *et al.*, 2016a, b). Remediation of the contaminated soil with the current technologies is costly, relatively slow and needs to be revamped urgently as the number of contaminated sites may increase worldwide in the near future (Abhilash *et al.*, 2013b). Microbial populations are known to affect heavy metals mobility and availability to the plant through release of chelating agents, acidification, phosphate solubilization, and redox changes (Abou-Shanab *et al.*, 2003) and play significant roles in recycling of plant nutrients, maintenance of soil structure, detoxification of noxious chemicals, and control of plant pests and plant growth (Giller *et al.*, 1998; Filip, 2002). The plants and the microbes have the capability to grow, uptake/degrade the pollutants from the contaminated environment, and this could be exploited for developing alternative cheap and efficient technologies for the clean-up of contaminated sites (Abhilash *et al.*, 2013a). Although the role of plant growth-promoting microorganisms have been extensively studied in various biotic and abiotic challenges, their importance in phyto/bioremediation is still underestimated (Fenner *et al.*, 2013; Tripathi *et al.*, 2016a,b). A better exploration of these microorganisms could be exploited for the sustainable clean-up of the contaminated sites (Kumar and Patra, 2013). The plant–microbe mediated remediation technology also improves the soil health by enhancing the content of soil organic carbon, micro and macro nutrients, soil porosity and permeability; however, productivity of such contaminated land is very low and could be improved further by efficient application of microbe-assisted phytoremediation.

5.3.1 Heavy metals

Heavy metals are one of main pollutants of the soil because of their toxicity, abundance in earth's crust, lower bioavailability and non-biodegradable nature, unlike their organic counterparts. So they can persist in the soil for longer periods, could be accumulated by the plant species growing in metalliferous soil and contaminate the food chain, and may leach out from the soil and contaminate groundwater as well (Glick, 2010; Ovečka and Takáč, 2014). This has led to a focus on development of more efficient, cost-effective heavy-metal remediation technologies. Plant–microbe interactions could provide helpful insight in removal of such pollutants from the soil. As toxicity and bioavailability of metals is the major problem affecting the phytoremediation process, application of heavy-metal-tolerant PGPR can improve the remediation period by promoting plant growth and enhancing metal uptake at a faster rate from soil (Patel and Patra, 2014; Sahay and Patra, 2014). Similarly the microbes could also produce chelating agents and biosurfactants which could further enhance bioavailability of metals in contaminated sites (Ma *et al.*, 2011). Plants with symbiotic association as in legumes could significantly benefit the process of phytoremediation if they show tolerance to metal contamination. This is the reason that legumes are getting significant attention for phytoremediation as they are also reported to grow profusely on heavy metal and other contaminated sites (Prasad and de Oliveira Freitas, 2003; Dary *et al.*, 2010). *Lupinus angustifolius,* a legume, phytostabilised the metal contaminated site when inoculated with triple consortia of heavy-metal-resistant PGPRs *Pseudomonas* sp. Az13, *Ochrobactrum cytisi* Azn6.2, and the nitrogen fixer *Bradyrhizobium* sp. 750. This could prevent food chain contamination and also be helpful in restoration of the soil fertility by increasing the nitrogen content of soil (Dary *et al.*, 2010). Similarly, *Bacillus subtilis* strain SJ-101 can enhance the phytobiomass production of *Brassica juncea* along with decreased Ni uptake in Ni contaminated soil by accumulating Ni and promoting growth of *Brassica juncea* (Zaidi *et al.*, 2006) (Table 5.2).Though these bacteria can support plant growth in metal-contaminated soil by virtue of their potential to promote plant growth and accumulate heavy metals, this technology might not be appropriate for remediation of the soil contaminated with multiple metals. Some microorganisms have the potential of remediating multiple metals from the soil. Mine tailings and fly ash are reported to be contaminated with multiple metals and are a potential source of environment contamination. Multiple-metal-tolerant bacteria like *Enterobacter intermedius* and *Pseudomonas* sp. can promote plant growth and increase the uptake and accumulation of Ni, Zn and Cd by plants, also protecting them against the inhibitory effect of heavy metals by producing phytostimulants to support plant health (Kumar and Patra, 2013; Płociniczak *et al.*, 2013).

Apart from bacteria various AMF have been reported to have the property of supporting plant growth in stressed environments such as heavy-metal-contaminated soil (Sahay and Patra, 2014). Some plants have the property to absorb high amounts of heavy metals and most of the plants were reported to have an AMF association (Cameron *et al.*, 2013). The AMF have the potential to promote plant growth and accumulate heavy metals in their hyphae and arbuscules, keeping their cytoplasm free from heavy metals thereby minimizing the heavy metal toxicity (Miransari, 2010). AMF could also regulate the genes involved in hyperaccumulation of heavy metals such as the metallothionein regulatory genes, and the metal transporter genes. The AMF symbiosis has been reported to help plants to grow in Zn, Cd, Cu, and Pb stressed environments (Cicatelli *et al.*, 2010; Miransari, 2010, 2011). Exploitation of AMF association with suitable hyperaccumulator host plants might be advised for remediation of the soils contaminated with heavy metals (Miransari, 2011).

Similarly, not only the association with AMF but the other symbiotic relationships between plant and plant growth-promoting endophytes (PGPE) could be used for bioaugmentation-assisted phytoremediation (Patel and Patra, 2014). Bioaugmentation of *Solanum nigrum* (a Cd hypreaccumulator) with *Pseudomonas* sp. Lk 9 assisted in the phytoremediation of not only Cd but of other heavy metals (Zn and Cu) as well. The bacterium also significantly promoted the plant growth and improved the soil health by producing siderophore, biosurfactants, acid phosphatases and increasing the status of the microbial biomass carbon of the soil (Chen *et al.*, 2014). Similarly the bacterium *Bacillus thuringiensis* GDB-1 isolated from the roots of *Pinus sylvestris* showed enhanced removal capacity of heavy metals such as As, Cu, Pb, Ni and Zn when applied with the native hyperaccumulator plant *Alanus firma* (Babu *et al.*, 2013). However, not all microorganisms have the capability to remove multiple metals from contaminated soils. Apart from this the response of the microorganisms also differs in field conditions from the laboratory condition (Glick, 2010). Thus they do not produce adequate results when applied in the field.

Genetic alteration of the microorganisms could be a possible solution for this limitation. On the other hand genetically engineered microorganisms could also be helpful in remediation of not only the multiple heavy metals but of complex contaminants such as organic pollutants and mixed heavy metals as well (Singh *et al.*, 2011). Genetically engineered strains of rhizobacteria *Pseudomonas* Pb2-1 and *Rhizobia* 10320D degraded trichloroethane and also showed an enhanced rate of heavy metal accumulation. Application of such genetically modified microbes could be helpful in restoration of lands contaminated with organic as well as inorganic pollutants (Lee *et al.*, 2006). In a study it was found that Cd-resistant strains of *Arthrobacter*, *Bacillus*, and *Pseudomonas* supported the degradation of 2,4-D by the cadmium-sensitive 2,4-D degrader *Ralstonia eutropha.* Thus dual bioaugmentation with metal detoxifying and organic pollutant degrading microbes provides a suitable approach for reclamation of multiple pollutant contaminated soils (Table 5.2).

Table 5.2. Role of PGPR in plant growth and environmental remediation (heavy metal/organic pollutant).

S.No.	Species	PGPR/ Environmental Remediation Attributes	Experimental Site/ Condition	Host Plant	Reference
	Pseudomonas sp.KS 51	PAH, Naphthalene (78.44 %) and Anthracene (63.53 %) degradation,IAA, hydrogen cyanide (HCN), siderophore and phosphate solubilization, shows biocontrol in response to phytopathogenic fungi.	*in vitro* Experiment	*Calotropis* R.Br.Combination of (*Loliumperenne*L.)	Shukla KP *et al.* (2012)
	Xanthobacter autotrophicus GJ10	Degradation of 1-2-dichloro ethane. Reclamation of halogenated aliphatic contaminated land.	*in vitro* experiment	*Nicotiana tabacum* L. 'Xanthi	Mena-Benitez *et al.* (2008)
	Comamonas sp.	Rhizoremediation of 4-chloronitrobenezene contaminated soil	Gnotobioticand outdoor pot experiments	Alfalfa (*Medicago sativa* L.)	Liu *et al.* (2007)
	Pseudomonas jesseniistrain PjM15, *Pseudomonas* sp. PsM6	IAA activity, inorganic phosphate,Ni, Cu solubilization and Zn sequestration	Pot experiment	*Ricinus communis* L.	Rajkumar and Freitas (2008)
	Consortium of *Bradyrhizobium* sp., *Pseudomonas* sp. and *Ochrobactrum cytisi*	nitrogen fixation, enhanced biomass, phytostablization of multimetal Cu, Zn, Cd contaminated land	Field experiments	*Lupinus luteus* L.	Dary *et al.* (2010)
	Paenibacillus sp., *Acinetobacter* sp.	Oil contaminated area restoration in mangrove forest has ability to produce IAA, siderophore, and phosphate solubilising activity	*in Vitro* experiment	*Avicennia schaueriana* Stapf and Leechm. ex Moldenke and *Laguncularia racemosa* C.F. Gaertn., *Rizophora mangle*	do Carmo *et al.* (2011)
	Agrobacterium radiobacter D14	Produces IAA and siderophore for plant growth promotion, helps in As bioremediation, bacterial inoculation, in the 300 mg kg^{-1} of contaminated soil removal efficiency 54 %	Pot experiment in greenhouse	*Populus deltoids* LH05-17	Wang *et al.* (2011)
	Bradyrhizobium sp. strain MRM6	Use for bioremediation of insecticides like fipronil and pyriproxyfen result in complete growth of plant biomass	Field experiments	*Vigna radiate* (L.) R.Wilczek	Ahemad *et al.* (2011a)

Continued

Table 5.2. Continued.

S.No.	Species	PGPR/ Environmental Remediation Attributes	Experimental Site/ Condition	Host Plant	Reference
	Cellulosimicrobium cellulans KUCr3	Used for bioremediation of Cr contaminated soil, also produces IAA, phosphate solubilizing activity.	*in vitro* experiment IPYG medium	Chilli (*Capsicum* L.) plants	Chatterjee *et al.* (2009)
	Gordonia sp. S2RP-17	Bioremediation of total petroleum hydrocarbon,TPH removal efficiencies was 95.8 % by the combination of *Zea mays* and *Gordonia* sp. S2RP-17 and enhances plant growth by ACC deaminase and siderophore	Mesocosm systems	*Zea mays* L.	Hong *et al.* (2011)
	*Azotobacter*sp.	Capable of degrading endosulfan suitable for plant growth promotion by producing IAA	*in Vitro*	Coffee farm	Castilloet *al.* (2011)
	Bacillus subtilis BS1	A good PGPR producing bio surfactant capable of removing PAH from soil	Pot experiment	*Zea mays* L.	Xiao *et al.* (2012)
	Bacillus subtilis, Sphingobacterium multivorum, Acinetobacter radioresistens, Rhodococcus erythropolis	Degradation of total petroleum hydrocarbon by using polyphenol oxidase activity and increases the plant biomass TPH removal rate was 58% after 162 days.	pot experiment	*Loliumperenne* L.	Tang *et al.* (2010)
	Bacillus pumilus ES4, *Bacillus pumilus* RIZO1, *A. brasilense*	Phytostabilization of mine tailings	Pot experiments in greenhouse	*Atriplex lentiformis* S.Watson	de-Bashan *et al.* (2010)
	Achromobacter xylosoxidans Ax10	Cu bioremediation and PGPR activity	Pot experiments	*Brassica juncea* (L.) Czern, *B. oxyrrhina*	Ma *et al.* (2009)
	Enterobacter intermedius MH8b	ACC deaminase, siderophore, IAA production, hydrocyanic acid for promotion of plant biomass and used for Zn (32 %), Cd (94 %) accumulation capacity, contaminated land reclamation	*in vitro* experiment	*Sinapis alba* L.	Płociniczak *et al.* (2013)
	Bacillus species PSB10	Reclamation of Cr contaminated land	Pot Experiments	*Cicer arietinum* L.	Wani and Khan (2010)

Serratia nematodiphila LRE07, *Enterobacter aerogenes* LRE17, *Enterobacter sp.* LSE04 *and Acineto-bacter sp.* LSE06.	IAA, Siderophores production, ACC diaminase, phosphate solubilizing activity helps in removal of Cd from soil.	Pot Experiments	*Solanum nigrum* L.	Chenet *al.* (2010)
Azotobacter chroococcum and Bacillus megaterium	adsorption of Pb^{2+} and Cd^{2+} in solution	*in vitro* experiment	Cd(NO3)2 and Pb(NO3)2 solutions were used	Wu *et al.* (2009b)
Rahnella aquatilis	Reclamation of Ni and Cd contaminated land	Pot Experiment	*Brassica juncea* (L.) Czern	Kumar *et al.* (2009)
Stenotrophomonas maltophilia, Stenotrophomonas rhizophila	Bioremediation of p-nitrophenol, 4-chlorophenol and 4-nitroaniline, nonylphenol, polypropylene glycols, herbicides 4-(2,4-dichlorophenoxy) butyric acid and 4-(4-chloro-2-methylphenoxy) butyric Acid, Cr (VI) resistance, also having PGPR activity like IAA, N fixation and biocontrol of fungal and bacterial pathogens	*in vitro* experiments	*Cucumissativus* L., *Brassica napus* L., *Solanum tuberosum, Fragaria xananassa, alfalfa Medicago sativa* L., *Helianthus annuus* L., *Zea mays* L., *Oryza sativa* L., *Triticum aestivum* L., *Salix herbacea* L., *Populus* L.	Ryan *et al.* (2009)
Pseudomonas putida W619-TCE	Reclamation of Ni and trichloroethylene contaminated land with increased phytoremediation efficiency and better Bioenergy production	Pot experiment	*Populus deltoids W. Bartram ex Marshall* x *trichocarpax-deltoides*	Weyens *et al.* (2009)

5.3.2 Organic contaminants

Soil contamination with organic pollutants has increased dramatically since the industrial revolution. The injudicious use of large amounts of pesticides, insecticides, chemical fertilizers, and other industrial and defence-related chemicals has resulted in severe and widespread contamination of the land with the toxic xenobiotic compounds, mostly organic in nature (Fenner *et al.*, 2013). The physicochemical methods for remediation of such polluted lands are costly, inefficient and environmentally destructive. Thus recently much attention has been paid to exploiting the plant–microbe association for removal of organic contaminants (Weyens *et al.*, 2009). Phytoremediation of the contaminated soil is also a low-input technology that could be applied for restoration of the degraded land; however it needs to address the toxicity of the harvested biomass (Dubey *et al.*, 2014). Various PGPRs are reported from the rhizosphere and the plant endosphere capable of degrading or modifying the organic pollutants (Castillo *et al.*, 2011). These microorganisms metabolize the organic pollutants for assimilating nutrients and generating energy. On the other hand plants can transform the toxic organic contaminants by the action of their wide spectrum enzymes (Fenner *et al.*, 2013). Plants can also draw the pollutants towards the rhizosphere via transpirational pull where the rhizospheric microorganisms can degrade these contaminants. Thus plant–microbe interaction could be applied as sustainable low-input biotechnology tools for remediation of organic contaminants from soil.

A number of sites are reported to be contaminated with hydrocarbon pollutants across the world. Hydrophobicity of the hydrocarbons reduces their availability for plants and microbes and adversely affects the health of plants (Jagtap *et al.*, 2014). Plant growth-promoting bacteria are reported to be helpful in phytoremediation of hydrocarbon-contaminated soil by promoting plant growth, degrading hydrocarbon, reducing its phytotoxicity and evapotranspiration (Kumari *et al.*, 2012). A total petroleum hydrocarbon (TPH) degrading thermophilic actinomycetes *Nocardia otitidiscaviarum* TSH1 was isolated from the soil which is capable of degrading polycyclic aromatic hydrocarbons and phenol as well. Thus it could be applied for remediation of a range of organic pollutants. Similarly, the highly lipophilic nature of the cell membrane of *Nocardia otitidiscaviarum* TSH1 helps in the uptake of hydrophobic molecules which have low bioavailability in soil (Zeinali *et al.*, 2007). *Nocardia otitidiscaviarum* TSH1 could not only help in remediation of oil-polluted soil but in remediation of other hydrophobic organic pollutants as well. Persistent organic pollutants (POPs) are another major group of organic pollutants posing serious threats to the environment due to their highly toxic, recalcitrant nature and ability to undergo long atmospheric transport. Successful remediation of POPs has been reported using the plant bacterial association (Abhilash *et al.*, 2013a). Bioaugmentation of *Cytisus striatus* with the endophyte *Rhodococcus erythropolis* ET54b and *Sphinigomonas* sp. D4 helped in enhanced phytoremediation of the persistent organic pollutant hexachlorocyclohexane (Becerra-Castro *et al.*, 2013). Similarly, an *Azotobacter* sp. isolated from coffee farm soil was found to be capable of degrading the POP endosulfan (Castillo *et al.*, 2011).Thus the plant–bacterial association could be successfully applied for phytoremediation of POPs.

Though bacteria are the most commonly reported microorganisms having the potential to remediate organic contaminants, fungi also play key roles in remediation of organic contaminants. Bioremediation potential of white rot fungi such as *Phanerochaete*, *Trametes*, *Bjerkandera* and *Pleurotus* is being extensively studied these days as they produce lignolytic enzymes which are non-specific in nature and degrade a series of recalcitrant hazardous chemicals (Hestbjerg *et al.*, 2003; Paszczynski *et al.*, 2008). For example, lignin peroxidase-producing fungi such as *Phanerochaete chrysosporium* and *Trametes hirsutus* degrade HCH (hexachlorocyclohexane) due to the non-specific nature of the lignin peroxidase enzyme. Xiao *et al.* (2012) reported that when inoculated together biocompatible strains of *Bacillus subtilis* BS1 and AMF *Glomus etunicatum* improved the mycoremediation of soils contaminated with phenanthrene, as the bacteria produced biosurfactant which increased the solubility and availability of phenanthrene. Sometimes a single microorganism is unable to effectively remediate the pollutant. A possible outcome for the same could be the use of microbial consortia or of a genetically modified microorganism capable of completely degrading the contaminant (Table 5.2).

5.4 Role of PGPR and PGPF in Biomass and Biofuel Production

Rapidly increasing energy scarcity stimulated many researches regarding biomass and biofuel production in recent years (Jagtap *et al.*, 2014; Tripathi *et al.*, 2016b). Plant growth-promoting microorganisms (PGPM) can provide an additional route for sustainable biofuel production. As PGPM exclusively serve as biofertilizers, they are therefore able to promote plant growth directly (Batty and Dolan, 2013). As a result, enhanced plant biomass can be utilized for biofuel production. Meanwhile, they have the potential to enhance the remediation processes. Recently, Jagtap *et al.* (2014), studied the enhanced phytoremediation potential and biomass production from *Pinus* sp., *Thuja* sp., and *Populus* sp. when inoculated with rhizospheric bacteria (Jagtap *et al.*, 2014). Therefore, bioremediation on marginal and contaminated lands could be integrated with biofuel production (Weyens *et al.*, 2009; Edrisi and Abhilash, 2016) (Table 5.3).These microorganisms basically include rhizobacteria, endophytes, and even microbes lying on the surface of the roots (rhizoplane). Previously, the studies for plant growth had been concentrated towards endophytes (Barac *et al.*, 2004; Zaidi *et al.*, 2006). But now, attention has been focused mostly on the plant growth-promoting capacity of rhizobacteria (Guo *et al.*, 2014; Werling *et al.*, 2014). Basically, a close association exists between endophytic bacteria and rhizobacteria. Even the bacteria from rhizoplanes are being involved in this sustainable process (Germaine *et al.*, 2006). The efficiency of PGPM in enhancing plant growth has been elucidated in several greenhouse and field studies of different plant species, such as *Lolium multiflorum* (Guo *et al.*, 2014), *Zea mays* (Couillerot *et al.*, 2013), *Saccharum officinarum* (Taulé *et al.*, 2012), *Glycine max* (Mishra *et al.*, 2009) and many others as mentioned in Table 5.3. These studies have depicted considerable enhancement in growth of biomass production. For example, increased efficiency of nodule-forming diazotrophs when combined with PGPR resulted in the enhanced growth of *G. max* plants (Mishra *et al.*, 2009) (Table 5.3). Poplars (*Populus* spp.) and willows (*Salix* spp.) are appropriate tree species for both phytoremediation and biomass production (Table 5.3). Also the growth of ryegrass was linked with the increased secretion of indole acetic acid, siderophores, 1-amino-cyclopropane-1-carboxylate daminase, soluble inorganic phosphate and metal-bearing minerals (Guo *et al.*, 2014). The above-mentioned studies also revealed that endophytic diazotrophs could help in providing nitrogen to these plant species under nitrogen-limiting conditions or in marginal lands. Furthermore, other rhizospheric microorganisms and PGPM including AMF have potential to enhance the phytoremediation process in contaminated sites as well as the biomass and biofuel production. Moreover, there are numerous options for use of plants for the production of biofuel with a sequential remediation of degraded or contaminated sites.

Furthermore, the enhanced biomass production of other bioenergy plants by using PGPR includes *Pinus densiflora* with *S. acidaminiphila* and *P. putida*, *Brassica napus* (Jagtap *et al.*, 2014), *Brassica oleracia* with *Enterobacter* and *Herbaspirillum* sp. (Ahmad *et al.*, 2013), *Jatropha curcas* with *Bacillus pumilus* (IM-3) (Sumarsih and Haryanto, 2012), *Prosopis julifora* with AMF (Solís-Domínguez *et al.*, 2011), *Lupinus luteus* with *Bradyrhizobium*, *Pseudomonas* and *Ochrobactrum* sp. (Dary *et al.*, 2010), *Medicago sativa* with a collection of different AMF and also with *Synorhizobium* and *Azotobacter* spp. (Gryndler *et al.*, 2008) (Table 5.3).

Table 5.3. Role of PGPR/PGPF in energy production and land reclamation.

S. No.	PGPR/PGPF Strains	Host Plant	PGPR/PGPF Attribute	Inhabiting Nature of Microbes	Energy Production	Land Suitability/ Reclamation	References
1	*Stenotrophomonas acidaminiphila* sp. nov. and *Pseudomonas putida* (Trevisan) Migula.	*Pinus densiflora* Siebold and Zucc, *Thuja orientalis* L. and *Populus tomentiglandulosa*	Degrades total petroleum hydrocarbons	Rhizospheric	Biomass, bioethanol	Diesel contaminated lands	Jagtap *et al.* (2014)
2	*Burkholderia* sp. D54	*Lolium multiflorum* Lam. (Ryegrass)	Ideal PGPR activity	Rhizospheric	Biomass	Zn, As, Cd and Pb contaminated lands	Guo *et al.* (2014)
3	*Azospirillum-Pseudomonas–Glomus* Consortia, *Azospirillum brasilense*	*Zea mays* L.	Promote shoot biomass by producing Indole-3-Acetic Acid	Rhizospheric	Biomass	Marginal land	Couillerot *et al.* (2013); Werling *et al.* (2014)
4	*Burkholderia* sp. SaZR4, *Burkholderia* sp. SaMR10, *Sphingomonas* sp. SaMR12, *Variovorax* sp. SaNR1and *Enterobacter* sp. SaCS20,	*Oryza sativa* L.	Ideal PGPR activity	Endophytic	Biomass	Can be used for Zn contaminated lands	Hiloidhari *et al.* (2012); Wang *et al.* (2014)
5	*Agrobacterium, Burkholderia, Enterobacter*	*Brassica napus* L.	IAA, Siderophore, Phosphate solubilizer, Nitrogen Fixer	Rhizospheric	Biofuel	Uptake of Cd, Pb, Zn contaminated lands	Farina *et al.* (2012); Jing *et al.* (2014)
6	*Stenotrophomonas* sp. *Pantoea* sp., *Achromobacter* sp.	*Saccharum officinarum L.*	Potential phosphate solubilizer	Endophytic	Bioethanol	Phosphate deprived land	Taulé *et al.* (2012)

7	*Enterobacter* sp., *Herbaspirillum* sp.	*Brassica oleracia*	Nitrogen fixing ability	Endophytic	Biodiesel	Sandy and marginal land	Zakria *et al.* (2008); Ahmad *et al.* (2013)
8	*Bacillus pumilus (IM-3)*	*Jatropha curcas* L.	siderophore, and ammonia production	Rhizospheric	Biodiesel	Degraded soil restoration	Sumarsih and Haryanto (2012)
9	*Pseudomonas fluorescens* and *P. putida*	*Jatropha curcas* L.	phosphate solubilization, IAA production	Rhizospheric	Biodiesel	Degraded soil restoration	Jha *et al.* (2010; 2012)
10	*Glomus intraradices* and a mix of *G. intraradices* and *G. deserticola*	*Prosopis julifora* (mesquite)	Supports plant growth and rhizosphere microbial community	Rhizospheric	Biomass	Reclamation of acidic Pb/Zn mine tailings	Solís-Domínguez *et al.* (2011)
11	*Bradyrhizobium* sp., *Pseudomonas* sp. and *Ochrobactrumcytisi*	*Lupinus luteus* L.	Phytostabilization and rhizoremediation	Rhizospheric	Improved Biomass	Remediation of Cu, Cd and Pb	Dary *et al.* (2010)
12	*Burkholderia, Rahnella, Sphingomonas* and *Acinetobacter*	*Populus trichocarpa* Torr. and A.Gray ex Hook. and *Salix sitchensis* Sanson ex Bong	Excellent PGPR activity	Endophytic	Biomass	Nitrogen deprived land	Doty *et al.* (2009)
13	*Bradyrhizobium japonicum*-SB1, *Bacillus thuringiensis*-KR1,	*Glycine max* (L.) Merr.	Nodule-forming diazotrophs, promote plant growth	Rhizospheric	Biodiesel	Nitrogen deprived land	Mishra *et al.* (2009)
14	Inocula of AMF [*Glomus intraradices* (BEG140), *G. claroideum* (BEG96) and *G. mosseae* (BEG95)] alone or with PGPR (*Synorhizobium* spp. and *Azotobacter* spp.)	*Medicago sativa* L. cv. Vlasta (alfalfa)	Enhances plant growth and enhances mycorrhizal colonization	Rhizospheric	Biomass	Reclamation of coal mine spoil banks	Gryndler *et al.* (2008)

Continued

Table 5.3. Continued.

S. No.	PGPR/PGPF Strains	Host Plant	PGPR/PGPF Attribute	Inhabiting Nature of Microbes	Energy Production	Land Suitability/ Reclamation	References
15	*Sinorhizobium* sp. and/or *Azotobacter* sp.	*Cannabis sativa* L. cv. Beniko and *Phalaris arundinacea* L.cv. Palaton S.	Enhances plant growth and naturally colonized by AMF	Rhizospheric	Biomass	Tolerate adverse conditions of spoil bank substrates	Gryndler *et al.* (2008)
16	*Bacillus subtilis* SJ-101	*Brassica juncea* L.	Potential phosphate solubilizer	Endophytic	Biomass	Phosphate deprived land	Zaidi *et al.* (2006)
17	*Pseudomonas putida* VM1450	*Pisum sativum* L.	Lowers phytotoxic effect	Rhizospheric, Rhizoplanic, Endophytic	Improved Biomass	Remediation of 2,4 D	Germaine *et al.* (2006)

PGPR also helps plant growth promotion passively by competing with the habitat and nutrients of pathogens. As a result, they suppress the proliferation and activity of these pathogens resulting in the enhanced growth of associated plants. Furthermore, many endophytes secrete several metabolites that considerably inhibits the growth of underlying pathogens (Brader *et al.*, 2014). For example, *Pseudomonas* sp. produce iron-binding siderophores under iron-deprived conditions (Barry and Challis, 2009). As a result, the siderophores act as iron scavenging molecules and create nutrient-limiting conditions for pathogens. Many reports suggest that endophytes also produce several molecules, antibiotics, chemicals and enzymes, that directly inhibit or even kill plant pathogens (Weyens *et al.*, 2009; Brader *et al.*, 2014). In this way these PGPM promote the plant growth and the biomass as well.

5.5 Role of PGPR and PGPF in Wasteland and Degraded Land Reclamation

Food, fuel and fibres are the primary products from plants; thus they have been exploited right from the beginning of human civilization. However, the plants and their associated microorganisms also provide ecological balance in the era of industrial pollution. Extensive land use or even the utilization of chemical fertilizers for agriculture has led the transition of productive lands into the marginal and further into the degraded or contaminated lands. Therefore the role of these novel organisms has attracted worldwide attention. As mentioned in the above section, biomass production can be efficiently enhanced by using PGPM. Thus, growing crops on the contaminated lands using PGPM can mutually enhance the biomass production as well as the reclamation of marginal lands, wastelands or contaminated lands (Table 5.3).

Developed technologies are in existence to remediate degraded lands. However, they need huge inputs, costs and maintenance, and as a result cause huge changes in soil physical and bio-chemical properties. Contrastingly, phytoremediation using PGPM can be a suitable and sustainable remediation process. Several studies have proven that these organisms have critical role in the reclamation of marginal, degraded, contaminated land or even waste land. There has been a study to increase the drought resilience of maize through endophytic colonization by *Burkholderia phytofirmans* PsJN and *Enterobacter* sp. FD17 (Naveed *et al.*, 2014). Furthermore, the establishment of switchgrass has been observed by inoculation with the strain of *Paenibacillus polymyxa* (Ker *et al.*, 2012). Moreover, it has also been found for the seeds of coriander, cumin and fennel that the thermotolerant bacteria *Bacillus* spp. and *Actinobacterium kocuria* sp. and the cynobacteria *Anabaena laxa* and *Calothrix elenkinii* have enhanced seed germination potential by around 25% from the control (Kumar *et al.*, 2013) and can be applied under marginal conditions for reclamation purposes. Similarly, the role of PGPR has also been observed in preventing soil erosion in arid regions by improving the growth of desert plants under reforestation scenarios (de-Bashan *et al.*, 2012). Further, it has also been found that endophytic actinobacteria that mostly belong to the *Streptomyces* genus have plant growth-promoting activity for *Jatropha curcas* L. growing in Panxi dry-hot valley soil. Hence, this has promising PGP attributes to be developed as biofertilizer to enhance soil fertility and promote plant growth (Qin *et al.*, 2015).

Since the process of phytoremediation is governed naturally through solar power and needs no maintenance, it has a high level of acceptance in society. Moreover, the role of PGPM is like a catalyst to enhance the remediation process efficiently (Bell *et al.*, 2014). Apart from these benefits, this sustainable process has some obstacles as well. These are phytotoxicity, evapotranspiration of volatile pollutants and degradation contaminant intermediates via the leaves (Weyens *et al.*, 2009). To overcome these lacunae, PGPM (Fester *et al.*, 2014), i.e. rhizobacteria (Nadeem *et al.*, 2014), endophytes (Brader *et al.*, 2014) and even AMF (Nadeem *et al.*, 2014) with specific characteristics can be used. The schematic representation (Fig. 5.1) clearly depicts how these microorganisms can play a vital role in the reclamation of degraded lands.

5.6 Role of Plant Growth-Promoting Microorganisms in Carbon Sequestration under Warming Climate

Soil microorganisms are the key players regulating the dynamics of soil carbon. On the other hand, plants have the capability to fix atmospheric carbon by reducing CO_2 for synthesising the photosynthates. The photosynthates are stored in plant tissues and also form exudate from roots called rhizodeposits (McNear, 2013). Exuded photosynthates are the source of energy and nutrients for the rhizospheric microbes (Singh *et al.*, 2004; Philippot *et al.*, 2013). Soil microorganisms utilise the rhizodeposits and incorporate the carbon and soil organic matter in their body (Wu *et al.*, 2009a; Stockmann *et al.*, 2013). However, the soil microorganisms also use the rhizodeposits as fresh carbon source to decompose the old recalcitrant carbon. Some microbes of the rhizospheric and the endophytic compartment benefit the plants by providing them with accessory nutrients and reducing stress (Weyens *et al.*, 2009, 2010; Glick, 2010; McNear, 2013). In this regard, the PGPRs can be used to enhance carbon sequestration by capturing the atmospheric carbon by plant growth promotion, soil aggregate formation and fixation of microbial carbon in soil systems. Thus PGPR can help in partial mitigation of the global climate change by increasing the C-sequestration from terrestrial ecosystems (Table 5.4).

Substantial amounts of soil carbon could be sequestered by manipulating the soil microorganisms. Among the soil microbiota fungi play the predominant role in nutrient dynamics. AMF are a known PGPR agent having the potential to enhance the soil carbon pool by improving soil aggregate formation. Thus, the alteration of the soil microbial community by increasing the proportion of AMF in the soil could help in higher carbon sequestration with increased soil aggregation (Zhang *et al.*, 2013). Individual inoculation of PGPR or AMF or its co-inoculation may also improve the soil C, N storage if able to synthesise glomalin and glomalin-related soil protein accordingly (Walley *et al.*, 2014). Recent reports show that *Pseudomonas fluorescens* inoculation increases plant productivity and is able to mitigate the positive feedback of elevated CO_2 by enhancing the C:N implants under elevated CO_2 (Nie *et al.*, 2015). A long-term study (Juwarkar *et al.*, 2010) also reported the successful reclamation of manganese mine land. They also found that application of plant growth-promoting bacteria along with site-specific multi-plant species increases the soil organic carbon pool as well. The study shows that the PGPRs have potential to sequester the carbon along with their attributes of plant growth promotion and degraded land reclamation. *Jatropha curcas*, a biofuel-producing crop, offers multiple benefits such as phytoremediation, reclamation of marginal lands by enhanced litter turnover and improved nutrient status of the soil (Abhilash *et al.*, 2013; Srivastava *et al.*, 2014; Edrisi *et al.*, 2015). All these attributes of *J.curcas* make it a suitable candidate for soil carbon sequestration. A bacterium *Enterobacter cancerogenus* MSA2 was reported to increase the plant growth of *J. curcas* (Jha *et al.*, 2012) and thus could be potentially used for enhancing the performance of *J. curcas* for soil carbon sequestration from marginal and degraded lands. Similarly, the plant growth-promoting endophytic bacterium *Enterobacter* sp. 638 can be used to enhance the carbon sequestration by poplar plants from marginal, non-agricultural soils by using this bacterium as a growth-promoting agent (Taghavi *et al.*, 2010).

It is predicted that in future a warmer climate may increase the labile carbon content of the soil by enhanced root exudation due to the fertilization effect of increased CO_2 levels on plants. This may increase microbial decomposition of the soil organic carbon due to increased microbial activity (Davidson and Janssens, 2006; Wieder *et al.*, 2013). Thus, long-term studies of physiology and adaptation of the microbes in a changing climate are essential for assessing the applicability

Table 5.4. Role of PGPR/PGPF in sustainable agriculture/carbon sequestration.

S.No.	Species	Host Plant	PGPR/PGPF attribute	Relationship to host	Sustainable Agriculture/Carbon sequestration	References
1	*Burkholderia, Rahnella, Sphingomonas,* and *Acinetobacter*	*Populus trichocarpa* Torr. and A.Gray ex Hook. and *Salix sitchensis* Sanson ex Bong	Having excellent PGPR activity	Endophyte	Higher biomass production can capture more carbon, phytoremediation of nitrogen-deprived marginal land and make it suitable for agriculture.	Doty *et al.* (2009)
2	*Bacillus subtilis* SJ-101	*Brassica juncea* L.	Potential phosphate solubilizer	Endophyte	Sustainable agriculture approach increases the availability of phosphate, crop and green carbon storage.	Zaidi *et al.* (2006)
3	*Stenotrophomonas* sp. *Pantoea* sp., *Achromobacter* sp.	*Saccharum officinarum*	Capable in N_2 fixation	Endophyte	Sustainable agronomic practices for increasing *Saccharum* production, and controling the phytopathogen	Taulé *et al.* (2011)
4	*Bradyrhizobium japonicum*-SB1, *Bacillus thuringiensis*-KR1, *Enterobacter* sp., *Herbaspirillum* sp.	*Glycine max* (L.) Merr., *Brassica oleracia*	Nodule forming diazotrophs, promote plant growth	Rhizospheric	Sustainable agriculture option increasing crop yield and reducing the negative impact of chemical fertilizer	Mishra *et al.* (2009) Zakria *et al.* (2008)
5	*Azospirillum–Pseudomonas–Glomus* Consortia, *Azospirillum brasilense*	*Zea mays* (L.) var. Costeño Mejorado	Promote shoot biomass by producing Indole-3-Acetic Acid	Rhizospheric	Sustainable agriculture option and reduce the N_2O induced GHGs emission	Couillerot *et al.* (2013)
6	*Enterobacter cancerogenus* MSA2, *Pseudomonas fluorescens* and *P. putida* *Bacillus pumilus* (IM-3)	*Jatropha curcas* L.	ACC deaminase, phytase, phosphate solubilization, IAA, siderophore, and ammonia production	Rhizospheric	Combination *Jatropha curcas* L. and its rhizospheric microbes can be used for carbon sequestration, degraded soil restoration, reduces desertification, deforestation and potent bioenergy crop	Jha *et al.* (2010 2012); Sumarsih and Haryanto (2012)
7	*Trichoderma* spp., *Pseudomonas* spp., *Actinomycetes* spp.	*Capsicum annuum* L., *Lycopersicum esculentum* Mill.	Good plant growth promotion activity	Rhizospheric	Increased the biomass and yield of the crop sustainably also act as a biocontrol agent	Graber *et al.* (2010)
8	*Agrobacterium, Burkholderia, Enterobacter*	*Brassica napus* L.	IAA, Siderophore, Phosphate solubilizer, Nitrogen Fixer	Rhizospheric	Used for increasing sustainable production of *Brassica napus.*	Farina *et al.* (2012)

Continued

Table 5.4. Continued.

S.No.	Species	Host Plant	PGPR/PGPF attribute	Relationship to host	Sustainable Agriculture/Carbon sequestration	References
9	*Bacillus* sp. +soil+ biochartreatment	*Phaseolus vulgaris* L.	Having good PGPR potential, increased the plant biomass	Rhizospheric	Used for the sustainable agricultural production of the crop, promoted the overall growth of the plant, can replace the use of chemical fertilizer	Saxena *et al.* (2013)
10	Arbuscular mycorrhizal fungi (AMF) *Acaulospora denticulata* *Scutellospora calospora*	*Pisum sativum* L. *Artemisia tridentate* Nutt. seedlings	Promoted plant biomass, help in macroaggregate formation	Rhizospheric	Sustainably enhances the plant biomass and yield of the crop. Capturing more carbon in (AMF biomass and Root biomass)	Pokharel *et al.* (2013)
11	Biofilm of (*Anabaena/ Trichoderma*) and (*Anabaena laxa* (T7) RP8/*Calothrix* sp.	*Vigna radiate* (L.) Wilczek and *Glycine max* (L.) Merr.	Excellent PGPR enhanced plant growth.	Rhizospheric	Open sustainable agriculture option and increases the crop yield and nutrient uptake efficiency	Prasanna *et al.* (2014)
12	Consortium of PGPR and Cyanobacterium (*Providenciasp.+ Anabaena laxa+ A. oscillarioides*) (*Providenciasp.+Brevundimonasdimunuta*)	*Oryza sativa* L.	Having plant growth promotion activity, like nitrogen fixation	Rhizospheric	Sustainably enhances the plant biomass, grain yield, minimizes the use of chemical fertilizer also enhances the rate of carbon sequestration in soil	Prasanna *et al.* (2012)
13	*Bacillus subtilis* and *Bacillus mucilaginosus* + vermicompost treatment	Tomato and spinach	Potential plant growth promotion, nutrients solubilization	Rhizospheric	Enhanced soil quality, microbial biomass in soil, crop yield, vitamin C content in tomato and soluble protein in spinach and reduces the risks of chemical fertilizer	Song *et al.* (2015)
14	*Rhodopseudomonas palustris* PP803 +Rice straw and rice husk ash (4:1 ratio)	*Oryza sativa* L. subsp. *indica*	Promote root and shoot length, having capacity to produce 5-aminolevulinic acid	Rhizospheric	Enhanced plant growth under salt stress, and reduces CH_4 and CO_2 emission by 100% and 47%	Kantha *et al.* (2015)

of plant-rhizosphere C sequestration potential of the microbes in order to prevent soil carbon loss and sequester more carbon in future. Again functional aspects of the terrestrial ecosystem and soil carbon sequestration potential will depend upon the aboveground and belowground responses in a changing climate and needs-integrated approches for detailed understanding (Abhilash and Dubey, 2014) (see Table 5.4).

5.7 Strategies for Enhancing the Performance of Plant Growth-Promoting Microorganisms

Plant–microbe interaction holds a key for the development of sustainable agricultural practices to meet the increasing human demand of food, fuel and fodder for livestock (Singh *et al.*, 2004; Weyens *et al.*, 2009) (Table 5.4). Soil microorganisms are key drivers of the organic mineralization and utilize different C sources of natural and xenobiotic nature. Plants may structure their rhizosphere and recruit the beneficial microbiota to their root supporting their growth. However, plant–microbe interactions are complex and not easy to decipher in the rhizosphere (Singh *et al.*, 2004). A number of plant growth-promoting rhizobacteria have been reported to date, but their successful utilization is yet to be fully exploited for increasing agricultural production and improving the remediation processes. A major limitation of the PGPRs is that while they may perform well in controlled conditions, the same results may not be reproduced during their field applications (Glick, 2010; Nadeem *et al.*, 2014). Their poor performance during field application might be due to a variety of processes like unsuccessful rhizospheric colonization, competition with other rhizospheric microbiota for resource utilization, lack of nutrients in the applied soil and spatio-temporal variations.

Furthermore, the change in atmospheric carbon and temperature will affect soil organic matter, carbon dynamics (Dennis *et al.*, 2010), nutrient cycling, and soil microbial biomass. However, we have no information about the effect of climate change on the rhizospheric plant–microbe interactions and how it will shape the process of plant growth promotion and microbe-assisted remediation of soil pollutants (Abhilash *et al.*, 2013; Abhilash and Dubey, 2014). It is believed that climate change may affect the process of root exudation. The increased or decreased root exudation will in turn affect the rhizospheric interactions, functions and the structure of the microbial community as well (Abhilash and Dubey, 2014). It may alter the behaviour and functions of the PGPRs, thus there is an urgent need to develop sustainable mechanisms for improving the efficiency of PGPRs to support plant growth and cope with different biotic and abiotic stresses in order to promote plant growth in field conditions as well (Fig. 5.5).

5.7.1 Agronomic practices

Adaptation of sustainable agronomic practices along with plant growth-promoting microorganisms could improve the performance of PGPR microorganisms in field conditions. The application of the conventional tillage practices in agriculture depletes the level of nutrients and organic matter in the soil. It also disturbs the structure of the soil microbial community (Kumar *et al.*, 2013). Application of reduced tillage practices such as minimum tillage, no tillage or zero tillage can improve soil structure and shape the structural and functional diversity of the soil and reduce the soil CO_2, NO and N_2O emissions (Lupwayi *et al.*, 1998; Marquina *et al.*, 2015). This will further enhance the diversity and function of the beneficial microorganisms in the soil supporting the microbial mediated processes of nutrient recycling, bioremediation of the xenobiotics, maintenance of soil structure, and aggregation.

Addition of biochar benefits the quality of the soil by improving its porosity, water holding capacity, labile carbon pool and

Fig. 5.5. Role of plant growth-promoting micro-organisms in sustainable agriculture. (A) Brinjal inoculated with the consortia of *Azotobacter chrococcum* MTCC-446 + *Pseudomonas aeruginosa* BHU PSB01+ *Trichoderma harzianum* + *B. megaterium* (BHU PSB14); (B) Pea grown in tetra-inoculation of *Rhizobium* sp. + *P. aeruginosa* + *Peribacillus polymyxia* BHU PSB17; (C) Chickpea supplemented with the microbial consortia of *Mesorhizobium* sp + *P. aeruginosa* + *T. harzianum*(D) Cauliflower + Spinach inoculated with *A. chrococcum* MTCC-446 + *P. aeruginosa* + *T. harzianum*; and (E) Tomato grown in tetra-inoculation of *A. chrococcum* MTCC-446 + *P. aeruginosa* + *T. harzianum* + *P. polymyxia* BHU PSB17.

other physicochemical properties (Jeffery *et al.*, 2015). Such changes in physicochemical properties of the soil will support the growth of the plant and reduce the N_2O emission from an agricultural field by tranfering electrons to the denitrifying microorgamism of the soil (Cayuela *et al.*, 2013). Increase in root growth and labile carbon portion will attract more microorganisms to the rhizosphere, also enhancing the number of PGPR in the rhizosphere and their synergistic effect of plant growth promotion (Kolton *et al.*, 2011). Application of biochar not only improves the soil quality (Roberts *et al.*, 2015) and attracts beneficial soil microbiota but also suppresses the growth of the phytopathogens. In a recent study, soil amendment with biochar was reported to suppress the growth of *Rhizoctonia solani* (Jaiswal *et al.*, 2014).

Inoculation of tomato seedlings with vesicular arbuscular mycorrhiza alone or in combination with PGPR and suitable irrigation practices may enhance the plant biomass, yield, water use efficiency and tolerance against biotic and abiotic stresses (Candido *et al.*, 2015). Agricultural wastes are often rich sources of organic matter and nutrients. Application of such agricultural wastes may increase the microbial activity in the soil. These microbes assist plant growth promotion in degraded soil and also help in degradation of toxic pollutants of the soil as well. Abhilash and Singh (2008) reported that application of sugarcane bagasse can accelerate lindane degradation by enhanced microbial activity. Similarly waste proteins such as blue algal sludge, rapeseed meal, poultry feathers, and chicken manure are also good sources of nitrogen. Solid state fermentation (SSF) of these protein sources helps in creation of a medium that supports higher biomass, lipopeptides and number of CFU for microorganisms. A SSF medium containing 7.61% rapeseed meal, 8.85% expanded feather meal, 6.47% dewatered blue algal sludge and 77.07 % chicken manure was established as a SSF substrate to get maximum SQR-9 biomass as $6.31 \pm 0.26 \times 10^8$ CFU/g Dry Weight (DW) and maximum amount of lipopeptides as 17.81 ± 0.72 mg/g DW. Such waste proteins could be used for value-added utilization for producing economical but high-quality bio-organic fertilizers (Huang *et al.*, 2015). Organic amendment can also improve the resource utilization ability of the PGPR microorganisms. Thus the application of sustainable agronomic practices may help in improvement of plant growth-promotion properties of the microorganisms. This will support development of PGPR-based sustainable agronomic practices to increase agricultural production and restore soil system function. In their study Chang *et al.* (2008) reported that *Pseudomonas putida* strains isolated from prolonged swine compost-treated soil have better polysaccharide utilization ability than *P. putida* strains isolated from an untreated control site. The *P. putida* strains isolated from the treated site were able to utilize specific polysaccharides such as L-rhamnnose and xylitol. However, the strains isolated from the control site had higher utilization ability for monosaccharides such as D-fructose, D-galactose, D-mannose and α-D-galactose. The addition of swine compost to trace-element-contaminated mine soil reduces the ecotoxicology of the soil due to decrease in mobility of the trace elements (Pardo *et al.*, 2014). Moreover, the amendment also increases activities of enzymes such as cellulase, β-galactosidase playing an essential role in organic matter mineralization and humification (Pardo *et al.*, 2014), and urease playing a role in the N cycle by catalyzing the urea degradation. All these enzymes are of microbial origin and very sensitive to toxicity of the trace elements. However, as amendment with swine compost immobilizes the trace elements, the microbial function is enhanced in such contaminated soil. Therefore, the amendment of compost may offer a suitable strategy for improving the function of PGPRs in contaminated soil sites (Zornoza *et al.*, 2012). Thus sustainable agricultural practices and biologically active agricutural and industrial waste could help in enhancing the activity of PGPM not ony in controlled environments but also in field conditions.

5.7.2 Rhizospheric engineering

PGPR microorganisms mostly thrive in the rhizosphere of plants. The rhizosphere is shaped by a variety of processes of solar-driven plant growth and root exudation. The rhizodeposits contain nutrient for driving the microbial functioning of the rhizosphere (McNear, 2013). The nature of the rhizodeposits changes with plant species and their spatio-temporal variation. The nature of the rhizospheric soil changes with the plant species, genotype and cultivar. The current monocropping system of modern agriculture has minimized the contribution of rhizospheric microorganisms for promoting plant growth and improving soil health (Philippot et al., 2013). An improvement of agricultural practices with mixed or intercropping systems can lead to more heterogeneous distribution of the plant root in soil, attracting more PGPRs to the plant root interface, i.e. the rhizosphere (Kumar et al., 2013). Identification of potential rhizospheric and endophytic microbes from rhizosphere regions of plant species under various stressed ecosystems and creating a consortia of compatible microbes for inducing multiple growth-enhancing effect in the plants is the prime objective in this venture. Plants' tolerance of various biotic and abiotic stresses can also be enhanced by detailed understanding of the rhizospheric semiochemicals and root exudation through rhizospheric engineering (Zhang et al., 2015).

The PGPRs and pollutant-degrading microbes colonize the rhizosphere in the same fashion as other microorganisms. In the rhizosphere, the PGPRs compete for the same resources with the other microorganisms (Hibbing *et al.*, 2010). Thus, resource limitation might be a factor affecting the functioning of the PGPRs during field application. Bioaugmentation of the PGPRs on the sterilized seed or root surface will increase the proliferation of the applied microbe in the rhizosphere and let them utilize rhizodeposits more efficiently. An increased number and activity of PGPRs in the rhizosphere will promote plant growth and support soil health (Segura and Ramos, 2013). Rhizobia are known to promote plant health by helping in nitrogen (N) fixation by leguminous plants; however, the N fixation process is sensitive to temperature and drought stress (Davies *et al.*, 2011). According to Davies *et al.* (2011), a better approach to improve the function of rhizobia would be its co-inoculation with PGPRs, having attributes to promote root branching and enhancing the secretion of root exudates. An increase in root branching will provide more sites for nodulation (Dardanelli *et al.*, 2008) and enhanced secretion of root exudates having more substrates for stimulation of *nod* genes. On the other hand, PGPRs also lower the ethylene level in the rhizosphere by utilizing the ACC exuded by the roots by their 1-aminocyclopropane-1-carboxylate deaminase (ACCD) activity (Davies *et al.*, 2011). Ethylene is a plant stress hormone which also inhibits the process of nodulation. A decreased level of ethylene will further help in the nodulation process (Davies *et al.*, 2011) and also enhance the root proliferation (Ryan *et al.*, 2009), thus improving the rhizobia-legume mediated N fixation. As the microbes compete for the resources in the rhizosphere, another approach could be the direct application of the microbial-derived compounds as soil additives. This technology is gaining popularity as fertilizers based on microbial products have a longer shelf life and are easier to apply. On the other hand, sometimes microbes do not perform optimally in field conditions; however, direct application of their products may provide better results. In a recent study Ali and McNear (2014) reported that the use of Soil Builder™ (containing products of the bacterial species – *bacillus*, *actimomyces* and proteobacteria – derived from bioreactors) increases the accumulation of flavonoids in the plants by enhancing the expression of the genes involved in the phenyl propanoid pathway. Phenylpropanoids not only help the plant species in combating stress but also assist the plants in their responses towards various biotic and abiotic stimulations (Vogt, 2010). Moreover phenylpropanoids also act as inducers for plant microbe symbioses and a phenylpropanoid-rich food provides additional health benefits to human beings (Singh *et al.*, 2014).

With the advent of the next generation genomic and transcriptomic analysis technologies, we have a much clearer picture of the catabolic pathways involved in the rhizospheric interactions (Benfey *et al.*, 2010; Lundberg *et al.*, 2012). Now we know that the plant species can programme the exudation process according to its need. A shift in the process of exudation changes the microbial community structure in the rhizosphere and induces the microbial genes involved in the utilization of the chemicals secreted by the root (Yergeau *et al.*, 2014). An insight into the maintenance and expression of the catabolic genes involved in plant growth promotion and remediation of the soil pollutants in the rhizosphere will hold a key to enhance plant growth and agricultural production in the near future (Badri *et al.*, 2009). Thus a further exploration of the plant genes involved in beneficial plant microbe interaction will provide the tool for the exploitation of PGPR as more efficient tools to improve plant health in agriculture (Benfey *et al.*, 2010).

5.7.3 Molecular approach

Nowadays there is an increased interest to find out the role of PGPRs in activities other than plant growth promotion as well. Though a number of microorganisms have been reported to have multiple attributes such as performing bioremediation, helping in carbon sequestration (Jagtap *et al.*, 2014), their successful application in field situations is still quite limited, owing to poor performance of the microbes as various biotic and abiotic stress factors limit their functioning in field conditions. Application of genetic modification and synthetic biology can alleviate these limitations and increase the advantage of using these microbes for sustainable agriculture. However, these genetic modification studies require a deep knowledge of the genetics of soil microbes, since changes or insertion of genes into a particular community may lead to environmental hazards if not carefully performed.

In this sense, study and genetics quantification of soil microbes is growing every day, making it possible to assess the impact that some activities cause on microbiota and how it responds to different situations. Techniques such as molecular quantification by qPCR (Real Time PCR) and metagenomics can make possible the association of genetic alteration with environment modification, also identification of the microbiota and its diversity with metabolism presented by the soil (Zhou *et al.*, 2012; Luo *et al.*, 2014). Another important factor that should be considered in relation to soil microbes is the natural physical processes such as heat and humidity that occur. These can interfere with the distribution of microorganisms present in the soil.

The evaluation of the nitrogen cycle and its genetic alterations can lead to imbalance between fixation, nitrification and denitrification. For example, this study can help you understand what happens in different variations of temperature, water and soil conditions. Another perspective is the identification of genes that degrade toxic compounds, and their possible future use in transgenesis or biomonitoring.

Effective colonization of the PGPRs in the rhizosphere is often the limiting constraint for their performance (Compant *et al.*, 2010). A recent study revealed that a green fluorescent protein (GFP) engineered bacterium *Bacillus amyloliquefaciens* FZB42 showed different colonizing patterns on the rhizoplanes of different plant species. Thus, GFP tagging is an excellent tool for studying the root colonization behaviour of different PGPRs in the diverse and competitive environmental rhizospheric niches (Fan *et al.*, 2011). Similarly, chemotaxis plays an important role in successful colonization in the rhizosphere. The root exudates act as chemoattractant and the microbes move towards the root exudate and get colonized in the rhizosphere or rhizoplane. A PGPR mutant of

P. fluorescens lacking the *che A* gene for chemotaxis showed inefficient colonization in tomato rhizosphere due to reduced movement of the bacterium towards the root exudates (de Weert *et al.*, 2002). Thus a detailed knowledge of the response genes such as *cheA, cheY,* and *pctA* involved in the process of chemotaxis (Compant *et al.*, 2010) will be helpful in modulating the regulation of these genes for efficient rhizospheric colonization of PGPRs through genetic engineering approaches.

PGPR have several plant-beneficial properties-contributing genes, and these genes could have been selected in these bacteria. Analysis of distribution of 25 genes among 25 proteobacterial PGPR and 279 other Alpha-, Beta- and Gammaproteobacteria representing various taxonomic groups and ecological status revealed that most of the 23 genes studied were also found in non-PGPR Proteobacteria and none of them were common to all 25 PGPR genomes studied. It suggests that cooperation interactions between Proteobacteria and plant roots might have established separately in taxonomically contrasted Proteobacteria (Bruto *et al.*, 2014). Further genome sequencing efforts targeting close relatives of these PGPR would be helpful in understanding the evolution of plant-beneficial traits among the PGPR and utilizing single genes or operons for enhancing plant health. The genome sequencing of the PGPR *Enterobacter* sp. *638* revealed the genes involved in the process of plant growth promotion-like uptake of nutrient, minimizing oxidative stress, phytohormones, siderophore and antimicrobial compound production, chemotaxis and colonization. Apart from the main chromosome, the bacterium also has a plasmid pENT638-1 responsible for endophytic colonization as it harbours the genes for the compounds that help in plant adhesion and colonization, e.g. hemagglutinin-related autotransporter (Taghavi *et al.*, 2010). The above plasmid could be used for moving the plant growth-promoting bacterium from the rhizosphere to the endophytic compartment which is much more favorable than the harsh and competitive environment of the rhizosphere. Similarly, the genome sequencing of the PGPR bacterium *Variovorax paradoxus* S110 revealed its dual survival as an individual and symbiont and its metabolic diversity of autotrophic and heterotrophic lifestyles (Han *et al.*, 2011). A more comprehensive analysis of the sequenced genomes will lead to characterization and identification of yet unknown genes involved in the process of root colonization, plant growth promotion and contaminant degradation. This will help to improve our understanding of the interaction of these beneficial microbes with plants and altering their biology for successful utilization in sustainable agriculture. A recent study shows that inoculation of *Bradyrhizobium diazoefficiens* with PGPR in *Glycin max* enhances the 9.7–43.6% seed yield per hectare. The molecular mechanism behind the yield enhancement was active nodulation, N fixation and *nifH*, bacteroid *dctA*, *phbC* and *otsA* gene expressions in the inoculated plants (Prakamhang *et al.*, 2015). Such types of molecular information may be utililized further for the development of genetically engineered legume or nonlegume crops.

Limited information is available on the plant growth-promoting attributes of the unculturable bacteria. The advent of next-generation sequencing technologies has revolutionized the field of metagenomics for the study of cultivation-independent microorganisms (Schenk *et al.*, 2012). Metagenomics analysis from different habitats might reveal novel PGPRs and their catabolic pathways involved in the process of plant growth promotion and in interaction with other microorganisms responsible for the functioning of the rhizospheric processes. However, rhizospheric interactions are very complex in nature and could not be easily defined (Philippot *et al.*, 2013). Systematic analysis of the preferential utilization of the root exudates by different microorganisms will be a key to finding out the functioning of the PGPRs in the rhizosphere and will help to further improve their performance for plant growth promotion.

5.8 Challenges and Future Research Perspectives

The problems of enhancing agricultural yield for a burgeoning population, alleviating hidden hunger and energy security are at its extremes. Solutions become more tedious when they utilize limited land without agricultural extensification practices. Plant growth-promoting microorganisms play a myriad of crucial roles in maintaining the soil system functioning and agroecosystem services. Beyond its extensive potential, major lacuna also exist on the field level application and global commercialization of these microbial inoculums. If we better understand soil microbiota, its operating mechanism and its interactions with plant growth it will be possible to perform better control and exploitation of existing resources. However, further investigation on the use of new technologies such as transgenics is mandatory while the consequences on soil are still unknown. In this reference, recent advances in genetics and genomics would unravel detailed biochemical processes, mechanisms and molecular understanding of the microbial functions and open a new horizon to sustainable agriculture for the completely sustainable development of the soil system.

Acknowledgements

PCA is thankful to DST, CSIR, INSA and RKD to CSIR for financial support (CSIR-SRF). AKS is thankful to DST for providing financial support under Women Scientist [SR/WOS-A/LS-176/2014(G)].

References

Abd-Alla, M.H. (2011) Nodulation and nitrogen fixation in interspecies grafts of soybean and common bean is controlled by isoflavanoid signal molecules translocated from shoot. *Plant Soil Environment* 57, 453–458.

Abd-Alla, M.H., El-Enany, A.W.E., Nafady, N.A., Khalaf, D.M. and Morsy, F.M. (2014) Synergistic interaction of *Rhizobium leguminosarum* bv. *viciae* and arbuscular mycorrhizal fungi as a plant growth promoting biofertilizers for faba bean (*Vicia faba* L.) in alkaline soil. *Microbiological Research* 169, 49–58.

Abhilash, P.C. and Dubey, R.K. (2014) Integrating aboveground–belowground responses to climate change. *Current Science* 1637, 12.

Abhilash, P.C. and Singh, N. (2008) Influence of the application of sugarcane bagasse on lindane (γ-HCH) mobility through soil column: implication for biotreatment. *Bioresource Technology* 99, 8961–8966.

Abhilash, P.C., Jamil, S. and Singh, N. (2009) Transgenic plants for enhanced biodegradation and phytoremediation of organic xenobiotics. *Biotechnology Advances* 27, 474–488.

Abhilash, P.C., Powell, J.R., Singh, H.B. and Singh, B.K. (2012) Plant–microbe interactions: novel applications for exploitation in multipurpose remediation technologies. *Trends in Biotechnology* 30, 416–420.

Abhilash, P.C., Dubey, R.K., Tripathi, V., Srivastava, P., Verma, J.P. and Singh, H.B. (2013a) Remediation and management of POPs-contaminated soils in a warming climate: challenges and perspectives. *Environmental Science and Pollution Research* 20, 5879–5885.

Abhilash, P.C., Dubey, R.K., Tripathi, V., Srivastava, P., Verma, J.P. and Singh, H.B. (2013b) Adaptive soil management. *Current Science* 104, 1275–1276.

Abhilash, P.C., Tripathi, V., Edrisi, S.A., Dubey, R.K., Bakshi, M. *et al.* (2016a) Sustainability of crop production from polluted lands. *Energy, Ecology and Environment* 1, 54–65.

Abhilash, P.C., Dubey, R.K., Tripathi, V., Gupta, V.K. and Singh, H.B. (2016b) Plant growth-promoting microorganisms for environmental sustainability. *Trends in Biotechnology* 34, 847–850.

Abou-Shanab, R.A., Angle, J.S., Delorme, T.A., Chaney, R.L., Van Berkum, P. *et al.* (2003) Rhizobacterial effects on nickel extraction from soil and uptake by *Alyssum murale*. *New Phytologist* 158, 219–224.

Adam, A. (2008) Elicitation of induced systemic resistance in tomato and cucumber and activation of the lipoxygenase pathway by non-pathogenic rhizobacteria. Doctoral dissertation, PhD Thesis, University of Liège, Liège, Belgium.

Adam, M., Heuer, H. and Hallmann, J. (2014) Bacterial antagonists of fungal pathogens also control root-knot nematodes by induced systemic resistance of tomato plants. *PloS One* 9, e90402.

Adesemoye, A.O., Torbert, H.A. and Kloepper, J.W. (2008) Enhanced plant nutrient use efficiency with PGPR and AMF in an integrated nutrient management system. *Canadian Journal of Microbiology* 54, 876–886.

Ahemad, M. and Khan, M.S. (2011a) Insecticide-tolerant and plant growth promoting *Bradyrhizobium* sp.(vigna) improves the growth and yield of greengram [*Vigna radiata* (L.) Wilczek] in insecticide-stressed soils. *Symbiosis* 54, 17–27.

Ahemad, M. and Khan, M.S. (2011b) Toxicological effects of selective herbicides on plant growth promoting activities of phosphate solubilizing *Klebsiella* sp. strain PS19. *Current Microbiology* 62, 532–538.

Ahmad, R., Lim, C.J. and Kwon, S.Y. (2013) Glycine betaine: a versatile compound with great potential for gene pyramiding to improve crop plant performance against environmental stresses. *Plant Biotechnology Reports* 7, 49–57.

Ahmed, E. and Holmström, S.J. (2014) Siderophores in environmental research: roles and applications. *Microbial Biotechnology* 7, 196–208.

Ali, M.B. and McNear, D.H. (2014) Induced transcriptional profiling of phenylpropanoid pathway genes increased flavonoid and lignin content in Arabidopsis leaves in response to microbial products. *BMC Plant Biology* 14, 1.

Armengaud, P., Breitling, R. and Amtmann, A. (2010) Coronatine-intensive 1 (COII) mediates transcriptional responses of *Arabidopsis thaliana* to external potassium supply. *Molecular Plant* 3, 390–405.

Asghar, H., Zahir, Z., Arshad, M. and Khaliq, A. (2002) Relationship between in vitro production of auxins by rhizobacteria and their growth-promoting activities in *Brassica juncea* L. *Biology and Fertility of Soils* 35, 231–237.

Babu, A.G., Kim, J.D. and Oh, B.T. (2013) Enhancement of heavy metal phytoremediation by *Alnus firma* with endophytic *Bacillus thuringiensis* GDB-1. *Journal of Hazardous Materials* 250, 477–483.

Badr, M.A. (2006). Efficiency of K-feldspar combined with organic materials and silicate dissolving bacteria on tomato yield. *Journal of Applied Sciences Research* 2, 1191–1198.

Badri, D.V., Weir, T.L., van der Lelie, D. and Vivanco, J.M. (2009) Rhizosphere chemical dialogues: plant–microbe interactions. *Current Opinion in Biotechnology* 20, 642–650.

Baggie, I., Rowell, D.L., Robinson, J.S. and Warren, G.P. (2005) Decomposition and phosphorus release from organic residues as affected by residue quality and added inorganic phosphorus. *Agroforestry Systems* 63, 125–131.

Bais, H.P., Weir, T.L., Perry, L.G., Gilroy, S. and Vivanco, J.M. (2006) The role of root exudates in rhizosphere interactions with plants and other organisms. *Annual Review of Plant Biology* 57, 233–266.

Banwart, S. (2011) Save our soils. *Nature* 474, 151–152.

Barac, T., Taghavi, S., Borremans, B., Provoost, A., Oeyen, L. *et al.* (2004) Engineered endophytic bacteria improve phytoremediation of water-soluble, volatile, organic pollutants. *Nature Biotechnology* 22, 583–588.

Barnard, A.M. and Salmond, G.P. (2007) Quorum sensing in *Erwinia* species. *Analytical and Bioanalytical Chemistry* 387, 415–423.

Barry, S.M. and Challis, G.L. (2009) Recent advances in siderophore biosynthesis. *Current Opinion in Chemical Biology* 13, 205–215.

Basak, B.B. and Biswas, D.R. (2009) Influence of potassium solubilizing microorganism (*Bacillus mucilaginosus*) and waste mica on potassium uptake dynamics by sudan grass (*Sorghum vulgare* Pers.) grown under two Alfisols. *Plant and Soil* 317, 235–255.

Batty, L.C. and Dolan, C. (2013) The potential use of phytoremediation for sites with mixed organic and inorganic contamination. *Critical Reviews in Environmental Science and Technology* 43, 217–259.

Beatty, P.H. and Good, A.G. (2011) Future prospects for cereals that fix nitrogen. *Science* 333, 416–417.

Becerra-Castro, C., Kidd, P.S., Rodríguez-Garrido, B., Monterroso, C., Santos-Ucha, P. and Prieto-Fernández, Á. (2013) Phytoremediation of hexachlorocyclohexane (HCH)-contaminated soils using *Cytisus striatus* and bacterial inoculants in soils with distinct organic matter content. *Environmental Pollution* 178, 202–210.

Bell, T.H., Joly, S., Pitre, F.E. and Yergeau, E. (2014) Increasing phytoremediation efficiency and reliability using novel omics approaches. *Trends in Biotechnology* 32, 271–280.

Benfey, P.N., Bennett, M. and Schiefelbein, J. (2010) Getting to the root of plant biology: impact of the Arabidopsis genome sequence on root research. *The Plant Journal* 61, 992–1000.

Berg, G. and Smalla, K. (2009) Plant species and soil type cooperatively shape the structure and function of microbial communities in the rhizosphere. *FEMS Microbiology Ecology* 68, 1–13.

Berta, G., Copetta, A., Gamalero, E., Bona, E., Cesaro, P. *et al.* (2014) Maize development and grain quality are differentially affected by mycorrhizal fungi and a growth-promoting pseudomonad in the field. *Mycorrhiza* 24, 161–170.

Bhattacharyya, P.N. and Jha, D.K. (2012) Plant growth-promoting rhizobacteria (PGPR): emergence in agriculture. *World Journal of Microbiology and Biotechnology* 28, 1327–1350.

Bisen, K., Keswani, C., Mishra, S., Saxena, A., Rakshit, A. and Singh, H.B. (2015) Unrealized potential of seed biopriming for versatile agriculture. In: Rakshit, A., Singh, H.B. and Sen, A. (eds) *Nutrient Use Efficiency: From Basics to Advances*. Springer, New Delhi, India, pp. 193–206.

Bisen, K., Keswani, C., Patel, J.S., Sarma, B.K. and Singh, H.B. (2016) *Trichoderma* spp.: efficient inducers of systemic resistance in plants. In: Chaudhary, D.K. and Verma, A. (eds) *Microbial-Mediated Induced Systemic Resistance in Plants*. Springer, Singapore, pp. 185–195.

Brader, G., Compant, S., Mitter, B., Trognitz, F. and Sessitsch, A. (2014) Metabolic potential of endophytic bacteria. *Current Opinion in Biotechnology* 27, 30–37.

Braun, V. and Hantke, K. (2011) Recent insights into iron import by bacteria. *Current Opinion in Chemical Biology* 15, 328–334.

Bresson, J., Varoquaux, F., Bontpart, T., Touraine, B. and Vile, D. (2013) The PGPR strain *Phyllobacterium brassicacearum* STM196 induces a reproductive delay and physiological changes that result in improved drought tolerance in *Arabidopsis*. *New Phytologist* 200, 558–569.

Bruto, M., Prigent-Combaret, C., Muller, D. and Moënne-Loccoz, Y. (2014) Analysis of genes contributing to plant-beneficial functions in plant growth-promoting rhizobacteria and related Proteobacteria. *Scientific Reports* 4.

Buee, M., Rossignol, M., Jauneau, A., Ranjeva, R. and Bécard, G. (2000) The pre-symbiotic growth of arbuscular mycorrhizal fungi is induced by a branching factor partially purified from plant root exudates. *Molecular Plant–microbe Interactions* 13, 693–698.

Bulgarelli, D., Schlaeppi, K., Spaepen, S., van Themaat, E.V.L., and Schulze-Lefert, P. (2013) Structure and functions of the bacterial microbiota of plants. *Annual Review of Plant Biology* 64, 807–838.

Cameron, D.D., Neal, A.L., van Wees, S.C. and Ton, J. (2013) Mycorrhiza-induced resistance: more than the sum of its parts? *Trends in Plant Science* 18, 539–545.

Candido, V., Campanelli, G., D'Addabbo, T., Castronuovo, D., Perniola, M. and Camele, I. (2015) Growth and yield promoting effect of artificial mycorrhization on field tomato at different irrigation regimes. *Scientia Horticulturae* 187, 35–43.

Castillo, J.M., Casas, J. and Romero, E. (2011) Isolation of an endosulfan-degrading bacterium from a coffee farm soil: persistence and inhibitory effect on its biological functions. *Science of the Total Environment* 412, 20–27.

Cavagnaro, T.R., Bender, S.F., Asghari, H.R. and van Der Heijden, M.G. (2015) The role of arbuscular mycorrhizas in reducing soil nutrient loss. *Trends in Plant Science* 20(5), 283–290.

Cayuela, M.L., Sánchez-Monedero, M.A., Roig, A., Hanley, K., Enders, A. and Lehmann, J. (2013) Biochar and denitrification in soils: when, how much and why does biochar reduce N_2O emissions? *Scientific Reports* 3.

Cesco, S., Mimmo, T., Tonon, G., Tomasi, N., Pinton, R. *et al.* (2012) Plant-borne flavonoids released into the rhizosphere: impact on soil bio-activities related to plant nutrition. A review. *Biology and Fertility of Soils* 48, 123–149.

Chang, C.Y., Chao, C.C. and Chao, W.L. (2008) Community structure and functional diversity of indigenous fluorescent *Pseudomonas* of long-term swine compost applied maize rhizosphere. *Soil Biology and Biochemistry* 40, 495–504.

Chatterjee, S., Sau, G.B. and Mukherjee, S.K. (2009) Plant growth promotion by a hexavalent chromium reducing bacterial strain, *Cellulosimicrobium cellulans* $KUCr_3$. *World Journal of Microbiology and Biotechnology* 25, 1829–1836.

Chen, L., Luo, S., Xiao, X., Guo, H., Chen, J. *et al.* (2010) Application of plant growth-promoting endophytes (PGPE) isolated from *Solanum nigrum* L. for phytoextraction of Cd-polluted soils. *Applied Soil Ecology* 46, 383–389.

Chen, L., Luo, S., Li, X., Wan, Y., Chen, J. and Liu, C. (2014) Interaction of Cd-hyperaccumulator *Solanum nigrum* L. and functional endophyte *Pseudomonas* sp. Lk9 on soil heavy metals uptake. *Soil Biology and Biochemistry* 68, 300–308.

Chisholm, S.T., Coaker, G., Day, B. and Staskawicz, B.J. (2006) Host-microbe interactions: shaping the evolution of the plant immune response. *Cell* 124, 803–814.

Cicatelli, A., Lingua, G., Todeschini, V., Biondi, S., Torrigiani, P. and Castiglione, S. (2010) Arbuscular mycorrhizal fungi restore normal growth in a white poplar clone grown on heavy metal-contaminated soil, and this is associated with upregulation of foliar metallothionein and polyamine biosynthetic gene expression. *Annals of Botany* 106, 791–802.

Cirou, A., Diallo, S., Kurt, C., Latour, X. and Faure, D. (2007) Growth promotion of quorum-quenching bacteria in the rhizosphere of *Solanum tuberosum*. *Environmental Microbiology* 9, 1511–1522.

Compant, S., van Der Heijden, M.G. and Sessitsch, A. (2010) Climate change effects on beneficial plant–microorganism interactions. *FEMS Microbiology Ecology* 73, 197–214.

Cornelis, P. (2010) Iron uptake and metabolism in pseudomonads. *Applied Microbiology and Biotechnology* 86, 1637–1645.

Couillerot, O., Ramírez-Trujillo, A., Walker, V., Von Felten, A., Jansa, J. *et al.* (2013). Comparison of prominent *Azospirillum* strains in *Azospirillum–Pseudomonas–Glomus* consortia for promotion of maize growth. *Applied Microbiology and Biotechnology* 97, 4639–4649.

Crépin, A., Barbey, C., Cirou, A., Tannières, M., Orange, N. *et al.* (2012) Biological control of pathogen communication in the rhizosphere: a novel approach applied to potato soft rot due to *Pectobacterium atrosepticum*. *Plant and Soil* 358, 27–37.

Crowley, D.E., Wang, Y.C., Reid, C.P.P. and Szaniszlo, P.J. (1991) Mechanisms of iron acquisition from siderophores by microorganisms and plants. *Plant and Soil* 130, 179–198.

Dardanelli, M.S., De Cordoba, F.J.F., Espuny, M.R., Carvajal, M.A.R., Díaz, M.E.S. *et al.* (2008) Effect of *Azospirillum brasilense* coinoculated with *Rhizobium* on *Phaseolus vulgaris* flavonoids and Nod factor production under salt stress. *Soil Biology and Biochemistry* 40, 2713–2721.

Dary, M., Chamber-Pérez, M.A., Palomares, A.J. and Pajuelo, E. (2010) "*In situ*" phytostabilisation of heavy metal polluted soils using *Lupinus luteus* inoculated with metal resistant plant-growth promoting rhizobacteria. *Journal of Hazardous Materials* 177, 323–330.

Davidson, E.A. and Janssens, I.A. (2006) Temperature sensitivity of soil carbon decomposition and feedbacks to climate change. *Nature* 440, 165–173.

Davies, W.J., Zhang, J., Yang, J. and Dodd, I.C. (2011) Novel crop science to improve yield and resource use efficiency in water-limited agriculture. *The Journal of Agricultural Science* 149, 123–131.

de-Bashan, L.E., Hernandez, J.P., Bashan, Y. and Maier, R.M. (2010) *Bacillus pumilus* ES4: candidate plant growth-promoting bacterium to enhance establishment of plants in mine tailings. *Environmental and Experimental Botany* 69, 343–352.

de-Bashan, L.E., Hernandez, J.P. and Bashan, Y. (2012) The potential contribution of plant growth-promoting bacteria to reduce environmental degradation – a comprehensive evaluation. *Applied Soil Ecology* 61, 171–189.

Dennis, P.G., Miller, A.J. and Hirsch, P.R. (2010) Are root exudates more important than other sources of rhizodeposits in structuring rhizosphere bacterial communities? *FEMS Microbiology Ecology* 72, 313–327.

de Oliveira Mendes, G., de Freitas, A.L.M., Pereira, O.L., da Silva, I.R., Vassilev, N.B. and Costa, M.D. (2014) Mechanisms of phosphate solubilization by fungal isolates when exposed to different P sources. *Annals of Microbiology* 64, 239–249.

de Souza, J.T., de Boer, M., de Waard, P., van Beek, T.A. and Raaijmakers, J.M. (2003) Biochemical, genetic, and zoosporicidal properties of cyclic lipopeptide surfactants produced by *Pseudomonas fluorescens*. *Applied and Environmental Microbiology* 69, 7161–7172.

de Weert, S., Vermeiren, H., Mulders, I.H., Kuiper, I., Hendrickx, N. *et al.* (2002) Flagella-driven chemotaxis towards exudate components is an important trait for tomato root colonization by *Pseudomonas fluorescens*. *Molecular Plant–microbe Interactions* 15, 1173–1180.

Dey, R., Pal, K.K., Bhatt, D.M. and Chauhan, S.M. (2004) Growth promotion and yield enhancement of peanut (*Arachis hypogaea* L.) by application of plant growth-promoting rhizobacteria. *Microbiological Research* 159, 371–394.

Dharni, S., Srivastava, A.K., Samad, A. and Patra, D.D. (2014a) Impact of plant growth promoting *Pseudomonas monteilii* PsF84 and *Pseudomonas plecoglossicida* PsF610 on metal uptake and production of secondary metabolite (monoterpenes) by rose-scented geranium (*Pelargonium graveolens* cv. *bourbon*) grown on tannery sludge amended soil. *Chemosphere* 117, 433–439.

Dharni, S., Maurya, A., Samad, A., Srivastava, S.K., Sharma, A. and Patra, D.D. (2014b) Purification, characterization, and in vitro activity of 2, 4-di-tert-butylphenol from *Pseudomonas monteilii* PsF84: conformational and molecular docking studies. *Journal of Agricultural and Food Chemistry* 62, 6138–6146.

Dixon, R. and Kahn, D. (2004) Genetic regulation of biological nitrogen fixation. *Nature Reviews Microbiology* 2, 621–631.

Do Carmo, F.L., Dos Santos, H.F., Martins, E.F., van Elsas, J.D., Rosado, A.S., and Peixoto, R.S. (2011) Bacterial structure and characterization of plant growth promoting and oil degrading bacteria from the rhizospheres of mangrove plants. *The Journal of Microbiology* 49, 535–543.

Doty, S.L., Oakley, B., Xin, G., Kang, J.W., Singleton, G. *et al.* (2009) Diazotrophic endophytes of native black cottonwood and willow. *Symbiosis* 47, 23–33.

Dubey, S.K. (1996) Combined effect of *Bradyrhizobium japonicum* and phosphate-solubilizing *Pseudomonas striata* on nodulation, yield attributes and yield of rainfed soybean (*Glycine max*) under different sources of phosphorus in vertisols. *Indian Journal of Agricultural Sciences* 66, 28–32.

Dubey, R.K., Tripathi, V., Singh, N. and Abhilash, P.C. (2014) Phytoextraction and dissipation of lindane by *Spinacia oleracea* L. *Ecotoxicology and Environmental Safety* 109, 22–26.

Dubey, R.K., Tripathi, V. and Abhilash, P.C. (2015) Book review: principles of plant–microbe interactions: microbes for sustainable agriculture. *Frontiers in Plant Science* 6, 986.

Dubey, R.K., Tripathi, V., Dubey, P.K., Singh, H.B. and Abhilash, P.C. (2016) Exploring rhizospheric interactions for agricultural sustainability: the need of integrative research on multi-trophic interactions. *Journal of Cleaner Production* 115, 362–365.

Duca, D., Lorv, J., Patten, C.L., Rose, D. and Glick, B.R. (2014) Indole-3-acetic acid in plant–microbe interactions. *Antonie Van Leeuwenhoek* 106, 85–125.

Edrisi, S.A., and Abhilash, P.C. (2016) Exploring marginal and degraded lands for biomass and bioenergy production: an Indian scenario. *Renewable and Sustainable Energy Reviews* 54, 1537–1551.

Edrisi, S.A., Dubey, R.K., Tripathi, V., Bakshi, M., Srivastava, P. *et al.* (2015) *Jatropha curcas* L.: a crucified plant waiting for resurgence. *Renewable and Sustainable Energy Reviews* 41, 855–862.

Egamberdieva, D. (2009) Alleviation of salt stress by plant growth regulators and IAA producing bacteria in wheat. *Acta Physiologiae Plantarum* 31, 861–864.

Estrada, G.A., Baldani, V.L.D., de Oliveira, D.M., Urquiaga, S. and Baldani, J.I. (2013) Selection of phosphate-solubilizing diazotrophic *Herbaspirillum* and *Burkholderia* strains and their effect on rice crop yield and nutrient uptake. *Plant and Soil* 369, 115–129.

Fan, B., Chen, X.H., Budiharjo, A., Bleiss, W., Vater, J. and Borriss, R. (2011) Efficient colonization of plant roots by the plant growth promoting bacterium *Bacillus amyloliquefaciens* FZB42, engineered to express green fluorescent protein. *Journal of Biotechnology* 151, 303–311.

FAO (2014) Hunger map. Available at: http://www.fao.org/hunger/en/ (accessed 6 July 2017).

Farina, R., Beneduzi, A., Ambrosini, A., de Campos, S.B., Lisboa, B.B. *et al.* (2012) Diversity of plant growth-promoting rhizobacteria communities associated with the stages of canola growth. *Applied Soil Ecology* 55, 44–52.

Fenner, K., Canonica, S., Wackett, L.P. and Elsner, M. (2013) Evaluating pesticide degradation in the environment: blind spots and emerging opportunities. *Science* 341, 752–758.

Fester, T., Giebler, J., Wick, L.Y., Schlosser, D. and Kästner, M. (2014) Plant–microbe interactions as drivers of ecosystem functions relevant for the biodegradation of organic contaminants. *Current Opinion in Biotechnology* 27, 168–175.

Filip, Z. (2002) International approach to assessing soil quality by ecologically-related biological parameters. *Agriculture, Ecosystems and Environment* 88, 169–174.

Frey-Klett, P., Burlinson, P., Deveau, A., Barret, M., Tarkka, M. and Sarniguet, A. (2011) Bacterial-fungal interactions: hyphens between agricultural, clinical, environmental, and food microbiologists. *Microbiology and Molecular Biology Reviews* 75, 583–609.

Germaine, K.J., Liu, X., Cabellos, G.G., Hogan, J.P., Ryan, D. and Dowling, D.N. (2006) Bacterial endophyte-enhanced phytoremediation of the organochlorine herbicide 2,4-dichlorophenoxyacetic acid. *FEMS Microbiology Ecology* 57, 302–310.

Geurts, R., Lillo, A. and Bisseling, T. (2012) Exploiting an ancient signalling machinery to enjoy a nitrogen fixing symbiosis. *Current Opinion in Plant Biology* 15, 438–443.

Ghosh, U.D., Saha, C., Maiti, M., Lahiri, S., Ghosh, S. *et al.* (2014) Root associated iron oxidizing bacteria increase phosphate nutrition and influence root to shoot partitioning of iron in tolerant plant *Typha angustifolia*. *Plant and Soil* 381, 279–295.

Giller, K.E., Witter, E. and Mcgrath, S.P. (1998) Toxicity of heavy metals to microorganisms and microbial processes in agricultural soils: a review. *Soil Biology and Biochemistry* 30, 1389–1414.

Glick, B.R. (2003). Phytoremediation: synergistic use of plants and bacteria to clean up the environment. *Biotechnology Advances* 21, 383–393.

Glick, B.R. (2010) Using soil bacteria to facilitate phytoremediation. *Biotechnology Advances* 28, 367–374.

Glick, B.R. (2012) Plant growth-promoting bacteria: mechanisms and applications. *Scientifica* 2012, 1–15.

Glick, B.R. (2014) Bacteria with ACC deaminase can promote plant growth and help to feed the world. *Microbiological Research* 169, 30–39.

Glick, B.R., Cheng, Z., Czarny, J. and Duan, J. (2007) Promotion of plant growth by ACC deaminase-producing soil bacteria. *European Journal of Plant Pathology* 119, 329–339.

Godfray, H.C.J., Beddington, J.R., Crute, I.R., Haddad, L., Lawrence, D. *et al.* (2010) Food security: the challenge of feeding 9 billion people. *Science* 327, 812–818.

Graber, E.R., Harel, Y.M., Kolton, M., Cytryn, E., Silber, A. *et al.* (2010) Biochar impact on development and productivity of pepper and tomato grown in fertigated soilless media. *Plant and Soil* 337, 481–496.

Gryndler, M., Sudová, R., Püschel, D., Rydlová, J., Janoušková, M. and Vosátka, M. (2008) Cultivation of high-biomass crops on coal mine spoil banks: can microbial inoculation compensate for high doses of organic matter? *Bioresource Technology* 99, 6391–6399.

Guerinot, M.L. and Yi, Y. (1994) Iron: nutritious, noxious, and not readily available. *Plant Physiology* 104, 815.

Güneş, A., Turan, M., Güllüce, M. and Şahin, F. (2014) Nutritional content analysis of plant growth-promoting rhizobacteria species. *European Journal of Soil Biology* 60, 88–97.

Guo, J., Feng, R., Ding, Y. and Wang, R. (2014) Applying carbon dioxide, plant growth-promoting rhizobacterium and EDTA can enhance the phytoremediation efficiency of ryegrass in a soil polluted with zinc, arsenic, cadmium and lead. *Journal of Environmental Management* 141, 1–8.

Guo, T. and Liao, M. (2014) Suppression of *Rhizoctonia solani* and induction of host plant resistance by *Paenibacillus kribbensis* PS04 towards controlling of rice sheath blight. *Biocontrol Science and Technology* 24, 116–121.

Habibi, S., Djedidi, S., Prongjunthuek, K., Mortuza, M.F., Ohkama-Ohtsu, N. *et al.* (2014) Physiological and genetic characterization of rice nitrogen fixer PGPR isolated from rhizosphere soils of different crops. *Plant and Soil* 379, 51–66.

Halbleib, C.M. and Ludden, P.W. (2000) Regulation of biological nitrogen fixation. *The Journal of Nutrition* 130, 1081–1084.

Han, H.S. and Lee, K.D. (2006) Effect of co-inoculation with phosphate and potassium solubilizing bacteria on mineral uptake and growth of pepper and cucumber. *Plant Soil and Environment* 52, 130.

Han, J.I., Choi, H.K., Lee, S.W., Orwin, P.M., Kim, J. *et al.* (2011) Complete genome sequence of the metabolically versatile plant growth-promoting endophyte *Variovorax paradoxus* S110. *Journal of Bacteriology* 193, 1183–1190.

Hartmann, A., Schmid, M., Van Tuinen, D. and Berg, G. (2009) Plant-driven selection of microbes. *Plant and Soil* 321, 235–257.

Hassan, S. and Mathesius, U. (2012) The role of flavonoids in root–rhizosphere signalling: opportunities and challenges for improving plant–microbe interactions. *Journal of Experimental Botany* 63, 3429–3444.

Hestbjerg, H., Willumsen, P.A., Christensen, M., Andersen, O. and Jacobsen, C.S. (2003) Bioaugmentation of tar-contaminated soils under field conditions using *Pleurotus ostreatus* refuse from commercial mushroom production. *Environmental Toxicology and Chemistry* 22, 692–698.

Hibbing, M.E., Fuqua, C., Parsek, M.R. and Peterson, S.B. (2010) Bacterial competition: surviving and thriving in the microbial jungle. *Nature Reviews Microbiology* 8, 15–25.

Hider, R.C. and Kong, X. (2010) Chemistry and biology of siderophores. *Natural Product Reports* 27, 637–657.

Hiloidhari, M., Baruah, D., Mahilary, H. and Baruah, D.C. (2012) GIS based assessment of rice (*Oryza sativa*) straw biomass as an alternative fuel for tea (*Camellia sinensis* L.) drying in Sonitpur district of Assam, India. *Biomass and Bioenergy* 44, 160–167.

Hodge, A., Campbell, C.D. and Fitter, A.H. (2001) An arbuscular mycorrhizal fungus accelerates decomposition and acquires nitrogen directly from organic material. *Nature* 413, 297–299.

Hong, S.H., Ryu, H.W., Kim, J. and Cho, K.S. (2011) Rhizoremediation of diesel-contaminated soil using the plant growth-promoting rhizobacterium *Gordonia* sp. S2 RP-17. *Biodegradation* 22, 593–601.

Hu, X., Chen, J. and Guo, J. (2006) Two phosphate- and potassium-solubilizing bacteria isolated from Tianmu Mountain, Zhejiang, China. *World Journal of Microbiology and Biotechnology* 22, 983–990.

Huang, Y., Sun, L., Zhao, J., Huang, R., Li, R. and Shen, Q. (2015) Utilization of different waste proteins to create a novel PGPR-containing bio-organic fertilizer. *Scientific Reports* 5, 7766.

Jagtap, S.S., Woo, S.M., Kim, T.S., Dhiman, S.S., Kim, D. and Lee, J.K. (2014) Phytoremediation of diesel-contaminated soil and saccharification of the resulting biomass. *Fuel* 116, 292–298.

Jain, A., Singh, A., Chaudhary, A., Singh, S. and Singh, H.B. (2014) Modulation of nutritional and antioxidant potential of seeds and pericarp of pea pods treated with microbial consortium. *Food Research International* 64, 275–282.

Jaiswal, A.K., Elad, Y., Graber, E.R. and Frenkel, O. (2014) *Rhizoctonia solani* suppression and plant growth promotion in cucumber as affected by biochar pyrolysis temperature, feedstock and concentration. *Soil Biology and Biochemistry* 69, 110–118.

Jechalke, S., Focks, A., Rosendahl, I., Groeneweg, J., Siemens, J. *et al.* (2014) Structural and functional response of the soil bacterial community to application of manure from difloxacin-treated pigs. *FEMS Microbiology Ecology* 87, 78–88.

Jeffery, S., Bezemer, T.M., Cornelissen, G., Kuyper, T.W., Lehmann, J. *et al.* (2015) The way forward in biochar research: targeting trade-offs between the potential wins. *GCB Bioenergy* 7, 1–13.

Jha, C.K., Patel, D., Rajendran, N. and Saraf, M. (2010) Combinatorial assessment on dominance and informative diversity of PGPR from rhizosphere of *Jatropha curcas* L. *Journal of Basic Microbiology* 50, 211–217.

Jha, C.K., Patel, B. and Saraf, M. (2012) Stimulation of the growth of *Jatropha curcas* by the plant growth promoting bacterium *Enterobacter cancerogenus* MSA2. *World Journal of Microbiology and Biotechnology* 28, 891–899.

Jing, Y.X., Yan, J.L., He, H.D., Yang, D.J., Xiao, L. *et al.* (2014) Characterization of bacteria in the rhizosphere soils of *Polygonum pubescens* and their potential in promoting growth and Cd, Pb, Zn uptake by *Brassica napus*. *International Journal of Phytoremediation* 16, 321–333.

Jourdan, E., Henry, G., Duby, F., Dommes, J., Barthelemy, J.P. *et al.* (2009) Insights into the defense-related events occurring in plant cells following perception of surfactin-type lipopeptide from *Bacillus subtilis*. *Molecular Plant–Microbe Interactions* 22, 456–468.

Juwarkar, A.A., Mehrotraa, K.L., Nair, R., Wanjari, T., Singh, S.K. and Chakrabarti, T. (2010) Carbon sequestration in reclaimed manganese mine land at Gumgaon, India. *Environmental Monitoring and Assessment* 160, 457–464.

Kanchiswamy, C.N., Malnoy, M. and Maffei, M.E. (2015) Bioprospecting bacterial and fungal volatiles for sustainable agriculture. *Trends in Plant Science* 20, 206–211.

Kantha, T., Kantachote, D. and Klongdee, N. (2015) Potential of biofertilizers from selected *Rhodopseudomonas palustris* strains to assist rice (*Oryza sativa* L. subsp. *indica*) growth under salt stress and to reduce greenhouse gas emissions. *Annals of Microbiology* 65, 2109–2118.

Karasu, A., Öz, M. and Dogan, R. (2009) The effect of bacterial inoculation and different nitrogen doses on yield and yield components of some chickpea genotypes (*Cicer arietinum* L.). *African Journal of Biotechnology* 8.

Kaur, G. and Reddy, M.S. (2014) Influence of P-solubilizing bacteria on crop yield and soil fertility at multilocational sites. *European Journal of Soil Biology* 61, 35–40.

Ker, K., Seguin, P., Driscoll, B.T., Fyles, J.W. and Smith, D.L. (2012) Switchgrass establishment and seeding year production can be improved by inoculation with rhizosphere endophytes. *Biomass and Bioenergy* 47, 295–301.

Keswani, C. (2015). Proteomics studies of thermotolerant strain of *Trichoderma* spp. Ph.D. Thesis, Banaras Hindu University, Varanasi, India, p. 126.

Keswani, C., Mishra, S., Sarma, B.K. and Singh, S.P. and Singh, H.B. (2014) Unraveling the efficient applications of secondary metabolites of various *Trichoderma* spp. *Applied Microbiology and Biotechnology* 98, 533–544.

Keswani, C., Bisen, K., Singh, V., Sarma, B.K. and Singh, H.B. (2016) Formulation technology of biocontrol agents: present status and future prospects. In: Arora, N.K., Mehnaz, S. and Balestrini, R. (eds) *Bioformulations: For Sustainable Agriculture*. Springer, New Delhi, India, pp. 35–52.

Khan, M.S., Zaidi, A. and Wani, P.A. (2007) Role of phosphate-solubilizing microorganisms in sustainable agriculture—a review. *Agronomy for Sustainable Development* 27, 29–43.

Köberl, M., Ramadan, E.M., Adam, M., Cardinale, M., Hallmann, J. *et al.* (2013) *Bacillus* and *Streptomyces* were selected as broad-spectrum antagonists against soilborne pathogens from arid areas in Egypt. *FEMS Microbiology Letters* 342, 168–178.

Kolton, M., Harel, Y.M., Pasternak, Z., Graber, E.R., Elad, Y. and Cytryn, E. (2011) Impact of biochar application to soil on the root-associated bacterial community structure of fully developed greenhouse pepper plants. *Applied and Environmental Microbiology* 77, 4924–4930.

Krewulak, K.D. and Vogel, H.J. (2008) Structural biology of bacterial iron uptake. *Biochimica et Biophysica Acta (BBA)-Biomembranes* 1778, 1781–1804.

Kumar, K.V. and Patra, D.D. (2013) Influence of nickel and cadmium resistant PGPB on metal accumulation and growth responses of *Lycopersicon esculentum* plants grown in fly ash amended soil. *Water, Air, and Soil Pollution* 224, 1–10.

Kumar, K.V., Srivastava, S., Singh, N. and Behl, H.M. (2009) Role of metal resistant plant growth promoting bacteria in ameliorating fly ash to the growth of *Brassica juncea*. *Journal of Hazardous Materials* 170, 51–57.

Kumar, M., Prasanna, R., Bidyarani, N., Babu, S., Mishra, B.K. *et al.* (2013) Evaluating the plant growth promoting ability of thermotolerant bacteria and cyanobacteria and their interactions with seed spice crops. *Scientia Horticulturae* 164, 94–101.

Kumari, B., Singh, S.N. and Singh, D.P. (2012) Characterization of two biosurfactant producing strains in crude oil degradation. *Process Biochemistry* 47, 2463–2471.

Lavakush, Yadav, J., Verma, J.P., Jaiswal, D.K. and Kumar, A. (2014) Evaluation of PGPR and different concentration of phosphorus level on plant growth, yield and nutrient content of rice (*Oryza sativa*). *Ecological Engineering*, 62, 123–128 .

Lazcano, C., Barrios-Masias, F.H. and Jackson, L.E. (2014) Arbuscular mycorrhizal effects on plant water relations and soil greenhouse gas emissions under changing moisture regimes. *Soil Biology and Biochemistry* 74, 184–192.

Lee, B., Lee, S. and Ryu, C.M. (2012) Foliar aphid feeding recruits rhizosphere bacteria and primes plant immunity against pathogenic and non-pathogenic bacteria in pepper. *Annals of Botany* 110, 281–290.

Lee, W., Wood, T.K. and Chen, W. (2006) Engineering TCE-degrading rhizobacteria for heavy metal accumulation and enhanced TCE degradation. *Biotechnology and Bioengineering* 95, 399–403.

Leeman, M., Van Pelt, J.A., Hendrickx, M.J., Scheffer, R.J., Bakker, P.A.H.M., and Schippers, B. (1995). Biocontrol of Fusarium wilt of radish in commercial greenhouse trials by seed treatment with *Pseudomonas fluorescens* WCS374. *Phytopathology* 85, 1301–1305.

Liu, A., Hamel, C., Hamilton, R.I., Ma, B.L. and Smith, D.L. (2000). Acquisition of Cu, Zn, Mn and Fe by mycorrhizal maize (*Zea mays* L.) grown in soil at different P and micronutrient levels. *Mycorrhiza* 9, 331–336.

Liu, L., Jiang, C.Y., Liu, X.Y., Wu, J.F., Han, J.G. and Liu, S.J. (2007) Plant–microbe association for rhizoremediation of chloronitroaromatic pollutants with *Comamonas* sp. strain CNB-1. *Environmental Microbiology* 9, 465–473.

Lugtenberg, B. and Kamilova, F. (2009) Plant-growth-promoting rhizobacteria. *Annual Review of Microbiology* 63, 541–556.

Lundberg, D.S., Lebeis, S.L., Paredes, S.H., Yourstone, S., Gehring, J. *et al.* (2012) Defining the core *Arabidopsis thaliana* root microbiome. *Nature* 488, 86–90.

Luo, J., Bai, Y., Liang, J. and Qu, J. (2014) Metagenomic approach reveals variation of microbes with arsenic and antimony metabolism genes from highly contaminated soil. *PLoS ONE* 9(10), e108185. doi.org/10.1371/journal.pone.0108185.

Lupwayi, N.Z., Rice, W.A. and Clayton, G.W. (1998) Soil microbial diversity and community structure under wheat as influenced by tillage and crop rotation. *Soil Biology and Biochemistry* 30, 1733–1741.

Ma, Y., Rajkumar, M. and Freitas, H. (2009) Improvement of plant growth and nickel uptake by nickel resistant-plant-growth promoting bacteria. *Journal of Hazardous Materials* 166, 1154–1161.

Ma, Y., Prasad, M.N.V., Rajkumar, M. and Freitas, H. (2011) Plant growth promoting rhizobacteria and endophytes accelerate phytoremediation of metalliferous soils. *Biotechnology Advances* 29, 248–258.

Maity, A., Pal, R.K., Chandra, R. and Singh, N.V. (2014) *Penicillium pinophilum*—A novel microorganism for nutrient management in pomegranate (*Punica granatum* L.). *Scientia Horticulturae* 169, 111–117.

Mardad, I., Serrano, A. and Soukri, A. (2013) Solubilization of inorganic phosphate and production of organic acids by bacteria isolated from a Moroccan mineral phosphate deposit. *African Journal Microbiological Research* 7, 626–635.

Marquina, S., Pérez, T., Donoso, L., Giuliante, A., Rasse, R. and Herrera, F. (2015) NO, N_2O and CO_2 soil emissions from Venezuelan corn fields under tillage and no-tillage agriculture. *Nutrient Cycling in Agroecosystems* 101, 123–137.

Marroquí, S., Zorreguieta, A., Santamaría, C., Temprano, F., Soberón, M. *et al.* (2001) Enhanced symbiotic performance by *Rhizobium tropici* glycogen synthase mutants. *Journal of Bacteriology* 183, 854–864.

Marschner, P., Crowley, D. and Rengel, Z. (2011) Rhizosphere interactions between microorganisms and plants govern iron and phosphorus acquisition along the root axis–model and research methods. *Soil Biology and Biochemistry* 43, 883–894.

Marx, J. (2004). The roots of plant–microbe collaborations. *Science* 304, 234–236.

Matilla, M.A., Ramos, J.L., Bakker, P.A., Doornbos, R., Badri, D.V. *et al.* (2010) *Pseudomonas putida* KT2440 causes induced systemic resistance and changes in Arabidopsis root exudation. *Environmental Microbiology Reports* 2, 381–388.

Matz, C. and Kjelleberg, S. (2005) Off the hook – how bacteria survive protozoan grazing. *Trends in Microbiology* 13, 302–307.

Matzanke, B.F. (1991) Structures, coordination chemistry and functions of microbial iron chelates. *CRC Handbook of Microbial Iron Chelates*, 15–64.

Mayak, S., Tirosh, T. and Glick, B.R. (2004) Plant growth-promoting bacteria that confer resistance to water stress in tomatoes and peppers. *Plant Science* 166, 525–530.

McNear Jr, D.H. (2013) The rhizosphere - roots, soil and everything in between. *Nature Education Knowledge* 4, 1.

Mena-Benitez, G.L., Gandia-Herrero, F., Graham, S., Larson, T.R., McQueen-Mason, S.J. *et al.* (2008) Engineering a catabolic pathway in plants for the degradation of 1,2-dichloroethane. *Plant Physiology* 147, 1192–1198.

Mendes, R., Kruijt, M., de Bruijn, I., Dekkers, E. and van der Voort, M. *et al.* (2011) Deciphering the rhizosphere microbiome for disease-suppressive bacteria. *Science* 332, 1097–2000.

Menezes-Blackburn, D., Jorquera, M.A., Greiner, R., Gianfreda, L. and de la Luz Mora, M. (2013). Phytases and phytase-labile organic phosphorus in manures and soils. *Critical Reviews in Environmental Science and Technology* 43, 916–954.

Miransari, M. (2010) Contribution of arbuscular mycorrhizal symbiosis to plant growth under different types of soil stress. *Plant Biology* 12, 563–569.

Miransari, M. (2011) Hyperaccumulators, arbuscular mycorrhizal fungi and stress of heavy metals. *Biotechnology Advances* 29, 645–653.

Mirza, M.S., Mehnaz, S., Normand, P., Prigent-Combaret, C., Moënne-Loccoz, Y. *et al.* (2006) Molecular characterization and PCR detection of a nitrogen-fixing *Pseudomonas* strain promoting rice growth. *Biology and Fertility of Soils* 43, 163–170.

Mishra, P.K., Mishra, S., Selvakumar, G., Kundu, S. and Shankar Gupta, H. (2009) Enhanced soybean (*Glycine max* L.) plant growth and nodulation by *Bradyrhizobium japonicum*-SB1 in presence of *Bacillus thuringiensis*-KR1. *Acta Agriculturae Scandinavica Section B–Soil and Plant Science* 59, 189–196.

Mishra, S., Singh, A., Keswani, C., Saxena, A., Sarma, B.K. and Singh, H.B. (2015) Harnessing plant–microbe interactions for enhanced protection against phytopathogens. In: Arora, N.K. (ed.) *Plant Microbes Symbiosis: Applied Facets*. Springer, New Delhi, India, pp. 111–125.

Mulligan, C.N. (2005) Environmental applications for biosurfactants. *Environmental Pollution* 133, 183–198.

Muzammil, S., Graillon, C., Saria, R., Mathieu, F., Lebrihi, A. and Compant, S. (2014) The Saharan isolate *Saccharothrix algeriensis* NRRL B-24137 induces systemic resistance in *Arabidopsis thaliana* seedlings against *Botrytis cinerea*. *Plant and Soil* 374, 423–434.

Nadeem, S.M., Ahmad, M., Zahir, Z.A., Javaid, A. and Ashraf, M. (2014) The role of mycorrhizae and plant growth promoting rhizobacteria (PGPR) in improving crop productivity under stressful environments. *Biotechnology Advances* 32, 429–448.

Naveed, M., Mitter, B., Reichenauer, T.G., Wieczorek, K. and Sessitsch, A. (2014) Increased drought stress resilience of maize through endophytic colonization by *Burkholderia phytofirmans* PsJN and *Enterobacter* sp. FD17. *Environmental and Experimental Botany* 97, 30–39.

Naznin, H.A., Kiyohara, D., Kimura, M., Miyazawa, M., Shimizu, M. and Hyakumachi, M. (2014) Systemic resistance induced by volatile organic compounds emitted by plant growth-promoting fungi in *Arabidopsis thaliana*. *PLoS One* 9, e86882.

Neal, A.L., Ahmad, S., Gordon-Weeks, R. and Ton, J. (2012) Benzoxazinoids in root exudates of maize attract *Pseudomonas putida* to the rhizosphere. *PLoS One* 7, e35498.

Nelson, E.B. (2004) Microbial dynamics and interactions in the spermosphere. *Annual Review of Phytopathology* 42, 271–309.

Nie, M., Bell, C., Wallenstein, M.D. and Pendall, E. (2015) Increased plant productivity and decreased microbial respiratory C loss by plant growth-promoting rhizobacteria under elevated CO_2. *Scientific Reports* 5, 9212.

Ongena, M., Jourdan, E., Adam, A., Paquot, M., Brans, A. *et al.* (2007). Surfactin and fengycin lipopeptides of *Bacillus subtilis* as elicitors of induced systemic resistance in plants. *Environmental Microbiology* 9, 1084–1090.

O'Rourke, J.A., Yang, S.S., Miller, S.S., Bucciarelli, B., Liu, J. *et al.* (2013) An RNA-Seq transcriptome analysis of orthophosphate-deficient white lupin reveals novel insights into phosphorus acclimation in plants. *Plant Physiology* 161, 705–724.

Ovečka, M. and Takáč, T. (2014) Managing heavy metal toxicity stress in plants: biological and biotechnological tools. *Biotechnology Advances* 32, 73–86.

Pardo, T., Bernal, M.P. and Clemente, R. (2014) Efficiency of soil organic and inorganic amendments on the remediation of a contaminated mine soil: I. Effects on trace elements and nutrients solubility and leaching risk. *Chemosphere* 107, 121–128.

Parra-Cota, F.I., Peña-Cabriales, J.J., de los Santos-Villalobos, S., Martínez-Gallardo, N.A. and Délano-Frier, J.P. (2014) *Burkholderia ambifaria* and *B. caribensis* promote growth and increase yield in grain amaranth (*Amaranthus cruentus* and *A. hypochondriacus*) by improving plant nitrogen uptake. *PloS One* 9, e88094.

Paszczynski, A. and Crawford, R.L. (1995) Potential for bioremediation of xenobiotic compounds by the white-rot fungus *Phanerochaete chrysosporium*. *Biotechnology Progress* 11, 368–379.

Patel, A. and Patra, D.D. (2014) Influence of heavy metal rich tannery sludge on soil enzymes vis-à-vis growth of *Tagetes minuta*, an essential oil bearing crop. *Chemosphere* 112, 323–332.

Patra, D.D. (2012) Production, purification, and characterization of antifungal metabolite from *Pseudomonas aeruginosa* SD12, a new strain obtained from tannery waste polluted soil. *Journal of Microbiology and Biotechnology* 22, 674–683.

Patra, D.D., Sachdev, M.S. and Subbiah, B.V. (1986) 15N studies on the transfer of legume-fixed nitrogen to associated cereals in intercropping systems. *Biology and Fertility of Soils* 2, 165–171.

Persello-Cartieaux, F., Nussaume, L. and Robaglia, C. (2003) Tales from the underground: molecular. *Plant, Cell and Environment* 26, 189–199.

Philippot, L., Raaijmakers, J.M., Lemanceau, P. and van der Putten, W.H. (2013) Going back to the roots: the microbial ecology of the rhizosphere. *Nature Reviews Microbiology* 11, 789–799.

Płociniczak, T., Sinkkonen, A., Romantschuk, M. and Piotrowska-Seget, Z. (2013) Characterization of *Enterobacter intermedius* MH8b and its use for the enhancement of heavy metals uptake by *Sinapisalba* L. *Applied Soil Ecology* 63, 1–7.

Pokharel, A.K., Jannoura, R., Heitkamp, F., Kleikamp, B., Wachendorf, C. *et al.* (2013) Development of aggregates after application of maize residues in the presence of mycorrhizal and non-mycorrhizal pea plants. *Geoderma* 202, 38–44.

Prakamhang, J., Tittabutr, P., Boonkerd, N., Teamtisong, K., Uchiumi, T. *et al.* (2015) Proposed some interactions at molecular level of PGPR coinoculated with *Bradyrhizobium diazoefficiens* USDA110 and *B. japonicum* THA6 on soybean symbiosis and its potential of field application. *Applied Soil Ecology* 85, 38–49.

Prasad, M.N.V. and de Oliveira Freitas, H.M. (2003) Metal hyperaccumulation in plants: biodiversity prospecting for phytoremediation technology. *Electronic Journal of Biotechnology* 6, 285–321.

Prasanna, R., Joshi, M., Rana, A., Shivay, Y.S. and Nain, L. (2012) Influence of co-inoculation of bacteria-cyanobacteria on crop yield and C–N sequestration in soil under rice crop. *World Journal of Microbiology and Biotechnology* 28, 1223–1235.

Prasanna, R., Triveni, S., Bidyarani, N., Babu, S., Yadav, K. *et al.* (2014) Evaluating the efficacy of cyanobacterial formulations and biofilmed inoculants for leguminous crops. *Archives of Agronomy and Soil Science* 60, 349–366.

Qin, S., Miao, Q., Feng, W.W., Wang, Y., Zhu, X. *et al.* (2015) Biodiversity and plant growth promoting traits of culturable endophytic actinobacteria associated with *Jatropha curcas* L. growing in Panxi dry-hot valley soil. *Applied Soil Ecology* 93, 47–55.

Raaijmakers, J.M., Sluis, L.V.D., Bakker, P.A., Schippers, B., Koster, M. and Weisbeek, P.J. (1995) Utilization of heterologous siderophores and rhizosphere competence of fluorescent *Pseudomonas* spp. *Canadian Journal of Microbiology* 41, 126–135.

Raaijmakers, J.M., De Bruijn, I., Nybroe, O. and Ongena, M. (2010) Natural functions of lipopeptides from *Bacillus* and *Pseudomonas*: more than surfactants and antibiotics. *FEMS Microbiology Reviews* 34, 1037–1062.

Radhakrishnan, R., Khan, A.L., Kang, S.M. and Lee, I.J. (2015) A comparative study of phosphate solubilization and the host plant growth promotion ability of *Fusarium verticillioides* RK01 and *Humicola* sp. KNU01 under salt stress. *Annals of Microbiology* 65, 585–593.

Rajkumar, M. and Freitas, H. (2008) Influence of metal resistant-plant growth-promoting bacteria on the growth of *Ricinus communis* in soil contaminated with heavy metals. *Chemosphere* 71, 834–842.

Ramírez, M., Valderrama, B., Arredondo-Peter, R., Soberón, M., Mora, J. and Hernández, G. (1999) *Rhizobium etli* genetically engineered for the heterologous expression of *Vitreoscilla* sp. hemoglobin: effects on free-living and symbiosis. *Molecular Plant–microbe Interactions* 12, 1008–1015.

Rana, A., Saharan, B., Nain, L., Prasanna, R. and Shivay, Y.S. (2012a) Enhancing micronutrient uptake and yield of wheat through bacterial PGPR consortia. *Soil Science and Plant Nutrition* 58, 573–582.

Rana, A., Joshi, M., Prasanna, R., Shivay, Y.S., Nain, L. (2012b) Biofortification of wheat through inoculation of plant growth promoting rhizobcateria and cyanobacteria. *European Journal of Soil Biology* 50, 118–126

Rathi, M.S., Paul, S. and Thakur, J.K. (2014) Response of wheat to inoculation with mycorrhizae alone and combined with selected rhizobacteria including *Flavobacterium* sp. as a potential bioinoculant. *Journal of Plant Nutrition* 37, 76–86.

Ray, S., Singh, S., Sarma, B.K. and Singh, H.B. (2016a) Endophytic *Alcaligenes* isolated from horticultural and medicinal crops promotes growth in okra (*Abelmoschus esculentus*). *Journal of Plant Growth Regulation* 35, 401–412.

Ray, S., Singh, V., Singh, S., Sarma, B.K. and Singh, H.B. (2016b) Biochemical and histochemical analyses revealing endophytic *Alcaligenes faecalis* mediated suppression of oxidative stress in *Abelmoschus esculentus* challenged with *Sclerotium rolfsii*. *Plant Physiology and Biochemistry* 109, 430–441.

Reay, D.S., Davidson, E.A., Smith, K.A., Smith, P., Melillo, J.M., Dentener, F. and Crutzen, P.J. (2012) Global agriculture and nitrous oxide emissions. *Nature Climate Change* 2, 410–416.

Requena, N., Jimenez, I., Toro, M. and Barea, J.M. (1997) Interactions between plant-growth-promoting rhizobacteria (PGPR), arbuscular mycorrhizal fungi and *Rhizobium* spp. in the rhizosphere of *Anthyllis cytisoides*, a model legume for revegetation in mediterranean semi-arid ecosystems. *New Phytologist* 136, 667–677.

Roberts, D.A., Paul, N.A., Dworjanyn, S.A., Bird, M.I. and de Nys, R. (2015) Biochar from commercially cultivated seaweed for soil amelioration. *Scientific Reports* 5, 9665.

Rodríguez, H. and Fraga, R. (1999) Phosphate solubilizing bacteria and their role in plant growth promotion. *Biotechnology Advances* 17, 319–339.

Rudrappa, T., Czymmek, K.J., Paré, P.W. and Bais, H.P. (2008) Root-secreted malic acid recruits beneficial soil bacteria. *Plant Physiology* 148, 1547–1556.

Ryan, R.P., Monchy, S., Cardinale, M., Taghavi, S., Crossman, L. *et al.* (2009) The versatility and adaptation of bacteria from the genus *Stenotrophomonas*. *Nature Reviews Microbiology* 7, 514–525.

Sahay, R. and Patra, D.D. (2014) Identification and performance of sodicity tolerant phosphate solubilizing bacterial isolates on *Ocimum basilicum* in sodic soil. *Ecological Engineering* 71, 639–643.

Sahni, S., Sarma, B.K., Singh, D.P., Singh, H.B. and Singh, K.P. (2008) Vermicompost enhances performance of plant growth-promoting rhizobacteria in *Cicer arietinum* rhizosphere against *Sclerotium rolfsii*. *Crop Protection* 27, 369–376.

Sarkar, A. and Reinhold-Hurek, B. (2014) Transcriptional profiling of nitrogen fixation and the role of NifA in the diazotrophic endophyte *Azoarcus* sp. strain BH72. *PloS One* 9(2) e86527.

Sarma, B.K., Singh, D.P., Mehta, S., Singh, H.B. and Singh, U.P. (2002) Plant growth-promoting rhizobacteria-elicited alterations in phenolic profile of chickpea (*Cicer arietinum*) infected by *Sclerotium rolfsii*. *Journal of Phytopathology* 150, 277–282.

Saxena, A., Raghuwanshi, R. and Singh, H.B. (2015) *Trichoderma* species mediated differential tolerance against biotic stress of phytopathogens in *Cicer arietinum* L. *Journal of Basic Microbiology* 55, 195–206.

Saxena, J., Rana, G. and Pandey, M. (2013) Impact of addition of biochar along with *Bacillus* sp. on growth and yield of French beans. *Scientia Horticulturae* 162, 351–356.

Schenk, P.M., Carvalhais, L.C. and Kazan, K. (2012) Unraveling plant–microbe interactions: can multi-species transcriptomics help? *Trends in Biotechnology* 30, 177–184.

Segura, A. and Ramos, J.L. (2013) Plant–bacteria interactions in the removal of pollutants. *Current Opinion in Biotechnology* 24, 467–473.

Sharma, A., Shankhdhar, D. and Shankhdhar, S.C. (2013) Enhancing grain iron content of rice by the application of plant growth promoting rhizobacteria. *Plant Soil and Environment* 59, 89–94.

Shen, J., Yuan, L., Zhang, J., Li, H., Bai, Z. *et al.* (2011) Phosphorus dynamics: from soil to plant. *Plant Physiology* 156, 997–1005.

Sheng, X.F., He, L.Y. and Huang, W.Y. (2002) The conditions of releasing potassium by a silicate-dissolving bacterial strain NBT. *Agricultural Sciences in China* 1, 662–666.

Shoresh, M., Harman, G.E. and Mastouri, F. (2010) Induced systemic resistance and plant responses to fungal biocontrol agents. *Annual Review of Phytopathology* 48, 21–43.

Shukla, K.P., Sharma, S., Singh, N.K. and Singh, V. (2012) Deciphering rhizosphere soil system for strains having plant growth promoting and bioremediation traits. *Agricultural Research* 1, 251–257.

Singh, A., Sarma, B.K., Upadhyay, R.S. and Singh, H.B. (2013) Compatible rhizosphere microbes mediated alleviation of biotic stress in chickpea through enhanced antioxidant and phenylpropanoid activities. *Microbiological Research* 168, 33–40.

Singh, A., Jain, A., Sarma, B.K., Upadhyay, R.S. and Singh, H.B. (2014) Beneficial compatible microbes enhance antioxidants in chickpea edible parts through synergistic interactions. *LWT-Food Science and Technology* 56, 390–397.

Singh, B.K., Millard, P., Whiteley, A.S. and Murrell, J.C. (2004) Unravelling rhizosphere–microbial interactions: opportunities and limitations. *Trends in Microbiology* 12, 386–393.

Singh, D.P., Prabha, R., Yandigeri, M.S. and Arora, D.K. (2011a) Cyanobacteria-mediated phenylpropanoids and phytohormones in rice (*Oryza sativa*) enhance plant growth and stress tolerance. *Antonie van Leeuwenhoek* 100, 557–568.

Singh, J.S., Abhilash, P.C., Singh, H.B., Singh, R.P. and Singh, D.P. (2011b) Genetically engineered bacteria: an emerging tool for environmental remediation and future research perspectives. *Gene* 480, 1–9.

Singh, S. and Kapoor, K.K. (1999). Inoculation with phosphate-solubilizing microorganisms and a vesicular-arbuscular mycorrhizal fungus improves dry matter yield and nutrient uptake by wheat grown in a sandy soil. *Biology and Fertility of Soils* 28, 139–144.

Singh, U.P., Sarma, B.K. and Singh, D.P. (2003) Effect of plant growth-promoting rhizobacteria and culture filtrate of *Sclerotium rolfsii* on phenolic and salicylic acid contents in chickpea (*Cicer arietinum*). *Current Microbiology* 46, 0131–0140.

Singh, V., Upadhyay, R.S., Sarma, B.K. and Singh, H.B. (2016a) *Trichoderma asperellum* spore dose depended modulation of plant growth in vegetable crops. *Microbiological Research* 193, 74–86.

Singh, V., Upadhyay, R.S., Sarma, B.K. and Singh, H.B. (2016b) Seed bio-priming with *Trichoderma asperellum* effectively modulate plant growth promotion in pea. *International Journal of Agriculture, Environment and Biotechnology* 9, 361–365.

Singhai, P.K., Sarma, B.K. and Srivastava, J.S. (2011) Biological management of common scab of potato through *Pseudomonas* species and vermicompost. *Biological Control* 57, 150–157.

Smil, V. (1999) Nitrogen in crop production: an account of global flows. *Global Biogeochemical Cycles* 13, 647–662.

Smil, V. (2000) Phosphorus in the environment: natural flows and human interferences. *Annual Review of Energy and the Environment* 25, 53–88.

Smith, S.E. and Read, D.J. (2008) *Mycorrhizal Symbiosis*. Academic Press, London.

Solís-Domínguez, F.A., Valentín-Vargas, A., Chorover, J. and Maier, R.M. (2011) Effect of arbuscular mycorrhizal fungi on plant biomass and the rhizosphere microbial community structure of mesquite grown in acidic lead/zinc mine tailings. *Science of the Total Environment* 409, 1009–1016.

Song, X., Liu, M., Wu, D., Griffiths, B.S., Jiao, J. *et al.* (2015) Interaction matters: synergy between vermicompost and PGPR agents improves soil quality, crop quality and crop yield in the field. *Applied Soil Ecology* 89, 25–34.

Spohn, M., Ermak, A. and Kuzyakov, Y. (2013) Microbial gross organic phosphorus mineralization can be stimulated by root exudates–A 33 P isotopic dilution study. *Soil Biology and Biochemistry* 65, 254–263.

Srivastava, P., Sharma, Y.K. and Singh, N. (2014) Soil carbon sequestration potential of *Jatropha curcas* L. growing in varying soil conditions. *Ecological Engineering* 68, 155–166.

Stewart, P.S. and Franklin, M.J. (2008) Physiological heterogeneity in biofilms. *Nature Reviews Microbiology* 6, 199–210.

Stockmann, U., Adams, M.A., Crawford, J.W., Field, D.J., Henakaarchchi, N. *et al.* (2013) The knowns, known unknowns and unknowns of sequestration of soil organic carbon. *Agriculture, Ecosystems and Environment* 164, 80–99.

Sumarsih, S. and Haryanto, D. (2012) *Pseudomonas fluorescens* and *Pseudomonas putida* for promoting growth of *Jatropha curcas* seedling root. *Journal of Tropical Life Science* 2, 53–57.

Taghavi, S., Van Der Lelie, D., Hoffman, A., Zhang, Y.B., Walla, M.D. *et al.* (2010) Genome sequence of the plant growth promoting endophytic bacterium *Enterobacter* sp. 638. *PLoS Genet* 6, e1000943.

Tang, J., Wang, R., Niu, X. and Zhou, Q. (2010) Enhancement of soil petroleum remediation by using a combination of ryegrass (*Lolium perenne*) and different microorganisms. *Soil and Tillage Research* 110, 87–93.

Tanimoto, E. (2005). Regulation of root growth by plant hormones—roles for auxin and gibberellin. *Critical Reviews in Plant Sciences* 24, 249–265.

Taulé, C., Mareque, C., Barlocco, C., Hackembruch, F., Reis, V.M. and Sicardi, M. (2011) The contribution of nitrogen fixation to sugarcane (*Saccharum officinarum* L.), and the identification and characterization of part of the associated diazotrophic bacterial community. *Plant Soil* 356, 35–49.

Taulé, C., Mareque, C., Barlocco, C., Hackembruch, F., Reis, V.M. *et al.* (2012) The contribution of nitrogen fixation to sugarcane (*Saccharum officinarum* L.), and the identification and characterization of part of the associated diazotrophic bacterial community. *Plant and Soil* 356, 35–49.

Tilman, D., Fargione, J., Wolff, B., D'Antonio, C., Dobson, A. *et al.* (2001). Forecasting agriculturally driven global environmental change. *Science* 292, 281–284.

Tilman, D., Cassman, K.G., Matson, P.A., Naylor, R. and Polasky, S. (2002) Agricultural sustainability and intensive production practices. *Nature* 418, 671–677.

Timmusk, S. and Wagner, E.G.H. (1999) The plant-growth-promoting rhizobacterium *Paenibacillus polymyxa* induces changes in *Arabidopsis thaliana* gene expression: a possible connection between biotic and abiotic stress responses. *Molecular Plant–microbe Interactions* 12, 951–959.

Tripathi, V., Dubey, R.K., Edrisi, S.A., Narain, K., Singh, H.B. *et al.* (2014a). Towards the ecological profiling of a pesticide contaminated soil site for remediation and management. *Ecological Engineering* 71, 318–325.

Tripathi, V., Dubey, R.K., Singh, H.B., Singh, N. and Abhilash, P.C. (2014b) Is *Vigna radiata* (L.) R. Wilczek a suitable crop for Lindane contaminated soil? *Ecological Engineering* 73, 219–223.

Tripathi, V., Abhilash, P.C., Singh, H.B., Singh, N. and Patra, D.D. (2015a) Effect of temperature variation on lindane dissipation and microbial activity in soil. *Ecological Engineering* 79, 54–59.

Tripathi, V., Fraceto, L.F. and Abhilash, P.C. (2015b) Sustainable clean-up technologies for soils contaminated with multiple pollutants: plant–microbe–pollutant and climate nexus. *Ecological Engineering* 82, 330–335.

Tripathi, V., Edrisi, S.A. and Abhilash, P.C. (2016a) Towards the coupling of phytoremediation with bioenergy production. *Renewable and Sustainable Energy Reviews* 57, 1386–1389.

Tripathi, V., Edrisi, S.A., O'Donovan, A., Gupta, V.K. and Abhilash, P.C. (2016b) Bioremediation for fueling the biobased economy. *Trends in Biotechnology* 34, 775–777.

Troufflard, S., Mullen, W., Larson, T.R., Graham, I.A., Crozier, A. *et al.* (2010) Potassium deficiency induces the biosynthesis of oxylipins and glucosinolates in *Arabidopsis thaliana*. *BMC Plant Biology* 10, 172.

Ullman, W.J., Kirchman, D.L., Welch, S.A. and Vandevivere, P. (1996) Laboratory evidence for microbially mediated silicate mineral dissolution in nature. *Chemical Geology* 132, 11–17.

Uroz, S., Dessaux, Y. and Oger, P. (2009) Quorum sensing and quorum quenching: the yin and yang of bacterial communication. *ChemBioChem* 10, 205–216.

Van der Helm, D. and Winkelmann, G. (1994) Hydroxamates and polycarboxylates as iron transport agents (siderophores) in fungi. In: Winkelmann, G. and Winge, D.R. (eds) *Metal Ions in Fungi*. Marcel Dekker, New York, pp. 39–98.

Vansuyt, G., Robin, A., Briat, J.F., Curie, C. and Lemanceau, P. (2007) Iron acquisition from Fe-pyoverdine by *Arabidopsis thaliana*. *Molecular Plant–microbe Interactions* 20, 441–447.

Verma, J.P., Yadav, J., Tiwari, N.K. and Jaiswal, D.K. (2014) Evaluation of plant growth promoting activities of microbial strains and their effect on growth and yield of chickpea (*Cicer arietinum* L.) in India. *Soil Biology and Biochemistry* 70, 33–37.

Vert, G., Grotz, N., Dédaldéchamp, F., Gaymard, F., Guerinot, M.L. *et al.* (2002) IRT1, an Arabidopsis transporter essential for iron uptake from the soil and for plant growth. *The Plant Cell* 14, 1223–1233.

Vogt, T. (2010) Phenylpropanoid biosynthesis. *Molecular Plant* 3, 2–20.

Vorholt, J.A. (2012) Microbial life in the phyllosphere. *Nature Reviews Microbiology* 10, 828–840.

Walker, V., Bertrand, C., Bellvert, F., Moënne-Loccoz, Y., Bally, R. and Comte, G. (2011) Host plant secondary metabolite profiling shows a complex, strain-dependent response of maize to plant growth-promoting rhizobacteria of the genus *Azospirillum*. *New Phytologist* 189, 494–506.

Walley, F.L., Gillespie, A.W., Adetona, A.B., Germida, J.J. and Farrell, R.E. (2014) Manipulation of rhizosphere organisms to enhance glomalin production and C sequestration: pitfalls and promises. *Canadian Journal of Plant Science* 94, 1025–1032.

Wang, Q., Xiong, D., Zhao, P., Yu, X., Tu, B. and Wang, G. (2011) Effect of applying an arsenic-resistant and plant growth–promoting rhizobacterium to enhance soil arsenic phytoremediation by *Populus deltoides* LH05-17. *Journal of Applied Microbiology* 111, 1065–1074.

Wang, Y., Yang, X., Zhang, X., Dong, L., Zhang, J. *et al.* (2014) Improved plant growth and Zn accumulation in grains of rice (*Oryza sativa* L.) by inoculation of endophytic microbes isolated from a Zn hyperaccumulator, *Sedum alfredii* H. *Journal of Agricultural and Food Chemistry* 62, 1783–1791.

Wani, P.A. and Khan, M.S. (2010) Bacillus species enhance growth parameters of chickpea (*Cicer arietinum* L.) in chromium stressed soils. *Food and Chemical Toxicology* 48, 3262–3267.

Waters, C.M. and Bassler, B.L. (2005). Quorum sensing: cell-to-cell communication in bacteria. *Annual Reviews Cell and Developmental Biology* 21, 319–346.

Webb, J., Ellis, S., Harrison, R. and Thorman, R. (2004) Measurement of N fluxes and soil N in two arable soils in the UK. *Plant Soil* 260, 253–270.

Weinert, N., Piceno, Y., Ding, G.C., Meincke, R., Heuer, H. *et al.* (2011) PhyloChip hybridization uncovered an enormous bacterial diversity in the rhizosphere of different potato cultivars: many common and few cultivar-dependent taxa. *FEMS Microbiology Ecology* 75, 497–506.

Welch, R.M. (1986) Effects of nutrient deficiencies on seed production and quality. *Advances in Plant Nutrition (USA)*.

Welch, R.M. and Graham, R.D. (2004) Breeding for micronutrients in staple food crops from a human nutrition perspective. *Journal of Experimental Botany* 55, 353–364.

Weller, D.M., Mavrodi, D.V., van Pelt, J.A., Pieterse, C.M., van Loon, L.C. and Bakker, P.A. (2012) Induced systemic resistance in *Arabidopsis thaliana* against *Pseudomonas syringae* pv. tomato by 2,4-diacetylphloroglucinol-producing *Pseudomonas fluorescens*. *Phytopathology* 102, 403–412.

Werling, B.P., Dickson, T.L., Isaacs, R., Gaines, H., Gratton, C. *et al.* (2014) Perennial grasslands enhance biodiversity and multiple ecosystem services in bioenergy landscapes. *Proceedings of the National Academy of Sciences of the United States of America* 111, 1652–1657.

Weyens, N., van der Lelie, D., Taghavi, S., Newman, L. and Vangronsveld, J. (2009) Exploiting plant–microbe partnerships to improve biomass production and remediation. *Trends in Biotechnology* 27, 591–598.

Weyens, N., Croes, S., Dupae, J., Newman, L., van der Lelie, D. *et al.* (2010) Endophytic bacteria improve phytoremediation of Ni and TCE co-contamination. *Environmental Pollution* 158, 2422–2427.

Wieder, W.R., Bonan, G.B. and Allison, S.D. (2013) Global soil carbon projections are improved by modelling microbial processes. *Nature Climate Change* 3, 909–912.

Winkelmann, G. (2007) Ecology of siderophores with special reference to the fungi. *Biometals* 20, 379–392.

Wu, C.H., Bernard, S.M., Andersen, G.L. and Chen, W. (2009a) Developing microbe–plant interactions for applications in plant-growth promotion and disease control, production of useful compounds, remediation and carbon sequestration. *Microbial Biotechnology* 2, 428–440.

Wu, S.C., Peng, X.L., Cheung, K.C., Liu, S.L. and Wong, M.H. (2009b) Adsorption kinetics of Pb and Cd by two plant growth promoting rhizobacteria. *Bioresource Technology* 100, 4559–4563.

Wu, S.H., Ho, C.T., Nah, S.L. and Chau, C.F. (2014) Global hunger: a challenge to agricultural, food, and nutritional sciences. *Critical Reviews in Food Science and Nutrition* 54, 151–162.

Xiao, X., Chen, H., Si, C. and Wu, L. (2012) Influence of biosurfactant-producing strain *Bacillus subtilis* BS1 on the mycoremediation of soils contaminated with phenanthrene. *International Biodeterioration and Biodegradation* 75, 36–42.

Yadav, J., Verma, J.P., Jaiswal, D.K. and Kumar, A. (2014) Evaluation of PGPR and different concentration of phosphorus level on plant growth, yield and nutrient content of rice (*Oryza sativa*). *Ecological Engineering* 62, 123–128.

Yang, M.M., Wen, S.S., Mavrodi, D.V., Mavrodi, O.V., von Wettstein, D. *et al.* (2014) Biological control of wheat root diseases by the CLP-producing strain *Pseudomonas fluorescens* HC1-07. *Phytopathology* 104, 248–256.

Yergeau, E., Sanschagrin, S., Maynard, C., St-Arnaud, M. and Greer, C.W. (2014) Microbial expression profiles in the rhizosphere of willows depend on soil contamination. *The ISME Journal* 8, 344–358.

Yuhashi, K.I., Ichikawa, N., Ezura, H., Akao, S., Minakawa, Y. *et al.* (2000) Rhizobitoxine production by *Bradyrhizobium elkanii* enhances nodulation and competitiveness on *Macroptilium atropurpureum*. *Applied and Environmental Microbiology* 66, 2658–2663.

Zaidi, S., Usmani, S., Singh, B.R. and Musarrat, J. (2006) Significance of *Bacillus subtilis* strain SJ-101 as a bioinoculant for concurrent plant growth promotion and nickel accumulation in *Brassica juncea*. *Chemosphere* 64, 991–997.

Zakria, M., Ohsako, A., Saeki, Y., Yamamoto, A. and Akao, S. (2008) Colonization and growth promotion characteristics of *Enterobacter* sp. and *Herbaspirillum* sp. on *Brassica oleracea*. *Soil Science and Plant Nutrition* 54, 507–516.

Zeinali, M., Vossoughi, M., Ardestani, S.K., Babanezhad, E. and Masoumian, M. (2007) Hydrocarbon degradation by thermophilic *Nocardia otitidiscaviarum* strain TSH1: physiological aspects. *Journal of Basic Microbiology* 47, 534–539.

Zhang, S., Li, Q., Lü, Y., Zhang, X. and Liang, W. (2013) Contributions of soil biota to C sequestration varied with aggregate fractions under different tillage systems. *Soil Biology and Biochemistry* 62, 147–156.

Zhang, Y., Ruyter-Spira, C. and Bouwmeester, H.J. (2015) Engineering the plant rhizosphere. *Current Opinion in Biotechnology* 32, 136–142.

Zhou, Y., Choi, Y.L., Sun, M. and Yu, Z. (2008) Novel roles of *Bacillus thuringiensis* to control plant diseases. *Applied Microbiology and Biotechnology* 80, 563–572.

Zhou, X., Wu, H., Koetz, E., Xua, Z. and Chen,C. (2012) Soil labile carbon and nitrogen pools and microbial metabolic diversity under winter crops in an arid environment. *Applied Soil Ecology* 53, 49–55.

Zornoza, R., Faz, A., Carmona, D.M., Martínez-Martínez, S. and Acosta, J.A. (2012) Plant cover and soil biochemical properties in a mine tailing pond five years after application of marble wastes and organic amendments. *Pedosphere* 22, 22–32.

6 *Pseudomonas* Communities in Soil Agroecosystems

Betina Cecilia Agaras,* Luis Gabriel Wall and Claudio Valverde
Laboratorio de Bioquímica, Microbiología e Interacciones Biológicas en el Suelo, Departamento de Ciencia y Tecnología, Universidad Nacional de Quilmes, Buenos Aires, Argentina

6.1 Introduction

Among all soil bacterial genera having a representative described as a plant-growth promoter, *Pseudomonas* comprise a wide variety of PGPR species, with different mechanisms of action (Lugtenberg and Kamilova, 2009). Several pseudomonads have demonstrated high rhizosphere competence, production of different kinds of secondary metabolites involved in antagonism, phytostimulation or fertilization, and an ability to degrade complex organic compounds, hence being able to contribute not only to plant health but also to bioremediation of soils (Lugtenberg and Dekkers, 1999; Haas and Défago, 2005; Tapadar and Jha, 2013; Agaras *et al.*, 2015; Mishra *et al.*, 2015; Kumar, 2016). Its physiological and genetic adaptability contribute to the widespread distribution of this genus in various ecosystems around the world (Stanier *et al.*, 1966; Palleroni and Moore, 2004; Silby *et al.*, 2011).

As members of the γ-Proteobacteria subphylum, which range from 1% to 34% of the abundance of total soil bacterial community of different environments (Aislabie and Deslippe, 2013), *Pseudomonas* are key members of the soil microbiome. Considered as copiotrophs, because they are specially present in areas where resource availability is high and carbon sources are simple (Fierer *et al.*, 2007), their remarkable nutritional versatility allows *Pseudomonas* to exploit diverse rhizosphere environments, where each plant exudes different kinds of organic compounds (Lugtenberg and Bloemberg, 2004).

In recent years, increasing efforts were made to characterize the bacterial community of natural and agricultural soils, trying to understand the different factors that shape the microbiome in each environment (Philippot *et al.*, 2013). Particularly, there has been an interest in studying the effect that agricultural practices have on microbial community structures (Cookson *et al.*, 2006; Costa *et al.*, 2006b; Picard *et al.*, 2008; Cycoń and Piotrowska-Seget, 2009; Figuerola *et al.*, 2012; Ding *et al.*, 2013; Agaras *et al.*, 2014; Figuerola *et al.*, 2015). This phenomenon is directly linked with increasing awareness about the development of more sustainable practices in cropping systems, looking for higher yields without depleting natural resources of soils (Cook, 2006). These include direct seeding (no-tillage), crop rotation, rational

*E-mail: betina_agaras@yahoo.com.ar

© CAB International 2017. *Advances in PGPR Research* (eds H.B. Singh, B.K. Sarma and C. Keswani)

use of agrochemicals and integrated pest management (FAO, 2003; AAPRESID, 2013a). In Argentina, almost 20×10^6 ha (78.5% of the arable land) are under no-till management (Albertengo *et al.*, 2014). It is known that conventional tillage destroys soil structure, and this disturbance affects both abiotic and biotic factors contributing to the soil ecosystem (Cook, 2006; Govaerts *et al.*, 2007). This is the main reason for the shift to no-tillage management in sustainable systems. Moreover, this dominant practice in our country is expanding towards marginal soils with more challenging climatic and nutritional conditions. When no-tillage is accompanied by crop rotation and rational use of chemical fertilizers/pesticides, soil quality is preserved over time, with high productivity levels (Derpsch *et al.*, 2010; Leoni Velazco, 2013). These are the so-called "Good Agricultural Practices" (FAO, 2003; AAPRESID, 2013a). By contrast, no-tillage associated with mono-cropping and misuse of fertilizers/pesticides results in soil quality decline (low crop yields, accumulation of chemical products in soil, higher soil erosion and higher incidence of plant diseases) (FAO, 2003; Raaijmakers *et al.*, 2009; AAPRESID, 2013b; Leoni Velazco, 2013), and in the loss of bacterial regional diversity (Figuerola *et al.*, 2015).

Since pseudomonads are key players in soil ecosystems, nutrient cycling and plant-growth promotion, there has been recent interest in studying the influence that tillage management, crop rotation, agrochemical applications and other agronomical practices may have on the abundance and community structure of *Pseudomonas* in the soil and/or the rhizosphere of cropped plants. In this chapter, we review relevant literature about the impact that various agricultural practices have in shaping the communities of *Pseudomonas* in soil agroecosystems.

6.2 Tillage Managements and Sustainable Agriculture Systems

Soil type is a main determinant factor of the microbiological community (Latour *et al.*, 1996; Cho and Tiedje, 2000; Girvan *et al.*, 2003; Garbeva *et al.*, 2004a; Berg and Smalla, 2009; Kuramae *et al.*, 2012). Nevertheless, agriculture is nowadays developed on different types of soils, and due to its versatility, *Pseudomonas* genus has been found worldwide (Weller *et al.*, 2002; De La Fuente *et al.*, 2006). Thus, studies has been in general focused on understanding the effect of human activities on *Pseudomonas* communities, regardless of soil type (Picard and Bosco, 2008; Adesemoye and Kloepper, 2009; Fischer *et al.*, 2010; Agaras *et al.*, 2014).

As mentioned before, tillage is a modulator factor of the soil microbial community across several geographical locations. Molecular techniques such as massive parallel sequencing, PCR-DGGE, qPCR, FISH, lipid profiles and enzymatic tests have allowed us to perform broad analysis of the whole microbial community structure and function in several soil types (Christensen *et al.*, 1999; Peixoto *et al.*, 2006; Govaerts *et al.*, 2007; Green *et al.*, 2007; Eickhorst and Tippkötter, 2008; Meriles *et al.*, 2009; Perez-Brandán *et al.*, 2012; Ding *et al.*, 2013; Dai *et al.*, 2015; Ferrari *et al.*, 2015; Figuerola *et al.*, 2015; Yan *et al.*, 2016), whereas individual microbial groups can similarly be studied with molecular methods but using oligonucleotides targeting taxon-specific genes (Wagner *et al.*, 1994; Zarda *et al.*, 1997; Hesselsøe *et al.*, 2001; Richardson *et al.*, 2002; Costa *et al.*, 2006a; Frapolli *et al.*, 2008; Mühling *et al.*, 2008; Rosa *et al.*, 2014). Nevertheless, culture-dependent techniques have also been useful tools for studying soil communities (Edwards *et al.*, 2001; Nesci *et al.*, 2006; San Miguel *et al.*, 2007; Montecchia *et al.*, 2011; Perez-Brandán *et al.*, 2012; López-Piñeiro *et al.*, 2013; Li *et al.*, 2013; Agaras *et al.*, 2014); this is notably the case of the genus *Pseudomonas*, which is easily cultured in formulated media with strong selective properties (Johnsen and Nielsen 1999; Johnsen *et al.*, 1999; Agaras *et al.*, 2012). With such an approach, in our laboratory we found that the abundance of culturable *Pseudomonas* was significantly higher in a no-tillage plot compared with a neighbouring plot with conventional tillage, specifically in the 5–10cm layer (Table 6.1).

Table 6.1. Effect of agricultural management on *Pseudomonas* abundance in different soil types.

Tillage system	Agricultural treatments							
	Extensive agriculture[1]	Vineyards[2]	Winter barley [3]	Maize[3]	Winter rye [3]	Maize mono-culture [4]	Wheat mono-culture [4]	Maize/ wheat rotation [5]
No tillage	2.0×10^4	2×10^7	4.5×10^6	3.4×10^6	4.4×10^6	7.8×10^3	1.0×10^4	0.0581
Conventional tillage	6.5×10^3	1×10^6	3.2×10^6	3.5×10^6	4.2×10^6	2.6×10^3	2.8×10^3	0.0174

[1] Values are CFU/g soil. Result obtained by plate counting in S1 Gould medium, from bulk soil samples of the 5–10cm layer of Funke experiment, with wheat as the previous crop (Bs. As., Argentina, unpublished data)
[2] Values are CFU/g soil. Estimated data from inter-row samples of the 0–10cm layer taken one week after harvest. Values obtained by plate counting on TSA and subsequent sequencing of 16S rDNA gene (López-Piñeiro *et al.*, 2013)
[3] Average values (CFU/g soil) from bulk soil samples of 0–15cm and 15–30cm layers, obtained by the analysis of the fatty-acid-methylester profile of colonies grown on glycerine-peptone agar (Höflich *et al.*, 1999)
[4] Values are CFU/g soil. Tillage systems differ mainly in the presence or absence of stubble. Fluorescent *Pseudomonas* abundance was estimated by the plate count method on King's B agar (Govaerts *et al.*, 2008)
[5] Values are relative proportions of *Pseudomonas* from total 16S rDNA sequences obtained after sequencing. Results correspond to NT with stubble and CT without stubble (tyical treatments) (Ceja-Navarro *et al.*, 2010)

Both plots were from Funke Village, in the southwest of the Pampean region of Argentina, where extensive agriculture is routinely in typical argiudoll loamy soils where precipitation is low and wind erosion is a real problem (Galantini *et al.*, 2013). This result is in agreement with data obtained from different crop systems: vineyards (López-Piñeiro *et al.*, 2013); wheat, maize, rye and barley systems (Höflich *et al.*, 1999); maize-wheat rotation, and wheat or maize mono-cropping (Govaerts *et al.*, 2008; Ceja-Navarro *et al.*, 2010) (Table 6.1).

In general, an increase in the abundance of *Pseudomonas* is correlated with the presence of residual crops on the surface of plots and with higher values of total soil organic carbon, both being conditions usually linked with no-tillage (Derpsch *et al.*, 2010). In the same experiment from Funke, we evaluated the proportion of culturable *Pseudomonas* community (total pseudomonads; TP) in the total culturable heterotrophic bacteria (TH), as well as the proportion of fluorescent *pseudomonas* (FP) among TP (Fig. 6.1), as we did before with samples from another agricultural treatment (Agaras *et al.*, 2014). Remarkably, we found that the TP/TH ratio is significantly higher in the 5–10cm layer (0.01±0.01) than in the 0–5cm layer (0.0003±0.0001), independent of management, whereas the FP/TP ratio is statistically higher in samples from the 0–5cm layer (0.54±0.28 versus 0.14±0.17), particularly for NT samples at this depth (Fig. 6.1). Therefore, although the abundance of culturable *Pseudomonas* in the 0–5cm layer is lower, this population seems to be mainly composed of fluorescent *Pseudomonas*, a subgroup that is intimately linked with disease suppression (Stutz *et al.*, 1986; Lemanceau and Alabouvette, 1993; Raaijmakers *et al.*, 1999).

In addition to the tillage regime, cropping management within the same tillage system also proved to alter the microbial community of soils. In a recent study of no-till plots located in a 400km west–east transect in the most productive region of Argentina (Wall, 2011), we found that Good Agricultural Practice (GAP) favoured the abundance of culturable pseudomonads in bulk soils. This higher abundance in GAP soils was also reflected in soybean rhizospheres, when we compared samples from GAP plots with samples from soybean monoculture plots. Moreover, we found that GAP increased the TP/TH ratio of bulk soils (Agaras *et al.*, 2014). Nevertheless, when we analyzed the community structure of *Pseudomonas* in each sampling location (as judged by PCR-RFLP of the genus-specific genes *oprF* and *gacA* (Bodilis *et al.*, 2006; Costa *et al.*, 2007; Agaras *et al.*, 2012)), we found that it was strongly influenced by the

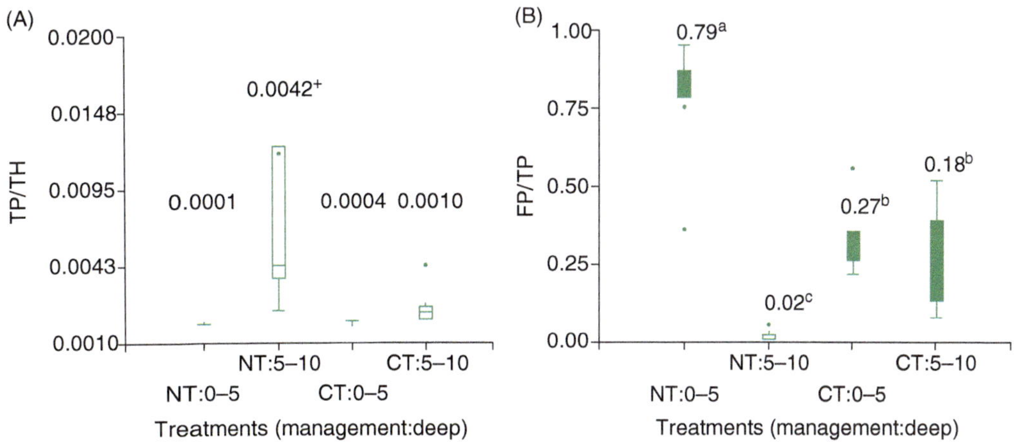

Fig. 6.1. Abundance proportion among culturable population of bacteria in no-tillage (NT) and conventional tillage (CT) management in 0–5 cm and 5–10 cm layers. Values were analyzed with ANOVA with Tukey test for comparison. For FP/TP values (B), different letters indicate significance difference among treatment ($p = 0.001$). This difference is also significant only for deep factor ($p < 0.001$), as we mentioned in the text. For TP/TH (A), the deep effect is statistically significant ($p = 0.005$), although NT: 0–5 showed an interesting significance of $p = 0.053$ (indicated by asterisk).

geographical location instead of the agricultural practice. Thus, these results are in agreement with the aforementioned work, which demonstrated the relevance of soil type on the microbiome (Agaras *et al.*, 2014).

Besides an influence of the cropping practice on abundance and of the geographical location on community structure, we found a strong seasonal effect on the soil pseudomonads community. On the one hand, the abundance was significantly higher in winter samples than in summer ones; on the other hand, the genetic relatedness of the most abundant culturable pseudomonads was higher for samples from the same sampling period (summer or winter) than those from either the same location or the same agricultural management (Agaras *et al.*, 2014). This seasonal shift, already reported for other bacterial groups (Cookson *et al.*, 2006; Prevost-Boure *et al.*, 2011; Rasche *et al.*, 2011), cannot be attributed to the absence of crops during winter because in this experiment GAP plots had winter cover crops that were absent in non-sustainable plots (Figuerola *et al.*, 2012). In the humid Pampa of Argentina, the summer air temperature tends to be 10°C higher than in winter periods (INDEC, 2013); as a consequence, winter soil samples contained c. 1.5% more moisture than those collected in summer. Therefore, the observed fluctuations in the abundance and community structure of pseudomonads across seasons matches previous studies that reported reduced survival of pseudomonads in drier soils and higher temperatures (Moffett *et al.*, 1983; O'Callaghan *et al.*, 2001).

In conclusion, the kind of soil management significantly influences *Pseudomonas* populations of bulk soils. As this bacterial group is intimately linked with plant-growth promotion, it may be important to promote its abundance in agroecosystems, since if *Pseudomonas* are present in soils they are able to subsequently colonize the rhizosphere of crop plants. In general, sustainable practices that include the maintenance of stubble, leading to more moistened soils with high organic carbon content, not only contribute to a better soil quality, but also favour the development of this bacterial genus from the intrinsic population of each geographical location.

6.3 Application of Agrochemicals

Another agricultural management practice widely applied is the use of herbicides for controlling weeds (Chauhan *et al.*, 2012).

Particularly after the development of resistant crops, such as soybean, maize, alfalfa, canola, sugar beet and cotton varieties, and the increasing acceptance of no-tillage systems, the use of glyphosate has been boosted remarkably (Dill *et al.*, 2008; Benbrook *et al.*, 2016). Moreover, this agrochemical is being employed in forest plantations, recreational areas and natural grasslands to remove exotic species or to promote winter forage species (Freedman, 1990; Thom *et al.*, 1993; Barnes, 2004; Powles, 2008). In particular, the application of glyphosate in native grasslands of the Flooding Pampa (Argentina) aims to increase winter productivity (Rodriguez and Jacobo, 2013), promoting forage production per hectare and thus allowing improvement of stocking rate and meat production in livestock systems (Bilello and Zeberio, 2002).

Effects of glyphosate application in various environments have been studied in the last decades, owing to the conflicts around its persistence and toxicity in soil and water ecosystems, especially in non-target organisms (Busse *et al.*, 2001; Ratcliff *et al.*, 2006; Zabaloy *et al.*, 2008; Zablotowicz *et al.*, 2009; Helander *et al.*, 2012; Ruiz-Toledo and Sánchez-Guillén, 2014; Druille *et al.*, 2016). Although some works have not found any change in soil or rhizosphere microbial community after glyphosate application (Ratcliff *et al.*, 2006; Weaver *et al.*, 2007; Barriuso *et al.*, 2011; Schafer *et al.*, 2014), or have even observed an enhancement in microbial biomass/respiration in rhizosphere samples immediately after the treatment (Haney *et al.*, 2002; Araújo *et al.*, 2003; Mijangos *et al.*, 2009; Nguyen *et al.*, 2016), there are reports that successive applications reduce soil microbial abundance and/or diversity (Zabaloy *et al.*, 2008; Druille *et al.*, 2015; Druille *et al.*, 2016). Interestingly, specific bacterial groups have been shown to become more abundant upon repetitive glyphosate treatment, suggesting an enrichment of potential glyphosate-degrading organisms (Lancaster *et al.*, 2010; Newman *et al.*, 2016). Members of *Pseudomonas* genus are within the latter (Jacob *et al.*, 1988; Selvapandiyan and Bhatnagar, 1994; Peñaloza-Vazquez *et al.*, 1995; Olawale and Akintobi, 2011). For example, Kuklinsky-Sobral and colleagues observed an enrichment of *P. oryzihabitans* in the endorhizosphere of soybean grown in a glyphosate-treated field, although they did not find any difference in the abundance of total endophytic bacteria (Kuklinsky-Sobral *et al.*, 2005). Also, Travaglia and colleagues have demonstrated that maize inoculation with a *Pseudomonas* sp. strain can improve germination, dry weight, leaf area, chlorophyll and carotene content, and phytohormone production of maize plants, and even reduce the glyphosate content of leaves and grains when this crop was grown in glyphosate-treated soils (Travaglia *et al.*, 2015).

In different soil types, Gimsing and colleagues demonstrated that the culturable population of *Pseudomonas* correlates positively with the mineralization rate of glyphosate in each soil (Gimsing *et al.*, 2004). However, when the effect of a single glyphosate application is considered for each soil type, the abundance of *Pseudomonas* in the soybean rhizosphere decreased after treatment with the herbicide (Zobiole *et al.*, 2011). In agreement with this result, we found a lower *Pseudomonas* abundance in grassland soils where commercial glyphosate was applied after 4 years of one annual application every late summer (Lorch *et al.*, 2016). On the basis of the well known *Pseudomonas* preference for rhizosphere environments, this effect could be linked with the reduced vegetation in glyphosate-treated plots compared with plots without applications.

Moreover, treatments with other herbicides, like glufosinate and metazachlor, have shown to negatively affect populations of *P. corrugata*, *P. tolaasii* and *P. fluorescens* groups, as well as to decrease the richness of the entire *Pseudomonas* population (Gyamfi *et al.*, 2002). The phenylurea herbicide linuron generated a selection of *P. mandelii* and *P. jesenni* groups and a subsequent enrichment in orchard soils after more than 10 years of soil treatment (El Fantroussi *et al.*, 1999). By contrast, application of the fungicide fenpropimorph did not modify *Pseudomonas* abundance in barley rhizosphere (Thirup *et al.*, 2000; Thirup *et al.*, 2001). Hence, it appears that the outcome of application of a pesticide on the *Pseudomonas* community mainly depends on the compound nature itself: in the presence of bacteria that could metabolize the organic molecule, the effect may be positive (i.e. enrichment of that sub-population); if there are no metabolizing species, the effect on the abundance of pseudomonads may be negative. This would be the reason why it has been possible to isolate different *Pseudomonas* strains from soils or rhizospheres contaminated with several pesticides. For instance, the nowadays restricted organochloride pesticides dichlorodiphenyltrichloroethane (DDT) and hexachlorocyclohexane (HCH) can be degraded by several species of *Pseudomonas*, including *P. aeruginosa* and *P. chlororaphis* (Sahu *et al.*, 1992; Nawab *et al.*, 2003; Kamanavalli and Ninnekar, 2004; Kumar *et al.*, 2005). Furthermore, Murthy and Manonmani could isolate from a polluted soil a HCH-degrading consortium formed of ten bacterial strains, among which seven individuals belonged to the *Pseudomonas* genus, including *P. putida, P. aeruginosa, P. fluorescens, P. diminuta, P. stutzeri* and *P. psedomallei* (Murthy and Manonmani, 2007). Similarly, the organophosphate insecticides diazinon and phorate can also select for several *Pseudomonas* species (Cycoń *et al.*, 2009; Jariyal *et al.*, 2014; Jariyal *et al.*, 2015) and the nematicide ethoprophos (O-ethyl S, S-dipropyl phosphorodithioate) is degraded by representatives of the *P. putida* group (Karpouzas and Walker, 2000). Altogether, these results show an enrichment of these bacterial species under the selection pressure of pesticides, suggesting that they can employ those molecules – or their metabolites – as energy sources.

In summary, the metabolic versatility of the *Pseudomonas* genus is a powerful advantage for their performance under different human-altered ecosystems. Furthermore, it could be a useful tool for the recovery of agricultural soils that are polluted with various agrochemicals or their residues, which persist in soils during decades (Mulbry and Kearney, 1991; Kannan *et al.*, 1997; Chowdhury *et al.*, 2008; Aktar *et al.*, 2009).

6.4 Crop Species

Plant species is a key factor modulating the *Pseudomonas* community structure (Berg and Smalla, 2009). For each plant species, its development stage, genotype, health conditions or fitness can influence the composition of its root microbiome (Lakshmanan, 2015). Rhizodepositions vary among host plants (Nguyen, 2003), thus selecting for a particular microflora (Paterson *et al.*, 2007), which should be able to employ those compounds as carbon and energy sources (Smalla *et al.*, 2001; Bais *et al.*, 2006). This phenomenon is known as the "rhizosphere effect" (Lynch and Whipps, 1990; Warembourg, 1997) and it is evident for members of the *Pseudomonas* genus colonizing diverse plant species (Garbeva *et al.*, 2004a; Costa *et al.*, 2006c; Fernández *et al.*, 2012). In contrast to elsewhere in the soil, there is a relatively high proportion of motile, rapidly growing bacteria in the rhizosphere: these are the so-called *r* strategists or copiotrophs, among which *Pseudomonas* spp. are found (Fierer *et al.*, 2007). Recent studies have demonstrated that plant selection is mainly based on functional traits instead of specific microbial species per se, and these functions seem to be important for the interaction with the plant: such functions include transporters, secretion systems, and metabolism of nitrogen, iron, phosphorus, hydrogen and potassium (Mendes *et al.*, 2014; Bulgarelli *et al.*, 2015; Yan *et al.*, 2016). In this regard, it has been recently demonstrated by using split-root assays that pathogen attack on one side of the barley root system resulted in a systemic enrichment of fluorescent *Pseudomonas* with genetic biocontrol traits on the other side (Dudenhöffer *et al.*, 2016). Furthermore, not only does the rhizosphere microbial community differ among plant species, but also bulk soil and endophytic communities are differentially shaped by this "plant effect". Even different cultivars of the same crop, as well as different transgenic lines, showed a marked effect on *Pseudomonas* community structure in bulk soil, rhizosphere soil and the endophytic compartment (Granér *et al.*, 2003; Mazzola *et al.*, 2004; Milling *et al.*, 2005; Marques *et al.*, 2014; Mendes *et al.*, 2014).

Pseudomonas spp. are widely recognized as preferential colonizers of the rhizosphere of field-grown soybean, maize and wheat (Kuklinsky-Sobral *et al.*, 2004; McSpadden Gardener *et al.*, 2005; Mittal and Johri, 2008;

Picard *et al.*, 2008; Von Felten *et al.*, 2010). In accordance, we observed that for soil samples under the same agricultural management, soybean and maize selected for a different culturable *Pseudomonas* community in their rhizospheres (i.e., higher relative abundance for soybean and differential genetic composition for both crops), independently of the geographical site (Agaras *et al.*, 2014). However, the TP/TH ratios were comparable, thus supporting the hypothesis that the higher pseudomonads abundance in soybean is due to a higher root surface exposed to soils, on a fresh weight basis, than that of the corn root system. For fluorescent pseudomonads, we found that the FP/TP ratio was statistically higher for the maize rhizosphere, suggesting that corn rhizodepositions are preferred by fluorescent *Pseudomonas* than soybean root exudates. In agreement with these results, McSpadden Gardener and collaborators have shown similar crop specific-effects in several fields over a 3-year period, with an enrichment of *phlD*$^{+}$*Pseudomonas*, a fluorescent subgroup with biocontrol potential (McSpadden Gardener *et al.*, 2005). Moreover, Latz and colleagues have provided evidence that plant identity, and especially the presence of *Lollium perenne*, can affect the expression of biocontrol-related genes of *P. protegens* CHA0 in microcosm experiments (Latz *et al.*, 2015).

Initially developed for studying the infection capacity of soils in N_2-fixing symbioses (Barnet *et al.*, 1985), the plant-trap method is also a useful tool to explore the impact of plant selection on soil and rhizosphere microbial communities (Haichar *et al.*, 2008; Marrero *et al.*, 2015). With this approach, we were able to detect different abundance and community composition of *Pseudomonas* in the rhizoplane of wheat, corn and soybean plantlets, upon introduction of surface-sterilized seeds into the same pristine soil having no previous record of agricultural intervention: total pseudomonads were more abundant in wheat than in soybean or maize roots, whereas genetically distinct pseudomonads could be isolated from every crop root sample (Marrero *et al.*, 2015). At the field scale, similar trends were observed in a single experimental field that was split into plots with different cropping regimes: maize mono-cropping, soybean mono-cropping, and two different crop rotation schemes. After only two cropping seasons, the full rotation scheme (maize/wheat/soybean/wheat) was able to drive a differential pseudomonads community structure in the bulk soil, and to increase the abundance of antibiotic producers (Agaras *et al.*, 2013). Furthermore, in grassland environments, it seems that plant diversity favours the abundance of biocontrol-related bacteria, i.e. *Pseudomonas* which can synthetize 2,4-diacetylphloroglucinol (DAPG) and pyrrolnitrin antibiotics, supporting the idea that mono-cropping is not a sustainable management for microflora diversity (Latz *et al.*, 2012).

As *Pseudomonas* is a ubiquitous bacterial genus in plant rhizospheres, crop species is a main factor that modulates its community, not only in rhizosphere environments but also in bulk soils after crop harvesting. In general, the bigger the root systems, the higher the *Pseudomonas* abundance. But this increase in population density does not imply a higher diversity. In contrast, plant species tend to select a singular bacterial group according to their needs. Owing to its wide range of plant growth-promoting activities, *Pseudomonas* is always a bacterial candidate to be recruited by plants.

6.5 Suppressive Soils and *Pseudomonas*: a Close Relationship

A singular case of study is the development of *suppressive soils*. These are ecosystems in which the plant pathogen: (i) does not establish or persist; (ii) establishes but causes little or no damage; (iii) establishes and causes disease for a while but thereafter the disease decreases its incidence. In all those circumstances, the causal agent persists in the soil but it is not able to express its pathogeny (Weller *et al.*, 2002). Suppressiveness can occur naturally (i.e. it is an inherent condition of soil), as described for soils suppressive to *F. oxysporum* species causing *Fusarium* wilt in several crops (Alabouvette, 1999) or to *Phytophthora cinnamomi*, the causal agent of root rot of many fruits and forest trees (Keen and Vancov, 2010). Besides, soil suppressiveness can be induced by the presence of the affected plant, the agriculture management or the inoculation of a pathogen, with the objective of favouring the abundance in soils of the involved biocontrol agent (Hornby, 1983). The latter group of soils includes most known cases of suppressiveness.

Take-all disease, caused by *Gaeumannomyces graminis* var. *tritici*, black root-rot of tobacco, originated by *Thielavopsis basicola*, and potato scab, which is produced by several *Streptomyces* species, are all diseases that declined after years of wheat, tobacco or potato mono-cropping, respectively (Gasser and Défago, 1981; Hornby, 1983; Meyer and Shew, 1991; Liu *et al.*, 1995; Weller *et al.*, 2002; Cook, 2003). For the pathogen *P. cinnamomi*, examples have also been described of induced suppressive soils after avocado, melon and watermelon mono-cropping (Keen and Vancov, 2010). Besides the control of fungal diseases, nematode suppressiveness has also been described: the ectoparasitic ring nematode, *Mesocriconema xenoplax*, causal agent of Peach tree short-life syndrome (PTSL), is controlled by the *P. synxantha* BG33 isolate obtained from the same peach orchards (Wechter *et al.*, 2001; Wechter *et al.*, 2002); the cereal cyst nematode *Heterodera avenae* Woll. has been effectively suppressed for more than 20 years in many soils in Northern Europe by several soil fungi (Kerry *et al.*, 1982), and the root-knot nematodes *Meloidogyne incognita* and *M. javanica* were suppressed in Florida soils by the parasitic bacteria *Pasteuria penetrans*, which increased its population after 7 years of tobacco monoculture (Weibelzahl-Fulton *et al.*, 1996), or by several fungi in California soils that have been planted with diverse perennial crops (Stirling *et al.*, 1979; Bent *et al.*, 2008). By contrast, continuous cropping of apple trees induced apple replant disease, which is triggered by a complex of fungi including *Cylindrocarpon destructans, Phytophthora cactorum, Pythium* spp. and *Rhizoctonia solani* (Mazzola, 1998), although it could be controlled by intercalating seasons of apple orchards with seasons of wheat, or by applying meal amendments (Mazzola and Gu, 2002; Mazzola, 2007).

In most of the aforementioned cases, monoculture was an effective management tool for the development of suppressiveness (Stirling *et al.*, 1979; Gasser and Défago, 1981; Hornby, 1983; Weller *et al.*, 2002; Cook, 2003), in contrast with the crop rotation recommended by manuals of sustainable agriculture (FAO, 2003; AAPRESID 2013a). It must be stressed here that those suppressive soils were all managed under conventional tillage. Nevertheless, it should be considered that mono-cropping is particularly detrimental under no-tillage managements, because the accumulation of crop residues increases the pathogen load in soils (Lin, 2011). On the other hand, the induction of disease-suppressive soils through crop monoculture or specific cropping sequences demonstrates the plants' role in building a disease-suppressive soil microbiome (Weller *et al.*, 2002; Peters *et al.*, 2003; Latz *et al.*, 2012). Notably, the suppressiveness of the soils described above is associated with the antagonistic potential of diverse *Pseudomonas* species (Stutz *et al.*, 1986; Lemanceau and Alabouvette, 1993; Duijff *et al.*, 1994; Mazzola, 2002; Mazzola *et al.*, 2004; Landa *et al.*, 2006; Mazzola, 2007; Weller, 2007; Weller *et al.*, 2007; Mazurier *et al.*, 2009; Hjort *et al.*, 2010; Mendes *et al.*, 2011; Michelsen *et al.*, 2015) (Table 6.2). For instance, there are wheat cultivars that are especially attractive for antibiotic-producing fluorescent *Pseudomonas* (Mazzola *et al.*, 2004; Landa *et al.*, 2006), thus underlying the fact that continuous wheat monoculture increases the abundance of the subgroup of DAPG producers in bulk soil to attain a population level that can synthesize enough antibiotic molecules to inhibit pathogen growth (Raaijmakers *et al.*, 1999). This phenomenon is the main reason of take-all decline (Cook, 2003). The same "wheat effect" on *Pseudomonas* population structure of bulk soils is a useful tool for suppressing apple replant disease: when apple orchard soils become conducive (i.e. non-suppressive), three cycles of wheat restored the *P. putida* level needed

Table 6.2. Suppressive soils with associated *Pseudomonas* community.

Disease	Pathogens	Susceptible crops	Location of the suppressive soils	*Pseudomonas* implied in suppressiveness	Mechanisms involved	References
Take-all	*Gaeumannomyces graminis* (several varieties)	Cereals (wheat, barley, oat, grass, maize)	Washington (USA)	Fluorescent pseudomonads, like:	Siderophores	Kloepper *et al.* (1980); Thomashow and Weller (1988); Weller (1988); Sarniguet *et al.* (1992); Raaijmakers and Weller (2001); Cook (2003)
			Horsham (Australia)	*P. fluorescens* Pf0-1	DAPG	
			The Netherlands	*P. fluorescens* B10	phenazine-1-carboxylic acid (PCA)	
			Rothamsted (UK)	*P. brassicasearum* Q8r1-96		
				P. chlororaphis subsp. *aureofaciens*30-84		
				P. fluorescens 2-79		
Fusarium-wilt	*Fusarium oxysporum* (several special forms)	Carnation	Chateaurenard (France)	*P. putida/P. fluorescens* groups, like:	Pseudobactin	Kloepper *et al.* (1980); Lemanceau and Alabouvette (1993); Duijff *et al.* (1994); Worku and Gerhardson (1996); Duijff *et al.* (1998); Chin-A-Woeng *et al.* (2001); Landa *et al.* (2006); Mazurier *et al.* (2009)
		Tomato	Salinas Valley, California (USA)	*P. putida* WCS358	Induced-systemic resistance (ISR)	
		Flax	Canary Islands (Spain)	*P. fluorescens* WCS417	Phenazines	
		Radish	Mt Vernon, Washington (USA)	*P. fluorescens* B10	Competition for niche and nutrients	
		Cucumber	Panama	*P. fluorescens* WCS374		
		Pea	Japan	*P. putida* A12		
		Spinach	Uppsala (Sweden)			
		Sweet potato	Fargo, North Dakota (USA)			
		Banana				
Potato scab	*Streptomyces scabies* Other *Streptomyces* species	Potato	Washington (USA)	Fluorescent pseudomonads	n.d.	Meng *et al.* (2012); Rosenzweig *et al.* (2012)
Stripe canker	*Phytophthora cinnamomi*	Several tree species (cinnamon, avocado, eucalyptus, jacaranda)	Queensland (Australia)	*P. putida/P. fluorescens* groups, like:	Hyphae, zoospores and sporangia lysis (particular mechanism not determined)	Broadbent *et al.* (1971); Stirling *et al.* (1992); Yang *et al.* (2001); Keen and Vancov (2010)
			California (USA)	*P. fluorescens* M24		
Root rot			South Africa	*P. fluorescens* 513		
Apple replant	*Cylindrocarpon destructans*	Apple	Washington (USA)	*P. putida* biotype A, like:	n.d.	Mazzola (1998); Mazzola (1999); Mazzola and Gu (2002)
	Phytophthora cactorum			*P. putida* 2C8		
	Pythium spp.					
	Rhizoctonia solani AG 5					

Stem rot and Tuber black scurf	*Rhizoctonia solani* AG 3	Potato	Inneruulalik (Greenland)	*P. fluorescens* In5	Nonribosomal peptides (NRP)	Garbeva *et al.* (2004b); Garbeva *et al.* (2004c); Michelsen *et al.* (2015)
			Bennekom (The Netherlands)	*P. putida/P. fluorescens* groups	Pyrrolnitrin	
Black root rot	*Thielaviopsis basicola*	Tobacco	Morens (Switzerland)	*P. protegens* CHA0	Siderophores Cyanic acid (HCN) Antibiotics (DAPG, pyrrolnitrin)	Ahl *et al.* (1986); Stutz *et al.* (1986); Voisard *et al.* (1989); Keel *et al.* (1992)
Damping-off	*Rhizoctonia solani* AG 2	Sugar beet	Hoeven (The Netherlands)	*P. protegens* Pf-5	Pyrrolnitrin	Howell and Stipanovic (1979); Mendes *et al.* (2011)
		Cotton		Fluorescent *Pseudomonas* haplotypes SH-A, SH-B and SH-C, like: *Pseudomonas sp.* strain SH-C52	A NRP (a putative chlorinated lipopeptide)	
Damping-off	*Phytium spp.*	Radish	Chinampa (Mexico)	Fluorescent pseudomonads	Antibiotics (pyoluteorin, DAPG)	Howell and Stipanovic (1980); Lumsden *et al.* (1987); Rezzonico *et al.* (2005); Rezzonico *et al.* (2007)
		Cotton	San Joaquín, California (USA)		Type III Secretion System	
Clubroot	*Plasmodiophora brassicae*	Cabbage	Uppsala (Sweden) Fukushima (Japan)	*Pseudomonas sp.*	Chitinase	Worku and Gerhardson (1996); Murakami *et al.* (2000); Hjort *et al.* (2007); Hjort *et al.* (2010)
Peach tree short life (PTSL)	*Mesocriconema (Criconemella) xenoplax*	Peach	Elgin, South Carolina (USA)	*P. synxantha* BG33	Egg-kill factor	Kluepfel *et al.* (1993); Wechter *et al.* (2001); Wechter *et al.* (2002)
Cyst nematodes	*Heterodera* spp.	Soybean Sugar beet	Heilongjiang (China) Riverside, California (USA)	*Pseudomonas sp.*	Invasion of cysts, but without details examined	Westphal and Becker (2001); Yin *et al.* (2003); Zhu *et al.* (2013)
Root-knot	*Meloidogyne hapla*	Tomato	Germany	*Pseudomonas kilonensis*	n.d.	Adam *et al.* (2014)

to antagonize *R. solani*, one of the pathogens involved in this disease, whilst decreasing *P. fluorescens* biovar. C abundance (not inhibitory to the replant pathogens), which have been promoted by apple roots (Mazzola and Gu, 2002; Mazzola *et al.*, 2004). By contrast, Almario and colleagues demonstrated for black root rot of tobacco that suppressiveness relies more on the presence of appropriate *phlD* genotypes and more favourable root conditions for the expression of DAPG synthesis than on a higher density of total fluorescent *Pseudomonas* in the tobacco root and rhizosphere (Almario *et al.*, 2013). In line with these observations, Gómez and colleagues found that the presence and previous adaptation of *P. fluorescens* SBW25 in a compost environment can modify the microbial community composition (Gómez *et al.*, 2016). These findings suggest that evolution occurring over ecological time scales can be a key driver of the structure of natural microbial communities, particularly in situations where some species have an evolutionary head start following large perturbations, such as in suppressive soils.

6.6 Perspectives and Future Directions

From its discovery, the *Pseudomonas* genus has been linked to plant-growth promotion, and this ability has been demonstrated for several plant species, including multiple crops of agronomical interest. Since pseudomonads are natural habitants of diverse soil environments, they are usually present as predominant bacterial groups of agroecosystems. Moreover, they are in general colonizing crop rhizospheres, and may be recruited by the plants themselves to help to improve their health. Due to the wide spectrum of probiotic properties shown by different *Pseudomonas* species, a deeper knowledge of their distribution in soils and rhizospheres, and a better understanding of factors altering those population dynamics, could help to look for mechanisms to improve probiotic pseudomonads communities in agricultural plots. Therefore, farmers would be able to reduce the employment of chemical inputs in agricultural systems, tending for more sustainable practices without affecting productivity, but managing natural resources of the ecosystem.

References

AAPRESID (2013a) *Manual de Buenas Prácticas Agrícolas e Indicadores de Gestión*. AAPRESID, Buenos Aires, Argentina.

AAPRESID (2013b) *Historia de la Siembra Directa*. Avalilable at: http://www.aapresid.org.ar/historia/ (accessed 7 July 2017).

Adam, M., Westphal, A., Hallmann, J. and Heuer, H. (2014) Specific microbial attachment to root knot nematodes in suppressive soil. *Applied and Environmental Microbiology* 80, 2679–2686.

Adesemoye, A.O. and Kloepper, J.W. (2009) Plant–microbes interactions in enhanced fertilizer-use efficiency. *Applied Microbiology and Biotechnology* 85, 1–12.

Agaras, B., Wall, L.G. and Valverde, C. (2012) Specific enumeration and analysis of the community structure of culturable pseudomonads in agricultural soils under no-till management in Argentina. *Applied Soil Ecology* 61, 305–319.

Agaras, B., Smalla, K., Wall, L.G. and Valverde, C. (2013) Short term impact of crop practice on the pseudomonads community structure of soil under no-tillage regime. In: Haas, D. (ed.) *14th International Conference on* Pseudomonas. UNIL, p. 303.

Agaras, B.C., Wall, L.G. and Valverde, C. (2014) Influence of agricultural practices and seasons on the abundance and community structure of culturable pseudomonads in soils under no-till management in Argentina. *Plant and Soil* 382, 117–131.

Agaras, B.C., Scandiani, M. and Luque, A. (2015) Quantification of the potential biocontrol and direct plant growth promotion abilities based on multiple biological traits distinguish different groups of *Pseudomonas* spp. isolates. *Biological Control* 90, 173–186.

Ahl, P., Voisard, C. and Defago, G. (1986) Iron bound-siderophores, cyanic acid, and antibiotics involved in suppression of *Thielaviopsis basicola* by a *Pseudomonas fluorescens* strain. *Phytopathology* 116, 121–134.

Aislabie, J. and Deslippe, J.R. (2013) Soil microbes and their contribution to soil services. In: Dymond, J.R. (ed.) *Ecosystems Services in New Zealand – Conditions and Trends*. Manaaki Whenua Press, Lincoln, NZ, pp 143–161.

Aktar, M.W., Sengupta, D. and Chowdhury, A. (2009) Impact of pesticides use in agriculture: their benefits and hazards. *Interdisciplinary Toxicology* 2, 1–12.

Alabouvette, C. (1999) *Fusarium* wilt suppressive soils: an example of disease-suppressive soils. *Australasian Plant Pathology* 28, 57–64.

Albertengo, J., Belloso, C. and Giraudo, M.B. (2014) Conservation agriculture in Argentina. In: Jat R.A., Sahrawat K.L. and Kassam A.H. (eds) *Conservation Agriculture: Global Prospects and Challenges*. CAB International, Wallingford, UK, pp 352–374.

Almario, J., Moënne-Loccoz, Y. and Muller, D. (2013) Monitoring of the relation between 2, 4-diacetylphloroglucinol-producing *Pseudomonas* and *Thielaviopsis basicola* populations by real-time PCR in tobacco black root-rot suppressive and conducive soils. *Soil Biology and Biochemistry* 57, 144–155.

Araújo, A.S.F., Monteiro, R.T.R. and Abarkeli, R.B. (2003) Effect of glyphosate on the microbial activity of two Brazilian soils. *Chemosphere* 52, 799–804.

Bais, H.P., Weir, T.L. and Perry, L.G. (2006) The role of root exudates in rhizosphere interactions with plants and other organisms. *Annual Review of Plant Biology* 57, 233–266.

Barnes, T.G. (2004) Strategies to convert exotic grass pastures to tall grass prairie communities. *Weed Technology* 18, 1364–1370.

Barnet, Y.M., Catt, P.C. and Hearne, D.H. (1985) Biological nitrogen fixation and root-nodule bacteria (*Rhizobium* sp. and *Bradyrhizobium* sp.) in two rehabilitating sand dune areas planted with *Acacia* spp. *Australian Journal of Botany* 33, 595–610.

Barriuso, J., Marín, S. and Mellado, R.P. (2011) Potential accumulative effect of the herbicide glyphosate on glyphosate-tolerant maize rhizobacterial communities over a three-year cultivation period. *Public Library of Science* 6, 27558.

Benbrook, C.M., Dill, G. and Sammons, R. (2016) Trends in glyphosate herbicide use in the United States and globally. *Environmental Sciences Europe* 28, 1.

Bent, E., Loffredo, A. and Mckenry, M.V. (2008) Detection and investigation of soil biological activity against *Meloidogyne incognita*. *Journal of Nematology* 40, 109–118.

Berg, G. and Smalla, K. (2009) Plant species and soil type cooperatively shape the structure and function of microbial communities in the rhizosphere. *FEMS Microbiology Ecology* 68, 1–13.

Bilello, G. and Zeberio, G. (2002) Incorporación tecnológica en explotaciones ganaderas de tipo familiar de la cuenca del salado. Control de paja colorada (*Paspalum quadrifarium*) y rejuvenecimiento de «ryegrass» en pastizales naturales. *Revista Facultad de Agronomía* 22, 107–120.

Bodilis, J., Hedde, M., Orange, N. and Barray, S. (2006) OprF polymorphism as a marker of ecological niche in *Pseudomonas*. *Environmental Microbiology* 8, 1544–1551.

Broadbent, P., Baker, K.F. and Waterworth, Y. (1971) Bacteria and actinomycetes antogonistic to fungal root pathogens in Australian soils. *Australian Journal of Biological Sciences* 24, 925–944.

Bulgarelli, D., Garrido-Oter, R. and Mnch, P.C. (2015) Structure and function of the bacterial root microbiota in wild and domesticated barley. *Cell Host & Microbe* 17, 392–403.

Busse, M.D., Ratcliff, A.W., Shestak,C.J. and Powers, R.F. (2001) Glyphosate toxicity and the effects of long-term vegetation control on soil microbial communities. *Soil Biology and Biochemistry* 33, 1777–1789.

Ceja-Navarro, J.A., Rivera-Orduna, F.N., Patino-Zúniga, L., Vila-Sanjurjo, A., Crossa, J. *et al.* (2010) Phylogenetic and multivariate analyses to determine the effects of different tillage and residue management practices on soil bacterial communities. *Applied and Environmental Microbiology* 76, 3685–3691.

Chauhan B.S., Singh R.G. and Mahajan, G. (2012) Ecology and management of weeds under conservation agriculture: a review. *Crop Protection* 38, 57–65. doi: 10.1016/j.cropro.2012.03.010.

Chin-A-Woeng, T.F., van den Broek, D., de Voer, G., van der Drift, K.M., Tuinman, S. *et al.* (2001) Phenazine-1-carboxamide production in the biocontrol strain *Pseudomonas chlororaphis* PCL1391 is regulated by multiple factors secreted into the growth medium. *Molecular Plant-Microbe Interactions* 14, 969–979.

Cho, J.C. and Tiedje, J.M. (2000) Biogeography and degree of endemicity of fluorescent *Pseudomonas* strains in soil. *Applied and Environmental Microbiology* 66, 5448–5456.

Chowdhury, A., Pradhan, S., Saha, M. and Sanyal, N. (2008) Impact of pesticides on soil microbiological parameters and possible bioremediation strategies. *Indian Journal of Microbiology* 48, 114–127.

Christensen, H., Hansen, M. and Sørensen, J. (1999) Counting and size classification of active soil bacteria by fluorescence *in situ* hybridization with an rRNA oligonucleotide probe. *Applied and Environmental Microbiology* 65, 1753–1761.

Cook, R.J. (2003) Take-all of wheat. *Physiological and Molecular Plant Pathology* 62, 73–86.

Cook, R.J. (2006) Toward cropping systems that enhance productivity and sustainability. *Proceedings of the National Academy of Sciences of the United States of America* 103, 18389–18394.

Cookson, W.R., Marschner, P., Clark, I.M., Milton, N., Smirk, M.N. *et al.* (2006). The influence of season, agricultural management, and soil properties on gross nitrogen transformations and bacterial community structure. *Soil Research* 44, 453–465.

Costa, R., Gomes, N.C.M. and Peixoto, R.S. (2006a) Diversity and antagonistic potential of *Pseudomonas* spp. associated to the rhizosphere of maize grown in a subtropical organic farm. *Soil Biology & Biochemistry*, 382434–382447.

Costa, R., Götz, M. and Mrotzek, N. (2006b) Effects of site and plant species on rhizosphere community structure as revealed by molecular analysis of microbial guilds. *FEMS Microbiology Ecology* 56, 236–49.

Costa, R., Salles, J.F., Berg, G. and Smalla, K. (2006c) Cultivation independent analysis of *Pseudomonas* species in soil and in the rhizosphere of field grown *Verticillium dahliae* host plants. *Environmental Microbiology* 8, 2136–2149.

Costa, R., Gomes, N., Krögerrecklenfort, E., Opelt, K., Berg, G. and Smalla, K. (2007) *Pseudomonas* community structure and antagonistic potential in the rhizosphere: insights gained by combining phylogenetic and functional gene-based analyses. *Environmental Microbiology* 9, 2260–2273.

Cycoń, M. and Piotrowska-Seget, Z. (2009) Changes in bacterial diversity and community structure following pesticides addition to soil estimated by cultivation technique. *Ecotoxicology* 18, 632–642.

Cycoń, M., Wójcik, M. and Piotrowska-Seget, Z. (2009) Biodegradation of the organophosphorus insecticide diazinon by *Serratia* sp. and *Pseudomonas* sp. and their use in bioremediation of contaminated soil. *Chemosphere* 76,494–501.

Dai, J., Hu, J., Zhu, A., Bai, J., Wang, J. and Lin, X. (2015) No tillage enhances arbuscular mycorrhizal fungal population, glomalin-related soil protein content, and organic carbon accumulation in soil macroaggregates. *Journal of Soils and Sediments* 15, 1055–1062.

De La Fuente, L., Mavrodi, D.V., Landa, B.B., Thomashow, L.S. and Weller, D.M. (2006) phlD-based genetic diversity and detection of genotypes of 2,4-diacetylphloroglucinol-producing *Pseudomonas fluorescens*. *FEMS Microbiology ecology* 56, 64–78.

Derpsch, R., Friedrich, T., Kassam, A. and Li, H. (2010) Current status of adoption of no-till farming in the world and some of its main benefits. *International Journal of Agricultural and Biological Engineering* 3, 1–25.

Dill, G.M., CaJacob, C.A. and Padgette, S.R. (2008) Glyphosate resistant crops: adoption, use and future considerations. *Pest Management Science* 64, 326–331.

Ding, G.C., Piceno, Y.M., Heuer, H., Weinert, N., Dohrmann, A.B. *et al.* (2013) Changes of soil bacterial diversity as a consequence of agricultural land use in a semi-arid ecosystem. *Public Library of Science* 8, 59497.

Druille, M., Cabello, M.N., Parisi, P.G., Golluscio, R.A. and Omacini, M. (2015) Glyphosate vulnerability explains changes in root-symbionts propagules viability in pampean grasslands. *Agriculture, Ecosystems & Environment* 202, 48–55.

Druille, M., García-Parisi, P.A., Golluscio, R.A., Cavagnaro, F.P. and Omacini, M. (2016) Repeated annual glyphosate applications may impair beneficial soil microorganisms in temperate grassland. *Agriculture, Ecosystems & Environment* 230, 184–190.

Dudenhöffer, J.H., Scheu, S. and Jousset, A. (2016) Systemic enrichment of antifungal traits in the rhizosphere microbiome after pathogen attack. *Journal of Ecology* 104, 1566–1575.

Duijff, B.J., Bakker, P.A. and Schippers, B. (1994) Suppression of fusarium wilt of carnation by *Pseudomonas putida* WCS358 at different levels of disease incidence and iron availability. *Biocontrol Science and Technology* 4, 279–288.

Duijff, B.J., Pouhair, D., Olivain, C., Alabouvette, C. and Lemanceau, P. (1998) Implication of systemic induced resistance in the suppression of fusarium wilt of tomato by *Pseudomonas fluorescens* WCS417r and by nonpathogenic *Fusarium oxysporum* Fo47. *European Journal of Plant Pathology* 104, 903–910.

Edwards, M.L., Lilley, A.K., Timms-Wilson, T.H., Thompson, I.P. and Cooper, I. (2001) Characterisation of the culturable heterotrophic bacterial community in a small eutrophic lake (Priest Pot). *FEMS Microbiology Ecology* 35, 295–304.

Eickhorst, T. and Tippkötter, R. (2008) Detection of microorganisms in undisturbed soil by combining fluorescence *in situ* hybridization (FISH) and micropedological methods. *Soil Biology and Biochemistry* 40, 1284–1293.

El Fantroussi, S., Verschuere, L., Verstraete, W. and Top, E. (1999) Effect of phenylurea herbicides on soil microbial communities estimated by analysis of 16S rRNA gene fingerprints and community-level physiological profiles. *Applied and Environmental Microbiology* 65, 982–988.

FAO (2003) Development of a Framework for Good Agricultural Practices. In: *Organization of the United Nations – Comm. Agric. – Seventeenth Session*. Available at: http://www.fao.org/prods/GAP/index_en.htm (accessed 7 July 2017).

Fernández, L., Agaras, B., Zalba, P., Wall, L.G. and Valverde, C. (2012) *Pseudomonas* spp. isolates with high phosphate-mobilizing potential and root colonization properties from agricultural bulk soils under no-till management. *Biology and Fertility of Soils* 48, 763–773.

Ferrari, A.E., Ravnskov, S., Larsen, J., Tønnersen, T., Maronna, R.A. and Wall, L.G. (2015) Crop rotation and seasonal effects on fatty acid profiles of neutral and phospholipids extracted from no-till agricultural soils. *Soil Use and Management* 31, 165–175.

Fierer, N., Bradford, M.A. and Jackson, R.B. (2007) Toward an ecological classification of soil bacteria. *Ecology* 88, 1354–1364.

Figuerola, E.L.M., Guerrero, L.D., Rosa, S.M., Simonetti, L., Duval, M.E. *et al.* (2012) Bacterial indicator of agricultural management for soil under no-till crop production. *PLoS One* 7, p. e51075.

Figuerola, E.L., Guerrero, L.D., Türkowsky, D., Wall, L.G. and Erijman, L. (2015) Crop monoculture rather than agriculture reduces the spatial turnover of soil bacterial communities at a regional scale. *Environmental Microbiology* 17, 678–688.

Fischer, S.E., Jofré, E.C., Cordero, P.V., Manero, F.J.G. and Mori, G.B. (2010) Survival of native *Pseudomonas* in soil and wheat rhizosphere and antagonist activity against plant pathogenic fungi. *Antonie van Leeuwenhoek* 97, 241–251.

Frapolli, M., Moënne-Loccoz, Y., Meyer, J. and Défago, G. (2008) A new DGGE protocol targeting 2, 4-diacetylphloroglucinol biosynthetic gene phlD from phylogenetically contrasted biocontrol pseudomonads for assessment of disease-suppressive soils. *FEMS Microbiology Ecology* 64, 468–481.

Freedman, B. (1990) Controversy over the use of herbicides in forestry, with particular reference to glyphosate usage. *Journal of Environmental Science & Health Part C* 8, 277–286.

Galantini. J.A., Duval, M., Iglesias, J. and Martinez, J. (2013) Efectos de largo plazo de la siembra directa en el SO Bonaerense: Dinámica y balance de carbono.

Garbeva, P., Van Veen, J.A. and Van Elsas, J.D. (2004a) Microbial diversity in soil: selection of microbial populations by plant and soil type and implications for disease suppressiveness. *Annual Review of Phytopathology* 42, 243–270.

Garbeva, P., van Veen, J.A. and van Elsas, J.D. (2004b) Assessment of the diversity, and antagonism towards Rhizoctonia solani AG3, of *Pseudomonas* species in soil from different agricultural regimes. *FEMS Microbiology Ecology* 47, 51–64.

Garbeva, P., Voesenek, K. and Van Elsas, J.D. (2004c) Quantitative detection and diversity of the pyrrolnitrin biosynthetic locus in soil under different treatments. *Soil Biology and Biochemistry* 36, 1453–1463.

Gasser, R. and Défago, G. (1981) Mise en évidence de la résistance de certaines terres à la pourriture noire des racines du tabac causée par le *Thielaviopsis basicola*. *Berichte der Schweizerischen Botanischen Gesellschaft Bulletin de la Societe Botanique Suisse* 75–80.

Gimsing, A.L., Borggaard, O.K., Jacobsen, O.S., Aamand, J. and Sørensen, J. (2004) Chemical and microbiological soil characteristics controlling glyphosate mineralisation in Danish surface soils. *Applied Soil Ecology* 27, 233–242.

Girvan, M.S., Bullimore, J., Pretty, J.N., Osborn, A.M. and Ball, A.S. (2003) Soil type is the primary determinant of the composition of the total and active bacterial communities in arable soils. *Applied and Environmental Microbiology* 69, 1800–1809.

Gómez, P., Paterson, S., De Meester, L., Liu, X., Lenzi, L. *et al.* (2016) Local adaptation of a bacterium is as important as its presence in structuring a natural microbial community. *Nature Communications* 7.

Govaerts, B., Mezzalama, M., Unno, Y., Sayre, K.D., Luna-Guido, M. *et al.* (2007) Influence of tillage, residue management, and crop rotation on soil microbial biomass and catabolic diversity. *Applied Soil Ecology* 37, 18–30.

Govaerts, B., Mezzalama, M., Sayre, K.D., Crossa, J., Lichter, K. *et al.* (2008) Long-term consequences of tillage, residue management, and crop rotation on selected soil micro-flora groups in the subtropical highlands. *Applied Soil Ecology* 38, 197–210.

Granér, G., Persson, P., Meijer, J. and Alström, S. (2003) A study on microbial diversity in different cultivars of *Brassica napus* in relation to its wilt pathogen, *Verticillium longisporum*. *FEMS Microbiology Letters* 224, 269–276.

Green, V.S., Stott, D.E., Cruz, J.C. and Curi, N. (2007) Tillage impacts on soil biological activity and aggregation in a Brazilian Cerrado Oxisol. *Soil and Tillage Research* 92, 114–121.

Gyamfi, S., Pfeifer, U., Stierschneider, M. and Sessitsch, A. (2002) Effects of transgenic glufosinate-tolerant oilseed rape (*Brassica napus*) and the associated herbicide application on eubacterial and *Pseudomonas* communities in the rhizosphere. *FEMS Microbiology Ecology* 41, 181–190.

Haas, D. and Défago, G. (2005) Biological control of soil-borne pathogens by fluorescent pseudomonads. *Nature Reviews Microbiology* 3, 307–319.

Haichar, F.E.Z., Marol, C., Berge, O., Rangel-Castro, J.I., Prosser, J.I. *et al.* (2008) Plant host habitat and root exudates shape soil bacterial community structure. *The ISME Journal* 2, 1221–1230.

Haney, R.L., Senseman, S.A. and Hons, F.M. (2002) Effect of Roundup Ultra on microbial activity and biomass from selected soils. *Journal of Environmental Quality* 31, 730–735.

Helander, M., Saloniemi, I. and Saikkonen, K. (2012) Glyphosate in northern ecosystems. *Trends in Plant Science* 17, 569–574.

Hesselsøe, M., Brandt, K.K. and Sørensen, J. (2001) Quantification of ammonia oxidizing bacteria in soil using microcolony technique combined with fluorescence *in situ* hybridization (MCFU–FISH). *FEMS Microbiology Ecology* 38, 87–95.

Hjort, K., Lembke, A., Speksnijder, A., Smalla, K. and Jansson, J.K. (2007) Community structure of actively growing bacterial populations in plant pathogen suppressive soil. *Microbial Ecology* 53, 399–413.

Hjort, K., Bergström, M., Adesina, M.F., Jansson, J.K., Smalla, K. and Sjöling, S. (2010) Chitinase genes revealed and compared in bacterial isolates, DNA extracts and a metagenomic library from a phytopathogen-suppressive soil. *FEMS Microbiology Ecology* 71, 197–207.

Höflich, G., Tauschke, M., Kühn, G., Werner, K., Frielinghaus, M. and Höhn, W. (1999) Influence of long-term conservation tillage on soil and rhizosphere microorganisms. *Biology and Fertility of Soils* 29, 81–86.

Hornby, D. (1983) Suppressive soils. *Annual Review Phytopathology* 21, 65–85.

Howell, C.R. and Stipanovic, R.D. (1979) Control of *Rhizoctonia solani* on cotton seedlings with *Pseudomonas fluorescens* and with an antibiotic produced by the bacterium by the soil tube method described previously. *Phytopathology* 69, 480–482.

Howell, C. and Stipanovic, R.D. (1980) Suppression of *Pythium ultimium* induced damping off of cotton seedlings by *Pseudomonas fluorescens* and its antibiotic pyoluterin. *Phytopathology* 70, 712–715.

INDEC (2013) Temperaturas medias mensuales por estación meteorológica. Año 2012. In: *INDEC – Geogr.* Available at: http://www.indec.gob.ar/nivel4_default.asp?id_tema_1=1&id_tema_2=15&id_tema_3=25 (accessed 8 July 2017).

Jacob, G.S., Garbow, J.R., Hallas, L.E., Kimack, N.M., Kishore, G.M. and Schaefer, J. (1988) Metabolism of glyphosate in *Pseudomonas* sp. strain LBr. *Applied and Environmental Microbiology* 54, 2953–2958.

Jariyal, M., Gupta, V.K., Mandal, K., Jindal, V., Banta, G. and Singh, B. (2014) Isolation and characterization of novel phorate-degrading bacterial species from agricultural soil. *Environmental Science and Pollution Research* 21, 2214–2222.

Jariyal, M., Gupta, V.K., Jindal, V. and Mandal, K. (2015) Isolation and evaluation of potent *Pseudomonas* species for bioremediation of phorate in amended soil. *Ecotoxicology and Environmental Safety* 122, 24–30.

Johnsen, K. and Nielsen, P. (1999) Diversity of *Pseudomonas* strains isolated with King's B and Gould's S1 agar determined by repetitive extragenic palindromic-polymerase chain reaction, 16S rDNA sequencing and Fourier transform infrared spectroscopy characterisation. *FEMS Microbiology Letters* 173, 155–162.

Johnsen, K., Enger, Ø., Jacobsen, C.S., Thirup, L. and Torsvik, V. (1999) Quantitative selective PCR of 16S ribosomal DNA correlates well with selective agar plating in describing population dynamics of indigenous *Pseudomonas* spp. in soil hot spots. *Applied and Environmental Microbiology* 65, 1786–1788.

Kamanavalli, C.M. and Ninnekar, H.Z. (2004) Biodegradation of DDT by a *Pseudomonas* species. *Current Microbiology* 48, 10–13.

Kannan, K., Tanabe, S., Giesy, J.P. and Tatsukawa, R. (1997) Organochlorine pesticides and polychlorinated biphenyls in foodstuffs from Asian and oceanic countries. In: *Reviews of Environmental Contamination and toxicology*. Springer, New York, 1–55.

Karpouzas, D.G. and Walker, A. (2000) Factors influencing the ability of *Pseudomonas putida* epl to degrade ethoprophos in soil. *Soil Biology and Biochemistry* 32, 1753–1762.

Keel, C., Schnider, U. and Maurhofer, M. (1992) Suppression of root disease by *Pseudomonas fluorescens* CHA0: Importance of the bacterial secondary metabolite 2,4-diacetylphloroglucinol. *Molecular Plant-Microbe Interactions* 5, 4–13.

Keen, B. and Vancov, T. (2010) *Phytophthora cinnamomi* suppressive soils. In: Méndez-Vilas, A. (ed.) *Current Research, Technology and Education Topics in Applied Microbiology and Microbial Biotechnology*. Formatex Research Center, Badajoz, Spain, pp 239–250.

Kerry, B.R., Crump, D.H. and Mullen, L.A. (1982) Natural control of the cereal cyst nematode, *Heterodera avenae* Woll., by soil fungi at three sites. *Crop Protection* 1, 99–109.

Kloepper, J.W., Leong, J., Teintze, M. and Schroth, M.N. (1980) *Pseudomonas* siderophores: a mechanism explaining disease-suppressive soils. *Current Microbiology* 4, 317–320.

Kluepfel, D.A., McInnis, T.M. and Zehr, E.I. (1993) Involvement of root-colonizing bacteria in peach orchard soils suppressive of the nematode *Criconemella xenoplax*. *Phytopathology* 83, 1240–1245.

Kuklinsky-Sobral, J., Araújo, W.L., Mendes, R., Geraldi, I.O., Pizzirani-Kleiner, A.A. and Azevedo, J.L. (2004) Isolation and characterization of soybean-associated bacteria and their potential for plant growth promotion. *Environmental Microbiology* 6, 1244–1251.

Kuklinsky-Sobral, J., Araujo, W.L., Mendes, R., Pizzirani-Kleiner, A.A. and Azevedo, J.L. (2005) Isolation and characterization of endophytic bacteria from soybean (*Glycine max*) grown in soil treated with glyphosate herbicide. *Plant and Soil* 273, 91–99.

Kumar, M., Chaudhary, P., Dwivedi, M., Kumar, R., Paul, D. *et al.* (2005) Enhanced biodegradation of β-and δ-hexachlorocyclohexane in the presence of α-and γ-isomers in contaminated soils. *Environmental Science & Technology* 39, 4005–4011.

Kumar, V.V. (2016) Plant growth-promoting microorganisms: interaction with plants and soil. In: Hakeem, K.R., Akhtar, M.S. and Abdullah S.N.A. (eds) *Plant, Soil and Microbes. Vol. 1: Implications in Crop Science*. Springer International Publishing, Cham, Switzerland, pp.1–16.

Kuramae, E.E., Yergeau, E., Wong, L.C., Pijl, A.S., van Veen, J.A. and Kowalchuk, G.A. (2012) Soil characteristics more strongly influence soil bacterial communities than land-use type. *FEMS Microbiology Ecology* 79, 12–24.

Lakshmanan, V. (2015) Root microbiome assemblage is modulated by plant host factors. *Advances in Botanical Research* 75, 57–79.

Lancaster, S.H., Hollister, E.B., Senseman, S.A. and Gentry, T.J. (2010) Effects of repeated glyphosate applications on soil microbial community composition and the mineralization of glyphosate. *Pest Management Science* 66, 59–64.

Landa, B.B., Mavrodi, O.V., Schroeder, K.L., Allende-Molar, R. and Weller, D.M. (2006) Enrichment and genotypic diversity of phlD-containing fluorescent *Pseudomonas* spp. in two soils after a century of wheat and flax monoculture. *FEMS Microbiology Ecology* 55, 351–368.

Latour, X., Corberand, T., Laguerre, G., Allard, F. and Lemanceau, P. (1996) The composition of fluorescent pseudomonad populations associated with roots is influenced by plant and soil type. *Applied and Environmental Microbiology* 62, 2449–2456.

Latz, E., Eisenhauer, N., Rall, B.C., Allan, E., Roscher, C. *et al.* (2012) Plant diversity improves protection against soil-borne pathogens by fostering antagonistic bacterial communities. *Journal of Ecology* 100, 597–604.

Latz, E., Eisenhauer, N., Scheu, S. and Jousset, A. (2015) Plant identity drives the expression of biocontrol factors in a rhizosphere bacterium across a plant diversity gradient. *Functional Ecology* 29, 1225–1234.

Lemanceau, P. and Alabouvette, C. (1993) Suppression of *Fusarium* wilts by fluorescent pseudomonads: mechanisms and applications. *Biocontrol Science and Technology* 3, 219–234.

Leoni Velazco, C. (2013) Crop rotation design in view of soilborne pathogen dynamics. A methodological approach illustrated with *Sclerotium rolfsii* and *Fusarium oxysporum* f.sp. cepae. PhD Thesis, Wageningen University, Wageningen, Netherlands.

Li, L., Al-Soud, W.A., Bergmark, L., Riber, L., Hansen, L.H. *et al.* (2013) Investigating the diversity of *Pseudomonas spp.* in soil using culture dependent and independent techniques. *Current Microbiology* 67, 423–430.

Lin, B.B. (2011) Resilience in agriculture through crop diversification: adaptive management for environmental change. *BioScience* 61, 183–193.

Liu, D., Anderson, N.A. and Kinkel, L.L. (1995) Biological control of potato scab in the field with antagonistic *Streptomyces scabies*. *Phytopathology* 85, 827–831.

López-Piñeiro, A., Muñoz, A., Zamora, E. and Ramírez, M. (2013) Influence of the management regime and phenological state of the vines on the physicochemical properties and the seasonal fluctuations of the microorganisms in a vineyard soil under semi-arid conditions. *Soil and Tillage Research* 126, 119–126.

Lorch, M., Agaras B.C., Omacini, M. *et al.* (2016) La aplicación reiterada del herbicida glifosato reduce la abundancia de Pseudomonas del suelo en un pastizal natural de la pampa deprimida. In: *XXIII Congreso ALAM XIV Congreso AAM*.

Lugtenberg, B.J.J. and Bloemberg, G.V. (2004) Life in the rhizosphere. In: Ramos, J.L. (ed.) *Pseudomonas: Genomics, Life Style and Molecular Architecture*. Kluwer Academic/Plenum Publishers, New York, pp. 403–430.

Lugtenberg, B.J. and Dekkers, L.C. (1999) What makes *Pseudomonas* bacteria rhizosphere competent. *Environment Microbiology* 1, 9–13.

Lugtenberg, B. and Kamilova, F. (2009) Plant-growth-promoting rhizobacteria. *Annual Review Microbiology* 63, 541–56.

Lumsden, R.D., García, R., Lewis, J. and Frías, G. (1987) Suppression of damping-off caused by *Pythium spp.* in soil from the indigenous mexican Chinampa agricultural system. *Soil Biology Biochemistry* 19, 501–508.

Lynch, J.M. and Whipps, J.M. (1990) Substrate flow in the rhizosphere. *Plant Soil* 129, 1–10.

Marques, J.M., da Silva, T.F., Vollu, R.E., Blank, A.F., Ding, G.C. *et al.* (2014) Plant age and genotype affect the bacterial community composition in the tuber rhizosphere of field-grown sweet potato plants. *FEMS Microbiology Ecology* 88, 424–435.

Marrero, M.A., Agaras, B., Wall, L.G. and Valverde, C. (2015) Enriquecimiento diferencial de *Pseudomonas spp.* en el rizoplano de distintas especies cultivadas. *Revista Argentina De Microbiología* 47, 32–137.

Mazurier, S., Corberand, T., Lemanceau, P. and Raaijmakers, J.M. (2009) Phenazine antibiotics produced by fluorescent pseudomonads contribute to natural soil suppressiveness to *Fusarium* wilt. *The ISME Journal* 3, 977–991.

Mazzola, M. (1998) Elucidation of the microbial complex having a causal role in the development of apple replant disease in Washington. *Phytopathology* 88, 930–938.

Mazzola, M. (1999) Control of replant disease of tree fruits with *Pseudomonas putida*. *U.S. Patent* 5, 948, 671.

Mazzola, M. (2002) Mechanisms of natural soil suppressiveness to soilborne diseases. *Antonie Van Leeuwenhoek* 81, 557–564.

Mazzola, M. (2007) Manipulation of rhizosphere bacterial communities to induce suppressive soils. *Journal of Nematology* 39, 213–220.

Mazzola, M. and Gu, Y-H. (2002) Wheat genotype-specific induction of soil microbial communities suppressive to disease incited by *Rhizoctonia solani* Anastomosis Group (AG)-5 and AG-8. *Phytopathology* 92, 1300–1307.

Mazzola, M., Funnell, D.L. and Raaijmakers, J.M. (2004) Wheat cultivar-specific selection of 2,4-diacetylphloroglucinol-producing fluorescent *Pseudomonas* species from resident soil populations. *Microbial Ecology* 48, 338–348.

McSpadden Gardener, B.B., Gutierrez, L.J. and Joshi, R. (2005) Distribution and biocontrol potential of *phlD*(+) *Pseudomonads* in corn and soybean fields. *Phytopathology* 95, 715–24.

Mendes, L.W., Kuramae, E.E. and Navarrete, A. (2014) Taxonomical and functional microbial community selection in soybean rhizosphere. *The ISME Journal* 8, 1577–1587.

Mendes, R., Kruijt, M. and de Bruijn, I. (2011) Deciphering the rhizosphere microbiome for disease-suppressive bacteria. *Science* 332, 1097–1100.

Meng, Q., Yin, J. and Rosenzweig, N. (2012) Culture-based assessment of microbial communities in soil suppressive to potato common scab. *Plant Disase* 96, 712–717.

Meriles, J.M., Gil, S.V. and Conforto, C. (2009) Soil microbial communities under different soybean cropping systems: characterization of microbial population dynamics, soil microbial activity, microbial biomass, and fatty acid profiles. *Soil Tillage Research* 103, 271–281.

Meyer, J.R. and Shew, H.D. (1991) Soils suppressive to black root rot of burley tobacco, caused by *Thielaviopsis basicola*. *Phytopathology* 81, 946–954.

Michelsen, C.F., Watrous, J., Glaring, M.A., Kersten, R., Koyama, N. *et al.* (2015) Nonribosomal peptides, key biocontrol components for *Pseudomonas fluorescens* In5, isolated from a Greenlandic suppressive soil. *MBio* 6, 1–9.

Mijangos, I., Becerril, J.M., Albizu, I., Epelde, L. and Garbisu, C. (2009) Effects of glyphosate on rhizosphere soil microbial communities under two different plant compositions by cultivation-dependent and-independent methodologies. *Soil Biology and Biochemistry* 41, 505–513.

Milling, A., Smalla, K., Maidl, F.X., Schloter, M. and Munch, J.C. (2005) Effects of transgenic potatoes with an altered starch composition on the diversity of soil and rhizosphere bacteria and fungi. *Plant and Soil* 266, 23–39.

Mishra, S., Singh, A., Keswani, C., Saxena, A., Sarma, B.K. and Singh, H.B. (2015) Harnessing plant-microbe interactions for enhanced protection against phytopathogens. In: Arora, N.K. (ed.) *Plant Microbes Symbiosis: Applied Facets*. Springer, New Delhi, India, pp. 111–125.

Mittal, S. and Johri, B.N. (2008) Influence of management practices on the diversity of pseudomonads in rhizosphere soil of wheat cropping system. *Biology and Fertility of Soils* 44, 823–831.

Moffett, M.L., Giles, J.E. and Wood, B.A. (1983) Survival of *Pseudomonas solanacearum* biovars 2 and 3 in soil: effect of moisture and soil type. *Soil Biology and Biochemistry* 15, 587–591.

Montecchia, M.S., Correa, O.S., Soria, M.A., Frey, S.D., García, A.F. and Garland, J.L. (2011) Multivariate approach to characterizing soil microbial communities in pristine and agricultural sites in Northwest Argentina. *Applied Soil Ecology* 47, 176–183.

Mühling, M., Woolven-Allen, J., Murrell, J.C. and Joint, I. (2008) Improved group-specific PCR primers for denaturing gradient gel electrophoresis analysis of the genetic diversity of complex microbial communities. *The ISME Journal* 2, 379–392.

Mulbry, W. and Kearney, P.C. (1991) Degradation of pesticides by micro-organisms and the potential for genetic manipulation. *Crop Protection* 10, 334–346.

Murakami, H., Tsushima, S. and Shishido, Y. (2000) Soil suppressiveness to clubroot disease of Chinese cabbage caused by *Plasmodiophora brassicae*. *Soil Biology and Biochemistry* 32, 1637–1642.

Murthy, H.R. and Manonmani, H.K. (2007) Aerobic degradation of technical hexachlorocyclohexane by a defined microbial consortium. *Journal of Hazardous Materials* 149, 18–25.

Nawab, A., Aleem, A. and Malik, A. (2003) Determination of organochlorine pesticides in agricultural soil with special reference to γ-HCH degradation by *Pseudomonas* strains. *Bioresource Technology* 88, 41–46.

Nesci, A., Barros, G., Castillo, C. and Etcheverry, M. (2006) Soil fungal population in preharvest maize ecosystem in different tillage practices in Argentina. *Soil and Tillage Research* 91, 143–149.

Newman, M.M., Hoilett, N., Lorenz, N., Dick, R.P., Liles, M.R. *et al.* (2016) Glyphosate effects on soil rhizosphere-associated bacterial communities. *Science of the Total Environment* 543, 155–160.

Nguyen, C. (2003) Rhizodeposition of organic C by plants: mechanisms and controls. *Agronomie* 23, 375–396.

Nguyen, D.B., Rose, M.T., Rose, T.J., Morris, S.G. and Van Zwieten, L. (2016) Impact of glyphosate on soil microbial biomass and respiration: a meta-analysis. *Soil Biology and Biochemistry* 92, 50–57.

O'Callaghan, M., Gerard, E.M. and Johnson, V.W. (2001) Effect of soil moisture and temperature on survival of microbial control agents. In: *Proceedings of The New Zealand Plant Protection Conference*, 128–135.

Olawale, A.K. and Akintobi, O.A. (2011) Biodegradation of glyphosate pesticide by bacteria isolated from agricultural soil. *Report and Opinion* 3, 124–128.

Palleroni N.J. and Moore, E. (2004) Taxonomy of pseudomonads: experimental approaches. In: Ramos, J.L. (ed.) *Pseudomonas: Genomics, Life Style and Molecular Architecture*. Kluwer Academic, New York, pp. 3–46.

Paterson, E., Gebbing, T., Abel, C., Sim, A. and Telfer, G. (2007) Rhizodeposition shapes rhizosphere microbial community structure in organic soil. *New Phytologist* 173, 600–610.

Peixoto, R.S., Coutinho, H.L.C., Madari, B., Machado, P.L.O.A., Rumjanek, N.G. *et al.* (2006) Soil aggregation and bacterial community structure as affected by tillage and cover cropping in the Brazilian Cerrados. *Soil and Tillage Research* 90, 16–28.

Peñaloza-Vazquez, A., Mena, G.L., Herrera-Estrella, L. and Bailey, A.M. (1995) Cloning and sequencing of the genes involved in glyphosate utilization by *Pseudomonas pseudomallei*. *Applied and Environmental Microbiology* 61, 538–543.

Perez-Brandán, C., Arzeno, J.L., Huidobro, J., Grümberg, B., Conforto, C. *et al.* (2012) Long-term effect of tillage systems on soil microbiological, chemical and physical parameters and the incidence of charcoal rot by *Macrophomina phaseolina* (Tassi) Goid in soybean. *Crop Protection* 40, 73–82.

Peters, R.D., Sturz, A.V., Carter, M.R. and Sanderson, J.B. (2003) Developing disease-suppressive soils through crop rotation and tillage management practices. *Soil and Tillage Research* 72, 181–192.

Philippot, L., Raaijmakers, J.M., Lemanceau, P. and van der Putten, W.H. (2013) Going back to the roots: the microbial ecology of the rhizosphere. *Nature Reviews Microbiology* 11, 789–799.

Picard, C. and Bosco, M. (2008) Genotypic and phenotypic diversity in populations of plant-probiotic *Pseudomonas spp.* colonizing roots. *Naturwissenschaften* 95, 1–16.

Picard, C., Baruffa, E. and Bosco, M. (2008) Enrichment and diversity of plant-probiotic microorganisms in the rhizosphere of hybrid maize during four growth cycles. *Soil Biology and Biochemistry* 40, 106–115.

Powles, S.B. (2008) Evolved glyphosate-resistant weeds around the world: lessons to be learnt. *Pest Management Science* 64, 360–365.

Prevost-Boure, N.C., Maron, P.A., Ranjard, L., Nowak, V., Dufrene, E. *et al.* (2011) Seasonal dynamics of the bacterial community in forest soils under different quantities of leaf litter. *Applied Soil Ecology* 47, 14–23.

Raaijmakers, J.M. and Weller, D.M. (2001) Exploiting genotypic diversity of 2,4-diacetylphloroglucinol-producing *Pseudomonas* spp.: characterization of superior root-colonizing *P. fluorescens* Strain Q8r1-96. *Applied and Environmental Microbiology* 67, 2545–2554.

Raaijmakers, J.M., Bonsall, R.F. and Weller, D.M. (1999) Effect of population density of *Pseudomonas fluorescens* on production of 2, 4-diacetylphloroglucinol in the rhizosphere of wheat. *Phytopathology* 89, 470–475.

Raaijmakers, J.M., Paulitz, T.C., Steinberg, C., Alabouvette, C. and Moënne-Loccoz, Y. (2009) The rhizosphere: a playground and battlefield for soilborne pathogens and beneficial microorganisms. *Plant and Soil* 321, 341–361.

Rasche, F., Knapp, D., Kaiser, C., Koranda, M., Kitzler, B. *et al.* (2011) Seasonality and resource availability control bacterial and archaeal communities in soils of a temperate beech forest. *The ISME Journal* 5, 389–402.

Ratcliff, A.W., Busse, M.D. and Shestak, C.J. (2006) Changes in microbial community structure following herbicide (glyphosate) additions to forest soils. *Applied Soil Ecology* 34, 114–124.

Rezzonico, F., Binder, C., Défago, G. and Moënne-Loccoz, Y. (2005) The type III secretion system of biocontrol *Pseudomonas fluorescens* KD targets the phytopathogenic *Chromista pythium ultimum* and promotes cucumber protection. *Molecular Plant-Microbe Interactions* 19, 991–1001.

Rezzonico, F., Zala, M., Keel, C., Duffy, B., Moënne-Loccoz, Y. and Défago, G. (2007) Is the ability of biocontrol fluorescent pseudomonads to produce the antifungal metabolite 2,4-diacetylphloroglucinol really synonymous with higher plant protection. *New Phytologist* 173, 861–872.

Richardson, R.E., Bhupathiraju, V.K., Song, D.L., Goulet, T.A. and Alvarez-Cohen, L. (2002) Phylogenetic characterization of microbial communities that reductively dechlorinate TCE based upon a combination of molecular techniques. *Environmental Science & Technology* 36, 2652–2662.

Rodriguez, A.M. and Jacobo, E.J. (2013) Glyphosate effects on seed bank and vegetation composition of temperate grasslands. *Applied Vegetation Science* 16, 51–62.

Rosa, S.M., Kraemer, F.B., Soria, M.A., Guerrero, L.D., Morrás, H.J. *et al.* (2014) The influence of soil properties on denitrifying bacterial communities and denitrification potential in no-till production farms under contrasting management in the Argentinean Pampas. *Applied Soil Ecology* 75, 172–180.

Rosenzweig, N., Tiedje, J.M., Quensen III, J.F., Meng, Q. and Hao, J.J. (2012) Microbial communities associated with potato common scab-suppressive soil determined by pyrosequencing analyses. *Plant Disease* 96, 718–725.

Ruiz-Toledo, J. and Sánchez-Guillén, D. (2014) Effect of the concentration of glyphosate present in body waters near transgenic soybean fields on the honeybee *Apis mellifera*, and the stingless bee *Tetragonisca angustula*. *Acta Zoológica Mexicana*, 30, 408–413.

Sahu, S.K., Patnaik, K.K. and Sethunathan, N. (1992) Dehydrochlorination of δ-isomer of hexachlorocyclohexane by a soil bacterium, *Pseudomonas* sp.. *Bulletin of Environmental Contamination and Toxicology* 48, 265–268.

San Miguel, C., Dulinski, M. and Tate, R.L. (2007) Direct comparison of individual substrate utilization from a CLPP study: a new analysis for metabolic diversity data. *Soil Biology and Biochemistry* 39, 1870–1877.

Sarniguet, A., Lucas, P. and Lucas, M. (1992) Relationships between take-all, soil conduciveness to the disease, populations of fluorescent pseudomonads and nitrogen fertilizers. *Plant and Soil* 145, 17–27.

Schafer, J.R., Hallett, S.G. and Johnson, W.G. (2014) Rhizosphere microbial community dynamics in glyphosate-treated susceptible and resistant biotypes of giant ragweed (*Ambrosia trifida*). *Weed Science* 62, 370–381.

Selvapandiyan, A. and Bhatnagar, R.K. (1994) Isolation of a glyphosate-metabolising *Pseudomonas*: detection, partial purification and localisation of carbon-phosphorus lyase. *Applied Microbiology And Biotechnology* 40, 876–882.

Silby, M.W., Winstanley, C., Godfrey, S.A., Levy, S.B. and Jackson, R.W. (2011) *Pseudomonas* genomes: diverse and adaptable. *FEMS Microbiology Reviews* 35, 652–680.

Smalla, K., Wieland, G., Buchner, A., Zock, A., Parzy, J. *et al.* (2001) Bulk and rhizosphere soil bacterial communities studied by denaturing gradient gel electrophoresis: plant-dependent enrichment and seasonal shifts revealed. *Applied and Environmental Microbiology* 67, 4742–4751.

Stanier, R.Y., Palleroni, N.J. and Doudoroff, M. (1966) The aerobic pseudomonads a taxonomic study. *Journal of General Microbiology* 43, 159–271.

Stirling, A.M., Hayward, A.C. and Pegg, K.G. (1992) Evaluation of the biological control potential of bacteria isolated from a soil suppressive to *Phytophthora cinnamomi*. *Australasian Plant Pathology* 21, 133–142.

Stirling, G.R., McKenry, M.V. and Mankau, R. (1979) Biological control of root-knot nematodes (*Meloidogyne spp.*) on peach. *Phytopathology*.

Stutz, E.W., Défago, G. and Kern, H. (1986) Naturally occurring fluorescent pseudomonads involved in suppression of black root rot of tobacco. *Phytopathology* 76, 181–185.

Tapadar, S.A. and Jha, D.K. (2013) Disease management in staple crops: a bacteriological approach. In: Maheshwari, D.K. (ed.) *Bacteria in Agrobiology: Disease Management*. Springer Berlin Heidelberg, Germany, pp. 111–152

Thirup, L., Ekelund, F., Johnsen, K. and Jacobsen, C.S. (2000) Population dynamics of the fast-growing sub-populations of *Pseudomonas* and total bacteria, and their protozoan grazers, revealed by fenpropimorph treatment. *Soil Biology and Biochemistry* 32, 1615–1623.

Thirup, L., Johnsen, K., Torsvik, V., Spliid, N.H. and Jacobsen, C.S. (2001) Effects of fenpropimorph on bacteria and fungi during decomposition of barley roots. *Soil Biology and Biochemistry* 33, 1517–1524.

Thom, E.R., Wildermoth, D.D. and Taylor, M.J. (1993) Growth and persistence of perennial ryegrass and white clover direct-drilled into a paspalum-dominant dairy pasture treated with glyphosate. *New Zealand Journal of Agricultural Research* 36, 197–207.

Thomashow, L.S. and Weller, D.M. (1988) Role of a phenazine antibiotic from *Pseudomonas fluorescens* in biological control of *Gaeumannomyces graminis* var. *tritici*. *Journal of Bacteriology* 170, 3499–3508.

Travaglia, C., Masciarelli, O., Fortuna, J., Marchetti, G., Cardozo, P. *et al.* (2015) Towards sustainable maize production: Glyphosate detoxification by *Azospirillum* sp. and *Pseudomonas* sp. *Crop Protection* 77, 102–109.

Voisard, C., Keel, C., Haas, D. and Dèfago, G. (1989) Cyanide production by *Pseudomonas fluorescens* helps suppress black root rot of tobacco under gnotobiotic conditions. *The EMBO Journal* 8, 351.

Von Felten, A., Défago, G. and Maurhofer, M. (2010) Quantification of *Pseudomonas fluorescens* strains F113, CHA0 and Pf153 in the rhizosphere of maize by strain-specific real-time PCR unaffected by the variability of DNA extraction efficiency. *Journal of Microbiological Methods* 81, 108–115.

Wagner, M., Erhart, R., Manz, W., Amann, R., Lemmer, H. *et al.* (1994) Development of an rRNA-targeted oligonucleotide probe specific for the genus *Acinetobacter* and its application for *in situ* monitoring in activated sludge. *Applied and Environmental Microbiology* 60, 792–800.

Wall, L.G. (2011) The BIOSPAS consortium: Soil biology and agricultural production. In: de Brujin, F. (ed.) *Handbook of Molecular Microbial Ecology I: Metagenomics and Complementary Approaches*. John Wiley & Sons, Chichester, UK, pp 299–306.

Warembourg, F.R. (1997) The "Rhizosphere Effect": a plant strategy for plants to exploit and colonize nutrient-limited habitats. *Bocconea* 7, 187–193.

Weaver, M.A., Krutz, L.J., Zablotowicz, R.M. and Reddy, K.N. (2007) Effects of glyphosate on soil microbial communities and its mineralization in a Mississippi soil. *Pest Management Science* 63, 388–393.

Wechter, W.P., Glandorf, D.C.M., Derrick, W.C., Leverentz, B. and Kluepfel, D.A. (2001) Identification of genetic loci in a rhizosphere-inhabiting, species of *Pseudomonas* involved in expression of a phytoparasitic nematode ovicidal factor. *Soil Biology and Biochemistry* 33, 1749–1758.

Wechter, W.P., Begum, D., Presting, G., Kim, J.J., Wing, R.A. and Kluepfel, D.A. (2002) Physical mapping, BAC-end sequence analysis, and marker tagging of the soilborne nematicidal bacterium, *Pseudomonas synxantha* BG33R. *Omics: A Journal of Integrative Biology* 6, 11–21.

Weibelzahl-Fulton, E., Dickson, D.W. and Whitty, E.B. (1996) Suppression of *Meloidogyne incognita* and *M. javanica* by *Pasteuria penetrans* in field soil. *Journal of Nematology* 28, 43.

Weller, D.M. (1988) Biological control of soilborne plant pathogens in the rhizosphere with bacteria. *Annual Review of Phytopathology* 26, 379–407.

Weller, D.M. (2007) *Pseudomonas* biocontrol agents of soilborne pathogens: looking back over 30 years. *Phytopathology* 97, 250–256.

Weller, D.M., Raaijmakers, J.M., Gardener, B.B.M. and Thomashow, L.S. (2002) Microbial populations responsible for specific soil suppressiveness to plant pathogens 1. *Annual Review of Phytopathology* 40, 309–348.

Weller, D.M., Landa, B.B., Mavrodi, O.V., Schroeder, K.L., De La Fuente, L. *et al.* (2007) Role of 2,4-diacetylphloroglucinol-producing fluorescent *Pseudomonas* spp. in the defense of plant roots. *Plant Biology* 9, 4–20.

Westphal, A. and Becker, J.O. (2001) Impact of soil suppressiveness on various population densities of *Heterodera schachtii*. *Annals of Applied Biology* 138, 371–376.

Worku, Y. and Gerhardson, B. (1996) Suppressiveness to clubroot, pea root rot and fusarium wilt in Swedish soils. *Journal of Phytopathology* 144, 143–146.

Yan, Y., Kuramae, E.E., de Hollander, M., Klinkhamer, P.G. and van Veen, J.A. (2016) Functional traits dominate the diversity-related selection of bacterial communities in the rhizosphere. *The ISME Journal*.

Yang, C.H., Crowley, D.E. and Menge, J.A. (2001) 16S rDNA fingerprinting of rhizosphere bacterial communities associated with healthy and Phytophthora infected avocado roots. *FEMS Microbiology Ecology* 35, 129–136.

Yin, B., Valinsky, L., Gao, X., Becker, J.O. and Borneman, J. (2003) Bacterial rRNA genes associated with soil suppressiveness against the plant-parasitic nematode Heterodera schachtii. *Applied and Environmental Microbiology* 6, 1573–1580.

Zabaloy, M.C., Garland, J.L. and Gómez, M.A. (2008) An integrated approach to evaluate the impacts of the herbicides glyphosate, 2, 4-D and metsulfuron-methyl on soil microbial communities in the Pampas region, Argentina. *Applied Soil Ecology* 40, 1–12.

Zablotowicz, R.M., Accinelli, C., Krutz, L.J. and Reddy, K.N. (2009) Soil depth and tillage effects on glyphosate degradation. *Journal of Agricultural and Food Chemistry* 57, 4867–4871.

Zarda, B., Hahn, D., Chatzinotas, A., Schönhuber, W., Neef, A. *et al.* (1997) Analysis of bacterial community structure in bulk soil by *in situ* hybridization. *Archives of Microbiology* 168, 185–192.

Zhu, Y., Shi, F., Tian, J., Liu, J., Chen, S. *et al.* (2013) Effect of soybean monoculture on the bacterial communities associated with cysts of *Heterodera glycines*. *Journal of Nematology* 45, 228.

Zobiole, L.H.S., Kremer, R.J., Oliveira, R.S. and Constantin, J. (2011) Glyphosate affects microorganisms in rhizospheres of glyphosate-resistant soybeans. *Journal of Applied Microbiology* 110, 118–127.

7 Management of Soilborne Plant Pathogens with Beneficial Root-Colonizing *Pseudomonas*

Dmitri V. Mavrodi,[1] Mingming Yang,[2]* Olga V. Mavrodi[1] and Shanshan Wen[2]
[1]*Department of Biological Sciences, University of Southern Mississippi, Hattiesburg, Mississippi, USA;* [2]*Department of Agronomy, Northwest A&F University, Yangling, Shaanxi, China*

7.1 Introduction

Soilborne plant pathogens are a significant constraint to crop production worldwide. There are no adequate seed treatments against many soilborne diseases, no resistant cultivars, and current trends towards reduced tillage and longer crop rotations favour the disease. Soilborne diseases reduce the quantity and quality of marketable yields, and their control adds considerably to the cost of production. Economic losses due to soilborne diseases in the United States alone are estimated at >$4 billion per year (Lumsden *et al.*, 1995). It has been estimated that from 2001 to 2003 an average of 7–15% of crop loss occurred on the main world crops due to soilborne fungi and oomycetes (*Gaeumannomyces graminis* var. *tritici*, *Fusarium oxysporum*, *Aphanomyces euteiches*, *Thielaviopsis basicola*, *Rhizoctonia solani*, *Phytophthora* and *Pythium* spp.), bacteria (*Streptomyces scabies* and *Ralstonia solanacearum*) and nematodes (*Heterodera*, *Meloidogyne* and *Criconemella xenoplax*) (Oerke, 2005; Singh *et al.*, 2014; Mishra *et al.*, 2015).

Fungi and oomycetes constitute the two most important classes of soilborne pathogens and share several features that make them particularly hard to control (Raaijmakers *et al.*, 2009). They are long-lived, persistent and produce highly mobile and resistant resting structures. Almost all soilborne fungi do not require a living host and obtain nutrients by killing the plant tissue with enzymes and toxins. Consequently, these pathogens can infect a broad range of crops via multiple attack strategies. Fungal pathogens often kill root tips and destroy fine feeder roots and root hairs, ultimately diminishing the ability of the plant to uptake nutrients and water. This leads to above-ground symptoms like reduced plant size, chlorosis, wilt or seedling damping-off. The below-ground symptoms include the rotting of roots, lesions, and loss of cortical tissue. The management of soilborne diseases with toxic, synthetic chemical pesticides imposes a burden on the environment and society that far exceeds the direct costs to growers and consumers. The long-term application of chemical pesticides strongly, and often permanently, alters the microbial community structure making sustainable agriculture impossible. Repeated use of the newer low-impact fungicides leads to the development of fungicide resistance in pathogens. Moreover, certain fumigants that

*E-mail: myang@nwsuaf.edu.cn

© CAB International 2017. *Advances in PGPR Research* (eds H.B. Singh, B.K. Sarma and C. Keswani)

have been traditionally used for the control of soilborne diseases deplete the ozone layer. Crop rotation can be used to mitigate some soilborne pathogens, but this approach is not always possible or desirable for economic reasons.

In the past decade, growing public awareness of the long-term impact of synthetic pesticides on human health, growing environmental concerns, and consumer demand for organically grown food have led to restrictions on the use of many synthetic pesticides in the developed countries (Glare *et al.*, 2012). This situation has led to a resurgence in the development, registration, and application of microbial biological control agents. Biological control represents a particularly attractive option for the management of soilborne diseases since plants naturally lack genetic resistance to most soilborne pathogens and instead rely on the stimulation and support of antagonistic rhizosphere microorganisms. Plant growth promoting rhizobacteria (PGPR) include *Acinetobacter, Agrobacterium, Arthrobacter, Azospirillum, Bacillus, Bradyrhizobium, Frankia, Pseudomonas, Rhizobium, Serratia, Thiobacillus* and many others. Among these, *Pseudomonas* spp. are particularly well adapted to the rhizospheric lifestyle and are ubiquitously distributed in agricultural soils worldwide. Pseudomonads are Gram-negative γ-proteobacteria known for their utilization of numerous organic compounds as energy sources, production of diverse secondary metabolites and resistance to antimicrobials. These bacteria can colonize eukaryotic hosts and include both commensals and economically important pathogens of plants and animals (Moore *et al.*, 2006; Schroth *et al.*, 2006; Keswani *et al.*, 2016). The genus *Pseudomonas* currently comprises >100 named species that have been separated based on multilocus sequence analysis into 14 species groups (Garrido-Sanz *et al.*, 2016). The *Pseudomonas fluorescens* group is the most diverse regarding both the genetic distances within it, the number of species and the large pan-genome that makes up >50% of the pan-genome of the genus as a whole (Loper *et al.*, 2012). The group also encompasses an unusually high proportion of strains that inhabit the plant rhizosphere and possess plant growth promoting and biological control properties. Such strains can rapidly colonize and multiply on plant roots and are capable of antagonizing soilborne pathogens through the production of various bioactive metabolites (i.e. antibiotics, siderophores and plant growth-promoting substances). This book chapter focuses on the prominent role of beneficial *Pseudomonas* spp. in the mitigation of soilborne disease by suppressive soils.

7.2 Rhizosphere Pseudomonads and Natural Suppression of Soilborne Plant Pathogens

Most crops lack genetic resistance to soilborne pathogens, and instead release complex mixtures of root exudates and secretions, lysates and mucilages (rhizodeposition) that shape and support a beneficial microbiome that serves as a first line of defence against pathogen attack (Cook *et al.*, 1995). Disease-suppressive soils represent the best example of indigenous microorganisms protecting plants against pathogens (Weller *et al.*, 2002; Weller *et al.*, 2007). These are soils in which the pathogen establishes and causes disease for a few seasons but thereafter the disease becomes less important, although the pathogen may persist in the soil (Baker and Cook, 1974). The suppressiveness of soils has been subdivided into "general" and "specific" suppression. General suppression is present in any soil and results from the collective competitive and antagonistic activity of the soil microbiome. In contrast, specific suppression is present only in certain soils and is due to development of a specific group of microorganisms capable of antagonizing a specific pathogen and often on a specific crop species. Specific suppression is highly effective and is superimposed over the background of general suppression. Soils that have developed specific suppressiveness are an environmentally sustainable resource for controlling soilborne pathogens. Once established, they require minimal inputs and may provide continuous control of a soilborne disease for decades. They are of special value for low-input and organic agroecosystems, which often have a limited supply of nutrients and higher disease pressure. Several well characterized disease-suppressive soils owe their activity to the build-up of distinct populations of biocontrol *Pseudomonas* spp. For example, the spontaneous decrease in the take-all disease of wheat, also known as take-all decline (TAD), is associated with the build-up of high populations of a distinct genotype of *Pseudomonas* spp. These bacteria actively colonize roots of wheat and suppress the take-all pathogen, *Gaeumannomyces graminis* var. *tritici*, via production of the antibiotic 2, 4-diacetylphloroglucinol (DAPG). Rhizosphere *Pseudomonas* that produce antibiotics also have been implicated as key antagonistic components of microbial communities from a Dutch soil suppressive to Rhizoctonia root rot of sugar beet (Mendes *et al.*, 2011), a Swiss soil suppressive to the black root rot of tobacco caused by *Thielaviopsis basicola* (Stutz *et al.*, 1986) and a French soil suppressive to Fusarium wilt of melon (Mazurier *et al.*, 2009).

7.2.1 Take-all decline

Take-all, caused by the soilborne fungus *G. graminis* var. *tritici* (Ggt), is one of the most important root diseases of wheat and is common throughout the world (Hornby, 1998; Freeman and Ward, 2004). Crop rotation and tillage are effective traditional approaches to manage take-all. However, many modern farming systems use reduced tillage and two or three crops of wheat before a break crop, which exacerbates take-all. Wheat cultivars lack resistance to take-all, and chemical controls, although available, have had only limited success in controlling the disease. TAD, the best-understood example of specific suppressiveness, is the spontaneous reduction in the incidence and severity of take-all and increase in yield occurring with continuous monoculture of wheat or barley following a severe attack of the disease (Hornby, 1998; Weller *et al.*, 2002). TAD is highly effective for management of take-all, and in the Pacific Northwest of the United States about 0.8 million ha of wheat suffer little damage from take-all, owing to TAD, even though the pathogen is still present in the soil (Cook, 2003). TAD suppressiveness is microbial in nature and results from the build-up of large populations (> 10^5 CFU g^{-1} root) of fluorescent *Pseudomonas* spp. that produce the antifungal metabolite DAPG. TAD is a field phenomenon that occurs globally and the DAPG-producing pseudomonads have been recovered from TAD soils studied in different parts of the world (Weller *et al.*, 2002). The genetic diversity of DAPG producers has been dissected by whole-cell repetitive sequence-based (rep)-PCR analysis, restriction fragment length polymorphism (RFLP), phylogenetic analysis of the DAPG biosynthesis gene *phlD* and whole genome sequencing (Landa *et al.*, 2002; De La Fuente *et al.*, 2006; Landa *et al.*, 2006; Weller *et al.*, 2007; Loper *et al.*, 2012). Several genotypes of DAPG-producing pseudomonads typically occur in a field, but usually one or two dominate on the roots of a crop grown in that soil. For example, 60–90% of the DAPG producers in TAD fields of the Pacific Northwest belong to the so called (rep)-PCR D-genotype (Weller *et al.*, 2002). These D-genotype strains, now classified as *Pseudomonas brassicacearum* on the basis of genomic sequence data, are primarily responsible for take-all suppression in the PNW and are exemplified by strain Q8r1-96, the focus of this proposal (Loper *et al.*, 2012).

DAPG plays a key role in the capacity of *P. brassicacearum* to suppress take-all. DAPG-nonproducing mutants do not control the disease (Kwak *et al.*, 2009), and the compound itself is highly inhibitory to Ggt *in vitro* (Raaijmakers and Weller, 1998; Raaijmakers *et al.*, 1999; Weller *et al.*, 2002; Weller *et al.*, 2007; Kwak *et al.*, 2009). Pathogen isolates do not differ significantly in antibiotic sensitivity, nor do they develop tolerance to DAPG in the field, probably because DAPG attacks multiple basic cellular pathways including membrane function, reactive oxygen regulation, and cell homeostasis (Kwak *et al.*, 2009; Kwak *et al.*, 2011; Kwak *et al.*, 2012). It has also been suggested that DAPG acts as a proton ionophore (Troppens *et al.*, 2013). In addition to suppressing Ggt, DAPG can exert a variety of effects on plants, including the induction of systemic resistance and promotion of amino acid exudation from roots (Iavicoli *et al.*, 2003; Phillips *et al.*, 2004; Weller *et al.*, 2012). Notably, DAPG also inhibits growth and seed germination in a variety of plants in a manner resembling the effects of the auxin-herbicide 2, 4-dichlorophenoxyacetic acid (2, 4-D) (Keel *et al.*, 1992; Kwak *et al.*, 2012). Brazelton *et al.* (2008) reported that DAPG altered tomato root architecture and interacted with an auxin-dependent signalling pathway in transgenic tobacco hypocotyl at concentrations comparable to those in the rhizosphere of plants harbouring DAPG-producing *Pseudomonas*.

The regulation of DAPG biosynthesis is complex and integrated into conserved global-regulatory circuits that enable cells of all *Pseudomonas* spp. to respond to external stimuli as well as to changes in the intracellular environment (Troppens *et al.*, 2013). At the top of the hierarchy is the critical Gac/Rsm two-component signal transduction pathway. During restricted growth and in response to an unknown signal, the phosphorylated sensor kinase GacS activates the response regulator GacA, which in turn induces expression of small RNAs that relieve translational repression of the DAPG biosynthesis operon mediated via the RNA-binding protein RsmA (Kay *et al.*, 2005). There is also an intracellular component to the regulatory system as TCA cycle intermediates, or imbalances in the cycle, also influence the GacA-dependent sRNAs, suggesting that the TCA cycle functions as a link between primary and secondary metabolism in *Pseudomonas* (Takeuchi *et al.*, 2009). Alternatively, TCA cycle intermediates, which are common constituents of plant root exudates, may be extracellular modulators of the Gac/Rsm regulon. Also, integrated with the Gac/Rsm system is the intracellular alarmone ppGpp, which recently was shown to be essential for epiphytic fitness and biocontrol activity in the DAPG-producing strain *P. protegens* CHA0 (Takeuchi *et al.*, 2012). Under stress conditions such as those in the rhizosphere, ppGpp activates the Gac/Rsm system, enabling cells to produce exometabolites including DAPG (Takeuchi *et al.*, 2012). Finally, DAPG is an autoinducer of its own synthesis and may also function as a signal that mediates interactions between DAPG-producing strains or between *Pseudomonas* and other organisms (Haas and Keel, 2003).

7.2.2 *Rhizoctonia*-suppressive soils

Rhizoctonia root rot, caused by *Rhizoctonia solani*, is a destructive soilborne disease of many economically important crops including wheat, rice and potato (Gonzalez *et al.*, 2006). The disease is hard to manage, as there are no resistant or tolerant adapted varieties, which is typical for generalist necrotrophic root pathogens such as *Rhizoctonia*. Fungicidal seed treatments are used but only give protection to the seeds and young seedlings and may have an adverse environmental impact. Thus, growers have to rely on cultural methods that often exacerbate soil erosion (deep tillage) or increase economical costs (crop rotation). As with take-all, soils harbouring microbial communities that function in the natural suppression of *Rhizoctonia* represent an attractive option of controlling this devastating pathogen. Several examples of *Rhizoctonia*-suppressive soils were described worldwide and in some cases the suppressiveness was associated with the presence of antagonistic *Pseudomonas* (MacNish, 1988; Roget, 1995; Mendes *et al.*, 2011; Yin *et al.*, 2013). The best-characterized *Rhizoctonia*-suppressive soil was described in The Netherlands by Mendes *et al.* (2011), who studied a field that became suppressive to the *Rhizoctonia* disease of sugar beet. The microbiological nature of the phenomenon was proved by the elimination of the suppressiveness by pasteurization and gamma irradiation. Furthermore, the mixing of suppressive soil with a conventional disease-conducive soil led to the partial protection of sugar beet from *Rhizoctonia*. The authors further embarked on a search for the key bacterial taxa and pathways involved in the pathogen suppression. The profiling of microbial DNA from the suppressive, pasteurized suppressive and conducive soils using a 16S rDNA PhyloChip microarray revealed the presence of over 33,000 species of bacteria and archaea. The compared soils harboured distinct microbial communities, and the level of disease suppressiveness positively correlated with the relative abundance of several bacterial taxa, including the Actinobacteria, Firmicutes (Lactobacillaceae) and Proteobacteria (Pseudomonadaceae, Burkholderiaceae, Xanthomonadales) (Mendes *et al.*, 2011). Culturing of bacteria from rhizospheres of sugar beet seedlings grown in the disease-suppressive soil produced a collection of isolates with antagonistic activity against *R. soliani.* The 16S rDNA-based analysis identified most of these isolates as members of the Pseudomonadaceae, which were further separated into three haplotype groups based on the results of DNA fingerprinting by BOX-PCR. Strains of the haplotype group II constituted the bulk of antagonistic isolates from the disease-suppressive soil and were subjected to random transposon mutagenesis. The functional analysis of strain SH-C52 yielded transposon mutants that lost the capacity to inhibit *R. solani* under *in vitro* conditions and protect sugar beet seedlings from the fungal infection in greenhouse assays. The analysis of genome regions affected by transposon insertions identified two clusters of genes involved in the synthesis of a nine amino acid chlorinated lipopeptide antibiotic. Such lipopeptides are synthesized in *Pseudomonas* by large enzymatic complexes known as non-ribosomal peptide synthetases (NRPSs) and have a broad range of activity against bacterial and fungal phytopathogens.

The second well characterized *Rhizoctonia*-suppressive soil was described by Yin *et al.* (2013) in Ritzville (Washington State, USA), at a cropping trial site that underwent a decline in Rhizoctonia patch disease of wheat. That study relied on the pyrosequencing analysis of 16S rDNA amplicons and revealed the abundance of the Acidobacteria, Gemmatimonadetes (*Gemmatimonas*), and certain Proteobacteria (*Dyella*) in the rhizosphere of wheat collected outside the patches of diseased plants or in recovered patches. Roots of diseased plants from inside the patches had higher abundance of the Bacteroidetes (*Chitinophaga, Pedobacter, Chryseobacterium*) and members of the Oxalobacteriaceae family (*Massilia* and *Duganella*). The authors further identified several strains of *Chryseobacterium soldanellicola* that antagonized *R. solani* AG-8 *in vitro* and reduced the Rhizoctonia disease of wheat in greenhouse tests. Interestingly, a significant fraction of 16S rDNA sequences recovered by Yin *et al.* (2013) from the *Rhizoctonia*-suppressive soil belonged to *Pseudomonas* spp., and pseudomonads producing the antibiotic phenazine-1-carboxylic acid (PCA) were found at high frequencies in samples collected at the studied site and neighbouring areas (Mavrodi *et al.*, 2012a,b; Parejko *et al.*, 2012) (see Section 7.4). PCA belongs to a large family of colourful, redox-active phenazine antibiotics produced by members of some fluorescent *Pseudomonas* and a few other bacterial genera (Turner and Messenger, 1986). In addition to the suppression of plant pathogens, phenazines can act as electron shuttles and contribute to the ecology, physiology, and morphology of the strains that produce them (Mazzola *et al.*, 1992; Chin-A-Woeng *et al.*, 2003; Hernandez *et al.*, 2004; Rabaey *et al.*, 2004; Rabaey *et al.*, 2005; Dietrich *et al.*, 2006; Maddula *et al.*, 2006; Mavrodi *et al.*, 2006; Price-Whelan *et al.*, 2006, 2007; Dietrich *et al.*, 2008; Maddula *et al.*, 2008; Pham *et al.*, 2008; Wang and Newman, 2008; Mentel *et al.*, 2009; Pierson and Pierson, 2010; Wang *et al.*, 2010).

Expression of the seven-gene phenazine biosynthesis (*phz*) operon is controlled in pseudomonads by the Gac/Rsm two-component signal transduction pathway and homoserine lactone (HSL)-mediated quorum sensing (Brint and Ohman, 1995; Latifi *et al.*, 1995; Wood and Pierson, 1996; Wood *et al.*, 1997; Chancey *et al.*, 1999; Khan *et al.*, 2005; Khan *et al.*, 2007). Phenazines and the quorum sensing are required for establishment and development of biofilms on surfaces of seeds and roots (Maddula *et al.*, 2006; Maddula *et al.*, 2008; Ramos *et al.*, 2010). In the rhizosphere, expression of *phz* genes can be induced by HSLs produced by heterologous isolates (Pierson *et al.*, 1998; Pierson and Pierson, 2007) or quenched by HSL-degrading rhizosphere inhabitants (Morello *et al.*, 2004). A series of independent studies revealed that these PCA-producing pseudomonads from Ritzville soils comprised at least four different species and could control *R. solani* AG-8 by producing high amounts of PCA in the rhizosphere of field-grown wheat (Mavrodi *et al.*, 2012a,b; Parejko *et al.*, 2012; Parejko *et al.*, 2013).

7.2.3 Soils suppressive to *Thielaviopsis basicola* and *Fusarium oxysporum*

Crop monoculture acts as a crucial factor in the establishment and maintenance of soils suppressive to *Rhizoctonia* and take-all disease of wheat (Kwak and Weller, 2013; Yin *et al.*, 2013). However, the monoculture plays a lesser role in natural or long-lasting suppressive soils. The two well studied examples of such naturally suppressive soils include those with suppressiveness to the black root rot of tobacco caused by *Thielaviopsis basicola* or *Fusarium oxysporum* mediated wilt of flax and other plants (Alabouvette, 1986; Stutz *et al.*, 1986). In both cases, antagonistic fluorescent pseudomonads were identified among the key taxa associated with the disease suppression.

The *T. basicola* suppressive soils from Morens (Switzerland) have been extensively studied for over three decades and contributed towards the progress in our understanding of plant protection mechanisms and rhizosphere ecology. *T. basicola* is a hemibiotrophic pathogen that infects a large number of crop plants and forms chlamydospores that can persist in soil for a long time (Huang and Kang, 2010). In the late 1960s, the black root rot of tobacco caused by *T. basicola* became a significant problem in Switzerland, but the disease did not affect several fields in the Morens region. The microbiological origin of the disease suppression in Morens was demonstrated by the loss of plant protection in heat-treated soil, and by the transfer of the suppressiveness to a conducive soil by mixing it with a small amount of soil from Morens (Stutz *et al.*, 1986).

The disease-suppressive soils of Morens and neighbouring conducive soils harbour an abundance of 2, 4-diacetylphloroglucinol (DAPG)-producing fluorescent *Pseudomonas* spp., many of which effectively protect tobacco from *T. basicola* (Stutz *et al.*, 1986). The genetic diversity and population levels of these pseudomonads were investigated using conventional microbiological techniques, 16S rRNA microarrays, quantitative PCR, and denaturing graduate gel electrophoresis (DGGE) of the DAPG biosynthesis gene *phlD* (Stutz *et al.*, 1986; Kyselkova *et al.*, 2009; Frapolli *et al.*, 2010; Almario *et al.*, 2013). Results of the DGGE-based profiling revealed a specific subset of DAPG-producing strains that were specifically enriched in the suppressive soil and produced, in addition to DAPG, the antimicrobial metabolites pyrrolnitrin, pyoluteorin and hydrogen cyanide. These strains were later assigned to a new species called *Pseudomonas protegens* and are exemplified by the model biocontrol strain CHA0. The key role of 2, 4-diacetylphloroglucinol in the ability of *P. protegens* to suppress *T. basicola* is supported by the sensitivity of this fungus to DAPG and failure of DAPG-nonproducing mutants to protect tobacco plants (Keel *et al.*, 1992; Haas and Defago, 2005). Other antimicrobials and the possible activation of the induced systemic resistance may further potentiate suppression of black root rot by CHA0-like strains (Almario *et al.*, 2013). In addition to DAPG-producing pseudomonads, the suppressive soils of Morens harboured high populations of *Azospirillum*, *Burkholderia*, *Comamonas*, *Gluconacetobacter*, *Herbaspirillum* and *Sphingomonadaceae*. Most of these taxa include species with documented plant growth promoting activity, but their exact contribution to the suppressiveness of *T. basicola*-mediated black root rot of tobacco remains to be characterized (Kyselkova *et al.*, 2009). The suppressive microflora seems to be supported by the unique composition of the Morens soils, which originate from morainic deposits and have high vermiculite content (Stutz *et al.*, 1989).

The Fusarium wilt suppressive soil from Chateaurenard (France) is another well characterized soil with long standing suppressiveness. This soil limits the incidence and severity of disease caused by *Fusarium oxysporum*, an important pathogen that causes devastating yield losses in many crops worldwide. The microbiological nature of the disease suppression in Chateaurenard was demonstrated in experiments that involved the elimination of soil suppressiveness by heat, fumigation or gamma irradiation (Alabouvette, 1986). The suppressiveness could also be transferred by mixing small amounts of the suppressive Chateaurenard soil to soils conducive to Fusarium wilt. The capacity of Chateaurenard soil to suppress Fusarium wilt was partially attributed to the abundance of non-pathogenic *F. oxysporum* that displaces the pathogen by competing for carbon sources (Steinberg *et al.*, 2007). The second component of the antagonistic microflora was represented by fluorescent pseudomonads that compete with the pathogen for iron by chelating it with siderophores. The antagonistic *Pseudomonas* also secretes antimicrobials that directly inhibit the growth of pathogenic *F. oxysporum*. Mazurier *et al.* (2009) identified and studied two distinct groups of antibiotic-producing pseudomonads in Chateaurenard soils. The first group included DAPG-producing strains that were present both in the Fusarium wilt suppressive Chateaurenard soil, as well as in the disease-conducive soil from Carquefou. In contrast, the second group of antagonistic *Pseudomonas* included phenazine-producing strains that were uniquely associated with the suppressive soil. These strains were capable of suppressing Fusarium wilt and acted synergistically with the beneficial non-pathogenic *F. oxysporum*. Further analyses identified these phenazine producers as *P. chlororaphis* and confirmed the critical role of phenazines in the ability of these organisms to control Fusarium wilt (Mazurier *et al.*, 2009; Mavrodi *et al.*, 2010). Collectively, these finding suggest that the Fusarium wilt suppressive soil of Chateaurenard harbours a unique consortium of antagonistic bacteria and fungi that control pathogens through a combination of antibiosis and competition for iron and carbon.

7.3 The emerging Role of Rhizodeposits in the Establishment and Performance of *Pseudomonas* Spp. in Suppressive Soils

Once established, induced suppressive soils remain active for decades, helping to manage soilborne diseases efficiently and with minimal environmental impact. Despite the economic and environmental benefits of such soils, however, they remain underutilized due to the lack of knowledge of the factors affecting the onset and robustness of suppressiveness. As mentioned above, sustained monoculture of a susceptible crop plays a crucial role in the selection and support of microbial communities with suppressiveness to a specific soilborne pathogen (Weller *et al.*, 2002). For example, the continuous monocropping of wheat is a fundamental prerequisite for the establishment of specific suppressiveness to Ggt in TAD soils. TAD suppressiveness is reduced or eliminated from the soil by a non-host crop and regained when wheat or barley is grown again (Weller *et al.*, 2002). The ability of *P. brassicacearum* to be the primary driver of TAD lies in the mutual affinity or preference of this bacterium and wheat. This affinity allows *P. brassicacearum* to colonize the wheat rhizosphere rapidly and maintain threshold population densities (> 10^5 CFU g^{-1} root) required for the suppression of take-all throughout the growing season (Raaijmakers and Weller, 2001; Weller *et al.*, 2007). The ability of crop monoculture to enrich for DAPG producers was also illustrated by two adjacent fields at Fargo, ND, each with greater than 100 years of continuous monoculture of wheat or flax. Finally, a pea monoculture field of 30 years' cultivation in Mt Vernon (Washington State, USA), which is suppressive to *F. oxysporum* f. sp. *pisi*, is enriched in DAPG producers (genotype P) that exhibit strong preference for pea (Landa *et al.*, 2002). Thus, there is mounting evidence that stressed plants actively recruit and shape their beneficial microbiome, but molecular details of this process are still very much a black box.

Plants are meta-organisms or holobionts that rely in part on their microbiome for specific functions and traits. Plant roots host distinct bacterial communities that profoundly affect plant health, development, vigour, disease resistance, and productivity. Beneficial root-colonizing bacteria supply plants with nutrients and defend them against soilborne pathogens. They also contribute to the ability of plants to survive under abiotic stress. In return, plants feed rhizosphere communities and influence their activity by depositing up to 40% of photosynthetically fixed carbon into the soil that directly surrounds plant roots. We hypothesize that this complex chemical cross-talk between microorganisms and plant roots forms a foundation for the crop-mediated selection of antagonistic rhizobacteria in suppressive soils. Primary root exudates include simple and complex sugars, amino acids, polypeptides, proteins, organic, aliphatic and fatty acids, sterols and phenolics (Nguyen, 2003; Badri *et al.*, 2009; Badri and Vivanco, 2009). These compounds serve as carbon and energy sources for rhizobacteria, and the presence of the intact corresponding catabolic pathways is essential for competitive colonization of roots and disease suppression (Lugtenberg *et al.*, 2001; Kamilova *et al.*, 2005; Lugtenberg and Kamilova, 2009). Root exudates also contain numerous signal molecules and secondary metabolites, the significance of which is only now emerging (Walker *et al.*, 2003; Bais *et al.*, 2005; Bais *et al.*, 2006). A handful of analyses of plant-induced gene expression by transcriptional profiling *in vitro* or in the rhizosphere (Silby and Levy, 2004; Mark *et al.*, 2005; Ramos-Gonzalez *et al.*, 2005; Matilla *et al.*, 2007; Barret *et al.*, 2009) have identified multiple genes that are differentially regulated by exposure to roots or root exudates. Bacterial pathways expressed during rhizosphere colonization control utilization of plant-derived metabolites, motility and chemotaxis, phase variation, outer membrane integrity, the ability to sequester limiting resources and resist environmental stresses (Raaijmakers *et al.*, 1995; Sarniguet *et al.*, 1995; Miller and Wood, 1996; Simons *et al.*, 1996, 1997; van Veen *et al.*, 1997; Dekkers *et al.*, 1998; Camacho-Carvajal, 2001; Schnider-Keel *et al.*, 2001; de Weert *et al.*, 2002; Sanchez-Contreras *et al.*, 2002; van den Broek *et al.*, 2005; de Weert *et al.*, 2006; Lugtenberg and Kamilova, 2009). In its spatial and temporal properties, root colonization resembles biofilm formation and biofilm-related pathways have also been implicated in adhesion to seeds and roots and rhizosphere colonization (Espinosa-Urgel *et al.*, 2000; Hinsa *et al.*, 2003; Yousef-Coronado *et al.*, 2008; Fuqua, 2010; Martinez-Gil *et al.*, 2010; Nielsen *et al.*, 2011).

Root exudates also strongly affect the expression of diverse plant growth promotion and biocontrol genes (Vacheron *et al.*, 2013). For example, there is considerable evidence that DAPG synthesis in the rhizosphere is regulated by plant-derived factors. Using DAPG-*lacZ* gene fusions, Notz *et al.* (2001) reported significantly greater expression in the rhizospheres of monocots as compared to dicots and differences in expression in response to variation of the maize host genotype. Root infection of maize by *Pythium ultimum* also stimulated gene expression independently of host or differences in rhizosphere colonization. Similarly, Jousset *et al.* (2010) showed that *Pythium* infection of barley resulted in increased expression of a DAPG reporter gene through a systemic mechanism manifested via increased exudation of diffusible molecules including vanillic, fumaric and *p*-coumaric acids. Very low concentrations of these organic acids also can induce DAPG production in vitro, suggesting that upon pathogen attack, plants launch a systemic response that can stimulate the antifungal activity in the rhizosphere microflora. De Werra *et al.* (2008) screened over 60 low molecular weight compounds, mostly of plant origin and found that many of them influenced the expression of a DAPG-GFP reporter construct. Despite obvious progress in understanding certain aspects of DAPG regulation, many of the steps linking the Gac/Rsm system to DAPG biosynthesis remain unknown. Also unknown are the external stimuli that control production of DAPG in the rhizosphere, and the broader biological role of DAPG in the rhizosphere settings.

7.4 Biocontrol *Pseudomonas* spp. as a Model for Climate-Driven Selection of Beneficial Microbiome

Recent studies revealed that the establishment of distinct groups of pseudomonads in suppressive soils is strongly influenced not only by the crop monoculture but also by the manner in which the crop is produced. This interesting topic was addressed in a series of studies focused on the low-precipitation zone of the Columbia Plateau (USA), which encompasses 1.6 million cropland hectares (Schillinger and Papendick, 2008). This unique agroecosystem is characterized by low annual precipitation (150 to 300 mm) and serves as a model to predict how global climate change could impact crops through changes in average temperatures, temperature extremes, wind erosion and moisture availability (Stockle *et al.*, 2010).

The survey of 61 commercial wheat fields scattered over 22,000 km^2 of the Columbia Plateau revealed that cereal crops grown in the low-precipitation zone support large populations (10^5–10^6 CFU g^{-1} root) of indigenous phenazine-producing rhizobacteria of the *P. fluorescens* complex (Mavrodi *et al.*, 2012a,b). These bacteria produce the broad-spectrum antibiotic phenazine-1-carboxylic acid (PCA) in the rhizosphere and control *R. solani* AG 8, a ubiquitous soilborne fungal pathogen (Bonsall *et al.*, 2012; Mavrodi *et al.*, 2012; Parejko *et al.*, 2012). Significantly, the phenazine producers were low or non-detectable in adjacent, irrigated wheat fields or neighbouring higher-precipitation areas, which were dominated by high populations of DAPG-producing pseudomonads. The DAPG-producing *P. brassicacearum* constitute the key group of antagonistic rhizobacteria in TAD suppressive soils (see Section 7.2.1). We correlated populations and frequencies of root systems colonized by PCA- and DAPG-producing rhizobacteria with agricultural practices, soil parameters and climatic variables and identified soil moisture (or the absence thereof) as a single major factor driving the development of antibiotic-producing *Pseudomonas* spp. Our findings represent the first example of selection of a phenotypically defined group of rhizobacteria that occurs on such a large scale in response to changes in soil moisture.

Interestingly, the differences in soil moisture also strongly influence the complex of necrotrophic soilborne fungal pathogens that pose significant yield constraints to cereal production across the Columbia Plateau. In the low-precipitation areas, crown rot caused by *Fusarium culmorum* and *Fusarium pseudograminearum* and root rots caused by *R. solani* AG-8 and *Rhizoctonia oryzae* are the most important diseases (Cook and Veseth, 1991). However, these pathogens are rarely seen in irrigated wheat fields, which are affected by take-all, caused by Ggt.

How soil moisture differentially affects beneficial *Pseudomonas* in the rhizosphere on wheat is currently unknown, but may result from several, non-mutually exclusive functional mechanisms. One possible explanation is that the shifts in the abundance of phenazine- and DAPG-producing pseudomonads involve interactions with indigenous microflora and/or changes in the predation by bacterivores. It is also possible that the two groups of pseudomonads differentially respond to variations in the amount and composition of root exudates at different soil moisture levels. Finally, phenazine-producing *Pseudomonas* may be better adapted for survival under conditions of water stress than their DAPG-producing counterparts. The latter hypothesis is supported by the fact that phenazine-producing pseudomonads thrived and produced PCA (up to 1.2 μg g^{-1} root) in the rhizosphere of wheat grown in arid conditions (<165 mm annual precipitation) (Mavrodi *et al.*, 2012a,b; Parejko *et al.*, 2012).

Rhizobacteria exist in an environment that regularly experiences dramatic changes in water activity, which can range from extremely hypotonic after a massive rainfall to extremely hypertonic during a prolonged drought. Bacteria use diverse physiological defensive mechanisms to cope with deleterious effects of water stress. Among rhizosphere pseudomonads, these mechanisms have been studied in considerable depth in *Pseudomonas putida*, which responds to water limitation by producing biofilms and accumulating compatible solutes (Potts, 1994; Elbein *et al.*, 2003; Fernandez-Aunion *et al.*, 2010). Bacterial biofilms are structurally complex assemblages of cells that are enclosed in an extracellular matrix comprising proteins, DNA, and exopolysaccharides (Watt *et al.*, 2006; Bloemberg, 2007; Borlee *et al.*, 2010; Zachow *et al.*, 2010). The exopolysaccharides (EPSs) can hold ten times their weight in water (Roberson and Firestone, 1992; Chenu and Roberson, 1996) and act as a major water-binding agent under water-limiting conditions (Sutherland, 2001). *P. putida* produces four different EPSs, i.e. alginate (*Alg*), putida exopolysaccharide A (*PeA*), putida exopolysaccharide b (*PeB*) and cellulose (*Bcs*) (Nielsen *et al.*, 2011; Nilsson *et al.*, 2011). Evaluation of the role of EPSs in saturated biofilms has revealed that alginate plays an important part in creating hydrated environments (Chang *et al.*, 2007; Mann and Wozniak, 2012), and that its biosynthesis genes are upregulated under conditions of water limitation (van de Mortel and Halverson, 2004; Nilsson *et al.*, 2011). Other EPSs also play important roles in the formation of rhizosphere biofilms by *P. putida*. Mutant testing has shown that Bcs and PeA contribute to hydration and that Bcs and alginate contribute to rhizosphere colonization in gnotobiotic assays (Nielsen *et al.*, 2011). A mutant devoid of all known EPS components produced biofilms similar in structure to those of the wild type *in vitro*, albeit with markedly reduced stability (Nilsson *et al.*, 2011). These results suggest that all four EPS components in *P. putida* contribute to biofilm integrity and highlight the importance of EPS in fitness under environmental stress. Although nearly all pseudomonads have the ability to produce alginate, individual species markedly differ in other forms of EPS (Mann and Wozniak, 2012) whose exact role and relative importance in biofilm formation, stability, and stress tolerance remains to be determined.

In addition to forming biofilms, most rhizobacteria respond to water stress by producing and/or taking up inert metabolites that help to balance the osmotic pressure across the cellular membrane without compromising protein folding or other cellular processes. These compounds are collectively known as compatible solutes, osmolytes or osmoprotectants and include certain polyols, sugars, amino acids, amino acid derivatives and peptides (Miller and Wood, 1996). When stressed in vitro osmotically, *P. putida* accumulates the osmoprotectants glycine betaine (GB), mannitol, glutamate, N-acetyl-glutaminylglutamine amide (NAGGN) and trehalose (Galvao *et al.*, 2006). Genes involved in the *de novo* synthesis and uptake of these osmoprotectants have been identified and characterized in *P. putida* and *P. syringae* (Chen and Beattie, 2008; Kurz *et al.*, 2010; Freeman *et al.*, 2013). Interestingly, all pseudomonads studied to date can utilize GB as an osmoprotectant, but at the same time lack genes for the *de novo* synthesis of this compound (Wargo, 2013). GB and its precursor choline are hypothesized to be ubiquitous and relatively abundant in plants, and most pseudomonads also are capable of converting choline to GB and have choline and GB transport systems (Storey and Wyn Jones, 1975; McNeil *et al.*, 1999). However, to the best of our knowledge, the exact role of GB and other osmolytes in rhizosphere settings has not been studied.

Despite recent advances in characterizing microbial biofilms and osmolytes, our understanding of the physiological responses to water stress in rhizobacteria remains incomplete. This is due to several factors. First of all, many rhizobacteria have large and highly plastic genomes, meaning that stress response traits are not universally shared. For example, PCA-producing strain *P. fluorescens* 2–79 has alginate genes but differs from *P. putida* in other structural EPSs, surface assemblages and adhesins. Second, the role of water stress response pathways to the rhizosphere fitness of bacteria is poorly understood because most of these traits were never tested under ecologically relevant conditions (i.e. in the presence of a plant host and indigenous soil microflora). Finally, the overwhelming majority of studies have been performed *in vitro* and the exchange of metabolites between rhizobacteria and a water-stressed plant has not been taken into account.

7.5 Conclusion

Soilborne plant pathogens are ubiquitous in agricultural soils worldwide, where they cause crop losses estimated in billions of dollars annually. Genetic resistance to many soilborne pathogens is rare and effective and affordable chemicals are often lacking. Instead, growers must rely on cultural practices and the antagonistic properties of the soil microbiome to reduce the impact of soilborne pathogens. The best examples of crop protection by indigenous rhizobacteria are disease suppressive soils in which the pathogen is held in check by antagonistic microorganisms.

Beneficial rhizosphere-dwelling *Pseudomonas* spp. play an essential role in supplying crop plants with nutrients and defending them against soilborne pathogens. They also contribute to the ability of plants to survive under abiotic stress. The Web of Science database (accessed on 30 September 2016) lists 3,409 papers dedicated to rhizosphere pseudomonads, and this vast body of literature highlights the broad scientific importance of these bacteria. It is now becoming apparent that such antagonistic *Pseudomonas* spp. differ significantly in their interactions with the host plant. Some are generalists: they readily colonize diverse plant species and produce an array of metabolites toxic to plant pathogens, but their populations soon dwindle to levels too low to suppress disease. In contrast, the most effective strains of pseudomonads are specialists: they produce a limited repertoire of antimicrobials but are extraordinarily competitive on the roots of particular host crops on which they provide extended disease suppression. These highly active biocontrol strains colonize roots actively and consistently, and provide extended protection of host plants from a variety of soil pathogens. Such indigenous pseudomonads are also key components of disease-suppressive soils where a specific pathogen does not persist despite favourable conditions (Baker and Cook, 1974).

There is mounting evidence that plants actively recruit their beneficial microbiome, but molecular details of this process are still poorly understood. The foundation for the differential affinity of rhizobacteria towards host plants is built upon complex chemical cross-talk between microorganisms and plant roots, which release photosynthetically fixed carbon in the form of exudates and other rhizodeposits. The molecular dialogue that occurs between rhizobacteria and plant roots is actively studied using state-of-the-art tools of functional genomics, bioinformatics, and metabolomics. The expected results will reveal the diversity and types of cellular pathways, physiological responses and selective forces that underlie the establishment of mutualistic interactions between beneficial rhizobacteria and the host plant. These studies will help to understand the molecular basis underlying suppressive soils and ultimately will provide a foundation for their more widespread integration into organic and conventional cropping systems.

References

Alabouvette, C. (1986) Fusarium wilt-suppressive soils from the Chateaurenard region: review of a 10-year study. *Agronomie* 6, 273–284.

Almario, J., Moënne-Loccoz, Y. and Muller, D. (2013) Monitoring of the relation between 2, 4-diacetylphloroglucinol producing *Pseudomonas* and *Thielaviopsis basicola* populations by real-time PCR in tobacco black root-rot suppressive and conducive soils. *Soil Biology and Biochemistry*, 144–155.

Badri, D.V. and Vivanco, J.M. (2009) Regulation and function of root exudates. *Plant, Cell and Environment* 32, 666–681.

Badri, D.V., Weir, T.L., van der Lelie, D. and Vivanco, J.M. (2009) Rhizosphere chemical dialogues: plant-microbe interactions. *Current Opinion in Biotechnology* 20, 642–650.

Bais, H.P., Prithiviraj, B., Jha, A.K., Ausubel, F.M. and Vivanco, J.M. (2005) Mediation of pathogen resistance by exudation of antimicrobials from roots. *Nature* 434, 217–221.

Bais, H.P., Weir, T.L., Perry, L.G., Gilroy, S. and Vivanco, J.M. (2006) The role of root exudates in rhizosphere interations with plants and other organisms. *Annual Review in Plant Biology* 57, 233–266.

Baker, K.F. and Cook, R.J. (1974) *Biological Control of Plant Pathogens.* W.H. Freeman, San Francisco, California.

Barret, M., Frey-Klett, P., Guillerm-Erckelboudt, A.Y., Boutin, M., Guernec, G. and Sarniguet, A. (2009) Effect of wheat roots infected with the pathogenic fungus *Gaeumannomyces graminis* var. *tritici* on gene expression of the biocontrol bacterium *Pseudomonas fluorescens* Pf29Arp. *Molecular Plant-Microbe Interactions* 22, 1611–1623.

Bloemberg, G.V. (2007) Microscopic analysis of plant-bacterium interactions using auto fluorescent proteins. *European Journal of Plant Pathology* 119, 301–309.

Borlee, B.R., Goldman, A.D., Murakami, K., Samudrala, R., Wozniak, D.J. and Parsek, M.R. (2010) *Pseudomonas aeruginosa* uses a cyclic-di-GMP-regulated adhesin to reinforce the biofilm extracellular matrix. *Molecular Microbiology* 75, 827–842.

Brazelton, J.N., Pfeufer, E.E., Sweat, T.A., Gardener, B.B. and Coenen, C. (2008) 2, 4-diacetylphloroglucinol alters plant root development. *Molecular Plant-Microbe Interactions* 21, 1349–1358.

Brint, J.M. and Ohman, D.E. (1995) Synthesis of multiple exoproducts in *Pseudomonas aeruginosa* is under the control of RhlR-RhlI, another set of regulators in strain PAO1 with homology to the autoinducer-responsive LuxR-LuxI family. *Journal of Bacteriology* 177, 7155–7163.

Camacho-Carvajal, M.M. (2001). Molecular characterization of the roles of type 4 pili, NDH-I and PyrR in rhizosphere colonization of *Pseudomonas fluorescens* WCS365. PhD thesis. University of Leiden, Leiden, Netherlands.

Chancey, S.T., Wood, D.W. and Pierson, L.S. (1999) Two-component transcriptional regulation of N-acyl-homoserine lactone production in *Pseudomonas aureofaciens. Applied and Environmental Microbiology* 65, 2294–2299.

Chang, W.S., van de Mortel, M., Nielsen, L., Nino de Guzman, G., Li, X. and Halverson, L.J. (2007) Alginate production by *Pseudomonas putida* creates a hydrated microenvironment and contributes to biofilm architecture and stress tolerance under water-limiting conditions. *Journal of Bacteriology* 189, 8290–8299.

Chen, C. and Beattie, G.A. (2008) *Pseudomonas syringae* BetT is a low-affinity choline transporter that is responsible for superior osmoprotection by choline over glycine betaine. *Journal of Bacteriology* 190, 2717–2725.

Chenu, C. and Roberson, E.B. (1996) Diffusion of glucose in microbial extracellular polysaccharide as affected by water potential. *Soil Biology and Biochemistry* 28, 877–884.

Chin-A-Woeng, T.F.C., Bloemberg, G.V. and Lugtenberg, B.J.J. (2003) Phenazines and their role in biocontrol by *Pseudomonas* bacteria. *New Phytologist* 157, 503–523.

Cook, R.J. (2003) Take-all of wheat. *Physiological and Molecular Plant Pathology* 62, 73–86.

Cook, R.J. and Veseth, R.J. (1991) *Wheat Health Management.* APS Press, St. Paul, Minnesota.

Cook, R.J., Thomashow, L.S., Weller, D.M., Fujimoto, D., Mazzola, M. *et al.* (1995) Molecular mechanisms of defense by rhizobacteria against root disease. *Proceedings of the National Academy of Sciences of the United States of America* 92, 4197–4201.

Dekkers, L.C., Phoelich, C.C., van der Fits, L. and Lugtenberg, B.J. (1998) A site-specific recombinase is required for competitive root colonization by *Pseudomonas fluorescens* WCS365. *Proceedings of the National Academy of Sciences of the United States of America* 95, 7051–7056.

De La Fuente, L., Mavrodi, D.V., Landa, B.B., Thomashow, L.S. and Weller, D.M. (2006) *phlD*-based genetic diversity and detection of genotypes of 2, 4-diacetylphlorogucinol-producing *Pseudomonas fluorescens. FEMS Microbiology Ecology* 56, 64–78.

de Weert, S., Vermeiren, H., Mulders, I.H., Kuiper, I., Hendrickx, N. *et al.* (2002) Flagella-driven chemotaxis towards exudate components is an important trait for tomato root colonization by *Pseudomonas fluorescens. Molecular Plant-Microbe Interactions* 15, 1173–1180.

de Weert, S., Dekkers, L., Bitter, W., Tuinman, S., Wijfjes, A. *et al.* (2006) The two-component *colR/S* system of *Pseudomonas fluorescens* WCS365 plays a role in rhizosphere competence through maintaining the structure and function of the outer membrane. *FEMS Microbiology Ecology* 58, 205–213.

de Werra, P., Baehler, E., Huser, A., Keel, C. and Maurhofer, M. (2008) Detection of plant-modulated alterations in antifungal gene expression in *Pseudomonas fluorescens* CHA0 on roots by flow cytometry. *Applied and Environmental Microbiology* 74, 1339–1349.

Dietrich, L.E.P., Price-Whelan, A., Petersen, A., Whiteley, M. and Newman, D.K. (2006) The phenazine pyocyanin is a terminal signalling factor in the quorum sensing network of *Pseudomonas aeruginosa*. *Molecular Microbiology* 61, 1308–1321.

Dietrich, L.E., Teal, T.K., Price-Whelan, A. and Newman, D.K. (2008) Redox-active antibiotics control gene expression and community behavior in divergent bacteria. *Science* 321, 1203–1206.

Elbein, A.D., Pan, Y.T., Pastuszak, I. and Carroll, D. (2003) New insights on trehalose: a multifunctional molecule. *Glycobiology* 13, 17R–27R.

Espinosa-Urgel, M., Salido, A. and Ramos, J.L. (2000) Genetic analysis of functions involved in adhesion of *Pseudomonas putida* to seeds. *Journal of Bacteriology* 182, 2363–2369.

Fernandez-Aunion, C., Hamouda, T.B., Iglesias-Guerra, F., Argandona, M., Reina-Bueno, M. *et al*. (2010) Biosynthesis of compatible solutes in rhizobial strains isolated from *Phaseolus vulgaris* nodules in Tunisian fields. *BMC Microbiology* 10, 192.

Frapolli, M., Défago, G. and Moënne-Loccoz, Y. (2010) Denaturing gradient gel electrophoretic analysis of dominant 2, 4-diacetylphloroglucinol biosynthetic *phlD* alleles in fluorescent *Pseudomonas* from soils suppressive or conducive to black root rot of tobacco. *Soil Biology and Biochemistry* 42, 649–656.

Freeman, B.C., Chen, C., Yu, X., Nielsen, L., Peterson, K. and Beattie, G.A. (2013) Physiological and transcriptional responses to osmotic stress of two *Pseudomonas syringae* strains that differ in epiphytic fitness and osmotolerance. *Journal of Bacteriology* 195, 4742–4752.

Freeman, J. and Ward, E. (2004) *Gaeumannomyces graminis*, the take-all fungus and its relatives. *Molecular Plant Pathology* 5, 235–252.

Fuqua, C. (2010) Passing the baton between laps: adhesion and cohesion in *Pseudomonas putida* biofilms. *Molecular Microbiology* 77, 533–536.

Galvao, T.C., de Lorenzo, V. and Canovas, D. (2006) Uncoupling of choline-O-sulphate utilization from osmoprotection in *Pseudomonas putida*. *Molecular Microbiology* 62, 1643–1654.

Garrido-Sanz, D., Meier-Kolthoff, J.P., Göker, M., Martín, M., Rivilla, R. and Redondo-Nieto, M. (2016) Genomic and genetic diversity within the *Pseudomonas fluorescens* complex. *PLoS One* 11, e0150183.

Glare, T., Caradus, J., Gelernter, W., Jackson, T., Keyhani, N. *et al*. (2012) Have biopesticides come of age? *Trends in Biotechnology* 30, 250–258.

Gonzalez, V., Portal, M.A. and Rubio, V. (2006) Review. Biology and systematics of the form genus *Rhizoctonia*. *Spanish Journal of Agricultural Research* 4, 55–79.

Haas, D. and Defago, G. (2005) Biological control of soil-borne pathogens by fluorescent pseudomonads. *Nature Reviews Microbiology* 3, 307–319.

Haas, D. and Keel, C. (2003) Regulation of antibiotic production in root-colonizing *Pseudomonas* spp. and relevance for biological control of plant disease. *Annual Reviews of Phytopathology* 41, 117–153.

Hernandez, M.E., Kappler, A. and Newman, D.K. (2004) Phenazines and other redox-active antibiotics promote microbial mineral reduction. *Applied and Environmental Microbiology* 70, 921–928.

Hinsa, S.M., Espinosa-Urgel, M., Ramos, J.L. and O'Toole, G.A. (2003) Transition from reversible to irreversible attachment during biofilm formation by *Pseudomonas fluorescens* WCS365 requires an ABC transporter and a large secreted protein. *Molecular Microbiology* 49, 905–918.

Hornby, D. (1998) *Take-All of Cereals: A Regional Perspective*. CAB International, Wallingford, UK.

Huang, J. and Kang, Z. (2010) Detection of *Thielaviopsis basicola* in soil with real-time quantitative PCR assays. *Microbiological Research* 165, 411–417.

Iavicoli, A., Boutet, E., Buchala, A. and Metraux, J.P. (2003) Induced systemic resistance in *Arabidopsis thaliana* in response to root inoculation with *Pseudomonas fluorescens* CHA0. *Molecular Plant-Microbe Interactions* 16, 851–858.

Jousset, A., Rochat, L., Scheu, S., Bonkowski, M. and Keel, C. (2010) Predator-prey chemical warfare determines the expression of biocontrol genes by rhizosphere-associated *Pseudomonas fluorescens*. *Applied and Environmental Microbiology* 76, 5263–5268.

Kamilova, F., Validov, S., Azarova, T., Mulders, I. and Lugtenberg, B. (2005) Enrichment for enhanced competitive plant root tip colonizers selects for a new class of biocontrol bacteria. *Environmental Microbiology* 7, 1809–1817.

Kay, E., Dubuis, C. and Haas, D. (2005) Three small RNAs jointly ensure secondary metabolism and biocontrol in *Pseudomonas fluorescens* CHA0. *Proceedings of the National Academy of Sciences of the United States of America* 102, 17136–17141.

Keel, C., Schnider, U., Maurhofer, M., Voisard, C., Laville, J. *et al.* (1992) Suppression of root diseases by *Pseudomonas fluorescens* CHA0 – importance of the bacterial secondary metabolite 2, 4-diacetylphloroglucinol. *Molecular Plant-Microbe Interactions* 5, 4–13.

Keswani, C., Bisen, K., Singh, V., Sarma, B.K. and Singh, H.B. (2016) Formulation technology of biocontrol agents: present status and future prospects. In: Arora, N.K., Mehnaz, S. and Balestrini, R. (eds) *Bioformulations: For Sustainable Agriculture*. Springer, New Delhi, India, pp. 35–52.

Khan, S.R., Mavrodi, D.V., Jog, G.J., Suga, H., Thomashow, L.S. and Farrand, S.K. (2005) Activation of the *phz* operon of *Pseudomonas fluorescens* 2–79 requires the LuxR homolog PhzR, *N*-(3-OH-hexanoyl)-L-homoserine lactone produced by the LuxI homolog PhzI, and a cis-acting *phz* box. *Journal of Bacteriology* 187, 6517–6527.

Khan, S.R., Herman, J., Krank, J., Serkova, N.J., Churchill, M.E. *et al.* (2007) *N*-(3-hydroxyhexanoyl)-l-homoserine lactone is the biologically relevant quormone that regulates the *phz* operon of *Pseudomonas chlororaphis* strain 30–84. *Applied and Environmental Microbiology* 73, 7443–7455.

Kurz, M., Burch, A.Y., Seip, B., Lindow, S.E. and Gross, H. (2010) Genome-driven investigation of compatible solute biosynthesis pathways of *Pseudomonas syringae* pv. *syringae* and their contribution to water stress tolerance. *Applied and Environmental Microbiology* 76, 5452–5462.

Kwak, Y.S. and Weller, D.M. (2013) Take-all of wheat and natural disease suppression: a review. *Plant Pathology Journal* 29, 125–135.

Kwak, Y.S., Bakker, P.A.H.M., Glandorf, D.C.M., Rice, J.T., Paulitz, T.C. and Weller, D.M. (2009) Diversity, virulence, and 2, 4-diacetylphloroglucinol sensitivity of *Gaeumannomyces graminis* var. *tritici* isolates from Washington State. *Phytopathology* 99, 472–479.

Kwak, Y.S., Han, S., Thomashow, L.S., Rice, J.T., Paulitz, T.C. *et al.* (2011) *Saccharomyces cerevisiae* genome-wide mutant screen for sensitivity to 2, 4-diacetylphloroglucinol, an antibiotic produced by *Pseudomonas fluorescens*. *Applied and Environmental Microbiology* 77, 1770–1776.

Kwak, Y.S., Bonsall, R.F., Okubara, P.A., Paulitz, T.C., Thomashow, L.S. and Weller, D.M. (2012) Factors impacting the activity of 2, 4-diacetylphloroglucinol-producing *Pseudomonas fluorescens* against take-all of wheat. *Soil Biology and Biochemistry* 54, 48–56.

Kyselkova, M., Kopecky, J., Frapolli, M., Defago, G., Sagova-Mareckova, M. *et al.* (2009) Comparison of rhizobacterial community composition in soil suppressive or conducive to tobacco black root rot disease. *ISME Journal* 3, 1127–1138.

Landa, B.B., Mavrodi, O.V., Raaijmakers, J.M., Gardener, B.B.M., Thomashow, L.S. and Weller, D.M. (2002) Differential ability of genotypes of 2, 4-diacetylphloroglucinol-producing *Pseudomonas fluorescens* strains to colonize the roots of pea plants. *Applied and Environmental Microbiology* 68, 3226–3237.

Landa, B.B., Mavrodi, O.V., Schroeder, K.L., Allende-Molar, R. and Weller, D.M. (2006) Enrichment and genotypic diversity of *phlD*-containing fluorescent *Pseudomonas* spp. in two soils after a century of wheat and flax monoculture. *FEMS Microbiology Ecology* 55, 351–368.

Latifi, A., Winson, M.K., Foglino, M., Bycroft, B.W., Stewart, G.S. *et al.* (1995) Multiple homologues of LuxR and LuxI control expression of virulence determinants and secondary metabolites through quorum sensing in *Pseudomonas aeruginosa* PAO1. *Molecular Microbiology* 17, 333–343.

Loper, J.E., Hassan, K.A., Mavrodi, D.V., Davis, E.W., Lim, C.K. *et al.* (2012) Comparative genomics of plant-associated *Pseudomonas* spp.: insights into diversity and inheritance of traits involved in multitrophic interactions. *PLoS Genetics* 8, e1002784.

Lugtenberg, B. and Kamilova, F. (2009) Plant-growth-promoting rhizobacteria. *Annual Reviews in Microbiology* 63, 541–556.

Lugtenberg, B.J., Dekkers, L. and Bloemberg, G.V. (2001) Molecular determinants of rhizosphere colonization by *Pseudomonas*. *Annual Reviews in Phytopathology* 39, 461–490.

Lumsden, R.D., Lewis, J.A. and Fravel, D.R. (1995). Formulation and delivery of biocontrol agents for use against soilborne plant pathogens. In: Hall, F.R., and Barry, J.W. (eds) *Biorational Pest Control Agents, Formulation and Delivery*. American Chemical Society, Washington, DC, pp. 166–182.

MacNish, G.C. (1988) Changes in take-all (*Gaeumannomyces graminis* var. *tritici*), rhizoctonia root rot (*Rhizoctonia solani*) and soil pH in continuous wheat with annual applications of nitrogenous fertilizer in Western Australia. *Australian Journal of Experimental Agriculture* 28, 333–341.

Maddula, V.S., Zhang, Z., Pierson, E.A. and Pierson, L.S. (2006) Quorum sensing and phenazines are involved in biofilm formation by *Pseudomonas chlororaphis (aureofaciens)* strain 30–84. *Microbial Ecology* 52, 289–301.

Maddula, V.S., Pierson, E.A. and Pierson, L.S. (2008) Altering the ratio of phenazines in *Pseudomonas chlororaphis (aureofaciens)* strain 30–84: effects on biofilm formation and pathogen inhibition. *Journal of Bacteriology* 190, 2759–2766.

Mann, E.E. and Wozniak, D.J. (2012) *Pseudomonas* biofilm matrix composition and niche biology. *FEMS Microbiology Reviews* 36, 893–916.

Mark, G.L., Dow, J.M., Kiely, P.D., Higgins, H., Haynes, J. *et al.* (2005) Transcriptome profiling of bacterial responses to root exudates identifies genes involved in microbe-plant interactions. *Proceedings of the National Academy of Sciences of the United States of America* 102, 17454–17459.

Martinez-Gil, M., Yousef-Coronado, F. and Espinosa-Urgel, M. (2010) LapF, the second largest *Pseudomonas putida* protein, contributes to plant root colonization and determines biofilm architecture. *Molecular Microbiology* 77, 549–561.

Matilla, M.A., Espinosa-Urgel, M., Rodriguez-Herva, J.J., Ramos, J.L. and Ramos-Gonzalez, M.I. (2007) Genomic analysis reveals the major driving forces of bacterial life in the rhizosphere. *Genome Biology* 8, R179.

Mavrodi, D.V., Blankenfeldt, W. and Thomashow, L.S. (2006) Phenazine compounds in fluorescent *Pseudomonas* spp.: biosynthesis and regulation. *Annual Reviews in Phytopathology* 44, 417–445.

Mavrodi, D.V., Peever, T.L., Mavrodi, O.V., Parejko, J.A., Raaijmakers, J.M. *et al.* (2010) Diversity and evolution of the phenazine biosynthesis pathway. *Applied and Environmental Microbiology* 76, 866–879.

Mavrodi, D.V., Mavrodi, O.V., Parejko, J.A., Bonsall, R.F., Kwak, Y.S. *et al.* (2012a) Accumulation of the antibiotic phenazine-1-carboxylic acid in the rhizosphere of dryland cereals. *Applied and Environmental Microbiology* 78, 804–812.

Mavrodi, O.V., Mavrodi, D.V., Parejko, J.A., Thomashow, L.S. and Weller, D.M. (2012b) Irrigation differentially impacts populations of indigenous antibiotic-producing *Pseudomonas* spp. in the rhizosphere of wheat. *Applied and Environmental Microbiology* 78, 3214–3220.

Mazurier, S., Corberand, T., Lemanceau, P. and Raaijmakers, J.M. (2009) Phenazine antibiotics produced by fluorescent pseudomonads contribute to natural soil suppressiveness to Fusarium wilt. *ISME Journal* 3, 977–991.

Mazzola, M., Cook, R.J., Thomashow, L.S., Weller, D.M. and Pierson, L.S. (1992) Contribution of phenazine antibiotic biosynthesis to the ecological competence of fluorescent pseudomonads in soil habitats. *Applied and Environmental Microbiology* 58, 2616–2624.

McNeil, S.D., Nuccio, M.L. and Hanson, A.D. (1999) Betaines and related osmoprotectants. Targets for metabolic engineering of stress resistance. *Plant Physiology* 120, 945–950.

Mendes, R., Kruijt, M., de Bruijn, I., Dekkers, E., van der Voort, M. *et al.* (2011) Deciphering the rhizosphere microbiome for disease-suppressive bacteria. *Science* 332, 1097–1100.

Mentel, M., Ahuja, E.G., Mavrodi, D.V., Breinbauer, R., Thomashow, L.S. and Blankenfeldt, W. (2009) Of two make one: the biosynthesis of phenazines. *Chembiochem* 10, 2295–2304.

Miller, K.J. and Wood, J.M. (1996) Osmoadaptation by rhizosphere bacteria. *Annual Reviews in Microbiology* 50, 101–136.

Mishra, S., Singh, A., Keswani, C., Saxena, A., Sarma, B.K. and Singh, H.B. (2015) Harnessing plant-microbe interactions for enhanced protection against phytopathogens. In: Arora, N.K. (ed.) *Plant Microbes Symbiosis: Applied Facets*. Springer, New Delhi, India, pp. 111–125.

Moore, E.R.B., Tindall, B.J., Martins Dos Santos, V.A.P., Pieper, D.H., Ramos, J.-L. and Palleroni, N.J. (2006) Nonmedical *Pseudomonas*. In: Dworkin, M., Falkow, S., Rosenberg, E., Schleifer, K.-H. and Stackebrandt, E. (eds) *The Prokaryotes*. Springer, New York, pp. 646–703.

Morello, J.E., Pierson, E.A. and Pierson, L.S., 3rd (2004) Negative cross-communication among wheat rhizosphere bacteria: effect on antibiotic production by the biological control bacterium *Pseudomonas aureofaciens* 30–84. *Applied and Environmental Microbiology* 70, 3103–3109.

Nguyen, C. (2003) Rhizodeposition of organic C by plants: mechanisms and controls. *Agronomie* 23, 375–396.

Nielsen, L., Li, X. and Halverson, L.J. (2011) Cell-cell and cell-surface interactions mediated by cellulose and a novel exopolysaccharide contribute to *Pseudomonas putida* biofilm formation and fitness under water-limiting conditions. *Environmental Microbiology* 13, 1342–1356.

Nilsson, M., Chiang, W.C., Fazli, M., Gjermansen, M., Givskov, M. and Tolker-Nielsen, T. (2011) Influence of putative exopolysaccharide genes on *Pseudomonas putida* KT2440 biofilm stability. *Environmental Microbiology* 13, 1357–1369.

Notz, R., Maurhofer, M., Schnider-Keel, U., Duffy, B., Haas, D. and Defago, G. (2001) Biotic factors affecting expression of the 2, 4-diacetylphloroglucinol biosynthesis gene *phlA* in *Pseudomonas fluorescens* biocontrol strain CHA0 in the rhizosphere. *Phytopathology* 91, 873–881.

Oerke, E.C. (2005) Crop losses to pests. *Journal of Agricultural Science* 144, 31–43.

Parejko, J.A., Mavrodi, D.V., Mavrodi, O.V., Weller, D.M. and Thomashow, L.S. (2012) Population structure and diversity of phenazine-1-carboxylic acid producing fluorescent *Pseudomonas* spp. from dryland cereal fields of central Washington state (USA) *Microbial Ecology* 63, 226–241.

Parejko, J.A., Mavrodi, D.V., Mavrodi, O.V., Weller, D.M. and Thomashow, L.S. (2013) Taxonomy and distribution of phenazine-producing *Pseudomonas* spp. in dryland agroecosystem of the Inland Pacific Northwest (U.S.). *Applied and Environmental Microbiology* 79, 3887–3891.

Pham, T.H., Boon, N., Aelterman, P., Clauwaert, P., De Schamphelaire, L. *et al.* (2008) Metabolites produced by *Pseudomonas* sp. enable a Gram-positive bacterium to achieve extracellular electron transfer. *Applied Microbiology and Biotechnology* 77, 1119–1129.

Phillips, D.A., Fox, T.C., King, M.D., Bhuvaneswari, T.V. and Teuber, L.R. (2004) Microbial products trigger amino acid exudation from plant roots. *Plant Physiology* 136, 2887–2894.

Pierson, E.A., Wood, D.W., Cannon, J.A., Blachere, F.M. and Pierson, L.S. (1998) Interpopulation signaling via *N*-acyl-homoserine lactones among bacteria in the wheat rhizosphere. *Molecular Plant-Microbe Interactions* 11, 1078–1084.

Pierson, L.S. and Pierson, E.A. (2007) Roles of diffusible signals in communication among plant-associated bacteria. *Phytopathology* 97, 227–232.

Pierson, L.S. 3rd, and Pierson, E.A. (2010) Metabolism and function of phenazines in bacteria: impacts on the behavior of bacteria in the environment and biotechnological processes. *Applied Microbiology and Biotechnology* 86, 1659–1670.

Potts, M. (1994) Desiccation tolerance of prokaryotes. *Microbiological Reviews* 58, 755–805.

Price-Whelan, A., Dietrich, L.E. and Newman, D.K. (2006) Rethinking 'secondary' metabolism: physiological roles for phenazine antibiotics. *Nature Chemical Biology* 2, 71–78.

Price-Whelan, A., Dietrich, L.E. and Newman, D.K. (2007) Pyocyanin alters redox homeostasis and carbon flux through central metabolic pathways in *Pseudomonas aeruginosa* PA14. *Journal of Bacteriology* 189, 6372–6381.

Raaijmakers, J.M. and Weller, D.M. (1998) Natural plant protection by 2, 4-diacetylphloroglucinol-producing *Pseudomonas* spp. in take-all decline soils. *Molecular Plant-Microbe Interactions* 11, 144–152.

Raaijmakers, J.M. and Weller, D.M. (2001) Exploiting genotypic diversity of 2, 4-diacetylphloroglucinol-producing *Pseudomonas* spp.: characterization of superior root-colonizing *P. fluorescens* strain Q8r1-96. *Applied and Environmental Microbiology* 67, 2545–2554.

Raaijmakers, J.M., Vandersluis, I., Koster, M., Bakker, P.A.H.M., Weisbeek, P.J. and Schippers, B. (1995) Utilization of heterologous siderophores and rhizosphere competence of fluorescent *Pseudomonas* spp. *Canadian Journal of Microbiology* 41, 126–135.

Raaijmakers, J.M., Bonsall, R.E. and Weller, D.M. (1999) Effect of population density of *Pseudomonas fluorescens* on production of 2, 4-diacetylphloroglucinol in the rhizosphere of wheat. *Phytopathology* 89, 470–475.

Raaijmakers, J.M., Paulitz, T.C., Steinberg, C., Alabouvette, C. and Moenne-Loccoz, Y. (2009) The rhizosphere: a playground and battlefield for soilborne pathogens and beneficial microorganisms. *Plant and Soil* 321, 341–361.

Rabaey, K., Boon, N., Siciliano, S.D., Verhaege, M. and Verstraete, W. (2004) Biofuel cells select for microbial consortia that self-mediate electron transfer. *Applied and Environmental Microbiology* 70, 5373–5382.

Rabaey, K., Boon, N., Hofte, M. and Verstraete, W. (2005) Microbial phenazine production enhances electron transfer in biofuel cells. *Environmental Science and Technology* 39, 3401–3408.

Ramos, I., Dietrich, L.E., Price-Whelan, A. and Newman, D.K. (2010) Phenazines affect biofilm formation by *Pseudomonas aeruginosa* in similar ways at various scales. *Research in Microbiology* 161, 187–191.

Ramos-Gonzalez, M.I., Campos, M.J. and Ramos, J.L. (2005) Analysis of *Pseudomonas putida* KT2440 gene expression in the maize rhizosphere: *in vivo* expression technology capture and identification of root-activated promoters. *Journal of Bacteriology* 187, 4033–4041.

Roberson, E.B. and Firestone, M.K. (1992) Relationship between desiccation and exopolysaccharide production in a soil *Pseudomonas* sp. *Applied and Environmental Microbiology* 58, 1284–1291.

Roget, D.K. (1995) Decline in root rot (*Rhizoctonia solani* AG-8) in wheat in a tillage and rotation experiment at Avon, South Australia. *Australian Journal of Experimental Agriculture* 35, 1009–1013.

Sanchez-Contreras, M., Martin, M., Villacieros, M., O'Gara, F., Bonilla, I. and Rivilla, R. (2002) Phenotypic selection and phase variation occur during alfalfa root colonization by *Pseudomonas fluorescens* F113. *Journal of Bacteriology* 184, 1587–1596.

Sarniguet, A., Kraus, J., Henkels, M.D., Muehlchen, A.M. and Loper, J.E. (1995) The sigma factor σ^s affects antibiotic production and biological control activity of *Pseudomonas fluorescens* Pf-5. *Proceedings of the National Academy of Sciences of the United States of America* 92, 12255–12259.

Schillinger, W.F. and Papendick, R.I. (2008) Then and now: 125 years of dryland wheat farming in the Inland Pacific Northwest. *Agronomy Journal* 100, S166–S182.

Schnider-Keel, U., Lejbolle, K.B., Baehler, E., Haas, D. and Keel, C. (2001) The sigma factor AlgU (AlgT) controls exopolysaccharide production and tolerance towards desiccation and osmotic stress in the biocontrol agent *Pseudomonas fluorescens* CHA0. *Applied and Environmental Microbiology* 67, 5683–5693.
Schroth, M.N., Hildebrand, D.C. and Panopoulos, N.J. (2006) Phytopathogenic pseudomonads and related plant-associated pseudomonads. In: Dworkin, M., Falkow, S., Rosenberg, E., Schleifer, K.-H. and Stackebrandt, E. (eds) *The Prokaryotes*. Springer, New York, pp. 714–740.
Silby, M.W. and Levy, S.B. (2004) Use of in vivo expression technology to identify genes important in growth and survival of *Pseudomonas fluorescens* Pf0-1 in soil: discovery of expressed sequences with novel genetic organization. *Journal of Bacteriology* 186, 7411–7419.
Simons, M., van der Bij, A.J., Brand, I., de Weger, L.A., Wijffelman, C.A. and Lugtenberg, B.J. (1996) Gnotobiotic system for studying rhizosphere colonization by plant growth-promoting *Pseudomonas* bacteria. *Molecular Plant-Microbe Interactions* 9, 600–607.
Simons, M., van der Bij, A.J., Brand, J., de Weger, L.A., Wijffelman, D.A. and Lugtenberg, B.J.J. (1997) Amino acid synthesis is necessary for tomato root colonization by *Pseudomonas fluorescens* strain WCS365. *Molecular Plant-Microbe Interactions* 10, 102–106.
Singh, H.B., Jain, A., Saxena, A., Singh, A., Keswani, C. *et al.* (2014) Deciphering the Pathogenic Behaviour of Phyto-Pathogens Using Molecular Tools. In: Sharma, N. (ed.) *Biological Controls for Preventing Food Deterioration: Strategies for Pre-and Postharvest Management*. John Wiley & Sons, Chichester, UK, pp. 377–408.
Steinberg, C., Edel-Hermann, V., Alabouvette, C. and Lemanceau, P. (2007) Soil suppressiveness to plant diseases. In: Van Elsas, D., Jansson, J.K., and Trevors, J.T. (eds) *Modern Soil Microbiology*, 2nd edn. CRC Press, Taylor & Francis Group, London, pp. 455–477.
Stockle, C.O., Nelson, R.L., Higgins, S., Brunner, J., Grove, G. *et al.* (2010) Assessment of climate change impact on Eastern Washington agriculture. *Climatic Change* 102, 77–102.
Storey, R. and Wyn Jones, R.G. (1975) Betaine and choline levels in plants and their relationship to NaCl stress. *Plant Science Letters* 4, 161–168.
Stutz, E.W., Defago, G. and Kern, H. (1986) Naturally occurring fluorescent pseudomonads involved in suppression of black root rot of tobacco. *Phytopathology* 76, 181–185.
Stutz, E., Kahr, G. and Défago, G. (1989) Clays involved in suppression of tobacco black root rot by a strain of *Pseudomonas fluorescens*. *Soil Biology and Biochemistry* 21, 361–366.
Sutherland, I. (2001) Biofilm exopolysaccharides: a strong and sticky framework. *Microbiology* 147, 3–9.
Takeuchi, K., Kiefer, P., Reimmann, C., Keel, C., Dubuis, C. *et al.* (2009) Small RNA-dependent expression of secondary metabolism is controlled by Krebs cycle function in *Pseudomonas fluorescens*. *Journal of Biological Chemistry* 284, 34976–34985.
Takeuchi, K., Yamada, K. and Haas, D. (2012) ppGpp controlled by the Gac/Rsm regulatory pathway sustains biocontrol activity in *Pseudomonas fluorescens* CHA0. *Molecular Plant-Microbe Interactions* 25, 1440–1449.
Troppens, D.M., Moynihan, J.A., Barret, M., O'Gara, F. and Morrissey, J.P. (2013) Genetics and evolution of 2, 4-diacetylphloroglucinol synthesis in *Pseudomonas fluorescens*. In: de Bruijn, F.J. (ed.) *Molecular Microbial Ecology of the Rhizosphere*. John Wiley and Sons, Hoboken, New Jersey, pp. 593–605.
Turner, J.M., and Messenger, A.J. (1986) Occurrence, biochemistry and physiology of phenazine pigment production. *Advances in Microbial Physiology* 27, 211–275.
Vacheron, J., Desbrosses, G., Bouffaud, M.L., Touraine, B., Moenne-Loccoz, Y. *et al.* (2013) Plant growth-promoting rhizobacteria and root system functioning. *Frontiers in Plant Science* 4, 356.
van de Mortel, M. and Halverson, L.J. (2004) Cell envelope components contributing to biofilm growth and survival of *Pseudomonas putida* in low-water-content habitats. *Molecular Microbiology* 52, 735–750.
van den Broek, D., Bloemberg, G.V. and Lugtenberg, B. (2005) The role of phenotypic variation in rhizosphere *Pseudomonas* bacteria. *Environmental Microbiology* 7, 1686–1697.
van Veen, J.A., van Overbeek, L.S. and van Elsas, J.D. (1997) Fate and activity of microorganisms introduced into soil. *Microbiology and Molecular Biology Reviews* 61, 121–135.
Walker, T.S., Bais, H.P., Halligan, K.M., Stermitz, F.R. and Vivanco, J.M. (2003) Metabolic profiling of root exudates of *Arabidopsis thaliana*. *Journal of Agricultural Food Chemistry* 51, 2548–2554.
Wang, Y. and Newman, D.K. (2008) Redox reactions of phenazine antibiotics with ferric (hydr)oxides and molecular oxygen. *Environmental Science and Technology* 42, 2380–2386.
Wang, Y., Kern, S.E. and Newman, D.K. (2010) Endogenous phenazine antibiotics promote anaerobic survival of *Pseudomonas aeruginosa* via extracellular electron transfer. *Journal of Bacteriology* 192, 365–369.
Wargo, M.J. (2013) Homeostasis and catabolism of choline and glycine betaine: lessons from *Pseudomonas aeruginosa*. *Applied and Environmental Microbiology* 79, 2112–2120.

Watt, M., Hugenholtz, P., White, R. and Vinall, K. (2006) Numbers and locations of native bacteria on field-grown wheat roots quantified by fluorescence *in situ* hybridization (FISH). *Environmental Microbiology* 8, 871–884.

Weller, D.M., Raaijmakers, J.M., Gardener, B.B.M. and Thomashow, L.S. (2002) Microbial populations responsible for specific soil suppressiveness to plant pathogens. *Annual Reviews in Phytopathology* 40, 309–348.

Weller, D.M., Landa, B.B., Mavrodi, O.V., Schroeder, K.L., De La Fuente, L. *et al.* (2007) Role of 2, 4-diacetylphloroglucinol-producing fluorescent *Pseudomonas* spp. in plant defense. *Plant Biology* 9, 4–20.

Weller, D.M., Mavrodi, D.V., van Pelt, J.A., Pieterse, C.M., van Loon, L.C. and Bakker, P.A. (2012) Induced systemic resistance in *Arabidopsis thaliana* against *Pseudomonas syringae* pv. *tomato* by 2, 4-diacetylphloroglucinol-producing *Pseudomonas fluorescens*. *Phytopathology* 102, 403–412.

Wood, D.W. and Pierson, L.S. 3rd (1996) The *phzI* gene of *Pseudomonas aureofaciens* 30–84 is responsible for the production of a diffusible signal required for phenazine antibiotic production. *Gene* 168, 49–53.

Wood, D.W., Gong, F., Daykin, M.M., Williams, P. and Pierson, L.S. (1997) N-acyl-homoserine lactone-mediated regulation of phenazine gene expression by *Pseudomonas aureofaciens* 30–84 in the wheat rhizosphere. *Journal of Bacteriology* 179, 7663–7670.

Yin, C., Hulbert, S., Schroeder, K., Mavrodi, O., Mavrodi, D. *et al.* (2013) The role of bacterial communities in the natural suppression of *Rhizoctonia solani* bare patch disease of wheat (*Triticum aestivum* L.). *Applied and Environmental Microbiology* 79, 7428–7438.

Yousef-Coronado, F., Travieso, M.L. and Espinosa-Urgel, M. (2008) Different, overlapping mechanisms for colonization of abiotic and plant surfaces by *Pseudomonas putida*. *FEMS Microbiology Letters* 288, 118–124.

Zachow, C., Fatehi, J., Cardinale, M., Tilcher, R. and Berg, G. (2010) Strain-specific colonization pattern of *Rhizoctonia* antagonists in the root system of sugar beet. *FEMS Microbiology Ecology* 74, 124–135.

8 Rhizosphere, Mycorrhizosphere and Hyphosphere as Unique Niches for Soil-Inhabiting Bacteria and Micromycetes

Elena Voronina* and Irina Sidorova
Lomonosov Moscow State University, Moscow, Russia

8.1 Introduction

Soil is a natural body abounding in diverse life forms which belong to all domains of life and to a range of functional groups. Soil heterogeneity at a fine scale provides numerous microhabitats and hosts a number of microbial communities different in size and composition, influenced by soil properties (Haq *et al.*, 2014). Vice versa, fungi, especially, symbiotic in mycorrhizas and bacteria, act as soil engineers, and it was revealed that more than 50% of the humus in boreal forest soil originated from roots and their microbe associates (Clemmensen *et al.*, 2013).

Soil microbial communities since the 20th century were known to play a key role in plant growth, health and productivity both at individual and ecosystem levels. Root microorganisms interfere with plant nutrition, attack the plant or protect it from attackers, and carry out multiple functions in plant life often based on intense interactions within the microbial community (Keswani *et al.*, 2013; Bisen *et al.*, 2015; Mishra *et al.*, 2015; Bisen *et al.*, 2016; Keswani *et al.*, 2016a, b). Understanding a root zone as a hot-spot of microorganisms' activities gave rise to the rhizosphere concept, which more recently was partly transformed to the mycorrhizosphere concept to emphasize the multidimensional roles and ubiquitous nature of mycorrhizal fungi that have accompanied plants since they emerged from the water in the Ordovician period and now inhabiting almost all plant communities (Smith and Read, 2008). The second half of the 20th century was marked by recognition of fungal mycelia's fundamental role in soil biochemistry and geocycling (Gadd, 2006) thus leading to concept of the hyphosphere as a soil zone modified by fungal influence and harbouring its own microbiota. All these "sphere" microniches stand apart of bulk soil in quantitative and qualitative composition of microbial communities they host and can serve as a reservoir of potential agents for bioremediation, biocontrol and sustainable agriculture "soil engineers". But delimitation within rhizo-, mycorrhizo- and hyphosphere is impaired by both natural continuity and overlapping of these zones and by different concepts applied to them by scientists.

Now known as microbiomes, microbial communities inhabiting such niches have recently gained substantial attention from researchers and were intensely studied by

*E-mail: mvsadnik@list.ru

© CAB International 2017. *Advances in PGPR Research* (eds H.B. Singh, B.K. Sarma and C. Keswani)

means of a plethora of modern molecular-based techniques allowing exploration *in situ* (Turner *et al.*, 2013), but there are lots of questions still waiting for the answers.

This chapter is aimed at summarizing concepts and experimental data on rhizo-, mycorrhizo- and hyphosphere with emphasis on causes for these zones to be unique microniches for soil bacteria and microfungi, and on the mechanisms of interactions between species leading to reciprocal influence and microhabitat transformation.

8.2 Historical Aspects of Rhizosphere, Mycorhizospere and Hyphosphere Study and Modern Research Approaches

8.2.1 The terms: brief background

Rhizosphere was recognized first of all mentioned above plant and fungal zones of influence on soil microbiota. The term was proposed as early as in 1904 by Dr Lorenz Hiltner, German plant scientist and soil microbiologist, the founding director of the Royal Agriculture–Botanical Institute in Munich (Hiltner, 1904; Sen, 2005). Working with different crop plants, Hiltner revealed the influence of root exudates on soil microorganisms' communities. He established the rhizosphere concept and described it as 'In soil influenced by roots, or within the "rhizosphere" as I will express myself further on, bacteria take up and immobilize the available nitrogen and thus support and enable the nitrogen fixation of the nodule and the enrichment of the soil with nitrogen. The rhizosphere creates the possibility that these useful activities develop' (Hartmann *et al.*, 2008). Hiltner for the first time formulated the main rhizosphere traits such as plant-species-specific effects (e.g. legumes are preferred by microorganisms instead of brassicaceae and some other crops), plant-determined selection of the most favorable bacteria in its own root-zone, and dependence of rhizosphere size on soil structure. Besides he supposed a beneficial role of rhizosphere bacteria in suppression of soilborne phytopathogens by excluding them from root vicinities (Hartmann *et al.*, 2008).

Hiltner's definition and description of this unique soil microhabitat remains topical even nowadays. A range of compendia of rhizosphere research were recently published providing a many-sided insight into the problem (Mukerji *et al.*, 2006; Cardon and Whitbeck, 2007; Varma *et al.*, 2008). Practical recommendations of Lorenz Hiltner on maintaining agricultural soils with the help of natural beneficial microorganisms, which with time are able to replace artificial fertilizers, transferred to the rather popular modern concept of low-input sustainable agriculture. The application of this idea became even more urgent in the 21st century because of the world-wide limitation and future deletion of ordinary fertilizers.

There is a rather unusual situation in the natural environment for the vast majority of plant species to exist in a non-symbiotic state. Mycorrhizal symbiosis is well acknowledged to be the most widespread type of fungal–plant interaction, involving the plant root system and some specific or non-specific mycobiont, predominantly presumed to be beneficial for both partners (Smith and Read, 2008). Thus as a rule the plant root's interactions with soil and its biota are directly (by external hyphal mantle at a root tip) or indirectly (via bioactive or signal compounds) mediated by mycorrhizal fungi. On the other hand, soil microorganisms can influence mycorrhizal establishment and development enhancing plant benefit from symbiosis or reducing it. Understanding the mycorrhizal biotic interaction's complexity and its crucial role for plants both at individual and community levels led to the emergence of the mycorrhizosphere concept which in the case of a majority of plant species should replace the rhizosphere one (Timonen and Marschner, 2006). The term mycorrhizosphere has a more intricate history than its partner term rhizosphere because several views on it appeared nearly at the same time. It was first mentioned as "mycorrhizasphere" by Rawlings (1958) to embrace all organic soil horizons influenced by mycorrhizal symbiosis including all the soil mass penetrated by mycobionts' hyphae. This term had not been widely acknowledged but some more recent researchers (Filion *et al.*, 1999) applied it *sensu*. Summerbell (2005a) proposed closely related "symbiorhizosphere" for a broad definition of all soil mass (including plant debris) influenced by mycorrhizal root system in total. A more widely accepted definition of the mycorrhizosphere as mycorrhizal root tips' only zone of influence was used by Foster and Marks (1967) and further became a common term (Summerbell, 2005a). However, in this *sensu stricto* mycorrhizosphere only those mycorrhizal types which imply formation of external hyphal mantle on a physiologically active root tip, have a direct contact zone between fungal symbiont and ambient soil with its biota. According to modern mycorrhizal symbioses classification provided by Brundrett (2004) these comprise ectomycorrhizas and partially arbutoid and monotropoid types. So-called endomycorrhizas, including such widespread and important examples as arbuscular, do not have any mycobiont-soil interface besides free extraradical hyphae.

The shaped concept of mycorrhizosphere was introduced by Angelo Rambelli in 1973 to depict a soil microhabitat influenced both by plant roots (via root exudates) and its mycobionts with its metabolites, which was in his opinion 'clearly delimitable' from other soil microhabitats (Rambelli, 1973). He summarized numerous experimental data and provided his own to demonstrate qualitative and quantitative changes of both bacterial and micromycete communities in the mycorrhizosphere and to propose mechanisms underlying biotic interactions. Furthermore the acknowledgement of this niche specificity was shared by a range of researchers and the data were summarized in reviews (Linderman, 1988; Timonen and Marschner, 2006). A very detailed and comprehensive overview of changing opinions and "white-outs" concerning ectomycorrhizosphere during the 20th century was provided by Summerbell (2005b).

Currently there is a global outburst of symbioses-related research, including mycorrhizas, and the term mycorrhizosphere has become very popular again. In a broad context it implies the rhizosphere of any mycorrhizal plant, in contrast to non-mycorrhizal ones. Anyone can prefer one or another term, but at present it is well acknowledged that mycorrhizal symbiosis is a multitrophic complex consisting of a host plant, its mycobiont and associated soil-borne microorganisms with a net of biotic interactions and reciprocal influences on each other, including both synergism and antagonism. Nowadays it is impossible to conduct a relevant study on mycorrhizas or plant root nutrition in the natural environment without paying attention to its microbial suite, for lots of "root" or "mycobiont" functions are fulfilled or facilitated by soil microorganisms (Timonen and Marschner, 2006).

For a long time free mycelium with hyphae penetrating substrates such as soil, litter, or wood debris and hidden in it got far less attention from scientists than it deserves for its key role in decomposition, nutrient cycling and plant community sustainability (Gadd, 2006). Partially this was due to absence of relevant techniques and approaches to mycelial study under field conditions. Apart from its key role in soil compounds transformation, free hyphae of fungi from different trophic guilds interact with a range of soil biota groups, such as bacteria and micromycetes, and have a profound impact both on its community structure and numbers (Boddy *et al.*, 2008; Zagryadskaya *et al.*, 2011; Sidorova *et al.*, 2017). These complex reciprocal interactions, including either stimulation or suppression, are known as the hyphosphere effect (Staněk, 1984), after the term hyphosphere, introduced for fungal hyphae surface and proximity by Royd Thornton in 1953 as a result of his observations of actinomycetes encrusting *Rhizoctonia solani* hyphae (Thornton, 1953). Another term applied to fungus-dependent microhabitats is mycosphere. According to some researchers (Staněk, 1984) it indicated more vast zone encircling substrate occupied by mycelium and influenced by fungal metabolites, and surface and vicinities of sporocarps. This definition implies hyphosphere to be a part of mycosphere, but a view of mycosphere as the synonym for hyphosphere 'the microhabitat that surrounds the fungal hyphae in soil' (Zhang *et al.*, 2014) is more widespread (Gilbert and Linderman, 1971; Warmink and van Elsas, 2008).

Summarizing the background, it is necessary to admit some confusion and ambiguity of terms described. This inconvenience is inevitable presuming the continuity and even overlapping of "spheres" niches which impede its clear and definite delimitation. Its size strongly depends on plant and/or fungus species, mycorrhizal type and a range of biotic and abiotic factors exerting in soil. In this chapter we accept rhizosphere as the non-mycorrhizal roots zone of influence (including non-mycorrhizal parts of root systems), mycorrhizosphere as its equivalent for mycorrhizal roots (excluding vast extraradical mycelial network), and hyphosphere as a zone of influence of different extraradical mycorrhizal or free-living substrate hyphal fungal structures, including hyphal cords, rhizomorphs, and sclerotia. Other terms will be avoided, if possible, in order to reduce confusion. Interrelations between the zones discussed above are outlined in Fig. 8.1.

8.2.2 Research approaches: some recent advances and classical techniques

The complexity of "sphere" habitats makes researching their structure and functions quite a challenge and requires cross-disciplinary approaches combining biology and soil science techniques. Classical microbiology methods widely applied to "spheres" microorganisms from Hiltner's time and throughout the 20th century were based on plate isolation on media and culture techniques. Allowing species identification and *in vitro* physiology and biochemistry study, these culture-dependent methods led to a vast data accumulation and have been applied successfully and are irreplaceable to date (Uroz and Frey-Klett, 2011). Their shortcoming is in missing a large body of unculturable organisms in the natural environment and providing no opportunity to delimit *in situ*

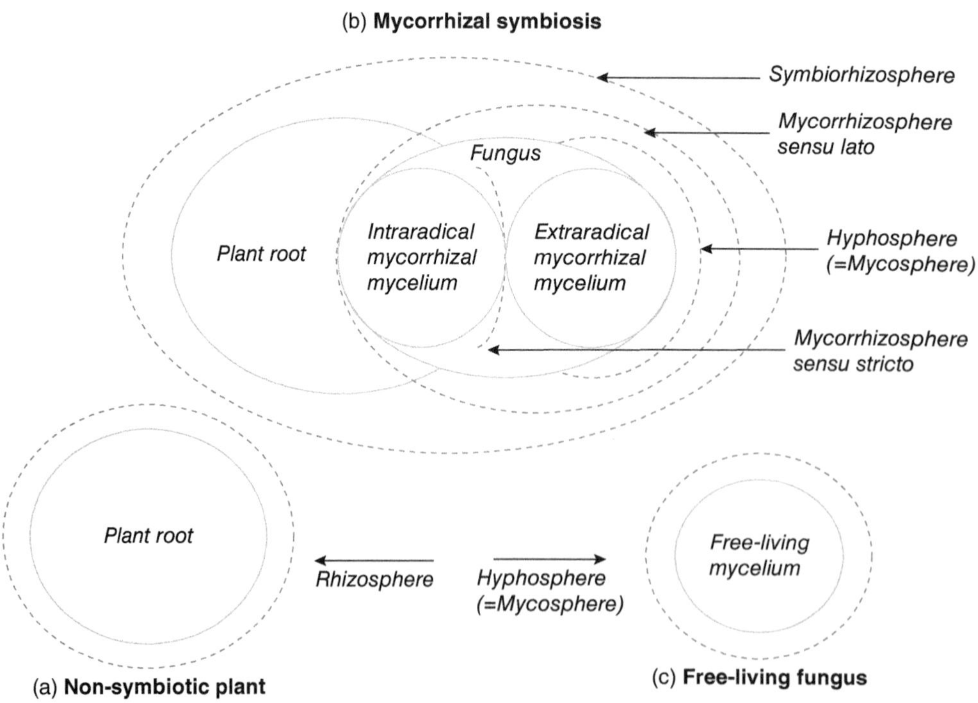

Fig. 8.1. Plant- and fungus-influenced microniches within the soil: different concepts outline.

active microorganisms from inactive ones. Numerous molecular techniques were developed for culture-independent microorganism identification *in situ* and the most common for prokaryote study is PCR-based amplification of 16S rRNA gene (Cardon and Whitbeck, 2007). Further high-throughput (next-generation) sequencing technologies applications allow the revealing of even minor components of the community (Bulgarelli *et al.*, 2012).

The common equivalent for eukaryotes, such as fungi, is ITS amplification (Turner *et al.*, 2013). This approach has a shortcoming especially critical in research on extremely multispecies environments like soil and its microniches. Restricted by primer design it allows detection of only the target organisms, inevitably narrowing and simplifying the output picture. To catch as much diversity as possible and to trace all possible interactions and impacts, such global analysis tools as metagenomics, metatranscriptomics and metaproteomics are applied to assess all life domains. Metagenomics represents functionality (amounts of genes taking part in metabolic processes), metatranscriptomics allows assessment of community-wide gene expression, and metaproteomics provides protein profiling (Philippot *et al.*, 2013; Turner *et al.*, 2013). Stable isotope probing can act as a complementary molecular technique for defining functionality of different groups of organisms by tracing a stable isotope from substrate to microbial cell. In particular it was applied for determination of main microorganism groups involved in soil carbon flow (Rangel-Castro *et al.*, 2005). Another way to detect activity and quantify active cells is the combination of fluorescence in situ hybridization (FISH) method with micro-autoradiography (Ladygina, 2009). FISH allows unculturable bacteria phylogenetic identification in natural environments using rRNA targeting fluorescent-specific phylogenetic probes and fluorescence microscopy.

In visualization techniques progress was made at the end of the 20th century with application of confocal laser scanning microscopy (CLSM) combining with fluorescent markers for observation of rhizosphere bacterial populations. To detect microbial activity along with the visualization, the combination of immunofluorescence and r-RNA-target probing proved to be efficient (Bloemberg and Lugtenberg, 2001).

Finally it should be noted that a holistic approach is required for microbial interrelations study instead of independent exploration of a rhizosphere plant or hyphosphere fungus separately from its microbial suite. As for techniques, only combining molecular and cultural methods can elucidate plant–microbe interactions at community level (Uroz and Frey-Klett, 2011; Turner *et al.*, 2013).

8.3 Rhizosphere, the Niche Influenced by Plant Roots

The Rhizosphere as a soil microniche differs drastically from bulk soil in pH, redox potential, water and nutrients concentration due to roots' uptake of water and mineral compounds and release of organic ones (Marschner, 1995). Increased carbon inflow supplied by root exudation and secretion (rhizodeposition), and autolysis of senescent parts of root systems makes this relatively small zone (only several millimetres around the root) a hot spot for soil biota of different groups (Girlanda and Perotto, 2005; Raaijmakers *et al.*, 2009). Roots release a range of compounds beneficial for microbiota which can be split into several groups: water-soluble exudates (sugars, organic acids, hormones and vitamins) leaking from roots independent of metabolic energy; products of secretion (polymeric carbohydrates, enzymes) dependent on metabolic energy; lysates of dying plant root cells; gases (ethylene, carbon dioxide, methane); and mucilage and mucigel (Ladygina, 2009; Dennis *et al.*, 2010). Rhizodeposition may vary widely from 10% to more than 40% of plant photosynthates (Grayston *et al.*, 1997). On the other hand, roots produce a number of metabolites with pronounced antifungal and antibacterial activities (Bais *et al.*, 2006) and thus engineer the microbial community by selection of forms tolerant, resistant or able to metabolize such compounds (Kowalchuk *et al.*, 2006).

Governed by root-derived compounds amounts and chemical composition, the rhizosphere provides an arena for a wide spectrum of intense biotic interactions from antagonistic to synergistic and represents one of the most dynamic biological interfaces. It hosts a variety of microorganisms which can be classified into deleterious (plant-pathogenic), beneficial, or neutral for the plant. Plant pathogenic bacteria and microfungi are often capable of germination and directional growth towards roots being chemically attracted by some components of rhizodeposits such as phenolic compounds (Philippot *et al.*, 2013). Beneficial soilborne organisms comprise four main functional groups: biofertilizers, phytostimulators, biocontrol agents and bioremediators. The first three groups are often considered as PGP (plant-growth-promoting) fungi or bacteria (Bloemberg and Lugtenberg, 2001). The main mechanism underlying the biofertilization effect is the ability of both fungi and bacteria to solubilize recalcitrant compounds and thus to free nutrients providing plant roots' access to them. Phosphorus, a crucial macronutrient, presents in soil, even rich in it, predominantly in the form of insoluble phosphates (P) inaccessible for plants itself. P-solubilizing rhizosphere microorganisms play a key role in plant mineral nutrition promoting plant growth (Whitelaw, 2000; Fomina *et al.*, 2006). Phytostimulators operate directly through phytohormones and biocontrol agents protect the root system by acting as natural antagonists of pathogens by direct competition (including bioactive compounds production) and parasitism or indirectly by plant defence systems induction (Haas and Défago, 2005; Raaijmakers *et al.*, 2009; van de Mortel *et al.*, 2012; Zamioudis and Pieterse, 2012). It was revealed recently that rhizosphere microorganisms play an important role in herbivore biocontrol too and that they are more efficient than carnivorous soil invertebrates, quite opposite to the accepted view on the problem (Piśkiewisz *et al.*, 2007). The fourth group, bioremediators, able to metabolize and remove toxic organic and inorganic pollutants, is rather promising for resolving the problem of soil contamination by the so called "rhizodegradation" or "rhizoremediation" method (Olson *et al.*, 2003; Chaudhry *et al.*, 2005).

Plant species along with the soil type are considered to be major factors influencing rhizosphere microbial communities (Berg and Smalla, 2009; Ladygina and Hedlund, 2010; Bulgarelli *et al.*, 2012; Lundberg *et al.*, 2012). It is obvious that multispecies natural phytocoenoses will provide more different rhizosphere microniches and thus host more diverse microorganism communities than monocrop agrocoenoses. Besides, even different cultivars of crops can have their unique rhizosphere microorganism suites. Presuming that crop plant breeding is conducted under artificial conditions in monoculture and with a minimal role of soil microbiota for plant growth and health, human selection can drive crops to the loss of some traits essential for hosting beneficial microorganisms in root systems (Yao and Wu, 2010; Bouffaud *et al.*, 2012). Summing up, plant and soil impact on rhizosphere microorganisms, the leading role of plant specifics in diverse natural phytocoenoses, underwent a long-term plant–microbial co-evolution which can be considered as shaping rhizosphere microbiomes in contrast to agrocoenoses where soil physico-chemical properties play the key role in this process (Philippot *et al.*, 2013).

A common point is the decrease of microbial diversity due to hard competition for root-derived organic compounds but it is difficult to depict rhizosphere biodiversity in total for there are discrepancies not only in environmental traits and plant species but in methods and research approaches as well (Philippot *et al.*, 2013). However, according to recent data, 'the diversity of microorganisms associated with the root system is enormous' and it leads to extension of plant functionality 'beyond imagination' (Bakker *et al.*, 2013). Depicting the rhizosphere in both spatial and temporal dimensions should be taken into account. At timescale the growth stage of a plant does matter. There is a range of research data showing the shifts in rhizosphere microbial communities during a plant life-cycle (van Overbeek and van Elsas, 2008). Spatial organization of the rhizosphere results in non-random and heterogenous distributions of microorganisms due to discrepancies in physiological activity of root systems in different parts and zones, and depends on root growth rate (Cardon and Whitbeck, 2007). For instance, bases and tips of roots show quite different proportions of fast-growing bacteria (Folman *et al.*, 2001). The most bacterial numbers were reported from the root elongation zone (Jaeger *et al.*, 1999).

8.3.1 Bacteria in rhizosphere

Numerous data demonstrate the dissimilarity of the rhizosphere and bulk soil in qualitative and quantitative microbial composition. In wild oat (*Avena fatua*) rhizosphere the richness of 147 from 1917 bacterial taxa was significantly different from those in surrounding soil (De Angelis *et al.*, 2009). According to many experimental data, dominant rhizosphere bacterial taxa include Proteobacteria (especially, Pseudomonadaceae and Burkholderiaceae), Actinobacteria and Firmicutes; less documented is the presence of Verrucomicrobia and Nitrospirae (De Angelis *et al.*, 2009; Mendes *et al.*, 2011; Philippot *et al.*, 2013; Turner *et al.*, 2013). Proteobacteria are known to be r-strategists able to consume a wide range of root-released organic compounds (Philippot *et al.*, 2013). It is considered that bacterial ability for organic acids (not sugars) utilization is critical for successful rhizosphere colonization (Bloemberg and Lugtenberg, 2001). Rhizosphere bacteria can influence root exudation as it was shown that axenically grown plants differ from non-axenical in root exudate composition (Micallef *et al.*, 2009).

The functional group of rhizosphere microorganisms most notorious and promising for application is PGPR (plant growth-promoting rhizobacteria), facilitating plant growth by direct or indirect influence. According to action mode they are split into fertilizers (diazotrophs providing plant with available nitrogen, phosphate-solubilizing bacteria), phytostimulators (directly enhancing plant growth or stress-tolerance by hormone production), and biocontrol agents (protecting plants from pathogens). The most efficient nitrogen fixation and biofertilization is provided by legume-associated root-nodulating bacteria (*Rhizobium*, *Sinorhizobium* and some others), but a range of free-living bacteria are capable of diazotrophic nutrition too. In the rhizosphere diazotrophs are represented predominantly by *Azospirillum* (Bloemberg and Lugtenberg, 2001). Another bacterial impact on soil fertility (though not restricted to rhizosphere and PGPR) is based on its ability to release nutrients from recalcitrant compounds for its own and plant nutrition causing bioweathering. Until recently weathering bacteria were detected in mycorrhizospheres only, but current data on its presence in mangrove trees and desert plants imply the possibility that the microorganisms could support plants in a stressful environment. Besides, some agents of weathering were revealed in soybean and temperate tree species rhizospheres. It was shown that weathering is more intense in root vicinities relative to bulk soil owing to the rhizosphere's different pH status and plant and microbial activities. The roots themselves can contribute to the process by releasing organic acids and mechanical disintegration of soil particles (Uroz *et al.*, 2009). The main substrates microbially weathered in the rhizosphere are phosphate (*Achromobacter*, *Acinetobacter*, *Agrobacterium*, *Azospirillum*, *Bacillus*, *Burkholderia*, *Citrobacter*, *Pseudomonas*, *Serratia*) and hydroxyapatite (*Rhanella*) (Uroz *et al.*, 2009). Mechanisms underlying the bacterial weathering are not always clear, they include oxidoreduction reactions, acidification and chelation. Amounts of macronutrients (carbon, nitrogen, phosphorus) available can influence weathering potential metabolites production by bacteria (Uroz *et al.*, 2009).

Some rhizosphere bacteria can facilitate plant iron nutrition by producing siderophores. Such an effect was demonstrated for fluorescent *Pseudomonas* species producing the high-affinity chelators, pyoverdines (Vansuyt *et al.*, 2007). Another PGP effect of iron chelation results in iron removal from the rhizosphere, making it unavailable to pathogens and thus protecting the plant (Alabouvette *et al.*, 2006).

Bacterial phytostimulation is based on direct influence. For instance, *Azospirillum*, mentioned as a diazotroph, secretes auxins and, in smaller amounts, cytokinins and gibberellins. It is considered that its auxin production contributes to the PGP effect rather than nitrogen fixation (Bloemberg and Lugtenberg, 2001).

PGPR can act as a first line of plant defense against pathogens. Several mechanisms can be involved: direct competition for nutrients, parasitism, niche exclusion, induction of systemic resistance or active (predominantly, against fungi) metabolites production (Bloemberg and Lugtenberg, 2001). The phenomenon was studied in detail for biocontrol strains of fluorescent pseudomonads (Haas and Défago, 2005). The most common classes of antifungal metabolites are phenazines, pyrrolnitrin, 2,4-diacetylphloroglucinol (DAPG), pyoluteorin and some lipopeptids, active against pathogen oomycete *Pythium* (Bloemberg and Lugtenberg, 2001). Rhizosphere non-pathogenic *Bacillus* and *Pseudomonas* spp. can induce a systemic resistance in a plant by priming it for activation of different protective responses and enhance production of defense secondary metabolites such as glucosinolates and some others of yet unknown structure, or facilitate cell wall reinforcement to protect the plant from pathogens or grazers. Bacterial compounds involved include siderophores and salicylic acid (van de Mortel *et al.*, 2012; Zamioudis and Pieterse, 2012). Rhizosphere bacteria cause the phenomenon of soil suppressiveness (control of plant diseases of fungal and bacterial origin) either by antifungal compound (such as DAPG) production, or by withdrawing pathogens from root-deriving carbon resources. Thus *Pseudomonas* spp. proved to be antagonistic to cereal-attacking *Gaeumannomyces graminis*. Proteobacteria, Firmicutes and Actinobacteria (notable for producing a wide array of compounds active against bacteria, viruses, fungi and invertebrates) are involved in suppression of root-rot causing *Rhizoctonia* (Mendes *et al.*, 2011; Turner *et al.*, 2013).

Synergism between PGPR can supply a plant with additional benefits. DAPG, an antifungal metabolite toxic for some nematodes, produced by *Pseudomonas* spp., was shown to increase gene expression in another PGPR *Azospirillum braziliense* leading to more effective root colonization and further plant growth promotion (Combes-Meynet *et al.*, 2011).

Another interesting mechanism of rhizosphere bacteria interactions is quorum quenching, antagonists' interference in cell to cell communication known as quorum sensing. This phenomenon can lead to expression either of beneficial or deleterious features in rhizosphere microorganisms. Quorum quenching can be an efficient tool for biocontrol, but pathogens possessing the same abilities can interfere with quorum-regulated bioactive compound synthesis thus competing with beneficial rhizosphere bacteria. Recently revealed mechanisms of quorum quenching comprise impairing or aborting of signal molecule production, signal molecule inactivation or signal perception disorder (Rasmussen and Givskov, 2006; Raaijmakers *et al.*, 2009).

8.3.2 Micromycetes in rhizosphere

Microfungal biota in the rhizosphere zone is studied far less than the bacterial, but there is obvious parallelism in plant-beneficial microfungal and bacterial functions. Namely, rhizosphere microfungi can also exert the PGP effect in bioweathering and biocontrol. Microfungi in rhizosphere contribute little to biofertilization compared to bacteria, lacking nitrogen-fixing ability and having only limited weathering capability relative to mycorrhizal fungi (Hoffland *et al.*, 2004). P-solubilizing activity was recognized in the genera *Aspergillus* and *Penicillium* (Whitelaw, 2000), ubiquitous soil inhabitants with some species showing preference for rhizosphere instead of bulk soil (Voronina, 2011).

More pronounced micromycetes' rhizosphere effects concern plant infection and, *vice versa*, biocontrol potential. Complex antagonistic interactions between microbial and fungal "plant attackers" and "plant protectors" are critical for plant performance both at individual and community level.

Contrary to bacteria, many fungal plant pathogens (considering both true fungi and oomycetes) are soilborne and their establishment in rhizosphere turns this niche into a "battlefield". Soilborne fungal pathogens (*Fusarium, Gaeumannomyces, Phytophthora, Pythium, Rhizoctonia, Verticillium*) are predominantly necrotrophic or semibiotrophic, and non-specific in contrast to aerial ones that underwent a co-evolution with their hosts, thereby plants have no specific resistance to them (Raaijmakers *et al.*, 2009).

The hypocrealean micromycete *Trichoderma* is the most noted and commercially applied rhizosphere fungal biocontrol agent. Its species is applied as a basis for biofungicides to control a range of both aerial and soilborne fungal phytopathogens. Strains of *Trichoderma* have a wide spectrum of antifungal activity from cell wall lytic enzyme production to induction of plant defense systems (Harman *et al.*, 2004). Some antagonistic strains also engaged in plant nutrition facilitation, thus having multiple impacts on plants (Girlanda and Perotto, 2005). Other common rhizosphere micromycetes with promising biocontrol abilities are *Gliocladium* and non-pathogenic *Fusarium oxysporum* strains (Raaijmakers *et al.*, 2009). A range of antimicrobial compounds produced by *Trichoderma* and *Gliocladium* participate in activity against pathogenic bacteria, fungi and nematodes (Kubicek *et al.*, 2001). It was revealed that in fungal cells phenazines affect the electron transport chain, hydrogen cyanide affects metalloenzymes, biosurfactants and DAPG break membrane integrity, but the mechanisms of action are not yet known for all active metabolites involved (Haas and Défago, 2005; Raaijmakers *et al.*, 2009). Hyperparasitism is another plant pathogen biocontrol manifestation. This mode of action was recognized in *Trichoderma* and *Gliocladium* against *Gaeumannomyces, Rhizoctonia, Sclerotinia,* and *Verticillium* (Harman *et al.*, 2004). *Trichoderma* species produce chitinases and cellulases, releasing signal molecules and triggering chemotropism to target pathogens. Hyphal contact between *Trichoderma* and its "prey" results in cell wall digestion of the latter and penetration hyphae by *Trichoderma* (Woo *et al.*, 2006). Endochitinases produced by *Gliocladium virens* cause cell wall damage in *Botrytis cinerea* thereby ensuring a biocontrol effect on the pathogen (Di Pietro *et al.*, 1993).

But the pathogen–plant protective microorganism duel does not always result in success for the latter. Antagonistic interactions between "plant attackers" and "plant protectors" can be based on interference with the biosynthesis of active compounds by beneficial microorganisms. For the first time it was shown for fluorescent *Pseudomonas*-DAPG producing strain. The biosynthesis of bacterial active metabolite was specifically repressed by fusaric acid produced by pathogenic *Fusarium oxysporum* f. sp. *radicis-lycopersici*. Further studies demonstrated variation in DAPG biosynthesis sensitivity to fusaric acid within *Pseudomonas* strains (Notz *et al.*, 2002; Duffy *et al.*, 2004; Raaijmakers *et al.*, 2009).

Antagonistic interactions between two fungal species can be illustrated by mycotoxin-producing *Fusarium* and mycoparasitic *Trichoderma* signal exchange. *Fusarium culmorum* and *F. graminearum* were shown to produce deoxynivalenol, a metabolite repressing chitinase production in *Trichoderma atroviride* affecting its *nag1* chitinase gene (Lutz *et al.*, 2003).

Pure rhizospheres' *sensu stricto* existence is not a common event in natural environments compared to mycorrhizospheres. Thus the data is restricted to a relatively narrow range of non-mycorrhizal or facultatively mycorrhizal plants, predominantly cultivated and genotypically far from the wild type and grown under artificial condition. Many modern researches on rhizosphere microbial composition and functionality were conducted with *Arabidopsis thaliana*, a well studied model plant lacking mycorrhiza and thus having a "pure" rhizosphere (Bulgarelli *et al.*, 2012; Lundberg *et al.*, 2012). That is why depicting this microniche from a "microbiocentric" view is far from complete and compels more research conducted in natural environments.

8.4 Mycorrhizosphere, the Niche Influenced both by Roots and Associated Mycobionts

The mycorrhizosphere is a more complex system than the rhizosphere owing to the presence of an additional key figure, a symbiotic fungus intimately interacting with the host root system. So, as a soil microhabitat, it undergoes not only the roots' but also its mycobionts' influence too. Up-to-date classification of mycorrhizae is based on symbiont interface ultrastructure and plant and/or fungus partner taxonomy, and recognizes seven types of symbiosis (Brundrett, 2004). Only two types, arbuscular and ectomycorrhizal, are within the scope of the chapter because of their highest environmental and application significance and, subsequently, the better state of research. At the first stage of mycorrhizal symbiosis exploration only two partners, plant and its mycobiont, were taken into account and attributed with all symbiotic-related beneficial functions. Now with a multitrophic concept of mycorrhiza establishment (Timonen and Marschner, 2006; Smith and Read, 2008) and progress in research techniques, the key role of associative mycorrhizosphere microbiota became obvious. Mycorrhizal symbiosis is extremely widespread in nearly all plant phyla and critically important for plant individuals and communities (Smith and Read, 2008). Plant beneficial mycorrhizal functions contributed by mycorrhizosphere organisms are summarized in Table 8.1.

Being analogous to the rhizosphere in its location close to roots with its exudates, the mycorrhizosphere should be considered as an independent type of soil microhabitat for the reason considered by Timonen and Marschner (2006) 'We may choose to call the mycorrhizosphere just rhizosphere with its associating microorganisms, these including the mycorrhizal fungi. However, this can easily be misleading in the case of mycorrhizal roots, which are an intimate relationship of a plant and a multicellular organism reaching far beyond the immediate vicinity of roots'. Apart from the influence of extraradical mycelial networks (here considered as hyphosphere), even the intraradical part of mycobiont contributes to mycorrhizosphere shaping. Rhizodeposition, a key point for root-associated microbiota, changed in mycorrhizosphere both quantitatively and

Table 8.1. Mycorrhizal functions contributed by associative microorganisms (based on data from Smith and Read, 2008).

Function	Mechanism	Associative microorganisms contribution
Individual plant level		
Nutrient (P, N) uptake enhancement	Depletion zone overstepping	–
	Increasing root-soil interface	–
	Recalcitrant substrates solubilizing	+
Resistance to metal ion toxicity enhancement	Sequestration in the fungal hyphae	–
	Chelating	+
Soil-borne pathogen protection	Root exudates sequestration	+
	Mechanic isolation of root tip (ectomycorrhiza)	–
	Antagonistic metabolites production	+
	Plant defense responses induction	+
Plant community level		
Competitive advantages	Host growth and health promotion	+
	Non-host suppression	+
Nutrient cycling enhancement	Nutrient supply to soil and further biochemical transformation	+
Soil detoxification	Metabolizing pollutants to non-toxic compounds	+

qualitatively due to partial sequestration by mycorrhizal fungi and partial replacement of rhizodeposits with mycobionts' exudates (Jones *et al.*, 2004). However, this effect turns out to be fungus specific: it was detected for basidiomycete mycobionts of ectomycorrhiza and was not revealed in glomalean arbuscular mycorrhiza (Rambelli, 1973; Laheurte *et al.*, 1990; Rillig, 2004). Mycorrhizal symbionts' nutrient exchange causes soil chemistry changes in the mycorrhizosphere compared to bulk soil. Nearly twice the increase of total carbon and nitrogen was detected in *Abies lasiocarpa* ectomycorrhizosphere in sub-boreal forest compared to surrounding soil (Arocena *et al.*, 1999).

The mycorrhizosphere harbours microbial communities initiating as rhizosphere ones but changing under direct influence of the mycobiont or indirect impact of the host plant's changed metabolism (Linderman, 1988). It provides more diverse microhabitats within, relative to a pure rhizosphere, including interface between symbionts and extraradical mycorrhizal fungus surface. Closeness of mycobiont may be unfavourable for soil microorganisms due to production of secondary metabolites but exudates and lysates from both symbiotic partners can provide an advantage. Besides trophic interactions, mycorrhizal root can provide microbiota with a compartment safe from draught and predators (Johansson *et al.*, 2004). Physical volume of the mycorrhizosphere often exceeds that of the rhizosphere because of intense lateral root branching caused in the plant by mycobiont-derived hormones (Gogala, 1991). All these factors contribute to a mycorrhizosphere complex microorganism community establishment conversely influencing mycorrhizal symbiosis in beneficial or antagonistic ways, determining its establishment and functioning.

8.4.1 Bacteria in mycorrhizosphere

Of a set of beneficial mycorrhizosphere microorganisms, mycorrhization helper bacteria (MHB) make up a group of extreme importance for plant-fungus symbiotic relationships. It was noted that some strains of *Bacillus* and *Pseudomonas* have a positive effect at ectomycorrhizal symbiosis formation and the term MHB was introduced (Garbaye, 1994). Now the MHB effect is shown for *Arthrobacter, Azotobacter, Bacillus, Burkholderia, Paenibacillus, Pseudomonas, Ralstonia, Streptomyces* and some others (Frey-Klett *et al.*, 2007; Kataoka and Futai, 2009). MHB display specificity both to pathogens and mycorrhizal fungi, enhancing mycorrhization with one species and inhibition with another. Even at the fungal strain level responses can be apparently different. Antibiotic production is a probable reason for sensitive fungal species inhibition by MHB strain *Streptomyces* sp. AcH 505 while promoting resistant ones (Frey-Klett *et al.*, 2007). However, a single strain of *Pseudomonas montelli* promoted both arbuscular and ectomycorrhiza formation in *Acacia holosericea*, and ectomycorrhizal symbiosis was successfully established with mycobionts from two genera (Duponnois and Plenchette, 2003). MHB are able to decrease amounts of antifungal metabolites in a mycorrhizosphere by direct antagonism against microorganisms deleterious for mycorrhizal fungi and influenced mycobionts at different stages of life cycle from spore germination until symbiosis establishment (Frey-Klett *et al.*, 2007). The MHB effects known by now include stimulation of spore germination, mycelial growth and extension, increasing symbionts' contact zone by root and hyphae branching stimulation and reducing negative environmental impact on mycorrhizal fungi. Each stage can be caused by several mechanisms involved. For instance, mycelial growth promotion can be achieved by growth factor supply, inhibition of antagonists or by detoxification of its active substances (Garbaye, 1994; Frey-Klett *et al.*, 2007). The group of helper bacteria plays multiple roles in the mycorrhizosphere acting as biofertilizers via nitrogen fixation and mineral solubilizing and biocontrol agents of root plant pathogens, once more proving multifaceted interaction within "sphere" habitats microbiomes.

Another interesting mycorrhizosphere trait is selection towards bacteria with high weathering potential. By combining soil analyses with cultivation-dependent techniques a significant enrichment in such microorganisms (*Burkholderia* was the most efficient) was detected in oak and beech mycorrhizospheres compared to bulk soil. It may indicate some indirect nutritional strategies possessed by trees to facilitate weathering by selecting causal microorganisms, though the mechanisms of such selection have not been elucidated yet (Calvaruso *et al.*, 2010).

Nitrogen-fixing bacteria stimulation in the mycorrhizosphere compared with other soil zones is argued because of a controversial data presence. Increased nitrogen fixation was demonstrated for the mycorrhizal root tip vicinity (Linderman, 1988) but it is a challenge to delimit free-living and associated diazotrophs in the natural environment. Data on fluorescent pseudomonads show no significant difference between bulk soil and mycorrhizosphere bacteria in nitrogenase activity (Frey-Klett *et al.*, 2005). Despite such contradictions, nitrogen-fixing microorganisms increase in the mycorrhizosphere is expected for root, its mycobiont and associated microbial complex respiration provide a low-oxygen microhabitat convenient for nitrogenase performance. Diazotroph association with ectomycorrhiza was revealed in *Pseudotsuga menziesii* – *Rhizopogon vinicolor* symbiosis (Li *et al.*, 1992). These interactions might be kinds of mutualistic ones with bacteria providing nitrogen and fungus carbon supply in the form of exudates.

It is clear that the mycorrhizosphere microbiome is dependent in its composition on the bulk soil microorganism reservoir and a range of abiotic factors. Non-specific soil inhabitants often present in the mycorrhizosphere, as well as in the pure rhizosphere, along with specific ones. Mycorrhizas of the same species have different microbial suites in different soil types while selective mycorrhizosphere function relative to bacteria plays its role (Bending *et al.*, 2002).

8.4.2 Micromycetes in mycorrhizosphere

Micromycetes colonizing the mycorrhizosphere were paid less attention compared to bacteria. These interactions were studied insufficiently, and data obtained under natural conditions *in situ* are lacking. The best studied aspect is microfungal antagonism with soilborne pathogens such as *Phytophthora, Pythium, Fusarium, and Cylindrocarpon* decreasing in the mycorrhizosphere compared with non-mycorrhizal roots (Timonen and Marschner, 2006). Mycorrhizal fungi are able to stimulate non-pathogenic micromycetes in contrast to deleterious ones. For example, *Glomus* extraction was shown to inhibit pathogenic *Fusarium oxysporum* conidia germination but to stimulate this process in the pathogen biocontrol agent *Trichoderma harzianum* (Filion *et al.*, 1999). Some ectomycorrhizal species (*Laccaria laccata*) *in vitro* demonstrated mycoparasitism against soil-borne microfungi, but it is not known if it can take place in the natural environment (Werner and Zadworny, 2003). Antagonistic effects of mycobionts on pathogenic micromycetes can exist due to: rhizodeposition sequestration aborting pathogen chemical attraction; exclusion of pathogens from root surface; production of fungicidal or fungistatic compounds; inducing plant defense responses (colonization by mycobiont as priming); and plant health promotion. The question on any micromycete species specificity to the mycorrhizosphere is still open. Data obtained from the natural environment by culture-dependent methods demonstrate some quantitative tendencies to restructuring of dominant and frequent species complex in the mycorrhizosphere compared to other microhabitats. No exclusively mycorrhizosphere inhabitants were detected within species with relatively high frequency (Voronina, 2011). Summerbell (2005a) reported a lack of micromycete high species specificity to an ectomycorrhizal mantle and postulated high levels of similarity between rhizosphere and mycorrhizosphere micromycete communities.

Considering the mycorrhizosphere in a narrow sense, devoid of free soil mycelia, here treated as hyphosphere, we assume that this microhabitat is more suitable for unique bacterial assemblage establishment, than for a microfungal one. But more data obtained with modern *in situ* approaches allowing unculturable organism detection are urgently needed to resolve this ambiguous case.

8.5 Hyphosphere, the Niche Influenced By Fungi

Nearly all types of soil host enormous amounts of fungal hyphae occupying a wide range of substrates due to diverse enzyme activity. Only ectomycorrhizal fungal free mycelium can contribute up to 40% of boreal forest soil microbial biomass (Högberg and Högberg, 2002). Turnover of its biomass is of great importance for forest carbon and nitrogen cycles and necromass decomposition potentially provides significant nutrient input in forest soil (Fernandez *et al.*, 2016). Mycelial modes of life provide ample possibilities for osmotrophic nutrition making fungi efficient in decomposing, bioweathering and nutrient turnover enhancing (Hoffland *et al.*, 2004; Smith and Read, 2008; Uroz *et al.*, 2009). The hyphal surface and a zone around it undergoing the influence of fungal metabolism create a specific microhabitat for soil microbiota named the hyphosphere. This niche varies according to fungal trophic guild and encircles either free-living saprotrophic organisms in total or a free-living extraradical mycelial part of the mycorrhizal symbiotrophs. While ectomycorrhizal fungi often develop vast mycelia, interconnecting trees in a single network with possibilities for common nutrient flow, arbuscular mycorrhizal Glomeromycota often have only a small extraradical part compared to the intraradical one (Leake *et al.*, 2004; Smith and Read, 2008). Different sources of carbon supply and discrepancy in decomposition activity give rise to specific traits of hyphospheres treated as a habitat for soil bacteria and micromycetes. To date nearly all hyphosphere-related research was focused on mycorrhizal fungal mycelia; data on saprobic species is lacking.

Mycelial metabolism drastically changes soil physico-chemical properties especially when hyphal aggregates such as perennial mats are forming. Ectomycorrhizal mat-forming species like *Hysterangium setchellii* and *Gautieria monticola* are known to provide access to nutrition for host plants by means of acidification-based bioweathering (Griffiths *et al.*, 1994). As the rhizosphere community goes round roots and rhizodeposits, the hyphosphere one is focused on hyphae exudates and secretion. Thus, in extraradical ectomycorrhizal mycelium plants sugars are transformed to specific fungal compounds such as trehalose and mannitol thus favoring microorganisms able to metabolize them (Söderström *et al.*, 1988). Fungal hyphae ability for excretion of oxalates, predominantly calcium oxalate, is well acknowledged. In the form of mono- or dihydrate calcium oxalate further crystallize on the mycelial surface or in hyphosphere stimulating oxalotrophs (Verrecchia *et al.*, 2006). Apart from direct biotic interactions, calcium oxalate formation by fungi contributed to soil geochemistry as calcium reservoir and important phosphate ability determinant (Fomina *et al.*, 2006).

The second dimension of the hyphosphere effect implies phoric rather than trophic interactions. Fast-growing hyphae, in addition capable of tunnelling not only soil aggregates but primary minerals as well are exploited by soil bacteria and sometimes another fungi too (Agerer, 2001; Hoffland *et al.*, 2004; Warmink and van Elsas, 2009; Warmink *et al.*, 2011; Nazir *et al.*, 2012a) as “highways” for “hitchhiking”.

8.5.1 Bacteria in hyphosphere

A wide range of bacterial genera is associated with fungal mycelium. It encompasses *Arthromonas, Burkholderia, Cellvibrio, Chondromyces, Chryseobacterium, Collimonas, Dyella, Flexibacter, Paenibacillus, Pseudomonas, Rhanella, Sphingomonas*, and *Streptomyces* (Haq et al., 2014). *Burkholderia* and *Pseudomonas* are the most diverse and abundant (De Boer *et al.*, 2005; Frey-Klett *et al.*, 2005). Similarly to the mycorrhizosphere, representatives of one genera can inhabit both the arbuscular and ectomycorrhizal fungal mycelial zone. Ability to consume fungal-derived compounds, especially low-molecular-weight organics is critical for hyphosphere bacteria; this can be a factor shaping microbial populations in the microniche. Fluorescent *Pseudomonas* strains from the hyphosphere of *Laccaria bicolor* were notable for preferential trehalose utilization thus differing from bulk soil matches (Frey *et al.*, 1997). In some cases production of such compounds is induced by a microorganism-consumer, as it was shown for glycerol-producing strain of saprobic fungus *Lyophyllum* and hyphosphere-specific bacterium *Burkholderia terrae* (Nazir *et al.*, 2012b). In arbuscular mycorrhizal fungi hyphosphere high abundance of bacteria of the Oxalobacteraceae family was detected which may denote some specific interactions between two groups of organisms, but the mechanism of microbial selection was not elucidated. Also fungal impact on bacterial attachment to hyphae was recognized (Scheublin *et al.*, 2010).

Mechanisms involved in fungus–bacteria interactions in soils include secretion systems for bacterial effector proteins (three and four types - T3SS and T4SS) (Warmink and van Elsas, 2008); chitinase production by bacteria to control or parasitize the fungus; biofilm formation genes (*B. terrae* can form biofilms along hyphae); and fungal-released compounds stimulating bacteria (Haq *et al.*, 2014).

The hyphosphere is considered to be a gene transfer arena where hyphae-derived metabolites direct bacterial growth promoting contacts through genetic interactions. A number of the genes migrate across the microbial community by means of plasmids which are essential for hyphosphere evolution. Recently evolutionarily important bacterium-to-fungus gene transfers were recognized in hyphosphere and, taking into account high probability of increase in recombination frequencies, this phenomenon can be a mechanism of beneficial microorganisms' local selection (Zhang *et al.*, 2014).

Microbial "hitchhikers" travelling in the soil with fungal growing hyphae is another specific hyphosphere effect (Warmink *et al.*, 2011; Nazir *et al.*, 2012a; Simon *et al.*, 2015). Bacterial migration within soil compartments is far more restricted compared to plant root and fungal hyphae directional growth. To arrive at a new habitat hyphosphere bacteria use fungal mycelium as transport (Warmink and van Elsas, 2009). This effect is a possible cue of particular bacterial groups' ecological success. With fungus "help" they can pass insurmountable barriers such as aerial gaps within soil particles. Such travelling bacteria have an ability to utilize fungal release substances near mycelial growth tips. "Hitchhiker" bacteria can be split into two groups. Single-strain migrators travel with the fungus as a single species (*B. terrae*) and co-migrators which can't move with the hyphae when present as a single strain (Warmink and van Elsas, 2009). Subsequently it was found out that capacity to travel with a hypha is widely distributed within four *Burkholderia* phylotypes isolated from soils with different properties (Nazir *et al.*, 2012a). The migration helper effect resulting in helpers (e.g. *B. terrae* BS001) promoting migration of the other within the soil was recently recognized. This effect seems to be specific and does not exist in some strain combinations. Probably, movement facilitation should be repaid by some other benefits (Warmink *et al.*, 2011). Thus migration ability of bacteria in a particular community is determined by abundance of migration-proficient helpers. Current concepts on fungal-bacterial interactions with mechanisms underlying the hyphosphere effect on bacteria are outlined in Haq *et al.* (2014).

8.5.2 Micromycetes in hyphosphere

Hyphosphere constitutes a microhabitat suitable for soilborne microfungi development, but there are only few research data on its numerical characteristic and community composition. Classical research by Staněk (1984) denotes mycopathogenic fungi and oomycetes (both necrotrophic and biotrophic), like *Didymella*, *Pythium*, and *Rhizoctonia*, and pointed out microbial antagonism against these disease agents. Saprotrophic anamorphic soil-borne fungi undergo profound influence of ectomycorrhizal basidiomycetes mycelia resulting both in quantitative and qualitative micromycete community shifts (Velikanov and Sidorova, 1997, 1998; Sidorova *et al.*, 2017). Species like *Cantharellus cibarius, Boletus edulis, Amanita phalloides, A. muscaria, and Lactarius mitissimus* were shown to decrease microfungi species diversity but effect was exerted with different intensity. Another type of effect was changing of dominant micromycete species cohort in the hyphosphere (Velikanov and Sidorova, 1998).

Mechanisms of interactions within the fungal hyphosphere community are not yet clear; an ability to consume hyphal exudates in microfungi, the same as for bacteria, and the subsequent antagonism with other guilds of soil biota can be proposed as probable causes.

The hyphosphere (treated as mycorrhizosphere by many authors in the case of mycorrhizal fungi mycelium) deserves further intense research with emphasis on such important though unclear trophic guilds as macrofungal saprotrophs. Some, yet not all, mechanisms of interaction were elucidated for bacteria–fungus interactions which can be applied in biocontrol or bioremediation programme development, but on fungus–fungus relationships information is lacking.

8.6 Conclusion

All three "sphere" microniches discussed reflect intricate webs of numerous trophic along with non-trophic interaction between the multiple dwellers. All of them are similar in the presence of a core: a root system, a mycorrhizal root, or a hypha, for rhizo-, mycorrhizo-, or hyphosphere, which subsequently engineers the zone and acts as an edificator for the consortium of associated organisms. The consortium feedback is essential; it can be expressed in an array of direct/indirect, synergistic/antagonistic, mutualistic/exploitive and other types of relationships, with mechanisms often obscure, within the community. All the niches contain "friends" and "foes", beneficial and deleterious for the core organisms and its mode of action is based on the same principles: signal and metabolite exchange, competition for nutrients, direct exclusion, etc. But it will be a mistake to mix these niches together for the striking discrepancy of its core, leading to multiple distinctions in physico-chemical characteristics of the niche and thereby shaping unique for each of "sphere" myco- and microbiome. Its uniqueness is more complicated and hard to depict simplistically, because specificity here is attributed not to species or other taxa, naturally originated from ambient soil or litter, but to the community as a whole. Species and locality specifics contribute to the problem too, but microorganism communities within and beyond the "sphere", as a rule demonstrate significant difference both in numbers and in composition. This had been stressed again and again. Graham (1988) pointed at physical, chemical and microbiological features sharply dissimilar in rhizosphere and mycorrhizosphere, which inevitably influenced microbial population numbers and diversity in a root zone. More recent authors defined mycosphere (hyphosphere) as this: 'The microhabitat provided by soil fungi allows soil bacteria to deal with unique resources provided by the fungal partner that would not be available in the bare soil. It is well possible that long-lasting bacterial–fungal associations are based on a "give and take" policy, ensuring that mutual benefits are warranted for both partners of the association. Here, we highlighted the mycosphere as a highly specific habitat of the soil and discussed how mycosphere-inhabiting organisms may adapt to the respective niches offered by the soil fungi, in addition to considerations as to what factors are primordial in their adaptation' (Haq *et al.*, 2014). Figure 8.2 illustrates microbial communities' dissimilarity assessed by frequency ranks between bulk soil, ectomycorrhizosphere and hyphosphere (ten agaricoid basidiomycete species). Each habitat formed a separate cluster without intermixing with others, with "spheres" related to each other closer than each of them to bulk soil.

There is a strong urge for integration of the experimental data obtained from different soil types, plant communities, etc., to enlighten this question, but it is a very hard task. Major obstacles arise due to natural continuity of soil and its compartments making delimiting difficult, and in the plethora of concepts existing and a confusion of terms and definitions.

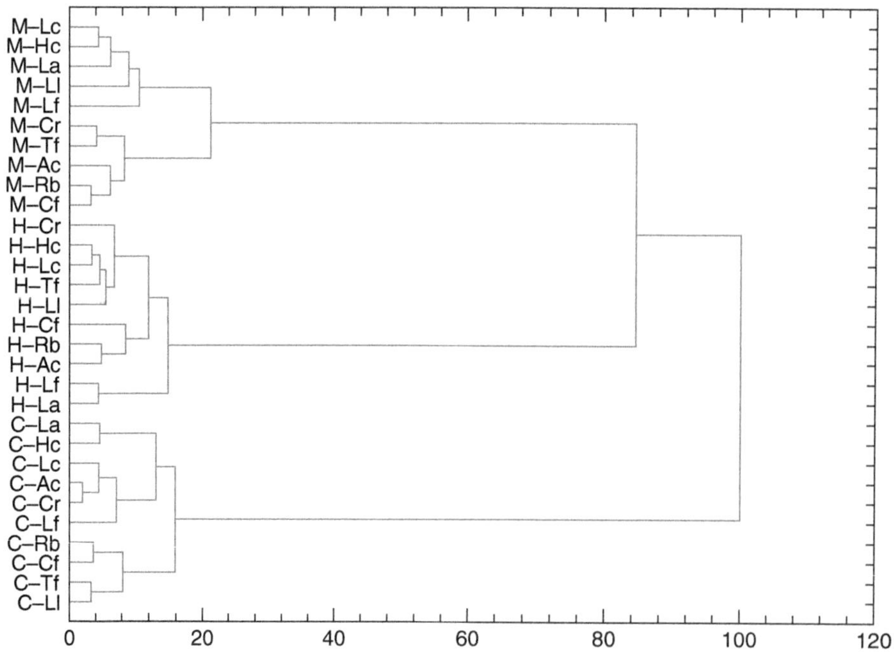

Fig. 8.2. Bacterial communities assessed by frequency ranks in bulk soil, ectomycorrhizosphere and hyphosphere (Boreal forest, Moscow Region, Russia): C – bulk soil, H – hyphosphere, M – ectomycorrhizosphere; agaricoid basidiomycete species: *Ac - Amanita citrina, Cf - Cortinarius flexipes, Cr - Cortinarius raphanoides, Hc - Hebeloma crustuliniforme, Ll - Laccaria laccata, La - Lactarius aurantiacus, Lc - Lactarius camphoratus, Lf - Lactarius flexuosus, Rb - Rhodocollybia butyracea, Tf - Tricholoma fulvum.*

By the way, lots of papers mentioning "rhizosphere" dealt with mycorrhizal plants and use the term as opposed to bulk soil. The word "mycorrhizosphere" has multiscale application, defining a space from a several millimetres long root tip vicinity to a hundreds of metres tree root system accompanied by an even more vast mycelial network. And a last but not least barrier is an absence of a single sampling and research protocol and a wide spectrum of techniques making comparative data analysis near impossible.

The soil microbiology paradigm is shifting again, from a "rhizo-" and "microbe-centric" view to "microbiome-" or even "interactome-centric". And a fourth recently introduced "sphere" is now emerging, the sapro-rhizosphere, a niche emphasizing the importance of saprotrophic fungi as food source for rhizosphere bacteria, for the former are active consumers of root exudates (Ballhausen and de Boer, 2016)

8.7 Future Trends and Perspectives

Advances in methods and techniques by the 21st century allowed us to use more ecologically relevant *in situ* approaches, and excellent reviews and experimental papers on the problem are numerous, but our competence on the nature and functions of the "spheres" is still outweighed by our ignorance. A holistic approach for understanding the mechanisms underlying interaction within the microbiome and integrative cross-disciplinary research are urgently needed.

Fundamental questions still waiting for answers are:

- What is the level of specificity of "spheres" biota (comprising both culturable and unculturable ones)?
- What are the key discriminators between rhizo-, mycorrhizo- and hyphosphere?

- What are the key biological processes taking place in "spheres" and what are their mechanisms?
- What is the role of signalling within "sphere" and beyond it?
- Is there a directed microbiota selection in "spheres" and what mechanisms are involved besides exudation?
- What roles play other biota groups apart from bacteria and some invertebrate groups in "sphere" communities?
- How stable these communities are and what is the direction of the succession?

Resolving these questions is necessary either for more large-scale and relevant practical application of "sphere" microbiota or screening new potential agents of biotechnology as well. Assuming the well acknowledged role of PGPR in sustainable low-input agriculture, contribution to promotion of valuable endangered plant species' growth and health, potential applications in crop breeding programmes, and the role of "sphere" microorganisms in bioremediation and restoration communities after disturbance, providing insight into mechanisms of its performance and interactions deserves much attention. No less attention should be paid in applied research to carefulness and caution. To sum up the discussion we make a quotation on the subject: 'Despite its complexity and dynamism, particularly in natural environments, it is important not to overlook the plant microbiome when interpreting experimental data, especially when it can lead to applications in the field. Genetic modification of plants, to resist disease for example, may have unforeseen consequences for the rest of the microbiome, which may or may not be physiologically relevant. The role of the microbiome and its relationship to plant health, productivity and biogeochemical cycles should be considered as much as the plant itself' (Turner *et al.*, 2013).

Acknowledgements

Financial support by the Russian Science Foundation (RSCF) to Elena Voronina (programme 14-50-00029) is gratefully acknowledged.

References

Agerer, R. (2001) Exploration types of ectomycorrhizae – A proposal to classify ectomycorrhizal mycelial systems according to their patterns of differentiation and putative ecological importance. *Mycorrhiza* 11, 107–114.

Alabouvette, C., Olivain, C. and Steinberg, C. (2006) Biological control of plant diseases: the European situation. *European Journal of Plant Pathology* 114, 329–341.

Arocena, J.M., Glowa, K.R., Massicotte, H.B. and Lavkulich, L. (1999) Chemical and mineral composition of ectomycorrhizosphere soils of subalpine fir (*Abies lasiocarpa* (Hook.) Nutt.) in the Ae horizon of a Luvisol. *Canadian Journal of Soil Science* 79, 25–35.

Bais, H., Weir, T., Perry, L., Gilroy, S. and Vivanco, J. (2006) The role of root exudates in rhizosphere interactions with plants and other organisms. *Annual Review of Plant Biology* 57, 233–266.

Bakker, P.A.H.M., Berendsen, R.L., Doornbos, R.F., Wintermans, P.C.A. and Pieterse, C.M.J. (2013) The rhizosphere revisited: root microbiomics. *Frontiers in Plant Science* 4, Article 165.

Ballhausen, M.B. and de Boer, W. (2016) The sapro-rhizosphere: carbon flow from saprotrophic fungi into fungus feeding bacteria. *Soil Biology and Biochemistry* 102, 14–17. Available at: http://dx.doi.org/10.1016/j.soilbio.2016.06.014 (accessed 8 July 2017).

Bending, G.D., Poole, E.J., Whipps, J.M. and Read, D.J. (2002) Characterisation of bacteria from Pinus sylvestris - Suillus luteus mycorrhizas and their effects on root-fungus interactions and plant growth. *FEMS Microbiology Ecology* 39, 219–227.

Berg, G. and Smalla, K. (2009) Plant species and soil types cooperatively shape the structure and function of microbial communities in the rhizosphere. *FEMS Microbiology Ecology* 68, 1–13.

Bisen, K., Keswani, C., Mishra, S., Saxena, A., Rakshit, A. and Singh, H.B. (2015) Unrealized potential of seed biopriming for versatile agriculture. In: Rakshit, A., Singh, H.B. and Sen, A. (ed.) *Nutrient Use Efficiency: From Basics to Advances*. Springer, New Delhi, India, pp. 193–206.

Bisen, K., Keswani, C., Patel, J.S., Sarma, B.K. and Singh, H.B. (2016) *Trichoderma* spp.: efficient inducers of systemic resistance in plants. In: Chaudhary, D.K. and Verma, A. (ed.) *Microbial-mediated Induced Systemic Resistance in Plants*. Springer, Singapore, pp. 185–195.

Bloemberg, G.V. and Lugtenberg, B.J.J. (2001) Molecular basis of plant growth promotion and biocontrol by rhizobacteria. *Current Opinion in Plant Biology* 4, 343–350.

Boddy, L., Frankland, J.C. and van West, P. (ed.) (2008) *Ecology of Saprotrophic Basidiomycetes. British Mycological Society Symposia Series, vol. 28*. Academic Press, London.

Bouffaud, M.L., Kyselková, M., Gouesnard, B., Grundmann, G., Muller, D. and Moënne-Loccoz, Y. (2012) Is diversification history of maize influencing selection of soil bacteria by roots? *Molecular Ecology* 21, 195–206.

Brundrett, M.C. (2004) Diversity and classification of mycorrhizal associations. *Biological Reviews of the Cambridge Philosophical Society* 79, 473–495.

Bulgarelli, D., Rott, M., Schlaepp, K., Ver, E., van Themaat, L. *et al.* (2012) Revealing structure and assembly cues for Arabidopsis root-inhabiting bacterial microbiota. *Nature* 488, 91–95.

Calvaruso, C., Turpault M.-P., Leclerc, E., Ranger, J. Garbaye, J. *et al.* (2010) Influence of forest trees on the distribution of mineral weathering-associated bacterial communities of the *Scleroderma citrinum* mycorrhizosphere. *Applied and Environmental Microbiology* 76, 4780–4787.

Cardon, Z.G. and Whitbeck, J.L. (eds) (2007) *The Rhizosphere: An Ecological Perspective*. Academic Press, Burlington, USA.

Chaudhry, Q., Blom-Zandstra, M., Gupta, S. and Joner, E.J. (2005) Utilising the synergy between plants and rhizosphere microorganisms to enhance breakdown of organic pollutants in the environment. *Environmental Science Pollution Research – International* 12, 34–48.

Clemmensen, K.E., Bahr, A., Ovaskainen, O., Dahlberg, A., Ekblad, A. *et al.* (2013) Roots and associated fungi drive long-term carbon sequestration in boreal forest. *Science* 339, 1615–1618.

Combes-Meynet, E., Pothier, J.F., Moenne-Loccoz, Y. and Prigent-Combaret, C. (2011) The Pseudomonas secondary metabolite 2,4-diacetylphloroglucinol is a signal inducing rhizoplane expression of *Azospirillum* genes involved in plant-growth promotion. *Molecular Plant-Microbe Interactions* 24, 271–284.

De Angelis, K.M., Brodie, E.L., De Santis, T.Z., Andersen, G.L., Lindow, S.E. and Firestone, M.K. (2009) Selective progressive response of soil microbial community to wild oat roots. *ISME Journal* 3, 168–178.

De Boer, W., Folman, L.B., Summerbell, R.C. and Boddy, L. (2005) Living in a fungal world: impact of fungi on soil bacterial niche development. *FEMS Microbiology Reviews* 29, 795–811.

Dennis, P.G., Miller, A.J. and Hirsch, P.R. (2010) Are root exudates more important than other sources of rhizodeposits in structuring rhizosphere bacterial communities? *FEMS Microbiology Ecology* 72, 313–327.

Di Pietro, A., Lorito, M., Hayes, C.K., Broadway, R.M. and Harman, G.E. (1993) Endochitinase from Gliocladium virens: isolation, characterization, and synergistic antifungal activity in combination with gliotoxin. *Phytopathology* 83, 308–313.

Duffy, B.K., Keel, C. and Défago, G. (2004) Potential role of pathogen signaling in multitrophic plant–microbe interactions involved in disease protection. *Applied and Environmental Microbiology* 70, 1836–1842.

Duponnois, R. and Plenchette, C. (2003) A mycorrhiza helper bacterium enhances ectomycorrhizal and endomycorrhizal symbiosis of Australian Acacia species. *Mycorrhiza* 13, 85–91.

Fernandez, C.W., Langley, J.A., Chapman, S., McCormack, M.L. and Koide, R.T. (2016) The decomposition of ectomycorrhizal fungal necromass. *Soil Biology and Biochemistry* 93, 38–49.

Filion, M., St-Arnaud, M. and Fortin, J.A. (1999) Direct interaction between arbuscular mycorrhizal fungus *Glomus intraradices* and different rhizosphere microorganisms. *New Phytologist* 141, 525–533.

Folman, L.B., Postma, J. and Van Veen, J.A. (2001) Ecophysiological characterization of rhizosphere bacterial communities at different root locations and plant developmental stages of cucumber grown on rockwool. *Microbial Ecology* 42, 586–597.

Fomina, M., Burford, E.P. and Gadd J.M. (2006) Fungal dissolution and transformation of minerals: significance for nutrient and metal mobility. In: Gadd, J.M. (ed.) *Fungi in Biogeochemical Cycles*. Cambridge University Press, New York, USA pp. 236–266.

Foster, R.C. and Marks, G.C. (1967) Observations on the mycorrhizas of forest trees. II. The rhizosphere of *Pinus radiata* D. Don. *Australian Journal of Biological Sciences* 20, 915–926.

Frey, P., Frey-Klett, P., Garbay,e J., Berge, O. and Heulin, T. (1997) Metabolic and genotypic fingerprinting of fluorescent pseudomonads associated with Douglas fir – Laccaria bicolor mycorrhizosphere. *Applied and Environmental Microbiology* 63, 1852–1860.

Frey-Klett, P., Chavatte, M., Clausse, M.-L., Courrier, S., Le Roux, C. *et al.* (2005) Ectomycorrhizal symbiosis affects functional diversity of rhizosphere fluorescent pseudomonads. *New Phytologist* 165, 317–328.

Frey-Klett, P., Garbaye, J. and Tarkka, M. (2007) Tansley review. The mycorrhiza helper bacteria revisited. *New Phytologist* 176, 22–36.

Gadd, J.M. (ed.) (2006) *Fungi in Biogeochemical Cycles*. Cambridge University Press, New York.

Garbaye, J. (1994) Tansley Review No. 76. Helper bacteria: a new dimension to the mycorrhizal symbiosis. *New Phytologist* 128, 197–210.

Gilbert, R.G. and Linderman, R.G. (1971) Increased activity of soil microorganisms near sclerotia of *Sclerotium rolfsii* in soil. *Canadian Journal of Microbiology* 17, 557–562.

Girlanda, M. and Perotto, S. (2005) A century of rhizosphere research: fungal interactions with plant's hidden half. *Mycological Research* 109, 1058–1061.

Gogala, N. (1991) Regulation of mycorrhizal infection by hormonal factors produced by hosts and fungi. *Experientia* 47, 331–340.

Graham, J.H. (1988) Interactions of mycorrhizal fungi with soilborne plant pathogens and other organisms. *Phytopathology* 78, 365–366.

Grayston, S.J., Vaughan, D. and Jones, D. (1997) Rhizosphere carbon flow in trees, in comparison with annual plants: the importance of root exudation and its impact on microbial activity and nutrient availability. *Applied Soil Ecology* 5, 29–56.

Griffiths, R.P., Baham, J.E. and Caldwell, B.A. (1994) Soil solution chemistry of ectomycorrhizal mats in forest soil. *Soil Biology and Biochemistry* 26, 331–337.

Haas, D. and Défago, G. (2005) Biological control of soil-borne pathogens by fluorescent pseudomonads. *Nature Reviews Microbiology* 3, 307–319.

Haq, I.U., Zhang, M., Yang, P. and van Elsas, J.D. (2014) The interactions of bacteria with fungi in soil: emerging concepts. *Advances in Applied Microbiology* 89, 185–216.

Harman, G.E., Howell, C.R., Viterbo, A., Chet, I. and Lorito, M. (2004) Trichoderma spp. – opportunistic avirulent plant symbionts. *Nature Reviews Microbiology* 2, 43–56.

Hartmann, A., Rothballer, M. and Schmid, M. (2008) Lorenz Hiltner, a pioneer in rhizosphere microbial ecology and soil bacteriology research. *Plant and Soil* 312, 7–14.

Hiltner, L. (1904) Über neurer erfahrungen und probleme auf dem gebiete der bodenbakteriologie unter besonderer berücksichtigung der gründüngung und brache. *Arbeiten der Deitschen Landwirtschaftlichen Gesellschaft* 98, 59–78.

Hoffland, E., Kuyper, T., Wallander, H., Plassard, C., Gorbushina, A. *et al.* (2004) The role of fungi in weathering. *Frontiers in Ecology and the Environment* 2, 258–264.

Högberg, M.N. and Högberg, P. (2002) Extramatrical ectomycorrhizal mycelium contributes one-third of microbial biomass and produces, together with associated roots, half the dissolved organic carbon in a forest soil. *New Phytologist* 154, 791–795.

Jaeger, C.H., Lindow, S.E. Miller, S. Clark, E. and Firestone, M.K. (1999) Mapping of sugar and amino acid availability in soil around roots with bacterial sensors of sucrose and tryptophan. *Applied and Environmental Microbiology* 65, 2685–2690.

Johansson, J.F., Paul, L.R. and Finlay, R.D. (2004) Microbial interactions in the mycorrhizosphere and their significance for sustainable agriculture. *FEMS Microbiology Ecology* 48, 1–13.

Jones, D.L., Hodge, A. and Kuzyakov, Y. (2004) Plant and mycorrhizal regulation of rhizodeposition. *New Phytologist* 163, 459–480.

Kataoka, R. and Futai, K. (2009) A new mycorrhizal helper bacterium, *Ralstonia* species, in the ectomycorrhizal symbiosis between *Pinus thunbergii* and *Suillus granulatus*. *Biology and Fertility of Soils* 45, 315–320.

Keswani, C., Singh, S.P. and Singh, H.B. (2013) A superstar in biocontrol enterprise: *Trichoderma* spp. *Biotech Today* 3, 27–30.

Keswani, C., Bisen, K., Singh, S.P., Sarma, B.K. and Singh, H.B. (2016a) A proteomic approach to understand the tripartite interactions between plant–*Trichoderma*–pathogen: investigating the potential for efficient biological control. In: Hakeem, K.R. and Akhtar, M.S. (eds) *Plant, Soil and Microbes Vol. 2. Mechanisms and Molecular Interactions*. Springer, Cham, Switzerland, pp. 79–93.

Keswani, C., Bisen, K., Singh, V., Sarma, B.K. and Singh, H.B. (2016b) Formulation technology of biocontrol agents: present status and future prospects. In: Arora, N.K., Mehnaz, S. and Balestrini, R. (eds) *Bioformulations: For Sustainable Agriculture*. Springer, New Delhi, India, pp. 35–52.

Kowalchuk, G., Hol, W. and van Veen, J. (2006) Rhizosphere fungal communities are influenced by *Senecio jacobaea* pyrrolizidine alkaloid content and composition. *Soil Biology and Biochemistry* 38, 2852–2859.

Kubicek, C.P., Mach, R.L., Peterbauer, C.K. and Lorito, M. (2001) Trichoderma: from genes to biocontrol. *Journal of Plant Pathology* 83, 11–23.

Ladygina, N. (2009) Indirect ecological interactions in the rhizosphere. Thesis, Lund University, Lund, Sweden.

Ladygina, N. and Hedlund, K. (2010) Plant species influence microbial diversity and carbon allocation in the rhizosphere. *Soil Biology and Biochemistry* 42, 162–168.

Laheurte, F., Leyval, C. and Berthelin, J. (1990) Root exudates of maize, pine and beech seedlings influenced by mycorrhizal and bacterial inoculation. *Symbiosis* 9, 111–116.

Leake, J.R., Johnson, D., Donnelly, D.P., Muckle, G.E., Boddy, L. and Read, D.J. (2004) Networks of power and influence: the role of mycorrhizal mycelium in controlling plant communities and agroecosystem functioning. *Canadian Journal of Botany* 82, 1016–1045.

Li, C.Y., Massicotte, H.B. and More, L.V.H. (1992) Nitrogen-fixing *Bacillus* sp. associated with Douglas-fir tuberculate ectomycorrhizae. *Plant and Soil* 140, 35–40.

Linderman, R.G. (1988) Mycorrhizal interactions with the rhizosphere microflora: the mycorrhizosphere effect. *Phytopathologist* 78, 366–371.

Lundberg, D.S., Lebeis, S.L., Herrera Paredes, S., Yourstone, S., Gehring, J. *et al.* (2012) Defining the core *Arabidopsis thaliana* root microbiome. *Nature* 488, 86–93.

Lutz, M., Feichtinger, G., Défago, G. and Duffy, B. (2003) Mycotoxigenic *Fusarium* and deoxynivalenol production repress chitinase gene expression in the biocontrol agent *Trichoderma atroviride* P1. *Applied and Environmental Microbiology* 69, 3077–3084.

Marschner, H (1995) *Mineral Nutrition of Higher Plants, 2nd edn*. Academic Press, London.

Mendes, R., Kruijt, M., de Bruijn, I., Dekkers, E., van der Voort, M. *et al.* (2011) Deciphering the rhizosphere microbiome for disease-suppressive bacteria. *Science* 332, 1097–1100.

Micallef, S.A., Shiaris, M.P. and Colon-Carmona, A. (2009) Influence of *Arabidopsis thaliana* accessions on rhizobacterial communities and natural variation in root exudates. *Journal of Experimental Botany* 60, 1729–1742.

Mishra, S., Singh, A., Keswani, C., Saxena, A., Sarma, B.K. and Singh, H.B. (2015) Harnessing plant-microbe interactions for enhanced protection against phytopathogens. In: Arora, N.K. (ed.) *Plant Microbes Symbiosis: Applied Facets*. Springer, New Delhi, India, pp. 111–125.

Mukerji, K.G., Manoharachary, C. and Singh, J. (ed.) (2006) *Microbial Activity in the Rhizosphere*. Springer Berlin Heidelberg, Germany.

Nazir, R., Zhang, M., de Boer, W. and van Elsas, J.D. (2012a) The capacity to comigrate with *Lyophyllum* sp. strain Karsten through different soils is spread among several phylogenetic groups within the genus *Burkholderia*. *Soil Biology and Biochemistry* 50, 221–233.

Nazir, R., Warmink, J.A., Voordes, D.C., van de Bovenkamp, H.H. and van Elsas, J.D. (2012b) Inhibition of mushroom formation and induction of glycerol release-ecological strategies of *Burkholderia terrae* BS001 to create a hospitable niche at the fungus *Lyophyllum* sp. strain Karsten. *Microbial Ecology* 65, 245–254.

Notz, R., Maurhofer, M., Dubach, H., Haas, D. and Défago, G. (2002) Fusaric acid-producing strains of *Fusarium oxysporum* alter 2,4-diacetylphloroglucinol biosynthetic gene expression in *Pseudomonas fluorescens* CHA0 in vitro and in the rhizosphere of wheat. *Applied and Environmental Microbiology* 68, 2229–2235.

Olson, P.E., Reardon, K.F. and Pilon-Smits, E.A.H. (2003) Ecology of rhizosphere bioremediation. In: McCutcheon, S.C. and Schnoor, J.L. (eds). *Phytoremediation: Transformation and Control of Contaminants*. Wiley, New York, pp. 317–355.

Philippot, L., Raaijmakers, J.M., Lemanceau, P. and van der Putten, W.H. (2013) Going back to the roots: the microbial ecology of the rhizosphere. *Nature Reviews Microbiology* 11, 789–799.

Piśkiewisz, A.M., Duyts, H., Berg, M.P., Costa, S.R. and van der Putten, W.H. (2007) Soil microorganisms control plant ectoparasitic nematodes in natural coastal foredunes. *Oecologia* 152, 505–514.

Raaijmakers, J.M., Paulitz, T.C., Steinberg, C., Alabouvette, C. and Moënne-Loccoz, Y. (2009) The rhizosphere: a playground and battlefield for soilborne pathogens and beneficial microorganisms. *Plant and Soil* 321, 341–361.

Rambelli, A. (1973) The rhizosphere of mycorrhizae. In: Marks, G.C. and Kozlowski T.T. (eds) *Ectomycorrhizae, Their Ecology and Physiology*. Academic Press, New York, pp. 299–349.

Rangel-Castro, J.I., Killham, K., Ostle, N., Nicol, G.W., Anderson, I.C. *et al.* (2005) Stable isotope probing analysis of the influence of liming on root exudate utilization by soil microorganisms. *Environmental Microbiology* 7, 828–838.

Rasmussen, T.B. and Givskov, M. (2006) Quorum sensing inhibitors: a bargain of effects. *Microbiology* 152, 895–904.

Rawlings, G.B. (1958) Some practical aspects of forest mycotrophy. *New Zealand Society of Soil Science Proceedings* 3, 41–44.

Rillig, M.C. (2004) Arbuscular mycorrhizae and terrestrial ecosystem processes. *Ecology Letters* 7, 740–754.

Scheublin, T.R., Sanders, I.R., Keel, C. and van der Meer, J.R. (2010) Characterisation of microbial communities colonising the hyphal surfaces of arbuscular mycorrhizal fungi. *ISME Journal* 4, 752–763.

Sen, R. (2005) Towards a multifunctional rhizosphere concept: back to the future? *New Phytologist* 168, 266–268.

Sidorova, I.I, Alexandrova, A.V. and Voronina, E.Yu. (2017) Hyphosphere of agaricomycetes with different trophic strategies: quantity of cultivated bacteria and micromycetes (in Russian). *Mikologiya I Fitopatologiya* 51(2), 78–89.

Simon, A., Bindschedler, S., Job, D., Wick, L.Y., Filippidou, S. *et al.* (2015) Exploiting the fungal highway: development of a novel tool for the *in situ* isolation of bacteria migrating along fungal mycelium. *FEMS Microbiology Ecology* 91, fiv 116.

Smith, S.E. and Read, D.J. (2008) *Mycorrhizal Symbiosis*, 3rd edn. Academic Press, New York.

Söderström, B., Finlay, R.D. and Read, D.J. (1988) The structure and function of the vegetative mycelium of ectomycorrhizal plants. IV. Qualitative analysis of carbohydrate contents of mycelium interconnecting host plants. *New Phytologist* 109, 163–166.

Staněk, M. (1984) Microorganisms in the hyphosphere of fungi. I. Introduction (in Czech). *Česká Mykologie* 38(1), 1–10.

Summerbell, R.C. (2005a) Root endophyte and mycorrhizosphere fungi of black spruce, *Picea mariana*, in a boreal forest habitat: influence of site factors on fungal distributions. *Studies in Mycology* 53, 121–145.

Summerbell, R.C. (2005b) From Lamarckian fertilizers to fungal castles: recapturing the pre-1985 literature on endophytic and saprotrophic fungi associated with ectomycorrhizal root systems. *Studies in Mycology* 53, 191–256.

Thornton, R.H. (1953) Features of growth of actinomyces in soil. *Research* 6, A1–A3.

Timonen, S. and Marschner, P. (2006) Mycorrhizosphere concept. In: Mukerji, K.G., Manoharachary, C. and Singh, J. (eds) *Microbial Activity in the Rhizosphere*. Springer Berlin Heidelberg, Germany, pp. 155–172.

Turner, T., James, E. and Poole, P. (2013) The plant microbiome. *Genome Biology* 14, 209.

Uroz, S. and Frey-Klett, P. (2011) Linking diversity to function: highlight on the mineral weathering bacteria. *Central European Journal of Biology* 6, 817–820.

Uroz, S., Calvaruso, C., Turpault, M.P. and Frey-Klett, P. (2009) Mineral weathering by bacteria: ecology, actors and mechanisms. *Trends in Microbiology* 17, 378–387.

van de Mortel, J.E., de Vos, R.C., Dekkers, E., Pineda, A., Guillod, L. *et al.* (2012) Metabolic and transcriptomic changes induced in Arabidopsis by the rhizobacterium *Pseudomonas fluorescens* SS101. *Plant Physiology* 160, 2173–2188.

van Overbeek, L. and van Elsas, J.D. (2008) Effects of plant genotype and growth stage on the structure of bacterial communities associated with potato (*Solanum tuberosum* L.). *FEMS Microbiology Ecology* 64, 283–296.

Vansuyt, G., Robin, A., Briat, J.F., Curie, C. and Lemanceau, P. (2007) Iron acquisition from Fe-pyoverdine by *Arabidopsis thaliana*. *Molecular Plant Microbe Interactions* 20, 441–447.

Varma, A., Abbott, L., Werner, D. and Hampp, R. (eds) (2008) *Plant Surface Microbiology*. Springer Berlin Heidelberg, Germany.

Velikanov, L.L. and Sidorova, I.I. (1997) Regulation of soil and litter myco- and microbiota by higher basidiomycetes in forest ecosystems. I. Basidiomycete effect on fungal and bacterial numbers (in Russian). *Mikologiya I Fitopatologiya* 31, 20–26.

Velikanov, L.L. and Sidorova, I.I. (1998) Regulation of soil and litter myco- and microbiota by higher basidiomycetes in forest ecosystems. II. Basidiomycete effect on soil micromycetes species diversity (in Russian). *Mikologiya I Fitopatologiya* 32, 33–36.

Verrecchia, E.P., Braissant, O. and Cailleau, G. (2006) The oxalate–carbonate pathway in soil carbon storage: the role of fungi and oxalotrophic bacteria. In: Gadd, J.M. (ed.) *Fungi in Biogeochemical Cycles*. Cambridge University Press, New York, pp. 289–310.

Voronina, E.Yu. (2011) Effect of the mycorrhizosphere on soil micromycete biodiversity and community structure and its relation to the rhizosphere and hyphosphere effects. *Microbiology* 80, 584–590.

Warmink, J.A. and van Elsas, J.D. (2008) Selection of bacterial populations in the mycosphere of *Laccaria proxima*: is type III secretion involved? *The ISME Journal* 2, 887–900.

Warmink, J.A. and van Elsas, J.D. (2009) Migratory response of soil bacteria to *Lyophyllum* sp. strain Karsten in soil microcosms. *Applied and Environmental Microbiology* 75, 2820–2830.

Warmink, J.A., Nazir, R., Corten, B. and van Elsas J.D. (2011) Hitchhikers on the fungal highway: the helper effect for bacterial migration via fungal hyphae. *Soil Biology and Biochemistry* 43, 760–765.

Werner A. and Zadworny M. (2003) In vitro evidence of mycoparasitism of the ectomycorrhizal fungus *Laccaria laccata* against *Mucor hiemalis* in the rhizosphere of *Pinus sylvestris*. *Mycorrhiza* 13, 41–47.

Whitelaw, M.A. (2000) Growth promotion of plants inoculated with phosphate-solubilizing fungi. *Advances in Agronomy* 69, 99–151.

Woo, S.L., Scala, F., Ruocco, M. and Lorito, M. (2006) The molecular biology of the interactions between *Trichoderma* spp., phytopathogenic fungi and plants. *Phytopathology* 96, 181–185.

Yao, H.Y. and Wu, F.Z. (2010) Soil microbial community structure in cucumber rhizosphere of different resistance cultivars to fusarium wilt. *FEMS Microbiology Ecology* 72, 456–463.

Zagryadskaya, Yu.A., Lysak, L.V., Lapygina, E.V., Voronina, E.Yu., Aleksandrova, A.V. and Sidorova I.I. (2011) The characteristics of bacterial communities in the hyphosphere of several basidial macromycetes. *Moscow University Soil Science Bulletin* 66, 129–133.

Zamioudis, C. and Pieterse, C.M.J. (2012) Modulation of host immunity by beneficial microbes. *Molecular Plant Microbe Interactions* 25, 139–150.

Zhang, M., de C. Pereira e Silva, M., De Mares Maryam, C. and van Elsas, J.D. (2014) The mycosphere constitutes an arena for horizontal gene transfer with strong evolutionary implications for bacterial-fungal interactions. *FEMS Microbiology Ecology* 89, 516–526.

9 The Rhizospheres of Arid and Semi-arid Ecosystems are a Source of Microorganisms with Growth-Promoting Potential

Fatima Berenice Salazar-Badillo,[1]* Silvia Salas-Muñoz,[2] Jorge Armando Mauricio-Castillo,[1] Jorge Sáenz-Mata,[3] Artemio Mendoza-Mendoza,[4] Maria Fernanda Nieto-Jacobo[4] and Johanna Steyaert[4]

[1]*Unidad Académica de Agronomía, Universidad Autónoma de Zacatecas, Cieneguillas, Zacatecas, México;* [2]*CONACYT-Campo Experimental Zacatecas, Instituto de Investigaciónes Forestales, Agrícolas y Pecuarias, Calera de V. R. Zacatecas, México;* [3]*Laboratorio de Ecología Microbiana, Facultad de Ciencias Biológicas, Universidad Juárez del Estado de Durango, Durango, México;* [4]*Bio-Protection Research Centre, Lincoln University, Canterbury, New Zealand*

9.1 Introduction

Approximately 47% of the earth's surface has been classified as arid lands (Fig. 9.1) (United Nations Environment Programme, 1992). In general terms an arid land is a region where the water supply and the values of precipitation and atmospheric moisture are lower than the annual global average (Rzedowski, 1968).

In Mexico, over 54% of the total area is classified as arid or semi-arid, corresponding to the Chihuahuan and Sonoran deserts. The Sonoran Desert covers the southwestern United States in Arizona and California, and northwestern Mexico in Sonora and the peninsula of Baja California. The Chihuahuan Desert is a large ecoregion, which comprises the states of Texas, New Mexico and Arizona in the United States, and on the Mexican side covers the states of Chihuahua, Coahuila, Durango, Zacatecas and small portions of Nuevo Leon and San Luis Potosí (Henrickson, 1974).

The weather conditions in these areas are characterized by low and erratic rainfall and marked fluctuations in temperature from 40°C in summer down to –10°C on winter nights. In addition, several factors such as the low availability of water, nitrogen and organic matter, low relative humidity and high salinity, among others, are considered major determinants for the growth and development of plants and microbes in these areas (Rodriguez-Valera, 1988). These extreme weather conditions have influenced the morphological, anatomical, physiological and molecular mechanisms of the organisms living in arid lands in order to survive. Microbial communities which survive in harsh environments such as arid ecosystems are called extremophiles (Ramírez *et al.*, 2006).

*E-mail: fatimiuxb681@gmail.com

© CAB International 2017. *Advances in PGPR Research* (eds H.B. Singh, B.K. Sarma and C. Keswani)

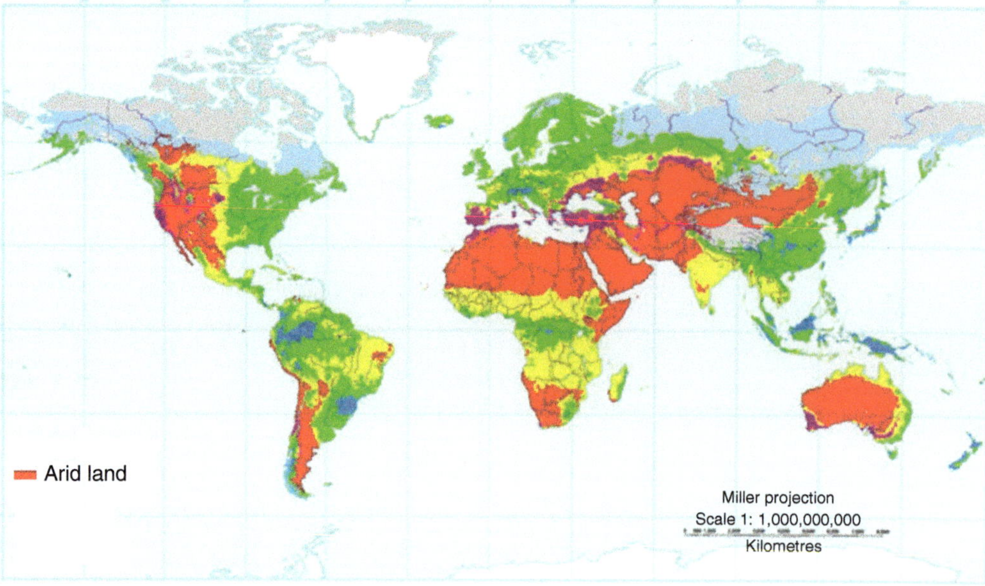

Fig. 9.1. Arid lands around the world. Arid lands around the world indicated in red (adapted from United States Department of Agriculture's Natural Resources.)

Desert soils are important reservoirs of extremophile microorganisms; the desert rhizospheres in natural ecosystems are suitable to find bacteria with high capacities to support and promote growth of plants.

Cost-efficient and easily adaptable technologies to mitigate the increasing problem of crop production loss due to climate-change-induced abiotic stress are currently being sought around the world (Venkateswarlu and Shanker, 2009; Bisen *et al.*, 2015; Keswani, 2015; Keswani *et al.*, 2016a, b). Extremophilic microorganisms have been receiving much attention recently in crop biotechnology as they are able to tolerate aggressive conditions and in many instances promote the growth of plants with which they associate (Van Den Burg, 2003).

9.2 Extremophile Microorganisms

Some microorganisms that are extremophiles have been shown to have a beneficial impact on growth promotion and abiotic stress tolerance induction in crops. In part, this is due to their ability to colonize the rhizosphere and/or endorhizosphere of plants. Interestingly, extremophile microbes have the ability to induce growth promotion by direct or indirect mechanisms in plants (Rodriguez-Valera, 1988; Ventosa *et al.*, 2008). Indirect mechanisms include suppression of plant pathogens by competitive production of antibiotics, siderophores and extracellular hydrolytic enzymes, stimulating the establishment of other beneficial microbes such as mycorrhizae and rhizobium, or/and removal of phytotoxic substances, allelopathy and competition with deleterious agents (Glick, 1995; Figueiredo *et al.*, 2008; Siddikee *et al.*, 2010; Bhattacharyya and Jha, 2012). On the other hand the direct mechanisms include production of growth regulators such as cytokinin, indoleacetic acid, and gibberellins; or the production of 1-aminocyclopropane-1-carboxylic acid deaminase, an enzyme that can lower plant ethylene levels that are typically increased by a wide variety of environmental stresses such as drought and salinity among other stresses. Other direct mechanisms are related to improved plant nutrition through phosphate and zinc solubilization, acquisition of iron by siderophores and nitrogen fixation (Bashan and de-Bashan, 2010; Hayat *et al.*, 2010; Siddikee *et al.*, 2010). The extremophiles are classified in five groups (Fig. 9.2): thermophiles (high temperatures), psychrophiles (low temperatures), halophiles (high salt content), alkaliphiles (growth in extreme pH conditions over 9) and acidophiles (growth in extreme pH conditions below 3) (Rodríguez-Valera F, 1988; Van Den Burg, 2003; Ramírez *et al.*, 2006).

9.2.1 Thermophiles

One of most prevalent problems due to increases in temperature is protein denaturation which leads to cellular damage in organisms. In nature, microorganisms have been found that grow naturally in the hottest places, such as semi-arid and arid places like the desert. These microorganisms are called thermophiles. An important feature of thermophiles is their optimal growth temperature, ranging from 45 to 80°C , with a maximum recorded temperature of 113°C (Van Den Burg, 2003). Survivability at high temperatures has resulted in two adaptations of particular interest:

1. High temperature metabolism: specifically the production and use of heat-stable enzymes. The most studied enzymes from thermophiles are the proteases, lipases, cellulases, chitinases and other polymer-degrading enzymes. The relevance of these enzymes is their capacity to improve the solubility of many polymeric substrates at elevated temperatures (Vieille and Zeikus, 2001). The laccases are among the enzymes produced by thermophilic microbes. These enzymes have been recognized by their capacity to accelerate the decomposition of lignin and catalyze the degradation of toxic chemicals in crop soil, resulting in increased organic matter and the improvement of water quality which, in turn, results in plant growth promotion (Chakroun *et al.*, 2010; Eichlerová *et al.*, 2012; Wong *et al.*, 2012).

2. Temperature-tolerant membrane: some thermophiles are able to synthesize heat shock proteins which are known for their ability to lessen and protect from cellular damage due to the elevated temperatures (Münchbach *et al.*, 1999; Grover *et al.*, 2011; Salas-Muñoz *et al.*, 2012). Interestingly, Ali *et al.* (2009) reported that *P. aeruginosa* AMK-P6 isolated from an arid environment in higher temperatures produces many heat shock proteins. In addition, some *Pseudomonas* strains are able to produce a biofilm of exopolysaccharides across plant roots which have a vital role in the stabilization

Fig. 9.2. Extremophile microorganisms in the desert. The root of the plants in arid and semiarid regions are reservoirs of diverse microorganism, which have the capacity to survive extreme conditions such as thermophiles (high temperatures), psychrophiles (low temperatures), halophiles (high salt content), alkaliphiles (growth in extreme pH conditions over 9) and acidophiles (growth in extreme pH conditions down to 5).

of soil aggregates, water flow and regulation of nutrients resulting in plant growth promotion (Tisdall and Oades, 1982; Roberson and Firestone, 1992). Thermophilic genera include: *Thermus, Cyanidium, Stetteria hydrogenophila, Methanothermus, Pyrobaculum, and Pseudomona* (Stetter, 1999).

9.2.2 Psychrophiles

Psychrophiles are able to grow in cold temperatures, ranging from −20°C to +10°C. Psychrophiles can adapt not only to low temperatures: in fact it has been reported that the variation of environmental conditions such as temperature, pH and salinity is vital for the growth and development of these microorganisms (Ramírez *et al.*, 2006). Among the microorganisms belonging to the psychrophiles are the genera *Bacillus, Rhizobium, Pseudomonas*, and *Listeria* (Fujii and Fulco, 1977; Cloutier *et al.*, 1992; Hebraud *et al.*, 1994; Mayr *et al.*, 1996). In agriculture these microorganisms play an important role in sustaining the production and productivity of crops. This is because their reported capability to solubilize nutrients and to fix nitrogen triggers plant growth promotion. In addition, they provide protection to the plant by the suppression of harmful pathogens and insects (Misaghi *et al.*, 1982; Volkmar and Bremer, 1998; Andrews and Harris, 2000; Katiyar and Goel, 2004).

Psychrophiles are also known for their capacity to produce cryoprotective proteins, phytohormones and induce the deamination of the precursor molecule of ethylene whose accumulation in root tissue is known to be detrimental to root growth and development (Glick *et al.*, 1998; Koda *et al.*, 2001; Mishra *et al.*, 2010). It is important to note that some enzymes such as proteases, amylases, cellulases, lipases and dehydrogenases from psychrophiles have been used in industry for the production of food, detergents and biosensors (Ramírez *et al.*, 2006).

9.2.3 Halophiles

Around the world, there exist many extreme habitats with high salt levels which combined with several other factors can limit the growth of some organisms. Microorganisms that thrive in high salt concentrations are called halophiles (Rodriguez-Valera, 1988). Halophiles are classified into five groups, depending on the salt concentration in which they can grow (Ventosa *et al.*, 2008): (1) halotolerant microbes are those that can grow in saturated concentrations; (2) extreme halophiles can grow in media containing above 20% salt concentration; (3) moderate halophiles are those that have the capacity to grow in media with no more than 20% salt concentration; (4) slight halophiles grow in media containing up to 10% salt; (5) nonhalophiles are microorganisms that require less than 1% of salt concentration in media for growth and development (Kushner and Kamekura, 1988; Ramírez *et al.*, 2006; Ventosa *et al.*, 2008). Siddikee *et al.* (2010) reported that 36 bacteria isolated from different soils with high concentrations of salt displayed a plant growth promoting activity by different mechanisms. The main mechanisms by which they acted were: nitrogen fixation, thiosulfate oxidation, production of indole acetic acid (IAA), ammonia, extracellular enzymes, phosphorus, and zinc production. In addition, 14 of these bacteria induced the amelioration of salt stress by increasing root length and dry weight in the plantlets inoculated with the halophile bacteria. Some of the most important microorganisms considered to be halophiles are the genera: *Halomonas, Volcaniella, Flavobacterium, Paracoccus, Pseudomonas, Bacillus, Brevibacterium, Planococcus, Zhihengliuella, Corynebacterium, Arthrobacter, Oceanimonas, Exibuobacterium, Micrococcus, Halovibrio, Chromobacterium, Natronobacterium, Dunaliella, Actinopolyspora, Actinopolyspora, Tetragenococcus* and *Azospirillum* (Das Sarma, 1995; Ramírez *et al.*, 2006; Nabti *et al.*, 2007; Ventosa *et al.*, 2008; Siddikee *et al.*, 2010).

9.2.4 Alkaliphiles

Alkaliphilic microorganisms have the ability to grow and develop under extreme pH conditions, i.e. above a pH of 9 (Horikoshi, 1999; Ramírez *et al.*, 2006; Godinho and Bhosle, 2013). Alkaliphiles can be aerobic or anaerobic including prokaryotes, eukaryotes, and archaea (Ramírez *et al.*, 2006; Godinho and Bhosle, 2013; Liu *et al.*, 2016). In order to avoid the possible damage caused by extracellular alkalinity and, to be able to grow and develop, these microorganisms have adapted to isolate the interior of the cell medium by diverse mechanisms such as:

1. Modification of internal pH: This is mainly due to various extremozymes produced in these organisms and located mainly in the cell wall, such as the α-galactosidase from *Micrococcus* sp. strain, proteases from *Brachybacterium sp.* LAP214, *Bacillus cohnii* LAP217, *Bacillus pseudofirmus* LAP220, *Brevibacterium casei* LAP223 and *Halomonas venusta* LAP515, among others (Horikoshi, 1999; Ramírez *et al.*, 2006; Rathod and Pathak, 2016).
2. Cell walls: It has been reported that alkaliphiles have the ability to modify the composition of their cell wall in order to avoid the damage caused by the extreme pH. The principal modification is in the peptidoglycans of the cell wall: variation in the amine content, an excess of diverse amino acids and the presence of glucosamine, muramic acid, D-glutamic acid, meso-diaminopimelic acid, acetic acid and D- and L-alanine (Horikoshi, 1999). Some alkaliphiles contain acidic polymers which induce the absorption of sodium and hydronium ions and discard hydroxide ions whereby the cell can grow in alkaline environments. Some of the acidic polymers are: galacturonic acid, gluconic acid, glutamic acid, aspartic acid and phosphoric acid (Aono and Horikoshi, 1983; Horikoshi, 1999; Ramírez *et al.*, 2006).
3. Membrane transport: they tightly regulate the concentration of Na^+ ions to maintain essential solute transport (Kitada and Horikoshi, 1977).

The most cited alkaliphiles are *Bacillus, Natrialba, Anaerobranca, Clostridium, Amphibacillus, Thermococcus, Tindallia, Atronobacterium*, Methanogens and diverse Cyanobacteria (Tanabe *et al.*, 1988; Boone *et al.*, 1993; Lodwick *et al.*, 1994; Cook *et al.*, 1996; Gerasimenko *et al.*, 1996; Desmarais *et al.*, 1997; Takeuchi *et al.*, 1997; Kevbrin *et al.*, 1998; Wiegel, 1998). Interestingly, it has been reported that some alkaliphiles, such as *Klebsiella* sp. D5A have the ability to promote plant growth by several mechanisms like the production of IAA, solubilization of phosphate, synthesis of siderophores, suppression of pathogenic fungi, resistance to abiotic stresses, etc. (Iniguez *et al.*, 2004; Pinto-Tomás *et al.*, 2009; Liu *et al.*, 2014; Wei *et al.*, 2014; Liu *et al.*, 2016).

9.2.5 Acidophiles

Little is known about microorganisms that can grow in extremely low pH environments and those that can thrive below pH 3 are considered acidophiles (Madigan *et al.*, 2003; Ramírez *et al.*, 2006; Baker-Austin and Dopson, 2007). Acidophiles are found mainly in inaccessible and isolated environments. The main mechanism by which they can grow and develop in these extreme environments is their capacity to induce a pH homeostasis in the cell; the mechanisms by which this occurs are: (1) the influx of protons produced by the F_0F_1 ATPase (Madshus, 1988; Baker-Austin and Dopson, 2007); (2) the cell membrane is highly impermeable to protons (Van de Vossenberg *et al.*, 1998; Konings *et al.*, 2002; Golyshina *et al.*, 2005); (3) the membrane channels have a reduced pore size (Amaro *et al.*, 1991; Guiliani and Jerez, 2000); (4) proton influx is inhibited by a chemiosmosis gradient (Hsung and Haug, 1977; Oshima *et al.*, 1977; Krulwich *et al.*, 1978; Cox *et al.*, 1979; Michels and Bakker, 1985; Goulbourne *et al.*, 1986; Krulwich and Guffanti, 1986; Suzuki *et al.*, 1999; She *et al.*, 2001; Fütterer *et al.*, 2004; Schäfer *et al.*, 2004); (5) excess protons are pumped out of the cell (Apel *et al.*, 1980; Michels and Bakker, 1985; Dopson *et al.*, 2002; Fütterer *et al.*, 2004; Tyson *et al.*, 2004; Golyshina and Timmis, 2005); (6) cytoplasmic buffering helps to maintain the intracellular pH (Zychlinsky and Matin, 1983; Goulbourne *et al.*, 1986; Krulwich and Guffanti, 1986; Castanie-Cornet *et al.*, 1999); (7) proton uncoupling by organic acids (Alexander *et al.*,

1987; Kishimoto *et al.*, 1990; Crossman *et al.*, 2004; Angelov and Liebl, 2006); (8) high expression of chaperones (Jerez *et al.*, 1988; Crossman *et al.*, 2004; Dopson *et al.*, 2005; Ram *et al.*, 2005; Dopson *et al.*, 2007); and (9) intracellular enzymes might be stabilized by 'iron rivets' (Nordstrom *et al.*, 2000; Golyshina *et al.*, 2006; Ferrer *et al.*, 2007).

Rani *et al.* (2009) reported that *Pseudomonas putida* 62BN, characterized as an acidophilic microbe, induced an increase in the root length, shoot length, wet weight, dry weight and chlorophyll in soybean plants growing in cadmium-contaminated soil. In fact, under these conditions it was demonstrated that when this microbe is used as a bioinoculant it can induce resistance against toxic contaminants in the plants (Rani *et al.*, 2009). Some of the most important microorganisms considered as acidophilic microbes are in the genera: *Thiobacillus*, *Leptospirillum* and *Acidiphilium*.

9.3 Concluding Remarks

In order to lessen the damage caused by increasing climate change and global warming, it is necessary to continue to explore the diversity of microorganisms present in areas with extreme conditions such as the arid and semi-arid lands. Extremophile microorganisms have recently drawn attention for their unique adaptability mechanisms to tolerate extreme environmental changes. Nevertheless a more comprehensive understanding of the genetics, biochemistry and physiology of these organisms is necessary to fully exploit their potential for bioremediating farmland, and promoting growth and reducing losses of crops.

References

Alexander, B., Leach, S. and Ingledew, W.J. (1987) The relationship between chemiosmotic parameters and sensitivity to anions and organic acids in the acidophile *Thiobacillus ferrooxidans*. *Journal of General Microbiology* 133, 1171–1179.

Ali, S.Z., Sandhya, V., Grover, M., Kishore, N., Rao, L.V. and Venkateswarlu, B. (2009) *Pseudomonas* sp. strain AKM-P6 enhances tolerance of sorghum seedlings to elevated temperatures. *Biology and Fertility of Soils* 46, 45–55.

Amaro, A.M., Chamorro, D., Seeger, M., Arredondo, R., Peirano, I. and Jerez, C.A. (1991) Effect of external pH perturbations on in vivo protein synthesis by the acidophilic bacterium *Thiobacillus ferrooxidans*. *Journal of Bacteriology* 173, 910–915.

Andrews, J.H. and Harris, R.F. (2000) The ecology and biogeography of microorganisms on plant surfaces. *Annual Review of Phytopathology* 38, 145–180.

Angelov, A. and Liebl, W. (2006) Insights into extreme thermoacidophily based on genome analysis of *Picrophilus torridus* and other thermoacidophilic archaea. *Journal of Biotechnology* 126, 3–10.

Aono, R. and Horikoshi, K. (1983) Chemical composition of cell walls of alkalophilic strains of *Bacillus*. *Journal of General Microbiology* 129, 1083–1087.

Apel, W.A., Dugan, P.R. and Tuttle, J.H. (1980) Adenosine 5′-triphosphate formation in *Thiobacillus ferrooxidans* vesicles by H^+ ion gradients comparable to those of environmental conditions. *Journal of Bacteriology* 142, 295–301.

Baker-Austin, C. and Dopson, M. (2007) Life in acid: pH homeostasis in acidophiles. *Trends in Microbiology* 15, 165–171.

Bashan, Y. and de-Bashan, L.E. (2010) Chapter two – How the plant growth-promoting bacterium *Azospirillum* promotes plant growth – A critical assessment. *Advances in Agronomy* 108, 77–136.

Bhattacharyya, P.N. and Jha, D.K. (2012) Plant growth-promoting rhizobacteria (PGPR): emergence in agriculture. *World Journal of Microbiology and Biotechnology* 28, 1327–1350.

Bisen, K., Keswani, C., Mishra, S., Saxena, A., Rakshit, A. and Singh, H.B. (2015) Unrealized potential of seed biopriming for versatile agriculture. In: Rakshit, A., Singh, H.B. and Sen, A. (eds) *Nutrient Use Efficiency: From Basics to Advances*. Springer, New Delhi, India, pp. 193–206.

Boone, D.R., Mathrani, I.M., Liu, Y., Menaia, J.A., Mah, R.A. and Boone, J.E. (1993) Isolation and characterization of *Methanohalophilus portucalensis* sp. nov and DNA re-association study of the genus *Methanohalophilus*. *International Journal of Systematic and Evolutionary Microbiology* 43, 430–437.

Castanie-Cornet, M.P., Penfound, T.A., Smith, D., Elliott, J.F. and Foster, J.W. (1999) Control of acid resistance in *Escherichia coli*. *Journal of Bacteriology* 181, 3525–3535.

Chakroun, H., Mechichi, T., Martinez, M.J., Dhouib, A. and Sayadi, S. (2010) Purification and characterization of a novel laccase from the ascomycete *Trichoderma atroviride*: application on bioremediation of phenolic compounds. *Process Biochemistry* 45, 507–513.

Cloutier, J., Prévost, D., Nadeau, P. and Antoun, H. (1992) Heat and cold shock protein synthesis in arctic and temperate strains of rhizobia. *Applied and Environmental Microbiology* 58, 2846–2853.

Cook, G.M., Russell, J.B., Reichert, A. and Wiegel, J. (1996) The Intracellular pH of *Clostridium paradoxum*, an anaerobic, alkaliphilic, and thermophilic bacterium. *Applied and Environmental Microbiology* 62, 4576–4579.

Cox, J.C., Nicholls, D.G. and Ingledew, W.J. (1979) Transmembrane electrical potential and transmembrane pH gradient in the acidophile *Thiobacillus ferrooxidans*. *Biochemical Journal* 178, 195–200.

Crossman, L., Holden, M., Pain, A. and Parkhill, J. (2004) Genomes beyond compare. *Nature Reviews Microbiology*, 2, 616–7.

Das Sarma, S. (1995) Halophilic archea - an overview. In: Robb, F.T. *et al* (eds), *Archea: A Laboratory Manual – Halophiles*. Cold Spring Harbor Laboratory Press, Cold Spring Harbor, NY, pp 3–11.

Desmarais, D., Jablonski, P.E., Fedarko, N.S. and Roberts, M.F. (1997) 2-Sulfotrehalose, a novel osmolyte in haloalkaliphilic archaea. *Journal of Bacteriology* 179, 3146–3153.

Dopson, M., Lindström, B.E. and Hallberg, K.B. (2002) ATP generation during reduced inorganic sulfur compound oxidation by *Acidithiobacillus caldus* is exclusively due to electron transport phosphorylation. *Extremophiles* 6, 123–129.

Dopson, M., Baker-Austin, C. and Bond, P.L. (2005) Analysis of differential protein expression during growth states of *Ferroplasma* strains and insights into electron transport for iron oxidation. *Microbiology* 151, 4127–4137.

Dopson, M., Baker-Austin, C. and Bond, P. (2007) Towards determining details of anaerobic growth coupled to ferric iron reduction by the acidophilic archaeon '*Ferroplasma acidarmanus*' Fer1. *Extremophiles* 11, 159–168.

Eichlerová, I., Šnajdr, J. and Baldrian, P. (2012) Laccase activity in soils: considerations for the measurement of enzyme activity. *Chemosphere* 88, 1154–1160.

Ferrer, M., Golyshina, O.V., Beloqui, A., Golyshin, P.N. and Timmis, K.N. (2007) The cellular machinery of *Ferroplasma acidiphilum* is iron-protein-dominated. *Nature* 445, 91–94.

Figueiredo, M.V.B., Martinez, C.R., Burity, H.A. and Chanway, C.P. (2008) Plant growth-promoting rhizobacteria for improving nodulation and nitrogen fixation in the common bean (*Phaseolus vulgaris* L.). *World Journal of Microbiology and Biotechnology* 24, 1187–1193.

Fujii, D.K. and Fulco, A.J. (1977) Biosynthesis of unsaturated fatty acids by bacilli. Hyperinduction and modulation of desaturase synthesis. *Journal of Biological Chemistry* 252, 3660–3670.

Fütterer, O., Angelov, A., Liesegang, H., Gottschalk, G., Schleper, C. *et al.* (2004) Genome sequence of *Picrophilus torridus* and its implications for life around pH 0. *Proceedings of the National Academy of Sciences of the United States of America* 101, 9091–9096.

Gerasimenko, L.M., Dubinin, A.V. and Zavarzin, G.A. (1996) Alkaliphilic cyanobacteria from soda lakes of Tuva and their ecophysiology. *Microbiology* 65, 736–740.

Glick, B.R. (1995) The enhancement of plant growth by free-living bacteria. *Canadian Journal of Microbiology* 41, 109–117.

Glick, B.R., Penrose, D.M. and Li, J. (1998) A model for the lowering of plant ethylene concentrations by plant growth-promoting bacteria. *Journal of Theoretical Biology* 190, 63–68.

Godinho, A. and Bhosle, S. (2013) Rhizosphere bacteria from coastal sand dunes and their applications in agriculture. In: Maheshwari, D.K., Saraf, M. and Aeron, A (eds) *Bacteria in Agrobiology: Crop Productivity*. Springer Berlin Heidelberg, Germany, pp. 77–96.

Golyshina, O.V. and Timmis, K.N. (2005) *Ferroplasma* and relatives, recently discovered cell wall-lacking archaea making a living in extremely acid, heavy metal-rich environments. *Environmental Microbiology* 7, 1277–1288.

Golyshina, O.V., Golyshin, P.N., Timmis, K.N. and Ferrer, M. (2006) The 'pH optimum anomaly' of intracellular enzymes of *Ferroplasma acidiphilum*. *Environmental Microbiology* 8, 416–425.

Goulbourne, E., Matin, M., Zychlinsky, E. and Matin, A. (1986) Mechanism of delta pH maintenance in active and inactive cells of an obligately acidophilic bacterium. *Journal of Bacteriology* 166, 59–65.

Grover, M., Ali, S.Z., Sandhya, V., Rasul, A. and Venkateswarlu, B. (2011) Role of microorganisms in adaptation of agriculture crops to abiotic stresses. *World Journal of Microbiology and Biotechnology* 27, 1231–1240.
Guiliani, N. and Jerez, C.A. (2000) Molecular cloning, sequencing, and expression of *omp-40,* the gene coding for the major outer membrane protein from the acidophilic bacterium *Thiobacillus ferrooxidans. Applied and Environmental Microbiology* 66, 2318–2324.
Hayat, R., Ali, S., Amara, U., Khalid, R. and Ahmed, I. (2010) Soil beneficial bacteria and their role in plant growth promotion: a review. *Annals of Microbiology* 60, 579–598.
Hebraud, M., Dubois, E., Potier, P. and Labadie, J. (1994) Effect of growth temperatures on the protein levels in a psychrotrophic bacterium, *Pseudomonas fragi. Journal of Bacteriology* 176, 4017–4024.
Henrickson, J. (1974) Saline habitats and halophytic vegetation of the Chihuahuan desert region. In: Wauer, R.H. and Riskind, D.H. (eds) *Transactions of the Symposium on the Biological Resources of the Chihuahuan Desert Region, United States and Mexico, Sul Ross State University, Alpine, TX, USA.* US Department of the Interior, Washington DC, pp. 289–314.
Horikoshi, K. (1999) Alkaliphiles: some applications of their products for biotechnology. *Microbiology and Molecular Biology Reviews* 63, 735–750.
Hsung, J.C. and Haug, A. (1977) Membrane potential of *Thermoplasma acidophila. FEBS Letters* 73, 47-50.
Iniguez, A.L., Dong, Y. and Triplett, E.W. (2004) Nitrogen fixation in wheat provided by *Klebsiella pneumoniae 342. Molecular Plant-Microbe Interactions* 17, 1078–1085.
Jerez, C., Chamorro, D., Peirano, I. and Toledo, H. (1988) Studies of the stress response in chemolithotrophic acidophilic bacteria. *Biochemical International* 17, 989–999.
Keswani, C. (2015). Proteomics studies of thermotolerant strain of *Trichoderma* spp. Ph.D. Thesis, Banaras Hindu University, Varanasi, India, p. 126.
Keswani, C., Bisen, K., Singh, S.P., Sarma, B.K. and Singh, H.B. (2016a) A proteomic approach to understand the tripartite interactions between plant-*Trichoderma*-pathogen: investigating the potential for efficient biological control. In: Hakeem, K.R. and Akhtar, M.S. (eds) *Plant, Soil and Microbes Vol. 2. Mechanisms and Molecular Interactions*. Springer, Switzerland, pp. 79–93.
Keswani, C., Bisen, K., Singh, V., Sarma, B.K. and Singh, H.B. (2016b) Formulation technology of biocontrol agents: present status and future prospects. In: Arora, N.K., Mehnaz, S. and Balestrini, R. (eds) *Bioformulations: For Sustainable Agriculture*. Springer, New Delhi, India, pp. 35–52.
Kevbrin, V.V., Zhilina, T.N., Rainey, F.A. and Zavarzin, G.A. (1998) *Tindallia magadii* gen. nov., sp. nov.: an alkaliphilic anaerobic ammonifier from soda lake deposits. *Current Microbiology* 37, 94–100.
Kishimoto, N., Inagaki, K., Sugio, T. and Tano, T. (1990) Growth inhibition of *Acidiphilium* species by organic acids contained in yeast extract. *Journal of Fermentation and Bioengineering* 70, 7–10.
Kitada, M. and Horikoshi, K. (1977) Sodium ion-stimulated alpha-[1-14C] aminoisobutyric acid uptake in alkalophilic *Bacillus* species. *Journal of Bacteriology* 131, 784–788.
Koda, N., Asaeda, T., Yamade, K., Kawahara, H. and Obata, H. (2001) A novel cryoprotective protein (CRP) with high activity from the ice-nucleating bacterium, *Pantoea agglomerans* IFO12686. *Bioscience, Biotechnology and Biochemistry* 65, 888–894.
Konings, W.N., Albers, S.V., Koning, S. and Driessen, A.J. (2002) The cell membrane plays a crucial role in survival of bacteria and archaea in extreme environments. *Antonie Van Leeuwenhoek* 81, 61–72.
Krulwich, T.A. and Guffanti, A.A. (1986) Regulation of internal pH in acidophilic and alkalophilic bacteria. *Methods in Enzymology* 125, 352–365.
Krulwich, T.A., Davidson, L.F., Filip, S.J., Zuckerman, R.S. and Guffanti, A.A. (1978) The proton motive force and betagalactoside transport in *Bacillus acidocaldarius. Journal of Biological Chemistry* 253, 4599–4603.
Kushner, D.J. and Kamekura, M. (1988) Physiology of halophilic eubacteria. *Halophilic Bacteria* 1, 109–140.
Liu, W., Hou, J., Wang, Q., Ding, L. and Luo, Y. (2014) Isolation and characterization of plant growth-promoting rhizobacteria and their effects on phytoremediation of petroleum-contaminated saline-alkali soil. *Chemosphere* 117, 303–308.
Liu, W., Wang, Q., Hou, J., Tu, C., Luo, Y. and Christie, P. (2016) Whole genome analysis of halotolerant and alkalotolerant plant growth-promoting rhizobacterium *Klebsiella* sp. D5A. *Scientific Reports*, 6, 26710; doi: 10.1038/srep26710
Lodwick, D., McGenity, T.J. and Grant, W.D. (1994) The phylogenetic position of the haloalkaliphilic archaeon *Natronobacterium magadii,* determined from its 23S ribosomal RNA sequence. *Systematic and Applied Microbiology* 17, 402–404.
Madigan, M.T., Martinko, J.M. and Parker, J. (2003) *Diversidad Procariótica: Archea. Brock Microbiología de los Microorganismos,* 10th edn. Pearson-Prentice Hall, Madrid.
Madshus, I.H. (1988) Regulation of intracellular pH in eukaryotic cells. *Biochemical Journal* 250, 1.

Mayr, B., Kaplan, T., Lechner, S. and Scherer, S. (1996) Identification and purification of a family of dimeric major cold shock protein homologs from the psychrotrophic *Bacillus cereus* WSBC 10201. *Journal of Bacteriology* 178, 2916–2925.

Michels, M. and Bakker, E.P. (1985) Generation of a large, protonophore-sensitive proton motive force and pH difference in the acidophilic bacteria *Thermoplasma acidophilum* and *Bacillus acidocaldarius*. *Journal of Bacteriology* 161, 231–237.

Misaghi, I.J., Stowell, L.J., Grogan, R.G. and Spearman, L.C. (1982) Fungistatic activity of water-soluble fluorescent pigments of fluorescent pseudomonads. *Phytopathology* 72(1), 33–36.

Mishra, P.K., Joshi, P., Bisht, S.C., Bisht, J.K. and Selvakumar, G. (2010) Cold-tolerant agriculturally important microorganisms. In: Maheshwari, D.K. (ed.) *Plant Growth and Health Promoting Bacteria*. Springer Berlin Heidelberg, Germany, pp. 273–296.

Münchbach, M., Nocker, A. and Narberhaus, F. (1999) Multiple small heat shock proteins in rhizobia. *Journal of Bacteriology* 181, 83–90.

Nabti, E., Sahnoune, M., Adjrad, S., Van Dommelen, A., Ghoul, M. *et al.* (2007) A halophilic and osmotolerant *Azospirillum brasilense* strain from Algerian soil restores wheat growth under saline conditions. *Engineering in Life Sciences* 7, 354–360.

Nordstrom, D.K., Alpers, C.N., Ptacek, C.J. and Blowes, D.W. (2000) Negative pH and extremely acidic mine waters from Iron Mountain, California. *Environmental Science & Technology* 34, 254–258.

Oshima, T., Arakawa, H. and Misako, B.A.B.A. (1977) Biochemical studies on an acidophilic, thermophilic bacterium, *Bacillus acidocaldarius*: isolation of bacteria, intracellular pH, and stabilities of biopolymers. *Journal of Biochemistry* 81, 1107–1113.

Pinto-Tomás, A.A., Anderson, M.A., Suen, G., Stevenson, D.M., Chu, F.S. *et al.* (2009) Symbiotic nitrogen fixation in the fungus gardens of leaf-cutter ants. *Science* 326, 1120–1123.

Ram, R.J., Ver Berkmoes, N.C., Thelen, M.P., Tyson, G.W., Baker, B.J. *et al.* (2005) Community proteomics of a natural microbial biofilm. *Science* 308, 1915–1920.

Ramirez, N., Serrano, J.A. and Sandoval, H. (2006) Microorganismos extremófilos. Actinomicetos halófilos en México. *Revista Mexicana de Ciencias Farmacéuticas* 37, 56–71.

Rani, A., Souche, Y.S. and Goel, R. (2009) Comparative assessment of *in situ* bioremediation potential of cadmium resistant acidophilic *Pseudomonas putida* 62BN and alkalophilic *Pseudomonas monteilli* 97AN strains on soybean. *International Biodeterioration & Biodegradation* 63(1), 62–66.

Rathod, M.G. and Pathak, A.P. (2016) Data on optimized production and characterization of alkaline proteases from newly isolated alkaliphiles from Lonar soda lake, India. *Data in Brief* 8, 863–866.

Roberson, E.B. and Firestone, M.K. (1992) Relationship between desiccation and exopolysaccharide production in a soil *Pseudomonas* sp. *Applied and Environmental Microbiology* 58(4), 1284–1291.

Rodríguez-Valera, F. (1988) Characteristics and microbial ecology of hypersaline environments. In: Rodriguez-Valera, F. (ed.) *Halophilic Bacteria*, vol. 1. CRC Press, Boca Raton, Florida, pp 3–30.

Rzedowski, J. (1968) Las principals zonas áridas de México y su vegetación. *Bios* 1, 4–24.

Salas-Muñoz, S., Gómez-Anduro, G., Delgado-Sánchez, P., Rodríguez-Kessler, M. and Jiménez-Bremont, J.F. (2012) The *Opuntia streptacantha OpsHSP18* gene confers salt and osmotic stress tolerance in *Arabidopsis thaliana*. *International Journal of Molecular Sciences* 13, 10154–10175.

Schäfer, K., Magnusson, U., Scheffel, F., Schiefner, A., Sandgren, M.O. *et al.* (2004) X-ray structures of the maltose–maltodextrin-binding protein of the thermoacidophilic bacterium *Alicyclobacillus acidocaldarius* provide insight into acid stability of proteins. *Journal of Molecular Biology* 335, 261–274.

She, Q., Singh, R.K., Confalonieri, F., Zivanovic, Y., Allard, G. *et al.* (2001) The complete genome of the crenarchaeon *Sulfolobus solfataricus* P2. *Proceedings of the National Academy of Sciences of the United States of America* 98, 7835–7840.

Siddikee, M.A., Chauhan, P.S., Anandham, R., Han, G.H. and Sa, T. (2010) Isolation, characterization, and use for plant growth promotion under salt stress, of ACC deaminase-producing halotolerant bacteria derived from coastal soil. *Journal of Microbiology and Biotechnology* 20(11), 1577–1584.

Stetter, K.O. (1999) Extremophiles and their adaptation to hot environments. *FEBS Letters* 452, 22–25.

Suzuki, I., Lee, D., Mackay, B., Harahuc, L. and Oh, J.K. (1999) Effect of various ions, pH, and osmotic pressure on oxidation of elemental sulfur by *Thiobacillus thiooxidans*. *Applied and Environmental Microbiology* 65(11), 5163–5168.

Takeuchi, M., Kaieda, N. and Koyama, N. (1997) Effect of amines on the (NH_4^+ + Na^+)-activated ATPase of an anaerobic alkaliphile, *Amphibacillus xylanus*. *Microbios* 90, 201–208.

Tanabe, H., Kobayashi, Y. and Akamatsu, I. (1988) Pretreatment of pectic wastewater with pectatelyase from an alkalophilic *Bacillus* sp. *Agricultural and Biological Chemistry* 52, 1855–1856.

Tisdall, J.M. and Oades, J. (1982) Organic matter and water-stable aggregates in soils. *Journal of Soil Science* 33, 141–163.

Tyson, G.W., Chapman, J., Hugenholtz, P., Allen, E.E., Ram, R.J. *et al.* (2004) Community structure and metabolism through reconstruction of microbial genomes from the environment. *Nature* 428, 37–43.

Van Den Burg, B. (2003) Extremophiles as a source for novel enzymes. *Current Opinion in Microbiology* 6(3), 213–218.

Van de Vossenberg, J.L., Driessen, A.J., Zillig, W. and Konings, W.N. (1998) Bioenergetics and cytoplasmic membrane stability of the extremely acidophilic, thermophilic archaeon *Picrophilus oshimae*. *Extremophiles* 2(2), 67–74.

Venkateswarlu, B. and Shanker, A.K. (2009) Climate change and agriculture: adaptation and mitigation stategies. *Indian Journal of Agronomy* 54(2), 226.

Ventosa, A., Mellado, E., Sanchez-Porro, C. and Marquez, M.C. (2008) Halophilic and halotolerant microorganisms from soils. In: Ventosa, A., Mellado, E., Sanchez-Porro, C. and Marquez, M.C. (eds) *Microbiology of Extreme Soils*, Springer Berlin Heidelberg, Germany, pp. 87–115.

Vieille, C. and Zeikus, G.J. (2001) Hyperthermophilic enzymes: sources, uses, and molecular mechanisms for thermostability. *Microbiology and Molecular Biology Reviews* 65, 1–43.

Volkmar, K.M. and Bremer, E. (1998) Effects of seed inoculation with a strain of *Pseudomonas fluorescens* on root growth and activity of wheat in well-watered and drought-stressed glass-fronted rhizotrons. *Canadian Journal of Plant Science* 78, 545–551.

Wei, C.Y., Lin, L., Luo, L.J., Xing, Y.X., Hu, C.J. *et al.* (2014) Endophytic nitrogen-fixing *Klebsiella variicola* strain DX120E promotes sugarcane growth. *Biology and Fertility of Soils* 50(4), 657–666.

Wiegel, J. (1998) Anaerobic alkalithermophiles, a novel group of extremophiles. *Extremophiles* 2, 257–267.

Wong, K.S., Huang, Q., Au, C.H., Wang, J. and Kwan, H.S. (2012) Biodegradation of dyes and polyaromatic hydrocarbons by two allelic forms of *Lentinula edodes* laccase expressed from *Pichia pastoris*. *Bioresource Technology* 104, 157–164.

Zychlinsky, E. and Matin, A. (1983) Cytoplasmic pH homeostasis in an acidophilic bacterium, *Thiobacillus acidophilus*. *Journal of Bacteriology* 156(3), 1352–1355.

10 Rhizosphere Colonization by Plant-Beneficial *Pseudomonas* spp.: Thriving in a Heterogeneous and Challenging Environment

Antoine Zboralski, Adrien Biessy and Martin Filion*
Université de Moncton, Département de biologie, Moncton, Canada

10.1 Introduction

Soils are the richest ecosystems on Earth in terms of biodiversity, as well as major components of agricultural systems (Hinsinger *et al.*, 2009). They are deeply involved in food webs, providing essential functions for sustaining life both below- and aboveground. However, soils are relatively poor in nutrients, except for some hotspots under the influence of living plant roots, a concept known as the rhizosphere. The rhizosphere is usually defined as the first 1–5 mm of soil surrounding plant roots (Bertin *et al.*, 2003; Angus and Hirsch, 2013; Prashar *et al.*, 2014). It is supplied in nutrients by plant roots through the release of 5% to 30% of the net carbon fixed by photosynthesis (Lynch and Whipps, 1990; Neumann, 2007; Uren, 2007). This in turn supports microbial growth to densities that are 10- to 1000-fold higher than those associated with the surrounding soil, called bulk soil (Lugtenberg and Bloemberg, 2004; Lugtenberg and Kamilova, 2009).

For the last decades, indigenous microorganisms inhabiting the rhizosphere have received more and more attention for their implication in nutrient uptake, plant growth promotion and disease suppression (Whipps, 2001; Lugtenberg and Kamilova, 2009; Brink, 2016). In this context, numerous plant-beneficial bacteria have been isolated from the rhizophere of different plant species (Antoun and Prévost, 2005; Weller, 2007; Lugtenberg and Kamilova, 2009; Mishra *et al.*, 2015). These bacteria have been named plant growth-promoting rhizobacteria, abbreviated "PGPR" (Kloepper *et al.*, 1980a). Different mechanisms are involved in the promotion of plant growth by PGPR, either directly or indirectly. Direct mechanisms of growth promotion have been associated with the production of plant hormones by PGPR like auxins, or with the improvement of nutrient availability for plants through soil nutrients solubilization (Kloepper *et al.*, 1980b; Lugtenberg *et al.*, 2002). Indirect mechanisms include biological control of pathogens by PGPR through competition for niches and nutrients, antibiotics production or the induction of plant defence mechanisms against pathogens, thus decreasing or preventing plant

*E-mail: martin.filion@umoncton.ca

© CAB International 2017. *Advances in PGPR Research* (eds H.B. Singh, B.K. Sarma and C. Keswani)

diseases development (Weller, 2007). Most PGPR strains discovered to date belong to the *Bacillus*, *Rhizobia* and *Pseudomonas* genera (Barriuso *et al.*, 2008), which are also some of the most common bacterial genera identified from the rhizosphere (Prashar *et al.*, 2014).

The *Pseudomonas* genus, one of the most important genera from which PGPR strains have been isolated and characterized, consists of rod-shaped motile non-sporulating Gram-negative bacteria (Peix *et al.*, 2009) and displays a wide distribution as well as a great ecological and metabolic diversity (Palleroni and Moore, 2004). Several plant-beneficial *Pseudomonas* spp., including *P. fluorescens*, *P. chlororaphis* and *P. putida* strains, have been identified from the rhizosphere of a wide variety of plants (Weller, 2007). Many display interesting plant growth-promotion abilities (Lugtenberg and Kamilova, 2009) and/or biocontrol traits against a wide range of plant pathogens (Haas and Défago, 2005; Weller, 2007).

The ability of plant-beneficial *Pseudomonas* spp. to promote plant growth strongly relies on their ability to colonize the rhizosphere (Kloepper *et al.*, 1980b; Lugtenberg *et al.*, 2001). Their population size has often been correlated with disease incidence reduction (Bull *et al.*, 1991; Raaijmakers *et al.*, 1995a) or with accumulation of antibiotics in the rhizosphere (Raaijmakers *et al.*, 1999; Mavrodi *et al.*, 2012). This ability to competitively colonize a plant rhizosphere and to persist, while maintaining a high population size throughout the growing season, has been called rhizocompetence (Weller, 1988; Raaijmakers *et al.*, 1995a). It consists of forging a lasting trophic relationship with the plant while competing with the indigenous microbiome for resources and space in a strongly heterogeneous environment that is the rhizosphere. In this chapter, we will focus on the establishment of this relationship with a special emphasis on the plant-driven nutrient heterogeneities of the rhizosphere and on some competitiveness-enhancing traits, involved in the success of several plant-beneficial *Pseudomonas* spp. strains.

10.2 The Rhizosphere: a Heterogeneous Environment Shaped by Plant Rhizodeposition

Plant roots grow into a spatially and temporally heterogeneous habitat both in structure and composition (Bardgett, 2005). These heterogeneities found in the rhizosphere are intensified by the plant, especially through a process called rhizodeposition (Hinsinger *et al.*, 2005). The rhizodeposition has been defined as the release of materials from plant roots into the rhizosphere, including soluble and insoluble exudates, lysates and gases such as carbon dioxide or ethylene (Shamoot *et al.*, 1968; Whipps and Lynch, 1985). It is mediated through several mechanisms and depends qualitatively and quantitatively on many biotic and abiotic factors, such as the plant species (Hütsch *et al.*, 2002; Nguyen, 2003), the plant physiological status (Neumann and Römheld, 2007), the rhizospheric microbiome characteristics, and the soil physical and chemical properties (Jones *et al.*, 2004; Prashar *et al.*, 2014). The main mechanisms involved in rhizodeposition are: (i) root exudation (Bais *et al.*, 2006), including mucilage secretion (Marschner and Marschner, 2012); and (ii) senescence of root outer cells, especially root hair and detached root-cap border cells (Nguyen, 2003; Vicré *et al.*, 2005; Hawes *et al.*, 2012) (Fig. 10.1).

10.2.1 Root exudation

Root exudation has originally been defined as the passive diffusion of low molecular weight compounds from root cells into the apoplasm or directly into the soil through the plasma membrane (Rovira *et al.*, 1979). However, active processes are now also known to be involved in the release of such compounds into the soil, which has led to a revision of its definition to include both passive and active processes (Nguyen, 2003). Some authors also include the release of high molecular weight compounds like mucilage in this definition (Bais *et al.*, 2006; Prashar *et al.*, 2014). Here, root exudation

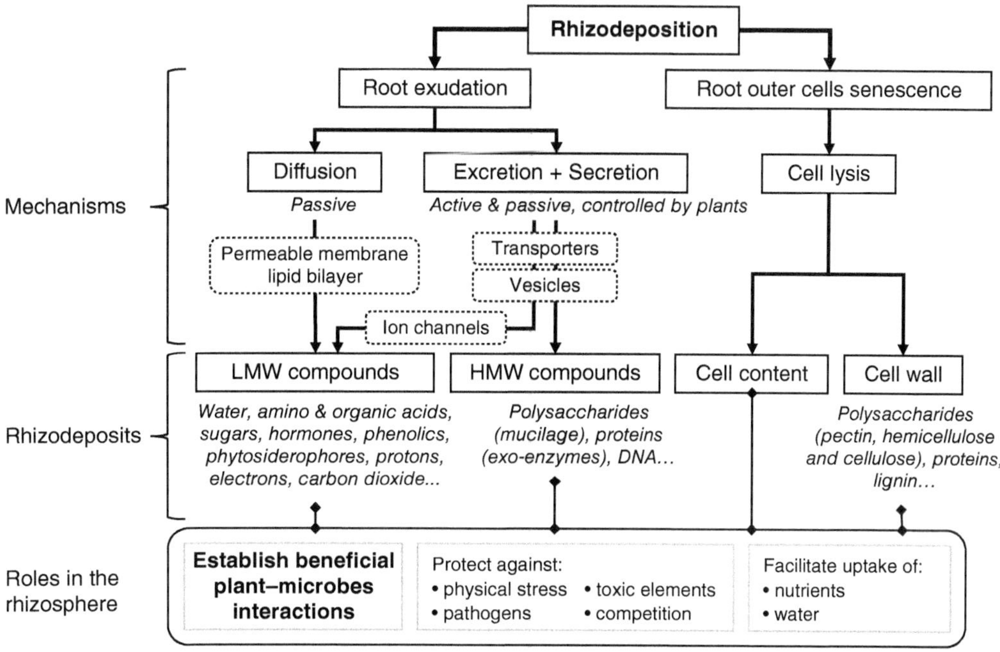

Fig. 10.1. Rhizodeposition of organic compounds: mechanisms, composition of rhizodeposits and roles in the rhizosphere (Neumann, 2007; Uren, 2007). LMW: low molecular weight; HMW: high molecular weight.

will be considered as the release of both low and high molecular weight (LMW and HMW, respectively) compounds by intact root cells into the soil through both passive and active mechanisms as illustrated in Fig. 10.1 (Rovira, 1969; Neumann, 2007).

Exudates display many plant-beneficial roles in the rhizosphere ecosystem, such as increasing nutrient availability (Neumann and Römheld, 2007), facilitating root growth through the soil (Marschner and Marschner, 2012), preventing soilborne plant diseases (Uren, 2007), or establishing profitable interactions with soil microbes (Badri *et al.*, 2009). These roles rely on the chemical diversity of the released compounds.

Composition

A great diversity of compounds are released through root exudation, especially organic ones (Uren, 2007). However, much of this diversity is displayed by LMW compounds (Bais *et al.*, 2006). The main LMW compounds are amino acids, organic acids, monosaccharides, oligosaccharides, plant hormones, phenolic compounds and various other secondary metabolites such as nucleotides or vitamins (Bertin *et al.*, 2003). Some inorganic LMW compounds are also released through exudation, such as ions, water, dioxygen, carbon dioxide, protons and electrons (Uren, 2007). As for HMW compounds, they are essentially represented by the mucilage, composed of polysaccharides and proteins (up to 6%), including enzymes (Nguyen, 2003; Walker *et al.*, 2003; Uren, 2007). Mucilage forms, in the presence of soil particles such as clays, a gelatinous and extremely water-absorbing layer called mucigel. This layer surrounds the root cap, facilitating the elongation of the root tip through the soil (Bertin *et al.*, 2003).

Mechanisms

Plant exudation occurs through diffusion, excretion and secretion (Neumann and Römheld, 2007; Uren, 2007). Diffusion through the plasma membrane is a passive mechanism over which the root exerts little direct control (Jones *et al.*, 2009). This mechanism is triggered by the sharp concentration gradient existing between the cytoplasm of root cells – mM – and the soil solution – μM (Farrar *et al.*, 2003; Neumann and Römheld, 2007). It relies on the membrane permeability, on the polarity of the diffusing compounds and on the cytosolic pH (Badri and Vivanco, 2009). Only LMW compounds are able to diffuse passively through the cell membrane (Nguyen, 2003). This process can be enhanced by stresses such as extreme temperature, nutrient deficiency or oxidative stress, directly altering membrane integrity (Neumann and Römheld, 2007).

Excretion and secretion are plant-controlled exudation mechanisms (Neumann and Römheld, 2007). They differ from each other according to a functional point of view (Bais *et al.*, 2006). On one side, excreted compounds are cell wastes, such as carbon dioxide produced through metabolic processes like respiration and released out of the cells to facilitate internal processes (Uren, 2007). On the other side, secreted compounds are produced in order to affect the surrounding environment of the root: they are directly involved in external processes, for example lubrication and plant–microbe signalling (Bais *et al.*, 2004). However, there seems to be an ambiguity in the literature regarding the energetic requirements of both processes: some authors indeed define them only as active ones (Bais *et al.*, 2006; Uren, 2007), while others include passive processes like diffusion through ion channels (Neumann, 2007). Here, we will consider that excretion and secretion can be driven by both active and passive processes, directly or indirectly controlled by plants, as suggested by Uren (2007). Compounds released by excretion or secretion can cross the plasmalemma through different cellular processes, especially exocytosis and membrane proteins-mediated exudation (Weston *et al.*, 2012). Exocytosis occurs mainly for secretion of HMW compounds such as polysaccharides, through the Golgi or endoplasmic reticulum vesicles (Battey and Blackbourn, 1993; Bertin *et al.*, 2003). It is also implicated in the release of some LMW compounds like phenolics and phytosiderophores (Negishi *et al.*, 2002; Neumann and Römheld, 2007). This vesicle-driven process greatly relies on the intra- and extracellular Ca^{2+} concentrations (Marschner and Marschner, 2012). As for the membrane transport proteins-mediated exudation, it relies on proteins belonging to diverse transporter families, such as ATP-binding cassette proteins (ABCs), multidrug and toxic compound extrusion proteins (MATEs), the major facilitator superfamily (MFS) or the aluminium-activated malate transporters family (ALMT) (Weston *et al.*, 2012). They are involved in the exudation of secondary metabolites such as flavonoids, phenolics, organic acids or phytosiderophores (Sugiyama *et al.*, 2007; Huang *et al.*, 2014). For example, under cytosolic pH (7.2–7.5), most organic acids, such as malic and citric acid, are negatively charged (Jones *et al.*, 2009), decreasing their permeability directly through the lipid bilayer of the membrane. However, an electrical gradient at the cell's plasma membrane is maintained through an active ATPase-mediated proton extrusion and a passive chemical gradient-mediated K^+ efflux (Neumann, 2007). The addition of both the concentration and electrical gradients drives LMW anionic organic acids out of the cell through ion channels like ALMTs (Ryan *et al.*, 2001; Roberts, 2006; Weston *et al.*, 2012).

Localization

Root exudation displays spatial heterogeneity along the root axis (Walker *et al.*, 2003; Compant *et al.*, 2010). Its localization strongly depends on the root system architecture, which relies on many biotic and abiotic factors such as the plant species and the soil composition (Badri and Vivanco, 2009). Currently, accurate knowledge about exudation localization seems to be lacking (Walker *et al.*, 2003; Badri and Vivanco, 2009), partly because of these multi-factor variabilities, except for mucilage secretion, which is known to be mainly released by root cap cells and border cells (Hawes *et al.*, 2003; Neumann, 2007; Haldar and Sengupta, 2015). Different exudation sites may display different exudate compositions (Frenzel, 1960; Badri and Vivanco, 2009), adding another layer of complexity to root exudation.

In general, the main exudation sites are the root apices, i.e. from the root hair zone to the apex (Uren, 2007), especially the root hair (Bertin *et al.*, 2003), the zone directly located above the root tip (Rovira, 1969; Jones *et al.*, 2009; Haldar and Sengupta, 2015) and the root cap (Hawes *et al.*, 2000). To a lesser extent, older root parts are also involved in exudation (Haichar *et al.*, 2014).

Regarding root hair, they are elongated epidermal unicellular structures located at a short distance above each root tip (Curl and Truelove, 1986). They represent 77% of the root surface of cultivated crops (Parker *et al.*, 2000). They are also involved in nutrient and water uptake (Marschner and Marschner, 2012). This is why root hair are often considered as the main point of contact between the roots and the rhizosphere (Haldar and Sengupta, 2015). In the *Proteaceae* family and in other species such as *Lupinus albus* or *Phaseolus vulgaris*, root hair have been shown to display an enhanced exudation of carboxylates into the soil (Lamont, 2003; Yan *et al.*, 2004). Moreover, mucilage has been detected in root hair zones, but it has not been clearly highlighted yet, whether or not it is directly released by root hair (Peterson and Farquhar, 1996).

The root mucilage is mainly produced by hypersecretory cells belonging to the root cap and called border cells (Hawes and Lin, 1990; Neumann and Römheld, 2007). These cells constantly detach from the root tip as it grows through the soil, but usually stay alive for a certain period of time. The majority of plant species exhibit border cells (Hawes *et al.*, 2000), or border-like cells such as in the case of the *Brassicaceae* family, including the model plant *Arabidopsis thaliana* (Driouich *et al.*, 2006). Up to several thousand cells are lost every day by the root cap (Nguyen, 2003), but remain alive up to several weeks after detachment from the root (Uren, 2007). Once detached, they even display an enhanced metabolic activity, and their transcriptomic and proteomic profiles differ clearly from those of cap cells (Hawes *et al.*, 2000). These observations have led to the confirmed assumption that border cells' roles in the rhizosphere go far beyond mucilage secretion for soil lubrication (Hawes *et al.*, 2003). These cells are indeed actively involved in root defence through secretion of antimicrobial compounds such as antibiotics, DNA, enzymes, and phytoalexins (Hawes *et al.*, 1998, 2012). When they die, these cells represent a valuable nutrient supply in the rhizosphere, even though they may represent only a small proportion of the carbon released by rhizodeposition (Jones *et al.*, 2009).

10.2.2 Senescence of root outer cells

Rhizodeposition through senescence of root outer cells – such as border cells and root hair – leads to the release of their content and of their cell wall into the rhizosphere (Lynch and Whipps, 1990; Nguyen, 2003). Little information about this process is available given the laboriousness of its study in the soil (Jones *et al.*, 2009; Dennis *et al.*, 2010). It occurs primarily for epidermal cells and for border cells (Marschner and Marschner, 2012) and may be enhanced by pathogen and mineral abrasion (Jones *et al.*, 2009). Epidermal cells include root hair, whose lifespan has been estimated around 3 days for some monocots of agricultural interest such as maize (Fusseder, 1987). Given this lifespan, Nguyen (2003) calculated that senescence of root hair accounted for negligible amounts in rhizodeposition in comparison to exudation, with a difference of three orders of magnitude between these processes. However, rhizodeposition through the release of root debris (not only root outer cells) and border cells seems to account for a relatively similar amount to exudates (Neumann, 2007; Uren, 2007).

10.2.3 Contributions of the rhizodeposition mechanisms

To understand the plant–microbes interactions at the root level, it is of great interest to determine the proportions of net photosynthetically fixed carbon released by rhizodeposition and the repartition of this carbon between the different release mechanisms. Many authors have reported on this information

(Lynch and Whipps, 1990; Whipps and Lynch, 1990; Nguyen, 2003; Neumann, 2007; Uren, 2007). However, the numbers obtained are highly variable, especially according to the plant species and their developmental stage, to the techniques used for measurements, but also to the definition ascribed to rhizodeposition and root exudation, leading to rough estimates (Uren, 2007) and, potentially, confusion. Furthermore, these numbers are not always calculated with the same units, complicating comparisons (Nguyen, 2003). According to the literature, the proportion of net fixed carbon released by rhizodeposition seems to range from 5 to 30% of total net fixed carbon (Lynch and Whipps, 1990; Neumann, 2007; Uren, 2007). As for the exudates, they may represent 5 to 20% of total fixed carbon (Jones *et al.*, 2004; Neumann, 2007). The border cells would account for a hundredth of what exudation provides in carbon (Neumann, 2007), and mucilage would represent between 2 and 12% of total rhizodeposition (Hirsch *et al.*, 2013).

Spatial, temporal and chemical variability of rhizodeposition mechanisms shape the rhizosphere into a highly heterogeneous environment compared to the bulk soil. This leads to a great diversity of ecological niches (Hawkes *et al.*, 2007) that can be colonized by microbes possibly interacting with the roots nearby. These interactions can be pathogenic, saprophytic or beneficial (Lugtenberg *et al.*, 2002). The latter is of great interest in agriculture, and has especially been studied for various strains belonging to the *Pseudomonas* genus in the hope of enhancing their efficiency in the field as PGPR (Bloemberg and Lugtenberg, 2001). How can these beneficial bacteria survive and thrive in such a heterogeneous environment as the rhizosphere of plants?

10.3 Beneficial *Pseudomonas* spp. Colonization of the Rhizosphere and Their Influence on the Plant Physiology

10.3.1 Rhizosphere colonization

Given its strong heterogeneity, the rhizosphere may be considered as a succession of favourable and less favourable ecological niches for plant-beneficial *Pseudomonas* spp. Favourable niches include junctions between epidermal root cells or sites of side roots appearances (Chin-A-Woeng *et al.*, 1997). Thus, only a small percent (~6%) of the root surface – the rhizoplane – is effectively colonized by plant-associated microbes (Lugtenberg and Bloemberg, 2004). The colonization of the rhizoplane by plant-beneficial *Pseudomonas* spp. is a dynamic process leading to the establishment of microcolonies on the surface of a continuously growing root. This process follows three steps: flagella-driven motility towards exudates, adhesion and biofilm formation.

The rhizosphere colonization starts with the perception of an exudate gradient by plant-beneficial *Pseudomonas* spp. and the resulting flagella-driven chemotaxis (Lugtenberg *et al.*, 2001; Molina *et al.*, 2003). Numerous exudates have been characterized as chemoattractants, including amino acids (Futamata *et al.*, 1998), organic acids (de Weert *et al.*, 2002) and secondary metabolites (Neal *et al.*, 2012). The perception of these compounds by specific sensors – some of which have already been characterized (Oku *et al.*, 2012, 2014) – leads to the migration of bacteria towards the vicinity of exudation sites. Interestingly, although flagella-driven motility is essential for chemotaxis, contrasting results were obtained regarding its involvement in rhizocompetence concerning the colonization of the rhizosphere of tomato, wheat and soybean (De Weger *et al.*, 1987; Howie *et al.*, 1987; Scher *et al.*, 1988). This suggests that other factors may be involved in bacterial transportation along the root, such as root growth (Lugtenberg *et al.*, 2001) or soil water flow (Trevors *et al.*, 1990).

Once bacteria reach the root, they attach to it. Several determinants implicated in the adhesion to the roots have been characterized. First, we can cite the hair-like structures called pili (Vesper, 1987) or a homolog to the OprF protein, which has been studied in the plant pathogen *Pseudomonas syringae* as a determinant in the attachment to the rhizoplane (De Mot *et al.*, 1992). The plant cells also possess proteins involved in plant–microorganism interactions, such as the glycoprotein called agglutinin. In *Pseudomonas putida*, a protein-coding gene, *aggA*, has been shown to be involved in the attachment to those glycoproteins (Anderson, 1983; Buell and Anderson, 1992), leading to the agglutination of bacteria to the root. A crucial step in the colonization of the rhizoplane is the shift from a transient adhesion (e.g. pili-mediated adhesion) to an irreversible attachment. A protein, the so-called Lap (large adhesion protein), has been shown to be involved in this process in *Pseudomonas putida* (Hinsa *et al.*, 2003). After irreversible attachment, bacteria multiply to reach a given population size and then form a biofilm.

Biofilms are multicellular aggregates of bacterial cells embedded in a complex matrix, mainly composed of extracellular polymeric substances (EPS) (Flemming and Wingender, 2010). EPS mainly constitute exopolysaccharides, extracellular proteins and extracellular DNA (eDNA) (Flemming and Wingender, 2010). Numerous functions have been associated with EPS including adhesion, cohesion of the biofilm, protection against high concentrations of toxic compounds and protection against desiccation (Danhorn and Fuqua, 2007; Flemming and Wingender, 2010). Due to the high density of bacterial cells, the threshold required to trigger quorum-sensing-regulated secondary metabolite synthesis is often reached in biofilms (Fuqua and Greenberg, 2002). This high bacterial density also enhances horizontal gene transfers (Madsen *et al.*, 2012).

10.3.2 *Pseudomonas* spp. toolbox to impact the plant

Once they have colonized the rhizosphere of a plant, many *Pseudomonas* spp. are able to use a variety of mechanisms to affect the plant's biological processes such as hormone signalling, nutrient uptake, immunity, and root exudation (Bakker *et al.*, 2007; Miller *et al.*, 2008; Höfte and Altier, 2010). Here, we will focus on some of these mechanisms, namely the disruption of plant hormone signalling, the alteration of root exudation, and the expression of the type III secretion system.

Disruption of plant hormone signalling

Several strains of *Pseudomonas* spp. have been shown to produce phytohormones, especially auxins and cytokinins (Miller *et al.*, 2008), thus promoting root growth, expanding the rhizosphere and increasing rhizodeposits available in the rhizosphere (Patten and Glick, 2002). These hormones are well known to be implicated in cell division and elongation, root initiation, apical dominance, delay of senescence, etc. (Spaepen, 2015). The main auxin synthesized by plants is the indole-3-acetic acid (IAA) (Woodward and Bartel, 2005). Many plant-pathogenic and beneficial *Pseudomonas* spp. have also been shown to produce it (Spaepen *et al.*, 2007; Miller *et al.*, 2008). The amount of IAA released by a bacterium determines its effects on a plant: the optimal concentration range for beneficial effects is very narrow, leading to deleterious effects when exceeded (Persello-Cartieaux *et al.*, 2003). Several IAA biosynthesis pathways have been reported in the *Pseudomonas* genus (Spaepen *et al.*, 2007). The main pathways are the indole-3-acetamide pathway and the indole-3-pyruvate pathway (Patten and Glick, 2002). The latter is particularly observed among beneficial *Pseudomonas* spp. (Miller *et al.*, 2008) and requires tryptophan as a precursor of IAA (Spaepen *et al.*, 2007). *Pseudomonas* spp. use this amino acid from root exudates to synthesize auxins (Lugtenberg and Kamilova, 2009). However, the exuded amounts vary greatly depending on the plant species (Kravchenko *et al.*, 2004), and the efficiency of IAA-producing *Pseudomonas* strains for root growth promotion may be correlated with these exuded amounts (Kamilova *et al.*, 2006; Kravchenko *et al.*, 2011).

Some *Pseudomonas* strains are also able to affect ethylene signalling within the plant by decreasing its concentration (Glick, 2014). Ethylene is a hormone produced by plants in response to a wide range of biotic and abiotic stresses (Morgan and Drew, 1997), leading to an inhibition of plant growth (Abeles *et al.*, 1992), but also to plant immunity responses (van Loon *et al.*, 2006). The mechanism by which some *Pseudomonas* spp. are able to reduce ethylene concentration within the plants relies on a particular enzyme: the 1-aminocyclopropane-1-carboxylate (ACC) deaminase (Glick *et al.*, 2007). This enzyme catalyzes the cleavage of ACC, which is the direct precursor of ethylene (Yang and Hoffman, 1984), into ammonia and α-ketobutyrate (Honma and Shimomura, 1978), thus limiting the biosynthesis of ethylene. This leads to a reduced impact of stress on plant development (Glick *et al.*, 2007). However, IAA produced by the plant and by some rhizospheric *Pseudomonas* spp. stimulates the plant production of ACC, potentially increasing the ethylene level (Glick, 2014). But the produced ethylene inhibits the IAA signal transduction in return, limiting the effect of IAA on ACC concentration and ethylene synthesis (Stearns *et al.*, 2012). Moreover, IAA is implicated in cell elongation, and especially in loosening the cell walls, increasing exudation of many compounds, including ACC (Glick, 2014). Thus, bacterial strains producing both IAA and ACC deaminase are able to stimulate ACC production and exudation from the plant and to metabolize it while lowering ethylene concentration in the plant, leading to an enhanced root growth (Glick, 2015).

Alteration of root exudation

As mentioned above, IAA-producing strains of *Pseudomonas* spp. are able to affect root exudation via cell wall loosening. Another mechanism allows *Pseudomonas* spp. to modulate exudation: the production of the antibiotic secondary metabolite 2,4-diacetylphloroglucinol (DAPG) (Miller *et al.*, 2008). Many *Pseudomonas* spp. are able to synthesize this compound (Mavrodi *et al.*, 2001), which has been shown to confer a wide-spectrum biocontrol activity (Weller *et al.*, 2002). It has been highlighted that DAPG may block amino acid uptake in plant roots, leading to an increase in net amino acid efflux from roots (Phillips, 2004). Even though this may result in an increase in rhizospheric amino acid availability, it has been shown, for at least one strain of *P. fluorescens*, that the ability to produce DAPG does not seem to affect its rhizocompetence, although the bacterial populations were only monitored for 30 days in the study concerned (Carroll *et al.*, 1995).

Other studies have shown that some *Pseudomonas* spp. are able to increase the soil carbon content (Naseby *et al.*, 1999) and to change the soil amino acid composition (Mozafar *et al.*, 1992), potentially by affecting root exudation. However, the particular processes involved in root exudation alteration by *Pseudomonas* spp. remain unclear and will require further research (Belimov *et al.*, 2015).

Type III secretion system

The Type III Secretion System (T3SS) is one of the main virulence factors of phytopathogenic bacteria such as *P. syringae*. It mediates translocation of virulence effector proteins via the hrp system (hypersensitive response and pathogenicity) into the host cells leading to plant disease (Alfano and Collmer, 2004). Interestingly, a heterologous system, called hrc (hypersensitive response and conserved), has been found in numerous genomes of plant-beneficial *Pseudomonas* spp. (Preston *et al.*, 2001; Loper *et al.*, 2012; Almario *et al.*, 2014) and a component of this system, hrcC, has been shown to be strongly induced in the rhizosphere of beet (Rainey, 1999) and wheat (Mavrodi *et al.*, 2011). The T3SS of several plant-beneficial *Pseudomonas* spp., including the SBW25 strain and the superior root colonizer Q8r1-96, has been shown to be fully functional – it can deliver effectors into plant cells – even though its inactivation does not alter the rhizocompetence of SDW25 nor Q8r1-96 (Preston *et al.*, 2001; Mavrodi *et al.*, 2011). Nonetheless, Q8r1-96 possesses Type III secreted effectors, which are injected in plant cells, leading to the suppression of the hypersensitive response and to the production of reactive oxygen species (Mavrodi *et al.*, 2011). These results suggest that a deeper relationship between plant-beneficial *Pseudomonas* spp. and their host likely occurs in the rhizosphere, although the underlying mechanisms of this interaction still remain unknown.

10.4 Competitiveness-Enhancing Traits Involved in *Pseudomonas* spp. Rhizosphere Colonization

When introduced into the rhizosphere, some plant-beneficial *Pseudomonas* spp. are poor rhizosphere colonizers whereas others are exceptional ones, able to maintain high population levels (10^5–10^7 CFU g^{-1}) during several crop cycles (Simons *et al.*, 1996; Raaijmakers and Weller, 2001; Ghirardi *et al.*, 2012). To achieve such success in colonization, plant-beneficial *Pseudomonas* spp. have to outmatch their indigenous competitors coveting the same nutrients and niches. Here, we will discuss five competitiveness-enhancing traits that have been shown to be involved in competitive rhizosphere colonization: root exudates utilization, siderophore production and uptake, nitrogen dissimilation, phase variation and phenazine production.

10.4.1 Root exudates utilization

Members of the genus *Pseudomonas* are able to use a wide variety of metabolites as a sole source of carbon and energy (Palleroni, 1984; Latour and Lemanceau, 1997) and are therefore adapted to the rhizosphere environment. Nonetheless, the capacity to use some root-released organic compounds, such as specific sugars (sucrose, trehalose and xylose), polyols (inositol and sorbitol) or amino acids (citrulline and trigonelline), has been more often observed in fluorescent *Pseudomonas* spp. (i.e. producing the fluorescent compound pyoverdine) from the rhizosphere than in fluorescent *Pseudomonas* spp. isolated from the bulk soil (Lemanceau *et al.*, 1995; Latour *et al.*, 2003). This suggests that the capacity to use some organic compounds as a source of carbon and energy is a competitiveness-enhancing trait involved in rhizosphere colonization. However, Lugtenberg *et al.* (1999) found no correlation between the colonizing ability of fluorescent *Pseudomonas* spp. towards tomato roots in gnotobiotic conditions and the capacity to use one of the major tomato root exudate sugars as the sole source of carbon and energy. Furthermore, despite hints suggesting that the superior root colonizing ability of the strain *Pseudomonas brassicacearum* Q8r1-96 in the wheat rhizosphere might come from its capacity to use trehalose, benzoate and valerate (Raaijmakers and Weller, 2001), a further study rejected this hypothesis: De La Fuente *et al.* (2007) analysed the ability of 55 DAPG-producing *Pseudomonas* spp., including Q8r1-96, to use those three compounds as sole source of carbon and energy as well as their growth rate when exposed to wheat and pea exudates, and found no differences between excellent and average root colonizers in terms of root exudates utilization.

Ghirardi *et al.* (2012) studied the ability of 23 strains of *Pseudomonas* sp. to survive in the rhizosphere of tomato seedlings grown in iron-limiting soil. They observed that an expansive substrate utilization profile plays a role in the rhizocompetence since strains included in two phenotypic clusters were significantly better colonizers than strains from other clusters. However, the authors suggested that other rhizosphere competence traits, such as the ability to efficiently scavenge ferric iron by siderophore uptake and the ability to use nitrogen oxides as final electron acceptor, were more important to discriminate between poor and good colonizers (Ghirardi *et al.*, 2012).

10.4.2 Siderophore production and uptake

Although being abundant in the soil, bioavailable iron compounds (Fe^{3+}) are scarce in the rhizosphere and in high demand. Hence, ferric iron is often the limiting factor to the growth of rhizospheric microorganisms (Loper and Buyer, 1991). In order to scavenge traces of bioavailable iron in the rhizosphere, plant-beneficial fluorescent *Pseudomonas* spp. produce and excrete high-affinity iron-chelating molecules called siderophores (Neilands, 1981). Once the released siderophores have chelated iron, they can be retrieved by the bacteria. This uptake is mediated by specific outer membrane receptors, most of which are TonB-dependent (Moeck and Coulton, 1998). The main siderophore produced by plant-beneficial fluorescent *Pseudomonas* spp. is pyoverdine, which is also the siderophore showing the highest affinity for Fe^{3+}. Pyoverdine production and utilization have been shown to be involved in competitive rhizosphere colonization of fluorescent *Pseudomonas* spp. As expected, mutants impaired in pyoverdine synthesis and uptake were less competitive in the rhizosphere than their parental strains (Höfte *et al.*, 1992; Mirleau *et al.*, 2000).

There is an important diversity within the pyoverdine family (Budzikiewicz, 2004), originating from the variability in the length and composition of the peptidic chain (Hohnadel and Meyer, 1988). The uptake of one kind of pyoverdine is mediated by a specific outer membrane receptor, which cannot be used for the uptake of another kind of pyoverdine. The nature of the pyoverdine produced by a strain, evaluated by siderotyping, has been shown to be correlated with its ability to colonize the rhizosphere (Ghirardi *et al.*, 2012).

In addition to the pyoverdine outer membrane receptors, plant-beneficial *Pseudomonas* spp. display an important diversity of outer membrane receptors. For example, 45 TonB-dependent outer membrane receptors have been found in the *Pseudomonas protegens* Pf5 genome (Paulsen *et al.*, 2005). Those outer membrane receptors have been shown to enable the uptake of heterologous siderophores (Hartney *et al.*, 2011), which are produced by other organisms. The capacity to use the siderophores produced by a competitor has been shown to confer a competitive advantage for the colonization of the rhizosphere of radish (Raaijmakers *et al.*, 1995b) and cucumber (Loper and Henkels, 1999).

10.4.3 Nitrogen dissimilation

Dioxygen sometimes represents a limiting factor for the growth of microorganisms in the rhizosphere when the demand from both the microorganisms and the plant roots is getting higher (Højberg and Sørensen, 1993). Soil aeration, which is directly linked to compaction or water content, can also influence the available dioxygen in the rhizosphere (Højberg *et al.*, 1999). Some strains of plant-beneficial *Pseudomonas* spp. are able to circumvent this low availability by using nitrogen oxide as an electron acceptor instead of dioxygen. The frequency of fluorescent *Pseudomonas* spp. able to reduce nitrogen oxide has been shown to be higher in the rhizosphere than in the bulk soil, suggesting that nitrogen reduction may be implicated in rhizosphere competence of indigenous fluorescent *Pseudomonas* spp. (von Berg and Bothe, 1992; Clays-Josserand *et al.*, 1995). Several studies have demonstrated the role of nitrate reduction in competitive rhizosphere colonization by plant-beneficial *Pseudomonas* spp. using isogenic mutants unable to synthesize nitrate reductases (Ghiglione *et al.*, 2000; Mirleau *et al.*, 2001). Isogenic mutants were impaired in competitive rhizosphere colonization compared to the wild types; the selective advantage given by nitrate reduction appearing to be even stronger in low dioxygen conditions (Mirleau *et al.*, 2001).

Nitrate reduction and total denitrification have to be distinguished from each other as their benefits to plant-beneficial *Pseudomonas* spp. differ significantly. The first consists of the reduction of nitrate (NO_3^-) to nitrite (NO_2^-) whereas the second leads to the production of dinitrogen (N_2) from a succession of reactions using nitrate (NO_3^-), nitrite (NO_2^-), nitric oxide (NO) and nitrous oxide (N_2O) as substrates. In the study of Ghirardi *et al.* (2012), the best colonizers all shared the ability to perform the complete denitrification cycle. Although nitrate reduction presents a higher energetic yield compared with the following reactions leading to total denitrification, no significant difference was found between nitrate reducers and non-denitrifers in this comparative analysis (Ghirardi *et al.*, 2012), suggesting that other determinants might be more important for the rhizosphere competence of fluorescent *Pseudomonas* spp.

10.4.4 Phase variation

Phase variation is a process used by plant-beneficial *Pseudomonas* spp. (among many other microorganisms) to adapt towards a changing environment by generating population diversity. It has been defined as a reversible, high-frequency phenotypic switching mediated by DNA mutation, reorganization or modification (Saunders, 2003; van den Broek *et al.*, 2005). The implication of phase variation in the rhizocompetence has been studied in many strains of plant-beneficial *Pseudomonas* spp., including *P. brassicacearum* NFM421 (Achouak *et al.*, 2004) and *P. fluorescens* F113 (Sanchez-Contreras *et al.*, 2002; Martínez-Granero *et al.*, 2005; Martínez-Granero *et al.*, 2006). During rhizosphere colonization, the authors noticed variants showing a different colony morphology and an increased motility. This increased mobility was correlated with an over-production of flagellin in the variants (Sanchez-Contreras *et al.*, 2002; Achouak *et al.*, 2004). Those variants efficiently colonized the distal parts of the roots, such as the root tips and newly forming roots, whereas the wild types were localized at the basal parts of the roots (Sanchez-Contreras *et al.*, 2002; Achouak *et al.*, 2004).

In *P. fluorescens* F113, most of the phenotypic variation is due to the activity of two site-specific recombinases of the λ integrase family, encoded by *xerC/sss* and *xerD* (Sanchez-Contreras *et al.*, 2002; Martínez-Granero *et al.*, 2005), which play a role in the rearrangement of the DNA (Sadowski, 1986). Interestingly, an isogenic mutant of *P. fluorescens* WCS365, affected in the production of the site-specific recombinase *sss*, was impaired in the colonization of the rhizosphere of potato, tomato, radish and wheat (Dekkers *et al.*, 1998). It has been suggested that the mutant was locked in a less favourable phenotypic configuration for rhizosphere colonization (Dekkers *et al.*, 1998). Nonetheless, a *sss*⁻ isogenic mutant of *P. brassicacearum* Q8r1-96 was not impaired in wheat root colonization, although it was less competitive than its parental strain when co-inoculated (Mavrodi *et al.*, 2006a). Interestingly, the insertion of the *sss* gene in two *Pseudomonas* spp. with contrasting rhizosphere colonization abilities (one good colonizer and one poor colonizer) was associated with an enhanced ability to colonize the root tip for both strains (Dekkers *et al.*, 2000).

10.4.5 Phenazine production

Phenazines are broad-spectrum antibiotics produced by some strains of fluorescent *Pseudomonas* spp. (Mavrodi *et al.*, 2006b). Phenazine derivatives have been shown to play a crucial role in the biocontrol of several plant pathogens including *Gaeumannomyces graminis* var. *tritici* (Thomashow and Weller, 1988), *Phythium* spp. (Gurusiddaiah *et al.*, 1986), *Fusarium oxysporum* f. sp. *radicis-lycopersici* (Chin-A-Woeng *et al.*, 1998) and *Streptomyces scabies* (St-Onge *et al.*, 2011; Arseneault *et al.*, 2013) and are frequently associated with disease-suppressiveness (Raaijmakers and Weller, 1998; Weller *et al.*, 2002; Mazurier *et al.*, 2009). Phenazine production is also involved in the rhizocompetence of several plant-beneficial *Pseudomonas* spp. strains (Mazzola *et al.*, 1992). Phenazine defective mutants of *Pseudomonas synxantha* 2–79 and *Pseudomonas chlororaphis* subsp. *aureofaciens* 30–84 were impaired in the colonization of the wheat rhizosphere grown in a non-pasteurized soil (Mazzola *et al.*, 1992): the populations of the two mutants were unable to maintain high levels and declined rapidly. In pasteurized soil, however, the mutants colonized the rhizosphere to the same extent the parental strains, suggesting that phenazine production is likely to enhance the ability to compete with indigenous microorganisms. However, despite its broad spectrum inhibition, several authors suggested that phenazine production by plant-beneficial *Pseudomonas* spp. does not impact their immediate competitors (Mavrodi *et al.*, 2006b; Pierson and Pierson, 2010) but could rather serve other purposes (Price-Whelan *et al.*, 2006).

Multiple studies have highlighted the beneficial effect of phenazine derivatives for their producers. Due to their high redox potential, phenazine derivatives may operate as electron shuttles in intracellular processes to maintain a high NADH/NAD+ ratio (Price-Whelan *et al.*, 2006), or increase the availability of ferric iron (Fe^{3+}) by reducing mineral iron (Hernandez *et al.*, 2004; Wang and Newman, 2008). Moreover, it has been suggested that the capacity of phenazine-producing *Pseudomonas* spp. to mobilize ferric iron may represent a decisive competitive advantage over other microorganisms under iron-limiting conditions (Mazurier *et al.*, 2009; Ghirardi *et al.*, 2012).

The production of phenazines has a strong impact on biofilm establishment (Maddula *et al.*, 2006) and the ratio between the different phenazine derivatives produced (Maddula *et al.*, 2008) strongly influences biofilm architecture. For example, 30–84 derivative 30-84O, which produces more 2-hydroxyphenazine-1-carboxylic acid (2-OH-PCA), has been shown to display an altered biofilm architecture compared to the wild type (Maddula *et al.*, 2008). Recently, this has been linked to the fact that 2-OH-PCA promotes the release of eDNA (Wang *et al.*, 2016), a structural component in biofilms. By promoting the construction of thicker and more robust biofilms, phenazines production might represent a crucial advantage in water-limited environments, where desiccation tolerance is essential. Interestingly, large indigenous communities of phenazine-producing *Pseudomonas* spp. have been sampled in dry lands (Mavrodi *et al.*, 2012), which attests their strong resistance against desiccation.

10.5 Conclusions and Future Prospects

The rhizosphere is a highly heterogeneous habitat rich in many organic substances released by living plants through their roots. This heterogeneity is exploited by a wide range of microorganisms that can be deleterious, neutral or beneficial for the plants, such as the bacteria belonging to the *Pseudomonas* genus. Plant-beneficial *Pseudomonas* spp. are of great interest in agriculture to protect crops against pathogens and to improve plant growth. However, their efficiency in the field essentially depends on their ability to aggressively colonize the rhizosphere. To thrive in this challenging environment, *Pseudomonas* spp. have many tools at their disposal, ranging from secondary metabolite biosynthesis to genetic plasticity, as well as enhanced nutrient metabolizing abilities. This diversity of bacterial mechanisms gives an insight into the complexity of the interactions occurring in the rhizosphere between plant-beneficial *Pseudomonas* spp., their plant hosts and their competitors.

Elucidating the competitiveness-enhancing traits of plant-beneficial *Pseudomonas* spp. remains challenging because of the overwhelming diversity of genetic determinants affecting the colonization of the rhizosphere. Comparative studies reporting differential colonization abilities between strains are often limited to a handful of genotypes (Landa *et al.*, 2002; Ghirardi *et al.*, 2012) and focus on a limited number of phenotypic attributes (De La Fuente *et al.*, 2007; Ghirardi *et al.*, 2012). Comparative analysis of the rhizocompetence of hundreds of plant-beneficial *Pseudomonas* spp. associated with whole genome sequencing should lead to the identification of new competitiveness-enhancing traits involved in rhizosphere colonization and may thus facilitate the screening of field-efficient PGPR strains.

From the plant point of view, as stated by Bais *et al.* (2006), roots are rhizosphere ambassadors of the plant, involved in interkingdom communication belowground. They shape the rhizosphere microbiome (Lareen *et al.*, 2016), impacting on inoculated PGPR such as *Pseudomonas* spp. Therefore, determining plant traits supporting bacterial colonization of the rhizosphere is as crucial as studying bacterial traits, and may lead to new plant varieties optimized for PGPR colonization. Large-scale development of efficient plant-beneficial *Pseudomonas* spp. will be achieved by embracing an integrated vision of the rhizosphere. This vision should include the bacterial rhizocompetence traits, as well as the plant's genetic determinants favourable to bacterial colonization, and the role of the indigenous microflora in the multitrophic interactions occurring within the rhizosphere.

References

Abeles, F.B., Morgan, P.W. and Saltveit, M.E. (1992) *Ethylene in Plant Biology*, 2nd edn., Academic Press, San Diego, California.

Achouak, W., Conrod, S., Cohen, V. and Heulin, T. (2004) Phenotypic variation of *Pseudomonas brassicacearum* as a plant root-colonization strategy. *Molecular Plant–Microbe Interactions* 17, 872–879.

Alfano, J.R. and Collmer, A. (2004) Type III secretion system effector proteins: double agents in bacterial disease and plant defense. *Annual Review of Phytopathology* 42, 385–414. doi:10.1146/annurev.phyto.42.040103.110731.

Almario, J., Gobbin, D., Défago, G., Moënne-Loccoz, Y. and Rezzonico, F. (2014) Prevalence of type III secretion system in effective biocontrol pseudomonads. *Research in Microbiology* 165, 300–304. doi:10.1016/j.resmic.2014.03.008.

Anderson, A.J. (1983) Isolation from root and shoot surfaces of agglutinins that show specificity for saprophytic pseudomonads. *Canadian Journal of Botany* 61, 3438–3443.

Angus, A.A. and Hirsch, A.M. (2013) Biofilm formation in the rhizosphere: multispecies interactions and implications for plant growth. In: de Bruijn, F.J. (ed.) *Molecular Microbial Ecology of the Rhizosphere: Volume 1 & 2*. Wiley-Blackwell, Hoboken, New Jersey, pp. 701–712. doi:10.1002/9781118297674.ch66.

Antoun, H. and Prévost, D.(2005) Ecology of plant growth promoting rhizobacteria. In: Siddiqui, Z.A. (ed.) *PGPR: Biocontrol and Biofertilization*. Springer, Dordrecht, Netherlands, pp. 1–38. doi:10.1007/1-4020-4152-7_1.

Arseneault, T., Goyer, C. and Filion, M. (2013) Phenazine production by *Pseudomonas* sp. LBUM223 contributes to the biological control of potato common scab. *Phytopathology* 103, 995–1000. doi:10.1094/PHYTO-01-13-0022-R.

Badri, D.V. and Vivanco, J.M. (2009) Regulation and function of root exudates. *Plant, Cell & Environment* 32, 666–681. doi:10.1111/j.1365-3040.2009.01926.x.

Badri, D.V., Weir, T.L., van der Lelie, D. and Vivanco, J.M. (2009) Rhizosphere chemical dialogues: plant–microbe interactions. *Current Opinion in Biotechnology* 20, 642–650. doi:10.1016/j.copbio.2009.09.014.

Bais, H.P., Park, S.-W., Weir, T.L., Callaway, R.M. and Vivanco, J.M. (2004) How plants communicate using the underground information superhighway. *Trends in Plant Science* 9, 26–32. doi:10.1016/j.tplants.2003.11.008.

Bais, H.P., Weir, T.L., Perry, L.G., Gilroy, S. and Vivanco, J.M.(2006) The role of root exudates in rhizosphere interactions with plants and other organisms. *Annual Review of Plant Biology* 57, 233–266. doi:10.1146/annurev.arplant.57.032905.105159.

Bakker, P.A.H.M., Pieterse, C.M.J. and van Loon, L.C. (2007) Induced systemic resistance by fluorescent *Pseudomonas* spp. *Phytopathology* 97, 239–243. doi:10.1094/PHYTO-97-2-0239.

Bardgett, R.D. (2005) *The Biology of Soil: A Community and Ecosystem Approach*. Oxford University Press, New York.

Barriuso, J., Solano, B.R., Lucas, J.A., Lobo, A.P., García-Villaraco, A. and Mañero, F.J.G. (2008) Ecology, genetic diversity and screening strategies of plant growth promoting rhizobacteria (PGPR). In: Ahmad, I., Pichtel, J. and Hayat, S. (eds) *Plant–Bacteria Interactions: Strategies and Techniques to Promote Plant Growth*. Wiley-VCH, Weinheim, Germany, pp. 1–17. doi: 10.1002/9783527621989.ch1.

Battey, N.H. and Blackbourn, H.D.(1993) The control of exocytosis in plant cells. *New Phytologist* 125, 307–338. doi:10.1111/j.1469-8137.1993.tb03883.x.

Belimov, A.A., Dodd, I.C., Safronova, V.I., Shaposhnikov, A.I., Azarova, T.S., *et al.* (2015) Rhizobacteria that produce auxins and contain 1-amino-cyclopropane-1-carboxylic acid deaminase decrease amino acid concentrations in the rhizosphere and improve growth and yield of well-watered and water-limited potato (*Solanum tuberosum*): Rhizobacterial effects on rhizosphere amino acid concentrations and potato growth. *Annals of Applied Biology* 167, 11–25. doi:10.1111/aab.12203.

Bertin, C., Yang, X. and Weston, L.A.(2003) The role of root exudates and allelochemicals in the rhizosphere. *Plant and Soil* 256, 67–83. doi:10.1023/A:1026290508166.

Bloemberg, G.V. and Lugtenberg, B.J. (2001) Molecular basis of plant growth promotion and biocontrol by rhizobacteria. *Current Opinion in Plant Biology* 4, 343–350. doi:10.1016/S1369-5266(00)00183-7.

Brink, S.C.(2016) Unlocking the secrets of the rhizosphere. *Trends in Plant Science* 21, 169–170. doi:10.1016/j.tplants.2016.01.020.

Budzikiewicz, H. (2004) Siderophores of the *Pseudomonadaceae sensu stricto* (fluorescent and non-fluorescent *Pseudomonas* spp.). In: Herz, W., Falk, H. and Kirby, G.W. (eds) *Progress in the Chemistry of Organic Natural Products*. Springer, Vienna, Austria, pp. 81–237. doi:10.1007/978-3-7091-0581-8_2.

Buell, C.R. and Anderson, A.J. (1992) Genetic analysis of the aggA locus involved in agglutination and adherence of *Pseudomonas putida*, a beneficial fluorescent pseudomonad. *Molecular Plant–Microbe Interactions* 5, 154–162. doi:10.1094/MPMI-5-154.

Bull, C.T., Weller, D.M. and Thomashow, L.S. (1991) Relationship between root colonization and suppression of *Gaeumannomyces graminis* var. *tritici* by *Pseudomonas fluorescens* strain 2-79. *Phytopathology* 81, 954–959.

Carroll, H., Moenne-Loccoz, Y., Dowling, D.N. and O'Gara, F.(1995) Mutational disruption of the biosynthesis genes coding for the antifungal metabolite 2,4-diacetylphloroglucinol does not influence the ecological fitness of *Pseudomonas fluorescens* F113 in the rhizosphere of sugarbeets. *Applied and Environmental Microbiology* 61, 3002–3007.

Chin-A-Woeng, T.F., de Priester, W., van der Bij, A.J. and Lugtenberg, B.J. (1997) Description of the colonization of a gnotobiotic tomato rhizosphere by *Pseudomonas fluorescens* biocontrol strain WCS365, using scanning electron microscopy. *Molecular Plant–Microbe Interactions* 10, 79–86. doi:10.1094/MPMI.1997.10.1.79.

Chin-A-Woeng, T.F., Bloemberg, G.V., van der Bij, A.J., van der Drift, K.M., Schripsema, J. *et al.* (1998) Biocontrol by phenazine-1-carboxamide-producing *Pseudomonas chlororaphis* PCL1391 of tomato root rot caused by *Fusarium oxysporum* f. sp. *radicis-lycopersici*. *Molecular Plant–Microbe Interactions* 11, 1069–1077. doi:10.1094/MPMI.1998.11.11.1069.

Clays-Josserand, A., Lemanceau, P., Philippot, L. and Lensi, R. (1995) Influence of two plant species (flax and tomato) on the distribution of nitrogen dissimilative abilities within fluorescent *Pseudomonas* spp. *Applied and Environmental Microbiology* 61, 1745–1749.

Compant, S., Clément, C. and Sessitsch, A. (2010) Plant growth-promoting bacteria in the rhizo- and endosphere of plants: their role, colonization, mechanisms involved and prospects for utilization. *Soil Biology and Biochemistry* 42, 669–678. doi:10.1016/j.soilbio.2009.11.024.

Curl, E.A. and Truelove, B. (1986) *The Rhizosphere*. Springer Berlin Heidelberg, Germany.

Danhorn, T. and Fuqua, C. (2007) Biofilm formation by plant-associated bacteria. *Annual Review of Microbiology* 61, 401–422. doi:10.1146/annurev.micro.61.080706.093316.

Dekkers, L.C., Phoelich, C.C., Van Der Fits, L. and Lugtenberg, B.J. (1998) A site-specific recombinase is required for competitive root colonization by *Pseudomonas fluorescens* WCS365. *Proceedings of the National Academy of Sciences of the United States of America* 95, 7051–7056.

Dekkers, L.C., Mulders, I.H., Phoelich, C.C., Chin-A-Woeng, T.F., Wijfjes, A.H. and Lugtenberg, B.J. (2000) The sss colonization gene of the tomato–*Fusarium oxysporum* f. sp. *radicis-lycopersici* biocontrol strain *Pseudomonas fluorescens* WCS365 can improve root colonization of other wild-type *Pseudomonas* spp. bacteria. *Molecular Plant–Microbe Interactions* 13, 1177–1183. doi:10.1094/MPMI.2000.13.11.1177.

De La Fuente, L., Mavrodi, D.V., Thomashow, L.S. and Weller, D.M.(2007) Utilization of trehalose, benzoate, valerate, and seed and root exudates by genotypes of 2,4-diacetylphloroglucinol producing *Pseudomonas fluorescens*. *Soil Biology and Biochemistry* 39, 2712–2722. doi:10.1016/j.soilbio.2007.04.029.

De Mot, R., Proost, P., Van Damme, J. and Vanderleyden, J. (1992) Homology of the root adhesin of *Pseudomonas fluorescens* OE 28.3 with porin F of *P. aeruginosa* and *P. syringae*. *Molecular and General Genetics* 231, 489–493. doi:10.1007/BF00292721.

Dennis, P.G., Miller, A.J. and Hirsch, P.R. (2010) Are root exudates more important than other sources of rhizodeposits in structuring rhizosphere bacterial communities? Root exudates and rhizosphere bacteria. *FEMS Microbiology Ecology* 72, 313–327. doi:10.1111/j.1574-6941.2010.00860.x.

De Weert, S., Vermeiren, H., Mulders, I.H., Kuiper, I., Hendrickx, N. *et al.* (2002) Flagella-driven chemotaxis towards exudate components is an important trait for tomato root colonization by *Pseudomonas fluorescens*. *Molecular Plant–Microbe Interactions* 15, 1173–1180. doi:10.1094/MPMI.2002.15.11.1173.

De Weger, L.A., Van der Vlugt, C.I., Wijfjes, A.H.M., Bakker, P.A., Schippers, B. and Lugtenberg, B. (1987) Flagella of a plant-growth-stimulating *Pseudomonas fluorescens* strain are required for colonization of potato roots. *Journal of Bacteriology* 169, 2769–2773.

Driouich, A., Durand, C. and Vicré-Gibouin, M. (2006) Formation and separation of root border cells. *Trends in Plant Science* 12, 14–19. doi:10.1016/j.tplants.2006.11.003.

Farrar, J., Hawes, M., Jones, D. and Lindow, S. (2003) How roots control the flux of carbon to the rhizosphere. *Ecology* 84, 827–837. doi:10.1890/0012-9658(2003)084[0827:HRCTFO]2.0.CO;2.

Flemming, H.-C. and Wingender, J. (2010) The biofilm matrix. *Nature Reviews Microbiology* 8, 623–633. doi:10.1038/nrmicro2415.

Frenzel, B. (1960) Zur Ätiologie der Anreicherung von Aminosäuren und Amiden im Wurzelraum von *Helianthus Annuus* L. *Planta* 55, 169–207. doi:10.1007/BF01914633.

Fuqua, C. and Greenberg, E.P. (2002) Listening in on bacteria: acyl-homoserine lactone signalling. *Nature Reviews Molecular Cell Biology* 3, 685–695. doi:10.1038/nrm907.

Fusseder, A. (1987) The longevity and activity of the primary root of maize. *Plant and Soil* 101, 257–265. doi:10.1007/BF02370653.

Futamata, H., Sakai, M., Ozawa, H., Urashima, Y., Sueguchi, T. and Matsuguchi, T. (1998) Chemotactic response to amino acids of fluorescent pseudomonads isolated from spinach roots grown in soils with different salinity levels. *Soil Science and Plant Nutrition* 44, 1–7. doi:10.1080/00380768.1998.10414421.

Ghiglione, J.-F., Gourbiere, F., Potier, P., Philippot, L. and Lensi, R. (2000) Role of respiratory nitrate reductase in ability of *Pseudomonas fluorescens* YT101 to colonize the rhizosphere of maize. *Applied and Environmental Microbiology* 66, 4012–4016. doi:10.1128/AEM.66.9.4012-4016.2000.

Ghirardi, S., Dessaint, F., Mazurier, S., Corberand, T., Raaijmakers, J.M., Meyer, J.-M. *et al.* (2012) Identification of traits shared by rhizosphere-competent strains of fluorescent Pseudomonads. *Microbial Ecology* 64, 725–737. doi:10.1007/s00248-012-0065-3.

Glick, B.R. (2014) Bacteria with ACC deaminase can promote plant growth and help to feed the world. *Microbiological Research* 169, 30–39. doi:10.1016/j.micres.2013.09.009.

Glick, B.R.(2015) Stress control and ACC deaminase. In: Lugtenberg, B. (ed.) *Principles of Plant–Microbe Interactions*. Springer International Publishing, Cham, Switzerland, pp. 257–264. doi:10.1007/978-3-319-08575-3_27.

Glick, B.R., Cheng, Z., Czarny, J. and Duan, J. (2007) Promotion of plant growth by ACC deaminase-producing soil bacteria. *European Journal of Plant Pathology* 119, 329–339. doi:10.1007/s10658-007-9162-4.

Gurusiddaiah, S., Weller, D.M., Sarkar, A. and Cook, R.J. (1986) Characterization of an antibiotic produced by a strain of *Pseudomonas fluorescens* inhibitory to *Gaeumannomyces graminis* var. *tritici* and *Pythium* spp. *Antimicrobial Agents and Chemotherapy* 29, 488–495. doi:10.1128/AAC.29.3.488.

Haas, D. and Défago, G. (2005) Biological control of soil-borne pathogens by fluorescent pseudomonads. *Nature Reviews Microbiology* 3, 307–319. doi:10.1038/nrmicro1129.

Haichar, F. el Z., Santaella, C., Heulin, T. and Achouak, W. (2014) Root exudates mediated interactions belowground. *Soil Biology and Biochemistry* 77, 69–80. doi:10.1016/j.soilbio.2014.06.017.

Haldar, S. and Sengupta, S. (2015) Plant-microbe cross-talk in the rhizosphere: insight and biotechnological potential. *The Open Microbiology Journal* 9, 1–7. doi:10.2174/1874285801509010001.

Hartney, S.L., Mazurier, S., Kidarsa, T.A., Quecine, M.C., Lemanceau, P. and Loper, J.E. (2011) TonB-dependent outer-membrane proteins and siderophore utilization in *Pseudomonas fluorescens* Pf-5. *Biometals* 24, 193–213. doi:10.1007/s10534-010-9385-2.

Hawes, M.C. and Lin, H.-J. (1990) Correlation of pectolytic enzyme activity with the programmed release of cells from root caps of pea (*Pisum sativum*). *Plant Physiology* 94, 1855–1859. doi:10.1104/pp.94.4.1855.

Hawes, M.C., Brigham, L.A., Wen, F., Woo, H.H. and Zhu, Y. (1998) Function of root border cells in plant health: pioneers in the rhizosphere. *Annual Review of Phytopathology* 36, 311–27. doi:10.1146/annurev.phyto.36.1.311.

Hawes, M.C., Gunawardena, U., Miyasaka, S. and Zhao, X. (2000) The role of root border cells in plant defense. *Trends in Plant Science* 5, 128–133. doi:10.1016/S1360-1385(00)01556-9.

Hawes, M.C., Bengough, G., Cassab, G. and Ponce, G. (2003) Root caps and rhizosphere. *Journal of Plant Growth Regulation* 21, 352–367. doi:10.1007/s00344-002-0035-y.

Hawes, M.C., Curlango-Rivera, G., Xiong, Z. and Kessler, J.O. (2012) Roles of root border cells in plant defense and regulation of rhizosphere microbial populations by extracellular DNA "trapping". *Plant and Soil* 355, 1–16. doi:10.1007/s11104-012-1218-3.

Hawkes, C.V., DeAngelis, K.M. and Firestone, M.K. (2007) Root interactions with soil microbial communities and processes. In: Cardon, Z.G. and Whitbeck, J.L. (eds) *The Rhizosphere: An Ecological Perspective*. Elsevier Academic Press, Amsterdam, pp. 1–29. doi:10.1016/B978-012088775-0/50003-3.

Hernandez, M.E., Kappler, A. and Newman, D.K. (2004) Phenazines and other redox-active antibiotics promote microbial mineral reduction. *Applied and Environmental Microbiology* 70, 921–928. doi:10.1128/AEM.70.2.921-928.2004.

Hinsa, S.M., Espinosa-Urgel, M., Ramos, J.L. and O'Toole, G.A. (2003) Transition from reversible to irreversible attachment during biofilm formation by *Pseudomonas fluorescens* WCS365 requires an ABC transporter and a large secreted protein: *P. fluorescens* biofilm formation. *Molecular Microbiology* 49, 905–918. doi:10.1046/j.1365-2958.2003.03615.x.

Hinsinger, P., Gobran, G.R., Gregory, P.J. and Wenzel, W.W. (2005) Rhizosphere geometry and heterogeneity arising from root-mediated physical and chemical processes: research review. *New Phytologist* 168, 293–303. doi:10.1111/j.1469-8137.2005.01512.x.

Hinsinger, P., Bengough, A.G., Vetterlein, D. and Young, I.M. (2009) Rhizosphere: biophysics, biogeochemistry and ecological relevance. *Plant and Soil* 321, 117–152. doi:10.1007/s11104-008-9885-9.

Hirsch, P.R., Miller, A.J. and Dennis, P.G. (2013) Do root exudates exert more influence on rhizosphere bacterial community structure than other rhizodeposits. In: de Bruijn, F.J. (ed.) *Molecular Microbial Ecology of the Rhizosphere: Volume 1 & 2*. Wiley-Blackwell, Hoboken, New Jersey, pp. 229–242. doi:10.1002/9781118297674.ch22.

Höfte, M. and Altier, N. (2010) Fluorescent pseudomonads as biocontrol agents for sustainable agricultural systems. *Research in Microbiology* 161, 464–471. doi:10.1016/j.resmic.2010.04.007.

Höfte, M., Boelens, J. and Verstraete, W. (1992) Survival and root colonization of mutants of plant growth-promoting pseudomonads affected in siderophore biosynthesis or regulation of siderophore production. *Journal of Plant Nutrition* 15, 2253–2262. doi:10.1080/01904169209364472.

Hohnadel, D. and Meyer, J.-M.(1988) Specificity of pyoverdine-mediated iron uptake among fluorescent *Pseudomonas* strains. *Journal of Bacteriology* 170, 4865–4873.

Højberg, O. and Sørensen, J. (1993) Microgradients of microbial oxygen consumption in a barley rhizosphere model system. *Applied and Environmental Microbiology* 59, 431–437.

Højberg, O., Schnider, U., Winteler, H.V., Sørensen, J. and Haas, D. (1999) Oxygen-sensing reporter strain of *Pseudomonas fluorescens* for monitoring the distribution of low-oxygen habitats in soil. *Applied and Environmental Microbiology* 65, 4085–4093.

Honma, M. and Shimomura, T. (1978) Metabolism of 1-aminocyclopropane-1-carboxylic acid. *Agricultural and Biological Chemistry* 42, 1825–1831. doi:10.1080/00021369.1978.10863261.

Howie, W.J., Cook, R.J. and Weller, D.M. (1987) Effects of soil matric potential and cell motility on wheat root colonization by fluorescent pseudomonads suppressive to take-all. *Phytopathology* 77, 286–292.

Huang, X.-F., Chaparro, J.M., Reardon, K.F., Zhang, R., Shen, Q. and Vivanco, J.M. (2014) Rhizosphere interactions: root exudates, microbes, and microbial communities. *Botany* 92, 267–275. doi:10.1139/cjb-2013-0225.

Hütsch, B.W., Augustin, J. and Merbach, W. (2002) Plant rhizodeposition-an important source for carbon turnover in soils. *Journal of Plant Nutrition and Soil Science* 165, 397. doi:10.1002/1522-2624(200208)165:4<397::AID-JPLN397>3.0.CO;2-C.

Jones, D.L., Hodge, A. and Kuzyakov, Y. (2004) Plant and mycorrhizal regulation of rhizodeposition. *New Phytologist* 163, 459–480. doi:10.1111/j.1469-8137.2004.01130.x.

Jones, D.L., Nguyen, C. and Finlay, R.D. (2009) Carbon flow in the rhizosphere: carbon trading at the soil–root interface. *Plant and Soil* 321, 5–33. doi:10.1007/s11104-009-9925-0.

Kamilova, F., Kravchenko, L.V., Shaposhnikov, A.I., Azarova, T., Makarova, N. and Lugtenberg, B. (2006) Organic acids, sugars, and L-tryptophane in exudates of vegetables growing on stonewool and their effects on activities of rhizosphere bacteria. *Molecular Plant-Microbe Interactions* 19, 250–256. doi:10.1094/MPMI-19-0250.

Kloepper, J.W., Leong, J., Teintze, M. and Schroth, M.N. (1980a) Enhanced plant growth by siderophores produced by plant growth-promoting rhizobacteria. *Nature* 286, 885–886. doi:10.1038/286885a0.

Kloepper, J.W., Schroth, M.N. and Miller, T.D. (1980b) Effects of rhizosphere colonization by plant growth-promoting rhizobacteria on potato plant development and yield. *Phytopathology* 70, 1078–1082. doi:10.1094/Phyto-70-1078.

Kravchenko, L.V., Azarova, T.S., Makarova, N.M. and Tikhonovich, I.A. (2004) The effect of tryptophan present in plant root exudates on the phytostimulating activity of rhizobacteria. *Microbiology* 73, 156–158. doi:10.1023/B:MICI.0000023982.76684.9d.

Kravchenko, L.V., Shapozhnikov, A.I., Makarova, N.M., Azarova, T.S., L'vova, K.A. *et al.* (2011) Exometabolites of bread wheat and tomato affecting the plant–microbe interactions in the rhizosphere. *Russian Journal of Plant Physiology* 58, 936–940. doi:10.1134/S1021443711050128.

Lamont, B.B. (2003) Structure, ecology and physiology of root clusters–a review. *Plant and Soil* 248, 1–19. doi:10.1023/A:1022314613217.

Landa, B.B., Mavrodi, O.V., Raaijmakers, J.M., McSpadden Gardener, B.B., Thomashow, L.S. and Weller, D.M. (2002) Differential ability of genotypes of 2, 4-diacetylphloroglucinol-producing *Pseudomonas fluorescens* strains to colonize the roots of pea plants. *Applied and Environmental Microbiology* 68, 3226–3237. doi:10.1128/AEM.68.7.3226-3237.2002.

Lareen, A., Burton, F. and Schäfer, P. (2016) Plant root-microbe communication in shaping root microbiomes. *Plant Molecular Biology* 90, 575–587. doi:10.1007/s11103-015-0417-8.

Latour, X. and Lemanceau, P. (1997) Métabolisme carboné et énergétique des *Pseudomonas* spp fluorescents saprophytes à oxydase positive. *Agronomie* 17, 427–443. doi:10.1051/agro:19970901.

Latour, X., Delorme, S., Mirleau, P. and Lemanceau, P. (2003) Identification of traits implicated in the rhizosphere competence of fluorescent pseudomonads: description of a strategy based on population and model strain studies. *Agronomie* 23, 397–405. doi:10.1051/agro:2003015.

Lemanceau, P., Corberand, T., Gardan, L., Latour, X., Laguerre, G. *et al.* (1995) Effect of two plant species, flax (*Linum usitatissinum* L.) and tomato (*Lycopersicon esculentum* Mill.), on the diversity of soilborne populations of fluorescent pseudomonads. *Applied and Environmental Microbiology* 61, 1004–1012.

Loper, J.E. and Buyer, J.S. (1991) Siderophores in microbial interactions on plant surfaces. *Molecular Plant–Microbe Interactions* 4, 5–13.

Loper, J.E. and Henkels, M.D. (1999) Utilization of heterologous siderophores enhances levels of iron available to *Pseudomonas putida* in the rhizosphere. *Applied and Environmental Microbiology* 65, 5357–5363.

Loper, J.E., Hassan, K.A., Mavrodi, D.V., Davis, E.W., Lim, C.K. *et al.* (2012) Comparative genomics of plant-associated *Pseudomonas* spp.: insights into diversity and inheritance of traits involved in multitrophic interactions. *PLoS Genetics* 8, e1002784. doi:10.1371/journal.pgen.1002784.

Lugtenberg, B.J. and Bloemberg, G.V. (2004) Life in the rhizosphere. In: Ramos, J.-L. (ed.) *Pseudomonas Vol. 1 – Genomics, Life Style and Molecular Architecture*. Springer, New York, pp. 403–430. doi:10.1007/978-1-4419-9086-0_13.

Lugtenberg, B. and Kamilova, F. (2009) Plant-growth-promoting rhizobacteria. *Annual Review of Microbiology* 63, 541–556. doi:10.1146/annurev.micro.62.081307.162918.

Lugtenberg, B.J., Kravchenko, L.V. and Simons, M. (1999) Tomato seed and root exudate sugars: composition, utilization by *Pseudomonas* biocontrol strains and role in rhizosphere colonization. *Environmental Microbiology* 1, 439–446. doi:10.1046/j.1462-2920.1999.00054.x.

Lugtenberg, B.J., Dekkers, L. and Bloemberg, G.V. (2001) Molecular determinants of rhizosphere colonization by *Pseudomonas*. *Annual Review of Phytopathology* 39, 461–490. doi:10.1146/annurev.phyto.39.1.461.

Lugtenberg, B.J., Chin-A-Woeng, T.F. and Bloemberg, G.V. (2002) Microbe–plant interactions: principles and mechanisms. *Antonie Van Leeuwenhoek* 81, 373–383. doi:10.1023/A:1020596903142.

Lynch, J.M. and Whipps, J.M. (1990) Substrate flow in the rhizosphere. *Plant and Soil* 129, 1–10. doi:10.1007/BF00011685.

Maddula, V.S.R.K., Zhang, Z., Pierson, E.A. and Pierson, L.S. (2006) Quorum sensing and phenazines are involved in biofilm formation by *Pseudomonas chlororaphis* (*aureofaciens*) Strain 30–84. *Microbial Ecology* 52, 289–301. doi:10.1007/s00248-006-9064-6.

Maddula, V.S.R.K., Pierson, E.A. and Pierson, L.S. (2008) Altering the ratio of phenazines in *Pseudomonas chlororaphis* (*aureofaciens*) Strain 30–84: effects on biofilm formation and pathogen inhibition. *Journal of Bacteriology* 190, 2759–2766. doi:10.1128/JB.01587-07.

Madsen, J.S., Burmølle, M., Hansen, L.H. and Sørensen, S.J. (2012) The interconnection between biofilm formation and horizontal gene transfer. *FEMS Immunology & Medical Microbiology* 65, 183–195. doi:10.1111/j.1574-695X.2012.00960.x.

Marschner, H. and Marschner, P. (2012) *Marschner's Mineral Nutrition of Higher Plants*, 3rd edn, Elsevier/Academic Press, London; Waltham, Massachusetts.

Martínez-Granero, F., Capdevila, S., Sánchez-Contreras, M., Martín, M. and Rivilla, R. (2005) Two site-specific recombinases are implicated in phenotypic variation and competitive rhizosphere colonization in *Pseudomonas fluorescens*. *Microbiology* 151, 975–983. doi:10.1099/mic.0.27583-0.

Martínez-Granero, F., Rivilla, R. and Martin, M. (2006) Rhizosphere selection of highly motile phenotypic variants of *Pseudomonas fluorescens* with enhanced competitive colonization ability. *Applied and Environmental Microbiology* 72, 3429–3434. doi:10.1128/AEM.72.5.3429-3434.2006.

Mavrodi, O.V., McSpadden Gardener, B.B., Mavrodi, D.V., Bonsall, R.F., Weller, D.M. and Thomashow, L.S. (2001). Genetic diversity of phlD from 2,4-diacetylphloroglucinol-producing *fluorescent Pseudomonas* spp. *Phytopathology* 91, 35–43. doi:10.1094/PHYTO.2001.91.1.35.

Mavrodi, O.V., Mavrodi, D.V., Weller, D.M. and Thomashow, L.S. (2006a) Role of ptsP, orfT, and sss recombinase genes in root colonization by *Pseudomonas fluorescens* Q8r1-96. *Applied and Environmental Microbiology* 72, 7111–7122. doi:10.1128/AEM.01215-06.

Mavrodi, D.V., Blankenfeldt, W. and Thomashow, L.S. (2006b) Phenazine compounds in fluorescent *Pseudomonas* spp. biosynthesis and regulation. *Annual Review of Phytopathology* 44, 417–445. doi:10.1146/annurev.phyto.44.013106.145710.

Mavrodi, D.V., Joe, A., Mavrodi, O.V., Hassan, K.A., Weller, D.M. *et al.* (2011) Structural and functional analysis of the type III secretion system from *Pseudomonas fluorescens* Q8r1-96. *Journal of Bacteriology* 193, 177–189. doi:10.1128/JB.00895-10.

Mavrodi, D.V., Mavrodi, O.V., Parejko, J.A., Bonsall, R.F., Kwak, Y.-S. *et al.* (2012) Accumulation of the antibiotic phenazine-1-carboxylic acid in the rhizosphere of dryland cereals. *Applied and Environmental Microbiology* 78, 804–812. doi:10.1128/AEM.06784-11.

Mazurier, S., Corberand, T., Lemanceau, P. and Raaijmakers, J.M. (2009) Phenazine antibiotics produced by fluorescent pseudomonads contribute to natural soil suppressiveness to Fusarium wilt. *The ISME journal* 3, 977–991. doi:10.1038/ismej.2009.33.

Mazzola, M., Cook, R.J., Thomashow, L.S., Weller, D.M. and Pierson, L.S. (1992) Contribution of phenazine antibiotic biosynthesis to the ecological competence of fluorescent pseudomonads in soil habitats. *Applied and Environmental Microbiology* 58, 2616–2624.

Miller, S.H., Mark, G.L., Franks, A. and O'Gara, F. (2008) *Pseudomonas*–plant interactions. In: Rehm, B.H. (ed.) *Pseudomonas: Model Organism, Pathogen, Cell Factory*. John Wiley & Sons, Weinheim, Germany, pp. 353–376. doi:10.1002/9783527622009.ch13.

Mirleau, P., Delorme, S., Philippot, L., Meyer, J.-M., Mazurier, S. and Lemanceau, P. (2000) Fitness in soil and rhizosphere of *Pseudomonas fluorescens* C7R12 compared with a C7R12 mutant affected in pyoverdine synthesis and uptake. *FEMS Microbiology Ecology* 34, 35–44. doi:10.1111/j.1574-6941.2000.tb00752.x.

Mirleau, P., Philippot, L., Corberand, T. and Lemanceau, P. (2001) Involvement of nitrate reductase and pyoverdine in competitiveness of *Pseudomonas fluorescens* strain C7R12 in soil. *Applied and Environmental Microbiology* 67, 2627–2635. doi:10.1128/AEM.67.6.2627-2635.2001.

Mishra, S., Singh, A., Keswani, C., Saxena, A., Sarma, B.K. and Singh, H.B. (2015) Harnessing plant-microbe interactions for enhanced protection against phytopathogens. In: Arora, N.K. (ed.) *Plant Microbes Symbiosis: Applied Facets*. Springer, New Delhi, India, pp. 111–125.

Moeck, G.S. and Coulton, J.W. (1998) TonB-dependent iron acquisition: mechanisms of siderophore-mediated active transport. *Molecular Microbiology* 28, 675–681. doi:10.1046/j.1365-2958.1998.00817.x.

Molina, M.A., Ramos, J.-L. and Espinosa-Urgel, M. (2003) Plant-associated biofilms. *Reviews in Environmental Science and Biotechnology* 2, 99–108. doi:10.1023/B:RESB.0000040458.35960.25.

Morgan, P.W. and Drew, M.C. (1997) Ethylene and plant responses to stress. *Physiologia Plantarum* 100, 620–630. doi:10.1111/j.1399-3054.1997.tb03068.x.

Mozafar, A., Duss, F. and Oertli, J.J. (1992) Effect of *Pseudomonas fluorescens* on the root exudates of two tomato mutants differently sensitive to Fe chlorosis. *Plant and Soil* 144, 167–176. doi:10.1007/BF00012873.

Naseby, D.C., Pascual, J.A. and Lynch, J.M. (1999) Carbon fractions in the rhizosphere of pea inoculated with 2,4 diacetylphloroglucinol producing and non-producing *Pseudomonas fluorescens* F113. *Journal of Applied Microbiology* 87, 173–181. doi:10.1046/j.1365-2672.1999.00809.x.

Neal, A.L., Ahmad, S., Gordon-Weeks, R. and Ton, J. (2012) Benzoxazinoids in root exudates of maize attract *Pseudomonas putida* to the rhizosphere. *PLoS ONE* 7, e35498. doi:10.1371/journal.pone.0035498.

Negishi, T., Nakanishi, H., Yazaki, J., Kishimoto, N., Fujii, F. *et al.* (2002) cDNA microarray analysis of gene expression during Fe-deficiency stress in barley suggests that polar transport of vesicles is implicated in phytosiderophore secretion in Fe-deficient barley roots. *The Plant Journal* 30, 83–94. doi:10.1046/j.1365-313X.2002.01270.x.

Neilands, J.B. (1981) Iron absorption and transport in microorganisms. *Annual Review of Nutrition* 1, 27–46. doi:10.1146/annurev.nu.01.070181.000331.

Neumann, G. (2007) Root exudates and nutrient cycling. In: Marschner, P. and Rengel, Z. (eds) *Nutrient Cycling in Terrestrial Ecosystems*. Springer Berlin Heidelberg, Germany, pp. 123–157. doi:10.1007/978-3-540-68027-7_5.

Neumann, G. and Römheld, V. (2007) The release of root exudates as affected by the plant physiological status. In: Pinton, R., Varanini, Z. and Nannipieri, P. (eds) *The Rhizosphere: Biochemistry and Organic Substances at the Soil-Plant Interface*. CRC Press, Boca Raton, Florida, pp. 23–72. Doi:10.1201/9781420005585.ch2.

Nguyen, C. (2003) Rhizodeposition of organic C by plants: mechanisms and controls. *Agronomie* 23, 375–396. Doi:10.1051/agro:2003011.

Oku, S., Komatsu, A., Tajima, T., Nakashimada, Y. and Kato, J. (2012) Identification of chemotaxis sensory proteins for amino acids in *Pseudomonas fluorescens* Pf0-1 and their involvement in chemotaxis to tomato root exudate and root colonization. *Microbes and Environments* 27, 462–469. Doi:10.1264/jsme2.ME12005.

Oku, S., Komatsu, A., Nakashimada, Y., Tajima, T. and Kato, J. (2014) Identification of *Pseudomonas fluorescens* chemotaxis sensory proteins for malate, succinate, and fumarate, and their involvement in root colonization. *Microbes and Environments* 29, 413–419. Doi:10.1264/jsme2.me14128.

Palleroni, N.J. (1984) *Pseudomonas*. In: Krieg, N.R. and Holt, J.C. (eds) *Bergey's Manual of Systematic Bacteriology*. Williams and Wilkins, Baltimore, Maryland.

Palleroni, N.J. and Moore, E.R. (2004) Taxonomy of pseudomonads: experimental approaches. In: Ramos, J.-L. (ed.) *Pseudomonas Vol. 1 – Genomics, Life Style and Molecular Architecture*. Springer, New York, pp. 3–44. Doi:10.1007/978-1-4419-9086-0_1.

Parker, J.S., Cavell, A.C., Dolan, L., Roberts, K. and Grierson, C.S. (2000) Genetic interactions during root hair morphogenesis in *Arabidopsis*. *The Plant Cell* 12, 1961–1974. Doi:10.1105/tpc.12.10.1961.

Patten, C.L. and Glick, B.R. (2002) Role of *Pseudomonas putida* indoleacetic acid in development of the host plant root system. *Applied and Environmental Microbiology* 68, 3795–3801. doi:10.1128/aem.68.8.3795-3801.2002.

Paulsen, I.T., Press, C.M., Ravel, J., Kobayashi, D.Y., Myers, G.S.A. *et al*. (2005) Complete genome sequence of the plant commensal *Pseudomonas fluorescens* Pf-5. *Nature Biotechnology* 23, 873–878. doi:10.1038/nbt1110.

Peix, A., Ramírez-Bahena, M.-H. and Velázquez, E. (2009) Historical evolution and current status of the taxonomy of genus *Pseudomonas*. *Infection, Genetics and Evolution* 9, 1132–1147. doi:10.1016/j.meegid.2009.08.001.

Persello-Cartieaux, F., Nussaume, L. and Robaglia, C. (2003) Tales from the underground: molecular plant–rhizobacteria interactions. *Plant, Cell & Environment* 26, 189–199. doi:10.1046/j.1365-3040.2003.00956.x.

Peterson, R.L. and Farquhar, M.L. (1996) Root hairs: specialized tubular cells extending root surfaces. *The Botanical Review* 62, 1–40. doi:10.1007/BF02868919.

Phillips, D.A. (2004) Microbial products trigger amino acid exudation from plant roots. *Plant Physiology* 136, 2887–2894. doi:10.1104/pp.104.044222.

Pierson, L.S. and Pierson, E.A. (2010) Metabolism and function of phenazines in bacteria: impacts on the behavior of bacteria in the environment and biotechnological processes. *Applied Microbiology and Biotechnology* 86, 1659–1670. doi:10.1007/s00253-010-2509-3.

Prashar, P., Kapoor, N. and Sachdeva, S. (2014) Rhizosphere: its structure, bacterial diversity and significance. *Reviews in Environmental Science and Biotechnology* 13, 63–77. doi:10.1007/s11157-013-9317-z.

Preston, G.M., Bertrand, N. and Rainey, P.B. (2001) Type III secretion in plant growth-promoting *Pseudomonas fluorescens* SBW25. *Molecular Microbiology* 41, 999–1014. doi:10.1046/j.1365-2958.2001.02560.x.

Price-Whelan, A., Dietrich, L.E.P. and Newman, D.K. (2006) Rethinking "secondary" metabolism: physiological roles for phenazine antibiotics. *Nature Chemical Biology* 2, 71–78. doi:10.1038/nchembio764.

Raaijmakers, J.M. and Weller, D.M. (1998) Natural plant protection by 2, 4-diacetylphloroglucinol-producing *Pseudomonas* spp. in take-all decline soils. *Molecular Plant-Microbe Interactions* 11, 144–152. doi:10.1094/mpmi.1998.11.2.144.

Raaijmakers, J.M. and Weller, D.M. (2001) Exploiting genotypic diversity of 2,4-Diacetylphloroglucinol-Producing *Pseudomonas* spp.: characterization of superior root-colonizing *P. fluorescens* Strain Q8r1-96. *Applied and Environmental Microbiology* 67, 2545–2554. doi:10.1128/aem.67.6.2545-2554.2001.

Raaijmakers, J.M., Leeman, M., Van Oorschot, M.M., Van der Sluis, I., Schippers, B. and Bakker, P. (1995a) Dose-response relationships in biological control of fusarium wilt of radish by *Pseudomonas* spp. *Phytopathology* 85, 1075–1080.

Raaijmakers, J.M., Sluis, L. van der, Bakker, P.A., Schippers, B., Koster, M. and Weisbeek, P.J. (1995b) Utilization of heterologous siderophores and rhizosphere competence of fluorescent *Pseudomonas* spp. *Canadian Journal of Microbiology* 41, 126–135. doi:10.1139/m95-017.

Rainey, P.B. (1999) Adaptation of *Pseudomonas fluorescens* to the plant rhizosphere. *Environmental Microbiology* 1, 243–257. doi:10.1046/j.1462-2920.1999.00040.x.

Roberts, S.K. (2006) Plasma membrane anion channels in higher plants and their putative functions in roots. *New Phytologist* 169, 647–666. doi:10.1111/j.1469-8137.2006.01639.x.

Rovira, A.D. (1969) Plant root exudates. *The Botanical Review* 35, 35–57. doi:10.1007/BF02859887.

Rovira, A.D., Foster, R.C. and Martin, J.K. (1979) Note on terminology: origin, nature and nomenclature of the organic materials in the rhizosphere. In: Harley, J.L. and Scott Russell, R. (eds) *The Soil–Root Interface*. Published under the aegis of the New Phytologist by Academic Press, London, pp. 1–4.

Ryan, P.R., Delhaize, E. and Jones, D.L. (2001) Function and mechanism of organic anion exudation from plant roots. *Annual Review of Plant Physiology and Plant Molecular Biology* 52, 527–560. doi:10.1146/annurev.arplant.52.1.527.

Sadowski, P. (1986) Site-specific recombinases: changing partners and doing the twist. *Journal of Bacteriology*, 165, 341.

Sanchez-Contreras, M., Martin, M., Villacieros, M., O'Gara, F., Bonilla, I. and Rivilla, R. (2002) Phenotypic selection and phase variation occur during alfalfa root colonization by *Pseudomonas fluorescens* F113. *Journal of Bacteriology* 184, 1587–1596. doi:10.1128/JB.184.6.1587-1596.2002.

Saunders, N.J. (2003) Mutation rates: estimating phase variation rates when fitness differences are present and their impact on population structure. *Microbiology* 149, 485–495. doi:10.1099/mic.0.25807-0.

Scher, F.M., Kloepper, J.W., Singleton, C., Zaleska, I. and Laliberte, M. (1988) Colonization of soybean roots by *Pseudomonas* and *Serratia* species: relationship to bacterial motility, chemotaxis, and generation time. *Phytopathology* 78, 1055–1059.

Shamoot, S., McDonald, L. and Bartholomew, W.V. (1968) Rhizo-deposition of organic debris in soil. *Soil Science Society of America Journal* 32, 817–820. doi:10.2136/sssaj1968.03615995003200060031x.

Simons, M., van der Bij, A.J., Brand, I., de Weger, L.A., Wijffelman, C.A. and Lugtenberg, B.J. (1996) Gnotobiotic system for studying rhizosphere colonization by plant growth-promoting *Pseudomonas* bacteria. *Molecular Plant–Microbe Interactions* 9, 600–607.

Spaepen, S. (2015) Plant hormones produced by microbes. In: Lugtenberg, B. (ed.) *Principles of Plant–Microbe Interactions*. Springer International Publishing, Cham, Switzerland, pp. 247–256. doi:10.1007/978-3-319-08575-3_26.

Spaepen, S., Vanderleyden, J. and Remans, R. (2007) Indole-3-acetic acid in microbial and microorganism-plant signaling. *FEMS Microbiology Reviews* 31, 425–448. doi:10.1111/j.1574-6976.2007.00072.x.

Stearns, J.C., Woody, O.Z., McConkey, B.J. and Glick, B.R. (2012) Effects of bacterial ACC deaminase on *Brassica napus* gene expression. *Molecular Plant-Microbe Interactions* 25, 668–676. doi:10.1094/mpmi-08-11-0213.

St-Onge, R., Gadkar, V.J., Arseneault, T., Goyer, C. and Filion, M. (2011) The ability of *Pseudomonas* sp. LBUM 223 to produce phenazine-1-carboxylic acid affects the growth of *Streptomyces scabies*, the expression of thaxtomin biosynthesis genes and the biological control potential against common scab of potato: Effect of PCA production ability on *S. scabies*. *FEMS Microbiology Ecology* 75, 173–183. doi:10.1111/j.1574-6941.2010.00992.x.

Sugiyama, A., Shitan, N. and Yazaki, K. (2007) Involvement of a soybean ATP-Binding cassette-type transporter in the secretion of genistein, a signal flavonoid in legume-rhizobium symbiosis. *Plant Physiology* 144, 2000–2008. doi:10.1104/pp.107.096727.

Thomashow, L.S. and Weller, D.M. (1988) Role of a phenazine antibiotic from *Pseudomonas fluorescens* in biological control of *Gaeumannomyces graminis* var. *tritici*. *Journal of Bacteriology* 170, 3499–3508.

Trevors, J.T., Van Elsas, J.D., Van Overbeek, L.S. and Starodub, M.E. (1990) Transport of a genetically engineered *Pseudomonas fluorescens* strain through a soil microcosm. *Applied and Environmental Microbiology* 56, 401–408.

Uren, N.C. (2007) Types, amounts, and possible functions of compounds released into the rhizosphere by soil-grown plants. In: Pinton, R., Varanini, Z. and Nannipieri, P. (eds) *The Rhizosphere: Biochemistry and Organic Substances at the Soil-Plant Interface*. CRC Press, Boca Raton, Florida, pp. 1–21. doi:10.1201/9781420005585.ch1.

van den Broek, D., Bloemberg, G.V. and Lugtenberg, B. (2005) The role of phenotypic variation in rhizosphere *Pseudomonas* bacteria. *Environmental Microbiology* 7, 1686–1697. doi:10.1111/j.1462-2920.2005.00912.x.

van Loon, L.C., Geraats, B.P.J. and Linthorst, H.J.M. (2006) Ethylene as a modulator of disease resistance in plants. *Trends in Plant Science* 11, 184–191. doi:10.1016/j.tplants.2006.02.005.

Vesper, S.J. (1987) Production of pili (fimbriae) by *Pseudomonas fluorescens* and correlation with attachment to corn roots. *Applied and Environmental Microbiology* 53, 1397–1405.

Vicré, M., Santaella, C., Blanchet, S., Gateau, A. and Driouich, A. (2005) Root border-like cells of *Arabidopsis*. Microscopical characterization and role in the interaction with rhizobacteria. *Plant Physiology* 138, 998–1008. doi:10.1104/pp.104.051813.

von Berg, K.-H.L. and Bothe, H. (1992) The distribution of denitrifying bacteria in soils monitored by DNA-probing. *FEMS Microbiology Letters* 86, 331–340. doi:10.1111/j.1574-6968.1992.tb04825.x.

Walker, T.S., Bais, H.P., Grotewold, E. and Vivanco, J.M. (2003) Root exudation and rhizosphere biology. *Plant Physiology* 132, 44–51. doi:10.1104/pp.102.019661.

Wang, D., Yu, J.M., Dorosky, R.J., Pierson III, L.S. and Pierson, E.A. (2016) The phenazine 2-hydroxy-phenazine-1-carboxylic acid promotes extracellular DNA release and has broad transcriptomic consequences in *Pseudomonas chlororaphis* 30–84. *PLoS ONE* 11, e0148003. doi:10.1371/journal.pone.0148003.

Wang, Y. and Newman, D.K. (2008) Redox reactions of phenazine antibiotics with ferric (hydr)oxides and molecular oxygen. *Environmental Science & Technology* 42, 2380–2386. doi:10.1021/es702290a.

Weller, D.M. (1988) Biological control of soilborne plant pathogens in the rhizosphere with bacteria. *Annual Review of Phytopathology* 26, 379–407. doi:10.1146/annurev.py.26.090188.002115.

Weller, D.M. (2007) *Pseudomonas* biocontrol agents of soilborne pathogens: looking back over 30 years. *Phytopathology* 97, 250–256. doi:10.1094/phyto-97-2-0250.

Weller, D.M., Raaijmakers, J.M., Gardener, B.B.M. and Thomashow, L.S. (2002) Microbial populations responsible for specific soil suppressiveness to plant pathogens. *Annual Review of Phytopathology* 40, 309–348. doi:10.1146/annurev.phyto.40.030402.110010.

Weston, L.A., Ryan, P.R. and Watt, M. (2012) Mechanisms for cellular transport and release of allelochemicals from plant roots into the rhizosphere. *Journal of Experimental Botany* 63, 3445–3454. doi:10.1093/jxb/ers054.

Whipps, J.M. (2001) Microbial interactions and biocontrol in the rhizosphere. *Journal of Experimental Botany* 52, 487–511. doi:10.1093/jexbot/52.suppl_1.487.

Whipps, J.M. and Lynch, J.M. (1985) Energy losses by the plant in rhizodeposition. In: Fuller, K.W. and Gallon, J.R. (eds) *Plant Products and the New Technology, Proceedings of the Phytochemical Society of Europe* 26. Oxford University Press, New York, pp. 59–71.

Whipps, J.M. and Lynch, J.M. (1990) Carbon economy. In: Lynch, J.M. and Leij, F. (eds) *The Rhizosphere*. John Wiley and Sons, Chichester, UK, 59–97.

Woodward, A.W. and Bartel, B. (2005) Auxin: regulation, action, and interaction. *Annals of Botany* 95, 707–735. doi:10.1093/aob/mci083.

Yan, X., Liao, H., Beebe, S.E., Blair, M.W. and Lynch, J.P. (2004) QTL mapping of root hair and acid exudation traits and their relationship to phosphorus uptake in common bean. *Plant and Soil* 265, 17–29. doi:10.1007/s11104-005-0693-1.

Yang, S.F. and Hoffman, N.E. (1984) Ethylene biosynthesis and its regulation in higher plants. *Annual Review of Plant Physiology* 35, 155–189. doi:10.1146/annurev.pp.35.060184.001103.

11 Endophytomicrobiont: A Multifaceted Beneficial Interaction

Shatrupa Ray,[1] Vivek Singh,[1] Kartikay Bisen,[2] Chetan Keswani,[3] Surendra Singh[1] and H.B. Singh[2]*

[1]Department of Botany, Institute of Science, Banaras Hindu University, Varanasi, India; [2]Department of Mycology and Plant Pathology, Institute of Agricultural Sciences, Banaras Hindu University, Varanasi, India; [3]Department of Biochemistry, Institute of Science, Banaras Hindu University, Varanasi, India

11.1 Introduction

Successful interaction between plants and beneficial microbes lays a foundation for improving plant growth and soil structure. However, several attempts to introduce beneficial bacteria into the rhizospheric region of agricultural plants have met with varying degrees of failure, particularly because of the huge competition posed by the pre-existing established rhizomicrobiota (Keswani *et al.*, 2013, 2014; Bisen *et al.*, 2015, 2016; Keswani, 2015; Keswani *et al.*, 2016a, b). Moreover, several reports claim loss of microbial bioactivity owing to long-term storage (Nautiyal, 1997). Considering the biodiversity and population density of indigenous soil microbiota, causing permanent structural changes to the rhizospheric microbiota may become quite hectic and cumbersome, or to be more succinct, impossible (Singh *et al.*, 2014; Mishra *et al.*, 2015; Singh *et al.*, 2016). Thus, a plausible shift in focus is directed towards promoting early establishment of selected communities of endophytic microorganisms within root systems. Though plant-associated bacteria generally trigger the notion of rhizospheric microbes benefitting from plant root exudates, a few bacteria and fungi are capable of entering the plants as endophytes and developing a permanent mutualistic establishment *in planta* without causing any disease symptoms.

Although Galippe (1887) had postulated that certain rhizospheric microbes may have the potential to enter and reside within plants, the prevailing general view understood any form of microbial occurrence within plant systems as directly corresponding to pathological conditions. However, Vogl (1898) reported for the first time an endophytic mycelium inhabiting the seeds of grass (*Lolium temulentum*). Subsequently, Perotti (1926) and Hennig and Villforth (1940) divulged the presence of bacteria in a large number of healthy plants thereby contradicting that presiding notion (Mano and Morisaki, 2007). Numerous reports have confirmed the presence of endophytes in a variety of plant species (Sturz *et al.*, 2000; Zinniel *et al.*, 2002).

Plants and their associated endophytes develop a mutualistic organization wherein the endophytic partner profits due to the enhanced availability of nutrients and

*E-mail: hbs1@rediffmail.com

© CAB International 2017. *Advances in PGPR Research* (eds H.B. Singh, B.K. Sarma and C. Keswani)

protection from various biotic and abiotic stresses, while the host benefits by growth enhancement and stress alleviation (Hardoim *et al.*, 2008). Bacterial endophytes can be transmitted by seeds or may be vegetatively propagated (Hallmann *et al.*, 1997). Seed transmitted and vegetatively propagated endophytes are transferred to the plants of the next generation without infection.

11.2 Endophytic Classification

The endophytic diversity within the host plant could be explained by their capabilities to survive within the host's internal environment (Compant *et al.*, 2010). The complexities of the endophytic community structures indicate a vigilant scrutiny by the host plants to select the section of rhizosphere microbiota that play essential roles in influencing the plant physiology to the extent of modulating the growth and development of the plant (Gaiero *et al.*, 2013). Hence, the populations of endophytic bacteria are found to be less diverse than the rhizospheric population. Podolich *et al.* (2015), proposed that endophytes are not independent 'players' on the 'plant field' but are closely manipulated by the host and the external environment. However, irrespective of the environmental changes or developmental stages of the plant host and plant signal compositions, the basic endophytic microbiome remains the same for a particular host genotype (Bulgarelli *et al.*, 2012).

Hardoim *et al.* (2008) classified bacterial endophytes as "obligate" and "facultative", describing them as culturable and non-culturable, respectively. The obligate endophytes exist in a dormant state wherein they remain alive but possess very low metabolic activity and an inability to divide (Podolich *et al.*, 2015). On the other hand, the prime reason for turning into non-culturable/facultative forms is nutrient deficiency or other stress conditions. Stress induces the production of phosphorylated GDP or GTP (ppGpp) via toxin-antitoxin system. Accumulation of ppGpp triggers the repression of genes for the normal metabolic pathway of endophytes which further leads to adaptation of endophytes to a non-growing state (Gaca *et al.*, 2013).

Other reports classify endophytes as commensals with unknown functions *in planta*, or they may be mutualists or antagonists, depending upon their expression of positive or negative impacts on hosts. Primarily, endophytic interactions with hosts are studied over a narrow habitat, i.e. within the host plant or taxonomically related hosts and rarely over a wide spectrum of taxonomically unrelated species. However, in the larger niche, reports suggest that inoculum-induced shifts in plant microbial community result in total replacement of harmful communities (Andreote *et al.*, 2009).

11.3 Recognition of Endophytic Status *In Planta*

Criteria to recognize the endophytic establishment of facultative endophytes include their isolation from surface-sterilized plant parts and their ability to re-colonize within the plant tissues as evidenced by the microscopic or *in vitro* examination of "tagged" bacteria within the plant tissues (Rosenblueth and Martínez-Romero, 2006; Botta *et al.*, 2013; Thomas and Reddy, 2013; Ray *et al.*, 2015). The obligate endophytes on the other hand, have mainly been evidenced by denaturing gradient gel electrophoresis profiles of 16S rRNA region amplified from the total plant genome (Araújo *et al.*, 2002). However, fortuitous amplification of plant small subunit (SSU) rDNA portion along with the endophytic SSU rDNA region may affect the authenticity of the above procedure. Thus, specific primers binding particularly to the SSU of the bacterial rDNA region without binding to the plant SSU rDNA has been found as an efficient alternative for identification of uncultivable endophytic strains (Chelius and Triplett, 2001; Sakai *et al.*, 2004; Sun *et al.*, 2008). Similarly, the ribosomal inter-genic spacer analysis (RISA) technique adopted by Ikeda *et al.* (2007) involves the amplification of the ribosomal inter-genic spacer region for the generation of plant-associated microbial profiles without the inclusion of plant DNA. Other similar metagenomic approaches have enabled a deeper delve into the uncultivable endophytic diversity (Manter *et al.*, 2010; Bulgarelli *et al.*, 2012; Sessitsch *et al.*, 2012; Bodenhausen *et al.*, 2013).

11.4 Plant Colonization by Endophytic Bacteria: the Complete Process

The site of origin of endophytic bacteria has for long remained a matter of debate. While the presence of root exudates and rhizodeposits in the rhizosphere region support endophytic colonization (Ray *et al.*, 2016), stem and leaf surfaces are also reported as producers of exudates that attract microorganisms (Mercado-Blanco and Prieto, 2012) (Table 11.1). However, abiotic stress factors including UV rays, heat and lack of nutrients reduce the possibilities of phyllosphere colonization, and only highly adaptable bacteria survive and enter through stomata or hydathodes (Compant *et al.*, 2010).

The primary events involved in endophytic colonization are similar to those adopted by rhizospheric bacteria. However, genes effective in the successful establishment of the rhizospheric bacteria within the plant contribute to its endophytic nature. For instance, the stress hormone ethylene produced within the plant endosphere has major consequences on the bacterial microbiota residing within. Here, bacteria producing 1-aminocyclopropane-1-carboxylate deaminase (ACC deaminase) would be sequestered from among the rhizomicrobiota by the plant as it would have the genetic machinery to ameliorate the stress response (Glick *et al.*, 1998). Moreover, 'rhizosphere competent endophytes' would be described as those which possess the requisite genetic machinery to colonize and carry on the endophytic lifestyle (Hardoim *et al.*, 2008). Figure 11.1 describes the series of events involved in successful establishment of bacterial endophytes within hosts.

Table 11.1. Endophytes in Plant Defence.

Endophytes	Disease	Causal organism	Reference
Streptomyces cyaneofuscatus JCM 4364 *Streptomyces mutabilis* NBRC 12800	Damping-off of Tomato	*Rhizoctonia solani*	Goudjal *et al.* (2014)
Bacillus subtilis	Anthracnose of bean	*Colletotrichum lindemuthianum*	Gholami *et al.* (2013)
Heteroconium chaetospira	Chinese cabbage	*Alternaria brassicicola, A. brassicae* and *A. raphani*	Morita *et al.* (2003)
Burkholderia cepacia Cs5	Grey mould of wine grapes	*Botrytis cinerea*	Kilani-Feki and Jaoua (2011)
*Verticillium*Vt305	Cauliflower verticillium wilt	*Verticillium longisporum*	Tyvaert *et al.* (2014)
Bacillus sp.EPCO102 *Bacillus* sp.EPCO16	Damping off of cotton	*Rhizoctonia solani*	Rajendran and Samiyappan (2008)
Streptomyces sp.	Powdery mildew of pea	*Erysiphepisi*	Sangmanee *et al.* (2009)
Bacillus subtilis EDR4	Stem rot of rape seed	*Sclerotinia sclerotiorum*	Chen *et al.* (2014)
Bacillus lentimorbus	Coffee leaf rust	*Hemileia vastatrix*	Shiomi *et al.* (2006)
Alcaligenes sp. *Pseudomonas aeruginosa*	Phytophthora leaf fall of rubber	*Phytophthora meadii*	Abraham *et al.* (2013)
Bacillus amyloliquefaciens BZ6-1	Peanut bacterial wilt	*Ralstonia solanacearum*	Wang and Liang (2014)

Fig. 11.1. Endophytic establishment commences with biofilm formation on host rhizoplane in presence of root exudates followed by intercellular colonization. Successful endophytic establishment leads to plant growth promotion and biocontrol against pathogenic microbes.

11.4.1 Chemotaxis

Photosynthetic products produced by plants are partly translocated to the root and secreted as root exudates. However, the concentration gradient of the secreted exudates formed in the rhizoplane region allows spatial differences in bacterial colonization (Mano and Morisaki, 2007). Apart from sugars and amino acids, flavonoids also play an essential role in plant–microbe communication. During plant–rhizobia symbiotic interactions, flavonoids and nod factors play key roles for endophytic symbiotic association. For instance, Webster *et al.* (1998) explained the function of flavonoid naringenin in stimulating the intercellular colonization of wheat roots by *Azorhizobium caulinodans*, while de Weert *et al.* (2002) suggested the role of organic acids in the directional motility of *Pseudomonas fluorescens* towards tomato root exudates. Bacilio-Jiménez *et al.* (2003) on the other hand suggested carbohydrates and amino acids as major chemo-attractants in the movement of *Corynebacterium flavescens* and *Bacillus pumilis* to rice.

Hardoim *et al.* (2015) suggested abundance of protein-encoding genes, such as aspartate/maltose (*Tar*) and dipeptides (*Tap*) among endophytes as compared to rhizospheric microbes. Moreover, response regulator proteins, such as CheBR, CheC as well as development of flagella were also reported to be more abundant in endophytes, which indicates utilization of aspartate and dipeptides present in root exudates by endophytes for getting attracted towards the host.

11.4.2 Biofilm formation: the basis of endophytism

In the presence of root exudates, the rhizosphere competent endophytes accelerate their metabolism to a physiological state which allows for optimal nutrient acquisition and

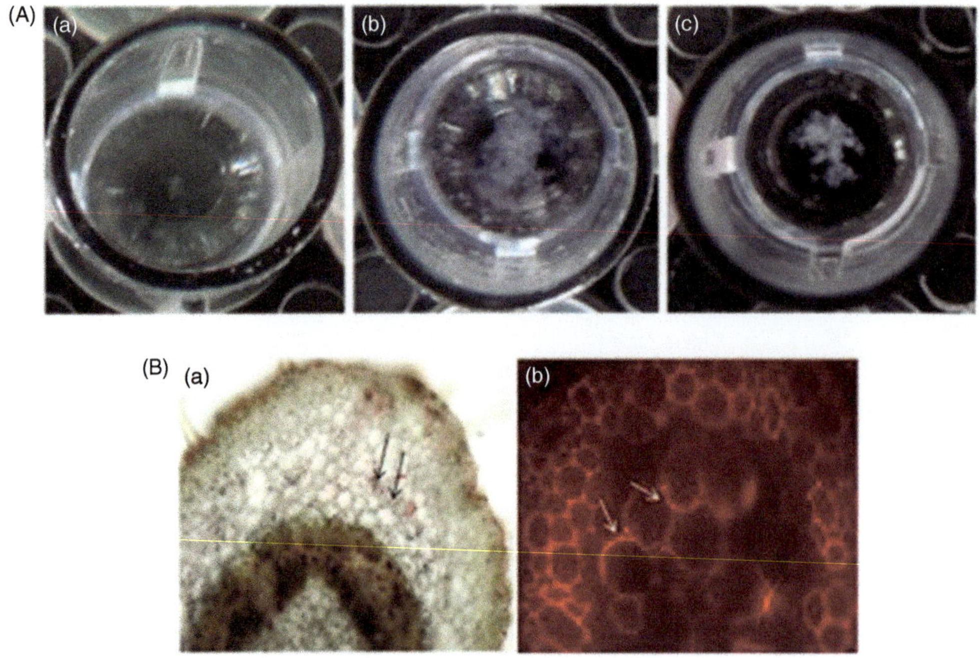

Fig. 11.2. (A) *In vitro* biofilm formation by endophytic *A. faecalis*as observed on microtitre plates in presence (b) and absence (c) of root exudates. (B) Endophytic *A. faecalis* colonies observed in the cortical (a) and vascular (b) regions of Okra plant.

growth (Fig. 11.2A). The primary features essential for root colonization include cyanide, pyoverdine and exoprotease production, which are regulated by the *gacA-gacS* (global antibiotic and cyanide control) regulatory system (Haas and Défago, 2005). Besides, the bacterial lipopolysaccharide (rhamnose) component, controlled by *RfbB* and *RfbC* also plays a key role in the attachment of the endophytic bacteria to the roots and colonization within the host plants (Balsanelli *et al.*, 2010) (Fig. 11.2B).

Biofilm formation on the surface of plant roots is a characteristic feature of endophytic colonization. Type IV pili has been reported as being important in biofilm formation and subsequent migration to the aerial regions though its presence can also be related to pathogenicity particularly in gram-negative bacteria (Reinhold-Hurek *et al.*, 2006). Type IV pili formation is regulated by the genes *pil A* and *pil T.* While *pilA* monitors the pilus formation, the characteristic twitching motility, a series of violent retractions is monitored by *pilT.* Consequently, *pil-T* mutants were reported to form pili but were incapable of twitching motility depriving them from endophytic colonization (Böhm *et al.*, 2007).

11.4.3 Tissue invasion for endophytic entry

Post establishment of the bacteria on the root surface as microcolonies, invasion of root tissues, particularly at the lateral root junctions might occur. In this process, the role of cell-wall degrading enzymes, such as cellulases, pectinases or endoglucanases comes into play for the degradation of plant cell envelopes and subsequent internal colonization (Compant *et al.*, 2005; Reinhold-Hurek *et al.*, 2006). Pérez-Donoso *et al.* (2010) reported endophytic *Xyllela fastidiosa* as a producer of endoglucanase and polygalacturonidase which aided in the enlargement of pore sizes of pit-membranes, thereby helping in its systemic colonization in grapevines. Alternatively, endophytic bacteria might enter plant tissues without cell-wall degradation, through spontaneously formed root-cracks between epidermal cells or wounds inflicted due to phytopathogens. In *Sesbania rostrata* and *Azorhizobium caulinodans,* invasion by plant rhizobia occurs through fissures in lateral root base, root cracks, etc. (Goormachtig *et al.*, 2004). Thus, the production of cell-wall degrading enzymes is not a mandatory feature for endophytic colonization. The synthesis of these degrading enzymes differentiates endophytic bacteria (produced in low levels) from other bacterial phytopathogens (produced in high levels) (Elbeltagy *et al.*, 2000).

In general, translocation of endophytic bacteria within plant tissues may be active or passive. While active translocation involves cell-wall degrading enzymes, passive translocation of endophytes can occur through ruptured endodermis or wounding caused by phytopathogens. From endodermis, bacteria move through the pericycle to finally reach the xylem vessels resulting in systemic translocation throughout host interior.

11.4.4 Plant defence genes involved in endophytic colonization

The defence response system in plants includes a two-way pathway. The first branch utilizes transmembrane pattern recognition receptors (PRRs) responding to microbial or pathogen-associated molecular patterns (MAMPs or PAMPs). This branch induces the activation of the salicylic acid defence response. The second pathway operates within the cell, using the polymorphic NB-LRR (nucleotide binding-leucine rich repeats) protein products that respond to pathogen virulence factors or effector proteins released into the cytoplasm. This pathway induces the activation of jasmonic acid and ethylene defence responses (Jones and Dangl, 2006; Reinhold-Hurek and Hurek, 2011).

Flagellar proteins are the basic MAMPSs that trigger the salicylic acid-dependent systemic acquired resistance (SAR pathway). Lack of the above structure might inhibit the defence response and enhance colonization by endophytic strains (Iniguez *et al.*, 2005). However, lack of flagellar structure displayed reduced colonization by *Salmonella enterica* within *Arabidopsis* (Iniguez *et al.*, 2005). *Burkholderia phytophormans* PsJN bioprimed grapevine plants were reported to express the induction of both SA- and JA-mediated pathways when challenged by the phytopathogen *Pseudomonas syringae* pv. *pisi.* However, the induced defence response was much lower in bioprimed plants as compared to the pathogen challenged plants (Reinhold-Hurek and Hurek, 2011). The SAR pathway is characterized by an early increase in the endogenously synthesized SA and subsequent activation of pathogenesis related proteins (PR proteins) (van Loon, 1997). Bacterial automation of induced systemic resistance requires activation of jasmonic acid and ethylene. Analogous to SAR, ISR depends upon *npr1* for the induction of further pathways (Fig. 11.3).

11.4.5 Entry and localization within plant tissues

Root cracks formed at the site of emergence of lateral roots or in the zone of elongation, emerging radicles and/or wounds inflicted by phytopathogens serve as plausible points for bacterial entry. Colonization of rhizospheric bacteria particularly at these points justifies the endophytic nature of the bacterial isolates, at least at a primary level (Dong *et al.*, 2003). A variety of detection techniques involving immunological detection (Quadt-Hallmann *et al.*, 1997), bright-field microscopy involving colouring stains; tagging with fluorescent protein-expressing genes, such as *dsRED*, *gfp*, etc.; antibiotic tagging, e.g. Rif-tagging; and microscopic techniques such as epifluorescence microscopy, confocal laser scanning microscopy and transmission electron microscopy, are common tools for the localization of endophytic bacteria within plant tissues. Some reports, however, contradict the use of *gfp* tagging as it negatively impacts the root development (van der Lelie *et al.*, 2009). Though several

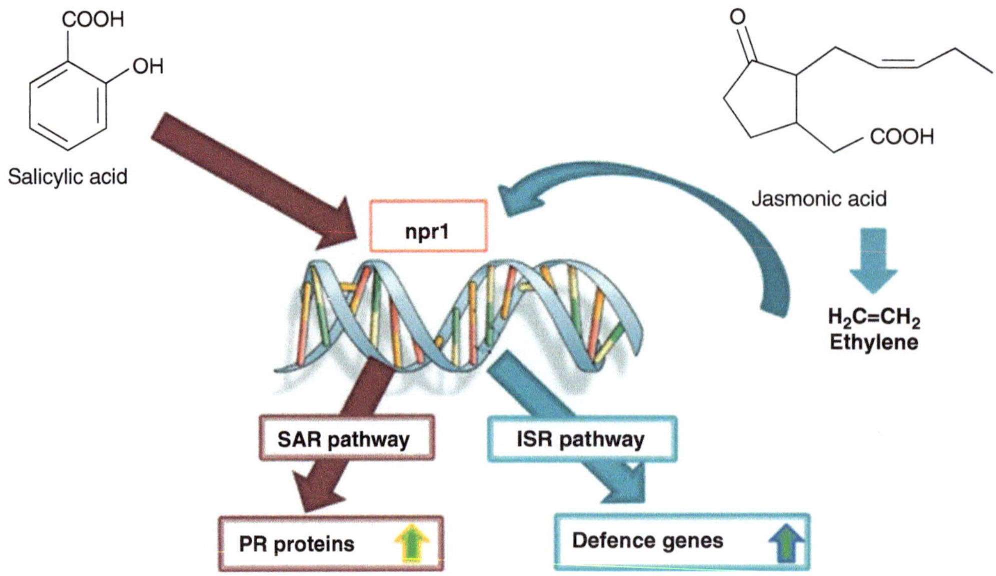

Fig. 11.3. Schematic representation of the two basic defence-related pathways in plants.

reports justify the intracellular presence of fungal endophytes, the bacterial endophytes, however, do not confirm such presence (Kogel *et al.*, 2006). Intercellular spaces between the cortical cells and the xylem vessels are the prime locations of the endophytic bacterial form. In most plants, roots have a higher diversity of endophytic microorganisms relative to other tissues (Compant *et al.*, 2010) (Table 11.2).

11.5 Multifaceted Benefits of Endophytic Bacteria

Benefits conferred by endophytic microbes vary from being beneficial to detrimental depending upon growth conditions of the host as well as the stage in the life cycle of the host (Hardoim *et al.*, 2015). For instance, Bacon *et al.* (2008) described *Fusarium verticillioides* playing dual function of a beneficial endophyte as well as a pathogen in maize, due to host genotype as well as the disturbance in endophytic balance within hosts caused by certain abiotic stress factors. Figure 11.4 elaborates the various beneficial attributes of endophytes which are further described below:

11.5.1 Plant growth promotion

Endophytic microbes play a significant role in enhancing plant growth and improving soil structure (Fig. 11.5). Microbial production of auxins is known to trigger increase in cell elongation, cell division and differentiation in various plants (Jain *et al.*, 2014; Keswani *et al.*, 2014; Bisen *et al.*, 2015; Jain *et al.*, 2015a,b; Patel *et al.*, 2015; Saxena *et al.*, 2015; Keswani *et al.*, 2016a; Singh *et al.*, 2016). Endophytic production of plant auxin (IAA) from tryptophan is reported to occur via three alternative pathways: (a) indolepyruvate and indole-3-acetaldehyde; (b) tryptamine and indole-3-acetaldehyde; and (c) indole-3-acetamide and indole-3-acetonitrile. Taghavi *et al.* (2009) reported the presence of all the three pathways in endophyte *Pseudomonas putida* W619.

Indole acetic acid produced by the bacterial cells behaves as a signalling compound for the plants and induces the ACS (1-aminocyclopropane-1-carboxylate synthase) and ACO (1-aminocyclopropane-1-carboxylate oxydase) multigene family. S-adenosylmethionine (SAM) is catalyzed to ACC by the phosphorylated form of ACS. The phosphorylation of ACS is catalyzed by a kinase

Table 11.2. Diversity of endophytic bacteria in different plant tissues.

Tissue	Endophytes	Reference
Root	*Azospirillum amazonense, A. brasilense, A. lipoferum, Bradyrhizobium* sp., *A. diazotrophicus, Rhizobium leguminosarum, Azoarcus* sp., *Burkholderia pickettii* and *Enterobacter* spp.	Reinhold-Hurek *et al.* (1993); McInroy and Kloepper (1995); Jiménez-Salgado *et al.* (1997); Yanni *et al.* (1997); Weber *et al.* (1999); Chaintreuil *et al.* (2000)
Stem	*Alcaligenes* sp., *Bacillus pumilus, B. cereus, Burkholderia cepacia, Curtobacterium flaccumfaciens, Enterobacter cloacae, Methylobacterium* spp., *Nocardia* sp., *Pantoea agglomerans, Streptomyces* sp., *Xanthomonas campestris, Methylobacterium extorquens, Pseudomonas synxantha, B. megaterium, Pantoea agglomerans, Enterobacter asburiae, Pseudomonas* sp., *Staphylococcus* sp., And *Agrobacterium* sp.	McInroy and Kloepper (1995); Araújo *et al.* (2002); Asis and Adachi (2003); Surette *et al.* (2003); Pirttilä *et al.* (2004)
Leaf	*H. seropedicae, B. cepacia, Herbaspirillum seropedicae, Herbaspirillum rubrisulbalbicans* and *Citrobacter* sp.	Olivares *et al.* (1996); Weber *et al.* (1999); Martínez *et al.* (2003)

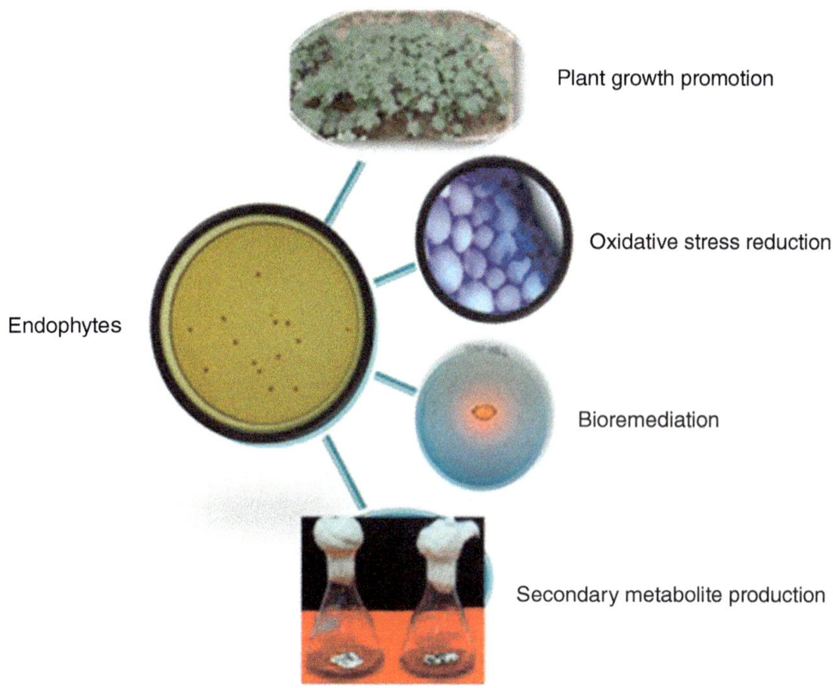

Fig. 11.4. Beneficial attributes of endophytic microbes.

induced by the bacterial IAA. Endophytic bacteria possessing ACC deaminase enzyme take up ACC prior to its oxidation to ethylene by ACC oxidase and cleave ACC into ammonia and α-ketobutyrate thereby modulating plant ethylene levels and defending the attack posed by the host cells (Hardoim *et al.*, 2008). Besides, ACC deaminase production by plant-associated bacteria promotes plant growth by regulating the synthesis of ethylene

Fig. 11.5. Plant growth promotional ability of endophytic *A. faecalis* as expressed on *Abelmoschus esculentus* under A) glass house and B) field conditions.

and thereby reducing its harmful effects (Glick *et al.*, 1998; Penrose and Glick, 2001; Mayak *et al.*, 2004; Madhaiyan *et al.*, 2006).

Volatile substances such as 2-3 butanediol and acetoin produced by rhizobacteria seem to be a newly discovered mechanism responsible for plant growth promotion (Ryu *et al.*, 2003). Acetolate synthase (AlsS) and acetolactate decarboxylase (AlsD) catalyze the switch from pyruvate to acetoin which further gets converted to 2,3-butanediol either by the host or the endophyte. AlsDS acetoin synthesis pathway has been reported in endophytic *Serratia proteamaculans* 568 and *Enterobacter* 638 (Taghavi *et al.*, 2009).

The microbial population also performs phosphate solubilization by the secretion of organic acids which convert insoluble phosphates into soluble monobasic and dibasic ions thereby making it water soluble (Taurian *et al.*, 2010). Siderophore-producing endophytic bacteria can restrict the growth of plant pathogens because of their strong affinity towards Fe (III) (Berg *et al.*, 2005). Gram-negative bacteria have Ton-B-dependent membrane receptors for specific uptake of ferric-siderophore complexes or other small molecules. This denotes the extreme scarcity of bioavailable iron in the endosphere of the root region, reinstating the role of potential endophytes for competitive uptake of ferrous ion in the soil region (Reinhold-Hurek and Hurek, 2011). The ability of siderophore production enables endophytic bacteria to behave as potential biocontrol agents.

11.5.2 Remediators of oxidative stress

Sudden colonization of aerobic microbes in host plant interior leads to an abrupt burst of reactive oxygen species (ROS) and reactive nitrogen species (RNS). Endophytes remain well armoured with enzymes possessing detoxification capacities. According to Hardoim *et al.* (2015), endophytes are well equipped with defence enzymes such as glutathione peroxidise (*btuE*), glutathione *S*-transferase (*gst*), catalase (*katE*) and nitricoxide reductase (*norR*). These defence enzymes enhance endophytes to cope with the plant oxidative burst pathway and also fortify the host to deal with any form of sudden stresses.

In this context, Ray *et al.* (2016) reported augmentation of the host defence system against attack of *Sclerotium rolfsii*. The phenylpropanoid pathway as well as the antioxidant pathway of host, *Abelmoschus esculentus* was found to be upregulated in response to attack by this deleterious fungus. Besides, a plethora of reports clarify the biocontrol potential of endophytes against several phytopathogens.

11.5.3 Bioremediation

Bioremediation exploits the metabolic property of microorganisms to degrade pollutants. Van Aken et al. (2004) reported nitroaromatic compounds degrading methylotrophic endophytic bacteria from poplar trees (Populus deltoides). Numerous endophytic bacteria including *Bacillus, Corynebacterium, Pseudomonas, Rhodococcus, Alcaligenes, Acinetobacter, Escherichia, Klebsiella, Microbacterium, Burkholderia, Achromobacter* and *Enterobacter* have been reported as potent hydrocarbon and polychlorinated biphenyl degraders as well as heavy metal reducers (Yousaf *et al.*, 2011; Joutey *et al.*, 2013). Germaine *et al.* (2006) reported endophytic *Pseudomonas putida* VM1450 as a successful organochloride herbicide degrader.

Heavy metals, on the other hand, are not degraded but are converted from one oxidation state to another. Endophytic bacteria efficiently carry out heavy metal bioremediation due to their ability to produce organic acids, such as 5-ketogluconic acids and chelators (Shin *et al.*, 2012) which leads to the effective sequestration of toxic metal ions from the soil, followed by their translocation and accumulation within the plant biomass (Weyens *et al.*, 2009). Madhaiyan *et al.* (2007) reported an endophytic strain of *Burkholderia* sp. as nickel and cadmium adsorbent in tomato plant. Phytoremediation of toxic metals by bacterial endophytes is enhanced due to their sequestration activities which affect translocation and accumulation of heavy metal in plant biomass. Chen *et al.* (2010) reported four heavy metal resistant endophytic bacteria, viz. *Serratia nematodophila* LRE07, *Enterobacter aerogenes* LRE17, *Enterobacter* sp. LSE04 and *Acinebacter* sp. LSE06 from *Solanum nigrum*.

Bioremediation of some ecologically toxic compounds by potential endophytic bacteria include: polychlorinated biphenyl and 1,2,3-trichloropropane (TCP) by *Herbaspirillum* sp. K1 in wheat (Männistö *et al.*, 2001); chlorobenzoic acid by *Pseudomonas aeruginosa* R75 and *P. savastanoi* CB35 in wild rye (Siciliano *et al.*, 1998); volatile organic compound (VOCs) and toluene by *Burkholderia cepacia* G4, *B. cepacia* Bu61, *Pseudomonas* sp. in poplar and yellow lupine (Barac *et al.*, 2004); and naphthalene by *P. putida* VM1441 in pea (Germaine *et al.*, 2009).

11.5.4 Antibiotic production

Endophytes are potentially active producers of these compounds having a wide diversity ranging from phytopathogens to human disease-causing bacteria, fungi, viruses and protozoa (Strobel and Daisy, 2003). Miller *et al.* (1998) reported endophytic *Pseudomonas viridiflava* as producers of ecomycins, a series of lipopeptides. Common amino acids involved in the structural biosynthesis of ecomycins include alanine, serine, threonine, glycine as well as some unusual amino acids, such as homoserine and β-hydroxyaspartic acid. The common targets of ecomycins include human pathogens *Cryptococcus neoformans* and *Candida albicans*.

Pseudomycins are another family of lipopeptide antibiotics, produced by endophytic *Pseudomonads* and active against *C. albicans, C. neoformans, Ceratocystis ulmi* and *Mycospharella fijiensis* (Strobel and Daisy, 2003). Common amino acids involved in the biosynthesis of pseudomycins include L-chlorothreonine, L-hydroxyaspartic acid and both D- and L- diaminobutyric acid.

Castillo *et al.* (2002) reported broad spectrum antibiotics munumbicins from *Streptomyces* sp. NRRL 30562, endophytic within *Kennedia nigriscans*. The munumbicins are peptide antibiotics with each molecule containing threonine, glutamic acid (or glutamine) and aspartic acid (or asparagine). The munumbicins are active against a wide spectrum of multidrug-resistant bacteria, including *Mycobacterium tuberculosis* and several highly pathogenic Gram-positive bacteria including *Bacillus anthracis*.

Kakadumycins, isolated from endophytic *Streptomyces* sp. NRRL 30566 growing within *Grevillea pteridifolia* are broad-spectrum antibiotics active especially against Gram-positive bacteria. Kakadumycins are peptide antibiotics having alanine, serine and an unknown amino acid (Castillo *et al.*, 2003; Strobel and Daisy, 2003).

11.6 Endophytes as Parasites

Endophytes and their hosts remain in mutualistic association which is in complete synchrony with the environment. However, due to sudden environmental changes, such as CO_2 accumulation, oxygen depletion, or presence of other microbial forms some endophytes may turn pathogenic (Sturz *et al.*, 1997).

Endophytic microbes, such as *Nocardia*, *B. cepacia*, *Salmonella*, etc. have been isolated from various plants that are closely linked to deleterious human pathogens (Guo *et al.*, 2002; Barac *et al.*, 2004; Pirttilä *et al.*, 2005). The potent biocontrol agents are aggressive nutrient consumers. In this process, they produce several secondary metabolites which may include antimicrobial compounds and they may themselves become resistant to several antibiotics (Parke and Gurian-Sherman, 2001). Hence, bacterial endophytes having related ancestry to human pathogens are strongly rejected for agricultural use, as the risk of horizontal gene transfer may render the opportunistic endophytic form to convert to pathogenic form (Rosenblueth and Martínez-Romero, 2006).

11.7 Conclusion and Future Prospects

Endophytic bacteria are plant-associated bacteria recognized for their capacity of plant growth promotion and stress amelioration. However, field application of the endophytic bacteria in several cases fails to bestow the benefits due to inefficient endosphere colonization. A complete understanding of the delivery methods for efficient colonization and the resulting plant–microbe interactions will enable a better use of these microbial inoculants in sustainable agriculture.

Another important issue addressed in this review is the *in planta* presence of several bacterial endophytes in their nonculturable state. Metagenomic approaches revealing more information about the beneficial attributes of those endophytic forms would further prove to be a milestone in endophytic research and organic farming.

Acknowledgements

SR is grateful to Banaras Hindu University for providing financial support via the RET-UGC fellowship. VS and HBS are thankful to DBT, New Delhi for awarding project grant (BT/PR5990/AGR/5/587/2012).

References

Abraham, A., Philip, S., Jacob, C.K. and Jayachandran, K. (2013) Novel bacterial endophytes from *Hevea brasiliensis* as biocontrol agent against Phytophthora leaf fall disease. *Biocontrol* 58, 675–684.

Araújo, W.L., Marcon, J., Maccheroni, W., van Elsas, J.D., van Vuurde, J.W. and Azevedo, J.L. (2002) Diversity of endophytic bacterial populations and their interaction with *Xylella fastidiosa* in citrus plants. *Applied and Environmental Microbiology* 68, 4906–4914.

Asis, C.A. and Adachi, K. (2003) Isolation of endophytic diazotroph *Pantoeaagglomerans* and nondiazotroph *Enterobacter asburiae* from sweet potato stem in Japan. *Letters in Applied Microbiology* 38, 19–23.

Bacilio-Jiménez, M., Aguilar-Flores, S., Ventura-Zapata, E., Pérez-Campos, E., Bouquelet, S. and Zenteno, E. (2003) Chemical characterization of root exudates from rice (*Oryzasativa*) and their effects on the chemotactic response of endophytic bacteria. *Plant and Soil* 249, 271–277.

Bacon, C.W., Glenn, A.E. and Yates, I.E. (2008) *Fusariumverticillioides*: managing the endophytic association with maize for reduced fumonisins accumulation. *Toxin Reviews* 27, 411–446.

Balsanelli, E., Serrato, R.V., De Baura, V.A., Sassaki, G., Yates, M.G. *et al.* (2010) *Herbaspirillum seropedicae* rfbB and rfbC genes are required for maize colonization. *Environmental Microbiology* 12, 2233–2244.

Barac, T., Taghavi, S., Borremans, B., Provoost, A., Oeyen, L. *et al.* (2004) Engineered endophytic bacteria improve phytoremediation of water-soluble, volatile, organic pollutants. *Nature Biotechnology* 22, 583–588.

Berg, G., Krechel, A., Ditz, M., Sikora, R.A., Ulrich, A. and Hallmann, J. (2005) Endophytic and ectophytic potato-associated bacterial communities differ in structure and antagonistic function against plant pathogenic fungi. *FEMS Microbiology Ecology* 51, 215–229.

Bisen, K., Keswani, C., Mishra, S., Saxena, A., Rakshit, A. and Singh, H.B. (2015) Unrealized potential of seed biopriming for versatile agriculture. In: Rakshit, A., Singh, H.B. and Sen, A. (eds) *Nutrient Use Efficiency: From Basics to Advances*. Springer, New Delhi, India, pp. 193–206.

Bisen, K., Keswani, C., Patel, J.S., Sarma, B.K. and Singh, H.B. (2016) *Trichoderma* spp.: efficient inducers of systemic resistance in plants. In: Chaudhary, D.K. and Verma, A. (eds) *Microbial-Mediated Induced Systemic Resistance in Plants*. Springer, Singapore, pp. 185–195.

Bodenhausen, N., Horton, M.W. and Bergelson, J. (2013) Bacterial communities associated with the leaves and the roots of *Arabidopsisthaliana*. *PLoS One* 8, e56329.

Böhm, M., Hurek, T. and Reinhold-Hurek, B. (2007) Twitching motility is essential for endophytic rice colonization by the N_2-fixing endophyte *Azoarcus* sp. strain BH72. *Molecular Plant-Microbe Interactions* 20, 526–533.

Botta, A.L., Santacecilia, A., Ercole, C., Cacchio, P. and Del Gallo, M. (2013) *In vitro* and *in vivo* inoculation of four endophytic bacteria on *Lycopersicon esculentum*. *Nature Biotechnology* 30, 666–674.

Bulgarelli, D., Rott, M., Schlaeppi, K., van Themaat, E.V.L., Ahmadinejad, N. *et al.* (2012) Revealing structure and assembly cues for *Arabidopsis* root-inhabiting bacterial microbiota. *Nature* 488, 91–95.

Castillo, U.F., Strobel, G.A., Ford, E.J., Hess, W.M., Porter, H. *et al.* (2002) Munumbicins, wide-spectrum antibiotics produced by *Streptomyces* NRRL 30562, endophytic on *Kennedian igriscans*. *Microbiology* 148, 2675–2685.

Castillo, U., Harper, J.K., Strobel, G.A., Sears, J., Alesi, K. *et al.* (2003) Kakadumycins, novel antibiotics from *Streptomyces* sp. NRRL 30566, an endophyte of *Grevillea pteridifolia*. *FEMS Microbiology Letters* 224, 183–190.

Chaintreuil, C., Giraud, E., Prin, Y., Lorquin, J., Ba, A. *et al.* (2000) Photosynthetic *Bradyrhizobia* are natural endophytes of the African wild rice *Oryza breviligulata*. *Applied and Environmental Microbiology* 66, 5437–5447.

Chelius, M.K., and Triplett, E.W. (2001) The diversity of archaea and bacteria in association with the roots of *Zea mays*. *Letters in Microbial Ecology* 41, 252–263.

Chen, L., Luo, S., Xiao, X., Guo, H., Chen, J. *et al.* (2010) Application of plant growth-promoting endophytes (PGPE) isolated from *Solanum nigrum* L. for phytoextraction of Cd-polluted soils. *Applied Soil Ecology* 46, 383–389.

Chen, Y., Gao, X., Chen, Y., Qin, H., Huang, L. and Han, Q. (2014) Inhibitory efficacy of endophytic *Bacillus subtilis* EDR4 against *Sclerotinia sclerotiorum* on rapeseed. *Biological Control* 78, 67–76.

Compant, S., Duffy, B., Nowak, J., Clément, C. and Barka, E.A. (2005) Use of plant growth-promoting bacteria for biocontrol of plant diseases: principles, mechanisms of action, and future prospects. *Applied and Environmental Microbiology* 71, 4951–4959.

Compant, S., Clément, C. and Sessitsch, A. (2010) Plant growth-promoting bacteria in the rhizo-and endosphere of plants: their role, colonization, mechanisms involved and prospects for utilization. *Soil Biology and Biochemistry* 42, 669–678.

de Weert, S., Vermeiren, H., Mulders, I.H., Kuiper, I., Hendrickx, N. *et al.* (2002) Flagella-driven chemotaxis towards exudate components is an important trait for tomato root colonization by *Pseudomonas fluorescens*. *Molecular Plant Microbe Interactions* 15, 1173–1180.

Dong, Y., Iniguez, A.L., Ahmer, B.M. and Triplett, E.W. (2003) Kinetics and strain specificity of rhizosphere and endophytic colonization by enteric bacteria on seedlings of *Medicago sativa* and *Medicago truncatula*. *Applied and Environmental Microbiology* 69, 1783–1790.

Elbeltagy, A., Nishioka, K., Suzuki, H., Sato, T., Sato, Y.I. *et al.* (2000) Isolation and characterization of endophytic bacteria from wild and traditionally cultivated rice varieties. *Soil Science and Plant Nutrition* 46, 617–629.

Gaca, A.O., Kajfasz, J.K., Miller, J.H., Liu, K., Wang, J.D. *et al.* (2013) Basal levels of (p) ppGpp in *Enterococcus faecalis*: the magic beyond the stringent response. *MBio* 4, e00646–13.

Gaiero, J.R., McCall, C.A., Thompson, K.A., Day, N.J., Best, A.S. and Dunfield, K.E. (2013) Inside the root microbiome: bacterial root endophytes and plant growth promotion. *American Journal of Botany* 100, 1738–1750.

Galippe, V. (1887) Note sur la présence de micro-organismesdans les tissusvégétaux (2nd note). *Comptes Rendu Hebdomadaires des Séances et Mémoires de la Société de Biologie* 39, 557–560.

Germaine, K.J., Liu, X., Cabellos, G.G., Hogan, J.P., Ryan, D. and Dowling, D.N. (2006) Bacterial endophyte-enhanced phytoremediation of the organochlorine herbicide 2, 4-dichlorophenoxyacetic acid. *FEMS Microbiology Ecology* 57, 302–310.

Germaine, K.J., Keogh, E., Ryan, D. and Dowling, D.N. (2009) Bacterial endophyte-mediated naphthalene phytoprotection and phytoremediation. *FEMS Microbiology Letters* 296, 226–234.

Gholami, M., Khakvar, R. and Aliasgar Zad, N. (2013) Application of endophytic bacteria for controlling anthracnose disease (*Colletotrichum lindemuthianum*) on bean plants. *Archives of Phytopathology* 46, 1831–1838.

Glick, B.R., Penrose, D.M. and Li, J. (1998) A model for the lowering of plant ethylene concentrations by plant growth-promoting bacteria. *Journal of Theoretical Biology* 190, 63–68.

Goormachtig, S., Capoen, W. and Holsters, M. (2004) Rhizobium infection: lessons from the versatile nodulation behaviour of water-tolerant legumes. *Trends in Plant Science* 9, 518–522.

Goudjal, Y., Toumatia, O., Yekkour, A., Sabaou, N., Mathieu, F. and Zitouni, A. (2014) Biocontrol of *Rhizoctonia solani* damping-off and promotion of tomato plant growth by endophytic actinomycetes isolated from native plants of Algerian Sahara. *Microbiological Research* 169, 59–65.

Guo, X., van Iersel, M.W., Chen, J., Brackett, R.E. and Beuchat, L.R. (2002) Evidence of association of *Salmonellae* with tomato plants grown hydroponically in inoculated nutrient solution. *Applied and Environmental Microbiology* 68, 3639–3643.

Haas, D. and Défago, G. (2005) Biological control of soil-borne pathogens by fluorescent pseudomonads. *Nature Reviews Microbiology* 3, 307–319.

Hallmann, J., Quadt-Hallmann, A., Mahaffee, W.F. and Kloepper, J.W. (1997) Bacterial endophytes in agricultural crops. *Canadian Journal of Microbiology* 43, 895–914.

Hardoim, P.R., van Overbeek, L.S. and van Elsas, J.D. (2008) Properties of bacterial endophytes and their proposed role in plant growth. *Trends in Microbiology* 16, 463–471.

Hardoim, P.R., van Overbeek, L.S., Berg, G., Pirttilä, A.M., Compant, S. *et al.* (2015) The hidden world within plants: ecological and evolutionary considerations for defining functioning of microbial endophytes. *Microbiology and Molecular Biology Reviews* 79, 293–320.

Hennig, K. and Villforth, F. (1940) Experimentelle untersuchungen zur frage der Bacteriesymbiose in höheren Pflanzen und ihre Beeinflussung durch 'Leitemente'. *Biochemische Zeitschrift* 305, 299–309.

Ikeda, S., Fuji, S.I., Sato, T., Furuya, H., Naito, H. *et al.* (2007) Microbial diversity in milled rice as revealed by ribosomal intergenic spacer analysis. *Microbes and Environments* 22, 165.

Iniguez, A.L., Dong, Y., Carter, H.D., Ahmer, B.M., Stone, J.M. and Triplett, E.W. (2005) Regulation of enteric endophytic bacterial colonization by plant defences. *Molecular Plant Microbe Interactions* 18,169–178.

Jain, A., Singh, A., Chaudhary, A., Singh, S. and Singh, H.B. (2014) Modulation of nutritional and antioxidant potential of seeds and pericarp of pea pods treated with microbial consortium. *Food Research International* 64, 275–282.

Jain, A., Singh, A., Singh, S., Sarma, B.K. and Singh, H.B. (2015a) Biocontrol agents-mediated suppression of oxalic acid induced cell death during *Sclerotinia sclerotiorum*–pea interaction. *Journal of Basic microbiology* 55, 601–606.

Jain, A., Singh, A., Singh, S. and Singh, H.B. (2015b). Biological management of Sclerotinia sclerotiorum in pea using plant growth promoting microbial consortium. *Journal of Basic Microbiology* 55, 961–972.

Jiménez Salgado, T., Fuentes-Ramirez, L.E., Tapia-Hernandez, A., Mascarua-Esparza, M.A., Martinez-Romero, E. and CaballeroMellado, J. (1997) *Coffea arabica* L., a new host plant for *Acetobacter diazotrophicus*, and isolation of other nitrogen-fixing acetobacteria. *Applied and Environmental Microbiology* 63, 3676–3683.

Jones, J.D. and Dangl, J.L. (2006) The plant immune system. *Nature* 444, 323–329.

Joutey, N.T., Bahafid, W., Sayel, H. and El Ghachtouli, N. (2013) Biodegradation: involved microorganisms and genetically engineered microorganisms. *Biodegradation-Life of Science*. InTech, Rijeka, Croatia pp. 289–320.

Keswani, C. (2015). Proteomics studies of thermotolerant strain of *Trichoderma* spp. Ph.D. Thesis, Banaras Hindu University, Varanasi, India, p. 126.

Keswani, C., Singh, S.P. and Singh, H.B. (2013) A superstar in biocontrol enterprise: *Trichoderma* spp. *Biotech Today* 3, 27–30.

Keswani, C., Mishra, S., Sarma, B.K., Singh, S.P. and Singh, H.B. (2014) Unraveling the efficient application of secondary metabolites of various *Trichoderma*. *Applied Microbiology and Biotechnology* 98, 533–544.

Keswani, C., Bisen, K., Singh, S.P., Sarma, B.K. and Singh, H.B. (2016a) A proteomic approach to understand the tripartite interactions between plant-*Trichoderma*-pathogen: investigating the potential for efficient biological control. In: Hakeem, K.R. and Akhtar, M.S. (eds) *Plant, Soil and Microbes Vol. 2. Mechanisms and Molecular Interactions*. Springer, Switzerland, pp. 79–93.

Keswani, C., Bisen, K., Singh, V., Sarma, B.K. and Singh, H.B. (2016b) Formulation technology of biocontrol agents: present status and future prospects. In: Arora, N.K., Mehnaz, S. and Balestrini, R. (eds) *Bioformulations: For Sustainable Agriculture*. Springer, New Delhi, India, pp. 35–52.

Kilani-Feki, O. and Jaoua, S. (2011) Biological control of *Botrytiscinerea* using the antagonistic and endophytic *Burkholderia cepacia* Cs5 for vine plantlet protection. *Canadian Journal of Microbiology* 57, 896–901.

Kogel, K.H., Franken, P. and Hückelhoven, R. (2006) Endophyte or parasite–what decides? *Current Opinion in Plant Biology* 9, 358–363.

Madhaiyan, M., Poonguzhali, S., Ryu, J. and Sa, T. (2006) Regulation of ethylene levels in canola (*Brassicacampestris*) by 1-aminocyclopropane-1-carboxylate deaminase-containing *Methylobacterium fujisawaense*. *Planta* 224, 268–278.

Madhaiyan, M., Poonguzhali, S. and Sa, T. (2007) Metal tolerating methylotrophic bacteria reduces nickel and cadmium toxicity and promotes plant growth of tomato (*Lycopersicon esculentum* L.). *Chemosphere* 69, 220–228.

Männistö, M.K., Tiirola, M.A. and Puhakka, J.A. (2001) Degradation of 2, 3, 4, 6-tetrachlorophenol at low temperature and low dioxygen concentrations by phylogenetically different groundwater and bioreactor bacteria. *Biodegradation* 12, 291–301.

Mano, H. and Morisaki, H. (2007) Endophytic bacteria in the rice plant. *Microbes and Environment* 23, 109–117.

Manter, D.K., Delgado, J.A., Holm, D.G. and Stong, R.A. (2010) Pyrosequencing reveals a highly diverse and cultivar-specific bacterial endophyte community in potato roots. *Microbial Ecology* 60, 157–166.

Martínez, L., Caballero, J., Orozco, J. and Martínez-Romero, E. (2003) Diazotrophic bacteria associated with banana (*Musa* spp). *Plant and Soil* 257, 35–47.

Mayak, S., Tirosh, T. and Glick, B.R. (2004) Plant growth-promoting bacteria that confer resistance to water stress in tomatoes and peppers. *Plant Science* 166, 525–530.

McInroy, J.A. and Kloepper, J.W. (1995) Survey of indigenous bacterial endophytes from cotton and sweet corn. *Plant and Soil* 173, 337–342.

Mercado-Blanco, J. and Prieto, P. (2012) Bacterial endophytes and root hairs. *Plant and Soil* 361, 301–306.

Miller, C.M., Miller, R.V., Garton-Kenny, D., Redgrave, B., Sears, J. *et al.* (1998) Ecomycins, unique antimycotics from *Pseudomonas viridiflava*. *Journal of Applied Microbiology* 84, 937–944.

Mishra, S., Singh, A., Keswani, C., Saxena, A., Sarma, B.K. and Singh, H.B. (2015) Harnessing plant-microbe interactions for enhanced protection against phytopathogens. In: Arora, N.K. (ed.) *Plant Microbes Symbiosis: Applied Facets*. Springer, New Delhi, India, pp. 111–125.

Morita, S., Azuma, M., Aoba, T., Satou, H., Narisawa, K. and Hashiba, T. (2003) Induced systemic resistance of Chinese cabbage to bacterial leaf spot and Alternaria leaf spot by the root endophytic fungus, *Heteroconium chaetospira*. *Journal of General Plant Pathology* 69, 71–75.

Nautiyal, C.S. (1997) Selection of Chickpea-Rhizosphere-Competent Pseudomonas fluorescens NBRI1303 Antagonistic to *Fusarium oxysporum* f. sp. *ciceri*, *Rhizoctonia bataticola* and *Pythium sp. Current Microbiology* 35, 52–58.

Olivares, F.L., Baldani, V.L.D., Reis, V.M., Baldani, J.I. and Dobereiner, J. (1996) Occurrence of the endophytic diazotrophs *Herbaspirillum* spp. in roots, stems, and leaves, predominantly of *Gramineae*. *Biology and Fertility of Soils* 21, 197–200.

Parke, J.L. and Gurian-Sherman, D. (2001) Diversity of the *Burkholderia cepacia* complex and implications for risk assessment of biological control strains. *Annual Reviews of Phytopathology* 39, 225–258.

Patel, J.S., Sarma, B.K., Singh, H.B., Upadhyay, R.S., Kharwar, R.N. and Ahmed, M. (2015). *Pseudomonas fluorescens* and *Trichoderma asperellum* enhance expression of Gα subunits of the pea heterotrimeric G-protein during *Erysiphe pisi* infection. *Frontiers in plant science*, 6.

Penrose, D.M. and Glick, B.R. (2001) Levels of ACC and related compounds in exudate and extracts of canola seeds treated with ACC deaminase-containing plant growth-promoting bacteria. *Canadian Journal of Microbiology* 47, 368–372.

Pérez-Donoso, A.G., Sun, Q., Roper, M.C., Greve, L.C., Kirkpatrick, B. and Labavitch, J.M. (2010) Cell wall-degrading enzymes enlarge the pore size of inter vessel pit membranes in healthy and *Xylella fastidiosa*-infected grapevines. *Plant Physiology* 152, 1748–1759.

Perotti, R. (1926) On the limits of biological enquiry in soil science. *Proceedings of International Society Soil Science* 2, 146–161.

Pirttilä, A.M., Joensuu, P., Pospiech, H., Jalonen, J. and Hohtola, A. (2004) Bud endophytes of Scots pine produce adenine derivatives and other compounds that affect morphology and mitigate browning of callus cultures. *Physiologia Plantarum* 121, 305–312.

Pirttilä, A.M., Pospiech, H., Laukkanen, H., Myllylä, R. and Hohtola, A. (2005) Seasonal variations in location and population structure of endophytes in buds of Scots pine. *Tree Physiology* 25, 289–297.

Podolich, O., Ardanov, P., Zaets, I., Pirttilä, A.M. and Kozyrovska, N. (2015) Reviving of the endophytic bacterial community as a putative mechanism of plant resistance. *Plant and Soil* 388, 367–377.

Quadt-Hallmann, A., Kloepper, J.W. and Benhamou, N. (1997) Bacterial endophytes in cotton: mechanisms of entering the plant. *Canadian Journal of Microbiology* 43, 577–582.

Rajendran, L. and Samiyappan, R. (2008) Endophytic *Bacillus* species confer increased resistance in cotton against damping off disease caused by *Rhizoctonia solani*. *Journal of Plant Pathology* 7, 1–12.

Ray, S., Singh, S., Sarma, B.K. and Singh, H.B. (2016a) Endophytic *Alcaligenes* isolated from horticultural and medicinal crops promotes growth in Okra *(Abelmoschus esculentus)*. *Journal of Plant Growth Regulation* 35, 401–412.

Ray, S., Singh, V., Singh, S., Sarma, B.K. and Singh, H.B. (2016b) Biochemical and histochemical analyses revealing endophytic *Alcaligenes faecalis* mediated suppression of oxidative stress in *Abelmoschus esculentus* challenged with *Sclerotium rolfsii*. *Plant Physiology and Biochemistry* 109, 430–441.

Reinhold-Hurek, B. and Hurek, T. (2011) Living inside plants: bacterial endophytes. *Current Opinion in Plant Biology* 14, 435–443.

Reinhold-Hurek, B., Hurek, T., Gillis, M., Hoste, B., Vancanneyt, M. *et al.* (1993) *Azoarcus* gen. nov., nitrogen-fixing proteobacteria associated with roots of Kallar grass (*Leptochloa fusca* (L.) Kunth) and description of two species, *Azoarcusindigens* sp. nov. And *Azoarcus communis* sp. nov. *International Journal of Systemic Bacteriology* 43, 574–584.

Reinhold-Hurek, B., Maes, T., Gemmer, S., Van Montagu, M. and Hurek, T. (2006) Anendoglucanase is involved in infection of rice roots by the not-cellulose-metabolizing endophyte *Azoarcus* sp. strain BH72. *Molecular Plant Microbe Interactions* 19, 181–188.

Rosenblueth, M. and Martínez-Romero, E. (2006) Bacterial endophytes and their interactions with hosts. *Molecular Plant Microbe Interactions* 19, 827–837.

Ryu, C.M., Farag, M.A., Hu, C.H., Reddy, M.S., Pare, P.W. and Kloepper, J.W. (2003) Volatiles produced by PGPR elicit plant growth promotion and induced resistance in *Arabidopsis*. *Proceedings of the Sixth International Workshop on Plant Growth Promoting Rhizobacteria*, Calicut, India, 436–443.

Sakai, M., Matsuka, A., Komura, T. and Kanazawa, S. (2004) Application of a new PCR primer for terminal restriction fragment length polymorphism analysis of the bacterial communities in plant roots. *Journal of Microbiological Methods* 59, 81–89.

Sangmanee, P., Bhromsiri, A. and Akarapisan, A. (2009) The potential of endophytic actinomycetes, (*Streptomyces* Sp.) for the biocontrol of powdery mildew disease in sweet pea (*Pisum sativum*). *Asian Journal of Food and AgroIndustry* 2 (Special Issue), S93–S98.

Saxena, A., Raghuwanshi, R. and Singh, H.B. (2015) *Trichoderma* species mediated differential tolerance against biotic stress of phytopathogens in *Cicerarietinum* L. *Journal of Basic Microbiology* 55, 195–206.

Sessitsch, A., Hardoim, P., Döring, J., Weilharter, A., Krause, A. *et al.* (2012) Functional characteristics of an endophyte community colonizing rice roots as revealed by metagenomic analysis. *Molecular Plant Microbe Interactions* 25, 28–36.

Shin, M.N., Shim, J., You, Y., Myung, H., Bang, K.S. *et al.* (2012) Characterization of lead resistant endophytic *Bacillus* sp. MN3-4 and its potential for promoting lead accumulation in metal hyperaccumulator *Alnusfirma*. *Journal of Hazardous Materials* 199, 314–320.

Shiomi, H.F., Silva, H.S.A., Melo, I.S.D., Nunes, F.V. and Bettiol, W. (2006) Bioprospecting endophytic bacteria for biological control of coffee leaf rust. *Scientia Agricola* 63(1), 32–39.

Siciliano, S.D., Goldie, H. and Germida, J.J. (1998) Enzymatic activity in root exudates of Dahurian wild rye (*Elymusdauricus*) that degrades 2-chlorobenzoic acid. *Journal of Agricultural and Food Chemistry* 46, 5–7.

Singh, V., Upadhyay, R.S., Sarma, B.K. and Singh, H.B. (2016) Seed bio-priming with *Trichoderma asperellum* effectively modulate plant growth promotion in pea. *International Journal of Agriculture, Environment and Biotechnology* 9, 361.

Strobel, G. and Daisy, B. (2003) Bioprospecting for microbial endophytes and their natural products. *Microbiology and Molecular Biology Reviews* 67, 491–502.

Sturz, A.V., Christie, B.R., Matheson, B.G. and Nowak, J. (1997) Biodiversity of endophytic bacteria which colonize red clover nodules, roots, stems and foliage and their influence on host growth. *Biology and Fertility of Soils* 25, 13–19.

Sturz, A.V., Christie, B.R. and Nowak, J. (2000) Bacterial endophytes: potential role in developing sustainable systems of crop production. *Critical Reviews in Plant Sciences* 19, 1–30.

Sun, L., Qiu, F., Zhang, X., Dai, X., Dong, X. and W. Song (2008) Endophytic bacterial diversity in rice (*Oryza Sativa* L.) roots estimated by 16S rDNA sequence analysis. *Microbial Ecology* 55, 415–424.

Surette, M.A., Sturz, A.V., Lada, R.R. and Nowak, J. (2003) Bacterial endophytes in processing carrots (*Daucuscarota* L. var. sativus): their localization, population density, biodiversity and their effects on plant growth. *Plant and Soil* 253, 381–390.

Taghavi, S., Garafola, C., Monchy, S., Newman, L., Hoffman, A. *et al.* (2009) Genome survey and characterization of endophytic bacteria exhibiting a beneficial effect on growth and development of poplar trees. *Applied and Environmental Microbiology* 75, 748–757.

Taurian, T., Anzuay, M.S., Angelini, J.G., Tonelli, M.L., Ludueña, L. *et al.* (2010) Phosphate-solubilizing peanut associated bacteria: screening for plant growth-promoting activities. *Plant and Soil* 329, 421–431.

Thomas, P. and Reddy, K.M. (2013) Microscopic elucidation of abundant endophytic bacteria colonizing the cell wall–plasma membrane peri-space in the shoot-tip tissue of banana. *AoB Plants* 5, plt011.

Tyvaert, L., França, S.C., Debode, J. and Höfte, M. (2014) The endophyte *Verticillium* Vt305 protects cauliflower against Verticillium wilt. *Journal of Applied Microbiology* 116, 1563–1571.

Van Aken, B., Peres, C.M., Doty, S.L., Yoon, J.M., and Schnoor, J.L. (2004) *Methylobacterium populi* sp. *nov.*, a novel aerobic, pink-pigmented, facultatively methylotrophic, methane-utilizing bacterium isolated from poplar trees (*Populusdeltoides× nigra* DN34). *International Journal of Systematic and Evolutionary Microbiology* 54, 1191–1196.

van der Lelie, D., Taghavi, S., Monchy, S., Schwender, J., Miller, L. *et al.* (2009) Poplar and its bacterial endophytes: coexistence and harmony. *Critical Reviews in Plant Sciences* 28, 346–358.

van Loon, L.C. (1997) Induced resistance in plants and the role of pathogenesis-related proteins. *European Journal of Plant Pathology* 103, 753–765.

Vogl, A. (1898) Mehl und die anderenMehlprodukte der Cerealien und Leguminosen. *NahrungsmUntersHygWarenk* 12, 25–29.

Wang, X. and Liang, G. (2014) Control efficacy of an endophytic *Bacillus amyloliquefaciens* Strain BZ6-1 against peanut bacterial wilt, *Ralstonia solanacearum*. *BioMed Research International* 2014.

Weber, O.B., Baldani, V.L.D., Teixeira, K.R.S., Kirchhof, G., Baldani, J.I. and Dobereiner, J. (1999) Isolation and characterization of diazotrophic bacteria from banana and pineapple plants. *Plant and Soil* 210, 103–113.

Webster, G., Jain, V., Davey, M.R., Gough, C., Vasse, J. *et al.* (1998) The flavonoid naringenin stimulates the intercellular colonization of wheat roots by *Azorhizobium caulinodans*. *Plant Cell and Environment* 21, 373–383.

Weyens, N., van der Lelie, D., Taghavi, S. and Vangronsveld, J. (2009) Phytoremediation: plant–endophyte partnerships take the challenge. *Current Opinion in Biotechnology* 20, 248–254.

Yanni, Y.G., Rizk, R.Y., Corich, V., Squartini, A., Ninke, K. *et al.* (1997) Natural endophytic association between *Rhizobium leguminosarum* bv. *trifolii* and rice roots and assessment of potential to promote rice growth. *Plant and Soil*, 194.

Yousaf, S., Afzal, M., Reichenauer, T.G., Brady, C.L. and Sessitsch, A. (2011) Hydrocarbon degradation, plant colonization and gene expression of alkane degradation genes by endophytic *Enterobacter ludwigii* strains. *Environmental Pollution* 159, 2675–2683.

Zinniel, D.K., Lambrecht, P., Harris, N.B., Feng, Z., Kuczmarski, D. *et al.* (2002) Isolation and characterization of endophytic colonizing bacteria from agronomic crops and prairie plants. *Applied and Environmental Microbiology* 68, 2198–2208.

12 Contribution of Plant Growth-Promoting Bacteria to the Maize Yield

Vivian Jaskiw Szilagyi Zecchin,[1]* Angela Cristina Ikeda[2] and Átila Francisco Mógor[1]
[1]*Federal University of Parana Department of Plant Science and Crop Protection, Curitiba, Brazil;* [2]*Federal University of Parana, Department of Forest Science, Curitiba, Brazil*

12.1 Introduction

Maize (*Zea mays* L.) is one of the most important cereal crops, revered widely for its high nutritional value and as a resource for animal feed and bioenergy. Responding to the increasing demands of society, modest initiatives, such as the development of plant production technologies promoting maize yield as well as reduction of synthetic inputs thereby improving farmers' profits, however, remain an unresolved issue. Fortunately, natural reservoirs, particularly rhizospheric and/or endophytic bacteria, can act as plausible alternative sources for plant nutrition or growth promotion (Barretti *et al.*, 2008).

Several reports suggest advantageous associations between maize and beneficial bacteria with effects on several aspects of plant growth promotion, such as biological nitrogen fixation (Dobbelaere *et al.*, 2003), phosphate solubilization (Campo and Hungria, 2007), siderophore production (Araújo *et al.*, 2010) and hormone production (Vessey, 2003).

Though several beneficial bacterial genera are reported to be associated to maize (Ikeda *et al.*, 2013), studies indicating this beneficial association under field conditions are as yet often not available.

To contribute to this field, we have here highlighted topics related to the biological and biochemical mechanisms that make the bacteria–plant interaction an efficient tool for maize yield improvements. We also present some representative studies which examined the three main bacterial genera associated with maize yield promotion: *Azospirillum*, *Pseudomonas* and *Azotobacter*; then we also look at three other genera with a smaller volume of work already published: *Serratia*, *Rhanella* and *Herbaspirillum*.

12.2 Bacteria and Maize

Biological inocula are utilized so as to improve the yield of a specific crop as well as enhancing its nutrient uptake, reducing its production costs and minimizing fertilizer addition—an efficient route towards environment protection (Conceição *et al.*, 2009). Advantageous associations between grass roots and soil bacteria include biological nitrogen-fixing and growth promoter groups. Similarly, maize (*Zea mays* L.) also shows great potential for all of these associations, which can yield major benefits such as yield increment.

*E-mail: vivian.szilagyi@gmail.com

© CAB International 2017. *Advances in PGPR Research* (eds H.B. Singh, B.K. Sarma and C. Keswani)

In Brazil, the first commercial inoculant for maize crops was developed using six strains of *Azospirillum brasilense* that achieved an average of 25% to 30% of increment in grain yield, meaning savings of U$ 1 billion per season, considering a cultivated area of 13 million hectares with an average yield of 3200 kg/ha (Hungria, 2011).

12.2.1 Endophytic bacteria

Microorganisms which are able to colonize internal tissues of a plant were called endophytes by De Bary in 1866 (Stone, 1988). Petrini (1991) proposed an expanded definition for this concept by including all microorganisms capable of colonizing internal tissues of plants without causing apparent damage. Initial reports considered them as contaminants resulting from insufficient surface disinfestation or weakly virulent pathogens. However, there were indications that these bacteria could influence plant growth and reduce disease symptoms in plants (Hallmann *et al.*, 1997).

Thus, on the whole, endophytic bacteria can be defined as those which colonize internal tissues of plants without causing apparent damage, regardless of the type of plant tissue, including leaves and roots. Furthermore, according to Hallmann (2001), endophytic microorganisms access nutrients and water more easily than those on the surface, and they are also better protected from fluctuations of environmental conditions.

Bacteria can penetrate a plant through lesions or emerging radicle root sites, and natural openings such as stomata, lenticels and hydathodes. Some endophytic bacteria release hydrolytic enzymes like cellulase or pectinase which help in penetration by the roots.

12.2.2 Rhizospheric bacteria

Plants can offer a wide range of habitats for bacterial growth, such as the surface of seeds, roots, leaves and fruits that can be a refuge for diverse microbial communities. Otherwise, flowers, stems, vascular tissue and intercellular spaces also support limited bacterial communities (Beattie, 2006).

Rhizobacteria primarily reside in and influence the rhizosphere community, which comprises the volume of soil around the roots and the rhizoplane, i.e. the surface of the plant roots and the strongly adhered soil particles (Kennedy, 2005). The rhizosphere is a highly nutritious region, since the roots transfer exudates containing compounds from cell lysis to the soil. Among the most important compounds are mucilage which contains hydrated polysaccharides, organic acids, vitamins and amino acids. This rhizodeposition is essential for microbial abundance and rhizospheric activity, forming an active microbial habitat in soil (Galvão *et al.*, 2010).

12.2.3 Plant growth-promoting bacteria

Bacteria that demonstrate any action of growth promotion in plants are called plant growth-promoting bacteria (PGPB) (Kloepper and Schroth, 1978). They can be directly related to increases in root length and number of root hairs, and these effects can be attributed to the production of plant growth promoting bioactive substances. A greater development in root system assists the plant to explore the soil and so improve water and nutrient uptake (Hungria, 2011; Singh *et al.*, 2014; Bisen *et al.*, 2015; Mishra *et al.*, 2015).

PGPBs may on the one hand directly promote growth in plants or may act indirectly by suppressing harmful microorganisms that inhibit plant growth. Biocontrol of phytopathogens by PGPBs involves substances produced by bacterial strains that cause antibiotic and antifungal effects (Kupper *et al.*, 2003; Keswani *et al.*, 2016). Furthermore, PGPBs can compete with phytopathogens for nutrients and colonization sites, as well as inducing systemic resistance mechanisms (Lugtenberg and Kamilova, 2009).

Plant growth promotion activity performed by endophytic bacteria and rhizobacteria in addition to biocontrol of diseases and stress protection in many plants has been reported (Peng *et al.*, 2009; Berg *et al.*, 2010). These accomplishments depend on the plant and bacterial genotypes, though the quality of root exudates can extensively determine preferences for certain bacterial species by plant cultivars (Nehl *et al.*, 1997; Coelho *et al.*, 2007).

Ikeda *et al.* (2013) established a collection of endophytic bacteria isolated from roots of different genotypes of maize (*Zea mays* L.). Morphophysiological and genetic characterization showed significant variability among isolated strains, and phylogenetic analysis revealed the isolates belonged to *Pantoea*, *Bacillus*, *Burkholderia* and *Klebsiella* genera, with characteristic nitrogen-fixing and plant growth promoting traits.

The *Azospirillum* spp. genus is described as one of the main groups associated to maize and comprises species that are associated with grasses, promoting plant growth and fixing atmospheric nitrogen in microaerobic conditions. Also in the maize crop can be found species such as diazotrophic rhizobacteria from *Klebsiella* spp., *Enterobacter* spp. and *Citrobacter* spp. genera and strains that are potentially plant growth promoters such as *Pantoea* spp. and *Serratia* spp. (Kennedy *et al.*, 2004; Hayat *et al.*, 2010).

Therefore, the discovery of bacteria colonizing plant tissues and potential plant growth promoters makes it possible to select strains efficiently to improve agricultural production.

12.3 Bacterial Mechanism of Plant Growth Promotion

Endophytic bacteria and rhizobacteria aid in plant growth promotion by several mechanisms, such as biological nitrogen fixing, phosphate solubilization, release of siderophores, and production of phytohormones like indole acetic acid.

12.3.1 Biological nitrogen fixation

Many free-living bacteria existing as endophytic and rhizosphere strains are described as diazotrophs or nitrogen fixers converting atmospheric nitrogen (N_2) into ammonia (NH_3) (Dobbelaere *et al.*, 2003). Diazotrophs are considered as direct plant growth promoters and non-associative bacteria, as they do not cause morphological changes in roots and interact with plants of C3 and C4 metabolism (i.e. rice, wheat, corn, sugarcane, cotton) thereby increasing their growth and yield (Hayat *et al.*, 2010).

Since *Beijerinckia fluminense* was isolated from sugarcane rhizosphere (Döbereiner and Ruschel, 1958) and *Azospirillum lipoferum* isolated from several grass roots was re-identified (Döbereiner and Day, 1976), new species of nitrogen-fixing bacteria have been isolated from root and leaf tissues of grasses. Asymbiotic association studies concerning nitrogen-fixing bacteria from maize roots have shown promising results since the 1970s. For example, Raju *et al.* (1972) published preliminary considerations of investigations in grasses, discussing aspects of microbiological activity in the rhizosphere of different species focusing on nitrogen-fixing activity, which was not well known at that time.

Presently many bacterial species are reported that have biological nitrogen-fixing (BNF) activity. *Herbaspirillum* and *Burkholderia* genera are known as producers of regulatory substances and act in banana growth promotion by fixing nitrogen as reported by Weber *et al.* (2000). Likewise, bacteria, such as *Herbaspirillum seropedicae* known for colonizing exclusively grasses, can be found in several crops such as common bean roots (Schmidt *et al.*, 2011).

BNF activity is carried out by nitrogenase enzyme. The *in vitro* nitrogenase test conducted inside liquid and semi-solid culture media allows the inoculation of bacteria that multiply in optimal conditions for nitrogen-fixing activity, i.e. under low oxygen tension that enable cell division without inhibiting nitrogenase activity (Döbereiner *et al.*, 1995). Nitrogenase enzyme has multiple subunits and their coding genes are used as molecular markers in studies of phylogeny, diversity and abundance of microorganisms (Gaby and Buckley, 2012).

One of the main genera of diazotrophic growth-promoting bacteria in maize is *Azospirillum* spp. that may be associated with increased absorptive surface of roots because of its endophytic condition. Other genera of endophytic bacteria in maize roots can also modify the morphology and diameter of roots due to production of growth promoting substances such as auxins and cytokinins (Cavallet *et al.*, 2000).

12.3.2 Phosphate solubilization

Bacteria can improve plant nutrition by processes that provide nutrients such as phosphorus, disposable by inorganic phosphate solubilization. Even in phosphorus-rich soils, this element can be found interacting with iron, calcium and aluminium in a complexed and immobilized form. Campo and Hungria (2007) found that 20–80% of phosphorus in soil is in organic form and its availability depends upon activity of microorganisms. The remaining inorganic form of phosphorus is not readily available to plants. Thus, phosphate-solubilizing microorganisms can mobilize soil nutrients, increasing yield and nutritional efficiency of plants even while remaining in association with the biological nitrogen-fixation process (Galvão *et al.*, 2010).

Phytate is a compound used by plants to store phosphorus and it is 20–50% of the organic phosphate from soil. The enzyme phytase hydrolyzes phytate and it has been reported in bacteria from *Bacillus* spp., *Enterobacter* spp., *Klebsiella* spp. and *Pseudomonas* spp. genera (Kerovuo *et al.*, 1998; Vohara and Satyanarayana, 2003).

12.3.3 Siderophore production

Bacteria have a developed iron-acquisition system involving production and secretion of siderophores, low molecular weight iron-complexing compounds produced under iron-limiting conditions (Araújo *et al.*, 2010). Siderophore complexation reduces free ions in the rhizosphere thereby affecting the development of phytopathogenic microorganisms which are less efficient concerning iron metabolism (Campo and Hungria, 2007). Molecules of siderophores capture and bind to Fe^{+3} ions from soil. The complex is recognized by the host plant and hence Fe^{+3} is transported through the vegetable cell membrane for use of the host cell or even by the microorganism (Buyer *et al.*, 1993). Iron is considered an important cofactor for enzymes that participate in many biochemical pathways of physiological processes in plants, such as photosynthesis and biological nitrogen-fixing activity. They act directly on iron availability and indirectly by antibiosis performance against phytopathogenic microorganisms which are less efficient in iron metabolism, thereby conferring a protective effect to the plant by siderophore production (Campo and Hungria, 2007; Galvão *et al.*, 2010).

12.3.4 Indole acetic acid production

Some bacteria also promote plant growth by producing substances such as plant hormones which enhance the plant root system. Synthesized by plants, the hormones act as messengers to regulate growth, development and cell and tissue differentiation. Vessey (2003) noted that some bacteria strains can promote plant growth due to production of phytohormones such as auxins, cytokinins and gibberellins. Auxins constitute one of the most important phytohormones aiding plant growth and indole acetic acid (IAA) is an excellent example of an auxin.

Auxin is produced in the top of the plant and is distributed by polar transport to the rest of the plant tissues. Many microorganisms are able to produce analogues of this phytohormone and also can colonize different plant parts. It means that they have beneficial effects in plant growing, such as improved seed germination rate, vegetable organ development, flower production and crop yield in greenhouse and field trails (Dey *et al.*, 2004).

According to Chaiharn and Lumyong (2011), the IAA produced by rhizobacteria can stimulate root elongation processes, cell division and cell differentiation and its production is associated to response of exudate produced by the plant rhizosphere. These exudates are rich sources of tryptophan, the amino acid precursor of IAA. These effects have been identified as significantly enhancing nutrient absorption when a bacterial strain that produces IAA is inoculated on maize. Therefore, bacteria are able to produce IAA with benefits for embryogenesis process, organ differentiation, roots and shoot architecture establishment, apical dominance and tropic responses (Spaepen *et al.*, 2007). The pathway of IAA production occurs mainly by indole-3-pyruvate, and this enzyme is described as functionally existing in bacteria like *Azospirillum* spp., *Bradyrhizobium* spp., *Enterobacter cloacae*, *Pantoeaag glomerans*, *Pseudomonas* spp. and *Rhizobium* spp. (Galvão *et al.*, 2010).

12.3.5 ACC-deaminase

Ethylene biosynthesis in plants uses methionine amino acid as biological precursor. First, a reaction occurs when *S*-adenosyl methionine (SAM) is converted into 1-aminocyclopropane-1-carboxylic acid (ACC). Then, the ACC is metabolized by ACC oxidase (ACCO) which uses oxygen (O_2) and iron in the presence of CO_2 to produce ethylene (Yang and Hoffman, 1984). Thus, it can be deduced that bacteria producing ACC are useful in ethylene metabolism.

The hormone ethylene has a wide range of biological activities and can affect plant growth and development in a large number of ways. Moreover, many studies reported the presence of this enzyme in plant growth promoting bacteria (PGPB) (Belimov *et al.*, 2001; Blaha *et al.*, 2006; Sgroy *et al.*, 2009). This enzyme produced by bacteria, facilitates plant growth when they colonize roots or seeds, and thus in response to tryptophan and other small molecules, the bacteria synthesize and secrete IAA (Patten and Glick, 1996). This bacterial IAA, together with endogenous plant IAA, can stimulate plant growth or induce the synthesis of the plant enzyme ACC synthase that converts the compound *S*-adenosyl methionine to ACC, the immediate precursor of ethylene in higher plants.

ACC degradation from the direct precursor of ethylene creates an ACC concentration gradient between the interior and the exterior of the plant, favouring its exudation and reducing the internal ethylene level. This, combining with auxins produced by the same microorganism, causes root system development, because of bacterial ACC deaminase competing with the plant's ACC oxidase. As a direct consequence of this enzyme's activity, the amount of ethylene produced by the plant is reduced. Therefore, root or seed colonization by bacteria that synthesize ACC deaminase prevents plant ethylene levels from becoming growth inhibitory (Glick, 1995; Glick *et al.*, 1998).

12.4 Maize Yield Improved by Bacteria in Field Trial

There are many bacterial genera associated with maize, among which many are beneficial. However, most research has been confined to greenhouse assays and *in vitro* screenings to assess potentialities of the bacterial strains. Assessment of the feasibility of beneficial bacteria under field conditions is still in preliminary stages which makes the choice of potent strain under field conditions quite difficult.

Below we present some representative studies in which were studied the three main bacterial genera associated with maize yield promotion: *Azospirillum, Pseudomonas* and *Azotobacter*. Then we present information on three other genera with a smaller volume of work already published: *Serratia*, *Rhanella* and *Herbaspirillum*.

12.4.1 *Azospirillum*

Study 1 (Fulchieri and Frioni, 1994)

The inoculation was made by pelleting an inoculant prepared by mixing cultures of three bacteria: *Azospirillum brasilense* (AZ 39); *A. lipoferum* (AZ 30) (AZ 39 and AZ 30 being obtained from INTA, Castelar, Argentina) and *A. brasilense* ATCC 29745 strain

Sp 7 (obtained from Embrapa, Rio de Janeiro, Brazil) to sterile peat. The concentration of the peat inoculum was 10^7 *Azospirillum* CFU per seed. The field trial was conducted in the University of Rio Cuarto, Cordoba, Argentina with a hybrid Cargill 155.

The number of seeds per year in the inoculated and fertilized treatments was about three times higher. The seed yield (kg ha^{-1}) in the inoculated treatment was 1.59 times higher. In the nitrogen treatment the seed yield was 1.48 times greater than the control. There were significant yield differences between inoculated and fertilized plots, meaning a saving of about 60 kg of fertilizer N ha^{-1}.

Study 2 (Swędrzyńska and Sawicka, 2000)

The strain 65B of *Azospirillum brasilense* used in this work originated from the Department of Microbiology, Institute of Soil Science and Plant Cultivation in Pulawy, Poland. Inoculation was performed before sowing: the inoculum was mixed with maize seeds and used for spraying the field. The spraying was done after emergence, at the developmental stage of two or three unfolded leaves, and applied under each plant in the vicinity of its roots. The bacterial inoculum amounted to 10^8–10^9 CFU. Field experiments were carried out in Zlotniki, at a site belonging to the Experimental and Didactic Station of August Cieszkowski Agricultural University of Poznań, Poland.

Inoculation contributed to the increase of average yield by 17%. The highest (27%) yield increase by inoculation was found in combinations not treated with a fungicide and without the application of nitrogen fertilizer. The yields from the inoculated plots with nitrogen fertilization and inoculation were 41% higher than control.

Study 3 (Hungria et al.*, 2010)*

Nine *Azospirillum* strains, isolated from maize plants, were evaluated after application to maize seeds. These strains have N_2-fixing capacity *in vitro* and produce indole acetic acid (IAA). These strains are deposited at the University Federal of Paraná, Department of Biochemistry and Molecular Biology, Curitiba, Brazil and also at the Culture Collection of Diazotrophic and Plant Growth Promoting Bacteria of Embrapa Soja, Londrina, Brazil.

Inoculants were prepared with *Azospirillum brasilense* strains at a concentration of 2×10^8 cells g^{-1} of peat or 3×10^8 ml^{-1} of liquid inoculum. The peat was applied at a rate of 250 g of inoculant 50 kg^{-1} of seeds, and to increase adhesion of the peat a solution of sucrose 10% (w/v) was used. The liquid inoculant was applied at a rate of 300 ml 50 kg^{-1} seeds.

Field experiments have to be performed in at least two different localities in the same State (Paraná) of Brazil (Londrina and Ponta Grossa cities). They represented the crop growing regions and were cultivated for two seasons. The experiments at Londrina were performed with Hybrid 9.486 and EMBRAPA-HD-28X, whereas at Ponta Grossa variety BR 201 and EMBRAPA-HD-28X were used. All experiments received 24 kg of N ha^{-1}, a low N fertilizer starter at sowing.

In the first set of experiments the *A. brasilense* strains Ab-V4, Ab-V5, Ab-V6 and Ab-V7 increased grain yields of maize by 662–823 kg ha^{-1}, or 24–30%, in relation to non-inoculated controls. In the second set of experiments combinations of Ab-V5 and Ab-V6 were used in a liquid and peat form; both proved to be effective for maize. On average Ab-V5 and Ab-V6 combined increased maize yield by 842 kg ha^{-1} or 27%. Effects of inoculation were attributed to general increases in uptake of P, K and Cu, not specifically to the biological nitrogen fixation.

Study 4 (Ferreira et al.*, 2013)*

A liquid inoculant with *A. brasilense* Ab-V5 (AZ) was used in two concentrations, the diluted form Azdil (0.53×10^6 cells ml^{-1}) and the concentrated form Azconc (1.4×10^{11} cells ml^{-1}). The preparation of the diluted form was done with 10% sucrose solution. An inoculum of 60 ml was used to inoculate 500 g of seed. Experimental assays were performed in Uberlândia, Minas Gerais State, Brazil.

The yield was increased by 29% in the inoculation treatment and nitrogen compared with nitrogen fertilization alone.

Study 5 (Morais et al., *2016)*

A commercial product was used that is based on *A. brasilense*, at a minimum concentration of 2 × 10^8 viable cells ml^{-1} composed of Ab-V5 and Ab-V6 strains. The experiment was carried out in the field, in soil of the Cerrado region of Brazil, in the municipality of Iraí de Minas, State of Minas Gerais. A Micron Combat spray was used attached to the seeder to inoculate the bacteria into the seed furrow at the time of seed distribution. The inoculation doses were 100 ml of commercial inoculant ha^{-1}, and increasing doses for 200, 300 and 400 ml ha^{-1}.

The dose of 200 ml inoculant ha^{-1} applied enhanced seed germination and early development of the maize, resulting in a greater number of plants ha^{-1}; the final stand was 4.7% higher compared with the treatment with no inoculation. The same dose was noteworthy for grain production, increasing around 610 kg ha^{-1}.

Study 6 (Müller et al., *2016)*

The inoculant was a commercial product composed of *A. brasilense* strains Ab-V5 and Ab-V6. Inoculation was performed with 100 ml ha^{-1} in seed treatment before planting and 300 ml ha^{-1} in planting furrow after planting. The application was done with backpack sprayers, with flat jet nozzles. The experiment was carried out in Guarapuava city, Paraná State, Brazil using the maize hybrid P30F53.

Inoculation with bacteria provided yield increase of 702 kg ha^{-1} for inoculation in seeding furrow and 432 kg ha^{-1} for inoculation in seed treatment, and both treatments did not differ between each other.

Study 7 (Fukami et al., *2016)*

The strains Ab-V5 and Ab-V6 of *A. brasilense* were used mixed on maize using different application modes: for seed inoculation and in-furrow 1.0 × 10^5 cells $seed^{-1}$ was applied, and for leaf spray and soil spray 1.0 × 10^5 cells $plant^{-1}$. In furrow inoculation, as well as of foliar and soil spray the inoculants were diluted with water to a final volume of 150 l ha^{-1}. Seed and in-furrow inoculations were performed at sowing, whereas leaf and soil spray inoculation took place when maize plants were at the V 2.5. These modes of inoculation were combined with levels of N fertilization.

Experiments were conducted in fields at the cities of: Cachoeira Dourada, Goiás State; Luis Eduardo Magalhães, Bahia State; and Ponta Grossa, Paraná State. All are located in Brazil's significant maize production areas. The hybrids used were 2B707 HX and P4285 H.

When inoculation was associated with 75% of the complementary dose of N, plant growth was increased compared to non-inoculated plants fertilized with 100% N. One of the best results was detected in Cachoeira Dourada, where the leaf spray inoculation and 75% N dose with the highest inoculant dose promoted an increase of 773 kg ha^{-1} in yield over the treatment that received the full dose of N fertilizer (100% N), with no inoculation.

12.4.2 *Pseudomonas*

Study 1 (Shaharoona et al., *2006)*

ACC-deaminase-containing *Pseudomonas fluorescens* biotype G strain N3, is a rhizobacteria isolate from the maize rhizosphere. For the inoculation, a bacteria suspension of 10^8–10^9 CFU ml^{-1} was injected into sterile peat (100 ml kg^{-1}, seed to peat ratio 1:1 w/w). Seed dressing was done with the inoculated peat mixed with 10% sugar solution. The work was conducted at the Institute of Soil and Environmental Sciences, University of Agriculture, Faisalabad, Pakistan.

P. fluorescens biotype G (N3) increased the grain yield 19.4% in the presence and 25.6% in the absence of N fertilizer, compared to their respective uninoculated controls. The results indicated that ACC-deaminase of the bacteria competes well with ACC-oxidase and thus eliminates the effect of NO_3^- induced ethylene on plant growth, if any. The effectiveness of N_3 might be related to its high root colonization ability and chitinase activity in addition to ACC-deaminase activity.

Study 2 (Hameeda et al., *2008)*

A P-solubilizing *Pseudomonas* sp. strain CDB 35 was used in a peat-based formulation at 150 g ha^{-1}; it was applied as seed coat with 1% CMC as adhesive. The number of viable cells was on the range of 106–107 CFU per seed. The experiments were conducted at International Crops Research Institute for the Semi-Arid Tropics (ICRISAT), Patancheru, India,

CDB 35 increased the grain yield of field-grown maize by 64% compared to the uninoculated control. The rise in P uptake of shoot and grain of maize was respectively 9 and 22 kg ha^{-1} with CDB 35 and the control was 5 and 11 kg ha^{-1}. N uptake was also enhanced for maize shoots (48 kg ha^{-1} with CDB 35), while the control was only 30 kg ha^{-1}. *Pseudomonas* sp. CDB 35 was an important tool for maize productivity and replacing P chemical fertilizer use.

Study 3 (Viruel et al., *2014)*

A bacterial culture (10^9 CFU ml^{-1}) of *Pseudomonas tolaasii* strain IEXb, a P-solubilizing (also IAA and siderophore-producing bacteria (Viruel *et al.*, 2011)) was used as bioinoculant at a final concentration of 50 ml kg^{-1} seed. For inoculation assays, seeds were soaked for 30 min in the bacterial suspension and then planted. Field assay was conducted at Instituto de Investigación Animal del Chaco Semiárido (IIACS), INTA Leales, Tucumán, Argentina, using the hybrid DK 390 MG RR2. The bacteria were evaluated in combination with triple superphosphate (TSP) as P fertilizer (50 kg P ha^{-1}).

The presence of IEXb stimulated seedling emergence (8%), grain yield (44%), 1000-grain weight (18%), and P content (56%) of maize plants. In general, *P. tolaasii* IEX was more efficient as a bioinoculant without P fertilizer than with triple superphosphate.

12.4.3 *Azotobacter*

Study 1 (Hussain et al., *1987)*

Seeds of maize variety UM-2 were inoculated by mixing in a bacterial suspension immediately before sowing. Experiments were conducted in the field of the Soil Science Department, University of Agriculture, Faisalabad, Pakistan.

Inoculated seeds of maize with 11 *Azotobacter* strains were sown in fields receiving no fertilizer and fertilizers (N and P). The effect was greater in unfertilized than in fertilized soil. The increase in yield due to fertilizers was 21.2% without inoculation and 37.1% with inoculation. Correlations between the total yield and N, P and K uptake did not show specific effects for any element. Possibly the increase in yield by inoculation was due not to the increase in N_2 fixation, but to growth promoting substances.

Study 2 (Pandey et al., *1998)*

Bacterial strain *Azotobacter chroococcum* W5 was obtained from the Division of Microbiology, Indian Agricultural Research Institute, New Delhi. It was isolated from wheat rhizosphere and was positive for the presence of the *Nif* gene, auxin and gibberellin producer. Maize seeds were inoculated with W5 at concentration of 10^6–10^7 CFU ml^{-1}. The experiment was carried out on a lower hill slope (Kamrang) under subtropical conditions.

Inoculation with W5 resulted in improved plant performance with yield enhancement of 1.15-fold over control. It also resulted in significantly higher values for nitrogen and phosphorus content of plants. The harvest index per plant and per unit area were two or three times higher in inoculated treatments than the control.

Study 3 (Hajnal-Jafari et al., *2012)*

The experiments were conducted in the experimental field of the Institute of Field and Vegetable Crops, Department of Organic Agriculture and Biodiversity, at the locality of Backi Petrovac, Serbia. Seed inoculation was performed introducing 100, 75 and 50 ml of *Azotobacter chroococcum* into 1000 grains of maize at 10^8 CFU ml^{-1}.

On average the inoculation promoted increases in grain yield of ZP 555 hybrid amounting to 1000 kg ha^{-1}, followed by hybrid 620k with 450 kg ha^{-1} and finally NS 6030 hybrid amounting to 280 kg ha^{-1}. *A. chroococcum* could increase maize yield by many stimulating processes such as seed germination, resistance of seedlings to stress conditions, nitrogen fixation and production of phytohormones.

12.4.4 *Serratia*

(Hameeda et al.*, 2008)*

The strain of *Serratia marcescens* EB 67 with P-solubilizing ability was tested in an experiment conducted at the International Crops Research Institute for the Semi-Arid Tropics (ICRISAT), Patancheru, India. A peat-based formulation was applied at 150 g ha^{-1} as a seed coat with 1% carboxymethyl cellulose as adhesive. The final concentration was 10^6–10^7 CFU per seed.

The seed treatment with EB 67 increased the grain yield 85% compared to the uninoculated control. The increases in P uptake of shoots and grain of maize were respectively 13 and 27 kg ha^{-1} in inoculated treatment, and 5 and 11 kg ha^{-1} control (without inoculation). The N uptake in shoots was 54 kg ha^{-1} with EB 67 and 30 kg ha^{-1} control. The use of *S. marcescens* was efficient to increase yield and can be used to reduce the use of P fertilizer.

12.4.5 *Rhanella*

(Montañez and Sicardi, 2013)

The bacterial strain *Rhanella* sp. EMA83 was isolated from maize (Montañez *et al.*, 2009) and this endophyte was characterized as nif*H* gene presence, IAA producer, and solubilizer of P. The field trial was conducted at Ombuesde Lavalle, Colonia, Uruguay, with two inoculation treatments, one onto the seed and the other in the soil. The dose for seed inoculation was 300 ml ha^{-1} while the dose for soil was 600 ml ha^{-1}at 1.5 × 10^9 CFU ml^{-1}. The maize cultivar NK900 was used.

In the field, maize grain yield (kg ha^{-1}) increased among treatments onto seed (15%) and soil (12%) with no N fertilization. At maximum fertilization rates (120 kg ha^{-1}) an increase was found only in soil application (16%) compared to the non-inoculated control.

12.4.6 *Herbaspirillum*

(Alves et al.*, 2015)*

Herbaspirillum seropedicae strain ZAE94, a diazotrophic bacterium from the collection of Embrapa Agrobiologia, was tested in field experiments. For the inoculation, 75 ml of inoculum containing 10^9 cells ml^{-1} was mixed with 175 g finely powdered, neutralized and autoclaved peat. The seeds were covered at 250 g peat inoculant per 10 kg of maize seed. The experiment was conducted in Seropédica city, Rio de Janeiro State, Brazil, with the hybrid SHS5050.

The *H. seropedicae* ZAE94 increased the maize yield up to 34%, depending on the plant genotype. The quantification of biological nitrogen fixation (BNF) revealed that 37% of the nitrogen on inoculated plants with this strain was BNF-derived.

12.5 Conclusion

The potentialities and the 'state of the art' on the use of bacteria to promote maize yield and reduce usage of chemical fertilizers were presented. The biological and biochemical mechanisms that make the bacteria and plant interactions an efficient tool on maize yield improvements were also presented and discussed. Therefore, widespread use of the above reported bacteria as a natural input on maize crop, could contribute significantly to the profitable sustainability of maize production worldwide.

Reference

Alves, G.C., Videira, S.S., Urquiaga, S. and Reis, V.M. (2015) Differential plant growth promotion and nitrogen fixation in two genotypes of maize by several *Herbaspirillum* inoculants. *Plant and Soil* 387, 307–321.

Araújo, W.L. de, Lacava, P.T., Marcon, J., Lima, A.O. de S., Sobral, J.K. *et al.* (2010) *Guia Prático: Isolamento e Caracterização de Microrganismos Endofíticos.* 2nd edn, CALO, Piracicaba, São Paulo, Brasil.

Barretti, P.B., Souza, R.M. and Pozza, E.A. (2008) Bactérias endofíticas como agentes promotores do crescimento de plantas de tomateiro e de inibição *in vitro* de *Ralstonia solanacearum*. *Ciência e Agrotecnologia* 3, 731–739.

Beattie, G. (2006) Plant associated bacteria: survey, molecular phylogeny, genomics and recent advances. *Biomedical and Life Sciences* 1, 1–56.

Belimov A.A., Safronova, V.I., Sergeyeva, T.A., Egorova, T.N., Matveyeva, V.A. *et al.* (2001) Characterization of plant growth promoting rhizobacteria isolated from polluted soils and containing 1-aminocyclopropane-1-carboxylate deaminase. *Canadian Journal of Microbiology* 47, 642–652.

Berg, G., Egamberdieva, D., Lugtenberg, B. and Hagemann, M. (2010) Symbiotic plant–microbe interactions: stress protection, plant growth promotion, and biocontrol by *Stenotrophomonas*. In: Seckbach, J. and Grube, M. (eds) *Symbioses and Stress (Cellular Origin, Life in Extreme Habitats and Astrobiology*, vol.17), Springer, Dordrecht, Netherlands, pp 445–460.

Bisen, K., Keswani, C., Mishra, S., Saxena, A., Rakshit, A. and Singh, H.B. (2015) Unrealized potential of seed biopriming for versatile agriculture. In: Rakshit, A., Singh, H.B. and Sen, A. (eds) *Nutrient Use Efficiency: From Basics to Advances*. Springer, New Delhi, India, pp. 193–206.

Blaha, D., Prigent-Combaret, C., Mirza, M.S. and Moënne-Loccoz, Y. (2006) Phylogeny of the 1-aminocyclopropane-1-carboxylic acid deaminase-encoding gene *acdS* in phytobeneficial and pathogenic *Proteobacteria* and relation with strain biogeography. *FEMS Microbiology Ecology* 56, 455–470.

Buyer, J.S., Kratzke, M.G. and Sikora, L.J. (1993) A method for detection of pseudobactin, the siderophore produced by a plant-growth-promoting *Pseudomonas* strain, in the barley rhizosphere. *Applied and Environmental Microbiology* 59, 677–681.

Campo, R.J. and Hungria, M. (2007) Anais da XIII RELARE. In: Campo R.J. and Hungria, M. (eds) *Reunião da Rede de Laboratórios para Recomendação, Padronização e Difusão de Tecnologia de Inoculantes Microbianos de Interesse Agrícola*. Embrapa Soja, Londrina, Parana, Brasil, p. 212.

Cavallet, L.H., Pessoa, A.C. dos S., Helmich, J.J., Helmich, P.R. and Ost, C.F. (2000) Produtividade do milho em resposta à aplicação de nitrogênio e inoculação das sementes com *Azospirillum* spp. *Revista Brasileira de Engenharia Agrícola e Ambiental* 4, 129–132.

Chaiharn, M. and Lumyong, S. (2011) Screening and optimization of indole-acetic-acid production and phosphate solubilization from rhizobacteria aimed at improving plant growth. *Current Microbiology* 62, 173–181.

Coelho, L.F., Freitas, S.S., Melo, A.M.T. and Ambrosano, G.M.B. (2007) Interação de bactérias fluorescentes do gênero *Pseudomonase* de *Bacillus* spp. com a rizosfera de diferentes plantas. *Revista Brasileira de Ciência do Solo* 31, 1413–1420.

Conceição, P.M., Vieira, H.D., Canellas, L.P., Olivares, F.L. and Conceição, P.S. (2009) Efeito dos ácidos húmicos na inoculação de bactérias diazotróficas endofíticas em sementes de milho. *Ciência Rural* 39, 1–4.

Dey, R., Pal, K.K., Bhatt, D.M. and Chauhan, S.M. (2004) Growth promotion and yield enhancement of peanut (*Arachis hypogaea* L.) by application of plant growth-promoting rhizobacteria. *Microbiological Research* 159, 371–394.

Dobbelaere, S., Vanderleyden, J. and Okon, Y. (2003) Plant growth-promoting effects of diazotrophs in the rhizosphere. *Critical Reviews in Plant Sciences* 22, 107–149.

Döbereiner, J. and Day, J.M. (1976) Associative symbioses in tropical grasses: characterization of microorganisms and nitrogen-fixing sites. In: Newton, W.E. and Nyman, C.J. (eds) *Proceedings of the 1st International Symposium on Nitrogen Fixation*. Washington State University Press, Pullmann, WA, pp. 518–538.

Döbereiner, J. and Ruschel, A.P. (1958) Uma nova espécie de *Beijerinckia*. *Revista de Biologia* 1, 261–272.

Döbereiner, J., Baldani, J.I. and Baldani, V.L.D. (1995) Como isolar e identificar bactérias diazotróficas de plantas não leguminosas. *Embrapa-CNPAB*, 60.

Ferreira, A.S., Pires, R.R., Rabelo, P.G., Oliveira, R.C., Luz, J.M.Q. and Brito, C.H. (2013) Implications of *Azospirillum brasilense* inoculation and nutrient addition on maize in soils of the Brazilian Cerrado under greenhouse and field conditions. *Applied Soil Ecology* 72, 103–108.

Fukami, J., Nogueira, M.A., Araujo, R.S. and Hungria, M. (2016) Accessing inoculation methods of maize and wheat with *Azospirillum brasilense*. *AMB Express* 6, 1–13.

Fulchieri, M. and Frioni, L. (1994) *Azospirillum* inoculation on maize (*Zea mays*): effect on yield in a field experiment in central Argentina. *Soil Biology and Biochemistry* 26, 921–923.

Gaby, J.C. and Buckley, D.H. (2012) A comprehensive evaluation of PCR primers to amplify the *nifH* gene of nitrogenase. *PLoS ONE* 7, 7.

Galvão, P.G., Urquiaga, S., Vidal, M.S. and Baldani, J.I. (2010) Interação entre plantas e bactérias promotoras do crescimento vegetal. In: Galvão, P.G., Urquiaga, S., Vidal, M.S. and Baldani, J.I. (eds) *Documentos 270*. Embrapa Agroecologia, Rio de Janeiro, Brasil, p. 63.

Glick, B.R. (1995) The enhancement of plant growth by free-living bacteria. *Canadian Journal of Microbiology* 41, 109–117.

Glick, B.R., Penrose, D.M. and Li, J. (1998) A model for the lowering of plant ethylene concentrations by plant growth-promoting bacteria. *Journal of Theoretical Biology* 190, 63–68.

Hajnal-Jafari, T., Latkovi´c, D., Duric, S., Mrkovacki, N. and Najdenovska, O. (2012) The use of *Azotobacter* in organic maize production. *Research Journal of Agricultural Science* 44, 28–32.

Hallmann, J. (2001) Plant interactions with endophytic bacteria. In: Jeger, M.J. and Spencer, N.J. (eds) *Biotic Interactions in Plant Pathogen Associations*. British Society for Plant Pathology, London, pp. 87–119.

Hallmann, J., Quadt-Hallmann, A., Mahaffee, W.F. and Kloepper, J.W. (1997) Bacterial endophytes in agricultural crops. *Canadian Journal of Microbiology* 43, 895–914.

Hameeda, B., Harini, G., Rupela, O.P., Wani, S.P. and Reddy, G. (2008) Growth promotion of maize by phosphate-solubilizing bacteria isolated from composts and macrofauna. *Microbiological Research* 163, 234–242.

Hayat, R., Ali, S., Amara, U., Khalid, R. and Ahmed, I. (2010) Soil beneficial bacteria and their role in plant growth promotion: a review. *American Microbiology* 60, 579–598.

Hungria, M. (2011) Inoculação com *Azospirillum brasiliense*: inovação em rendimento a baixo custo. In: Hungria, M. (ed.) *Documentos 325*. Embrapa Soja, Londrina, Parana, Brasil, p. 36.

Hungria, M., Campo, R.J., Souza, E.M. and Pedrosa, F.O. (2010). Inoculation with selected strains of *Azospirillum brasilense* and *A. lipoferum* improves yields of maize and wheat in Brazil. *Plant and Soil* 331(1–2), 413–425.

Hussain, A., Arshad, M. and Hussain, E. (1987) Response of maize (*Zea mays*) to *Azotobacter* inoculation under fertilized and unfertilized conditions. *Biology and Fertility of Soils* 4, 73–77.

Ikeda, A.C., Bassani, L.L., Adamoski, D., Stringari, D., Cordeiro, V.K. *et al.* (2013) Morphological and genetic characterization of endophytic bacteria isolated from roots of different maize genotypes. *Microbial Ecology* 65, 154–160.

Kennedy, A.C. (2005) Rhizosphere. In: Sylvia, D.M. (ed.) *Principles and Applications of Soil Microbiology*. Pearson, Prentice Hall, New Jersey, pp. 357–384.

Kennedy, I.R., Choudhury, A.I.M.A. and Keeskes, M.L. (2004) Symbiotic bacterial diazotrophs in crop-farming systems: can their potential for plant growth promotion be better exploited? *Soil Biology and Biochemistry* 6, 1229–1244.

Kerovuo, J., Lauraeus, M., Nurminen, P., Kalkkinen, N. and Apajalahti, J. (1998) Isolation, characterization, molecular gene cloning, and sequencing of a novel phytase from *Bacillus subtilis*. *Applied and Environmental Microbiology* 64, 2079–2085.

Keswani, C., Bisen, K., Singh, V., Sarma, B.K. and Singh, H.B. (2016) Formulation technology of biocontrol agents: present status and future prospects. In: Arora, N.K., Mehnaz, S. and Balestrini, R. (eds) *Bioformulations: For Sustainable Agriculture*. Springer, New Delhi, India, pp. 35–52.

Kloepper, J.W. and Schroth, M.N. (1978) Plant growth promoting rhizobacteria on radishes. *Proceedings of the Fourth International Conference on Plant Pathogen Bacteria, Angers, France* INRA Angers, France, 2, 879–882.

Kupper, K.C., Gimenes-Fernandes, N. and Goes, A. de. (2003) Controle biológico de *Colletotrichum acutatum*, agente causal da queda prematura dos frutos cítricos. *Fitopatologia Brasileira* 28, 251–257.

Lugtenberg, B. and Kamilova, F. (2009) Plant-growth-promoting rhizobacteria. *Annual Review of Microbiology* 63, 541–556.

Mishra, S., Singh, A., Keswani, C., Saxena, A., Sarma, B.K. and Singh, H.B. (2015) Harnessing plant–microbe interactions for enhanced protection against phytopathogens. In: Arora, N.K. (ed.) *Plant Microbes Symbiosis: Applied Facets*. Springer, New Delhi, India, pp. 111–125.

Montañez, A. and Sicardi, M. (2013) Effects of inoculation on growth promotion and biological nitrogen fixation in maize (*Zea mays* L.) under greenhouse and field conditions. *Basic Research Journals of Agricultural Science and Review* 2, 102–110.

Morais, T.P.D., Brito, C.H.D., Brandão, A.M. and Rezende, W.S. (2016) Inoculation of maize with *Azospirillum brasilense* in the seed furrow. *Revista Ciência Agronômica* 47, 290–298.

Müller, T.M., Sandini, I.E., Rodrigues, J.D., Novakowiski, J.H., Basi, S. and Kaminski, T.H. (2016) Combination of inoculation methods of *Azospirillum brasilense* with broadcasting of nitrogen fertilizer increases corn yield. *Ciência Rural* 46, 210–215.

Nehl, D.B., Allem, S.J. and Brown, J.F. (1997) Deleterious rhizosphere bacteria: an integrating perspective. *Applied Soil Ecology* 5, 1–20.

Pandey, A., Sharma, E. and Palni, L.M.S. (1998) Influence of bacterial inoculation on maize in upland farming systems of the Sikkim Himalaya. *Soil Biology and Biochemistry* 30, 379–384.

Patten, C.L. and Glick, B.R. (1996) Bacterial biosynthesis of indole-3-acetic acid. *Canadian Journal of Microbiology* 42, 207–220.

Peng, G., Zhang, W., Luo, H., Xie, H., Lai, W. and Tan, Z. (2009) *Enterobacter oryzae* sp. nov., a nitrogen-fixing bacterium isolated from the wild rice species *Oryza latifolia*. *International Journal of Systematic and Evolutionary Microbiology* 59, 1650–1655.

Petrini, O. (1991) Fungal endophytes of tree leaves. In: Andrews, J.A. and Hirano, S.S. (eds) *Microbial Ecology of Leaves*. Springer, New York, 179–197.

Raju, P.N., Evans, H.J. and Seidler, R.J. (1972) An asymbiotic nitrogen-fixing bacterium from the root environment of corn. *Proceedings of the National Academy of Science of the United States of America* 69, 3474–3478.

Schmidt, M.A., Souza, E.M., Baura, V., Wassem, R., Yates, M.G. *et al.* (2011) Evidence for the endophytic colonization of *Phaseolus vulgaris* (common bean) roots by the diazotroph *Herbaspirillum seropedicae*. *Brazilian Journal of Medical and Biological Research* 44, 182–185.

Sgroy, V., Cassán, F., Masciarelli, O., Del Papa, M.F., Lagares, A. and Luna, V. (2009) Isolation and characterization of endophytic plant growth-promoting (PGPB) or stress homeostasis-regulating (PSHB) bacteria associated to the halophyte *Prosopis strombulifera*. *Applied Microbiology and Biotechnology* 85, 371–381.

Shaharoona, B., Arshad, M., Zahir, Z.A. and Khalid, A. (2006) Performance of *Pseudomonas* spp. containing ACC-deaminase for improving growth and yield of maize (*Zea mays* L.) in the presence of nitrogenous fertilizer. *Soil Biology and Biochemistry* 38, 2971–2975.

Singh, H.B., Jain, A., Saxena, A., Singh, A., Keswani, C. *et al.* (2014) Deciphering the pathogenic behaviour of phyto-pathogens using molecular tools. In: Sharma, N. (ed.) *Biological Controls for Preventing Food Deterioration: Strategies for Pre-and Postharvest Management*. John Wiley & Sons, Chichester, UK, pp. 377–408.

Spaepen, S., Vanderleyden, J. and Remans, R. (2007) Indole-3-acetic acid in microbial and microorganism-plant signaling. *FEMS Microbiology Reviews* 31, 425–448.

Stone, J.K. (1988) Fine structure of latent infections by *Rhabdocline parkeri* on Douglas-fir with observations on uninfected epidermal cells. *Canadian Journal of Botany* 66, 45–54.

Swędrzyńska, D. and Sawicka, A. (2000) Effect of inoculation with *Azospirillum brasilense* on development and yielding of maize (*Zea mays* ssp. *saccharata* L.) under different cultivation conditions. *Polish Journal of Environmental Studies* 9, 505–509.

Vessey, J.K. (2003) Plant growth promoting rhizosphere as biofertilisers. *Plant and Soil* 255, 571–586.

Viruel, E., Lucca, M.E. and Siñeriz, F. (2011). Plant growth promotion traits of phosphobacteria isolated from Puna, Argentina. *Archives of Microbiology* 193, 489-496.

Viruel, E., Erazzú, L.E., Martínez-Calsina, L., Ferrero, M.A., Lucca, M.E. and Siñeriz, F. (2014) Inoculation of maize with phosphate solubilizing bacteria: effect on plant growth and yield. *Journal of Soil Science and Plant Nutrition* 14, 819–831.

Vohara, A. and Satyanarayana, T. (2003) Pitases: microbial sources, production, purification and potential biotechnological applications. *Critical Reviews in Biotechnology* 23, 29–60.

Weber, O.B., Baldani, J.I. and Döbereiner, J. (2000) Bactérias diazotróficas em mudas de bananeira. *Pesquisa Agropecuária Brasileira* 35, 2277–2285.

Yang, S.F. and Hoffman, N.E. (1984) Ethylene biosynthesis and its regulation in higher plants. *Annual Review of Plant Physiology* 35, 155–189.

13 The Potential of Mycorrhiza Helper Bacteria as PGPR

Marieta Marin Bruzos

Belowground Ecosystem Group, Department of Forest and Conservation Sciences, University of British Columbia, Vancouver, British Columbia, Canada

13.1 Introduction

The belowground environment is an active space where living organisms and plant roots interact among themselves and with the soil components. As a consequence, the root system of many crops in different ecosystems lives in a mutualistic interaction with mycorrhiza-forming fungi. The resulting association benefits the plants by improving their nutrients uptake and increasing the resistance against soilborne pathogens and abiotic stresses (Finlay, 2008). While this symbiosis is generally considered a dual plant–fungus interaction, other microorganisms like bacteria and yeasts are also closely related (Frey-Klett and Garbaye, 2005).

Foster and Marks (1967) introduced the definition of "mycorrhizosphere" as the soil area influenced by the mycorrhizal roots and peripheral fungal mycelium. Some of the bacterial groups living within the mycorrhizosphere are able to stimulate the mycorrhiza development. Bacterial strains showing this property were named mycorrhiza helper bacteria (MHB) by Garbaye (1994). Since then, different studies have been performed to evaluate the combined effect of MHB and their associated fungi on the plant growth, especially in the enhancement of nutrient acquisition (Frey-Klett *et al.*, 2007). However, with the development of molecular and genomics techniques new knowledge has been added for a better understanding of this tripartite association.

Several MHB are currently considered as plant growth-promoting rhizobacteria (PGPR), such as some isolates of *Pseudomonas* spp. and *Bacillus* (Probanza *et al.*, 2001; Jaleel *et al.*, 2007; Bisen *et al.*, 2015; Keswani *et al.*, 2016). Because of the similarity of species present in both groups, this classification can be overlapping. An additional issue that makes the division between the PGPR and MHB group difficult is that in general experiments performed with PGPR the competence of the isolates to enhance the mycorrhiza development is not usually studied (Rigamonte *et al.*, 2010). However, some fungal metabolic pathways are commonly regulated by different rhizospheric bacteria, whereas other signalling molecules are particular to the MHB group (Deveau *et al.*, 2007).

In this chapter we review some of the early works that contributed to the finding of MHB, the proposed mechanisms of action that allow this group of bacteria to interact

E-mail: mmarinb@mail.ubc.ca

© CAB International 2017. *Advances in PGPR Research* (eds H.B. Singh, B.K. Sarma and C. Keswani)

with the mycorrhizal fungi, the recent discoveries on the MHB topic with the use of transcriptomic and genomic techniques, and the potential of this bacterial group to be used as PGPR.

13.2 Early Findings

According to different fossil records, mycorrhizal associations developed around 50 million years ago; however, there are data that suggest the existence of symbiotic structures over 180 million years ago (LePage *et al.*, 1997). However, the occurrence of bacteria directly involved in the mycorrhizal establishment was described by Ridge and Theodorou (1972) who found that fumigation with methyl bromide enhanced the infection of *Rhizopogon luteolus* in *Pinus radiate* in one nursery soil but reduced it in another one. So they concluded that this could be correlated to different microorganisms recolonizing the soils.

In following works the occurrence of bacteria able to promote mycorrhiza formation was suggested by other researchers: in *Pisolithus tinctorius* by Marx *et al.* (1989) and De Oliveira (1988) in beech seedlings interacting with *Hebeloma crustuliniforme*. Accordingly with those findings Garbaye and Bowen (1989) hypothesized that some helper bacteria must be adapted to live in cooperation with the fungi; consequently, if they were to be found in soil, they were probably more numerous in the close vicinity of the fungus. Thus, these authors isolated bacteria from surface-sterilized ectomycorrhiza formed by *Rhizopogon luteolus* in the root system of *Pinus radiata* and studied their effect on the formation of mycorrhizal structures with the same symbiotic partners in soil previously sterilized. They determined about 10^6 colony-forming units per gram (fresh weight) of mycorrhiza: 80% of them displayed a significant helper effect on mycorrhiza formation while 20% were neutral or inhibitory. Finally, Garbaye (1994) reviewed the previous work on this topic and proposed the term MHB for this bacterial group. Table 13.1 summarizes some of the early experiments done in the MHB topic considered as milestones.

13.3 Proposed Helper Mechanisms

The mycorrhiza establishment is determined by the interactions among biotic and abiotic environmental factors, the physiology of the fungus and the plant root susceptibility to colonization. According to Frey-Klett *et al.* (2007), MHB may stimulate the development of mycorrhiza at different phases during the bacteria–fungus–root interaction. Figure 13.1 shows the different sites of action of MHB.

13.3.1 Promoted germination of fungal propagules

The secondary metabolites of MHB can stimulate the germination of fungal spores. When the roots of sea oats (*Unicola paniculata*) were inoculated with *Klebsiella pneumonia* the spore germination rate increased as well as a faster elongation of *Glomus deserticola* mycelium (Will and Sylvia, 1990). *Pseudomonas fluorescens* was able to promote root colonization by *Glomus mosseae* (Pivato *et al.*, 2009). In the same way Hidayat *et al.* (2013) found that *Pseudomonas diminuta* enhanced spore germination percentage and hyphal length of *Glomus* sp. as much as 224% and 330% respectively than control.

13.3.2 Promoted mycelial growth

Co-cultures of fungus and bacteria are easily implemented *in vitro*, and consequently are often used as a first step when searching for MHB isolates that promote hyphal growth. If the inoculation of helper bacteria is able to increase the mycelial biomass in the rhizosphere, then the frequency of root–fungus encounters should increase too, resulting in a faster mycorrhizal establishment (Brulé *et al.*, 2001). The growth of the fungal

Table 13.1. Early works in the Mycorrhizal Helper Bacteria research.

Findings	Fungus	Tree	References
Fumigation with methyl bromide enhanced infection in one nursery soil but reduced it in another one.	*Rhizopogon luteolus*	*Pinus radiate*	Ridge and Theodorou (1972)
Several bacterial isolates from forest soil stimulated mycorrhiza formation under controlled conditions.	*Hebeloma crustuliniforme*	Beech seedlings	De Oliveira (1988)
Vegetative inoculum contaminated by fungi and bacteria was sometimes more efficient for mycorrhiza formation than non-contaminated inoculum when inoculating pine seedlings in fumigated nursery soils.	Not specified	*Pisolithus tinctorius*	Marx *et al.* (1989)
True helper bacteria should be adapted to live in cooperation with the fungus or in a soil more frequent in the close vicinity of the fungus.	*Rhizopagon luteolus*	*Pinus radiata*	Garbaye and Bowen (1989)

Adapted from: Garbaye (1994)

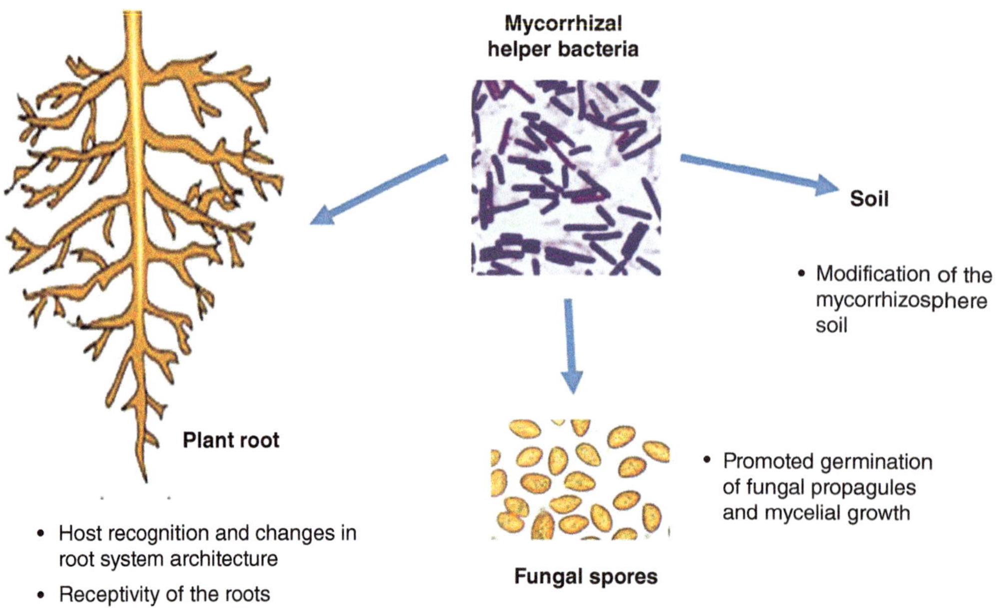

Fig. 13.1. The sites of action for Mycorrhiza Helper Bacteria.

mycelium is stimulated by secondary metabolic compounds produced by the MHB. Those metabolites can affect the fungal metabolism and modulate gene expression. Some of these compounds have been identified as gases that increase the radial mycelial growth of the fungus *Pisolithus albu* when it is growing on tryptic soy broth agar or on a minimal medium amended with trehalose (Duponnois and Kisa, 2006); similarly identified was auxofuran, so named owing to the relationship of its chemical structure with the auxins (Riedlinger *et al.*, 2006). With the use of microarray methodology it was possible to identify different genes of early response involved in the priming helper effect of the bacterial strain *P. fluorescens* BBc6R8 on the development and morphological structure of its mycorrhizal fungal partner *Laccaria bicolor* A238N (Deveau *et al.*, 2007).

13.3.3 Modification of the mycorrhizosphere soil

A great number of soil microorganisms, including mycorrhizal fungi, can produce toxic compounds to inhibit the growth of competitors. Duponnois and Garbaye (1990) evaluated how the MHB modulated the concentrations of antagonistic metabolites produced by mycorrhizal fungi. They reported that the bacteria were able to detoxify the liquid media from the inhibitory fungal metabolites. Helper bacteria could perhaps also suppress the production of toxic substances by soil microbes (Rigamonte *et al.*, 2010). Some MHB strains while colonizing the mycorrhizosphere can compete for nutrients with bacteria that inhibit mycorrhization and therefore reduce the concentration of anti-fungal metabolites in the rhizosphere environment (Garbaye, 1994).

In experiments performed on Cd or Zn contaminated substrates, the AM (arbuscular mycorrhizal) colonization and the growth of extra radical mycelium in plants colonized by *Glomus mosseae* was observed to be higher when *Brevibacillus brevis* was inoculated (Vivas *et al.*, 2003a,b). In the same way, Malekzadeh *et al.* (2011) found that *Micrococcus roseus*, a strain tolerant to Cd, improved the nutrient uptake by the plants and the mycorrhiza formation between *G. mosseae* and *Zea mays* in polluted soils.

13.3.4 Host recognition and modifications in root system architecture

Another possible mechanism that has been studied in MHB is their competence to stimulate the development of lateral roots in mycorrhizal plants probably due to the production of auxins or auxin-related compounds by the bacteria. Taking into account that MHB can also promote fungal growth, there can be a dual effect that increases the number of potential interaction sites between the plant root system and the fungus (Schrey *et al.*, 2005), and subsequently stimulate a higher plant mycorrhization rate by the mycobiont. In addition, it is possible that different helper strains may develop other MHB traits, even for the same pair of mycorrhizal symbionts (Rigamonte *et al.*, 2010). In this way, it was found that the helper strain *Burkholderia* sp. EJP67 isolated from the *Pinus sylvestris-Lactarius rufus* ectomycorrhizae promoted both first- and second-order mycorrhizal roots, while *Paenibacillus* sp. EJP73 isolated from the same ectomycorrhizae only stimulated the development of secondary mycorrhizal roots (Poole *et al.*, 2001).

The morphological structure of fungal mycelia during the mycorrhization process, when helper bacteria are involved, has not been well studied. However, Deveau *et al.* (2007) observed substantial morphological variations of the hyphal apex density and branching angles in co-cultures of *L. bicolor* S238N with MHB bacterial strains, which depended on the bacterial strains.

13.3.5 Receptivity of the roots

According to the last mechanism proposed, the bacterium allows the colonization of the plant root while growing in the rhizospheric soil earlier than the interaction among the mycorrhizal fungus and the host plant. This could take place through controlled excretion of cell wall digesting enzymes by the MHB-like cellulases, permitting the infiltration of the roots by the fungal hyphae and facilitating their extent inside the root tissues. An additional fact that could also help the mycorrhization is the inhibition of the plant defense response by the MHB prior to the fungal colonization process.

Aspray *et al.* (2006) proved that physical contact among the helper bacteria cells and the symbionts is required to carry on the stimulatory effect. The MHB can increase the nutrient uptake of the fungus, providing nitrogen in the case of diazotrophic bacteria, contributing to the solubilization of minerals like phosphate or iron by the secretion of protons and complexing agents such as organic acids of low molecular weight or siderophores (Rigamonte *et al.*, 2010). It is probably as well that the MHB increase the production of hypaphorine, a phenolic fungal compound, enhancing the aggressiveness of the mycobiont (Duponnois and Plenchette, 2003).

13.4 Genomic Approaches

The progress in molecular and genomic techniques has brought new insights into the MHB research. The development of real-time quantitative PCR (qPCR) techniques has provided a tool to detect and quantify directly the helper bacteria strains and mycorrhizal fungi in the plant rhizosphere. The data from qPCR showed that the existence of microorganisms and tree roots can together affect the nature of the interaction between the MHB *Streptomyces* sp. AcH 505 and *Piloderma cruceum*, and this mycorrhizal fungus may increase MHB growth (Kurth *et al.*, 2013). Similarly, this technique has been used to correlate the responses among bacterial strains and the expression of seven target/reporter genes from *L. bicolor* S238N mycelium in pairwise analyses (Labbé *et al.*, 2014).

Soil metagenomics methods have been implemented to study rhizobacterial communities associated with mycorrhizae (Daniel, 2005). Performing molecular screening of 16S rDNA libraries, researchers have found several bacterial taxa with a predominance of species from the genera *Pseudomonas*, *Burkholderia* and *Bacillus* (De Boer *et al.*, 2005). In the same way, Streptomycetes isolates associated with ectomycorrhizal fungi have been reported as modulators of plant symbiosis (Schrey and Takka, 2008), whereas Archaebacteria populations present in the rhizosphere of mycorrhizal fungi have been found only in boreal regions (Bomberg and Timonen, 2007).

DNA-based techniques have also expanded the identification of MHB to non-culturable species. Combining microscopic techniques and 16S rRNA molecular analysis, bacteria related to *Burkholderia* species have been reported living in symbiosis with members of the gigasporaceae. These MHB were grouped in a new taxon phylogenetically close to *Burkholderia* but named "*Candidatus* Glomeribacter gigasporarum" due to their unculturability (Bianciotto *et al.*, 2003). *Gigaspora margarita* BEG34 and its associated endobacterium "*Candidatus* Glomeribacter gigasporarum" are now employed as a model system to study endobacteria–AM fungi associations (Salvioli *et al.*, 2010).

Recently, the complete genome of *Pseudomonas fluorescens* BBc6R8 was sequenced (Deveau *et al.*, 2014) and that data was combined with transcriptomic and mutagenesis approaches to reveal molecular determining factors of the helper effect. The *in vitro* result suggests that the production of helper molecules appears to be constitutive. The helper effect seems to be pleiotropic and to depend on trophic interactions (Galet *et al.*, 2015). In the same way, Kurth *et al.* (2015) used large transcriptomic analysis to study the interaction between the mycorrhiza helper bacteria *Streptomyces* sp. AcH 505 and pedunculated oak *Quercus robur*. They concluded that the treatment with AcH 505 induced and sustained the expression of signalling genes that encode candidate receptor protein kinases and transcription factors, and thus leads to differential expression of genes related to cell wall variation in pedunculate oak microcuttings. Gene expression responses to the inoculation with *Streptomyces* sp. AcH 505 alone and in combination with *P. croceum* were more evident in root resting stages, probably because non-growing roots re-direct their metabolic activity towards plant defense instead of growth.

13.5 Potential use of MHB as PGPR

Taking into account the positive impact on mycorrhiza formation, improvement of plant nutrition, growth and control of soil phytopathogens by the helper bacteria the following potential applications of MHB as PGPR are listed:

- Use of helper bacteria to stimulate plant growth: it has been proved that many MHB can stimulate plant development through different mechanisms such as phytohormone production and nitrogen fixation. For example, the MHB *Pseudomonas fluorescens* 92 stimulates the growth of cucumber plants, and *Arthrobacter* sp BB1 stimulated the growth and mycorrhization rates of *Pinus pinea* (Gamalero *et al.*, 2003; Barriuso *et al.*, 2008).

- Increased plant survival in contaminated soils: MHB increasing plant tolerance to Cd or Zn was described by Vivas *et al.* (2003a, b). Khan (2005) studied the role of symbiotic AMF and helper bacteria in sustainable plant growth on nutrient-poor soils contaminated with heavy metals and reported that the plants surviving on such sites were living in association with MHB and showed higher arbuscular mycorrhizal infection.
- Use to protect plants against soil pathogens and controlled mycorrhization: the fungus specificity among the MHB shows that MHB could be used for a simultaneous promotion of certain symbiotic fungi and for the biocontrol of plant pathogenic fungi. Antagonism against different soil phytopathogens has been often determined during *in vitro* assays with MHB (Maier *et al.*, 2004). On the other hand, the inoculation of MHB could be very advantageous when implementing techniques of controlled mycorrhization in forest management, through its application to soil in nurseries. The inoculation together with the mycorrhizal-forming fungi allows a more effective use of fungal inoculum while increasing the quality of the mycorrhizal association in early stages of the plants (Garbaye, 1994).

13.6 Future Challenges in MHB Research

The understanding of biological interactions among MHB and mycorrhizal fungi is still at a juvenile stage and substantial investigation is needed to comprehend them completely. According to Frey-Klett *et al.* (2007) there are some research priorities to realize this goal that almost ten years later are still significant and need to be covered:

- Search for MHB in a broader range of mycorrhizal association must be done for a better understanding of their specificity. The implementation of molecular and metagenomics techniques may help the identification and characterization of mycorrhiza-associated bacteria, but culture-based assays are still needed to evaluate the helper effect and study mechanism of action.
- Identification of functional specific genes for the helper effect must be done in fungi and bacteria (Schrey *et al.*, 2005; Deveau *et al.*, 2007). This will provide shortcuts when searching for helper strains in new mycorrhizal systems. This task will be simplified by genome sequencing of mycorrhizal fungi and some MHB strains (Frey-Klett and Garbaye, 2005).
- Microscopic techniques must be employed to localize specific bacterial cells and their activities related to the helper effect, as well as studying the metabolic variations during fungal–bacterial cometabolism. This understanding will increase our comprehension of the helper mechanisms.
- The influence of mycorrhiza helper bacteria on mycorrhizal functions such as nutrition improvement, water acquisition, control of soil pathogens and plant provision with growth factors must be studied. This will provide new elements to the evolutionary biology, physiology and ecology of mycorrhizal symbioses.
- Finally, the mycorrhization practices in agriculture and forestry must be reconsidered: MHB may increase the effectiveness of fungal inocula with a little extra cost since bacteria are cheaper to grow in commercial amounts than mycorrhizal fungi. Thus, more MHB research should be dedicated to mycorrhizal fungi of commercial importance. In addition, emergent concern about heavy-metal contamination of the soils together with the needs of organic practices, should lead to adoption of more

eco-friendly methods such as controlled mycorrhization or microbial bioremediation by using mycorrhizal fungi as carriers of depolluting bacteria.

We can conclude that the joining together of scientific, agricultural-related interests and ecological needs, sustained by the advance of new genomic techniques, can represent an exceptional prospect to place MHB as PGPR at the forefront of upcoming mycorrhiza research and to enhance the field of plant–microbe collaborations for sustainable agriculture.

References

Aspray, T., Jones, E., Whipps, J. and Bending, G.D. (2006) Importance of mycorrhization helper bacteria cell density and metabolite localization for the *Pinus sylvestris–Lactarius rufus* symbiosis. *FEMS Microbiology Ecology* 56, 25–33.

Barriuso, J., Solano, R., Santamaria, C., Daza, A. and Manero, F. (2008) Effect of inoculation with putative plant growth-promoting rhizobacteria isolated from *Pinus* spp. on *Pinus pinea* growth, mycorrhization and rhizosphere microbial communities. *Journal of Applied Microbiology* 105, 1298–1309.

Bianciotto, V., Lumini, E., Bonfante, P. and Vandamme, P. (2003) '*Candidatus* Glomeribacter gigasporarum', an endosymbiont of arbuscular mycorrhizal fungi. *International Journal of Systematic and Evolutionary Microbiology* 53, 121–124.

Bisen, K., Keswani, C., Mishra, S., Saxena, A., Rakshit, A. and Singh, H.B. (2015) Unrealized potential of seed biopriming for versatile agriculture. In: Rakshit, A., Singh, H.B. and Sen, A. (eds) *Nutrient Use Efficiency: From Basics to Advances*. Springer, New Delhi, India, pp. 193–206.

Bomberg, M. and Timonen, S. (2007) Distribution of Cren- and Euryarchaeota in Scots pine mycorrhizospheres and boreal forest humus. *Microbiology Ecology* 54, 406–416.

Brulé, C., Frey-Klett, P., Pierrat, J.C., Courrier, S., Gérard, F. *et al.* (2001) Survival in the soil of the ectomycorrhizal fungus *Laccaria bicolor* and effect of a mycorrhiza helper *Pseudomonas fluorescens*. *Soil Biology and Biochemistry* 33, 1683–1694.

Daniel, R. (2005) The metagenomics of soil. *Nature Reviews Microbiology* 3, 470–78.

De Boer, W., Folman, L.B., Summerbell, R.C. and Boddy, L. (2005) Living in a fungal world: impact of fungi on soil bacterial niche development. *FEMS Microbiology Reviews* 29, 795–811.

De Oliveira, V.L. (1988) Interactions entre les microorganismes du sol et l'etablissement de la symbiose ectomycorhizienne chez le hêter (*Fagus silvatica* L.) avec *Hebeloma crustuliniforme* et *Paxillus involutus*. Ph.D. thesis, University of Nancy, Nancy, France.

Deveau, A., Palin, B., Delaruelle, C., Peter M. Kohler, A., Pierrat, J.C. *et al.* (2007) The mycorrhiza helper *Pseudomonas fluorescens* BBc6R8 has a specific priming effect on the growth, morphology and gene expression of the ectomycorrhizal fungus *Laccaria bicolor* S238N. *New Phytologist* 175, 743–755.

Deveau, A., Gross, H., Morin, E., Karpinets, T., Utturkar, S. *et al.* (2014). Genome sequence of the mycorrhizal helper bacterium *Pseudomonas fluorescens* BBc6R8. *Genome Announcements* 2(1):e01152-13. doi:10.1128/genomeA.01152-13.

Duponnois, R. and Garbaye, J. (1990) Some mechanisms involved in growth stimulation of ectomycorrhizal fungi by bacteria. *Canadian Journal of Botany* 68, 2148–2152.

Duponnois, R. and Kisa, M. (2006) The possible role of trehalose in the mycorrhiza helper effect. *Canadian Journal of Botany* 84, 1005–1008.

Duponnois, R. and Plenchette, C. (2003) A mycorrhiza helper bacterium enhances ectomycorrhizal and endomycorrhizal symbiosis of Australian *Acacia* species. *Mycorrhiza* 13, 85–91.

Finlay, R. (2008) Ecological aspects of mycorrhizal symbiosis: with special emphasis on the functional diversity of interactions involving the extraradical mycelium. *Journal of Experimental Botany* 59, 1115–1126.

Fitter, A.H. and Garbaye, J. (1994) Interactions between mycorrhizal fungi and other soil organisms. *Plant Soil* 159, 123–132.

Foster, R. and Marks, G. (1967) Observations on the mycorrhizas of forest trees. II. The rhizosphere of *Pinus radiata* D. Don. *Australian Journal of Biological Sciences* 20, 915–926.

Frey-Klett, P. and Garbaye J. (2005) Mycorrhiza helper bacteria: a promising model for the genomic analysis of fungal-bacterial interactions. *New Phytologist* 168, 4–8.

Frey-Klett, P., Garbaye, J. and Tarkka, M. (2007) The mycorrhiza helper bacteria revisited. *New Phytologist* 176, 22–36.

Galet, J., Deveau, A., Hôtel, L. and Frey-Klett, P. (2015) *Pseudomonas fluorescens* pirates both ferrioxamine and ferricoelichelin siderophores from *Streptomyces ambofaciens*. *Applied and Environmental Microbiology* 81(9), 3132–3141.

Gamalero, E., Fracchia, L., Cavaletto, M., Garbaye, J., Frey-Klett, P. *et al.* (2003) Characterization of functional traits of two fluorescent pseudomonads isolated from basidiomes of ectomycorrhizal fungi. *Soil Biology and Biochemistry* 35, 55–65.

Garbaye, J. (1994) Helper bacteria: a new dimension to the mycorrhizal symbiosis. *New Phytologist* 128, 197–210.

Garbaye, J. and Bowen, G.D. (1989) Stimulation of mycorrhizal infection of *Pinus radiata* by some microorganisms associated with the mantle of ectomycorrhizas. *The New Phytologist* 112, 383–388.

Garbaye, J. and Duponnois, R. (1992) Specificity and function of mycorrhization helper bacteria (MHB) associated with the *Pseudotsuga menziesii–Laccaria laccats* symbiosis. *Symbiosis* 14, 335–344.

Hidayat, C., Arief, D., Nurbaity, A. and Jajang, S. (2013) Rhizobacteria selection to enhance spore germination and hyphal length of arbuscular mycorrhizal fungi *in vitro*. *Asian Journal of Agriculture and Rural Development* 3(4), 199–204.

Jaleel, C.A., Manivannan, P., Sankar, B., Kishorekumar, A., Gopi, R. *et al.* (2007) *Pseudomonas fluorescens* enhances biomass yield and ajmalicine production in *Catharanthus roseus* under water deficit stress. *Colloids and Surfaces B: Biointerfaces* 60, 7–11.

Keswani, C., Bisen, K., Singh, V., Sarma, B.K. and Singh, H.B. (2016) Formulation technology of biocontrol agents: present status and future prospects. In: Arora, N.K., Mehnaz, S. and Balestrini, R. (eds) *Bioformulations: For Sustainable Agriculture*. Springer, New Delhi, India, pp. 35–52.

Khan, A.G. (2005) Role of soil microbes in the rhizospheres of plants growing on trace element contaminated soils in phytoremediation. *Journal of Trace Elements in Medicine and Biology* 18, 355–364.

Kurth, F., Zeitler, K., Feldhahn, L., Neu, T. and Weber, T. (2013) Detection and quantification of a mycorrhization helper bacterium and a mycorrhizal fungus in plant-soil microcosms at different levels of complexity. *BMC Microbiology* 13, 205–211.

Kurth, F., Feldhahn, L., Bonn, M., Buscott, F. and Tarkka, M. (2015) Large scale transcriptome analysis reveals interplay between development of forest trees and a beneficial mycorrhiza helper bacterium. *BMC Genomics* 16, 658–670.

Labbé, J.L., Weston, D.J., Dunkirk, N., Pelletier, D.A. and Tuskan, G.A. (2014) Newly identified helper bacteria stimulate ectomycorrhizal formation in *Populus*. *Frontiers in Plant Science* 5, 579.

LePage, B.A., Currah, R.S., Stockey, R.A. and Rothwell, G.W. (1997) Fossil ectomycorrhizae from the middle Eocene. *American Journal of Botany* 84, 410–412.

Maier, A., Riedlinger, J., Fiedler, H.P. and Hampp, R. (2004) Actinomycetales bacteria from a spruce stand: characterization and effects on growth of root symbiotic, and plant parasitic soil fungi in dual culture. *Mycological Progress* 3, 129–136.

Malekzadeh, E., Alikhani, H., Savaghebi-Firoozabadi, G. and Zarei, M. (2011) Influence of arbuscular mycorrhizal fungi and an improving growth bacterium on Cd uptake and maize growth in Cd-polluted soils. *Spanish Journal of Agricultural Research* 9, 1213–1223.

Marx, D.H., Cordell, C.E., Maul, S.B. and Ruehle, J.L. (1989) Ectomycorrhizal development on pine by *Pisolithus tinctorius* in bare-root and container seedling nurseries. *New Forests* 3, 45–56.

Pivato, B., Offre, P., Marchelli, S., Barbonaglia, B., Mougel, C. and Lemanceau, P. (2009) Bacterial effects on arbuscular mycorrhizal fungi and mycorrhiza development as influenced by the bacteria, fungi, and host plant. *Mycorrhiza* 19, 81–90.

Poole, E.J., Bending, G.D., Whipps, J.M. and Read D.J. (2001) Bacteria associated with *Pinus sylvestris–Lactarius rufus* ectomycorrhizas and their effects on mycorrhiza formation *in vitro*. *New Phytologist* 151, 743–751.

Probanza, A., Mateos, J.L., Lucas, J.A., Ramos, B., De Felipe, M.R. and Gutiérrez Mañero, F.J. (2001) Effects of inoculation with PGPR *Bacillus* and *Pisolitus tinctorius* on *Pinus pinea* L. growth, bacterial rhizosphere colonization and mycorrhizal infection. *Microbiology Ecology* 41, 140–148.

Ridge, E.H. and Theodorou, C. (1972) The effect of soil fumigation on microbial recolonization and mycorrhizal infection. *Soil Biology and Biochemistry* 4, 295–305.

Riedlinger, J., Schrey, S.D., Tarkka, M.T., Hampp, R., Kapur, M. and Fiedler, H.P. (2006) Auxofuran, a novel substance stimulating growth of fly agaric, produced by the mycorrhiza helper bacterium *Streptomyces* AcH 505. *Applied and Environmental Microbiology* 72, 3550–3557.

Rigamonte, T.A., Pylro, V.S., and Duarte, G.F. (2010) The role of mycorrhization helper bacteria in the establishment and action of ectomycorrhizae associations. *Brazilian Journal of Microbiology* 41, 832–840.

Salvioli, A., Chiapello, M., Fontaine, J., Lounes, A., Graandmougin-Ferjani, A. *et al.* (2010) Endobacteria affect the metabolic profile of their host *Gigaspora margarita,* an arbuscular mycorrhizal fungus. *Environmental Microbiology* 12 (8), 2083–2095.

Schrey, S.D. and Takka, M.T. (2008) Friends and foes: streptomycetes as modulators of plant disease and symbiosis. *Antoine van Leeuwenhoek* 94, 11–19.

Schrey, S.D., Schellhammer, M., Ecke, M., Hampp, R. and Tarkka, M.T. (2005) Mycorrhiza helper bacterium *Streptomyces* AcH 505 induces differential gene expression in the ectomycorrhizal fungus *Amanita muscaria, New Phytologist* 168, 205–216.

Vivas, A., Azcón, R., Biró, B., Barea, J.M. and Ruiz-Lozano, J.M.. (2003a). Influence of bacterial strains isolated from lead-polluted soil and their interactions with arbuscular mycorrhizae on the growth of *Trifolium pretense* L. under lead toxicity. *Canadian Journal of Microbiology* 49, 577–588.

Vivas, A., Marulanda, A., Gomez, M., Barea, J.M. and Azcon, R. (2003b) Physiological characteristics (SDH and ALP activities) of arbuscular mycorrhizal colonization as affected by *Bacillus thuringiensis* inoculation under two phosphorus levels. *Soil Biology Biochemistry* 35, 987–996.

Will, M. and Sylvia, D. (1990) Interaction of rhizosphere bacteria, fertilizer, and vesicular arbuscular mycorrhizal fungi with sea oats. *Applied and Environmental Microbiology* 56, 2073–2079.

14 Methods for Evaluating Plant Growth-Promoting Rhizobacteria Traits

Antonio Castellano-Hinojosa*[1,2] and E.J. Bedmar[2]
[1]*Department of Soil Microbiology and Symbiotic Systems, Estación Experimental del Zaidín-CSIC. E-419, Granada, Spain;* [2]*Department of Microbiology, Faculty of Pharmacy, Campus of Cartuja, University of Granada, Granada, Spain*

14.1 Introduction

14.1.1 Plant growth-promoting rhizobacteria

The presence of microorganisms, bacteria, fungi, actinomycetes, protozoa and algae is critical to the maintenance and health of soil function, in both natural and managed agricultural soils. This is due to their involvement in key processes such as soil structure formation, decomposition of organic matter, toxin removal, suppression of plant disease and, overall, the cycling of carbon, nitrogen, phosphorus and sulphur (Doran *et al.*, 1996; van Elsas *et al.*, 1997; Mishra *et al.*, 2015; Keswani *et al.*, 2016). Bacteria are the most common of those microorganisms reaching 10^8 to 10^{10} cells gm^{-1} of agricultural, non-stressed soils. The presence of bacteria in the plant rhizosphere is typically higher than in bulk soils because of the presence of sugars, amino acids, organic acids and other small molecules from plant root exudates that favor bacterial nutrition and growth. Within this soil microbiota, some bacterial populations are able to competitively colonize plant roots and stimulate growth, thereby reducing the incidence of plant diseases, and are now best known as rhizobacteria for which the term plant growth-promoting rhizobacteria was coined, and are commonly recognized by the initials PGPR (Kloepper and Schroth, 1978; Kloepper and Schroth, 1981).

14.1.2 Types of PGPR

Following its degree of interaction, positive, negative or neutral, between rhizobacteria and their host plants (Whipps, 2001), most of the PGPR positively influence plant growth and have been divided into two groups: symbiotic bacteria and free-living rhizobacteria (Khan, 2005). They can also be categorized into two groups according to their residing sites: (a) intracellular plant growth-promoting rhizobacteria (iPGPR) (symbiotic bacteria) live inside symbiosis-specific plant organs called nodules where dinitrogen (N_2) fixation takes place; and (b) extracellular plant growth-promoting rhizobacteria (ePGPR) (free-living rhizobacteria), which may exist in the rhizosphere, the rhizoplane or in the space between the cells of the root cortex and do not produce

*E-mail: ach@ugr.es

© CAB International 2017. *Advances in PGPR Research* (eds H.B. Singh, B.K. Sarma and C. Keswani)

nodules (Gray and Smith, 2005; Martínez-Viveros *et al.*, 2010).

The iPGPR includes members of the Alpha-proteobacteria collectively called rhizobia as well as the Betaproteobacteria *Burkholderia* and *Cupriavidus*, and the actinomycetes *Frankia*, (Peix *et al.*, 2015; Trujillo *et al.*, 2015). In this paper we focus on ePGPR (from henceforth PGPR), which comprise a high number of phylogenetically unrelated genera within different phyla (Rodríguez-Díaz *et al.*, 2008; Babalola, 2010; Hayat *et al.*, 2010; Bhattacharyya and Jha, 2012; Bulgarelli *et al.*, 2013; Ahemad and Kibret, 2014).

14.1.3 Mechanisms of action

In general, the potential functions of PGPR involved in plant growth promotion include direct and indirect mechanisms. The best studied direct mechanisms include: (a) production of indole acetic acid (IAA) or IAA-like compounds; (b) facilitating the acquisition of resources/nutrients that plants lack, such as fixed nitrogen, iron, phosphorus, etc.; and (c) production of the enzyme 1-aminocyclopropane-1-carboxylate (ACC deaminase) involved in reduction of plant ethylene. Among the indirect mechanisms those associated with enhancement of plant growth are (a) production of antibiotics and lytic enzymes, (b) production of siderophores, (c) competition for sites in the plant roots, (d) induction of systemic resistance, etc. Comprehensive reviews covering mechanisms related to plant promotion by PGPR have been published (Lugtenberg and Kamilova, 2009; Compant *et al.*, 2010; Hayat *et al.*, 2010; Gamalero and Glick, 2011; Saharan and Nehra, 2011; Bhattacharyya and Jha, 2012; Glick, 2012; Bisen *et al.* 2015; Tkacz and Poole, *et al.*, 2015). PGPR may use various mechanisms, whether direct or indirect, which may take place simultaneously or sequentially at different plant growth stages (Berg *et al.*, 2002; Mantelin and Touraine, 2004; Haas and Défago, 2005; Bais *et al.*, 2006; Berg and Smalla, 2009; Müller *et al.*, 2009; Compant *et al.*, 2010; Keswani, 2015).

14.1.4 The need for PGPR utilization in agricultural practice

Since the mid-sixties and the emergence of the green revolution, the application of excess N- and P-based fertilizers has led to an unprecedented contamination of soils and waters, leading to harm to ecosystems, causing pollution, and spreading disease; nutrient depletion, soil acidification and eutrophication are also common consequences of inadequate soil management (Hungria and Vargas, 2000; Gerhardson, 2002). In addition to nitrate entering into the soil from biological nitrogen fixation, industrial manufacture of ammonium by the Haber–Bosch process contributes to a considerable increase in soil nitrate concentration. This excess nitrate cannot be removed by denitrification, the biological process by which nitrate (NO_3^-) is reduced to N_2 via the formation of nitrite (NO_2^-), nitric oxide (NO) and nitrous oxide (N_2O), which results in the accumulation of nitrate in soils, waters, and sediments.

More than half the fertilizer applied to soils ends up in rivers, lakes and seas. This large increase in N load in the environment, in turn leads to serious alterations in the cycling of N contributing to eutrophication and massive growth of algae on offshore continental platforms (Diaz and Rosenberg, 2008) and causing severe damage to environmental services at local, regional and global scales (Galloway *et al.*, 2008).

Because most bacteria do not carry out each one of the reduction steps that comprise denitrification, their gaseous intermediates NO and N_2O can be released into the atmosphere. Hence, excess nitrates affects not only terrestrial and marine ecosystems, but also contributes to the release of greenhouse gases involved in climate change (Sutton *et al.*, 2011).

The Earth's natural resources are being overexploited to attend the nutrient demands for an increasingly growing human population, currently ~7 billion and estimated to reach ~8 billion by 2020 (Glick, 2012).

In the current scenario, fertilizers are normally overused in developed countries so that plants are able to reach their current yield potential. Considering that crop yield per hectare has to be doubled by 2020 to try to meet the population's demand for dietary proteins (Ray *et al.*, 2013; Robinson *et al.*, 2014), it is a paradox that utilization of fertilizers cannot continuously increase if the wealth of the planet has to be preserved.

Because PGPR enhance plant growth, their use as biofertilizers, rhizoremediators or phytostimulators could reduce the use of chemical fertilizers and support ecofriendly sustainable food production conditions (van Loon, 2007; Lugtenberg and Kamilova, 2009; Babalola, 2010; Hayat *et al.*, 2010; Bhattacharyya and Jha, 2012; Glick, 2012; Bulgarelli *et al.*, 2013; Hardoim *et al.*, 2015). Although understanding of the PGPR–plant interactions are not yet well resolved, many bacterial species are used commercially as adjuncts to agricultural practice (Glick, 1995, 2012; Podile and Kishore, 2006; Dardanelli *et al.*, 2009; Babalola, 2010; Hayat *et al.* 2010; Bhattacharyya and Jha, 2012; Singh *et al.*, 2014). In this regard, identification and characterization of bacterial traits involved in plant promotion is of interest to continue searching for PGPR. In this chapter we cover methodologies used to qualitatively and quantitatively analyze the common traits that characterize PGPR.

14.2 Determination of PGPR Properties

14.2.1 Nitrogen fixation

Nitrogen (N) is the fourth most abundant element in the biomass, where it is part of essential compounds such as proteins, nucleic acids, hormones, etc. With the exception of water, N is the most common limiting element in agriculture. Although up to about 80% of the air around the earth's surface is N, most of it is found as dinitrogen gas (N_2), a form which is not suitable for plant and animal consumption, thus representing a major constraint to life on our planet. Diazotrophic microorganisms, mainly bacteria, contain the enzyme nitrogenase, which converts inert N_2 gas to bio-available ammonia (NH_4^+). This process is called biological nitrogen fixation and initiates the N cycle in the biosphere. Besides N_2 to NH_4^+, the nitrogenase complex also reduces acetylene, azide, cyanide, nitrous oxide and protons. In most cases, activity of nitrogenase is detected using the acetylene-dependent ethylene production (acetylene reduction activity, ARA) assay (Hardy *et al.*, 1973). Nitrogenase biosynthesis is encoded by the *nif* genes, which in many diazotrophic bacteria are arranged in a single cluster of approximately 20–24 kb with seven separate operons that together encode 20 distinct proteins. The *nifHDK* genes encode the structural components of the molybdenum nitrogenase enzyme complex. The *nifH* gene is responsible for the synthesis of the NifH protein, also designated the Fe-protein of the nitrogenase enzyme and the *nifDK* gene codes for the FeMo protein (Rubio and Ludden, 2008; Curatti and Rubio, 2014). PCR amplification of the *nifH* gene is widely used to assess the capability of a bacterium to fix N_2.

Following the above considerations, the activity of nitrogenase can be determined using three different assays: (1) growth of bacterial cells in N-free media, (2) the ARA assay, and (3) amplification of genes involved in nitrogenase synthesis.

Growth in solid N-free media

Theoretically considered, N_2-fixing bacteria can grow in laboratory media lacking N in their composition. Different media have been described with different nutrient compositions, among them the Burk medium (Wilson and Knight, 1952), Nfb medium, JMV medium, and LIGP are generally used (Döbereiner and Day, 1976; Reis *et al.*, 1994; Reis *et al.*, 2004).

PROCEDURE:

1. Grow the bacterial cells in the medium (e.g. ~2.5 mL) in which they are usually cultured until an optical cell density of ~0.5 is reached. Wash the cells by centrifugation and resuspension of the pellet several times. For that purpose, a microfuge (12,000 rpm for 3 min) can be used. Saline (0.9% NaCl) or phosphate-buffered (5 mM of K_2HPO_4, 5 mM of KH_2PO_4, 150 mM of NaCl, pH 7.0) solutions can be used to wash the cells.

2. Take aliquots (e.g. 100 µL) of the cell suspension to inoculate Petri dishes containing N-free media. Different media can be used to facilitate bacterial growth. Incubate at 30 °C until appearance of colony forming units (CFUs).

3. Check for purity of a CFU by repeated striking onto the same N-free medium they grew first.

NOTES:

All media must be prepared with high-quality products to prevent N contamination so that traces of N cannot be used by N-scavenging bacteria. Because nitrogenase activity is inhibited by oxygen, growth of bacterial cells can be carried out placing the plates into an anaerobic jar and making the internal ambient anoxic either by flushing with N_2, or more expensively Ar or He, for 5–10 min. The commercial reagent Anaerocult, a commercial registered product of Merck, can also be used.

Growth in semisolid N-free media

As an alternative to the above methodology, semisolid N-free media can be used.

PROCEDURE:

The procedure is similar to that above, except that after washing the cells:

1. Prepare dilutions (1:150, 1:250, 1:500 and 1:1000; v:v) to inoculate flasks filled with about 2/3 of a semisolid, usually containing 0.3–0.5% purified agar, N-free medium.

2. Close the flasks hermetically (e.g. by using rubber septa). Incubate at 30°C until a dense cellular film is observed in the subsurface of the medium.

3. Remove the culture medium with a sterile spoon and transfer the bacterial layer to tubes containing sterile saline and 0.5 cm diameter glass beads.

4. Homogenize the cell suspension by vortexing for 1 minute. Then centrifuge in a microfuge (12,000 rpm, 3 min). Supernatants are serially diluted and used for inoculation of Petri dishes containing the different solid N-free medium.

As an alternative to this protocol, after inoculation, the flasks can be flushed with either N_2, Ar or He to establish anoxic conditions from the beginning of the cultures. Flushing time is dependent on the internal volume of the flask; 5–10 min is generally used. The commercial reagent Anaerocult, a commercial registered product of Merck, can also be used.

NOTE:

Growth of a bacterium on solid/semisolid N-free media should be considered as an initial test and never as a definite proof for N_2 fixation.

Acetylene (C_2H_2)-dependent ethylene (C_2H_4) production (acetylene reduction activity (ARA) assay

Besides N_2 to NH_4^+, the nitrogenase complex also reduces acetylene, azide, cyanide, nitrous oxide and protons, yet the ARA assay is the method most used to analyze N_2-fixation capability.

PROCEDURE:

1. Grow the bacterial cells in the medium (e.g. ~2.5 mL) in which they are usually cultured until an optical cell density of ~0.5 is reached. Wash the cells by centrifugation and resuspension of the pellet several times. For that purpose, a microfuge (12,000 rpm for 3 min) can be used. Saline (0.9% NaCl) or phosphate-buffered (5 mM of K_2HPO_4, 5 mM of KH_2PO_4, 150 mM of NaCl, pH 7.0) solutions can be used to wash the cells.

2. Take aliquots (e.g.100 µL) of the cell suspension to inoculate flasks filled with about 2/3 of their internal volume with liquid N-free medium (see previous procedure). Different media can be used to facilitate bacterial growth. Close the flasks hermetically. Volume of the samples and volume of the flasks may vary to optimize the assay. Septa must allow injection and withdrawal of the internal atmosphere of the flasks. As a blank, include non-inoculated flasks.

3. Close the flasks and incubate the cultures at 30°C, with agitation (e.g. 120–150 rpm), until a cell density of about 0.3–0.5 at 600 nm is reached. Lower cell densities can be used.

4. At that time, replace 10% of the internal atmosphere of the flasks by the same volume of acetylene (C_2H_2). Mix well by shaking. Commercial acetylene, or that obtained by mixing calcium carbide (CaC_2) and water (1:15 w:v), can be used.
5. Incubate cultures at 30°C. Incubation time depends on the acetylene reduction activity of the samples.
6. Take gas aliquots from the headspace of the flasks for injection onto a gas chromatograph. Aliquots from 100 to 1000 μL can be used.

NOTES:
Alternatively, after closing the flasks, it is possible to flush the headspace with either N_2, Ar or He to create anoxic conditions. Flushing time is dependent on the internal volume of the flask; 5-10 min is generally used. In these conditions, as a starter to favour bacterial growth, the N-free medium can be supplemented with 0.5-1% fixed nitrogen (e.g. yeast extract).

Kinetic of ethylene (C_2H_4) production can be followed by taking samples over time. Because assays are run in closed systems, gas accumulation may cause feedback inhibition of the enzymatic activity. Accordingly, gas samples must be taken during the exponential phase of gas production.

Ethylene can be determined by gas chromatography using a flame ionization detector (FID). The chromatograph is usually provided with N_2 as a carrier gas, and H_2 and synthetic air to make up the flame. Gas fluxes through the chromatograph as well as through the oven; injector and detector temperatures may vary depending on each chromatograph commercial brand and the type of column used for the chromatography. Concentration of ethylene in each sample can be calculated from standards of pure ethylene. A correction for dissolved ethylene in water (Bunsen solubility coefficient) has to be considered. Values are usually expressed as mol C_2H_4 produced × mg protein × h^{-1}. Gas tight syringes should be used.

PCR amplification of nif *genes*

Because the presence of *nif* genes in the bacterial genome is indicative of the ability to fix N_2, amplification of any nitrogen fixation-related gene can be used to assess this capacity. *nifH* is widely used for that purpose because it is evolutionarily conserved and, thus, of great value for detection and identification of diazotrophs by cultivation-independent methods. Several primers and PCR conditions have been described for PCR amplification of the *nifH* gene; here we will refer to amplification of the gene as described by Gaby and Buckley (2012).

PROCEDURE:
In a microtube, mix 2.5 μL of PCR buffer 1X, 2.5 mM $MgCl_2$, 0.2 mM dNTPs (a mixture of dATP, dCTP, dGTP and dTTP), 0.05% BSA, 1 mM of each primer (forward and reverse), 1 to 5 ng of genomic DNA, 2.5 U (unit) high fidelity Taq polymerase, and complete up to 50 μL with ultrapure water.

NOTES:
Separate the PCR products by electrophoresis on 1% agarose gels in 0.5X TBE buffer (Trizma Base, 10g/L; boric acid, 5.5 g/L; EDTA, 0.9 g/L; pH 8.5) at 80V. Samples (4 μL) can be supplemented with 1 μL loading buffer (40% sucrose and 0.25% bromophenol blue) before loading the samples on the gel. After electrophoresis, DNA is stained with either ethidium bromide, GelRed, Sybr Green I, etc., and visualized under UV light.

Readers may refer to Ueda *et al.* (1995) and Widmer *et al.* (1999) to learn more on primers and reaction mixtures and conditions used for PCR-amplification of *nif* genes.

14.2.2 Phosphate solubilization

Phosphorus (P) is one of the most essential plant nutrients and profoundly affects the overall growth of plants (Wang *et al.*, 2009) by influencing various key metabolic processes such as cell division and development, energy transport, signal transduction, macromolecular biosynthesis, photosynthesis and respiration of plants (Ahemad *et al.*, 2009; Khan *et al.*, 2009). In general, the concentration of available P in soil is very low, typically at levels ranging between 5 and 30 mg/kg, due to the fact that soluble P reacts with ions such as calcium, iron or aluminium causing precipitation or fixation and, consequently, reducing its availability to plants (Peix *et al.*, 2003; Vyas and Gulati, 2009). Inorganic phosphates, which are applied as chemical fertilizers, are also immobilized on the soil matrix and, therefore, cannot be used for crops (Atlas and Bartha, 1997; Rizvi *et al.*, 2014). Of the various strategies adopted by microbes to cause the solubilization of insoluble P, the involvement of low molecular mass organic acids secreted by microorganisms has been a well recognized and widely accepted theory, and there are also reports which suggest that insoluble P could be transformed into soluble forms by chelation and reduction processes (Asea *et al.*, 1988; Illmer and Schinner, 1992; Altomare *et al.*, 1999; Chen *et al.*, 2006). Whatever the mechanism, microorganisms capable of solubilizing phosphates play a key role in optimizing the availability of P for the plant, which could result in increased crop yields in non-stressed and stressed plants (Rodríguez and Fraga 1999; Igual *et al.*, 2001; Chen *et al.*, 2006; Qureshi *et al.*, 2012; Panwar *et al.*, 2014). One of the important attributes of PGPR is phosphate solubilization and the group of microorganisms capable of converting inorganic P into soluble forms is known as P-solubilizing microorganisms (Khan *et al.*, 2007). Both qualitative and quantitative assays can be performed to study bacterial capability to solubilize phosphates.

Qualitative assay

PROCEDURE:

1. Grow the bacterial cells in the medium (eg. ~2.5 mL) in which they are usually cultured until an optical cell density of ~ 0.5 is reached. Wash the cells by centrifugation and resuspension of the pellet several times. For that purpose, a microfuge (12,000 rpm for 3 min) can be used. Saline (0.9% NaCl) or phosphate-buffered (5 mM of K_2HPO_4, 5 mM of KH_2PO_4, 150 mM of NaCl, pH 7.0) solutions can be used to wash the cells.

2. Take aliquots (2.5 μL–10 μL) of the cell suspension and inoculate Petri dishes containing any of the following media: PVK (Pikovskaya, 1948), NBRIP (Mehta and Nautiyal, 2001), YED-P (Peix *et al.*, 2001), Ashby (Abdel-Malek and Ishac, 1968), and yeast extract mannitol (YEM) medium (Vincent, 1970) supplemented with 0.2% of $Ca_3(PO_4)_2$. Volume of the samples may vary to optimize the assay. Rock phosphate (0.2%) can be used instead of $Ca_3(PO_4)_2$.

3. Incubate at 30°C. Appearance of a clear halo around the UFC indicates solubilization of phosphate.

4. Measure the diameter of the halo during the incubation time. Allow the halo to develop until it comes to a stop.

5. A solubilization index (SI) and the solubilizing efficiency (SE) can be calculated according to Premono *et al.* (1996):

SI = (diameter of the colony + diameter of the halo)/diameter of the colony

SE = (diameter of the halo/diameter of the colony) × 100

NOTE:

Avoid bubble formation during pouring of the medium into the Petri dishes. Carefully shake the plate until complete gelation of the medium to avoid phosphate precipitation.

Quantitative assay

1. Grow the bacterial cells in the medium (e.g. ~2.5 mL) in which they are usually cultured until an optical cell density of ~0.5 is reached. Wash the cells by centrifugation and resuspension of the pellet several times. For that purpose, a microfuge (12,000 rpm for 3 min) can be used. Saline (0.9% NaCl) or phosphate buffered (5 mM of K_2HPO_4, 5 mM of KH_2PO_4, 150 mM of NaCl, pH 7.0) solutions can be used to wash the cells.

2. Inoculate (1:100; v:v) flasks containing either PVK, NBRIP, YED-P, Ashby or YEM medium supplemented with 0.2% of $Ca_3(PO_4)_2$ or rock phosphate. Volume of the aliquots and volume of the flasks may vary to optimize the assay. Growth media can be supplemented with a colorimetric pH change indicator (e.g. 0.025% bromophenol blue).

3. Incubate the cultures at 30°C with agitation (120–150 rpm). Incubation time can be extended as required.
4. Take aliquots (e.g. 2–3 mL) every 1–3 d and centrifuge them. For that purpose a microfuge can be used (12,000 rpm for 3 min). Sampling time is dependent on the growth rate of the bacterium to be analyzed.
5. Recover the supernatant.
6. Measure the pH of the supernatant using a pH meter.

Because the pH is proportional to the amount of phosphate that has been solubilized, a standard curve relating pH and phosphate content has to be constructed.

Determination of phosphate concentration can be done according to Murphy and Riley (1962) as follows:

Solution A: Weigh 5 g of $[(NH_4)_6Mo_7O_{24} \times 4H_2O]$ and add distilled water to complete up to 50 mL.
Solution B: Add 14 mL of concentrated H_2SO_4 (density = 1.82) to 90 mL of distilled water.
Solution C: Weigh 2.7 g of $C_6H_8O_6$ (ascorbic acid) and add distilled water to complete up to 50 mL.
Solution D: Weigh 0.034g of $C_4H_2KO_6Sb_6 \times 1.5H_2O$ (antimony potassium tartrate) and add distilled water to complete up to 25 mL.
Solution E: Mix 10 mL of solution A, 25 mL of solution B, 10 mL of solution C and 5 mL of solution D. Mix 30 μL of solution E with 250 μL of aliquots taken from the bacterial supernatants in the above point 5.

Incubate 10 min at room temperature. Determine the absorbance at 665 nm.

Phosphate content is estimated by comparison with a standard curve prepared using serial dilutions (e.g. 1:150, 1:250, 1:500 and 1:1000; v:v) of 6 mM KH_2PO_4 (250 μL) mixed with solution E (30 μL).

14.2.3 Siderophores production

Iron is a vital nutrient for almost all forms of life as it plays a key role in cellular processes such as electron transport, activation of oxygen, peroxide reduction, amino acid and nucleoside synthesis, DNA synthesis, photosynthesis, etc. (Neilands, 1995). Iron requirement for the optimal growth of plants is about 10^{-9} M and that for microbes is in the range of 10^{-7} to10^{-5} M (Raymond *et al.*, 2003), both of which are far greater than the biological availability which is 10^{-17} M at physiological pH 7.0. Despite its abundance, most iron in soils is found in the form of ferro-magnesium silicates or iron oxides and hydroxides, which are not readily utilizable by microorganisms and plants (Rajkumar *et al.*, 2010). Also, alkaline conditions prevent iron dissolution in the soil water and render it unavailable to plants and microorganisms. Therefore, there is always a kind of iron-stressed condition prevalent in most soils (Desai and Archana, 2011). To overcome the low bioavailability of iron, microorganisms, mainly bacteria and fungi, can synthesize and secrete low molecular weight iron-specific chelators known as siderophores. Different microorganisms produce different types of siderophores which are characterized by possessing iron-chelating functional groups with high affinity for ferric iron. This, in turn, results in formation of complexes with great thermodynamic stability. Production of siderophores by PGPR can enhance plant growth by controlling growth of phytopathogens, as chelation of iron in the rhizosphere makes it unavailable to pathogenic bacteria (Desai and Archana, 2011; Panwar *et al.*, 2014). The types and chemistry of siderophores and their role in crop improvement have been comprehensively reviewed in Desai and Archana (2011).

Because detection of all possible known forms of siderophores would require numerous assays, a universal methodology was developed by Schwyn and Neilands (1987) based on utilization of chrome azurol S (CAS) and hexadecyl-trimethyl-ammonium bromide (HDTMA) as indicators with the modification of Alexander and Zuberer (1991) and Payne (1994).The CAS-HDTMA mixture forms a tight complex with ferric iron to produce a blue color. When a strong iron chelator such as a siderophore removes iron from the dye complex, the color changes from blue to orange.

Qualitative assay

PROCEDURE:

1. Preparation of the CAS agar medium.

The CAS agar medium is formed by a mixture of 4 solutions:

Solution MM9: Weigh 1 g of NH_4Cl, 0.5g of NaCl, 0.3 g of KH_2PO_4, 0.25 g of $MgSO_4$, 0.011 g of $CaCl_2$ and add distilled water to complete up to 1L. Remove any traces of iron in the MM9 solution as follows: mix 1 L of solution MM9 with 1g of 8-hydroxyquinoline diluted in 10 mL of chloroform and stir (120–150 rpm) the mixture for 36 h at 4°C. Then, add 20 mL of chloroform and mix again by vigorous stirring. Take the aqueous phase and wash it with 20 mL of chloroform. Finally, remove the residual chloroform by heating at 45°C with stirring for 3 min.

Solution 1: To 50 mL of solution MM9, add 5 g of $C_5H_9NO_4$ (glutamic acid) and 3 g of casamino acids, and complete with distilled water up to 100 mL. Remove residual iron as indicated above. Adjust the pH to 6.5. Sterilize by filtration.

Solution 2: CAS solution: Mix 10 mL of $FeCl_3 \times 6H_2O$ diluted in HCl 10 mM, 50 mL of chrome azurol S (CAS) (1.21 mg/mL), and 40 mL of $C_{19}H_{42}BrN$ (HDTMA) (1.82 mg/mL). The components of solution 2 must be mixed in the order shown above. The HDTMA solution must be added very slowly, under continuous stirring, to prevent formation of aggregates. Sterilize by autoclave (120 °C for 20 min).

Solution 3: To 50 mL of solution MM9 add 30.24 g of piperazine-N,N'-bis(2-ethanesulfonic acid) (PIPES), 15 g of agar and complete with distilled water up to 800 L. Adjust the pH to 6.5. Sterilize by autoclave.

To prepare the CAS agar medium, mix (v:v) solution 1 with solution 3. Then slowly add solution 2. The appearance of a blue–gray color should be noticed after addition of solution 2.

2. Pour the CAS agar medium into Petri dishes carefully to avoid bubble formation.

3. Grow the bacterial cells in the medium (e.g. ~2.5 mL) in which they are usually cultured until an optical cell density of ~0.5 is reached. Wash the cells by centrifugation and resuspension of the pellet several times. For that purpose, a microfuge (12,000 rpm for 3 min) can be used. Saline (0.9% NaCl) or phosphate-buffered (5 mM of K_2HPO_4, 5 mM of KH_2PO_4, 150 mM of NaCl, pH 7.0) solutions can be used to wash the cells.

4. Take aliquots (2.5–10 μL) of the cell suspension and deposit them as drops onto Petri dishes containing CAS agar medium.

5. Incubate at 30°C until appearance of a yellow–orange halo around the bacterial UFC.

6. Measure the diameter of the halo during the incubation time. Allow the halo to develop until it comes to a stop.

7. A siderophore production index (SI) and siderophore production efficiency (SE) can be calculated according to Premono *et al.* (1996) (see the qualitative assay in §14.2.2 above).

NOTES:

HDTMA can be toxic for Gram-positive bacteria. Because it is possible that a given bacterium is unable to grow in CAS medium, other modified CAS assays have been developed which allow bacterial growth (Ames-Gottfred *et al.*, 1989; Machuca and Milagres, 2003; Pérez-Miranda *et al.*, 2007; Castellano-Hinojosa *et al.*, 2015). The O-CAS method is widely used in which bacterial cells are grown on the solid media on which they are usually cultured and then overlaid with a modified CAS solution (Pérez-Miranda *et al.*, 2007).

Quantitative assay

PROCEDURE:

1. Grow the bacterial cells in the medium (e.g. ~2.5 mL) in which they are usually cultured until an optical cell density of ~0.5 is reached. Wash the cells by centrifugation and resuspension of the pellet several times. For that purpose, a microfuge (12,000 rpm for 3 min) can be used. Saline (0.9% NaCl) or phosphate-buffered (5 mM of K_2HPO_4, 5 mM of KH_2PO_4, 150 mM of NaCl, pH 7.0) solutions can be used to wash the cells. Other defined media can be used to grow the cells, among them M9 (Miller, 1972), SM (Meyer

and Abdallah, 1978), Bergersen (Bergersen, 1961) and Modi (Modi *et al.*, 1985). Excess phosphate or other weak iron chelators should be avoided in the composition of the media as they interfere with the reaction.

2. Take aliquots of the supernatants and mix (1:1; v:v) with solution 2. Mix by vortexing for 20 s.

3. Incubate the mixture at room temperature, in the dark, for 20 min. Before incubation of the mixtures, a solution of 0.2 M 5-sulfosalicylic acid can be added (10:1; v:v) to facilitate transfer of iron from the solution 2 to bacterial siderophores.

4. Measure the absorbance at 630 nm. A mixture (1:1; v:v) of solution 2 and the medium used for bacterial growth can be used as a reference.

Siderophore production can be estimated as indicated earlier by Castellano-Hinojosa *et al.* (2015):

$$\% \text{ siderophores units} = [(A_r - A_s)/A_r]/100$$

where A_r = absorbance of the reference solution, and A_s = absorbance of the sample.

14.2.4 Indole acetic acid production

Tien *et al.* (1979) first suggested that rhizobacteria could enhance plant growth by phytohormones excretion. Among them, auxin is the generic name given to represent a group of chemical compounds associated in plants with processes such as gravitropism and phototropism, vascular tissue differentiation, apical dominance, lateral and adventitious root initiation, and stimulation of cell division and stem and root elongation (Ross *et al.*, 2000). In addition to indole acetic acid (IAA) other auxin-like molecules have been described as a product of bacterial metabolism such as indole-3-butyric acid (IBA) (Costacurta *et al.*, 1994; Costacurta and Vanderleyden, 1995), indole-3-lactic acid (ILA) (Crozier *et al.*, 1988), indole-3-acetamide (IAM) (Hartmann *et al.*, 1983), indole-3-acetaldehyde (Costacurta *et al.*, 1994), indole-3-ethanol and indole-3-methanol (Crozier *et al.*, 1988), tryptamine, anthranilate and other yet uncharacterized indolic compounds whose physiological function remains unknown (for a review see Cassán *et al.*, 2011; Khan *et al.*, 2014)

Quantitative assay of intrinsic IAA and IAA-related compounds

IAA and IAA-related compounds produced by PGPR can be quantitatively evaluated by the method of Gordon and Weber (1951), later modified by Bric *et al.* (1991) and Gravel *et al.* (2007).

PROCEDURE:

1. Grow the bacterial cells in the medium (e.g. ~2.5 mL) in which they are usually cultured until an optical cell density of ~0.5 is reached. Wash the cells by centrifugation and resuspension of the pellet several times. For that purpose, a microfuge (12,000 rpm for 3 min) can be used. Saline (0.9% NaCl) or phosphate-buffered (5 mM of K_2HPO_4, 5 mM of KH_2PO_4, 150 mM of NaCl, pH 7.0) solutions can be used to wash the cells.

2. Inoculate (1:100; v:v) flasks containing a defined medium for bacterial growth. In the absence of an appropriate growth medium, cells can be grown in complete media. Volume of the aliquots and volume of the flasks may vary to optimize the assay. Defined media M9, SM, Bergersen or Modi can also be used for bacterial growth.

3. Incubate at 30°C with agitation (120–150 rpm) for 14 d. Incubation time can be extended as required.

4. Take aliquots (e.g. 2–3 mL) of the cultures and centrifuge them. For that purpose a microfuge (12,000 rpm for 2 min) can be used. Sampling time is dependent on the growth rate of the bacterium to be analyzed.

5. Recover the supernatant.

6. Mix the supernatant (1:2; v:v) with Salkowski reagent (150 mL of $HClO_4$, 250 mL of distilled water and 7.5 mL of 0.5M $FeCl_3 \times 6H_2O$. Incubate at room temperature in darkness for 20 min.

7. Measure the absorbance at 535 nm.

8. Auxin content is estimated as IAA equivalents by comparison with a standard curve prepared using serial dilutions of 50 µg/mL IAA.

Quantitative determination of potential IAA and IAA-related compounds

Because intrinsic production of IAA and IAA-related compounds can be very low, the capacity of a strain to produce auxin-like compounds can be analyzed studying the effect of L-tryptophan on IAA production. The procedure to follow is as indicated for the quantitative assay above, except that IAA production is evaluated after addition of L-tryptophan to the growth medium (e.g. 50, 100, 200, 400, and 600 mg/mL).

14.2.5 1-aminocyclopropane-1-carboxylate (ACC) deaminase activity

Ethylene is a plant hormone which plays an important role in root initiation and elongation, abscission and ripening, senescence, legume nodulation and in stress signalling (Glick *et al.*, 2007). Exogenous application of ethylene causes adventitious root formation and root hair initiation as well as fruit ripening, flower wilting and leaf senescence. Regarding the production of endogenous ethylene during the plant developmental processes, ethylene is involved in xylem formation and flowering in some plants; it also induces fruit ripening and flower wilting. In plants subjected to stress conditions, production of ethylene inhibits root elongation, nodulation and auxin transport; it also causes hypertrophies, accelerates aging and provokes senescence and abscission.

Many PGPR may promote plant growth by lowering the levels of ethylene in plants (Glick *et al.*, 2007). This is attributed to the activity of the enzyme 1-aminocyclopropane-1-carboxylate (ACC) deaminase, which hydrolyzes ACC, the immediate biosynthesis precursor of ethylene in plants (Yang and Hoffman, 1984; Glick, 1995). Bacteria that possess this enzyme can cleave ACC to ammonia and α-ketobutyrate which, in turn, can be used by bacteria as a source of nitrogen and carbon for growth (Klee *et al.*, 1991). In this sense, PGPR can act as a sink for ACC by lowering ethylene levels in plants and, consequently, preventing ethylene accumulation and its toxic effects (Glick *et al.*, 1998, 2007; Steenhoudt and Vanderleyden, 2000; Saleem *et al.*, 2007).

Qualitative assay

Ability of bacterial strains to hydrolyze ACC is usually assayed following the methodology described by Honma and Shimomura (1978) with the modifications introduced by Penrose and Glick (2003).

PROCEDURE:

1. Grow the bacterial cells in the medium (e.g. ~2.5 mL) in which they are usually cultured until an optical cell density of ~0.5 is reached. Wash the cells by centrifugation and resuspension of the pellet several times. For that purpose, a microfuge (12,000 rpm for 3 min) can be used. Saline (0.9% NaCl) or phosphate-buffered (5 mM of K_2HPO_4, 5 mM of KH_2PO_4, 150 mM of NaCl, pH 7.0) solutions can be used to wash the cells.

2. Take aliquots (e.g. 2.5–5 µL) and deposit them as a drop onto Petri dishes containing DF salts (Dworkin and Foster, 1958) defined medium supplemented with 3 mM ACC as the sole nitrogen source. Other defined media can be used to grow the cells, among them M9, SM, Bergersen and Modi supplemented with 3 mM ACC.

3. Incubate at 30°C. Appearance of bacterial growth is indicative of ACC deaminase activity.

4. Take a CFU with an inoculating loop and inoculate flasks containing DF salts medium (e.g. 2.5 mL) supplemented with 3 mM ACC.

5. Incubate at 30°C, with agitation (120–150 rpm) until they reach a 0.3–0.5 optical density at 600 nm. Then, inoculate (1:100; v:v) flasks containing 50 mL of DF salts medium supplemented with 3 mM ACC and incubate again.

6. After growth, centrifuge the flask in a microfuge (6000 rpm for 10 min) at 4°C.

7. Remove the supernatant and wash the cells three times using 5 mL of 0.1 M Tris-HCl, pH.

8. Centrifuge the flask in a microfuge (6000 rpm for 10 min) at 4°C .

9. Remove the supernatant and keep the flask containing the cell pellet at –20°C until use.

Quantitative assay

1. Resuspend the cell pellet kept at –20°C in 1 mL of 0.1 M Tris-HCl, pH 7.6 and transfer the cell suspension to a new microtube.
2. Centrifuge at 12,000 rpm for 2 min. Remove the supernatant and resuspend the cell pellet in 600 μL of 0.1M Tris-HCl, pH 8.5.
3. Add 30 μL of toluene and mix by vortexing for 30 s at maximum frequency. Then, transfer 200 μL toluenized cells to a new microtube.
4. Add 20 μL of filter-sterilized 0.5 M ACC and mix by vortexing. Incubate the cell suspensions at 30°C for 15 min. Keep aliquots (e.g. 100 μL) of the toluenized cells to determine protein concentration. Protein content can be determined according to Lowry *et al.* (1951) or Bradford (1976).
5. Add 1 mL of 0.56 M HCl, mix briefly by vortex and centrifuge at 13,000 rpm for 3 min. Take 1 mL of supernatant and add 800 μL of HCl 0.56 M. Shake manually by inversion.
6. Add 300 μL of 2,4-dinitrophenylhydrazine (0.2% of 2,4-dinitrophenylhydrazine in 2M of HCl). Incubate at 30°C for 30 min.
7. Add 2 mL of 2N NaOH to the microtubes (including the blank, see below), mix well by vortex and let the mixtures stand for 30 min.
8. Measure the absorbance of the mixtures at 540 nm. A mixture made of 600 μL of 0.1 M Tris-HCl, pH 8.5, 30 μL of toluene, 1 mL of 0.56 M HCl and 300 μL of 2,4-dinitrophenylhydrazine can be used as a reference.

ACC activity is estimated by comparison with a standard curve prepared using serial dilutions (e.g. ranging from 0.1 to 1 mmol) of 10 mM α-ketobutyrate prepared in 0.1 M Tris-HCl, pH 8.5, and stored at 4°C. Activity can be expressed as mol of α-ketobutyrate $mg^{-1} h^{-1}$.

14.2.6 PGPR as biocontrol agent

Plant diseases are responsible for high-value economic losses. Because the use of agrochemicals is negatively perceived by consumers and is increasingly banned by governmental policies, the use of microbes is an environment-friendly approach in the fight against plant diseases. Biological control, or biocontrol, is the process of suppressing pathogenic living organisms by using other living organisms, and is extensively studied under laboratory and field conditions leading to formulation of commercial products. Competition for nutrients, niche exclusion, induced systemic resistance, lytic enzyme production and production of antimicrobial, generally fungi, compounds are main mechanisms involved in biocontrol activity (Bloemberg and Lugtenberg, 2001). Bacteria involved in biocontrol, their mechanism of action and applications have been dealt with in Chernin and Chet (2002), Compant *et al.* (2005), Lugtenberg and Kamilova (2009), Babalola (2010), Hayat *et al.* (2010), Saharan and Nehra (2011), Bhattacharyya and Jha (2012), Glick (2012), and Bulgarelli *et al.* (2013). Here we present protocols to assay some traits involved in biocontrol such as antagonism, antibiosis, hydrogen cyanide production, exo-polysaccharides production and lytic enzymes (cellulase, pectinase and chitinase production).

Antagonism

PROCEDURE:

1. Grow the fungal pathogen under study in the appropriate culture medium (e.g. potato dextrose agar, PDA) at the corresponding temperature, usually 25°C.
2. Grow the bacterial cells in the medium (e.g. ~2.5 mL) in which they are usually cultured until an optical cell density of ~0.5 is reached. Wash the cells by centrifugation and resuspension of the pellet several times. For that purpose, a microfuge (12,000 rpm for 3 min) can be used. Saline (0.9% NaCl) or phosphate-buffered (5 mM of K_2HPO_4, 5 mM of KH_2PO_4, 150 mM of NaCl, pH 7.0) solutions can be used to wash the cells.
3. Fill Petri dishes with PDA medium. With the help of a Drigalsky's spatula, spread (e.g. 100 μL) the bacterial cells on the plates. Incubate at 30°C until bacterial growth is apparent.
4. Using a sterile cork borer take 1 cm diameter agar plug from the PDA medium containing the fungus and place it in the

middle of the plate inoculated the day before with the bacterial strain. As a control, use PDA plates inoculated only with the fungus. Incubate at the appropriate temperature for fungal growth.

5. Measure the diameter of the fungal mycelium zone every 2 d. Continue incubation until growth of the halo comes to a stop. Percentage of inhibition can be determined according to Landa *et al.* (1997):

$$\% \text{ Inhibition} = [(G_c - G_s)/G_c] \times 100$$

where G_c = diameter of the fungal mycelium in plates not inoculated with the bacterial culture; G_s = diameter of the fungal mycelium in plates inoculated with the bacterial culture.

Antibiosis

PROCEDURE:

1. Grow the fungal pathogen under study in the appropriate culture medium (e.g. potato dextrose agar, PDA) at the corresponding temperature, usually 25°C.

2. Grow the bacterial cells in the medium (eg. ~2.5 mL) in which they are usually cultured until an optical cell density of ~0.5 is reached. Wash the cells by centrifugation and resuspension of the pellet several times. For that purpose, a microfuge (12,000 rpm for 3 min) can be used. Saline (0.9% NaCl) or phosphate-buffered (5 mM of K_2HPO_4, 5 mM of KH_2PO_4, 150 mM of NaCl, pH 7.0) solutions can be used to wash the cells.

3. Fill Petri dishes with solid PDA medium.

4. Using a sterile cork borer, take 1 cm diameter agar plug from the PDA medium containing the fungus and place it near the border of the plate. Then, with the help of an inoculating loop, spread the bacterial cell suspension onto the opposite side of the plate following a straight line. As a control, use PDA plates inoculated only with the fungus. Incubate at the appropriate temperature for fungal growth.

5. Measure the diameter of the fungal mycelium every 2 d. Continue incubation until growth of the mycelium comes to a stop. The inhibitory effect of the bacterial strain on fungal growth can be determined as above (see Antagonism procedure).

Hydrogen cyanide production

Hydrogen cyanide (HCN) is a gas known to negatively affect root metabolism and root growth by inhibiting cytochrome oxidase respiration (Schippers *et al.*, 1990). This volatile metabolite can inhibit growth of plant pathogens (Voisard *et al.*, 1989) and thereby suppress diseases (Glick, 1995). HCN can be produced during the early stationary growth phase (Knowles and Bunch, 1986) by several bacteria by oxidative decarboxylation pathway using glycine, glutamate or methionine as precursors (Curl and Truelove, 1986). HCN production can be determined according to Bakker and Schipper (1987).

PROCEDURE:

1. Grow the bacterial cells in the solid medium in which they are usually cultured, supplemented with 0.4% of glycine, until appearance of bacterial UFC.

2. Dip a disc of sterile Whatman filter paper no. 1 into 0.5% picric acid supplemented with 2% of Na_2CO_3.

3. Place the disc on the lid of the Petri dishes and seal them with parafilm paper.

4. Incubate at 30°C. A change from white to orange–brown colour of the filter paper is indicative of HCN production.

NOTE:

Avoid direct contact between the filter paper and the bacterial culture.

Exo-polysaccharides production

Exo-polysaccharides (EPS) are carbohydrate polymers secreted by a wide variety of bacteria. They can remain associated to the cell wall to form a bound capsule layer or can be released to the cell surroundings as extracellular slime (Glick *et al.*, 1999). EPS have vital roles in a variety of processes such as biofilm formation (Bhaskar and Bhosle, 2005), protection of bacterial cell from desiccation (Pal *et al.*, 1999), maintenance of primary cellular functions, antibacterial activity against predators, gelling ability and pollutant degradation kinetics (Fusconi and Godinho, 2002), and bioremediation activity and plasma substituting capacity (Allison, 1998). EPS production can be determined according to Mody and Modi (1987).

PROCEDURE:

1. Grow the bacterial cells in the medium (e.g. ~5–10 mL) in which they are usually cultured supplemented with 5% of sucrose until an optical cell density of ~0.5 is reached. Incubate at 30°C with agitation (e.g. 120–150 rpm).
2. Centrifuge the cells at 12,000 rpm for 3 min. Transfer the supernatant to a new microtube.
3. Mix the supernatant with chilled acetone (CH_3COCH_3) (1:3; v:v) and incubate overnight at 4°C.
4. Centrifuge the cells at 12,000 rpm for 30 min and remove the supernatant.
5. Wash the precipitated EPS repeatedly three times alternately with distilled water and acetone.
6. Transfer the precipitated EPS to a Whatman filter paper no. 1, dry it overnight at room temperature and finally weigh the filter paper.

NOTES:

Other organic solvents such as absolute ethanol, propanol and isopropanol can also be used for EPS precipitation. After obtaining the precipitated EPS, the carbohydrate content can be estimated using the phenol–sulfuric acid method (Dubois *et al.*, 1956).

Lytic enzyme production

PGPRs produce enzymes, among them cellulase, pectinases and chitinase, that can lyse the cell walls of many pathogenic microorganisms. By hydrolyzing cellulose, pectins and chitin, they play a pivotal role in suppression of plant pathogens (Chernin and Chet, 2002; Kamensky *et al.*, 2003; Ovadis *et al.*, 2004; Kim *et al.*, 2008).

Cellulase. The methodology described by Kasana *et al.* (2008) is widely used.

PROCEDURE:

1. Grow the bacterial cells in the medium (e.g. ~2.5 mL) in which they are usually cultured until an optical cell density of ~0.5 is reached. Wash the cells by centrifugation and resuspension of the pellet several times. For that purpose, a microfuge (12,000 rpm for 3 min) can be used. Saline (0.9% NaCl) or phosphate-buffered (5 mM of K_2HPO_4, 5 mM of KH_2PO_4, 150 mM of NaCl, pH 7.0) solutions can be used to wash the cells.
2. Take aliquots (e.g. 5–10μL) and deposit them in the middle of a Petri dish containing carboxy methyl cellulose (CMC) agar medium (Kasana *et al.*, 2008). Incubate the cultures at 30°C until appearance of bacterial UFC. Because it is possible that a given bacterium is unable to grow in CMC medium, other modified CMC media have been developed to allow bacterial growth (Hankin and Anagnostakis, 1977).
3. Three alternatives can be used to detect the production of cellulose. (a) Flood the Petri dish with 1% of hexadecyltrimethyl ammonium bromide (HAB). Incubate at room temperature for 30 min. (b) Flood the Petri dish with 0.1% of Congo red. Incubate 20 min at room temperature. Remove the liquid with a pipette and flood the dish with 1 M of NaCl. Incubate at room temperature for 20 min. (c) Flood the Petri dish with iodine solution (2 g KI and 1 g iodine in 300 mL of distilled water). Incubate at room temperature for 5 min. Regardless of the method used, remove liquids with a pipette.
4. Appearance of unstained areas indicates degradation of CMC due to cellulose production by the bacterial strain.
5. Measure the diameter of the clear halo. A cellulase production index (SI) and cellulase production efficiency (SE) can be calculated according to Premono *et al.* (1996) (see the qualitative assay in §14.2.2 above).

Pectinase. Pectinase activity can be assayed as indicated by Yogesh *et al.* (2009)

PROCEDURE:

1. Grow the bacterial cells in the medium (e.g. ~2.5 mL) in which they are usually cultured

until an optical cell density of ~ 0.5 is reached. Wash the cells by centrifugation and resuspension of the pellet several times. For that purpose, a microfuge (12,000 rpm for 3 min) can be used. Saline (0.9% NaCl) or phosphate-buffered (5 mM of K_2HPO_4, 5 mM of KH_2PO_4, 150 mM of NaCl, pH 7.0) solutions can be used to wash the cells.

2. Dip a disc of sterile Whatman filter paper no. 1 into the microbial culture and place it onto a Petri dish containing pectinase agar (PEC) medium (Yogesh *et al.*, 2009). Incubate the cultures at 30°C until appearance of bacterial UFC. Because it is possible that a given bacterium is unable to grow in PEC medium, other media have been developed to allow bacterial growth, among them, the MS (Gerhardt *et al.*, 1994) and PSAM (Yogesh *et al.*, 2009) media.

3. Flood the plate with 50 mM of iodine solution (as in the third part of the procedure for cellulose set out above). Incubate at room temperature for 15 min.

4. Remove the liquid with a pipette.

5. Appearance of clear halos indicates the ability of the strain to produce pectinase.

6. Measure the diameter of the clear halo. A pectinase production index (SI) and pectinase production efficiency (SE) can be calculated according to Premono *et al.* (1996) (see the qualitative assay in §14.2.2 above).

Chitinase. Chitinase production can be performed following the protocol published by Dunne *et al.* (1997).

PROCEDURE:

1. Grow the bacterial cells in the medium (e.g. ~2.5 mL) in which they are usually cultured until an optical cell density of ~0.5 is reached. Wash the cells by centrifugation and resuspension of the pellet several times. For that purpose, a microfuge (12,000 rpm for 3 min) can be used. Saline (0.9% NaCl) or phosphate-buffered (5 mM of K_2HPO_4, 5 mM of KH_2PO_4, 150 mM of NaCl, pH 7.0) solutions can be used to wash the cells.

2. Take aliquots (e.g. 5–10 μL) and deposit them in the middle of a Petri dish containing solid chitin minimal agar (CMM) medium supplemented with 1.6% of colloidal chitin as the sole source of carbon (Dunne *et al.*, 1997). Incubate the cultures at 30°C until appearance of a clear halo around the bacterial cells. Because it is possible that a given bacterium is unable to grow in CMM medium, other media have been developed to allow bacterial growth, among them, CMM modified medium is commonly used (Kuddus and Ahmad, 2013).

3. Measure the diameter of the clear halo. A chitinase production index (SI) and chitinase production efficiency (SE) can be calculated according to Premono *et al.* (1996) (see the qualitative assay in §14.2.2 above).

14.3 Conclusion

Utilization of reference strains during assays to analyze traits involved in PGPR activity is recommended. *Brevibacillus brevis* strain BEA1 and *Azospirillum brasilense* strain C16 (Moreno *et al.*, 2009; Cárdenas *et al.*, 2010) are usually employed as models when assessing IAA/IAA-related compounds and phosphate solubilization, respectively. To date, however, there is no general agreement about the bacterium that should be used as the type strain for a given PGPR property. Efforts should be made to select reference strains to be used worldwide.

Acknowledgements

This study was supported by the ERDF-co-financed grants AGR2012-1968 from Consejería de Economía y Conocimiento de la Junta de Andalucía, Spain, and the MINECO-CSIC Agreement for the RECUPERA 2020 Project.

References

Abdel-Malek, Y. and Ishac, Y.Z. (1968) Evaluation of methods used in counting *Azotobacter*. *Journal of Applied Microbiology* 31, 269–275.

Ahemad, M. and Kibret, M. (2014) Mechanisms and applications of plant growth promoting rhizobacteria: current perspective. *Journal of King Saud University – Science* 26, 1–20.

Ahemad, M., Zaidi, A., Khan, M.S. and Oves, M. (2009) Biological importance of phosphorus and phosphate solubilizing microbes. In: Khan, M.S. and Zaidi, A. (ed.) *Phosphate Solubilizing Microbes for Crop Improvement*. Nova Science Publishers, New York, pp. 1–14.

Alexander, D.B. and Zuberer, D.A. (1991) Use of chrome azurol S reagents to evaluate siderophore production by rhizosphere bacteria. *Biology and Fertility of Soils* 12, 39–45.

Allison, A.G. (1998) Exopolysaccharide production in bacterial biofilm. *Biofilm Journal* 3, 1–19.

Altomare, C., Norvell, W.A., Borjkman, T. and Harman, G.E. (1999) Solubilization of phosphates and micronutrients by the plant growth promoting and biocontrol fungus *Trichoderma harzianum* Rifai. *Applied and Environmental Microbiology* 65, 2926–2933.

Ames-Gottfred, N.P., Christie, B.R. and Jordan, D.C. (1989) Use of the Chrome Azurol S agar plate technique to differentiate strains and field isolates of *Rhizobium leguminosarum* biovar *trifolii*. *Applied and Environmental Microbiology* 55, 707–710.

Asea, P.E.A., Kucey, R.M.N. and Stewart, J.W.B. (1988) Inorganic phosphate solubilization by two *Penicillium* species in solution culture and soil. *Soil Biology and Biochemistry* 20, 459–464.

Atlas, R. and Bartha, R. (1997) *Microbial Ecology*. Addison Wesley Longman, New York.

Babalola, O.O. (2010) Beneficial bacteria of agricultural importance. *Biotechnology Letters* 32, 1559–1570.

Bais, H.P., Weir, T.L., Perry, L.G., Gilroy, S. and Vivanco, J.M. (2006) The role of root exudates in rhizosphere interactions with plants and other organisms. *Annual Review Plant Biology* 57, 234–266.

Bakker, A.W. and Schipper, B. (1987) Microbial cyanide production in the rhizosphere in relation to potato yield reduction and *Pseudomonas* spp. mediated plant growth stimulation. *Soil Biology and Biochemistry* 19, 451–457.

Berg, G. and Smalla, K. (2009) Plant species and soil type cooperatively shape the structure and function of microbial communities in the rhizosphere. *FEMS Microbiology Ecology* 68, 1–13.

Berg, G., Roskot, N., Steidle, A., Eberl, L., Zock, A. and Smalla, K. (2002) Plant-dependent genotypic and phenotypic diversity of antagonistic rhizobacteria isolated from different *Verticillium* host plants. *Applied and Environmental Microbiology* 68, 3328–3338.

Bergersen, F.J. (1961) Nitrate reductase in soybean root nodules. *Biochimica et Biophysica Acta* 52, 206–207.

Bhaskar, P. and Bhosle, N.B. (2005) Microbial extracellular polymeric substances in marine biogeochemical processes. *Current Science* 88, 45–53.

Bhattacharyya, P.N. and Jha, D.K. (2012) Plant growth-promoting rhizobacteria (PGPR): emergence in agriculture. *World Journal of Microbiology and Biotechnology* 28, 1327–1350.

Bisen, K., Keswani, C., Mishra, S., Saxena, A., Rakshit, A. and Singh, H.B. (2015) Unrealized potential of seed biopriming for versatile agriculture. In: Rakshit, A., Singh, H.B. and Sen, A. (eds) *Nutrient Use Efficiency: From Basics to Advances*. Springer, New Delhi, India, pp. 193–206.

Bloemberg, G.V. and Lugtenberg, B.J. (2001) Molecular basis of plant growth promotion and biocontrol by rhizobacteria. *Current Opinion in Plant Biology* 4, 343–350.

Bradford, M.M. (1976) A rapid and sensitive method for the quantitation of microgram quantities of protein utilizing the principle of protein-dye binding. *Analytical Biochemistry* 72, 248–254.

Bric, J.M., Bostock, R.M. and Silverstone, S.E. (1991) Rapid in situ assay for indole acetic acid production by bacteria immobilized on nitrocellulose membrane. *Applied and Environmental Microbiology* 57, 535–538.

Bulgarelli, D., Schlaeppi, K., Spaepen, S., Ver Loren van Themaat, E. and Schulze-Lefert, P. (2013) Structure and functions of the bacterial microbiota of plants. *Annual Review of Plant Biology* 64, 807–838.

Cárdenas, D.M., Garrido, M.F., Bonilla, R. and Baldani, V.L. (2010) Isolation and identification of *Azospirillum* sp. in Guinea grass (*Panicum maximum* Jacq.) of the Valle del César. *Pastos y Forrajes* 33, 285–300.

Cassán, F., Perrig, D., Sgroy, V. and Luna, V. (2011) Basic and technological aspects of phytohormone production by microorganisms: *Azospirillum* sp. as a model of plant growth promoting rhizobacteria. In: Maheshwari, D.K. (ed.) *Bacteria in Agrobiology: Plant Nutrient Management*. Springer Berlin Heidelberg, Germany, pp. 141–182.

Castellano-Hinojosa, A., Correa-Galeote, D., Palau, J. and Bedmar, E.J. (2015) Isolation of N_2-fixing rhizobacteria from *Lolium perenne* and evaluating their plant growth promoting traits. *Journal of Basic Microbiology* 55, 1–7.

Chen, Y.P., Rekha, P.D., Arun, A.B., Shen, F.T., Lai, W.A. and Young, C.C. (2006) Phosphate solubilizing bacteria from subtropical soil and their tricalcium phosphate solubilizing abilities. *Applied Soil Ecology* 34, 33–41.

Chernin, L. and Chet, I. (2002) Microbial enzymes in biocontrol of plant pathogens and pests. In: Burns, R.G. and Dick, R.P. (eds) *Enzymes in the Environment: Activity, Ecology, and Applications*. Marcel Dekker, New York, pp. 171–225.

Compant, S., Reiter, B., Sessitsch, A., Nowak, J., Clément, C. and Barka, E.A. (2005) Endophytic colonization of *Vitis vinifera* L. by plant growth-promoting bacterium *Burkholderia* sp. strain PsJN. *Applied and Environmental Microbiology* 71, 1685–1693.

Compant, S., Clément, C. and Sessitsch, A. (2010) Plant growth-promoting bacteria in the rhizo- and endosphere of plants: their role, colonization, mechanisms involved and prospects for utilization. *Soil Biology and Biochemistry* 42, 669–678.

Costacurta, A. and Vanderleyden, J. (1995) Synthesis of phytohormones by plant-associated bacteria. *Critical Reviews in Microbiology* 21, 1–18.

Costacurta, A., Keijers, V. and Vanderleyden, J. (1994) Molecular cloning and sequence analysis of an *Azospirillum brasilense* indole-3-pyruvate decarboxylase. *Molecular Genetics and Genomics* 243, 463–472.

Crozier, A., Arruda, P., Jasmim, J.M., Monteiro, A.M. and Sandberg, G. (1988) Analysis of indole-3-acetic acid and related indoles in culture medium from *Azospirillum lipoferum* and *Azospirillum brasilense*. *Applied and Environmental Microbiology* 54, 2833–2837.

Curatti, L. and Rubio, L.M. (2014) Challenges to develop nitrogen-fixing cereals by direct *nif* gene transfer. *Plant Science* 225, 130–137.

Curl, E.A. and Truelove, B. (1986) *The Rhizosphere*. Springer Berlin Heidelberg, Germany.

Dardanelli, M.S., Manyani, H., Gonzalez-Barroso, S., Rodrıguez-Carvajal, M.A., Gil-Serrano, A.M. *et al.* (2009) Effect of the presence of the plant growth promoting rhizobacterium (PGPR) *Chryseobacterium balustinum* Aur9 and salt stress in the pattern of flavonoids exuded by soybean roots. *Plant and Soil* 328, 483–493.

Desai, A. and Archana, G. (2011) Role of siderophores in crop improvement. In: Maheshwari, D.K. (ed.) *Bacteria in Agrobiology: Plant Nutrient Management*. Springer Berlin Heidelberg, Germany, pp. 109–139.

Diaz, R.J. and Rosenberg, R. (2008) Spreading dead zones and consequences for marine ecosystems. *Science* 321, 926–928.

Döbereiner, J. and Day, J.M. (1976) Physiological aspects of N_2 fixation by a spirillum from *Digitaria* roots. *Soil Biology and Biochemistry* 8, 45–50.

Doran, J.W., Sarrantonio, M. and Liebig, M.A. (1996) Soil health and sustainability. *Advances in Agronomy* 56, 2–54.

Dubois, M., Gilles, K.A., Hamilton, J.K., Rebers, P.A. and Smith, F. (1956) Colorimetric method for determination of sugar and relayed substances. *Analytical Chemistry* 28, 350–356.

Dunne, C., Delany, I., Fenton, A. and O'Gara, F. (1997) Mechanisms involved in biocontrol by microbial inoculants. *Agronomie* 16, 721–729.

Dworkin, M. and Foster, J. (1958) Experiments with some microorganisms which utilize ethane and hydrogen. *Journal of Bacteriology* 75, 592–601.

Fusconi, R. and Godinho, M.J.L. (2002) Screening for exopolysaccharide-producing bacteria from sub-tropical polluted groundwater. *Brazilian Journal of Biology* 62, 363–369.

Gaby, J.C. and Buckley, D.H. (2012) A comprehensive evaluation of PCR primers to amplify the *nifH* gene of nitrogenase. *PLoS ONE* 7, e42149.

Galloway, J.N., Townsend, A.R., Erisman, J.W., Bekunda, M., Cai, Z. *et al.* (2008) Transformation of the nitrogen cycle: recent trends, questions and potential solutions. *Science* 320, 889–892.

Gamalero, E. and Glick, B.R. (2011) Mechanisms used by plant growth-promoting bacteria. In: Maheshwari, D.K. (ed.) *Bacteria in Agrobiology: Plant Nutrient Management*. Springer Berlin Heidelberg, Germany.

Gerhardson, B. (2002) Biological substitutes for pesticides. *Trends in Biotechnology* 20, 338–343.

Gerhardt, P., Murray, R.G.E., Wood, W.A. and Kreig, N.R. (1994) *Methods for General and Molecular Bacteriology*. ASM, Washington.

Glick, B.R. (1995) The enhancement of plant growth by free–living bacteria. *Canadian Journal of Microbiology* 41, 109–117.

Glick, B.R. (2012) Plant growth-promoting bacteria: mechanism and applications. *Scientifica* 2012, 1–15.

Glick, B., Shah, S., Li, J., Penrose, D. and Moffatt, B. (1998) Isolation and characterization of ACC deaminase genes from two different plant growth-promoting rhizobacteria. *Canadian Journal of Microbiology* 44, 833–843.

Glick, B.R., Patten, C.L., Holguin, G. and Penrose, D.M. (1999) *Biochemical and Genetic Mechanisms Used by Plant Growth-promoting Bacteria*. Imperial College Press, London.

Glick, B.R., Todorovic, B., Czarny, J., Cheng, Z., Duan, J. and McConkey, B. (2007) Promotion of plant growth by bacterial ACC deaminase. *Critical Reviews in Plant Sciences* 26, 227–242.

Gordon, S.A. and Weber, R.A. (1951) Colorimetric estimation of indole acetic acid. *Plant Physiology* 26, 192–195.

Gravel, V., Antoun, H. and Tweddell, R.J. (2007) Growth stimulation and fruit yield improvement of greenhouse tomato plants by inoculation with *Pseudomonas putida* or *Trichoderma atroviride*: possible role of indole acetic acid (IAA). *Soil Biology and Biochemistry* 39, 1968–1977.

Gray, E.J. and Smith, D.L. (2005) Intracellular and extracellular PGPR: commonalities and distinctions in the plant–bacterium signaling processes. *Soil Biology and Biochemistry* 37, 395–412.

Haas, D. and Défago, G. (2005) Biological control of soil borne pathogens by fluorescent pseudomonads. *Nature Review Microbiology* 3, 307–319.

Hardoim, P., van Overbeek, L., Berg, G., Pirttilä, A.M., Compant, S. *et al.* (2015) The hidden world within plants: ecological and evolutionary considerations for defining functioning of microbial endophytes. *Microbiology and Molecular Biology Reviews* 79, 293–320.

Hardy, R.W.F., Burns, R.C. and Holsten, R.D. (1973) Applications of the acetylene-ethylene assay for measurement of nitrogen fixation. *Soil Biology and Biochemistry* 5, 47–81.

Hartmann, A., Singh, M. and Klingmüller, W. (1983) Isolation and characterization of *Azospirillum* mutants excreting high amounts of indoleacetic acid. *Canadian Journal of Microbiology* 29, 916–923.

Hayat, R., Ali, S., Amara, U., Khalid, R. and Ahmed, I. (2010) Soil beneficial bacteria and their role in plant growth promotion: a review. *Annals Microbiology* 60, 579–598.

Honma, M. and Shimomura, T. (1978) Metabolism of 1-aminocyclopropane-1-carboxylic acid. *Agricultural and Biological Chemistry* 43, 1825–1831.

Hungria, M. and Vargas, M.A.T. (2000) Environmental factors affecting N_2 fixation in grain legumes in tropics with an emphasis on Brazil. *Field Crops Research* 65, 151–154.

Igual, J.M., Valverde, A., Cervantes, E. and Velázquez, E. (2001) Phosphate-solubilizing bacteria as inoculants for agriculture: use of updated molecular techniques in their study. *Agronomie* 21, 561–568.

Illmer, P.A. and Schinner, F. (1992) Solubilization of inorganic phosphates by microorganisms isolated from forest soil. *Soil Biology and Biochemistry* 24, 389–395.

Kamensky, M., Ovadis, M., Chet, I. and Chernin, L. (2003) Soil-borne strain IC14 of *Serratia plymuthica* with multiple mechanisms of antifungal activity provides biocontrol of *Botrytis cinerea* and *Sclerotinia sclerotiorum* diseases. *Soil Biology and Biochemistry* 35, 323–331.

Kasana, R.C., Salwan, R., Dhar, H., Dutt, S. and Gulati, A. (2008) A rapid and easy method for the detection of microbial cellulases on agar plates using Gram's iodine. *Current Microbiology* 55, 503–507.

Keswani, C. (2015) Proteomics studies of thermotolerant strain of *Trichoderma* spp. Ph.D. Thesis, Banaras Hindu University, Varanasi, India, p. 126.

Keswani, C., Bisen, K., Singh, V., Sarma, B.K. and Singh, H.B. (2016) Formulation technology of biocontrol agents: present status and future prospects. In: Arora, N.K., Mehnaz, S. and Balestrini, R. (eds) *Bioformulations: For Sustainable Agriculture*. Springer, New Delhi, India, pp. 35–52.

Khan, A.G. (2005) Role of soil microbes in the rhizospheres of plants growing on trace metal contaminated soils in phytoremediation. *Journal of Trace Elements in Medicine and Biology* 18, 355–364.

Khan, M.S., Zaidi, A. and Wani, P.A. (2007) Role of phosphate solubilizing microorganisms in sustainable agriculture-a review. *Agronomy for Sustainable Development* 27, 29–43.

Khan, M.S., Zaidi, A., Wani, P.A., Ahemad, M. and Oves, M. (2009) Functional diversity among plant growth-promoting rhizobacteria. In: Khan, M.S., Zaidi, A. and Musarrat, J. (eds) *Microbial Strategies for Crop Improvement*. Springer Berlin Heidelberg, Germany, pp. 105–132.

Khan, M.S., Zaidi, A. and Ahmad, E. (2014) Mechanism of phosphate solubilization and physiological functions of phosphate-solubilizing microorganisms. In: Khan, M.S., Zaidi, A. and Musarrat, J. (eds) *Phosphate Solubilizing Microorganisms*. Springer, Cham, Switzerland, pp. 31–62.

Kim, Y.C., Jung, H., Kim, K.Y. and Park, S.K. (2008) An effective biocontrol bioformulation against *Phytophthora* blight of pepper using growth mixtures of combined chitinolytic bacteria under different field conditions. *European Journal of Plant Pathology* 120, 373–382.

Klee, H.J., Hayford, M.B., Kretzmer, K.A., Barry, G.F. and Kishore, G.M. (1991) Control of ethylene synthesis by expression of a bacterial enzyme in transgenic tomato plants. *The Plant Cell* 3, 1187–1193.

Kloepper, J.W. and Schroth, M.N. (1978) Plant growth promoting rhizobacteria on radishes. *Proceedings of the 4th International conference on pathogenic bacteria (Vol II)*. Station de Pathologie Végétale et Phytobacteriologie, INRA, Angers, France, pp. 879–882.

Kloepper, J.W. and Schroth, M.N. (1981) Plant growth promoting rhizobacteria and plant growth under gnotobiotic conditions. *Phytopathology* 71, 642–644.

Knowles, C.J. and Bunch, A.W. (1986) Microbial cyanide metabolism. *Advances in Microbial Physiology* 27, 73–111.

Kuddus, S.M. and Ahmad, R.I.Z. (2013) Isolation of novel chitinolytic bacteria and production optimization of extracellular chitinase. *Journal of Genetic Engineering and Biotechnology* 11, 39–46.
Landa, B.B., Hervá, S.A., Bettiol, W. and Jiménez-Diaz, R.M. (1997) Antagonistic activity of bacteria from the chickpea rhizosphere against *Fusarium oxysporum* sp. *ciceris*. *Phytoparasitica* 25, 413–434.
Lowry, O.H., Rosebrough, N.J., Farr, A.L. and Randall, R.J. (1951) Protein measurement with the Folin Phenol reagents. *Journal of Biological Chemistry* 193, 265–275.
Lugtenberg, B. and Kamilova, F. (2009) Plant-growth-promoting rhizobacteria. *Annual Review Microbiology* 63, 541–556.
Machuca, A. and Milagres, A.M. (2003) Use of CAS-agar plate modified to study the effect of different variables on the siderophore production by *Aspergillus*. *Letters in Applied Microbiology* 36, 177–181.
Mantelin, S. and Touraine, B. (2004) Plant growth-promoting bacteria and nitrate availability: impacts on root development and nitrate uptake. *Journal of Experimental Botany* 55, 27–34.
Martínez-Viveros, O., Jorquera, M.A., Crowley, D.E., Gajardo, G. and Mora, M.L. (2010) Mechanisms and practical considerations involved in plant growth promotion by rhizobacteria. *Journal of Soil Science and Plant Nutrition* 10, 293–319.
Mehta, S. and Nautiyal, C.S. (2001) An efficient method for qualitative screening of phosphate-solubilizing bacteria. *Current Microbiology* 43, 51–56.
Meyer, J.M. and Abdallah, M.A. (1978) The fluorescent pigment of *Pseudomonas fluorescens*: biosynthesis, purification and physicochemical properties. *Journal of General Microbiology* 107, 319–328.
Miller, J.H. (1972) (ed.) Formulas and recipes. In: *Experiments in Molecular Genetics*. Cold Spring Harbor Laboratory, New York, pp. 431–433.
Mishra, S., Singh, A., Keswani, C., Saxena, A., Sarma, B.K. and Singh, H.B. (2015) Harnessing plant-microbe interactions for enhanced protection against phytopathogens. In: Arora, N.K. (ed.) *Plant Microbes Symbiosis: Applied Facets*. Springer, New Delhi, India, pp. 111–125.
Modi, M., Shah, K.S. and Modi, V.V. (1985) Isolation and characterization of catechol-like siderophore from cowpea *Rhizobium* RA-1. *Archives of Microbiology* 141, 156–158.
Mody, B.R. and Modi, V.V. (1987) Peanut agglutinin induced alteration in capsular and extracellular polysaccharide synthesis and explant nitrogenase activity of cowpea rhizobia. *Journal of Biosciences* 12, 289–296.
Moreno, B., Vivas, A., Nogales, R. and Benitez, E. (2009) Solvent tolerance acquired by *Brevibacillus brevis* during an olive waste vermicomposting process. *Ecotoxicology and Environmental Safety* 72, 2109–2114.
Müller, H., Westendorf, C., Leitner, E., Chernin, L., Riedel, K. *et al*. (2009) Quorum-sensing effects in the antagonistic rhizosphere bacterium *Serratia plymuthica* HRO-C48. *FEMS Microbiology and Ecology* 67, 468–478.
Murphy, J. and Riley, I.P. (1962) A modified single solution method for the determination of phosphate in natural waters. *Analytica Chimica Acta* 27, 31–36.
Neilands, J.B. (1995) Siderophores: structure and function of microbial iron transport compounds. *Journal of Biological Chemistry* 270, 26723–26726.
Ovadis, M., Liu, X., Gavriel, S., Ismailov, Z., Chet, I. and Chernin, L. (2004) The global regulator genes from biocontrol strain *Serratia plymuthica* IC1270: cloning, sequencing, and functional studies. *Journal of Bacteriology* 186, 4986–4993.
Pal, S., Manna, A. and Paul, A.K. (1999) Production of poly (β-hydroxy butyricacid) and exopolysaccharide by *Azotobacter beijerinckii* WDN - 01. *World Journal of Microbiology and Biotechnology* 15, 15–21.
Panwar, M., Tewari, R. and Nayyar, H. (2014) Microbial consortium of plant growth-promoting rhizobacteria improves the performance of plants growing in stressed soils: An overview In: Khan, M.S., Zaidi, A. and Musarrat, J. (eds) *Phosphate Solubilizing Microorganisms, Principles and Application of Microphos Technology*. Springer Berlin Heidelberg, Germany, pp. 257–285.
Payne, S.M. (1994) Detection, isolation and characterization of siderophores. *Methods in Enzymology* 235, 329–344.
Peix, A., Rivas-Boyero, A.A., Mateos, P.F., Rodríguez-Barrueco, C., Martínez-Molina, E. and Velázquez, E. (2001) Growth promotion of chickpea and barley by a phosphate solubilizing strain of *Mesorhizobium mediterraneum* under growth chamber conditions. *Soil Biology and Biochemistry* 33, 103–110.
Peix, A., Rivas, R., Mateos, P.F., Martínez-Molina, E., Rodríguez-Barrueco, C. and Velázquez, E. (2003) *Pseudomonas rhizosphaerae* sp. nov., a novel species that actively solubilizes phosphate *in vitro*. *International Journal of Systematic and Evolutionary Microbiology* 53, 2067–2072.
Peix, A., Ramírez-Bahena, M.H., Velázquez, E. and Bedmar, E.J. (2015) Bacterial associations with legumes. *Critical Reviews in Plant Sciences* 34, 1–3.

Penrose, D.M. and Glick, B.R. (2003) Methods for isolating and characterizing ACC deaminase containing plant growth-promoting rhizobacteria. *Physiologia Plantarum* 118, 10–15.

Pérez-Miranda, S., Cabirol, N., George-Téllez, R., Zamudio-Rivera, L.S. and Fernández, F.J. (2007) O-CAS, a fast and universal method for siderophore detection. *Journal of Microbiological Methods* 70, 127–131.

Pikovskaya, R.I. (1948) Mobilization of phosphorus in soil connection with the vital activity of some microbial species. *Microbiologiya* 17, 362–370.

Podile, A.R. and Kishore, G.K. (2006) Plant growth-promoting rhizobacteria. In: Gnanamanickam S.S. (ed.) *Plant Associated Bacteria*. Springer, Dordrecht, Netherlands, pp. 195–230.

Premono, M.E., Moawad, A.M. and Vlek, P.L.G. (1996) Effect of phosphate-solubilizing *Pseudomonas putida* on the growth of maize and its survival in the rhizosphere. *Indonesian Journal of Crop Science* 11, 13–23.

Qureshi, M.A., Iqbal, A., Akhtar, N., Shakir, M.A. and Khan, A. (2012) Co-inoculation of phosphate solubilizing bacteria and rhizobia in the presence of L-tryptophan for the promotion of mashbean (*Vigna mungo* L.). *Soil and Environment* 31, 47–54.

Rajkumar, M.A.N., Prasad, M.N.V. and Freitas, H. (2010) Potential of siderophore-producing bacteria for improving heavy metal phytoextraction. *Trends in Biotechnology* 28, 142–149.

Ray, D.K., Mueller, N.D., West, P.C. and Foley, J.A. (2013) Yield trends are insufficient to double global crop production by 2050. *PLoS One* 8, e66428.

Raymond, K.N., Dertz, E.A. and Kim, S.S. (2003) Enterobactin: an archetype for microbial iron transport. *Proceedings of the National Academy of Sciences of the United States of America* 100, 3584–3588.

Reis, V.M., Olivares, F.L. and Döbereiner, J. (1994) Improved methodology for isolation of *Acetobacter diazotrophicus* and confirmation of its endophytic habitat. *World Journal of Microbiology and Biotechnology* 10, 101–104.

Reis, V.M., Estrada de los Santos, P., Tenorio-Salgado, S., Vogel, J., Stofels, M. *et al.* (2004) *Burkholderia tropioca* sp *nov*, a novel nitrogen-fixing, plant associated bacterium. *International Journal of Systematic and Evolutionary Microbiology* 54, 2155–2162.

Rizvi, A., Khan, M.S. and Ahmad, E. (2014) Inoculation impact of phosphate-solubilizing microorganisms on growth and development of vegetable crops. In: Khan, M.S., Zaidi, A. and Musarrat, J. (eds) *Phosphate Solubilizing Microorganisms*. Springer, Cham, Switzerland, pp. 31–62.

Robinson, T.P., Wint, G.R., Conchedda, G., Van Boeckel, T.P., Ercoli, V. *et al.* (2014) Mapping the global distribution of livestock. *PLoS One* 9, e96084.

Rodríguez, H. and Fraga, R. (1999) Phosphate solubilizing bacteria and their role in plant growth promotion. *Biotechnology Advances* 17, 319–339.

Rodríguez-Díaz, M., Gonzalés, B.R., Clemente, C.P., Martínez-Toledo, M.V. and González-López, J. (2008) A review on the taxonomy and possible screening traits of plant growth promoting rhizobacteria. In: Ahmad, I., Pichtel, J. and Hayat, S. (eds) *Plant-Bacteria Interactions, Strategies and Techniques to Promote Plant Growth*. Wiley VCH., Verlag, Germany, pp. 55–80.

Ross, J., O'Neill, D., Smith, J. and Kerckhoffs, E.R. (2000) Evidence that auxin promotes the gibberellin A1 biosynthesis in pea. *The Plant Journal* 21, 547–552.

Rubio, L.M. and Ludden, P.W. (2008) Biosynthesis of the iron–molybdenum cofactor of nitrogenase. *Annual Review Microbiology* 62, 93–111.

Saharan, B. and Nehra, V. (2011) Plant growth promoting rhizobacteria: a critical review. *Life Sciences and Medicine Research* 21, 1–30.

Saleem, M., Arshad, M., Hussain, S. and Bhatti, A.S. (2007) Perspective of plant growth promoting rhizobacteria (PGPR) containing ACC deaminase in stress agriculture. *Journal of Industrial Microbiology and Biotechnology* 34, 635–648.

Schippers, B., Bakker, A.W., Bakker, P.A.H.M. and van Peer, R. (1990). Beneficial and deleterious effects of HCN-producing pseudomonads on rhizosphere interactions. *Plant and Soil* 129, 75–83.

Schwyn, B. and Neilands, J.B. (1987) Universal chemical assay for the detection and determination of siderophores. *Analytical Biochemistry* 160, 47–56.

Singh, H.B., Jain, A., Saxena, A., Singh, A., Keswani, C. *et al.* (2014) Deciphering the pathogenic behaviour of phyto-pathogens using molecular tools. In: Neeta, S. (ed.) *Biological Controls for Preventing Food Deterioration: Strategies for Pre-and Postharvest Management*. John Wiley & Sons, Chichester, UK, pp. 377–408.

Steenhoudt, O. and Vanderleyden, J. (2000) *Azospirillum*, a free-living nitrogen-fixing bacterium closely associated with grasses: genetic, biochemical and ecological aspects. *FEMS Microbiology Reviews* 24, 487–506.

Tien, T., Gaskins, M. and Hubbell, D. (1979) Plant growth substances produced by *Azospirillum brasilense* and their effect on the growth of pearl millet (*Pennisetum americanum* L). *Applied and Environmental Microbiology* 37, 1016–1024.

Tkacz, A. and Poole, P. (2015) Role of root microbiota in plant productivity. *Journal of Experimental Botany* 66, 2167–2175.

Trujillo, M.E., Riesco, R., Benito, P. and Carro, L. (2015) Endophytic actinobacteria and the interaction of *Micromonospora* and nitrogen fixing plants. *Frontiers in Microbiology* 6, 1341.

Ueda, T., Suga, Y., Yahiro, N. and Matsuguchi, T. (1995) Remarkable N_2-fixing bacterial diversity detected in rice roots by molecular evolutionary analysis of *nifH* gene sequences. *Journal of Bacteriology* 177, 1414–1417.

van Elsas, J.D., Trevors, J.Y. and Wellington, E.M.H. (1997) *Modern Soil Microbiology*. Marcel Dekker, New York.

van Loon, L.C. (2007) Plant responses to plant growth promoting rhizobacteria. *European Journal of Plant Pathology* 119, 243–254.

Vincent, J.M. (1970) *A Manual for the Practical Study of the Root Nodule Bacteria*. Blackwell Scientific (ed.) International Biological Programme, Oxford.

Voisard, C., Keel, C., Haas, D., and Defago, G. (1989) Cyanide production by *Pseudomonas fluorescens* helps suppress black root rot of tobacco under gnotobiotic conditions. *EMBO Journal* 8, 351–358.

Vyas, P. and Gulati, A. (2009) Organic acid production *in vitro* and plant growth promotion in maize under controlled environment by phosphate-solubilizing fluorescent *Pseudomonas*. *BMC Microbiology* 9, 174–188.

Wang, X., Wang, Y., Tian, J., Lim, B.L., Yan, X. and Liao, H. (2009) Overexpressing AtPAP15 enhances phosphorus efficiency in soybean. *Plant Physiology* 151, 233–240.

Whipps, J.M. (2001) Microbial interactions and biocontrol in the rhizosphere. *Journal of Experimental Botany* 52, 487–511.

Widmer, F., Shaffer, B.T., Porteous, L.A. and Seidler, R.J. (1999) Analysis of *nifH* gene pool complexity in soil and litter at a Douglas Fir forest site in the Oregon Cascade Mountain Range. *Applied of Environmental Microbiology* 65, 374–380.

Wilson, P. and Knight, S.G. (1952) *Experiments in Bacterial Physiology*. Burgess Publishing, Minneapolis, USA.

Yang, S.F. and Hoffman, N.E. (1984) Ethylene biosynthesis and its regulation in higher plants. *Annual Review of Plant Biology* 35, 155–189.

Yogesh, K., Vamsi, K.K., Amol, B., Nikhil, G., Soham, T. *et al.* (2009) Study of pectinase production in submerged fermentation using different strains of *Aspergillus niger*. *International Journal of Microbiology Research* 1, 13–17.

15 The Rhizosphere Microbial Community and Methods of its Analysis

Mukesh Meena,* Manish Kumar Dubey, Prashant Swapnil, Andleeb Zehra, Shalini Singh, Punam Kumari and R.S. Upadhyay
Department of Botany, Institute of Science, Banaras Hindu University, Varanasi, India

15.1 Introduction

The rhizosphere is the narrow zone of soil surrounding a root wherein the biological, chemical and physical parameters of soil are influenced by the living plant root. The rhizosphere supports a favourable environment for the multiplication of diverse, microbial population, which has a significant role in the organic matter transformation and biogeochemical cycles of the essential nutrients of plant (Bisen *et al.*, 2015; Lagos *et al.*, 2015; Keswani *et al.*, 2016a, b). The components of root exudates act as chemotactic attractants for microbes, where they flourish in a carbon-rich environment (Lugtenberg and Kamilova, 2009; Philippot *et al.*, 2013).

The rhizosphere of actively growing plants and their root exudates play an important role in plant–microbe interaction (Badri and Vivanco, 2009). Various compounds of root exudation are sugars, organic acid anions and amino acids which are released within proximity of the roots, provide nutrients and support to numerous microorganisms for their robust growth and activity (Mendes *et al.*, 2013). The rhizosphere microbiota includes bacteria, fungi, nematodes, viruses, protozoa, and algae inhabiting the rhizosphere in diverse forms; they may interact with the host plant either independently, mutualistically or antagonistically (Raaijmakers *et al.*, 2009; Mendes *et al.*, 2013; Mishra *et al.*, 2015; Bisen *et al.*, 2016) resulting in either beneficial, deleterious or neutral effects on plant growth (Bonkowski *et al.*, 2009; Buée *et al.*, 2009; Raaijmakers *et al.*, 2009; Raaijmakers and Mazzola, 2012) (Fig. 15.1).

Plant physiology and development are influenced by rhizospheric microbial communities. The most abundant and active microorganisms in the rhizosphere include fungi and bacteria which can competitively colonize plant roots and stimulate plant growth. They have an important role in biogeochemical cycling of organic matter, mineral nutrients and therapeutics for several diseases, and in abiotic and biotic stress tolerance (Haney *et al.*, 2015; Breidenbach *et al.*, 2016). These are collectively called plant growth-promoting microorganisms (PGPM) (Nivedhitha *et al.*, 2008). Other non-symbiotic microbes that inhabit the rhizosphere can serve in an antagonistic relationship; this activity of some microbes may cause a reduction in plant growth which results in specific diseases in crops and leads to loss of crop yields (Burr *et al.*, 1978). Therefore, it is essential to develop

*E-mail: mukeshmeenabhu@gmail.com

© CAB International 2017. *Advances in PGPR Research* (eds H.B. Singh, B.K. Sarma and C. Keswani)

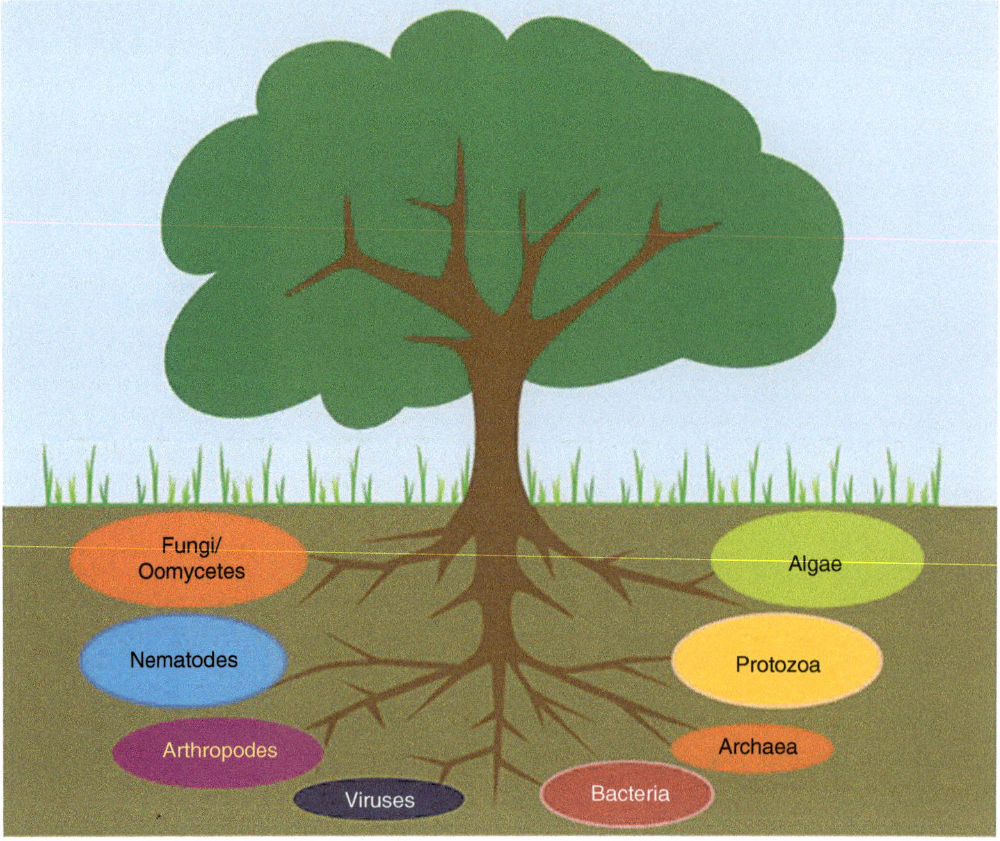

Fig. 15.1. Schematic presentation of the rhizospheric microbial communities associated with plants.

an effective method to assess the composition, diversity, distribution and behaviour of microbes in the rhizosphere for better understanding and characterization of soil health.

Researchers have developed two methods for analysing the composition, functional diversity, structure, and microbial activity in soil: (1) culture-dependent and (2) culture-independent method. The culture-based method is the conventional approach and uses a variety of culture media such as Luria–Bertani medium, Nutrient Agar, and Tryptic Soy Agar (Kirk *et al.*, 2004) for isolation and characterization of diverse microbial groups. However, only a small portion (less than 1%) of the total microbial population is cultured by using this technique (Vartoukian *et al.*, 2010). To overcome these problems, researchers prioritised the need for culture-independent methods which rely on a nucleic acid approach and include analyses of whole genomes or selected genes like 16S rRNA. Over the last few decades different techniques have been developed to describe and characterize the phylogenetic and functional diversity of microorganisms after the direct cloning of environmental DNA which was proposed by Pace *et al.* (1985). These techniques are grouped into two classes: (1) partial-community analysis approaches and (2) whole-community analysis approaches. PCR-based methods are used for the partial community analysis, where the environmental sample was used for the total DNA/RNA extraction and non-PCR based methods like estimation of G+C content, hybridization, and whole genome sequencing are used for the whole-community analysis (Sharma *et al.*, 2014).

Soil microbiologists face the complicated task of attempting to recognize and characterize microorganisms and their role. This chapter summarizes some of the most common biochemical (culture-dependent) and molecular (culture-independent) methods used to study rhizospheric soil microbial communities. Although molecular methods have the benefit of obtaining information regarding non-culturable organisms, they also have boundaries that cannot be ignored.

15.2 Rhizospheric Microbial Communities

Plant growth and productivity are highly influenced by the intensive interactions between plant root surfaces and the microbial communities within the soil. The nutrient-rich plant rhizosphere is naturally colonized by many microorganisms such as bacteria, fungi, nematodes, protozoa, algae and microarthropods that may have positive (beneficial), negative (harmful) or no visible (neutral) effect on its growth, development and productivity (Bais *et al.*, 2006; Raaijmakers *et al.*, 2009; Keswani, 2015). For the promotion of plant growth activity, it is essential to have information about the microbial community colonized around the rhizospheric region and also their ecological niche (functional role). However, still, for the vast majority of plant-associated microorganisms, there is limited knowledge of their impact on plant growth and wellness.

The rhizosphere microbiome has been well studied over the years for beneficial effects on plant growth and health; it may harbour nitrogen-fixing symbionts, mycorrhizal fungi, plant growth-promoting rhizobacteria (PGPR), endophytes, biocontrol microorganisms, mycoparasitic fungi and protozoa. These rhizospheric microbes have been shown to have many positive impacts on plants through a variety of mechanisms, including biological nitrogen fixation, increased nutrient availability and uptake (Morrissey *et al.*, 2004), averting plant diseases by suppressing or killing the phytopathogens (Mendes *et al.*, 2011), enhanced resistance to abiotic stresses, such as extreme temperatures, heavy metals and salinity (Selvakumar *et al.*, 2012; Zolla *et al.*, 2013), and biotic stresses (Zamioudis and Pieterse, 2012; Badri *et al.*, 2013) that ultimately lead to enhanced growth and plant productivity (Berg, 2009; Huang *et al.*, 2014) (Fig. 15.2). On the basis of the above-mentioned primary effect, i.e. their most well known beneficial effect on the plant, the beneficial microbes

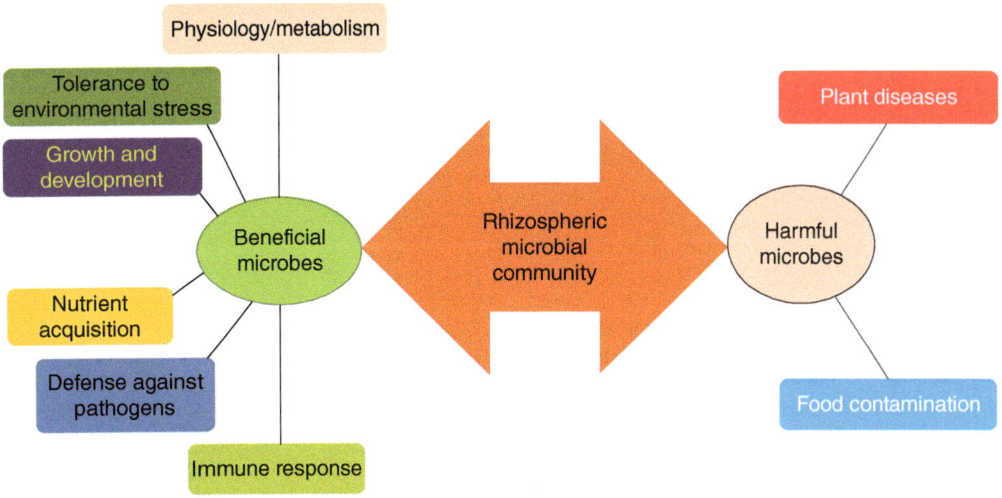

Fig. 15.2. Schematic diagram of the function and impact of beneficial and harmful rhizosphere microorganisms on the plant.

are generally classified into two broad groups (i) microorganisms with direct effects on plant growth promotion (plant growth-promoting microorganisms (PGPM)) and (ii) biological control agents (BCA) that indirectly assist with plant productivity through the control of plant pathogens (Avis *et al.*, 2008; Keswani *et al.*, 2014). Plant growth promotion and development may be facilitated both directly and indirectly by PGPM and BCA (van der Heijden *et al.*, 2008; Schnitzer *et al.*, 2011).

Mostly PGPM induced plant growth directly by the production of plant growth regulators (e.g., auxins, gibberellins, cytokinins and ethylene), providing biologically fixed nitrogen, siderophores, increasing phosphorus uptake by solubilising inorganic phosphates, potassium and zinc, and alleviating the various stress responses by secreting ACC (1-aminocyclopropane-1-carboxylate) deaminase enzyme, while indirect mechanisms involve suppression of bacterial, fungal, viral, and nematode pathogens (Barea *et al.*, 2005). These mechanisms are well documented for rhizobacteria belonging to the *Proteobacteria* and *Firmicutes*, i.e. *Pseudomonas* and *Bacillus* (Morgan *et al.*, 2005; Kogel *et al.*, 2006). In the rhizosphere, bacteria are the most abundant of all the micro-organisms and those which positively affect the plant growth have been called plant growth-promoting rhizobacteria (PGPR) (Kloepper and Schroth, 1978). Several bacteria of various genera such as *Arthrobacter*, *Azospirillium*, *Azotobacter*, *Serratia*, *Azoarcus*, *Pseudomonas*, *Bacillus*, *Burkholderia*, *Enterobacter*, *Erwinia*, *Gluconacetobacter*, *Klebsiella*, *Pseudomonas*, *Beijerinckia*, *Rhizobium*, etc., are well recognized as PGPR (Murphy *et al.*, 2003; Esitken *et al.*, 2006).

The benefits of plant–PGPR interaction include increased seed germination rate, seedling vigour, emergence, root growth, shoot growth, yield, seed weight, leaf area, chlorophyll content, nutrient uptake, protein content, early flowering, hydraulic activity, tolerance to abiotic stress, the total biomass of plant, biocontrol, and delayed senescence (Adesemoye and Kloepper, 2009; Compant *et al.*, 2010). In addition, some PGPR have also shown potential as antagonists and biocontrol agents (Beneduzi *et al.*, 2012). Various beneficial rhizobacteria like *Azotobacter* and *Gluconacetobacter* have been reported for their antagonistic effect on a variety of plant parasitic nematodes including *Meloidogyne incognita* (Bansal *et al.*, 2005). Some bacteria can also positively interact with plants by producing protective biofilms or by degradation by plant- and microbially-produced compounds in the soil that would otherwise be allelopathic or even autotoxic. The fungi from the Deuteromycetes, that is, *Trichoderma* and *Gliocladium*, and from the Sebacinales order, that is, *Piriformospora*, are most commonly investigated for plant growth promotion (Qiang *et al.*, 2012; Singh *et al.*, 2016). However, the best-known example is the mycorrhizal fungi that form a close and mutually beneficial symbiotic relationship with approximately 80% of all terrestrial plant species by translocating nutrients and minerals for the plants in return for photosynthates (Morgan *et al.*, 2005; Salvioli *et al.*, 2016). The beneficial effects of mycorrhizal fungi can be best observed especially in poor soils and under stress conditions like drought, metal stress, or nutrient deficiency. Thus, vesicular arbuscular mycorrhiza (VAM) is widely used as biofertilizer for various crops (Behl *et al.*, 2007).

Beneficial interactions also occur between nitrogen-fixing symbionts belonging to *Rhizobium*, *Bradyrhizobium*, *Sinorhizobium*, *Mesorhizobium*, *Azorhizobium* and *Photorhizobium* genera with legume plants (Narula *et al.*, 2009). The filamentous actinobacterial genus *Frankia* was also reported to form an intracellular nitrogen-fixing symbiosis with over 200 angiosperm species (Daniel *et al.*, 2007). There are also certain blue–green algae that possess the ability for symbiotic associations with some other beneficial microorganisms such as fungi, mosses, liverworts, and aquatic ferns (*Azolla*) (Bhattacharyya *et al.*, 2016). Most of these symbionts are used as biofertilizer for most of the crops worldwide (Deaker *et al.*, 2004). Several asymbiotic, associative associations such as *Azospirillum* with grass family crops (barley, sorghum, wheat and barley), *Acetobacter* with sugarcane or sweet potato, and *Achromobacter* with rice have gained much attention in recent years because of their effect on enhancement and health of crops. Recently, some microbes such as *Rhizobium*, *Bacillus*, *Pseudomonas* and *Glomus* spp. have been shown to also play a role in reducing disease (indirect effect) apart from their role in plant growth promotion (Avis *et al.*, 2008; Kumar, 2012; Gopalakrishnan *et al.*, 2015).

The BCA exerts indirect effects on plant growth by suppressing or preventing the growth of bacterial, fungal, viral, and nematode pathogens through production of antibiotics, siderophores, toxins, hydrogen cyanide (HCN), hydrolytic enzymes (chitinase, protease, lipase, etc.), competition for nutrients and suitable colonization of niches at the root surface or stimulation of the plant systemic resistance (Compant *et al.*, 2005; Maksimov *et al.*, 2011). Therefore, BCAs could be considered as an alternative to chemical pesticides. Among all the existing BCAs, *Trichoderma* spp. are probably the most studied for their effects on reducing plant diseases. The mycorrhizal symbiosis can also lead to a reduction in symptoms and lower susceptibility of the plant to pathogenic microorganisms. The role of VAM fungus (*Glomus mosseae*) as BCA of flag smut (*Urocystis agropyri*) of wheat was already well established (Chhabra *et al.*, 1996). Bacteria such as *Pseudomonas* species and *Bacillus* species, as well as fungal species such as *Trichoderma*, *Gliocladium*, *Ampelomyces* and *Chaetomium* produce antibiotics as a normal part of their self-protective arsenals and therefore, these organisms have great potential as a BCA (Pereg and McMillan, 2015). Many different bacterial genera produce HCN, including *Alcaligenes*, *Aeromonas*, *Bacillus*, *Rhizobium* and *Pseudomonas* spp. (Ahmad *et al.*, 2008). Several studies showed that inoculation with these BCA reduces the development of various plant pathogens (Siddiqui *et al.*, 2006; Lanteigne *et al.*, 2012). Several other microorganisms are also known to be capable of providing disease control; among them bacterial genera including *Streptomyces*, *Agrobacterium*, *Enterobacter*, *Erwinia*, *Serratia*, and *Azotobacter* strains are common (Narula *et al.*, 2009). In addition, the mycorrhizal symbiosis can also lead to a reduction in symptoms and lower susceptibility of the plant to pathogenic microorganisms. The mode of action of BCA has been multifaceted including parasitism, competition, antibiosis, and induction of the plant's defence mechanisms. Conversely, some BCA such as *Trichoderma* and *Pseudomonas* spp. have also been demonstrated to stimulate the plant growth (direct effect) in the absence of a pathogen.

Although most rhizospheric microbes appear to be benign, deleterious microorganisms affect the growth of the plants in a negative manner by inducing disease, production of phytotoxins, competition for nutrients, and removal of essential substances from the soil or even plant death (Morgan *et al.*, 2005; Nihorimbere *et al.*, 2011; Mihalache *et al.*, 2015) (Fig. 15.2). The most important rhizosphere plant pathogens are fungi and the fungal-like oomycetes, followed by bacteria and viruses (Lugtenberg *et al.*, 2002; Mendes *et al.*, 2013). Many plant pathogens, bacteria as well as fungi, have co-evolved with plants and show a high degree of host specificity (Raaijmakers *et al.*, 2009). The root exudates of plants are known to influence the proliferation of many pathogenic fungi. The common fungal and oomycete phytopathogens include members of *Fusarium*, *Phytophthora*, *Sclerotium*, *Aphanomyces*, *Pythium*, *Rhizopus*, *Rhizoctonia*, *Verticillium*, *Heterobasidion* and *Armillariella* (Asiegbu and Nahalkova, 2005; Tournas and Katsoudas, 2005; Narula *et al.*, 2009). However, only some bacteria have detrimental effects on plant health and survival through pathogenic or parasitic infection such as *Ralstonia solanacearum* which can cause bacterial wilt of tomato, *Agrobacterium tumefaciens* known as crown gall agent, *Pantoaea stewartii* – cause of Stewart's wilt of corn, and *Xanthomonas campestris* – a vascular pathogen that causes black rot of cabbage and other cruciferous plants, etc. (Von Bodman *et al.*, 2003; Mansfield *et al.*, 2012). Other common and well investigated bacterial agents include *Erwinia carotovora*, *Pseudomonas* and *Streptomyces scabies* (Nihorimbere *et al.*, 2011). The mechanisms by which these rhizobacteria affect the plant growth relate to the production of phytotoxins and phytohormones, competition for nutrients, and inhibition of mycorrhizal fungi (Morgan *et al.*, 2005). Plant pathogenicity factors that have been identified in bacteria refer to type III effectors also known as TTSS and toxins (Abramovitch and Martin, 2004). Other important bacterial virulence factors include phytotoxins such as coronatine toxin which mimics jasmonic acid and interferes with salicylic acid (Chisholm *et al.*, 2006) while syringomycin or pectate lyases act through the formation of ion channels in plant plasma membranes which lead to a cascade of intercellular signalling events (Lugtenberg *et al.*, 2002). Bacterial auxin synthesis is sometimes associated with pathogenesis as it can enhance the bacterial gall formation. Bacteria such as *Agrobacterium stumefaciens*, *A. rhizogenes*, *Pseudomonas savastanoi* and *Pantoaea agglomerans* pv. *gypsophilae* possess the indole-3-acetamide (IAM) pathways involved in indole-3-acetic acid (IAA) synthesis and pathogenesis (Mihalache *et al.*, 2015).

15.3 Methods for Microbial Community Analysis

The diversity of species comprises several aspects: species richness, species evenness, the number of total species present, and distribution of species. There are various methods to measure rhizospheric microbial diversity, which can be divided into two groups; biochemical-based (culture-dependent) methods and molecular-based (cultural-independent) techniques (Fig. 15.3). Classically, diversity studies comprise the comparative diversities of communities across a rise of stress, interruption or other biotic and abiotic variation. However, it is complicated with present techniques to study diversity accurately, since we do not identify what is present and we have no way to calculate the accuracy of extraction and detection methods.

15.3.1 Culture-dependent methods

The culture-based method is a traditional method used to measure the microbial diversity of natural and anthropogenically affected environmental samples, but they are able to recover only a small fraction of microbial species (Rastogi and Sani, 2011). This method includes isolation and characterization of microbes using different types of commercial growth media like Luria–Bertani medium, Nutrient Agar, and Tryptic Soy Agar (Kirk *et al.*, 2004). The culture-dependent methods mainly used are dilution plating and culturing methods; another is the community-level physiological profile to study microbial diversity.

Dilution plating and culturing methods

Traditionally, these techniques were used to analyse diversity by using selective plating and direct viable counts. There are various reports where this technique was employed for assessment of various soil quality parameters like disease suppression and organic matter decomposition (Wertz *et al.*, 2006; Gil *et al.*, 2009; Bonanomi *et al.*, 2011). In this technique, different types of culture media are used to recover a maximum number of different microbial populations from soil (Hill *et al.*, 2000). The major limitation of this technique is that less than 0.1% of soil microorganisms can be cultured (Torsvik *et al.*, 1998). In addition, this technique is biased towards fast-growing microbes and fungal species that produce a large number of spores (Dix and Webster, 1995); it also requires specific conditions like temperature, pH and light for the growth of the microbes.

Community-level physiological profiles

Community-level physiological profiling (CLPP) is a frequently used culture-dependent method; it is a rapid and inexpensive approach to assessing the functional diversity of soil microbes (Campbell *et al.*, 1997; Garland, 1997; Staddon *et al.*, 1998). This technique is based on bacterial species being identified through sole source carbon utilization (SSCU) patterns. Community-level physiological profiles uses the BIOLOG system, a suite of 95 different carbon sources (Garland and Mills, 1991). By using this method information can be obtained about

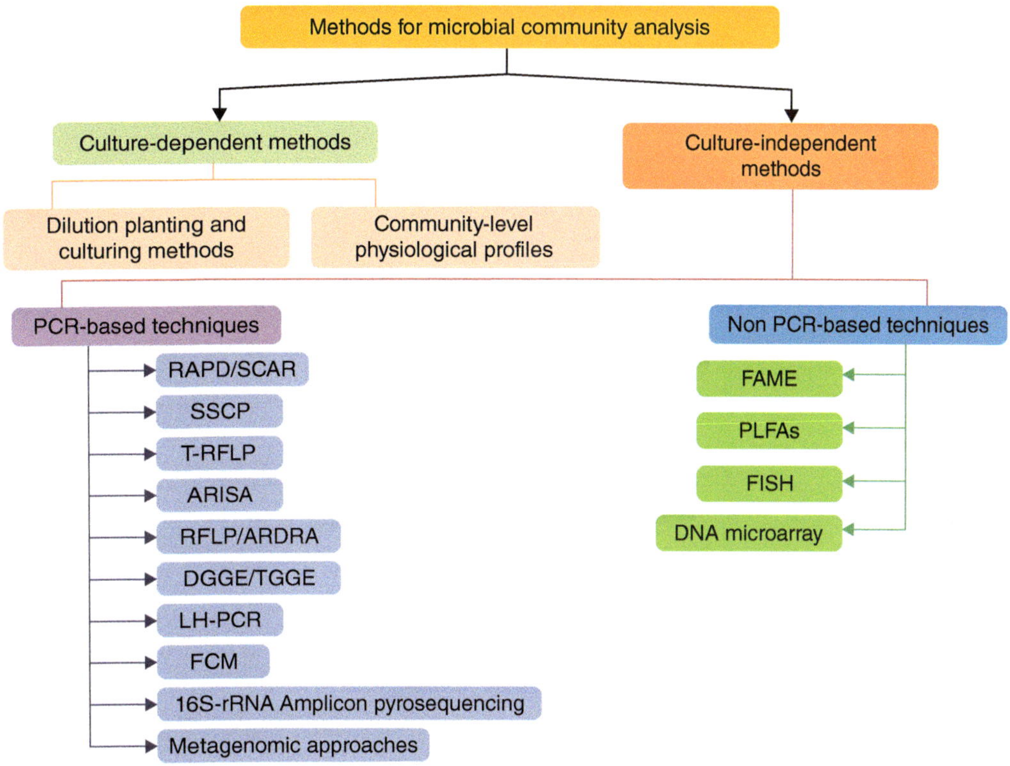

Fig. 15.3. Different methods used to study microbial community analysis in the rhizospheric soil.

metabolic abilities of soil microbial communities to utilize specific carbon sources. Differences in utilization of sole carbon sources can be used for comparison among microbes as an indication of differences in the physiological functions of microbial communities (Garland, 1997). Although CLPP technique has become popular for analysis of microbial community functional diversity, it has several limitations: for example, only culturable microbes can be assessed by the BIOLOG system, whereas slow-growing bacteria and soil fungi are left untouched. The wide range of single carbon sources in BIOLOG sole C-source test plates and triphenyl tetrazolium chloride (TTC) are buffered at nearly neutral pH, thereby presenting a different environmental condition which is therefore unlike the natural environment for those microorganisms which are well adapted to acidic or alkaline soils (Sharma *et al.*, 2014). Thus, the BIOLOG response provides a very biased representation of the functional/metabolic capabilities of the soil community (Bossio *et al.*, 1998).

15.3.2 Culture-independent techniques

Culture-independent techniques play an important role in the analysis of microbial communities. Ogram *et al.* (1987) showed that lipid and nucleic acid analysis is the most commonly used technique, while phospholipid fatty acids (PLFAs) also proved to be a useful method for the study of soil and aquatic microbial communities (Vestal and White, 1989). Many molecular and biochemical methods have been developed nowadays and are useful for the study of the great diversity of soil microorganisms.

Fatty acid methyl ester analysis (FAME)

Grouping of the fatty acids and the changes in the microorganisms due to toxic substances can be best studied by fatty acid methyl ester analysis. In this process, fatty acids are extracted from soil, methylated and analyzed by using gas chromatography (Ibekwe and Kennedy, 1999). Cluster and principal components analyses were used to recognize similarities and differences amid soil microbial communities described by using FAME profiles (Cavigelli *et al.*, 1995). Some of the species such as *Bacillus*, *Pseudomonas*, Gram-positive cocci and rods, Gram-negative non-fermenters and environmental organisms found in pharmaceutical facilities were also discriminated by this technique (Sutton and Cundell, 2004). Anaerobic, aerobic and/or facultatively aerobic bacteria, present in wastewater treatment systems were characterized by this technique. In a forensic investigation, FAME profiling is a useful technique for the study of the spore production methods of *Bacillus* organisms (Ehrhardt *et al.*, 2010). This tool also acts as a biomarker for spore discrimination, and highlights natural changes in microbial communities to provide soil profiles and patterns for locating clandestine graves (Breton *et al.*, 2015).

Phospholipid fatty acids (PLFAs)

PLFAs are the most important components of microbial cell membranes. PLFAs in extracted soil examination can provide much more information such as soil quality, quantitative indicators of soil responses to land management and further, environmental stressors regarding the whole structure of microbial communities. Further, PLFAs also act as great potential biomarkers for the study of the important attributes of microbial communities such as viable biomass, nutritional status and structure of the microorganisms (White and Ringelberg, 1997). In aquifer environment, PLFA was generally preferred for the study of microbial diversity (Pratt *et al.*, 2012). On the other hand, cyclopropane fatty acids produced from the genus *Alicyclobacillus* during peat heating were demonstrated by PLFA profiles (Ranneklev and Bååth, 2003). PLFA analysis in combination with DNA techniques highlighted the diversity of bacteria in the polyhumic lake Mekkojarvi (Taipale *et al.*, 2009). In the pelagic zone, this tool revealed the presence of *Cytophaga-Flavobacteria*, diatoms, green algae and dinoflagellates (Pace *et al.*, 2007). Microbial dynamics associated with rhizosphere carbon cycling can also be revealed by the PLFA tool coupled with ^{13}C pulse-chase labeling and proves to be a most effective approach (Butler *et al.*, 2003). PLFA coupled with the terminal restriction fragment length polymorphism (T-RFLP) of 16S rDNA analyzed genes for any changes in the microbial community composition in Illinois River (Baniulyte *et al.*, 2009). On the other hand, PLFA analysis of soil samples via the Sherlock PLFA analysis software and Agilent GC provides an automated and comprehensive method for analyzing PLFAs from soil microbiota. Coupled with a high-throughput extraction method, the (whole-cell fatty acid) MIDI-PLFA solution results in a standardized PLFA protocol that can be implemented for detailed study of the soil microbiota (Buyer and Sasser, 2012; Fernandes *et al.*, 2013).

Fluorescent in situ hybridization (FISH)

Fluorescent *in situ* hybridization allows the cell visualization using an epifluorescence or confocal laser scanning microscope. CARD-FISH an improved FISH method is the catalysed reporter deposition fluorescence *in situ* hybridization (Pernthaler *et al.*, 2002). This tool can be used for the detection of rRNA, mRNA and genes encoded on chromosomes in microorganisms (Kubota, 2013). Pathogens in biofilms to be penetrated through extracellular polymeric substance matrix can be studied by a useful technique called high-affinity peptide nucleic acid (PNA)-FISH (Lehtola *et al.*, 2006). Many scientists have characterized the microscale spatial structure of microbial communities in a miscellaneous range of ecosystems with the help of combined FISH-based techniques and microscopy, including those present in the soil microbiome, mammalian intestine, and on marine snow (Berlemont and Martiny, 2013; Thiele *et al.*, 2015; Cordero and Datta, 2016). Recently, Doi *et al.* (2007) analyzed the soil bacterial community in cultivated fields of rice using PCR-DGGE and FISH. The results revealed that these two methods, more than conventional methods, provided a rapid and simple analysis of rhizosphere bacteria. FISH analysis indicated the predominance of Gram-positive low GC bacteria in rhizospheres and a higher proportion of *Clostridium* spp. in lowland fields, which is consistent with results of PCR-DGGE. The study also suggests that the applications of PCR-DGGE and FISH to agricultural field experiments are reliable methods to evaluate the rhizosphere bacteria in the soil. Further, FISH method could be used to study plant–microbial interactions, with a few modifications. Pulse-labelling of plants with $^{14}CO_2$, followed by FISH–microautoradiography analyses of the plant rhizospheric soil, could reveal the identity of bacteria utilizing root exudates (Singh *et al.*, 2004).

Flow cytometry (FCM)

Flow cytometry is also the most reliable technique to analyse the bacterial community in the soil and sediment samples. Scientists used a time-efficient flow-cytometric (FCM) counting process involving cell detachment and separation from matrix particles by centrifugation in tubes receiving sample suspensions and histodenz solution (Frossard *et al.*, 2016). This technique is used to assess bacterial abundances in various soils (natural and agricultural), sediments (streams and lakes) and sludge from sand-filters in a drinking water treatment plant. The use of microbial flow cytometry allows the analysis of physiological heterogeneity amid speed and precision for complex biological populations (Whiteley *et al.*, 2003). Total cells were purified from soil cores and intact extractable cell counts assessed by SYBR Green II fluorescence, while vigorous cell counts were determined by 5-cyano-2,3-ditolyl tetrazolium chloride reduction (CTC-positive cells). Shamir *et al.* (2009) reported the effect of *Tamarix aphylla* on the live/dead bacterial population ratio on a spatial and temporal range and concluded that the effect of abiotic factors, which changed on spatial as well as temporal scales, and also the size of the active soil bacterial community, which fluctuated between 1.44% and 25.4% in summer and winter, respectively.

Automated ribosomal intergenic spacer analysis (ARISA)

ARISA provides broad and fast snapshots of microbial diversity and has proved to be one of the best fingerprinting techniques for the purpose of comparison. This tool proved to be significant for the characterization of length variability and nucleotide sequence of benthic microbes based upon 16S and 23S rRNA genes in the rRNA operon (Daffonchio *et al.*, 2003). In environmental samples, this tool is also well suited to know the variability of *Methylobacterium* communities (Knief *et al.*, 2008). In *Eneoelia* and *Armillaria*, ARISA profiles were used to evaluate the between-site variation (Ranjard *et al.*, 2001). ARISA fingerprints also demonstrated the effect of *Azospirillum lipoferum* on the structure of rhizobacterium in maize (Baudoin *et al.*, 2009). An *in silico* method is a useful approach to obtain a dataset of bacterial 16S-23S spacers which simulate ARISA profiles to assess species richness in low-diversity ecosystems (Kovacs et al., 2010). Bacterial communities isolated from four types of soil conflicting in geographic origins, vegetation cover and physicochemical properties were evaluated by the ARISA technique, which showed discrete and contained numerous analytical peaks with respect to size and intensity (Ranjard *et al.*, 2001).

16S rRNA amplicon pyrosequencing

Amplicon sequencing is a widely applied technique for the study of the composition, organization, spatial and temporal patterns of microbial communities (Olsen *et al.*, 1986). Recently, Sinclair *et al.* (2015) reported the reliability of performing bacterial 16S rRNA gene amplicon sequencing on the MiSeq technology. 16S rRNA pyro sequencing was also used to identify the microbial communities in rice roots and rhizosphere soil (Hernández *et al.*, 2015). Recently, Pascual *et al.* (2016) assessed the bacterial diversity in the rhizosphere of *Thymus zygis* growing in the Sierra Nevada National Park (Spain) through16S rRNA pyro sequencing approach, while Sun *et al.* (2014) used the Illumina amplicon sequencing of 16S rRNA approaches to reveal and characterize the bacterial community development in the rhizosphere of apple nurseries.

Denaturing gradient gel electrophoresis (DGGE) and temperature gradient gel electrophoresis (TGGE)

Denaturing gradient gel electrophoresis (DGGE) and temperature gradient gel electrophoresis (TGGE) are PCR-dependent techniques used to study microbial genetic diversity based on the difference in base composition and secondary structure of fragments of the 16S rDNA molecule. These techniques detect point mutations in DNA sequences and also are used to assess the diversity of bacteria and fungi (Smalla *et al.*, 2001) caused by nutrition changes and by the addition of chemicals (Whiteley and Bailey, 2000). Zhou *et al.* (2012) also analyzed changes in bacterial and fungal communities under different concentrations of the autotoxin. TGGE technique demonstrated the change in a genetically modified microorganism and their impact on eubacteria, α and β–proteobacteria, actinobacteria and acidobacteria (Carter *et al.*, 2007). The analysis of amplified 16S or 18S rRNA genes by denaturing or temperature gradient gel electrophoresis (DGGE/TGGE) have been frequently used to examine the microbial diversity of rhizosphere soil and to reveal the changes in microbial communities (Lin and Pan, 2010; Soni *et al.*, 2010; Cleary *et al.*, 2012; Pires *et al.*, 2012). These two approaches can be useful in exploration of microbial communities of different environments (Campbell *et al.*, 2009; Frerichs *et al.*, 2012). TTGE exploits the principle on which DGGE is based, without requiring a chemical denaturing gradient. DGGE for microbial community analysis is now common (Nakatsu *et al.*, 2000; Chong *et al.*, 2009; Ning *et al.*, 2009), but TTGE has not been explored much (Rincon-Florez *et al.*, 2013). However, DGGE studies also used well conserved protein-coding genes, such as a *nifD* gene, (*nifH*) gene, (*rpoB*) gene and (*dsrB*) gene (Geets *et al.*, 2006). In an anoxygenic phototroph, *pufM*-based DGGE analysis have been reported (Karr *et al.*, 2003).

Restriction fragment length polymorphism (RFLP)/amplified ribosomal DNA restriction analysis (ARDRA)

RFLP is a tool to study soil microorganism diversity that relies on DNA polymorphisms. In this method, PCR-amplified rDNA is digested with the help of restriction enzymes and different fragment sizes are detected by using agarose or non-denaturing polyacrylamide gel electrophoresis in the case of microbial community analysis (Liu *et al.*, 1997; Tiedje *et al.*, 1999). RFLP banding patterns can be used to screen clones or used to evaluate bacterial community organization (Massol-Deya *et al.*, 1995; Pace, 1996). This method is helpful for detecting structural changes in microbial communities but not as a measure of diversity or detection of accurate phylogenetic groups (Liu *et al.*, 1997). Banding patterns in different communities become too complex to analyze using RFLP while a single species could have 4-6 restriction fragments (Tiedje *et al.*, 1999). Possibly by using a six-base cutting enzyme, the total of restriction fragments per species could be reduced, thus gaining an improvement in this method. Poly *et al.* (2001) reported that three restriction enzymes, *Hae*III, *Nde*II and *Mnl*I, selected for RFLP analyses, were the most selective for the study of *nifH* gene diversity, which was applied to assess the genetic range of the *nifH* gene pool in soil.

Terminal restriction fragment length polymorphism (T-RFLP)

T-RFLP with improved 16S rRNA gene sequence is a rapid method which provides accurate information about community structure and dynamics at low cost. T-RFLP is not completely obsolete and proves to be valuable in microbial ecology (Prakash *et al.*, 2014). In a metal-contaminated soil, T-RFLP has demonstrated the spatial and temporal changes in bacterial communities, monitored populations and assessed the diversity in *Viola calaminaria* (Tonin *et al.*, 2001). T-RFLP approaches provide changes in the structure and composition of soil communities and also investigate rhizosphere microbial communities associated with the dwarf shrubs *Calluna vulgaris* and *Vaccinium myrtillus* (Singh *et al.*, 2006).

Single-strand conformation polymorphism (SSCP)

Single-strand conformation polymorphism (SSCP) is a technique which is adapted for the analysis of microbial communities. This technique is based on electrophoretic separation of single-strand DNA fragments differing in dissimilar length. The diversity of polyhydroxyalkanoate-producing bacteria used single-strand conformation polymorphism as a culture-independent approach and confirms that rhizosphere is the main reservoir of polyhydroxyalkanoate bacteria (Gasser *et al.*, 2006). Bouasria *et al.* (2012) reported that bacterial and fungal communities' diversities were evaluated by using SSCP technique. SSCP analysis was also used to assess *Trichoderma*-specific communities with low diversity on the Canary Islands (Zachow *et al.*, 2008). Recently, the SSCP technique has been used for the quick profiling of soil microbial communities (Stefanis *et al.*, 2013) and phylogenetic studies (Badin *et al.*, 2012). Schwieger and Tebbe (1998) also revealed that SSCP of DNA method is widely used in the analysis and differentiation of cultivated pure-culture soil microbes and noncultivated rhizosphere microbial community (Schwieger and Tebbe 1998). This technique has also been used for detection of genetic mutations in bacteriological and viral systems (Fujita *et al.*, 1992).

Amplicon length-heterogeneity PCR (LH-PCR)

This method is based on the natural differences between lengths of amplified gene fragments and is analogous to ARISA. In LH-PCR, a fluorescently labelled primer is used to establish the relative amounts of amplified sequences originating from various rhizospheric soils. These labelled fragments are separated by gel electrophoresis and detected by laser-induced fluorescence with an automated gene sequencer (Ritchie *et al.*, 2000). This tool provides insight into the community structure without the construction of clone libraries (Mills *et al.*, 2007). LH-PCR technique is a monitoring tool to enhance microbial ecology, assess differences in soil bacterial community and characterizes phylotypes in soil fungal communities. On the other hand, during the production of H_2, the LH-PCR profiling associated with the 16S rRNA genes, sequencing was used for the characterization of the bacterial community (Bonito et al., 2013). It is reported that different soil microbial communities can be altered during the time period which is determined by the use of the LH-PCR technique (Moreno *et al.*, 2011). Researchers found that the LH-PCR method is proficient, consistent, and an extremely reproducible method that should be a useful tool in future assessments of soil microbial composition (Ritchie *et al.*, 2000).

Random amplified polymorphic DNA (RAPD) and sequence-characterized amplified region (SCAR) technique

RAPD-derived molecular markers are used to locate random segments of genomic DNA and also revealed polymorphism. Randomly amplified polymorphic DNA (RAPD) is a commonly used effective technique to assess the diversity of soil microbial communities. RAPD has been widely used in species classification and phylogenetic analysis, species identification, and genetic analysis of soil microbial populations. Several studies applied RAPD technique for soil microbial community analysis (Yang *et al.*, 2000; Yao *et al.*, 2006; Ranjan *et al.*, 2013). Recently, Li *et al.* (2014) monitored the bacterial community and dynamics of dominant bacterial species in ginseng rhizosphere soil during the growth of *Panax ginseng* by using random amplified polymorphic DNA (RAPD), and amplified ribosomal DNA restriction analysis (ARDRA).

The SCAR based markers analyses the inherent genomic strain variability from others which enable the rapid detection and identification of it in a complex sample. This approach has already been used for several soil bacteria, such as *Azospirillum* (Felici *et al.*, 2008; Couillerot *et al.*, 2010; Priya *et al.*, 2016), *Bacillus subtilis* (Felici *et al.*, 2008), *Pseudomonas fluorescens* (Von Felten *et al.*, 2010), and fungal microbes such as *Colletotrichum coccodes* (Dauch *et al.*, 2003), *Beauvaria bassiana* (Castrillo *et al.*, 2003), and *Trichoderma* spp. (Savazzini *et al.*, 2009; Pérez *et al.*, 2014). All these studies used a SCAR marker to monitor the fate and behaviour of the strain in the soil, which is essential in order to assess their potential spread and impact. In lettuce, RAPD-SCAR marker demonstrated downy mildew resistance genes (Dahlberg *et al.*, 2002). SCARs have the advantages over RAPD markers in having additional specificity and reproducibility. RAPD-SCAR markers have also been utilized in detection of *Fusarium oxysporum* f. sp. *ciceris* and *cubense*, for the selection of *Saccharomyces cerevisiae* and for the identification of powdery mildew resistance genes (Bhagyawant, 2016).

DNA-Microarray

Microarray genotyping covers an entire set of the genes of an organism and can also be used to measure the expression levels of large numbers of genes simultaneously. Microarray technology is much faster and plays a significant role in the analysis of microorganisms in different environmental samples (Asuming-Brempong, 2012). DNA microarray can represent large spans of genomic DNA for comparative genomic hybridization (CGH) analysis and also monitor pre-mRNA splicing on a genomic scale. The PhyloChip (made on the slide which attached thousands of oligonucleotide probes of the 16S rRNA genes) microarray allows the biologist to examine the levels of 16S rRNA genes for analyzing soil microbial community. This technique can allow the comparable detection of up to several thousand microbial strains, species, genera or advanced taxonomic groups in a single experiment (Peplies *et al.*, 2003). Kim *et al.* (2008) reported that bacterial communities present in the soil were analyzed by 16S rRNA gene sequences from representative clones from the microarray and established the phylogenetic assignments provided by the higher taxon probes. Cong *et al.* (2015) analyzed soil microbial functional gene diversity and causative factors in a tropical rainforest with the help of microarray-based metagenomic techniques and found that high microbial functional gene diversity and various soil microbial metabolic potential for biogeochemical processes were measured to exist in a tropical rainforest.

Metagenomic approaches

Metagenomics is a non-culture based approach to comprehensively analyse microbial communities (including viruses, bacteria, archaea, fungi, and protists) in different ecosystems ranging from oceans to the human microbiome by applying several bioinformatics approaches that include taxonomic systems, sequence databases, and sequence-alignment tools (Neelakanta and Hameeda, 2013). The composition and size of bacterial communities in two rhizosphere soil samples of plants (*Ramonda serbica* and *Ramonda nathaliae*) were analyzed using a metagenomic approach like fluorescent *in situ* hybridization (FISH), amplification of 16S rDNA genes from metagenomic DNAs, and restriction fragment length polymorphism (RLFP) (Lidija *et al.*, 2010). Fractioning of the metagenomic DNA as a function of (i) vertical soil sampling, (ii) density gradients (cell separation), (iii) cell lysis stringency, and (iv) DNA fragment size distribution, is used for analysis of unique genetic diversity, which was based on ribosomal intergenic spacer analysis [RISA] fingerprinting and phylochips (Delmont *et al.*, 2011). Analysis of DNA or RNA from soil provides information about microbial diversity and polymerase chain reaction-denaturing gradient gel electrophoresis (PCR-DGGE) is one such metagenomic approach that is used in monitoring the microbial diversity in the soil affected by certain chemicals or fertilizer (Fujii *et al.*, 2009). Moreover, the metagenomic approach also characterizes the taxonomic and functional diversity of bacterial and fungal communities present in soil by shotgun sequencing (Castañeda and Barbosa, 2016) and T-RFLP approach (Castañeda *et al.*, 2015). Thus the overall study revealed that metagenomic approach gives a more accurate overview of soil microbial diversity and community composition compared to other methods.

15.4 Conclusions and Future Perspectives

Rhizospheric microbial diversity is important not only for fundamental scientific research but also to understand the connection between diversity and community function. Microbial diversity could be adversely affected by human influences such as pollution, agricultural and chemical applications. Ecosystem sustainability is important to understanding the connection between diversity and role of the ecosystem. Our information on soil microbial diversity is incomplete by our inability to study soil microbes. For example about 1% of the soil bacterial population can be cultured by standard laboratory practices. An approximate 1,500,000 species of fungi survive in the earth (Giller *et al.*, 1997), but these cannot be cultured by present standard laboratory methods (van Elsas *et al.*, 2000). There are many problems associated with soil microbial diversity. These occur not only from methodological restrictions, but also from lack of taxonomic information.

Microbial community composition analysis allows a comprehensive insight into the diversity and possible environmental role and fosters a perspective of multifaceted microbial processes (Vanwonterghem *et al.*, 2014). In recent years, a healthy rise has been seen in sequencing approaches targeting microbial communities such as amplicon sequencing and metatranscriptomic approaches (Caporaso *et al.*, 2012; Grosskopf and Soyer, 2014; Fischer *et al.*, 2016). In this chapter, we have described various methods for studying rhizospheric soil microbial diversity. Although, molecular methods have the advantage of obtaining information concerning non-culturable organisms, there remain several limitations which cannot be ignored.

Study of soil microbial diversity is most challenging for the soil microbiologists due to lack of accuracy of soil microbial diversity analysis techniques. It is very difficult to conclude what is present in one gram of soil and one technique could not be sufficient to understand the microbial diversity of the soil. Therefore, we need a variety of techniques with different end points and degrees of resolution to acquire the widest possible and most authentic results regarding the community of microbes in the rhizospheric soil. We need to expand our knowledge to understand the links between structural diversity and function of below- and above-ground ecosystems which is influenced by biological, chemical and physical factors.

References

Abramovitch, R.B. and Martin, G.B. (2004) Strategies used by bacterial pathogens to suppress plant defenses. *Current Opinion in Plant Biology* 7, 356–364.

Adesemoye, A. and Kloepper, J. (2009) Plant–microbes interactions in enhanced fertilizer-use efficiency. *Applied Microbiology and Biotechnology* 85, 1–12.

Ahmad, F., Ahmad, I. and Khan, M.S. (2008) Screening of free-living rhizospheric bacteria for their multiple plant growth promoting activities. *Microbiology Research* 163, 232–238.

Asiegbu, F.O. and Nahalkova, J.L.G. (2005) Pathogen inducible cDNAs from the interaction of the root rot fungus *Heterobasidion annosum* with Scots pine (*Pinus sylvestris* L.). *Plant Science* 168, 365–372.

Asuming-Brempong, S. (2012) Microarray technology and its applicability in soil science – a short review. *Open Journal of Soil Science* 2, 333–340.

Avis, T.J., Gravel, V., Antoun, H. and Tweddell, R.J. (2008) Multifaceted beneficial effects of rhizosphere microorganisms on plant health and productivity. *Soil Biology and Biochemistry* 40, 1733–1740.

Badin, A-L., Mustafa, T., Bertrand, C., Monier, A., Delolme, C. *et al.* (2012) Microbial communities of urban stormwater: the phylogentic structure of bacterial communities varies with porosity. *FEMS Microbiology Ecology* 81, 324–338.

Badri, D.V. and Vivanco, J.M. (2009) Regulation and function of root exudates. *Plant, Cell and Environment* 32, 666–681.

Badri, D.V., Zolla, G., Bakker, M.G., Manter, D.K. and Vivanco, J.M. (2013) Potential impact of soil microbiomes on the leaf metabolome and on herbivore feeding behavior. *New Phytologist* 198, 264–273.

Bais, H.P., Weir, T.L., Perry, L.G., Gilroy, S. and Vivanco, J.M. (2006) The role of root exudates in rhizosphere interactions with plants and other organisms. *Annual Review of Plant Biology* 57, 233–266.

Baniulyte, D., Favila, E. and Kellyl, J.J. (2009) Shifts in microbial community composition following surface application of dredged river sediments. *Microbial Ecology* 57, 160–169.

Bansal, R.K., Dahiya, R.S., Narula, N. and Jain, R.K. (2005) Management of *Meloidogyne incognitain* cotton, using strains of bacterium *Gluconacetobacter diazotrophicus*. *Nematologia Mediterranea* 33, 103–105.

Barea, J.M., Pozo, M.J., Azcon, R. and Azcon-Aguilar, C. (2005) Microbial cooperation in the rhizosphere. *Journal of Experimental Botany* 56, 1761–1778.

Baudoin, E., Nazaret, S., Mougel, C., Ranjard, L. and Moënne-Loccoz, Y. (2009) Impact of inoculation with the phytostimulatory PGPR *Azospirillum lipoferum* CRT1 on the genetic structure of the rhizobial community of field-grown maize. *Soil Biology and Biochemistry* 41, 409–413.

Behl, R.K., Ruppel, S., Kothe, E. and Narula, N. (2007) Wheat x Azotobacter x VA Mycorrhiza interactions towards plant nutrition and growth – a review. *Journal of Applied Botany and Food Quality* 81, 95–109.

Beneduzi, A., Ambrosini, A. and Passaglia, M.P. (2012) Plant growth-promoting rhizobacteria (PGPR): Their potential as antagonists and biocontrol agents. *Genetics and Molecular Biology* 35, 1044–1051.

Berg, G. (2009) Plant–microbe interactions promoting plant growth and health: perspectives for controlled use of microorganisms in agriculture. *Applied Microbiology and Biotechnology* 84, 11–18.

Berlemont, R. and Martiny, A.C. (2013) Phylogenetic distribution of potential cellulases in bacteria. *Applied and Environmental Microbiology* 79, 1545–1554.

Bhagyawant, S.S. (2016) RAPD-SCAR markers: an interface tool for authentication of traits. *Journal of Biosciences and Medicines* 4, 1–9.

Bhattacharyya, P.N., Goswami, M.P. and Bhattacharyya, L.H. (2016) Perspective of beneficial microbes in agriculture under changing climatic scenario: a review. *Journal of Phytology* 8, 26–41.

Bisen, K., Keswani, C., Mishra, S., Saxena, A., Rakshit, A. and Singh, H.B. (2015) Unrealized potential of seed biopriming for versatile agriculture. In: Rakshit, A., Singh, H.B. and Sen, A. (eds) *Nutrient Use Efficiency: From Basics to Advances*. Springer, New Delhi, India, pp. 193–206.

Bisen, K., Keswani, C., Patel, J.S., Sarma, B.K. and Singh, H.B. (2016) *Trichoderma* spp.: efficient inducers of systemic resistance in plants. In: Chaudhary, D.K. and Verma, A. (eds) *Microbial-mediated Induced Systemic Resistance in Plants*. Springer, Singapore, pp. 185–195.

Bonanomi, G., D'Ascoli, R., Antignani, V., Capodilupo, M., Cozzolino, L. *et al.* (2011) Assessing soil quality under intensive cultivation and tree orchards in Southern Italy. *Applied Soil Ecology* 47, 184–194.

Bonito, G., Gryganskyi, A., Schadt, C., Pelletier, D., Schaefer, A. *et al.* (2013) Genomic analysis of *Mortierella elongata* and its endosymbiotic bacterium. *27th Fungal Genetics Conference* Asilomar, CA.

Bonkowski, M., Villenave, C. and Griffiths, B. (2009) Rhizosphere fauna: the functional and structural diversity of intimate interactions of soil fauna with plant roots. *Plant and Soil* 321, 213–233.

Bossio, D.A., Scow, K.M., Gunapala, N. and Graham, K.J. (1998) Determinants of soil microbial communities: effects of agricultural management, season, and soil type on phospholipid fatty acid profiles. *Microbial Ecology* 36, 1–12.

Bouasria, A., Mustafa, T., De Bello, F., Zinger, L., Lemperiere, G. *et al.* (2012) Changes in root-associated microbial communities are determined by species specific plant growth responses to stress and disturbance. *European Journal of Soil Biology* 52, 59–66.

Breidenbach, B., Pump, J. and Dumont, M.G. (2016) Microbial community structure in the rhizosphere of rice. *Frontiers in Microbiology* 6, 01537.

Breton, G., Lee, J., Zhou, Y.J., Schreiber, J.J., Keler, T. *et al.* (2015) Circulating precursors of human CD1c^{+} and CD141^{+} dendritic cells. *Journal of Experimental Medicine* 212, 401–413.

Buée, M., De Boer, W., Martin, F., van Overbeek, L. and Jurkevitch, E. (2009) The rhizosphere zoo: an overview of plant-associated communities of microorganisms, including phages, bacteria, archaea, and fungi, and of some of their structuring factors. *Plant and Soil* 321, 189–212.

Burr, T.J., Schroth, M.N. and Suslow, T. (1978) Increased potato yields by treatment of seed pieces with specific strains of *Pseudomonas fluorescens* and *P. putida*. *Phytopathology* 68, 1377–1383.

Butler, J.L., Williams, M.A., Bottomley, P.J. and Myrold, D.D. (2003) Microbial community dynamics associated with rhizosphere carbon flow. *Applied and Environmental Microbiology* 69, 6793–6800.

Buyer, J.S. and Sasser, M. (2012) High throughput phospholipid fatty acid analysis of soils. *Applied Soil Ecology* 61, 127–130.

Campbell, C.D., Grayston, S.J. and Hirst, D.J. (1997) Use of rhizosphere carbon sources in sole carbon source tests to discriminate soil microbial communities. *Journal of Microbiological Methods* 30, 33–41.

Campbell, J.H., Clark, J.S. and Zak, J.C. (2009) PCR-DGGE comparison of bacterial community structure in fresh and archived soils sampled along a Chihuahuan Desert elevational gradient. *Microbial Ecology* 57, 261–266.

Caporaso, J.G., Lauber, C.L., Walters, W.A., Berg-Lyons, D., Huntley, J. *et al.* (2012) Ultra-high-throughput microbial community analysis on the Illumina HiSeq and MiSeq platforms. *ISME Journal* 6, 1621–1624.

Carter, G.W., Prinz, S., Neou, C., Shelby, J.P., Marzolf, B. *et al.* (2007) Prediction of phenotype and gene expression for combinations of mutations. *Molecular Systems Biology* 3, 96.

Castañeda, L.E. and Barbosa, O. (2016) Metagenomic analysis exploring taxonomic and functional diversity of soil microbial communities in Chilean vineyards and surrounding native forests. *PeerJ PrePrints.* (Available at: https://doi.org/10.7287/peerj.preprints.1661v1, accessed 10 July 2017.)

Castañeda, L.E., Manzano, M., Godoy, K., Marquet, P.A. and Barbosa, O. (2015) Comparative study between soil microbial structure communities from Vineyards and *Sclerophyllous* forest in Central Chile. *Ecology and Evolution*, doi:10.1002/ece3.1652.

Castrillo, L.A., Vandenberg, J.D. and Wraight, S.P. (2003) Strain-specific detection of introduced *Beauveria bassianain* agricultural fields by use of sequence-characterized amplified region markers. *Journal of Invertebrate Pathology* 82, 75–83.

Cavigelli, M.A., Robertson, G.P. and Klug, M.J. (1995) Fatty acid methyl ester (FAME) profiles as measures of soil microbial community structure. *Plant and Soil* 170, 99–113.

Chhabra, M.L., Beniwal, M.S., Karwasra, S.S., Chand, H. and Jalali, B.L. (1996) VAM: a biological tool for control of flag smut of wheat under field conditions. *Proceedings of National Symposium on Molecular Approaches in Plant Disease Management*, Shimla, India.

Chisholm, S.T., Coaker, G., Day, B. and Staskawicz, B.J. (2006) Host-microbe interactions: Shaping the evolution of the plant immune response. *Cell* 124(4), 803–814.

Chong, C.W., Tan, G.Y.A., Wong, R.C.S., Riddle, M.J. and Tan, I.K.P. (2009) DGGE fingerprinting of bacteria in soils from eight ecologically different sites around Casey station, Antarctica. *Polar Biology* 32, 853–860.

Cleary, D.F.R., Smalla, K., Mendonca-Hagler, L.C.S. and Gomes, N.C.M. (2012) Assessment of variation in bacterial composition among microhabitats in a mangrove environment using DGGE fingerprints and barcoded pyrosequencing. *PLoS One* 7, e29380.

Compant, S., Duffy, B., Nowak, J., Clement, C. and Barka, E.A. (2005) Use of plant growth promoting bacteria for biocontrol of plant diseases: Principles, mechanisms of action and future prospects. *Applied and Environmental Microbiology* 71, 4951–4959.

Compant, S., Clément, C. and Sessitsch, A. (2010) Plant growth-promoting bacteria in the rhizo- and endosphere of plants: Their role, colonization, mechanisms involved and prospects for utilization. *Soil Biology and Biochemistry* 42, 669–678.

Cong, J., Liu, X., Lu, H., Xu, H., Li, Y. *et al.* (2015) Analyses of the influencing factors of soil microbial functional gene diversity in tropical rainforest based on GeoChip 5.0. *Genomic Data* 5, 397–398.

Cordero, O.X. and Datta, M.S. (2016) Microbial interactions and community assembly at microscales. *Current Opinion in Microbiology* 31, 227–234.

Couillerot, O., Bouffaud, M-L., Baudoin, E., Muller, D., Caballero-Mellado, J. and Moënne-Loccoz, Y. (2010) Development of a real-time PCR method to quantify the PGPR strain *Azospirillum lipoferum* CRT1 on maize seedlings. *Soil Biology and Biochemistry* 42, 2298–2305.

Daffonchio, D., Cherif, A., Brusetti, L., Rizzi, A., Mora, D. *et al.* (2003) Nature of polymorphisms in 16s–23s rRNA gene intergenic transcribed spacer fingerprinting of *Bacillus* and related genera. *Applied and Environmental Microbiology* 69, 5128–5137.

Dahlberg, J.A., Zhang, X., Hart, G.E. and Mullet, J.E. (2002) Comparative assessment of variation among sorghum germplasm accessions using seed morphology and RAPD measurements. *Crop Science* 42, 291–296.

Daniel, G., Jaffré, T. and Prin, Y. (2007) Abundance of *Frankia* from *Gymnostoma* spp. in the rhizosphere of *Alphitonia neocaledonica*, a non-nodulated *Rhamnaceae* endemic to New Caledonia. *European Journal of Soil Biology* 36, 169–175.

Dauch, A.L., Watson, A.K. and Jabaji-Hare, S.H. (2003) Detection of the biocontrol agent *Colletotrichum coccodes* (183088) from the target weed velvetleaf and from soil by strain-specific PCR markers. *Journal of Microbiological Methods* 55, 51–64.

Deaker, R., Roughley, R.J. and Kennedy, I.R. (2004) Legume seed inoculation technology – a review. *Soil Biology and Biochemistry* 36, 1275–1288.

Delmont, T.O., Patrick, R., Sébastien, C., Ian, M.C., Florentin, C. *et al.* (2011) Accessing the soil metagenome for studies of microbial diversity. *Applied and Environmental Microbiology* 77, 1315–1324.

Dix, N.J. and Webster, J. (eds) (1995) *Fungal Ecology*. Chapman & Hall, London.
Doi, T., Hagiwara, Y., Abe, J. and Morita, S. (2007) Analysis of rhizosphere bacteria of rice cultivated in Andosol lowland and upland fields using molecular biological methods. *Plant Root* 1, 66–74.
Ehrhardt, C.J., Chu, V., Brown, T., Simmons, T.L., Swan, B.K. *et al.* (2010) Use of fatty acid methyl ester profiles for discrimination of *Bacillus cereus* T-strain spores grown on different media. *Applied and Environmental Microbiology* 76, 1902–1912.
Esitken, A., Pirlak, L., Turan, M. and Sahin, F. (2006) Effects of floral and foliar application of plant growth promoting rhizobacteria (PGPR) on yield, growth and nutrition of sweet cherry. *Scientia Horticulturae* 110, 324–327.
Felici, C., Vettori, L., Giraldi, E., Forino, L.M.C., Toffanin, A. *et al.* (2008) Single and co-inoculation of *Bacillus subtilis* and *Azospirillum brasilense* on *Lycopersicon esculentum*: effects on plant growth and rhizosphere microbial community. *Soil Biology and Biochemistry* 40, 260–270.
Fernandes, M.F., Saxena, J. and Dick, R.P. (2013) Comparison of whole-cell fatty acid (MIDI) or phospholipid fatty acid (PLFA) extractants as biomarkers to profile soil microbial communities. *Microbial Ecology* 66, 145–157.
Fischer, M.A., Güllert, S., Neulinger, S.C., Streit, W.R. and Schmitz, R.A. (2016) Evaluation of 16S rRNA gene primer pairs for monitoring microbial community structures showed high reproducibility datasets generated with multiple archaeal and bacterial primer pairs. *Frontiers in Microbiology* 7, 01297, doi: 10.3389/fmicb.2016.01297.
Frerichs, J., Oppermann, B.I., Gwosdz, S., Möller, I., Herrmann, M. and Krüger, M. (2012) Microbial community changes at a terrestrial volcanic CO_2 vent induced by soil acidification and anaerobic microhabitats within the soil column. *FEMS Microbiology Ecology* 84, 60–74.
Frossard, A., Hammes, F. and Gessner, M.O. (2016) Flow cytometric assessment of bacterial abundance in soils, sediments and sludge. *Frontiers in Microbiology* 7, 903, doi: 10.3389/fmicb.2016.00903.
Fujii, T., Morimoto, S., Hoshino, Y.T., Okada, H., Wang, Y. *et al.* (2009) Studies of diversity and functions of soil microbes and nematodes in NIAES using nucleic acids extracted from soil. *Macro Symposium* Workshop 5.
Fujita, M., Inoue, M., Tanizawa, O., Iwamoto, S. and Enomoto, T. (1992) Alterations of the *p53* gene in human primary cervical carcinoma with and without human papillomavirus infection. *Cancer Research* 52, 5323–5328.
Garland, J.L. (1997) Analysis and interpretation of community-level physiological profiles in microbial ecology. *FEMS Microbiology Ecology* 24, 289–300.
Garland, J.L. and Mills, A.L. (1991) Classification and characterization of heterotrophic microbial communities on the basis of patterns of community-level sole-carbon-source utilization. *Applied and Environmental Microbiology* 57, 2351–2359.
Gasser, R.B., Hu, M., Chilton, N.B., Campbell, B.E., Jex, A.J. *et al.* (2006) Single-strand conformation polymorphism (SSCP) for the analysis of genetic variation. *Nature Protocols* 1, 3121–3128.
Geets, J., Borremans, B., Diels, L., Springael, D., Vangronsveld, J. *et al.* (2006) *DsrB* gene-based DGGE for community and diversity surveys of sulfate-reducing bacteria. *Journal of Microbiological Methods* 66, 194–205.
Gil, S.V., Meriles, J., Conforto, C., Figoni, G., Basanta, M. *et al.* (2009) Field assessment of soil biological and chemical quality in response to crop management practices. *World Journal of Microbiology and Biotechnolology* 25, 439–448.
Giller, K.E., Beare, M.H., Lavelle, P., Izac, A.-M.N. and Swift, M.J. (1997) Agricultural intensification, soil biodiversity and agroecosystem function. *Applied Soil Ecology* 6, 3–16.
Gopalakrishnan, S., Sathya, A., Vijayabharathi, R., Varshney, R.K., Gowda, C.L.L. and Krishnamurthy, L. (2015) Plant growth promoting rhizobia: Challenges and opportunities. *3 Biotech* 5, 355–377.
Grosskopf, T. and Soyer, O.S. (2014) Synthetic microbial communities. *Current Opinion in Microbiology* 18, 72–77.
Haney, C.H., Samuel, B.S., Bush, J. and Ausubel, F.M. (2015) Associations with rhizosphere bacteria can confer an adaptive advantage to plants. *Nature Plants* 1, 15051, doi: 10.1038/nplants.2015.51.
Hernández, M., Dumont, M.G., Yuan, Q. and Conrad, R. (2015) Different bacterial populations associated with the roots and rhizosphere of rice incorporate plant derived carbon. *Applied and Environmental Microbiology* 81, 2244–2253.
Hill, G.T., Mitkowski, N.A., Aldrich-Wolfe, L., Emele, L.R., Jurkonie, D.D. *et al.* (2000) Methods for assessing the composition and diversity of soil microbial communities. *Applied Soil Ecology* 15, 25–36.
Huang, X-F., Chaparro, J.M., Reardon, K.F., Zhang, R., Shen, Q. and Vivanco, J.M. (2014) Rhizosphere interactions: root exudates, microbes, and microbial communities. *Botany* 92, 267–275.

Ibekwe, A. and Kennedy, A. (1999) Fatty acid methyl ester (FAME) profiles as a tool to investigate community structure of two agricultural soils. *Plant Soil* 206, 151–161.

Karr, E.A., Sattley, W.M., Jung, D.O., Madigan, M.T. and Achenbach, L.A. (2003) Remarkable diversity of phototrophic purple bacteria in a permanently frozen Antarctic lake. *Applied and Environmental Microbiology* 69, 4910–4914.

Kim, J-S., Dungan, R.S. and Crowley, D. (2008) Microarray analysis of bacterial diversity and distribution in aggregates from a desert agricultural soil. *Biology and Fertility of Soils* 44, 1003–1011.

Kirk, J.L., Beaudette, L.A., Hart, M., Moutoglis, P., Klironomos, J.N. *et al.* (2004) Methods of studying soil microbial diversity. *Journal of Microbiological Methods* 58, 169–188.

Keswani, C. (2015) Proteomics studies of thermotolerant strain of *Trichoderma* spp. Ph.D. Thesis, Banaras Hindu University, Varanasi, India, p. 126.

Keswani, C., Mishra, S., Sarma, B.K., Singh, S.P. and Singh, H.B. (2014) Unraveling the efficient application of secondary metabolites of various *Trichoderma*. *Applied Microbiology and Biotechnology* 98, 533–544.

Keswani, C., Bisen, K., Singh, S.P., Sarma, B.K. and Singh, H.B. (2016a) A proteomic approach to understand the tripartite interactions between plant-*Trichoderma*-pathogen: investigating the potential for efficient biological control. In: Hakeem, K.R. and Akhtar, M.S. (eds) *Plant, Soil and Microbes Vol. 2. Mechanisms and Molecular Interactions*. Springer, Cham, Switzerland, pp. 79–93.

Keswani, C., Bisen, K., Singh, V., Sarma, B.K. and Singh, H.B. (2016b) Formulation technology of biocontrol agents: present status and future prospects. In: Arora, N.K., Mehnaz, S. and Balestrini, R. (eds) *Bioformulations: For Sustainable Agriculture*. Springer, New Delhi, India, pp. 35–52.

Kloepper, J.W. and Schroth, M.N. (1978) Plant-growth promoting rhizobacteria on radishes. In: *Proceedings of the 4th International Conference on Plant Pathogenic Bacteria* 2, 879–882.

Knief, C., Frances, L., Cantet, F. and Vorholt, J.A. (2008) Cultivation-independent characterization of *Methylobacterium* populations in the plant phyllosphere by automated ribosomal intergenic spacer analysis. *Applied and Environmental Microbiology* 74, 2218–2222.

Kogel, K.H., Franken, P. and Hückelhoven, R. (2006) Endophyte or parasite–what decides? *Current Opinion in Plant Biology* 9, 358–363.

Kovacs, A., Yacoby, K. and Gophna, U. (2010) A systematic assessment of automated ribosomal intergenic spacer analysis (ARISA) as a tool for estimating bacterial richness. *Research in Microbiology* 161, 192–197.

Kubota, K. (2013) CARD-FISH for environmental microorganisms: technical advancement and future applications. *Microbes and Environments* 28, 3–12.

Kumar, P. (2012) Ph.D. thesis, Gurukul Kangri University, Haridwar, India.

Lagos, M.L., Maruyama, F., Nannipieri, P., Mora, M.L., Ogram, A. and Jorquera, M.A. (2015) Current overview on the study of bacteria in the rhizosphere by modern molecular techniques: a mini–review. *Journal of Soil Science and Plant Nutrition* 15, 504–523.

Lanteigne, C., Gadkar, V., Wallon, T., Novinscak, A. and Filion, M. (2012) Production of DAPG and HCN by *Pseudomonas* sp. LBUM300 contributes to the biological control of bacterial canker of tomato. *Phytopathology* 102, 967–973.

Lehtola, M.J., Laxander, M., Miettinen, I.T., Hirvonen, A., Vartiainen, T. and Martikainen, P.J. (2006) The effects of changing water flow velocity on the formation of biofilms and water quality in pilot distribution system consisting of copper or polyethylene pipes. *Water Research* 40, 2151–2060.

Li, Y., Ying, Y.X. and Ding, W.L. (2014) Dynamics of *Panax ginseng* rhizospheric soil microbial community and their metabolic function. *Evidence-Based Complementary and Alternative Medicine*, Article ID 160373, 6, doi: 10.1155/2014/160373.

Lidija, Đ., Savić, M., Tanja, N. and Branka, V. (2010) Metagenomic analysis of soil microbial communities. *Archives of Biological Sciences* (Serbia) 62, 559–564.

Lin, C-H. and Pan, T-M. (2010) PCR-Denaturing gradient gel electrophoresis analysis to assess the effects of a genetically modified *cucumber mosaic virus*-resistant tomato plant on soil microbial communities. *Applied and Environmental Microbiology* 76, 3370–3373.

Liu, W.T., Marsh, T.L., Cheng, H. and Forney, L.J. (1997) Characterization of microbial diversity by determining terminal restriction fragment length polymorphisms of genes encoding 16S rRNA. *Applied and Environmental Microbiology* 63, 4516–4522.

Lugtenberg, B. and Kamilova, F. (2009) Plant-growth-promoting-rhizobacteria. *Annual Review of Microbiology* 63, 541–556.

Lugtenberg, B.J., Chin-A-Woeng, T.F. and Bloemberg, G. (2002) Microbe–plant interactions: principles and mechanisms. *Antonie van Leeuwenhoek* 81, 373–383.

Maksimov, I.V., Abizgil'dina, R.R. and Pusenkova, L.I. (2011) Plant growth promoting rhizobacteria as alternative to chemical crop protectors from pathogens (Review). *Applied Biochemistry and Microbiology* 47, 333–345.

Mansfield, J., Genin, S., Magori, S., Citovsky, V., Sriariyanum, M. *et al.* (2012) Top 10 plant pathogenic bacteria in molecular plant pathology. *Molecular Plant Pathology* 13, 614–629.

Massol-Deya, A.A., Odelson, D.A., Hickey, R.F. and Tiedje, J.M. (1995) Bacterial community fingerprinting of amplified 16S and 16S-23S ribosomal DNA gene sequences and restriction endonuclease analysis (ARDRA). In: Akkerman, A.D.L., van Elsas, J.D. and De-Briijn, F.J. (eds) *Molecular Microbial Ecology Manual 3.3.2*. Kluwer, Boston, pp. 1–8.

Mendes, R., Kruijt, M., de Bruijn, I., Dekkers, E., van der Voort, M. *et al.* (2011) Deciphering the rhizosphere microbiome for disease-suppressive bacteria. *Science* 332, 1097–1100.

Mendes, R., Garbeva, P. and Raaijmakers, J.M. (2013) The rhizosphere microbiome: significance of plant beneficial, plant pathogenic, and human pathogenic microorganisms. *FEMS Microbiology Reviews* 37, 634–663.

Mihalache, G., Zamfirache, M-M. and Ştefan, M. (2015) Root associated bacteria – friends or enemies? a review. *Memoirs of the Scientific Sections of the Romanian Academy* XXXVIII, 27–54.

Mills, D.K., Entry, J.A. and Gillevet, P.M. (2007) Assessing microbial community diversity using amplicon length heterogeneity polymerase chain reaction. *Soil Science Society of America Journal* 71, 572–578.

Mishra, S., Singh, A., Keswani, C., Saxena, A., Sarma, B.K. and Singh, H.B. (2015) Harnessing plant-microbe interactions for enhanced protection against phytopathogens. In: Arora N.K. (ed.) *Plant Microbes Symbiosis: Applied Facets*. Springer, New Delhi, pp. 111–125.

Moreno, L.I., Mills, D., Fetscher, J., John-Williams, K., Meadows-Jantz, L. and McCord, B. (2011) The application of amplicon length heterogeneity PCR (LH-PCR) for monitoring the dynamics of soil microbial communities associated with cadaver decomposition. *Journal of Microbiological Methods* 84, 388–393.

Morgan, J.A.W., Bending, G.D. and White, P.J. (2005) Biological costs and benefits to plant–microbe interactions in the rhizosphere. *Journal of Experimental Botany* 56, 1729–1739.

Morrissey, J.P., Dow, J.M., Mark, G.L., and O'Gara, F. (2004) Are microbes at the root of a solution to world food production? *EMBO Reports* 5, 922–926.

Murphy, J.F., Reddy, M.S., Ryu, C.M., Kloepper, J.W. and Li, R.H. (2003) Rhizobacteria-mediated growth promotion of tomato leads to protection against *Cucumber mosaic virus*. *Phytopathology* 93, 1301–1307.

Nakatsu, C.H., Torsvik, V. and Øvreås, L. (2000) Soil community analysis using DGGE of 16s rDNA polymerase chain reaction product. *Soil Science Society of America Journal* 64, 1382–1388.

Narula, N., Kothe, E. and Behl, R.K. (2009) Role of root exudates in plant-microbe interactions. *Journal of Applied Botany and Food Quality* 82, 122–130.

Neelakanta, G. and Hameeda, S. (2013) The use of metagenomic approaches to analyze changes in microbial communities. *Microbiology Insights* 6, 37–48.

Nihorimbere, V., Ongena, M., Smargiassi, M. and Thonart, P. (2011) Beneficial effect of the rhizosphere microbial community for plant growth and health. *Biotechnology, Agronomy, Society and Environment* 15, 327–337.

Ning, J., Liebich, J., Kästner, M., Zhou, J., Schäffer, A. and Burauel, P. (2009) Different influences of DNA purity indices and quantity on PCR-based dgge and functional gene microarray in soil microbial community study. *Applied Microbiology and Biotechnology* 82, 983–993.

Nivedhitha, V.R., Shwetha, B., Deepa, D.D., Manojkumer, N. and Raghavendra, R.B. (2008) Plant growth promoting microorganisms (PGPMs) from bamboo rhizosphere. *Journal of Advanced Biotechnology* 7, 33–35.

Ogram, A., Sayler, G.S. and Barkay, T. (1987) The extraction and purification of microbial DNA from sediments. *Journal of Microbiology Methods* 7, 57–66.

Olsen, G.J., Lane, E.J., Giovannoni, S.J., Pace, N.R. and Stahl, D.A. (1986) Microbial ecology and evolution: a ribosomal RNA approach. *Annual Review of Microbiology* 40, 337–355.

Pace, M.L., Carpenter, S.R., Cole, J.J., Coloso, J.J., Kitchell, J.F. *et al.* (2007) Does terrestrial organic carbon subsidize the planktonic food web in a clear-water lake? *Limnology and Oceanography* 52, 2177–2189.

Pace, N.R. (1996) New perspective on the natural microbial world: molecular microbial ecology. *ASM News* 62, 463–470.

Pace, N.R., Stahl, D.J., Lane, D.J. and Olsen, G.J. (1985) Analyzing natural microbial populations by rRNA sequences. *ASM News* 51, 4–12.

Pascual, J., Blanco, S., García-López, M., García-Salamanca, A., Bursakov, S.A. *et al.* (2016) Assessing bacterial diversity in the rhizosphere of *Thymus zygis* growing in the Sierra Nevada National Park (Spain) through culture-dependent and independent approaches. *PLoS One* 11(1), e0146558. doi:10.1371/journal.pone.0146558.

Peplies, J., Glockner, O.F. and Amann, R. (2003) Optimization strategies for DNA microarray-based detection of bacteria with 16S rRNA targeting oligonucleotide probes. *Applied and Environmental Microbiology* 69, 1397–1407.

Pereg, L. and McMillan, M. (2015) Scoping the potential uses of beneficial microorganisms for increasing productivity in cotton cropping systems. *Soil Biology and Biochemistry* 80, 349–358.

Pérez, G., Verdejo, V., Godim-Porto, C., Orlando, J. and Carú, M. (2014) Designing a SCAR molecular marker for monitoring *Trichoderma* cf. *harzianum* in experimental communities. *Journal of Zhejiang University-Sciences B (Biomedicine and Biotechnology)* 15, 966–978.

Pernthaler, A., Pernthaler, J. and Amann, R. (2002) Fluorescence in situ hybridization and catalyzed reporter deposition for the identification of marine bacteria. *Applied and Environmental Microbiology* 68, 3094–3101.

Philippot, L., Raaijmakers, J.M., Lemanceau, P. and van der Putten, W.H. (2013) Going back to the roots: the microbial ecology of the rhizosphere. *Nature Review Microbiology* 11, 789–799.

Pires, A.C.C., Cleary, D.F.R., Almeida, A., Almeida, A., Cunha, A. *et al.* (2012) Denaturing gradient gel electrophoresis and barcoded pyrosequencing reveal unprecedented archaeal diversity in mangrove sediment and rhizosphere samples. *Applied and Environmental Microbiology* 78, 5520–5528.

Poly, F., Monrozier, L.J. and Bally, R. (2001) Improvement in the RFLP procedure for studying the diversity of *nifH* genes in communities of nitrogen fixers in soil. *Research in Microbiology* 152, 95–103.

Prakash, O., Pandey, P.K., Kulkarni, G.J., Mahale, K.N. and Shouche, Y.S. (2014) Technicalities and glitches of terminal-restriction fragment length polymorphism (T-RFLP). *Indian Journal of Microbiology* 54, 255–261.

Pratt, B., Riesen, R. and Johnston, C.G. (2012) PLFA analyses of microbial communities associated with PAH-contaminated riverbank sediment. *Microbial Ecology* 64, 680–691.

Priya, P.R., Antony, R.S., Gopalaswamy, G. and Balachandar, D. (2016) Development of sequence-characterized amplified region (SCAR) markers as a quality standard of inoculants based on *Azospirillum*. *Archives of Microbiology* 198, 257–267.

Qiang, X., Weiss, M., Kogel, K.H. and Schafer, P. (2012) *Piriformospora indica* a mutualistic basidiomycete with an exceptionally large plant host range. *Molecular Plant Pathology* l13, 508–518.

Raaijmakers, J. and Mazzola, M. (2012) Diversity and natural functions of antibiotics produced by beneficial and pathogenic soil bacteria. *Annual Review of Phytopatholgy* 50, 403–424.

Raaijmakers, J.M., Paulitz, T.C., Steinberg, C., Alabouvette, C. and Moënne-Loccoz, Y. (2009) The rhizosphere: a playground and battlefield for soilborne pathogens and beneficial microorganisms. *Plant and Soil* 321, 341–361.

Ranjan, A., Mahalakshmi, M.R. and Sridevi, M. (2013) Isolation and characterization of phosphate solubilizing bacteria, bacterial species from different crop fields of Salem, Tamil Nadu, India. *International Journal of Food Science and Nutrition* 3, 29–33.

Ranjard, L., Poly, F., Lata, J.C., Mougel, C., Thioulouse, J. and Nazaret, S. (2001) Characterization of bacterial and fungal soil communities by automated ribosomal intergenic spacer analysis fingerprints: biological and methodological variability. *Applied and Environmental Microbiology* 67, 4479–4487.

Ranneklev, S.B. and Bååth, E. (2003) Use of phospholipid fatty acids to detect previous self-heating events in stored peat. *Applied and Environmental Microbiology* 69, 3532–3539.

Rastogi, G. and Sani, R.K. (2011) molecular techniques to assess microbial community structure, function, and dynamics in the environment. In: Ahmad, I., Ahmad, F. and Pichtel. J. (eds) *Microbes and Microbial Technology: Agricultural and Environmental Applications*. doi: 10.1007/987-1-4419-7931-5_2.

Rincon-Florez, V.A., Carvalhais, L.C. and Schenk, P.M. (2013) Culture-independent molecular tools for soil and rhizosphere microbiology. *Diversity* 5, 581–612.

Ritchie, N.J., Schutter, M.E., Dick, R.P. and Myrold, D.D. (2000) Use of length heterogeneity PCR and Fatty acid methyl ester profiles to characterize microbial communities in soil. *Applied and Environmental Microbiology* 66, 1668–1675.

Salvioli, A., Ghignone, S., Novero, M., Navazio, L., Venice, F. *et al.* (2016) Symbiosis with an endobacterium increases the fitness of a mycorrhizal fungus, raising its bioenergetic potential. *The ISME Journal* 10, 130–144.

Savazzini, F., Longa, C.M.O. and Pertot, I. (2009) Impact of the biocontrol agent *Trichoderma atroviride* SC1 on soil microbial communities of a Vineyard in northern Italy. *Soil Biology and Biochemistry* 41, 1457–1465.

Schnitzer, S.A., Klironomos, J.N., Hillerislambers, J., Kinkel, L.L., Reich, P.B. *et al.* (2011) Soil microbes drive the classic plant diversity-productivity pattern. *Ecology* 92, 296–303.

Schwieger, F. and Tebbe, C.C. (1998) A new approach to utilize PCR-single-strand-conformation polymorphism for 16S rRNA gene-based microbial community analysis. *Applied and Environmental Microbiology* 64, 4870–4876.

Selvakumar, G., Panneerselvam, P. and Ganeshamurthy, A.N. (2012) Bacterial mediated alleviation of abiotic stress in crops. In: Maheshwari, D.K. (ed.) *Bacteria in Agrobiology: Stress Management.* Springer Berlin Heidelberg, Germany, pp. 205–224.

Shamir, I., Zahavy, E. and Steinberger, Y. (2009) Bacterial viability assessment by flow cytometry analysis in soil. *Frontiers of Biology in China* 4, 424–435.

Sharma, B., Narzary, D. and Jha, D.K. (2014) Cultural independent diversity analysis of soil microbial community and their significance. In: Maheswari, D.K. (ed.) *Bacterial Diversity in Sustainable Agricultural. Sustainable Development and Biodiversity 1.* Springer, Cham, Switzerland, doi: 10.1007/978-3-319-05936-5_12.

Siddiqui, I.A., Shaukat, S.S., Sheikh, I.H. and Khan, A. (2006) Role of cyanide production by *Pseudomonas fluorescens* CHAO in the suppression of root-knot nematode, *Meloidogyne javanica* in tomato. *World Journal of Microbiology and Biotechnology* 22, 641–650.

Sinclair, L., Osman, O.A., Bertilsson, S. and Eiler, A. (2015) Microbial community composition and diversity via 16S rRNA gene amplicons: evaluating the illumina platform. *PLoS One* 10(2), e0116955.

Singh, B.K., Millard, P., Whiteley, A.S. and Murrell, J.C. (2004) Unravelling rhizosphere–microbial interactions: opportunities and limitations. *Trends in Microbiology* 12, 386–393.

Singh, B.K., Munro, S., Reid, E., Ord, B., Potts, J.M. *et al.* (2006) Investigating microbial community structure in soils by physiological, biochemical, and molecular fingerprinting methods. *European Journal of Soil Science* 57, 72–82.

Singh, V., Upadhyay, R.S., Sarma, B.K. and Singh, H.B. (2016) Seed biopriming with *Trichoderma asperellum* effectively modulate plant growth promotion in pea. *International Journal of Agriculture, Environment and Biotechnology* 9, 361–365.

Smalla, K., Wieland, G., Buchner, A., Zock, A., Parzy, J. *et al.* (2001) Bulk and rhizosphere soil bacterial communities studied by denaturing gradient gel electrophoresis: plant-dependent enrichment and seasonal shifts revealed. *Applied and Environmental Microbiology* 67, 4742–4751.

Soni, R., Saluja, B. and Goel, R. (2010) Bacterial community analysis using temporal temperature gradient gel electrophoresis (TTGE) of 16S rDNA PCR products of soil metagenome. *EKOLOGIJA* 56, 94–98.

Staddon, W.J., Duchesne, L.C. and Trevors, J.T. (1998) Acid phosphatase alkaline phosphatase and arylsulfatase activities in soils from a jack pine (*Pinus banksiana* Lamb.) ecosysystem after clear–cutting prescribed burning, and scarification. *Biology and Fertility of Soils* 27, 1–4.

Stefanis, C., Alexopoulos, A., Voidarou, C., Vavias, S. and Bezirtzoglou, E. (2013) Principal methods for isolation and identification of soil microbial communities. *Folia Microbiologica* 58, 61–68.

Sun, J., Zhang, Q., Zhou, J. and Wei, Q. (2014) Illumina amplicon sequencing of 16s rRNA tag reveals bacterial community development in the rhizosphere of apple nurseries at a replant disease site and a new planting site. *PLoS One* 9(10), e111744.

Sutton, S.V.W. and Cundell, A.M. (2004) Microbial identification in the pharmaceutical industry. *Pharmacopeial Forum* 30, 1884–1894.

Taipale, S., Jones, R.I. and Tiirola, M. (2009) Vertical diversity of bacteria in an oxygen-stratified humic lake, evaluated using DNA and phospholipid analyses. *Aquatic Microbial Ecology* 55, 1–16.

Thiele, S., Fuchs, B.M., Amann, R. and Iversen, M.H. (2015) Colonization in the photic zone and subsequent changes during sinking determine bacterial community composition in marine snow. *Applied and Environmental Microbiology* 81, 1463–1471.

Tiedje, J.M., Asuming-Brempong, S., Nusslein, K., Marsh, T.L. and Flynn, S.J. (1999) Opening the black box of soil microbial diversity. *Applied Soil Ecology* 13, 109–122.

Tonin, C., Vandenkoornhuyse, P., Joner, E.J., Straczek, J. and Leyval, C. (2001) Assessment of arbuscular mycorrhizal fungi diversity in the rhizosphere of *Viola calaminaria* and effect of these fungi on heavy metal uptake by clover. *Mycorrhiza* 10, 161–168.

Torsvik, V., Daae, F.L., Sandaa, R.A. and Overeas, L. (1998) Novel techniques for analyzing microbial diversity in natural and perturbed environments. *Journal of Biotechnology* 64, 53–62.

Tournas, V.H. and Katsoudas, E. (2005) Mould and yeast flora in fresh berries, grapes and citrus fruits. *International Journal of Food Microbiology* 105, 11–17.

van der Heijden, M.G.A., Bardgett, R.D. and van Straalen, N.M. (2008) The unseen majority: soil microbes as drivers of plant diversity and productivity in terrestrial ecosystems. *Ecology Letters* 11, 296–310.

van Elsas, J.D., Frois-Duarte, G., Keijzer-Wolters, A. and Smit, E. (2000) Analysis of the dynamics of fungal communities in soil via fungal-specific PCR of soil DNA followed by denaturing gradient gel electrophoresis. *Journal of Microbiological Methods* 43, 133–151.

Vanwonterghem, I., Jensen, P.D., Dennis, P.G., Hugenholtz, P., Rabaey, K. and Tyson, G.W. (2014) Deterministic processes guide long-term synchronised population dynamics in replicate anaerobic digesters. *ISME Journal* 8, 2015–2028.

Vartoukian, S.R., Palmer, R.M. and Wade, W.G. (2010) Cultivation of a synergistetes strain representing a previously uncultivated lineage. *Environmental Microbiology* 12, 916–928.

Vestal, J.R. and White, D.C. (1989) Lipid analysis in microbial ecology: quantitative approaches to the study of microbial communities. *Bioscience* 39, 535–541.

Von Bodman, S.B., Bauer, W.D. and Coplin, D.L. (2003) Quorum sensing in plant-pathogenic bacteria. *Annual Review of Phytopathology* 41, 455–482.

Von Felten, A., Défago, G. and Maurhofer, M. (2010) Quantification of *Pseudomonas fluorescens* strains F113, CHA0 and Pf153 in the rhizosphere of maize by strain-specific real-time PCR unaffected by the variability of DNA extraction efficiency. *Journal of Microbiological Methods* 81, 108–115.

Wertz, S., Degrange, V., Prosser, J.I., Poly, F., Commeaux, C. *et al.* (2006) Maintenance of soil functioning following erosion of microbial diversity. *Environmental Microbiology* 8, 2162–2169.

White, D.C. and Ringelberg, D.B. (1997) Utility of the signature lipid biomarker analysis in determining *in situ* microbial biomass, community structure and nutritional/physiological status of deep subsurface microbiota. In: Amy, P.S. and Haldeman, D.L. (eds) *The Microbiology of the Terrestrial Deep Subsurface*. Lewis Publishers, Boca Raton, Fla, pp. 119–136.

Whiteley, A.S. and Bailey, M.J. (2000) Bacterial community structure and physiological state within an industrial phenol bioremediation system. *Applied and Environmental Microbiology* 66, 2400–2407.

Whiteley, A.S., Griffiths, R.I. and Bailey, M.J. (2003) Analysis of the microbial functional diversity within water-stressed soil communities by flow cytometric analysis and CTC+ cell sorting. *Journal of Microbiological Methods* 54, 257–267.

Yang, Y.H., Yao, J., Hu, S. and Qi, Y. (2000) Effects of agricultural chemicals on DNA sequence diversity of soil microbial community: a study with RAPD marker. *Microbial Ecology* 39, 72–79.

Yao, H.Y., Jiao, X.D. and Wu, F.Z. (2006) Effects of continuous cucumber cropping and alternative rotations under protected cultivation on soil microbial community diversity. *Plant and Soil* 284, 195–203.

Zachow, C., Tilcher, R. and Berg, G. (2008) Sugar beet-associated bacterial and fungal communities show a high indigenous antagonistic potential against plant pathogens. *Microbial Ecology* 55, 119–129.

Zamioudis, C. and Pieterse, C.M. (2012) Modulation of host immunity by beneficial microbes. *Molecular Plant Microbe Interactions* 25, 139–150.

Zhou, X., Yu, G. and Wu, F. (2012) Responses of soil microbial communities in the rhizosphere of cucumber (*Cucumis sativus* L.) to exogenously applied p-hydroxybenzoic acid. *Journal of Chemical Ecology* 38, 975–983.

Zolla, G., Badri, D.V., Bakker, M.G., Manter, D.K. and Vivanco, J.M. (2013) Soil microbiomes vary in their ability to confer drought tolerance to *Arabidopsis*. *Applied Soil Ecology* 68, 1–9.

16 Improving Crop Performance under Heat Stress using Thermotolerant Agriculturally Important Microorganisms

M.K. Chitara,[1] Chetan Keswani,[2]* Kartikay Bisen,[1] Vivek Singh,[3] S.P. Singh,[2] B.K. Sarma[1] and H.B. Singh[1]

[1]Department of Mycology and Plant Pathology, Institute of Agricultural Sciences, Banaras Hindu University, Varanasi, India; [2]Department of Biochemistry, Institute of Science, Banaras Hindu University, Varanasi, India; [3]Department of Botany, Institute of Science, Banaras Hindu University, Varanasi, India

16.1 Introduction

Agriculture, particularly in tropical regions, is considered as a sector highly prone to climate-change and crop production, due to incessant stresses caused by natural and anthropogenic factors. Increasing incidences of biotic and abiotic stresses have become a major cause for decline in productivity of crops. Global climate change, with a predicted 1.5–5.8°C rise in temperatures by 2100, is imposing a great risk to agricultural production (Rosenzweig *et al.*, 2001).

Average temperature on the Indian subcontinent has increased by 0.57°C in the last 100 years and models project that it is likely to increase by a maximum of 2.5°C by 2050 and 5.8°C by 2100 (Kumar *et al.*, 2006). High temperatures may cause severe cellular injury and cell death may occur within a short time, thereby leading to a catastrophic collapse of cellular organization (Schoffl *et al.*, 1999). Heat stress severely affects plant metabolism, thereby adversely altering growth, development, physiological processes and yield (Hasanuzzaman *et al.*, 2012, 2013) (Fig. 16.1). One of the major effects of high temperature (HT) stress is the excess generation of reactive oxygen species (ROS), which leads to oxidative stress (Hasanuzzaman *et al.*, 2012, 2013). Effects of high temperature can be seen at different levels of plant behaviour, i.e. morphological, physiological (photosynthesis, respiration), and biochemical/molecular changes, in growth as well as in developmental changes resulting in altered life cycle duration. The basic properties of cellular organelles, such as strength of cell membrane, thylakoid structures, cell size, and stomatal regulation are adversely affected. Moreover, high temperature alters the degree of cellular hydration and programmed cell death thereby promoting production of reactive oxygen species. Other similar deleterious effects include osmotic damage, alteration in primary and secondary metabolite profiles, water and ion uptake or movement and altered hormone concentrations.

In the present context heat stress has become a serious problem throughout the

*E-mail: chetankeswani@rediffmail.com

© CAB International 2017. *Advances in PGPR Research* (eds H.B. Singh, B.K. Sarma and C. Keswani)

Fig. 16.1. Major effects of high temperature on plants.

world in crop production (Hall, 2001, 1992). In several crop species, the impact of high temperature stress plays a more significant role in reproductive development than in vegetative growth. Rising temperatures inactivate the enzymatic processes and cause pollen infertility of plants, which leads to yield losses (Young *et al.*, 2004; Zinn *et al.*, 2010). Day by day, increasing demand for food from the emerging economic countries such as India and China is posing an arduous challenge to plant breeders and farmers, and it is expected that by 2050, the expected decline per capita caloric availability will aggravate malnutrition in children by 20% (Nelson, 2009; Chhetri and Chaudhary, 2011). Human survivability depends on food availability irrespective of climate change. Thus there arises an immediate need to develop heat-stress-tolerant crop varieties acclimatized to stress conditions, possibly by incorporating heat-stress-tolerant genes, from such microorganisms in host plants.

In this chapter, we will discuss the thermotolerant agriculturally important microorganisms which play a pivotal role in improving crop performance under thermal stress conditions and diminish the effect of heat stresses on plants through production of exopolysaccharates and biofilm formation.

16.2 What is Heat Stress?

The condition where high temperature causes physiological or biochemical functional changes in plants is aptly referred to as heat/thermal stress. In addition, high temperature increases the rate of reproductive development, thereby shortening the time for photosynthesis and contributing to fruit or seed production. The above-mentioned effect is also a consequence of the heat-stress effect, even though no permanent damage to plant development is caused. The plant intrinsically nurses certain mechanisms to tolerate effects of the heat stress, viz. long-term evolutionary phenological and morphological adaptations and short-term avoidance or acclimatization mechanisms, such as altered leaf orientation, transpirational cooling, or alteration of membrane lipid compositions, etc. (Wahid and Close, 2007).

16.3 Effects of High Temperature on Plants

16.3.1 Seed germination and emergence

Seedling vigour and seed germination are important characteristics for obtaining a good crop stand and high yield. Soil temperature is the major environmental factor that not only affects the proportion of germinated seeds, but also the rate of emergence and subsequent establishment, even under optimum soil and irrigation conditions (Prasad *et al.*, 2006).

The effects of HT on germination were investigated in several crops and serious effects on seed germination were observed. Temperatures above 45°C do not allow a proper rate of germination due to cell death and embryo damage.

16.3.2 Growth and morphology

Due to heat stress, retardation of growth is observed in plants. Heat stress, along with drought stress, induces more harmful effects on growth, yield and productivity of crops than when induced individually (Prasad *et al.*, 2008). Heat stress in higher plants significantly alters cell division and cell elongation rates thereby affecting leaf size and weight. Heat stress may also decrease the stem growth, resulting in reduced plant height (Prasad *et al.*, 2006).

16.3.3 Physiological effects

An alteration in environmental temperature generally affects the physiological processes of plants. The ability of plants to survive in high-temperature conditions is a complex process, determined by environmental factors and also by the genetic capability of the plant. Increase of temperature causes decline in plant growth, photosynthesis rate, respiration rate and enzyme activity of the plants.

16.3.4 Photosynthesis

Generally there is a positive correlation between change in temperature and photosynthesis. But when temperature increases above the normal growing range (15°C to 45°C) of plants, heat injury is caused, resulting in damage of enzymes responsible for photosynthesis. Under HT condition, the activity of stomatal conductance of guard cells (g_s) also decreases significantly (Tan *et al.*, 2011). Deactivation of RuBisCO is the major cause associated with lowering of photosynthesis under HT. Reports suggest that the heat-induced deactivation of RuBisCO is the primary constraint on photosynthesis at moderately HT. However, contradictorily, Chl fluorescence signals from PSII that cause significant deactivation of RuBisCO are not affected by elevated temperatures (Haldimann and Feller, 2004)

16.3.5 Water relations

Plant water status is considered as the most important variable under ambient temperature ranges (Mazorra *et al.*, 2002). HT induces reduction in leaf water level, thereby causing reduction in hydraulic conductance, finally leading to a decline in water absorption (Morales *et al.*, 2003)

16.3.6 Dry matter partitioning

Temperature plays a significant role in dry matter (DM) partitioning of various crops. Stresses like heat and water deficiency lower the assimilation process and mineral uptake during the grain/pod filling period. Sometimes, HT causes harm to sink activity due to earlier panicle senescence, whereas the source activity still exists as leaf senescence does not occur (Morita *et al.*, 2004; Kim *et al.*, 2011). In those cases, grain filling is terminated earlier than complete leaf senescence. It is reported that relevant HT increases the rates of grain filling, fraction of DM partitioning to panicle, and leaf senescence.

16.3.7 Reproductive development

Reproductive development of plants is more sensitive to HT because plant fertility is reduced when temperatures increase (Mckee and Richards, 1998).

16.3.8 Yield

Reduction in crop yield under HT varies with temperature and genotypes of the crop (Table 16.1). HT can reduce crop yield by affecting both source and sink of assimilates (Mendham and Salsbury, 1995).

16.4 Heat Stress Tolerance in Plants

Plants utilize various types of mechanisms for surviving under high temperatures, together with prolonged evolutionary, morphological and phenological adaptations, and temporary avoidance or acclimatization mechanisms such as cooling of plant canopy by transpiration, alteration of membrane lipid compositions or changing leaf orientation. Plants suffer from various stresses at developmental stages and generally try to alleviate the stress by adopting various types of response mechanisms at the tissue level (Queitsch *et al.*, 2000). Stress signals may be received as a change in membrane fluidity or osmotic changes in cells which result in triggering of downstream signalling mechanisms which activate stress-responsive processes to re-establish homeostasis and to protect and repair damaged membranes and proteins.

Up-regulation of several genes has been reported as helping the plant to survive under stress (Tuteja, 2009). Stressed plants receive external and internal signals through different interlinked or independent pathways which are used to adjust various responses for its tolerance development (Kaur and Gupta, 2005). Plant reactions to stress are linked with more than one pathway. To produce a response in specific cellular tissues against a certain stimulus, interactions of signalling molecules and cofactors are desired. Signalling molecules are the results of stress-responsive genes. There are different kinds of signal transduction molecules related to stress-responsive gene activation depending upon the type of plant stress. Some major groups of these are the mitogen-activated protein kinase (MAPK/MPKs), Ca-dependent protein kinases (CDPKs), NO, phytohormones, and sugar (as signalling molecule) (Ahmad *et al.*, 2012). These molecules bind with transcription factors and activate the stress-responsive genes.

After activation of stress-responsive genes, detoxification of reactive oxygen species (ROS) by activating detoxifying enzymes and free radical scavengers occurs; also reactivation of structural proteins and essential enzymes (Ciarmiello *et al.*, 2011) which help to maintain the cellular homeostasis takes place. The underlying signalling mechanisms under HT stress involve the basic helix-loop-helix (bHLH) transcription factor, phytochrome interacting factor 4 (PIF4), whose orthologs have been identified in several crop species (Proveniers and van Zanten, 2013).

16.5 Role of Microorganisms to Improve Crop Performance under Stress

Microorganisms play a crucial role in adjustment and increase of tolerance to abiotic stresses in agricultural plants. Plant growth-promoting rhizobacteria (PGPR) are associated with plant roots and effectively diminish the harmful effects of abiotic stresses (high temperatures, low temperature, drought, salinity and metal toxicity) on plants through production of exo-polysaccharides and biofilm. When plants confront stress conditions, rhizospheric microorganisms interfere with plant cells by different mechanisms such as induction of osmoprotectors and heat-shock proteins, etc.

Table 16.1. Effects of high temperature stress in different crop species.

Crop	Heat treatment	Growth stage	Major effects	Reference
Chili pepper (*Capsicum annuum*)	38/30°C (day/night)	Reproductive, maturity and harvesting stage	Reduced fruit width and fruit weight, increased proportion of abnormal seeds per fruit	Cao *et al.* (2009)
Maize (*Zea mays*)	33–40°C, 15 days	During pre-anthesis and silking onwards	Severe effect on plant and ear growth rates	Zhang *et al.* (2013)
	35/27°C (day/night), 14 days	Reproductive stage	Reduced ear expansion, particularly suppression of cob extensibility by impairing hemicellulose and cellulose synthesis through reduction of photosynthate supply	Yin *et al.* (2010)
Rice (*Oryza sativa*)	Above 33°C, 10 days	Heading stage	Reduced rates of pollen and spikelet fertility.	Hurkman *et al.* (2009)
	25–42.5°C	Vegetative growth stage	Decrease in the CO_2 assimilation rate	Djanaguiraman *et al.* (2011)
	32°C (night temperature)	Reproductive stage	Decreased yield, increased spikelet sterility, decreased grain length, width and weight	Suwa *et al.* (2010)
Sorghum (*Hordeum vulgare*)	40/30°C (day/night)	65 DAS to maturity stage	Decreased chlorophyll (chl) content, chla fluorescence, decreased photosystem II (PSII) photochemistry, P_n and antioxidant enzyme activity and increased ROS content, thylakoid membrane damage, reduced yield	Mohammed and Tarpley (2010)
Soybean (*Glycine max*)	38/28°C (day/night), 14 days	Flowering stage	Decreased leaf P_n and stomatal conductance (g_s), increased thicknesses of the palisade and spongy layers, damaged plasma membrane, chloroplast membrane, and thylakoid membranes, distorted mitochondrial membranes, cristae and matrix	Tan *et al.* (2011)
Tobacco (*Nicotiana tabacum*)	43°C, 2 h	Early growth stage	Decreased net photosynthetic rate (P_n), stomatal conductance as well as the apparent quantum yield (AQY) and carboxylation efficiency (CE) of photosynthesis. Reduced activities of antioxidant enzymes	Gunawardhana *et al.* (2011)
	32 and 34°C	Throughout the growing period	Reduced yield, damages in pod quality parameters such as fibre content and breakdown of the Ca-pectate	Edreira and Otegui (2012)
Wheat (*Triticum aestivum*)	38°C, 24 and 48 h	Seedling stage	Decreased Chl and relative water content (RWC); diminished antioxidative capacity	Hasanuzzaman *et al.* (2013)
	32/24°C (day/night), 24 h	At the end of spikelet initiation stage	Spikelet sterility, reduced grain yield	Saitoh (2008)
	37/28°C (day/night), 20 days	Grain filling and maturity stage	Shortened duration of grain filling and maturity, decrease in kernel weight and yield	Rahman *et al.* (2009)
	30/25°C day/night	From 60 DAS to maturity stage	Reduced leaf size, shortened period for days to booting, heading, anthesis and maturity, drastic reduction in number of grains/spike, smaller grain size and reduced yield	Djanaguiraman *et al.* (2010)

16.5.1 Adaptation of microorganisms as a response to abiotic stress

Soil microflora is influenced by several environmental factors. Some factors are referred to as *modulators* (Balser *et al.*, 2002), e.g., pH, soil temperature, salinity, and water potential (as distinct from factors such as carbon and nitrogen which are better considered as resources for the growth and development of microbial communities). Plant and microbial flora change in response to stress conditions and develop new, tolerant communities, adapted through complex regulatory processes involving many genes (Milosevic and Marinkovic, 2011). Change in the environment alters the biomass and composition of a microbial community. All microorganisms have a set of optimal environmental conditions, which secure their optimal growth (Pettersson, 2004).

The capacity of microorganisms to survive under high temperatures depends on the temperature's duration and intensity as well as microbial adaptation to stress, which is a complex regulatory process involving a number of genes (Tobor-Kapłon *et al.*, 2006; Grover *et al.*, 2011). Certain microbial species live in extreme habitats (halophytes and thermophiles) and they use different mechanisms to reduce stress (Madigen, 1999; Grover *et al.*, 2010; Mishra *et al.*, 2015). In stress conditions, most of the rhizobacteria produce osmoprotectors (K^+, glutamate, trehalose, proline, glycine, and polysaccharides).

Rhizosphere-competent fungi of the genus *Trichoderma* are widely used as biofertilizers and biopesticides in commercial formulations because of their significant beneficial role on plant growth and disease resistance (Tucci *et al.*, 2011). The antifungal properties of *Trichoderma* spp. have been associated with different mechanisms of action, such as the production of antibiotics (Vinale *et al.*, 2008; Keswani *et al.*, 2014) or/and hydrolytic enzymes (Benítez *et al.*, 2004), and competition for nutrients and space (Elad, 2000), and also showing mycoparasitism on pathogens. Abiotic stresses continuously affect the growth and productivity of major crops reducing economic masses to less than half of that possible under ideal growing conditions (Boyer, 1982). It is also known that *Trichoderma* spp. have the capacity to induce resistance to abiotic and biotic stress in plants and promote plant growth and also alleviate the stress (Kuc, 2001; Harman *et al.*, 2004). It was also reported that *Trichoderma harzianum* T22 treated seeds expressed higher germination than untreated seeds whether the stress applied was osmotic, salt, or suboptimal temperatures (Mastouri *et al.*, 2010). The ability of *Trichoderma* spp. to overcome extreme environments facilitates their existence in very varied geographical locations, from Caribbean countries to Antarctica (Hermosa *et al.*, 2004). The T22 strain of *Trichoderma* has improved the tolerance to water loss of tomato seedlings by activating the antioxidant defence mechanism and activity of ascorbate and glutathione-recycling enzymes (Mastouri *et al.*, 2012).

Hence, these studies point to the possibility of employing thermotolerant strains of agriculturally important microorganisms in alleviation of heat stress in crops by relocating them in rhizosphere for modulation of the oxidative and defence network of the host.

16.5.2 PGPR-mediated alleviation of abiotic stress

Reports suggest that certain microorganisms enhance plant tolerance to abiotic stresses such as heat, drought, salinity, nutrient deficiency or excess, etc. (Yang *et al.*, 2008), and high contents of heavy metals (Rajapaksha *et al.*, 2004; Grover *et al.*, 2010; Milošević and Marinković, 2011; Keswani *et al.*, 2013, 2014, 2016a, b; Bisen *et al.*, 2015, 2016; Keswani, 2015). Basically, rhizospheric microorganisms have the greatest role in the tolerance of agricultural plants to abiotic stresses. Near the rhizosphere soil microorganisms trigger various mechanisms that play an eminent role in affecting plant tolerance to stress. It produces some hormones like IAA (indole acetic acid), gibberellins, and other substances that promote growth of root hairs and increase total root area, which in turn facilitates nutrients uptake by plants. Plant growth-promoting rhizobacteria (PGPR), which live in association with plant roots, elicit the largest influence on plants, affecting their productivity and immunity. PGPR inhabit the rhizosphere of many agricultural plants and participate in increasing plant growth and reducing the diseases caused by pathogenic fungi, bacteria, nematodes, and viruses (Klopper *et al.*, 2004; Yang *et al.*, 2008). The mechanism of induced systemic tolerance (IST) causes defence responses via. physical and chemical changes in plants, which can help the plant to alleviate abiotic stresses.

Generally reports suggest production of the enzyme 1-aminocyclopropane-1-carboxylate (ACC) deaminase by bacteria that aid in stimulation of plant growth. Under stress conditions, the bacterial enzyme facilitates the growth of plants by decomposing plant ACC (ethylene precursor in plants). Saleem *et al.* (2007) described the role of ACC deaminase-containing PGPRs in crop production. By reducing the level of ethylene, the plant becomes more resistant to stress conditions in the environment (Glick, 1999).

AM (Arbuscular mycorrhiza) fungi alleviate the effects of stresses (drought and salinity), osmoregulation, and proline accumulation. *Glomus intraradices* increases the tolerance of *Pterocarpus officinalis* to excessive moisture (Grover *et al.*, 2010). In addition, dual symbiotic systems tend to mitigate the effect of abiotic stress on plants. The endophytic fungus *Cuvularia* sp. has been isolated from *Dichanthelium lanuginosum* growing on geothermal soil and reported to be thermotolerant at temperatures ranging from 50°C to 65°C, while when the plant and the fungus grow separately, they do not tolerate temperatures above 38°C (Redman et al., 2002).

16.6 Conclusion

It is well documented that the incessant global rise in temperature affects plant performance, which exhibits a variety of responses such as qualitative and quantitative changes in growth and morphology. Plants can cope with various abiotic stresses and adjust themselves at different growth and developmental stages, while on the other hand microorganisms also help agricultural crop plants to alleviate the abiotic and biotic stress. Some strains of microorganism play a vital role in enabling plants to tolerate heat stress by adopting several mechanisms such as root colonization, association and mutual interaction.

Acknowledgement

CK is grateful to DST-PURSE for providing financial support.

References

Ahmad, P., Bhardwaj, R. and Tuteja, N. (2012) Plant signaling under abiotic stress environment. In: Ahmad, P. and Prasad, M.N.V. (eds) *Environmental Adaptations and Stress Tolerance of Plants in the Era of Climate Change*. Springer, New York, pp. 297–324.

Balser, T.C., Kinzig, A.P. and Firestone, M.K. (2002) Linking soil microbial communities and ecosystem functioning. In: Kinzig, A.P., Pacala, S.W. and Tilman, D. *The Functional Consequences of Biodiversity: Empirical Progress and Theoretical Extensions*. Princeton University Press, Princeton, NJ, pp. 265–293.

Benítez, T., Rincón, A.M., Limón, M.C. and Codón, A.C. (2004) Biocontrol mechanisms of *Trichoderma* strains. *International Microbiology* 7(4):249-60.

Bisen, K., Keswani, C., Mishra, S., Saxena, A., Rakshit, A. and Singh, H.B. (2015) Unrealized potential of seed biopriming for versatile agriculture. In: Rakshit, A., Singh, H.B., Sen, A. (eds) *Nutrient Use Efficiency: From Basics to Advances*. Springer, New Delhi, India, pp. 193–206.

Bisen, K., Keswani, C., Patel, J.S., Sarma, B.K. and Singh, H.B. (2016) *Trichoderma* spp.: efficient inducers of systemic resistance in plants. In: Chaudhary, D.K. and Verma, A. (eds) *Microbial-mediated Induced Systemic Resistance in Plants*. Springer, Singapore, pp. 185–195.

Chhetri, N. and Chaudhary, P. (2011) Green revolution: pathways to food security in an era of climate variability and change. *Journal of Disaster Research* 6, 486–497.

Ciarmiello, L.F., Woodrow, P., Fuggi, A., Pontecorvo, G. and Carillo, P. (2011) Plant genes for abiotic stress. In: Shanker, A.K. and Venkateswarlu, B. (eds) *Abiotic Stress in Plants—Mechanisms and Adaptations*. INTECH Open Access Publisher, Croatia, pp. 283–308.

De Silva, C.S. and Gunawardhana, M.D.M. (2011) Impact of temperature and water stress on growth yield and related biochemical parameters of okra. *Tropical Agricultural Research* 23, 77–83.

Djanaguiraman, M., Prasad, P.V.V. and Seppanen, M. (2010) Selenium protects sorghum leaves from oxidative damage under high temperature stress by enhancing antioxidant defense system. *Plant Physiology and Biochemistry* 48, 999–1007.

Djanaguiraman, M., Prasad, P.V.V. and Al-Khatib, K. (2011) Ethylene perception inhibitor 1-MCP decreases oxidative damage of leaves through enhanced antioxidant defense mechanisms in soybean plants grown under high temperature stress. *Environmental and Experimental Botany* 71, 215–223.

Edreira, J.I.R. and Otegui, M.E. (2012) Heat stress in temperate and tropical maize hybrids: Differences in crop growth, biomass partitioning and reserves use. *Field Crops Research* 130, 87–98.

Elad, Y. (2000) Biological control of foliar pathogens by means of *Trichoderma harzianum* and potential modes of action. *Crop Protection* 19, 709–714.

Glick, B.R. (2005) Modulation of plant ethylene levels by the bacterial enzyme ACC deaminase. *FEMS Microbiology Letters* 251, 1–7.

Grover, M., Ali, S.Z., Sandhya, V., Rasul, A. and Venkateswarlu, B. (2011) Role of microorganisms in adaptation of agriculture crops to abiotic stresses. *World Journal of Microbiology and Biotechnology* 27, 1231–1240.

Haldimann, P. and Feller, U. (2004) Inhibition of photosynthesis by high temperature in oak (*Quercus pubescens* L.) leaves grown under natural conditions closely correlates with a reversible heat-dependent reduction of the activation state of ribulose-1,5-bisphosphate carboxylase/oxygenase. *Plant, Cell & Environment* 27, 1169–1183.

Hall, A.E. (1992) Breeding for heat tolerance. *Plant Breeding Reviews* 10, 129–168.

Hall, A.E. (2001) *Crop Responses to Environment*. CRC Press, Boca Raton, FL.

Harman, G.E., Howell, C.R., Viterbo, A., Chet, I. and Lorito, M. (2004) *Trichoderma* species—opportunistic, avirulent plant symbionts. *Nature Reviews Microbiology* 2, 43–56.

Hasanuzzaman, M., Hossain, M.A., da Silva, J.A.T. and Fujita, M. (2012) Plant response and tolerance to abiotic oxidative stress: antioxidant defense is a key factor. In: Venkateswarlu, B., Shanker, A.K., Shanker, C. and Maheswari, M. (eds) *Crop Stress and Its Management: Perspectives and Strategies*. Springer, Netherlands, pp. 261–315.

Hasanuzzaman, M., Nahar, K. and Fujita, M. (2013a) Plant response to salt stress and role of exogenous protectants to mitigate salt-induced damages. In: Ahmad, P., Azooz, M.M. and Prasad, M.N.V. (eds) *Ecophysiology and Responses of Plants under Salt Stress*. Springer, New York, pp. 25–87.

Hasanuzzaman, M., Nahar, K. and Fujita, M. (2013b) Extreme temperatures, oxidative stress and antioxidant defense in plants. In: Vahdati, K. and Leslie, C. (eds) *Abiotic Stress—Plant Responses and Applications in Agriculture*. INTECH Open Access Publisher, Croatia, pp. 169–205.

Hurkman, W.J., Vensel, W.H., Tanaka, C.K., Whitehand, L. and Altenbach, S.B. (2009) Effect of high temperature on albumin and globulin accumulation in the endosperm proteome of the developing wheat grain. *Journal of Cereal Science* 49, 12–23.

Kaur, N. and Gupta, A.K. (2005) Signal transduction pathways under abiotic stresses in plants. *Current Science* 88, 1771–1780.

Keswani, C. (2015) Proteomics studies of thermotolerant strain of *Trichoderma* spp. Ph.D. Thesis, Banaras Hindu University, Varanasi, India, p. 126.

Keswani, C., Singh, S.P. and Singh, H.B. (2013) A superstar in biocontrol enterprise: *Trichoderma* spp. *Biotech Today* 3, 27–30.

Keswani, C., Mishra, S., Sarma, B.K., Singh, S.P. and Singh, H.B. (2014) Unraveling the efficient application of secondary metabolites of various *Trichoderma*. *Applied Microbiology and Biotechnology* 98, 533–544.

Keswani, C., Bisen, K., Singh, S.P., Sarma, B.K. and Singh, H.B. (2016a) A proteomic approach to understand the tripartite interactions between plant–*Trichoderma*–pathogen: investigating the potential for efficient biological control. In: Hakeem, K.R. and Akhtar, M.S. (eds) *Plant, Soil and Microbes Vol. 2. Mechanisms and Molecular Interactions*. Springer, Switzerland, pp. 79–93.

Keswani, C., Bisen, K., Singh, V., Sarma, B.K. and Singh, H.B. (2016b) Formulation technology of biocontrol agents: present status and future prospects. In: Arora, N.K., Mehnaz, S. and Balestrini, R. (eds) *Bioformulations: For Sustainable Agriculture*. Springer, New Delhi, India, pp. 35–52.

Kim, J., Shon, J., Lee, C.K., Yang, W., Yoon, Y. *et al.* (2011) Relationship between grain filling duration and leaf senescence of temperate rice under high temperature. *Field Crops Research* 122, 207–213.

Mastouri, F., Björkman, T. and Harman, G.E. (2012) *Trichoderma harzianum* enhances antioxidant defense of tomato seedlings and resistance to water deficit. *Molecular Plant-Microbe Interactions* 25, 1264–1271.

Mazorra, L.M., Nunez, M., Hechavarria, M., Coll, F. and Sánchez-Blanco, M.J. (2002) Influence of brassinosteroids on antioxidant enzymes activity in tomato under different temperatures. *Biologia Plantarum* 45, 593–596.

McKee, J. and Richards, A.J. (1998) The effect of temperature on reproduction in five *Primula* species. *Annals of Botany* 82, 359–374.

Mendham, N.J. and Salsbury, P.A. (1995) Physiology, crop development, growth and yield. In: Kimber, D.S. and McGregor, D.I. (eds) *Brassica Oilseeds: Production and Utilization*. CAB International, Wallingford, UK, pp. 11–64.

Mishra, S., Singh, A., Keswani, C., Saxena, A., Sarma, B.K. and Singh, H.B. (2015) Harnessing plant–microbe interactions for enhanced protection against phytopathogens. In: Arora, N.K. (ed.) *Plant Microbes Symbiosis: Applied Facets*. Springer, New Delhi, India, pp. 111–125.

Mohammed, A.R. and Tarpley, L. (2010) Effects of high night temperature and spikelet position on yield-related parameters of rice (*Oryza sativa* L.) plants. *European Journal of Agronomy* 33, 117–123.

Morales, D., Rodríguez, P., Dell'Amico, J., Nicolas, E., Torrecillas, A. and Sánchez-Blanco, M.J. (2003) High-temperature preconditioning and thermal shock imposition affects water relations, gas exchange and root hydraulic conductivity in tomato. *Biologia Plantarum* 47, 203–208.

Morita, S., Shiratsuchi, H., Takahashi, J. and Fujita, K. (2004) Effect of high temperature on grain ripening in rice plants: analysis of the effects of high night and high day temperatures applied to the panicle and other parts of the plant. *Japanese Journal of Crop Science* 73, 77–83.

Pettersson, M. (2004) *Factors affecting rates of change in soil bacterial communities*. Marie Pettersson, Högbovägen 1F, 227 31 Lund, Sweden.

Prasad, P.V., Boote, K.J. and Allen, L.H. (2006) Adverse high temperature effects on pollen viability, seed-set, seed yield and harvest index of grain-sorghum [*Sorghum bicolor* (L.) Moench] are more severe at elevated carbon dioxide due to higher tissue temperatures. *Agricultural and Forest Meteorology* 139, 237–251.

Prasad, P.V.V., Staggenborg, S.A. and Ristic, Z., (2008) Impacts of drought and/or heat stress on physiological, developmental, growth, and yield processes of crop plants. In: Ahuja, L.H. and Saseendran, S.A. (eds) *Response of Crops to Limited Water: Understanding and Modeling Water Stress Effects on Plant Growth Processes. Advances in Agricultural Systems Modeling Series, Vol. 1*. ASA-CSSA, Madison, WI, USA, pp. 301–355.

Proveniers, M.C. and van Zanten, M. (2013) High temperature acclimation through PIF4 signaling. *Trends in Plant Science* 18, 59–64.

Queitsch, C., Hong, S.W., Vierling, E. and Lindquist, S. (2000) Heat shock protein 101 plays a crucial role in thermotolerance in *Arabidopsis*. *The Plant Cell* 12, 479–492.

Rahman, M.A., Chikushi, J., Yoshida, S. and Karim, A.J.M.S. (2009) Growth and yield components of wheat genotypes exposed to high temperature stress under control environment. *Bangladesh Journal of Agricultural Research* 34, 360–372.

Rajapaksha, R.M.C.P., Tobor-Kapłon, M.A. and Bååth, E. (2004) Metal toxicity affects fungal and bacterial activities in soil differently. *Applied and Environmental Microbiology* 70, 2966–2973.

Redman, R.S., Sheehan, K.B., Stout, R.G., Rodriguez, R.J. and Henson, J.M. (2002) Thermotolerance generated by plant/fungal symbiosis. *Science* 298, 1581–1581.

Rosenzweig, C., Iglesias, A., Yang, X.B., Epstein, P.R. and Chivian, E. (2001) Climate change and extreme weather events; implications for food production, plant diseases, and pests. *Global Change & Human Health* 2, 90–104.

Saitoh, H. (2008) *Ecological and Physiology of Vegetable*. Nousangyoson Bunka Kyoukai, Tokyo.

Saleem, M., Arshad, M., Hussain, S. and Bhatti, A.S. (2007) Perspective of plant growth promoting rhizobacteria (PGPR) containing ACC deaminase in stress agriculture. *Journal of Industrial Microbiology & Biotechnology* 34, 635–648.

Suwa, R., Hakata, H., Hara, H., El-Shemy, H.A., Adu-Gyamfi, J.J. *et al.* (2010) High temperature effects on photosynthate partitioning and sugar metabolism during ear expansion in maize (*Zea mays* L.) genotypes. *Plant Physiology and Biochemistry* 48, 124–130.

Tan, W., wei Meng, Q., Brestic, M., Olsovska, K. and Yang, X. (2011) Photosynthesis is improved by exogenous calcium in heat-stressed tobacco plants. *Journal of Plant Physiology* 16, 2063–2071.

Tobor-Kapłon, M.A., Bloem, J. and De Ruiter, P.C. (2006) Functional stability of microbial communities from longterm stressed soils to additional disturbance. *Environmental Toxicology and Chemistry* 25, 1993–1999.

Tucci, M., Ruocco, M., De Masi, L., De Palma, M. and Lorito, M. (2011) The beneficial effect of *Trichoderma* spp. on tomato is modulated by the plant genotype. *Molecular Plant Pathology* 12, 341–354.

Tuteja, N. (2009) Integrated calcium signaling in plants. In: Baluska, F. and Mancuso, S. (eds) *Signaling in Plants*. Springer Berlin Heidelberg, Germany, pp. 29–49.

Vinale, F., Sivasithamparam, K., Ghisalberti, E.L., Marra, R., Woo, S.L. and Lorito, M. (2008) *Trichoderma*–plant–pathogen interactions. *Soil Biology and Biochemistry* 40, 1–10.

Wahid, A. and Close, T.J. (2007) Expression of dehydrins under heat stress and their relationship with water relations of sugarcane leaves. *Biologia Plantarum* 51, 104–109.

Yang, J., Kloepper, J.W. and Ryu, C.M. (2009) Rhizosphere bacteria help plants tolerate abiotic stress. *Trends in Plant Science* 14, 1–4.

Yin, Y., Li, S., Liao, W., Lu, Q., Wen, X. and Lu, C. (2010) Photosystem II photochemistry, photoinhibition, and the xanthophyll cycle in heat-stressed rice leaves. *Journal of Plant Physiology* 167, 959–966.

Young, L.W., Wilen, R.W. and Bonham-Smith, P.C. (2004) High temperature stress of *Brassica napus* during flowering reduces micro- and megagametophyte fertility, induces fruit abortion, and disrupts seed production. *Journal of Experimental Botany* 55, 485–495.

Yun-Ying, C., Han, D., Li-Nian, Y., Zhi-Qin, W., Li-Jun, L. and Jian-Chang, Y. (2009) Effect of high temperature during heading and early grain filling on grain yield of *indica* rice cultivars differing in heat-tolerance and its physiological mechanism. *Acta Agronomica Sinica* 35, 512–521.

Zhang, X., Cai, J., Wollenweber, B., Liu, F., Dai, T. *et al.* (2013) Multiple heat and drought events affect grain yield and accumulations of high molecular weight glutenin subunits and glutenin macropolymers in wheat. *Journal of Cereal Science* 57, 134–140.

Zinn, K.E., Tunc-Ozdemir, M. and Harper, J.F. (2010) Temperature stress and plant sexual reproduction: uncovering the weakest links. *Journal of Experimental Botany* 53, 1–10.

17 Phytoremediation and the Key Role of PGPR

Elisabetta Franchi[1]* and Giannantonio Petruzzelli[2]
[1]*Eni S.p.A, Renewable Energy & Environmental R&D, S. Donato Milanese, Italy;*
[2]*Institute of Ecosystem Study, National Council of Research, Pisa, Italy*

17.1 Phytoremediation

The original concept of phytoremediation is derived from studies on plants which can uptake and tolerate extremely high levels of heavy metals. These plants were defined hyperaccumulators (Brooks *et al.*, 1977) and these studies originated from an article (Minguzzi and Vergnano, 1948), describing the ability of *Alyssum bertolonii* to accumulate very high amounts of nickel. Brooks (1998) underlined the seminal importance of this article for the development of phytoremediation: 'a small perennial shrub in Tuscany, Italy, was destined to lead the way to a whole range of new technologies and discoveries'. Nowadays, phytoremediation identifies a series of plant-based technologies that can be applied to a wide range of organic and inorganic contaminants for remediating polluted soil, water and sediments, by exploiting the multiple properties of plants, which can be used in different specific processes.

In soil remediation, some fundamental strategies, which have been extensively described (ITRC, 2009; Samasrdjieva *et al.*, 2011), can be briefly summarized as follows:

- Phytoextraction, which involves the cultivation of plant species able to uptake and accumulate the contaminants in the aerial part and the subsequent removal of the vegetable biomass, enriched in contaminants.
- Phytostabilization, which is based on the ability of plants to immobilize the contaminants in the root zone preventing the leaching of dissolved contaminants while stabilizing the soil, thus reducing the aerial dispersion of contaminated soil particles.
- Phytodegradation, which exploits the ability of plants both to absorb and degrade organic contaminants and to increase the microbial activity promoting the biodegradation of the pollutants.
- Phytovolatilization, which involves the use of plants for the transfer of contaminants from the polluted media into the atmosphere through the process of transpiration. It has been proposed for volatile organic compounds and mercury contamination. However, this procedure comprises some risks due to the toxicity of the volatilized contaminants.
- Rhizofiltration is the main technology to decontaminate polluted water. It is

*E-mail: elisabetta.franchi@eni.com

© CAB International 2017. *Advances in PGPR Research* (eds H.B. Singh, B.K. Sarma and C. Keswani)

based on aquatic plants either floating or submerged which uptake and concentrate, by their roots, the contaminants, removing them from the aqueous environment.

Phytotechnologies, in addition to economic advantage over conventional remediation techniques, promote beneficial side effects, such as carbon sequestration, soil erosion control, improvement of soil quality and landscape image. Thus, these green remediation approaches have achieved an increasing interest from the stakeholders and public opinion.

17.1.1 Phytoremediation mechanisms

Plants act on organic contaminants by means of different mechanisms: direct absorption of the contaminant and subsequent accumulation of metabolites in plant tissues, or release of compounds that, stimulating the microbial activity of the soil, promotes the degradation of organic molecules (US EPA, 2001).

Direct absorption of the contaminants through the root system depends on several factors such as the concentration of contaminants in the soil solution, the transpiration rate, the chemical species of the contaminants and soil moisture conditions. This mechanism is very effective for the remediation of sites polluted by moderately hydrophobic compounds (defined by a low octanol–water partition coefficient: log $K_{ow} < 3.5$), such as BTEX (Benzene, Toluene, Ethylbenzene and Xylenes), chlorinated solvents and low molecular weight aliphatic compounds (Boonsaner *et al.*, 2011; Collins *et al.*, 2011). The hydrophobic compounds (log $K_{ow} > 3.5$) are strongly retained on the root surfaces and cannot be translocated in the plant.

For the degradation of organic molecules, the best effective application of technology stems from the increase in degradation processes in the rhizosphere deriving from the interactions of plant and microorganisms (US EPA 2006) (Chaudhry *et al.*, 2005; Fester *et al.*, 2014).

The use of vegetation increases the content of humic substances, promotes the activity of bacteria and fungi and has a positive influence on all those factors favouring the degradation of the organic compounds in the soil. Through the root exudates, substances such as sugars, alcohols and organic acids are released, with a consequent stimulating effect on microbial capability to degrade organic molecules. The symbiotic relationship between plants and microorganisms is thus responsible for a more rapid degradation of the soil contaminants (Azaizeh *et al.*, 2011). Moreover, plants can secrete several enzymes capable of degrading organic contaminants such as explosives, chlorinated compounds, herbicides, polyaromatic hydrocarbons, etc. (Alkorta and Garbisu, 2001). The smaller molecules derived from the degradation process are generally less toxic and less persistent than the originals (Chhikara *et al.*, 2010).

Phytostabilization and phytoextraction are the most used strategies for the remediation of soils contaminated with inorganics, such as heavy metals. Phytostabilization processes use plants to immobilize contaminants in the soil by adsorption on the roots and/or immobilization in the rhizosphere. These processes reducing the mobility of contaminants prevent their migration into groundwater and decrease their bioavailability. This technique does not involve the definitive removal of the pollutants that, once immobilized, remain on the site (Alkorta *et al.*, 2010). Phytostabilization is particularly appropriate when the contaminants' concentration is so high that the phytoextraction would take too long to achieve the remediation targets, as, for example, in mining sites. Suitable plant species must grow vigorously to exert a hydraulic control in addition to a proper action of contaminant immobilization. Plants growing during the phytostabilization process improve the structural stability of soil and reduce the risk of erosive processes.

Phytoextraction is the most attractive phytotechnology since it enables the use of a biological technique to remove non-biodegradable pollutants, such as heavy metals, from a contaminated site. The phytoextraction technology was originally designed for hyperaccumulator plants, capable of absorbing huge amounts of metals from the soil, accumulating them in the above-ground biomass and allowing the subsequent removal from the site by harvesting.

Hyperaccumulators have often been identified as plants naturally evolved on soils with high specific metal content. Thus, it is often difficult to use these plants to clean up polluted soil where metals have a quite different origin or several elements are present at the same time. Moreover, hyperaccumulators are often characterized by a reduced biomass production and a lack of commercially available seeds. Then, the use of other species, including herbaceous field crops is considered a viable alternative to hyperaccumulators for removing trace metals, since these species may compensate their lower metal uptake by a greater biomass yield.

Thus, phytoextraction has essentially followed two strategies: continuous or natural phytoextraction, in which hyperaccumulator species are involved (Ghaderian *et al.*, 2007) and assisted phytoextraction, in which additives are used to release the metals from solid phases into soil solution, thus increasing the bioavailable amounts of metals promoting a greater uptake by tolerant crop plants. Both strategies depend on biomass production and the amount of contaminants absorbed by plants. These variables are strictly dependent on the plants' ability to grow in contaminated soils as well as contaminants' bioavailability.

17.1.2 Focus on bioavailability

The efficiency of all *in situ* technologies is closely dependent on soil properties that determine the distribution of contaminants between the different soil phases (solid, liquid and gaseous). This is particularly important for phytoextraction, since plants uptake the contaminants, which are, or become, bioavailable only if they are in the liquid phase of the soil (the soil solution). The assessment of contaminants' bioavailability is therefore essential for a successful application of technology (Barbafieri *et al.*, 2013).

In soil, bioavailability is the result of complex mass transfer and adsorption/release mechanisms, which depend on the contaminant properties, the chemical-physical characteristics of the soil and the biology of the organisms involved that, in this case, are the plants (NRC, 2002). The first critical step for increasing bioavailability is the transfer of the contaminants from the solid phase to the soil solution: only after being released in the aqueous phase can a contaminant move towards the roots of the plants to be absorbed.

With regard to heavy metals, the most important class of inorganic contaminants, their chemical form in the soil is crucial for the possible use of phytoextraction, and thus for the success of remediation (Pedron *et al.*, 2009, 2013). Heavy metals' bioavailability depends on their chemical properties and the soil features regulating the processes of adsorption/desorption, precipitation/dissolution, complexation/complex dissociation, which regulate their distribution between the solid phase and the soil solution (Peijnenburg *et al.*, 2007). In the soil environment, these several processes are interdependent, and the soil solution is the focal point of metal reactivity, since the liquid phase surrounding soil particles and root surfaces is the "continuum" from which plants and microorganisms can absorb nutrients and contaminants. In the soil solution, elements in soluble forms are in equilibrium with those adsorbed on soil surfaces. Ion removal from solution promotes their desorption from exchange sites.

These reactions determine contaminants' mobility and bioavailability for plant uptake, which is largely dependent on soil properties, in particular pH, organic matter, clay content, cation exchange capacity (CEC), and redox potential (Pezzarossa *et al.*, 2011).

In soils characterized by high content of humic acids or with a significant presence of clays, metals are strongly held by these components, with the consequent reduction of their availability for the plants (Abdullah and Sarem, 2010; Wanga *et al.*, 2010). Even more important is the soil pH, which determines the precipitation, or solubilization of the metals (Li *et al.*, 2003; Chaney *et al.*, 2005). Redox potential affects both directly the bioavailability of some metals, such as Hg, As and Cr, and indirectly, by influencing the reactivity of iron and manganese oxyhydroxides, having a high adsorption capacity for all metals (Cherlatchka and Cambier, 2000). Moreover, the establishment of stronger bonds between metals and the soil surfaces tends to increase the persistence in soil, and to reduce the bioavailability with a consequent decrease in their phytoextraction (Shelmerdine *et al.*, 2009).

Thus, to increase the efficiency of this technology it is necessary to promote metal bioavailability in soils. This can be achieved with amendments, such as chelating agents, promoting the desorption of metals from the solid phase, increasing their concentration in the soil solution and therefore plant uptake. Assisted phytoextraction, as previously stated, is based on this procedure.

Many soil conditioners have been used and several promising results are ascribable to the increase of metal solubility, particularly at laboratory or greenhouse scale. Organic acids such as ethylenediaminetetraacetic acid (EDTA), N-(hydroxyethyl)-ethylenediaminetriacetic acid (HEDTA), and diethylenetriaminepentaacetic acid (DTPA) were the most commonly used additives. EDTA has been repeatedly used since it can complex many heavy metals (Seth *et al.*, 2011). These ligands generally increase the metals' transfer from soil to the roots, but their high mobilizing capacity could exceed the bioavailable quantity that plants are able to uptake. Thus, with their long persistence in the soil, an increase of metal concentration in the soil liquid phase (Luo *et al.*, 2005; Santos *et al.*, 2006), could lead to a potential risk of leaching into the groundwater and this side effect should always be considered.

To provide metal chelators which are less phytotoxic and more readily biodegradable, several low molecular weight organic acids, such as ethylenediamine-N,N'-disuccinic acid (EDDS), have been used as alternatives to EDTA (Luo *et al.*, 2005; Doumett *et al.*, 2011, Pedron *et al.*, 2014). Furthermore, easily biodegradable organic acids, being compounds similar to those naturally produced by plants in root exudates, can positively influence the microbial activity in the rhizosphere.

The use of additives is particularly interesting when, with a single product, it is possible to increase, at the same time, the bioavailability of more than one element. Interesting results have recently been obtained (Petruzzelli *et al.*, 2014) with the addition of ammonium thiosulfate, a common fertilizer, to a soil contaminated by mercury (Hg) and arsenic (As). Arsenic and mercury, which are non-essential elements for plants, are characterized, in the soil, by very different chemical properties. In many contaminated sites, arsenic and mercury are simultaneously present and, generally, different clean-up procedures are applied. Hg phytoextraction is often based on the use of a thiosulphate salt (Moreno *et al.*, 2004; Pedron *et al.*, 2013), whereas for As, a phosphate salt (Tassi *et al.*, 2004) is considered the most efficient additive. In this experiment with *B. juncea* and *L. albus* on a multi-contaminated industrial soil, the addition of ammonium thiosulfate greatly promoted the uptake and translocation of both Hg and As in the aboveground parts of the plants. The use of the same additive able to increase the plant uptake of both contaminants can greatly reduce both time and costs of remediation.

The contaminants' bioavailability in the soil is also important for organics' phytoremediation, even if methods may be greatly different in relation to the chemical characteristics of the various organic compounds. Moreover, the various differences of both analytical procedures and associated measurement are often a source of uncertainty (Cui *et al.*, 2013). Considering, for example, an important class of organic contaminants, the polycyclic aromatic hydrocarbons, it is possible to identify some general aspects that can be extended to other classes of organic compounds. The absorption through the root system and the distribution inside the plant are much reduced due to the low solubility in water and to the adsorption process of hydrophobic substances on soil organic matter. The absorption by the roots is possible for the hydrocarbons with three or four aromatic rings, but it is very unlikely for more complex molecular structures, and, an adsorption of hydrocarbons on the roots surface, without their transfer within the plant, often occurs (Meng *et al.*, 2011).

Phytoremediation offers many advantages over other technologies since the costs can be from 40% to 90% lower compared to *ex situ* technologies; it is applicable simultaneously in the presence of different classes of contaminants and after remediation the quality of the soil is maintained or even improved. However, like all remediation technologies, phytoremediation has some drawbacks and cannot be applied to all sites. The major limitations of the technology are related to plant features, such as biomass production, bioaccumulation capacity, and the volume of soil explored by the root system. Nevertheless, it is important to emphasize that the phytoremediation efficiency depends not only on plant-related factors, but basically on aspects related to soil characteristics controlling contaminant mobility and bioavailability. Applying phytotechnologies entails specific operations for each contaminated site. The selection of plant species, sowing and harvesting must be carefully planned according to the precise properties of the contaminated soil.

Since the amount of biomass that can be produced is a critical aspect, it is essential to provide measures of fertilization and irrigation adapted to the climatic conditions of the site, promoting all the strategies able to increase plant growth, such as the addition of PGPR. The use of PGPR has been shown to positively influence the efficiency of phytoremediation both for PAHs and heavy metals (Franchi *et al.*, 2016a, b).

17.2 Significance of PGPR for an Effective Phytoremediation

Many plants are able to hyperaccumulate metals or degrade organic molecules, but these environmental pollutants often induce a stressful situation limiting plants' growth and hence their phytoremediation performance. Hyperaccumulator plants (e.g. *Pteris vittata* for As; *Alyssum bertolonii* for Ni; *Thlaspi caerulescens* for Cd, Zn, Pb and Mn) are the most effective at removing metals from the soil, but, in general, are small and slow-growing, thus reducing their potential for metal phytoextraction (Rascio and Navari-Izzo, 2011). Plant growth and sometimes metal phytoextraction can be helped by soil microorganisms living in close association with plant roots (Glick, 2014). Root exudates promote the proliferation of specific groups of microorganisms able to aggressively colonize the root surface, affecting plant growth and often enhancing the biodegradation of organic compounds in the soil (Vacheron *et al.*, 2013). Some rhizosphere microorganisms can metabolize the organic pollutants by their own degradative capabilities (rhizodegradation), while some soil bacteria positively affect plants by improving growth and health, enhancing root development and plant tolerance to various environmental stresses (Fig. 17.1). Certainly, the phytoremediation of both organic and inorganic contaminants will produce better results if plants are larger and in good health (Glick, 2010).

It has long been known (Chakrabarty, 1981) that some soil microorganisms can efficiently degrade several toxic organic compounds and along with this finding, microbial degradation provides a practical and effective resource. Hydrocarbons derived from petroleum industries and accidental leakages during transport and storage of oil derivatives have an impact on the environment strong enough to be considered the main xenobiotic organic molecules. Hydrocarbon-degrading bacteria have developed efficient biodegradation strategies to transform the hydrocarbons to more easily metabolizable substrates (Prince *et al.*, 2010; Nie *et al.*, 2014) and, for the most part, they belong to the phyla of Proteobacteria, Actinobacteria and Firmicutes with a predominance of Gamma proteobacteria (*Pseudomonas* spp.). But, during a biodegradation process, with the depletion of the pollutants the bacterial community often changes, and a modification of the dominant phylotypes can occur (Militon *et al.*, 2010). *Pseudomonas* genus is naturally widespread in the

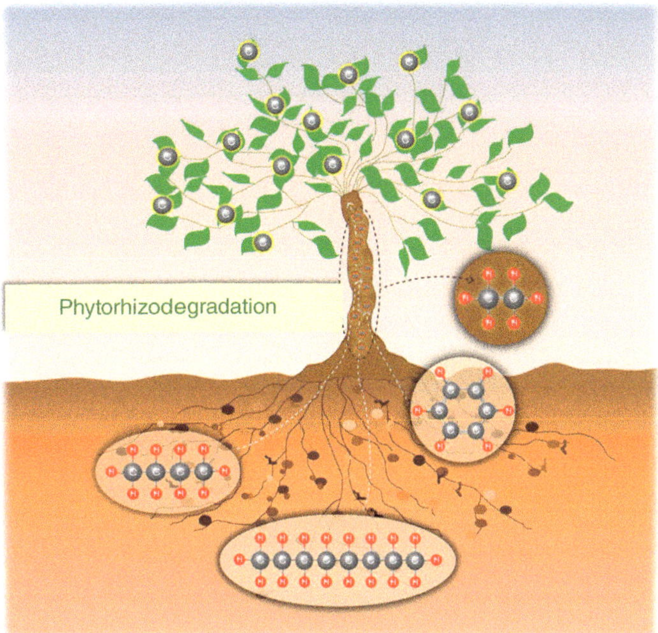

Fig. 17.1. Schematic overview of phytorhizodegradation.

environment and its versatility is well known. Most *Pseudomonas* spp. are able to synthesize rhamnolipids (Meliani and Bensoltane, 2014; Silva *et al.*, 2016), biosurfactant molecules causing larger dispersion of water-insoluble n-alkanes in the aqueous phase. This interaction among cells and the smaller solubilized hydrocarbons led to a fast uptake of hydrocarbon into bacterial cells (Das and Chandran, 2011).

The addition of hydrocarbon-degrading bacteria to contaminated soils is frequently effective at a laboratory scale, but in the field the biodegradation rate may be too slow and a successful bioremediation is often more difficult to realize. Nevertheless, the addition of plants to contaminated soils for metabolizing and removing toxic compounds could be equally difficult since, even if hypertolerant, plants' growth is usually considerably reduced and the biomass is not enough to allow an efficient degradation of the contaminants within a reasonable time frame. The use of hydrocarbon-degrading bacteria, together with plants, is the best strategy to overcome the restrictions of both conventional bioremediation and phytoremediation (Glick, 2010). These biodegrading bacteria efficiently stick to the plant roots and some of them (endophytes) are able to enter inside the plant cells (Lumactud *et al.*, 2016). Root endophytic communities are generally thought to be a subgroup of the rhizospheric bacteria (Weyens *et al.*, 2009a) although a recent report (Gottel *et al.*, 2011) showed that, in poplar trees, root endophytic communities are distinct groups rather than opportunistic subsets of the rhizosphere community (Germaine *et al.*, 2013). In addition, some of these bacteria possess plant growth-promoting, as well as biodegradative, activities. Most of the published studies have therefore been done under controlled laboratory conditions, growth chamber or greenhouse conditions, and few studies include field tests. Over the past twenty years, several combined approaches with plants and biodegradative bacteria have been applied to remove total petroleum hydrocarbons (TPH), polycyclic aromatic hydrocarbons (PAH) and several halogenated compounds (Nakamura *et al.*, 2004; Escalante-Espinosa *et al.*, 2005; Radwan *et al.*, 2005; Germaine *et al.*, 2006; Kim *et al.*, 2006; Leigh *et al.*, 2006; Liu *et al.*, 2007; Alarcón *et al.*, 2008; Al-Awadhi *et al.*, 2009; Germaine *et al.*, 2009; Uhlik *et al.*, 2009; Barrutia *et al.*, 2011; Slater *et al.*, 2011; Leewis *et al.*, 2013; Bramley-Alves *et al.*, 2014; Budhadev *et al.*, 2014; Wei *et al.*, 2014; Xu *et al.*, 2014; Yergeau *et al.*, 2014; Pagé *et al.*, 2015; McIntosh *et al.*, 2016; Leewis *et al.*, 2016) from contaminated soils.

One of the most important properties of hydrocarbon-degrading bacteria, is definitely the production of biosurfactants, and this feature makes polycyclic aromatic hydrocarbons, relatively insoluble compounds, more bioavailable. Biosurfactants are small molecules affecting a decrease in interfacial tension, but also amphiphilic macromolecules stabilizing the emulsion (Menezes Bento *et al.*, 2005). These molecules reduce surface and interfacial tensions in both aqueous solutions and hydrocarbon mixtures making them potential agents for bioremediation (Banat *et al.*, 2000; Mnif *et al.*, 2015; Vijayakumar and Saravanan, 2015; Das and Kumar, 2016). Biosurfactants increase the surface area of hydrophobic water-insoluble substrates and consequently their bioavailability. The emulsification created by the biosurfactant molecules enhances the growth of bacteria and hence the rate of bioremediation. Very often, rhizospheric bacteria showing PGP traits are also biosurfactant producers, in particular *Pseudomonas* and *Bacillus* spp. (Kumar *et al.*, 2014).

The phytoremediation of heavy metals is, in general, technically more difficult than that of organic compounds. While hydrocarbons can be broken down *in situ* either in plants or in the soil, metals cannot be degraded and have to be removed from the soils. The main limitations of most metal phytoextraction processes are the bioavailability of the target metal(s) and the ability of various plants to accumulate metals within their aboveground biomass (Fig. 17.2). As previously described, metal bioavailability can be increased through the addition of various chelating agents, a strategy working well at a laboratory scale but often much less effective in the field. Several vegetable species have been tested for their ability to take up high levels of metals and to translocate them from roots to leaves and shoots, but, many of the so-called hyperaccumulating plants do not produce sufficient biomass

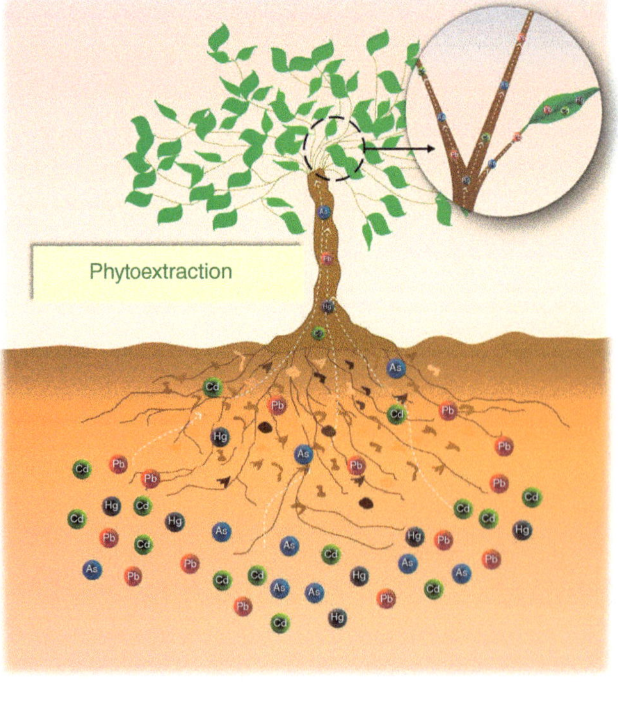

Fig. 17.2. Schematic overview of phytoextraction.

to make this process efficient in the field (Raskin and Ensley, 1999). The use of soil bacteria (PGPR) as helpers in metal phytoremediation can significantly facilitate the growth of plants in the presence of high and occasionally inhibitory levels of metals, but usually do little or nothing to increase metal bioavailability. These soil bacteria are typically selected for resistance to the target toxic metal(s) and then verified for the presence of the most important PGPR traits such as the occurrence of the enzyme ACCD (1-aminocyclopropane-1-carboxylate deaminase) and the ability to synthesize IAA (3-indoleacetic acid) and siderophores. Even if the precise mechanisms of plant growth promotion in the presence of metals are not definitively recognized, most published data are consistent with the involvement of IAA, siderophores and ACCD. Experimental data have demonstrated that the presence of all or even some of these activities is sufficient to help and enhance plant growth (Tak *et al.*, 2013), suggesting that IAA promotes plant growth (Patten and Glick, 2002), ACCD prevents ethylene stress, reducing the inhibitory effect on plant growth (Glick *et al.*, 2007) and siderophores help plants to assimilate the required iron in the presence of great amounts of other metals potentially competing for the uptake (Burd *et al.*, 2000; Rajkumar *et al.*, 2010; Bisen *et al.*, 2015; Mishra *et al.*, 2015; Keswani *et al.*, 2016).

Besides ACCD activity, IAA and siderophores production, many other bacterial features may assist metal phytoremediation; also, genetically engineered bacteria producing various metal-binding peptides making some metals more bioavailable, have been used in phytoremediation approaches (Wu *et al.*, 2006; Ike *et al.*, 2007). Moreover, bacteria having an active inorganic phosphate solubilization system, seem to facilitate phytoremediation by assisting metal uptake (Ma *et al.*, 2011), and the bacterial biosurfactants, establishing complexes with heavy metals at the soil interface, desorb metals from soil matrix, thus increasing metal solubility and bioavailability in the soil solution (Rajkumar *et al.*, 2012).

The last decade has been important to a better understanding of the various bacterial contributions to phytoremediation. The efficacy of bacterially assisted phytoremediation has been mainly demonstrated under laboratory conditions, but, especially for organic contaminants, this approach has already been found to be effective in the field and actually field trials are rapidly growing. To achieve efficient phytoremediation of metal-contaminated soils is, although crucial and still under investigation, the problem of metal bioavailability.

During a phytoremediation experiment, a large number of variables come into play, such as: the plant type (allochthonous or native), the physical-chemical features of the soil (pH, cation exchange capacity (CEC), soil texture, total organic and inorganic percentages (TOC, TIC)), the indigenous microbial community, and the kind and the amount of contaminants. Despite the complexity of the variables involved, published data allow us to extrapolate some general suggestions that can be applied to facilitate the phytoremediation of different contaminated soils. Essential requirements are certainly the ability to degrade soil contaminants, plant growth-promoting capability based on ACCD activity, the synthesis of IAA and the endophytic bacteria capable of colonizing the internal tissues of the plant (Weyens *et al.*, 2009a). As already said, for a successful phytoremediation of metals it is crucial to find practical solutions to enhance the bioavailability of many metals. Finally, the simple strategy of adding PGPR (preferably endophytes) able to reduce plant ethylene levels by ACCD activity and with the ability to synthesize the phytohormone IAA can significantly (and often dramatically) increase both plant growth and phytoremediation activity (Glick and Stearns, 2011).

17.3 PGPR Effect on Metals Phytoextraction

The majority of metals comes from mining, smelting, fertilizers, pesticides, coal combustion, medical waste, combustion of leaded petrol, and batteries (Wuana and Okieimen, 2011), and can contaminate soils, plants, sediments, and surface water (Ullah *et al.*, 2015). Common toxic metals are mercury (Hg), lead (Pb) and cadmium (Cd), while others, such as copper (Cu), chromium (Cr), manganese (Mn), zinc (Zn), and nickel (Ni) are essential microelements which become toxic at high concentrations. To these heavy metals should be added aluminium (Al) and two metalloids, equally toxic, antimony (Sb) and arsenic (As) (Duruibe *et al.*, 2007). These elements have severe effects on terrestrial and aquatic ecosystems and are extremely hazardous to humans since they can enter the body through food, water, air and contact with the skin (Tchounwou *et al.*, 2012). It is therefore crucial to decrease these health threats, eliminating them from the environment. Remediation of heavy metals is needed to protect humans, plants and animals from their toxic effects, thus saving the environment for future generations (Glick, 2010).

Metalliferous plants can grow on metal-enriched soils without any symptoms of toxicity. These plants can be categorized into three groups: indicators (Smith, 2013), excluders and hyperaccumulators (Baker, 1981; Baker and Brooks, 1989). Indicator plants reflect soil metals' concentration in their shoots. Excluder plants are able to grow on metalliferous soils, but metals are largely excluded from uptake into plant tissues and hence, their concentrations in the shoots are lower than those in the roots. Hyperaccumulator plants are able to accumulate large amounts of metals in their aboveground tissues (Sessitsch *et al.*, 2013).

Three basic strategies of metal phytoextraction have been developed: (i) natural phytoextraction using hyperaccumulators; (ii) natural phytoextraction using fast-growing and high biomass plants; and (iii) chemically assisted phytoextraction adding soil additives to increase metal mobility (solubility) in the soil (Vangronsveld *et al.*, 2009).

Plant-associated bacteria can contribute to enhance metal uptake by plants and thus the efficiency and rate of phytoextraction (Kuffner *et al.*, 2008; 2010; Sessitsch and Puschenreiter, 2008; Sheng *et al.*, 2008; Rajkumar *et al.*, 2009; Weyens *et al.*, 2009b; Glick, 2010). Also, an improved biomass production can increase the efficiency of the metal phytoextraction.

Trace metal biogeochemistry is deeply affected by microorganisms that can influence metal speciation facilitating mobilization or immobilization mechanisms, often altering the equilibrium of metal species between soluble and insoluble phases. Mobilization of metals can be achieved by protonation, chelation, or chemical transformation. Immobilization can instead take place by precipitation of insoluble organic or inorganic compounds but also by sorption, uptake or intracellular sequestration (Gadd, 2004). These sorbed, precipitated or occluded metals can then be solubilized by acidification, chelation and ligand-induced dissolution. Protons, exported by bacteria, replace metal cation adsorption sites, dissolving trace elements containing minerals such as phosphates. After chelation, organic chelator compounds scavenge metals from sorption sites protecting them from resorption processes. Bacterially produced natural chelators are carboxylic acid anions and siderophores (Gadd, 2004).

Bacteria producing chelating organic acids, such as citric, oxalic or acetic have been shown to mobilize various metals in soil (Li *et al.*, 2009). Acid-producing rhizosphere bacteria, able to release phosphorus from insoluble metal phosphate species, are often referred to as phosphate solubilizers and hence PGPR (Gupta *et al.*, 2002). Increased metal uptake in various plants after inoculation with acid producers or phosphate solubilizers has been reported (Ma *et al.*, 2011).

In addition, some bacteria thriving in metalliferous soils and producing siderophore molecules, carboxylic acids and proteins, such as phytochelatins, metallothioneins and metallohistins have a probable role in trace element complexation in the rhizosphere (Haferburg and Kothe, 2010).

Metal-resistant PGPR can improve plant growth under stress conditions due to toxic trace element concentrations. The positive effect on plant growth and biomass is largely due to the production of phytohormones (such as IAA), suppression of stress ethylene production (due to ACCD activity), or improvement in nutritional status by the presence of N_2 fixers, PO_4-solubilisers, or siderophore-producers. Doubtless, a simple improvement in biomass results in an increase in the overall yield of phytoextracted metals; thus, it can be said that plant growth promotion plays a key role in the extraction and removal of metals (Sessitsch *et al.*, 2013).

With the aim to identify useful strains for phytoextraction purposes, several rhizospheric and endophytic bacteria associated with metal-tolerant plants have been isolated. Most of these microorganisms have shown plant growth promotion features and, when reinoculated, the growth of the host plants was normally enhanced (Ma *et al.*, 2011).

In 1998, a spill occurred at the Aznalcóllar mine that released over 5000 tons of sludge and acidic waters contaminated with extremely high concentrations of heavy metals and metalloids along the Guadiamar river, 20 km from the city of Sevilla (Grimalt *et al.*, 1999). The accident is considered one of the greatest environmental tragedies to have happened in Europe and severe residual contamination by As, Cd, Cu, Pb and Zn has been reported (Galán *et al.*, 2002). Legume plants have been found among the first colonizers after the toxic spill (Prasad and Freitas, 2003; Carrasco *et al.*, 2005). Besides their capacity to tolerate heavy metals, legumes are able to establish symbiotic interaction with rhizobia, being a source of combined nitrogen for the biosphere and a model for microbe–plant interaction studies (Graham and Vance, 2003). In particular, *Lupinus* species have been proposed for phytoremediation of metals (Vázquez *et al.*, 2008). Dary *et al.* (2010) showed, during field experiments carried out in the zone affected by the toxic spill of the Aznalcóllar mine, that *Lupinus luteus* is adequate for metal stabilization of soils with a moderate level of heavy metal pollution. Yellow lupines accumulated heavy metals mainly in roots (Cu, Cd and especially Pb were poorly translocated to shoots), demonstrating a potential use in metal phytostabilization. Co-inoculation of lupines with a consortium of metal-resistant PGPR (including *Bradyrhizobium* sp., *Pseudomonas* sp. and *Ochrobactrum cytisi*) produced an enhancement of plant biomass and a decrease in metal accumulation in shoots and roots, probably due to a protective effect on the rhizosphere.

Several mechanisms of action on phytoremediation by PGPR can occur and most probably are both plant- and substrate-dependent (Grandlic *et al.*, 2008). Environmental conditions to which an inoculant is exposed will undoubtedly impact on the activation of certain plant growth-promotion traits (Becerra-Castro *et al.*, 2012). These processes may be delayed by the high concentrations of metals in the soils (Dell'Amico *et al.*, 2005), as suggested by Marques *et al.* (2013), assessing the effects of inoculating metal resistant PGPR on the growth of *Helianthus annuus* grown in Zn and Cd spiked soils. PGPR strains, *Ralstonia eutropha* and *Chrysiobacterium humi,* produced modifications in metal bioaccumulation and bioconcentration, reducing losses of weight in metal-exposed plants. They also observed that bacterial community diversity decreased with increasing metal levels in the soil, while, after inoculation with PGPR, a higher bacterial diversity in rhizospheric soil of plants was maintained throughout the experimental period. Inoculation of sunflower with *Chrysiobacterium humi* seems to be a good method to enhance the short-term stabilization potential of the plant in metal-contaminated soil, reducing losses in plant biomass and aboveground tissue contamination.

Zinc is an essential trace element but, at millimolar levels, may be toxic to organisms through soil or water contamination. Zinc toxicity limits are ranging from 150 to 300 mg kg^{-1}, depending upon the growth stage and plant species (Nagajyoti *et al.*, 2010; Yadav, 2010). Greenhouse experiments with *Brassica juncea* plants exposed to 400 mg Zn kg^{-1} investigated the capabilities of *Pseudomonas brassicacearum* (strain DBK11) and *Rhizobium leguminosarum* (strain WSM1325) to promote growth (Adediran *et al.*, 2015). Reduced growth in non-inoculated plants was ascribed to accumulation of Zn oxalate and Zn sulphate in roots. *P. brassicacearum* displayed a modest plant growth promoting ability, while *R. leguminosarum* alone and also in combination with *P. brassicacearum* showed a greater effect on plant growth and Zn phytoextraction. The improved growth, together with the increased metal accumulation detected in inoculated plants, were attributed to the storage of Zn as Zn phytate and Zn cysteine in the root. Thus, since both bacteria do not statistically improve *B. juncea* growth in the absence of Zn, the authors suggest that this bacteria-induced metal chelation could represent the main mechanism of plant growth-promoting bacteria in toxicity attenuation and microbial-assisted phytoremediation. The authors, however, point out that more studies are needed, by using different forms of Zn, in order to simulate natural conditions in real-life metal-contaminated soils.

Chen *et al.* (2014) examine the effects of inoculation with the endophytic bacterium *Sphingomonas* SaMR12, on plant growth, root morphology, and root exudates. SaMR12 was isolated from *Sedum alfredii* which showed heavy metal (in particular zinc) resistance and the ability to efficiently transport the metals from the roots to the shoots. Organic acids, such as malic, oxalic and tartaric, mainly produced by *S. alfredii* roots, are probably involved in increasing heavy metal bioavailability (Li *et al.*, 2013). Through hydroponic experiments, SaMR12 inoculation considerably improved the efficiency of zinc phytoextraction by increasing the biomass, zinc absorption, root morphology, and root exudates.

Under iron-deficient conditions, rhizobacteria produce siderophores. The activity of *Pseudomonas aeruginosa* ATCC 15692, a well-known siderophore-producing rhizobacterium was assayed with *Pteris cretica L.* in an As-contaminated soil (Jeong *et al.*, 2014). *P. cretica* grown in the siderophore-amended soil showed a higher As uptake than the plant grown in the EDTA-treated soil, and As, taken up by roots in the presence of siderophores, seemed to be favourably translocated to shoots. Lampis and colleagues (2015) carried out a greenhouse pot experiment to assess the efficiency of arsenic phytoextraction by another As-hyperaccumulating fern, *Pteris vittata*, growing in a soil contaminated with arsenopyrite cinders, with the help of selected rhizobacteria isolated from the polluted soil. The bacteria (*Pseudomonas* sp., *Delftia* sp., *Bacillus* sp., *Variovorax* sp. and *Pseudoxanthomonas* sp.) were selected for multiple beneficial traits such as the production of IAA and siderophores and the capability to reduce arsenate to arsenite. The inoculation of contaminated soil with these best-performing strains increased plant biomass achieving an eight-fold increase in the arsenic BCF (bioconcentration factor) and a three-fold increase in PE (phytoremediation efficiency) compared to non-inoculated plants. These results demonstrate that the inoculation of the hyperaccumulator fern species *P. vittata* with bacteria selected for their plant growth-promoting features can significantly enhance arsenic phytoextraction from highly contaminated environmental matrices. Due to its high As tolerance, the rhizosphere of *Pteris vittata* is a valuable source of endophytic microorganisms (Zhu *et al.*, 2014; Tiwari *et al.*, 2016; Xu *et al.*, 2016).

Cadmium is a metal with a small biological requirement and is marginally degradable by abiotic and biotic mechanisms. It may come from soil erosion and human activities such as mining operations, but also from agricultural practices causing adverse effects in humans upon consumption of contaminated vegetables (Tchounwou *et al.*, 2012). Some plants (tobacco, rice, other cereal grains, potatoes and other vegetables) take up cadmium from the soil more readily than other heavy metals such as lead and mercury (Satarug *et al.*, 2010). *Micrococcus* sp. strain MU1 and *Klebsiella* sp. strain BAM1, cadmium-resistant PGPR, effectively increased cadmium solubilization in cadmium-supplemented soil and promoted *Helianthus annuus* root elongation at toxic concentrations of cadmium (Prapagdee *et al.*, 2013). After inoculation with *Micrococcus* MU1, an increase in plant dry weight and cadmium accumulation in *H. annuus* was observed. Moreover, *Klebsiella* BAM1 promoted the translocation of cadmium from the roots to the shoots demonstrating that these plant–microbe interactions were efficacious in promoting plant growth and cadmium phytoextraction of *H. annuus* planted in polluted soils.

Very often, in soils, sediments and waters, not only one heavy metal is present, but several different trace elements characterize the contamination. Therefore, rhizospheric and endophytic bacteria usually show more than one resistance. Indigenous As- and Hg-tolerant bacteria isolated and selected from a multi-contaminated soil (belonging to *Proteobacteria*, *Firmicutes* and *Actinobacteria* subdivisions) showed *in vitro* properties known to have possible positive influence on plant growth. The addition of this PGPR consortium to *Brassica juncea* and *Lupinus albus* plant roots in microcosms led to a meaningful increase of the biomass and a moderate increase in the concentration of metals in shoots, thus supporting and enhancing the effect on phytoextraction by the use of the common fertilizer thiosulphate (Franchi *et al.*, 2016a).

A recent study (Álvarez-López *et al.*, 2016) evaluated different bacterial inoculation methods (seed inoculation, soil inoculation, dual soil inoculation event, and seed+soil inoculation) of tobacco plants growing in a mine-soil contaminated with Pb, Zn, and Cd. The inoculation with IAA-producing *Rhodococcus erytropolis* strain P30 positively affected the phytoextraction process of the three metals. The most pronounced effect was observed in plant biomass production and, to a lesser extent, in the shoot metal accumulation and plant nutritional status. Unexpectedly, a single soil inoculation event led to the best results together with the lower bacterial density (10^6 CFU mL^{-1} vs 10^8 CFU mL^{-1}). These outcomes pointed out the importance of the inoculation method but also the significance of the cellular density of the inoculum, as both can modify the results in terms of plant performance and soil metal removal.

In the perspective of a post-process biomass to energy conversion, Janssen *et al.* (2015) propose a remediation strategy for metal-contaminated land through the exploitation of short-rotation coppicing of willow and poplar. Since metal phytoextraction looked insufficient to obtain a rapid reduction of soil metal contents, two strategies were suggested: (i) *in situ* selection of the best performing clones and (ii) bioaugmentation of these clones with beneficial plant-associated bacteria. Several cultivable bacterial populations were isolated from the rhizosphere, roots and twigs of two *Salix* clones (*S. viminalis* and *S. alba x alba*), selected on the basis of field data. Compared to the best performing commercial clones, considerable increases (up to 74% for Cd and 91% for Zn) in stem metal extraction were achieved. Two *Salix* clones grown in Cd-Zn-Pb contaminated soil were then inoculated with five bacterial strains, selected for their plant growth and metal uptake-promoting features. However, although the selected strains used for inoculation, showed *in vitro* beneficial characters (such as ACCD activity and IAA production), increases in biomass and metals extraction were not always observed. These data clearly indicate that predicting the *in vivo* effect of a bacterial strain on plant growth or metal uptake, based only on phenotypic characteristics expressed *in vitro*, is not as straightforward as desired (Weyens *et al.*, 2013). An appropriate plant colonization is a crucial step required to obtain valuable effects inside the plant, and the possible competition with trillions of indigenous bacteria is also to be taken into account (Lugtenberg and Kamilova, 2009).

17.4 Rhizoremediation of Organic Contaminants

Microorganisms can use organic contaminants as carbon source and electron reservoir for respiration (Germida *et al.*, 2002). Soil conditions are essential for hydrocarbon degradation by microorganisms, and the following levels are considered optimal: soil moisture at 30% of water-holding capacity, soil pH between 6.5 and 8, oxygen content between 10% and 40%, and low clay or silt content for soil type (Das and Chandran, 2011). Since most individual bacterial species do not own the whole metabolic pathway, degradation is generally achieved via a consortium of microorganisms with various enzyme systems (Chaudhry *et al.*, 2005). The fastest and greatest degradation of most organic contaminants occurs under aerobic conditions, and several aerobic bacterial genera such as *Pseudomonas*, *Alcaligenes*, *Sphingomonas*, *Rhodococcus* and *Arthrobacter* have been reported to degrade alkanes and polyaromatic compounds (Thavamani *et al.*, 2012; Kuppusamy *et al.*, 2016). Rhizospheric bacteria and fungi generally live under conditions of "nutrient starvation" and are, thus, continuously looking for nutrients (Rohrbacher and St-Arnaud, 2016). Root exudates are high in organic acids (citric, malic, succinic, oxalic, pyruvic), carbohydrates (glucose, xylose, fructose, maltose, sucrose, ribose), amino acids, fatty acids, proteins, enzymes, nucleotides and vitamins (Badri and Vivanco, 2009; Ben, 2015). Microorganisms have developed sensory systems (chemotaxis) driving them to these root-secreted molecules just for getting the necessary nutrients and energy for their survival and reproduction (Gao and Zhu, 2005). Diversity in root exudates creates different microbial communities, specific to each plant species (Huang *et al.*, 2014). Therefore, root exudates are crucial in determining the composition and diversity of the rhizosphere microbial community (Johnson *et al.*, 2005; Rentz *et al.*, 2005; Singh *et al.*, 2007; Mathesius and Watt, 2010; Badri *et al.*, 2013).

Also, root exudates can significantly affect the communication among bacteria, with molecules mimicking bacterial signals of quorum sensing (Mangwani *et al.*, 2015). Exudates can similarly facilitate plant–microbe and microbe–microbe interactions by recruiting beneficial specific microorganisms such as PGPR, mycorrhizal fungi or nitrogen-fixing bacteria (Baudoin *et al.*, 2003; Berendsen *et al.*, 2012; Huang *et al.*, 2014). Indeed, the release of exudates (in particular sugars and amino acids) was shown to attract PGPR (Somers *et al.*, 2004; Huang *et al.*, 2014). *Pseudomonas* spp. hold chemotactic proteins for malic acid, citric acid, and amino acids (especially leucine) assisting colonization of tomato roots (Oku *et al.*, 2012) and released malic acid allows the enrolment of the PGPR *Bacillus subtilis* (Rudrappa *et al.*, 2008). The phenolic compounds in root exudates can act as specific substrates and signalling molecules, playing roles in rhizospheric plant–microbe interactions, such as legume–rhizobia symbioses (Mandal *et al.*, 2010; Michalet *et al.*, 2013). Actually, legume plants are able to secrete phenolic compounds attracting and inducing the chemotaxis of *Rhizobium* species (Mandal *et al.*, 2009). Then, root exudates play a pivotal role in biodegradation providing carbon source and energy to hydrocarbon-degrading microorganisms and improving the hydrocarbon degradation in the rhizosphere (Dzantor, 2007; Gao *et al.*, 2011; Phillips *et al.*, 2012).

Land farming and bioremediation are the major practices used for remediation of soils contaminated with petroleum and polycyclic aromatic hydrocarbons (PAH). However, their efficacy in the removal of persistent and highly hydrophobic hydrocarbons is restricted. Huang and colleagues (2004) evaluated a multi-process phytoremediation approach to remove PAH from contaminated soils. In real contaminated sites, mixtures of chemical pollutants are generally very complex, and a multi-process system could be necessary. Land farming increases the oxidative potential of the soil, enhancing physical volatilization and photochemical oxidation, and at the same time, improves environmental conditions for soil microorganisms, promoting the biodegradation activity. Inoculation of plants with PAH-degrading bacteria promotes a microbial degradation process that is not limited by the availability of degradation bacteria in the soil. Further, the addition of PGPR provides better plant growth by increasing plant tolerance to contaminants in the soil. This multiple approach greatly accelerated the remediation process since, although land farming, bioremediation, and phytoremediation have some efficacy in remediation of persistent PAHs, the success of each method alone is limited. The combination of these processes, together with the inoculation of plants with PGPR, can overcome the limitations of the individual methods and the effectiveness of this multi-process remediation system to eliminate persistent contaminants has been demonstrated.

A similar approach was applied to a multi-contaminated soil by using hydrocarbon-degrading indigenous bacteria (Franchi *et al.*,

2016b). The phytoremediation described in this paper was made possible only after a land farming pre-treatment stimulating and supporting the microbial activity. Phytoremediation was further sustained by the addition of a microbial consortium made up of indigenous bacteria showing *in vitro* PGP features.

Radwan *et al.* (2007) reported that nodule bacteria (*Rhizobium leguminosarum* and *Bradyrhizobium japonicum*) and PGPR (*Pseudomonas* spp. and *Flavobacterium* spp.) isolated from the roots of the legume *Vicia faba* were able to grow on crude oil and individual pure hydrocarbons (n-alkanes with chain lengths from C9 to C40 and aromatic hydrocarbons: benzene, biphenyl, naphthalene, phenanthrene, and toluene) as sole carbon and energy sources. Quantitative hydrocarbon analysis confirmed that both nodule bacteria and PGPR were active in hydrocarbon consumption. Moreover, intact nodules of *V. faba* containing bacteria reduced hydrocarbon levels in a medium in which those nodules were shaken. Legume crops are therefore suitable phytoremediation tools for oily soil, since they enrich such soils with fixed nitrogen, and also with hydrocarbon-utilizing microorganisms.

A three-year field test of a PGPR-enhanced phytoremediation (PEP) system at a contaminated land farm led to a successful remediation of TPH (Gurska *et al.*, 2009). The remediation strategy consisted of physical manipulation of the soil performed through tilling, sunlight exposure (aeration/photo-oxidation) and plant growth with PGPR. The inoculation of PGPR (*Pseudomonas* spp.) led to an extensive development of the root system enhancing contaminant degradation and supporting an active rhizosphere that effectively promoted TPH degradation, including high molecular weight petroleum fractions often resistant to remediation.

Hong *et al.* (2011) studied the effects of the inoculation of *Gordonia* sp. S2RP-17 on the growth of *Zea mays* and on diesel removal in diesel-contaminated soil, using mesocoms systems. *Gordonia* sp. S2RP-17 was isolated from the rhizoplane of *Equisetum arvense* that had inhabited diesel-contaminated soil for a long period and was verified to have ACCD activity and siderophore synthesizing ability. Results show that this bacterial isolate can actually enhance remediation efficiency in diesel-contaminated soil, also by promoting the growth of *Zea mays*.

Very often, due to the poor hydrocarbon accessibility, bioremediation strategies are limited. This low solubility can be overcome by the use of biosurfactants. Two *Pseudomonas* strains (RK4 and RK3) isolated from oil-contaminated soils (Kumar *et al.*, 2014) were discovered to have plant growth-promoting features as well as biosurfactant properties. The influence of the inoculum of these strains and the interaction with *Withania somnifera,* in oil-contaminated soil, reveal that biosurfactants (rhamnolipids) secreted by *Pseudomonas* strains were able to help lowering oil hydrocarbon toxicity, and plant growth-promoting features improved both the growth and the antioxidant activity of *W. somnifera*. Consortia of both strains showed better results with respect to the individual strains, suggesting beneficial synergistic interactions.

Festuca arundinacea L. (tall fescue) is a perennial species with a highly branched fine fibrous root system that could meaningfully increase the efficiency of hydrocarbon degradation in the soil. During a pot experiment (Hou *et al.*, 2015) designed to study the micro-ecological mechanism of PGPR enhancing phytoremediation, tall fescue biomass increased by PGPR addition. Furthermore, petroleum hydrocarbons, primarily C21-C34 fractions of aliphatic hydrocarbons (AHs) and polycyclic aromatic hydrocarbons (PAHs), were removed, at the highest level, by two PGPR inoculation treatments. The authors demonstrated that the removal efficiency was not related to bacterial diversity but to the selective effect of phytoremediation on specific bacterial communities. RDA (Redundance Analysis, depicting the relationship between petroleum fractions and bacterial diversity) revealed that *Lysobacter*, *Pseudoxanthomonas*, *Planctomyces*, *Nocardioides*, *Hydrogenophaga* and *Ohtaekwangia* genera were positively correlated with high molecular weight petroleum hydrocarbons (C21–C34 AHs and PAHs).

Rhizoremediation comprises the use of PGPR to remove organic pollutants. The efficiency of rhizoremediation by inoculation of the common and cost-effective plant Alfalfa (*Medicago sativa*) with hydrocarbon-degrading bacteria, was evaluated (Bano *et al.*, 2015). Among the different strains isolated from oily sludge, having different capabilities to enhance biodegradation, *Bacillus altitudinis* (KF859970) showed an accelerated rate of degradation of n-alkanes and some methyl branched hydrocarbons, with respect to non-inoculated soils, proving to degrade straight-chain hydrocarbons faster than any other strains. Even if the other bacterial strains showed interaction with alfalfa and biodegradation of oily sludge, the degradation rate was slower than that obtained with *B. altitudinis*.

The same isolated strains from oily sludge were used to evaluated the role of a PGPR consortium on the physiology of maize (*Zea mays*) grown under oily sludge stress environment (Shahzad *et al.*, 2016). This is the first study reporting the fate of bacterial consortium and fertilizers on maize antioxidant defence system estimated by peroxidase (POD) and superoxide dismutase (SOD) content, under oily sludge stress environment. POD and SOD content was greater in maize plants when grown on 30% oily sludge contaminated soil, suggesting that maize is tolerant to 30% of oily sludge and the antioxidant defence system works accurately under an oily sludge stress environment. Combined application of consortium and ammonium nitrate and ammonium phosphate as fertilizers improved the germination percentage, protein and proline content in maize plants, and decreased SOD and POD of the maize leaves grown in oily sludge. Degradation of total petroleum hydrocarbon (TPHs) was 59% higher with the combined addition of consortium and fertilizer than untreated maize. The bacterial consortium is therefore able to enhance maize tolerance to oily sludge promoting TPH degradation (Liao *et al.*, 2015).

References

Abdullah, S. and Sarem, S.M. (2010) The potential of chrysanthemum and pelargonium for phytoextraction of lead-contaminated soils. *Journal of Civil Engineering* 4, 409–416.

Adediran, G.A., Ngwenya, B.T., Mosselmans, J.F.W., Heal, K.V. and Harvie, B.A. (2015) Mechanisms behind bacteria induced plant growth promotion and Zn accumulation in *Brassica juncea*. *Journal of Hazardous Materials* 283, 490–499, DOI:10.1016/j.jhazmat.2014.09.064.

Alarcón, A., Davies, F.T. Jr, Autenrieth, R.L. and Zuberer, D.A. (2008) Arbuscular mycorrhiza and petroleum-degrading microorganisms enhance phytoremediation of petroleum contaminated soil. *International Journal of Phytoremediation* 10, 251–263, DOI: 10.1080/15226510802096002.

Al-Awadhi, H., El-Nemr, I., Mahmoud, H., Sorkhoh, N.A. and Radwan, S. (2009) Plant-associated bacteria as tools for the phytoremediation of oily nitrogen-poor soils. *International Journal of Phytoremediation* 11, 11–27, DOI:10.1080/15226510802363261.

Alkorta, I. and Garbisu, C. (2001) Phytoremediation of organic contaminants in soils. *Bioresource Technology* 79, 273–276, DOI: 10.1016/S0960-8524(01)00016-5.

Alkorta, I., Becerril, J.M. and Garbisu, C. (2010) Phytostabilization of metal contaminated soils. *Reviews on Environmental Health* 25(2), 135–146, DOI: 10.1515/REVEH.2010.25.2.135.

Álvarez-López, V., Prieto-Fernández, A., Janssen, J., Herzig, R., Vangronsveld, J. and Kidd, P.S. (2016) Inoculation methods using *Rhodococcus erythropolis* strain P30 affects bacterial assisted phytoextraction capacity of *Nicotiana tabacum*. *International Journal of Phytoremediation* 18(4), 406–415, DOI:10.1080/15226514.2015.1109600.

Azaizeh, H., Castro, P.M.L. and Kidd, P. (2011) Biodegradation and rhizoremediation of xenobiotics using rhizospheric bacteria. In: Schröder, P. and Collins, C.D. (eds) *Organic Xenobiotics and Plants: From Mode of Actions to Ecophysiology*. Springer, Dordrecht, Netherlands, pp. 191–216, DOI: 10.1007/978-90-481-9852-8.

Badri, D.V. and Vivanco, J.M. (2009) Regulation and function of root exudates. *Plant Cell Environment* 32, 666–681, DOI: 10.1111/j.1365-3040.2008.01926.x.

Badri, D.V., Chaparro, J.M., Zhang, R.F., Shen, Q.R. and Vivanco, J.M. (2013) Application of natural blends of phytochemicals derived from the root exudates of arabidopsis to the soil reveal that phenolic-related compounds predominantly modulate the soil microbiome. *Journal of Biological Chemistry* 288, 4502–4512, DOI: 10.1074/jbc.M112.433300.

Baker, A.J.M. (1981) Accumulators and excluders - strategies in the response of plants to heavy metals. *Journal of Plant Nutrition* 3, 643–654.

Baker, A.J.M. and Brooks, R.R. (1989) Terrestrial higher plants which accumulate metallic elements: a review of their distribution, ecology and phytochemistry. *Biorecovery* 1, 81–126.

Banat, I.M., Makkar, R.S. and Cameotra, S.S. (2000) Potential commercial applications of microbial surfactants. *Applied Microbiology and Biotechnology* 53, 495–508.

Bano, A., Shahzad, A. and Siddiqui, S. (2015) Rhizodegradation of hydrocarbon from oily sludge. *Journal of Bioremediation & Biodegradation* 6, 289, DOI:10.4172/2155-6199.1000289.

Barbafieri, M., Japenga, J., Romkens, P., Petruzzelli, G. and Pedron, F. (2013) Protocols for applying phytotechnologies in metal-contaminated soils. In: Gupta, D.K. (ed.) *Plant-Based Remediation Processes, Soil Biology 35*, Springer Berlin Heidelberg, Germany, pp. 19–37, DOI:10.1007/978-3-642-35564-6_2.

Barrutia, O., Garbisu, C., Epelde, L., Sampedro, M.C., Goicolea, M.A. and Becerril, J.M. (2011) Plant tolerance to diesel minimizes its impact on soil microbial characteristics during rhizoremediation of diesel-contaminated soils. *Science of the Total Environment* 409(19), 4087–4093, DOI: 10.1016/j.scitotenv.2011.06.025.

Baudoin, E., Benizri, E. and Guckert, A. (2003) Impact of artificial root exudates on the bacterial community structure in bulk soil and maize rhizosphere. *Soil Biology & Biochemistry* 35, 1183–1192, DOI: 10.1016/S0038-0717(03)00179-2.

Becerra-Castro, C., Monterroso, C., Prieto-Fernandez, A., Rodriguez-Lamas, L., Loureiro-Vinas, M. *et al.* (2012) Pseudometallophytes colonising Pb/Zn mine tailings: a description of the plant microorganism rhizosphere soil system and isolation of metal tolerant bacteria. *Journal of Hazardous Materials* 217–218, 350–359, DOI: 10.1016/j.jhazmat.2012.03.039.

Ben, L. (2015) Life of microbes in the rhizosphere. In: Lugtenberg, B. (ed.) *Principles of Plant–Microbe Interactions – Microbes for Sustainable Agriculture*. Springer, Berlin, pp. 7–15, DOI: 10.1007/978-3-319-08575-3_3.

Berendsen, R.L., Pieterse, C.M.J. and Bakker, P.A.H.M. (2012) The rhizosphere microbiome and plant health. *Trends in Plant Sciences* 17, 478–486, DOI: 10.1016/j.tplants.2012.04.001.

Bisen, K., Keswani, C., Mishra, S., Saxena, A., Rakshit, A. and Singh, H.B. (2015) Unrealized potential of seed biopriming for versatile agriculture. In: Rakshit, A., Singh, H.B. and Sen, A. (eds) *Nutrient Use Efficiency: From Basics to Advances*. Springer, New Delhi, India, pp. 193–206.

Boonsaner, M., Borrirukwisitsak, S. and Boonsaner, A. (2011) Phytoremediation of BTEX contaminated soil by *Cannax generalis*. *Ecotoxicology and Environmental Safety* 74, 1700–1707, DOI: 10.1016/j.ecoenv.2011.04.011.

Bramley-Alves, J., Wasley, J., King, C.K., Powell, S. and Robinson, S.A. (2014) Phytoremediation of hydrocarbon contaminants in subantarctic soils: an effective management option. *Journal of Environmental Management* 1(142), 60–69, DOI: 10.1016/j.jenvman.2014.04.019.

Brooks, R.R. (1998) *Plants that Hyperaccumulate Heavy Metals*. CAB International, Wallingford, UK.

Brooks, R.R., Lee, J., Reeves, R.D. and Jaffre, T. (1977) Detection of nickeliferous rocks by analysis of herbarium specimens of indicator plants. *Journal of Geochemical Exploration* 7, 49–57, DOI:10.1016/0375-6742(77)90074-7.

Budhadev, B., Rubul, S., Sabitry, B. and Hari Prasad, S. (2014) Phytoremediation of petroleum hydrocarbon (PHC) contaminated soil by using *Mimosa pudica* L. *Journal of Environmental Science and Engineering* 56(3), 327–332.

Burd, G.I., Dixon, D.G. and Glick, B.R. (2000) Plant growth-promoting bacteria that decrease heavy metal toxicity in plants. *Canadian Journal of Microbiology* 46, 237–245.

Carrasco, J.A., Armario, P., Pajuelo, E., Burgos, A., Caviedes, M.A. *et al.* (2005) Isolation and characterisation of symbiotically effective *Rhizobium* resistant to arsenic and heavy metals after the toxic spill at the Aznalcóllar pyrite mine. *Soil Biology and Biochemistry* 37, 1131–1140, DOI:10.1016/j.soilbio.2004.11.015.

Chakrabarty, A.M. (1981) Microorganisms having multiple compatible degradative energy-generating plasmids and preparation thereof. US Patent 4(259), 444.

Chaney, R.L., Angle, J.S., McIntosh, M.S., Reeves, R.D., Li, Y.M. *et al.* (2005) Using hyperaccumulator plants to phytoextract soil Ni and Cd. *Zeitschrift für Naturforschung* C 60, 190–198.

Chaudhry, Q., Blom-Zandstra, M., Gupta, S. and Joner, E.J. (2005) Utilizing the synergy between plants and rhizosphere microorganisms to enhance break down of organic pollutants in the environment. *Journal of Environmental Science and Pollution Research International* 12, 34–48, DOI:10.1065/espr2004.08.213.

Chen, B., Shen, J., Zhang, X., Pan, F., Yang, X., and Feng, Y. (2014) The endophytic bacterium, sphingomonas SaMR12, improves the potential for zinc phytoremediation by its host, *Sedum alfredii*. *PLoS ONE* 9, 9, DOI:10.1371/journal.pone.0106826.

Cherlatchka, R. and Cambier, P. (2000) Influence of reducing conditions on solubility of trace metals in contaminated soils. *Water, Air, and Soil Pollution* 118, 143–167, DOI: 10.1023/A:1005195920876.

Chhikara, S., Paulose, B., White, J.C. and Dhankher, O.P. (2010) Understanding the physiological and molecular mechanism of persistent organic pollutant uptake and detoxification in cucurbit species (Zucchini and Squash). *Environmental Science and Technology* 44, 7295–7301, DOI: 10.1021/es100116t.

Collins, C.D., Martin, I. and Doucette, W. (2011) Plant uptake of xenobiotics. In: Schroeder, P. and Collins, C.D. (eds) *Organic Xenobiotics and Plants: From Mode of Actions to Ecophysiology*. Springer Science and Business Media, Dordrecht, Netherlands pp. 3–16.

Cui, X., Mayer, P. and Ganc, J. (2013) Methods to assess bioavailability of hydrophobic organic contaminants: principles, operations, and limitations. *Environmental Pollution* 172, 223–234, DOI: 10.1016/j.envpol.2012.09.013.

Dary, M., Chamber-Pérez, M.A., Palomares, A.J. and Pajuelo, E. (2010) "In situ" phytostabilisation of heavy metal polluted soils using *Lupinus luteus* inoculated with metal resistant plant-growth promoting rhizobacteria. *Journal of Hazardous Materials* 177, 323–330, DOI:10.1016/j.jhazmat.2009.12.035.

Das, A.J. and Kumar, R. (2016) Bioremediation of petroleum contaminated soil to combat toxicity on *Withania somnifera* through seed priming with biosurfactant producing plant growth promoting rhizobacteria. *Journal of Environmental Management* 174, 79–86, DOI: 10.1016/j.jenvman.2016.01.031.

Das, N. and Chandran, P. (2011) Microbial degradation of petroleum hydrocarbon contaminants: an overview. *Biotechnology Research International*, DOI: 10.4061/2011/941810.

Dell'Amico, E., Cavalca, L. and Andreoni, V. (2005) Analysis of rhizobacterial communities in perennial Graminaceae from polluted water meadow soil, and screening of metal resistant potentially plant growth promoting bacteria. *FEMS Microbiology Ecology* 52, 153–162, DOI: 10.1016/j.femsec.2004.11.005.

Doumett, S., Fibbi, D., Azzarello, E., Mancuso, S., Mugnai, S. *et al.* (2011) Influence of the application renewal of glutamate and tartrate on Cd, Cu, Pb and Zn distribution between contaminated soil and *Paulownia tomentosa* in a pilot-scale assisted phytoremediation study. *International Journal of Phytoremediation* 13, 1–17, DOI: 10.1080/15226514.2011.605304.

Duruibe, J.O., Ogwuegbuand, M.O.C. and Egwurugwu, J.N. (2007) Heavy metal pollution and human biotoxic effects. *International Journal of Physical Sciences* 2, 112–118.

Dzantor, E.K. (2007) Phytoremediation: the state of rhizosphere 'engineering' for accelerated rhizodegradation of xenobiotic contaminants. *Journal of Chemical Technology and Biotechnology* 82, 228–232, DOI: 10.1002/jctb.1662.

Escalante-Espinosa, E., Gallegos-Martínez, M.E., Favela-Torres, E. and Gutiérrez-Rojas, M. (2005) Improvement of the hydrocarbon phytoremdiation rate by *Cyperus laxus* Lam. inoculated with a microbial consortiumin a model system. *Chemosphere* 59, 405–413, DOI: 10.1016/j.chemosphere.2004.10.034.

Fester, T., Giebler, J., Wick, LY., Schlosser, D. and Kastner, M. (2014) Plant-microbe interactions as drivers of ecosystem functions relevant for the biodegradation of organic contaminants. *Current Opinion in Biotechnology* 27, 168–175, DOI: 10.1016/j.copbio.2014.01.017.

Franchi, E., Rolli, E., Marasco, R., Agazzi, G., Borin, S. *et al.* (2016a) Phytoremediation of a multi contaminated soil: mercury and arsenic phytoextraction assisted by mobilizing agent and plant growth promoting bacteria. *Journal of Soils and Sediments* 1–13, DOI: 10.1007/s11368-015-1346-5.

Franchi, E., Agazzi, G., Rolli, E., Marasco, R., Chiaberge, S. *et al.* (2016b) Exploiting hydrocarbon-degrading indigenous bacteria for bioremediation and phytoremediation of a multicontaminated soil. *Chemical Engineering Technology* 39, 1676–1684, DOI: 10.1002/ceat.201500573.

Gadd, G.M. (2004) Microbial influence on metal mobility and application for bioremediation. *Geoderma* 122, 109–119, DOI:10.1016/j.geoderma.2004.01.002.

Galán, E., González, I. and Fernández-Caliani, J.C. (2002) Residual pollution load of soil impacted by the Aznalcóllar (Spain) mining spill after cleanup operations. *Science of the Total Environment* 286, 167–179, DOI:10.1016/S0048-9697(01)00974-3.

Gao, Y.Z. and Zhu, L.Z. (2005) Phytoremediation for phenanthrene and pyrene contaminated soils. *Journal of Environmental Sciences (China)* 17(1), 14–18.

Gao, Y., Yang, Y., Ling, W., Kong, H. and Zhu, X. (2011) Gradient distribution of root exudates and polycyclic aromatic hydrocarbons in rhizosphere soil. *Soil Science Society of America Journal* 75, 1694–1703, DOI:10.2136/sssaj2010.0244.

Germaine, K.J., Liu, X., Cabellos, G.G., Hogan, J.P., Ryan, D. and Dowling, D.N. (2006) Bacterial endophyte enhanced phytoremediation of the organochlorine herbicide 2,4-dichlorophenoxyacetic acid. *FEMS Microbiology Ecology* 57, 302–310, DOI: 10.1111/j.1574-6941.2006.00121.x.

Germaine, K.J., Keogh, K., Ryan, D. and Dowling, D.N. (2009) Bacterial endophyte-mediated naphthalene phytoprotection and phytoremediation. *FEMS Microbiology Letters* 296, 226–234, DOI: 10.1111/j.1574-6968.2009.01637.x.

Germaine, K.J., McGuinness, M. and Dowling, D.N. (2013) Improving phytoremediation through plant-associated bacteria. In: de Bruijn, F.J. (ed.) *Molecular Microbial Ecology of the Rhizosphere*. John Wiley, Chichester, UK, Vol.2, Cap. 91.

Germida, J.J., Frick, C.M. and Farrell, R.E. (2002) Phytoremediation of oil-contaminated soils. In: Violante, A., Huang, P.M., Bollag, J.M. and Gianfreda, L. (eds) *(Soil Mineral-Organic Matter-Microorganism Interactions and Ecosystem Health). Developments in Soil Science* 28, (2), 169–186. DOI: 10.1016/S0166-2481(02)80015-0.

Ghaderian, S.M., Mohtadi, A., Rahiminejad, R., Reeves, R.D. and Baker, A.J.M. (2007) Hyperaccumulation of nickel by two *Alyssum* species from the serpentine soils of Iran. *Plant and Soil* 293, 91–97, DOI: 10.1007/s11104-007-9221-9.

Glick, B.R. (2010) Using soil bacteria to facilitate phytoremediation. *Biotechnology Advances* 28, 367–374, DOI: 10.1016/j.biotechadv.2010.02.001.

Glick, B.R. (2014) Bacteria with ACC deaminase can promote plant growth and help to feed the world. *Microbiological Research* 169, 30–39, DOI:10.1016/j.micres.2013.09.009.

Glick, B.R. and Stearns, J.C. (2011) Making phytoremediation work better: maximizing a plant's growth potential in the midst of adversity. *International Journal of Phytoremediation* 13(S1), 4–16, DOI: 10.1080/15226514.2011.568533.

Glick, B.R., Cheng, Z., Czarny, J. and Duan, J. (2007) Promotion of plant growth by ACC deaminase containing soil bacteria. *European Journal of Plant Pathology* 119, 329–339, DOI:10.1016/j.micres.2013.09.009.

Gottel, N.R., Castro, H.F., Kerley, M, Yang, Z., Pelletier, D.A. *et al.* (2011) Distinct microbial communities within the endosphere and rhizosphere of *Populus deltoides* roots across contrasting soil types. *Applied Environmental Microbiology* 77, 934, DOI: 10.1128/AEM.05255-11.

Graham, P.H. and Vance, C.P. (2003) Legumes: importance and constraints to greater use. *Plant Physiology* 131, 872–877, DOI:10.1104/pp.017004.

Grandlic, C.J., Mendez, M.O., Chorover, J., Machado, B. and Maier, R.M. (2008) Plant growth promoting bacteria for phytostabilisation of mine tailings. *Environmental Science & Technology* 42, 2079–2084, DOI: 10.1021/es072013j.

Grimalt, J.O., Ferrer, M. and Macpherson, E. (1999) The mine tailing accident in Aznalcóllar. *Science of the Total Environment* 242, 3–12, DOI:10.1016/S0048-9697(99)00372-1.

Gupta, A., Meyer, J.M. and Goel, R. (2002) Development of heavy metal-resistant mutants of phosphate-solubilizing *Pseudomonas* sp. NBRI 4014 and their characterization. *Current Microbiology* 45, 323–327, DOI: 10.1007/s00284-002-3762-1.

Gurska, J., Wang, W., Gerhardt, K.E., Khalid, A.M., Isherwood, D.M. *et al.* (2009) Three year field test of a plant growth promoting rhizobacteria enhanced phytoremediation system at a land farm for treatment of hydrocarbon waste. *Environmental Science and Technology* 43, 12, 4472–4479.

Haferburg, G. and Kothe, E. (2010) Metallomics: lessons for metalliferous soil remediation. *Applied Microbiology and Biotechnology* 87, 1271–1280, DOI: 10.1007/s00253-010-2695-z.

Hong, S.H., Ryu, H., Kim, J. and Cho, K.S. (2011) Rhizoremediation of diesel-contaminated soil using the plant growth-promoting rhizobacterium *Gordonia* sp. S2RP-17. *Biodegradation* 22(3), 593–601, DOI: 10.1007/s10532-010-9432-2.

Hou, J., Liu, W., Wang, B., Wang, Q., Luo, Y. and Franks, A.E. (2015) PGPR enhanced phytoremediation of petroleum contaminated soil and rhizosphere microbial community response. *Chemosphere* 138, 592–598, DOI: 10.1016/j.chemosphere.2015.07.025.

Huang, X-D., El-Alawi, Y., Penrose, D.M., Glick, B.R. and Greenberg, B.M. (2004) A multi-process phytoremediation system for removal of polycyclic aromatic hydrocarbons from contaminated soils. *Environmental Pollution* 130(3), 465–476, DOI: 10.1016/j.envpol.2003.09.031.

Huang, X.F., Chaparro, J.M., Reardon, K.F., Zhang, R.F., Shen, Q.R. and Vivanco, J.M. (2014) Rhizosphere interactions: Root exudates, microbes, and microbial communities. *Botany* 92, 267–275, DOI: 10.1139/cjb-2013-0225.

Ike, A., Sriprang, R., Ono, H., Murooka, Y. and Yamashita, M. (2007) Bioremediation of cadmium contaminated soil using symbiosis between leguminous plant and recombinant rhizobia with the MTL4 and the PCS genes. *Chemosphere* 66, 1670–1676, DOI: 10.1016/j.chemosphere.2006.07.058.

ITRC (Interstate Technology & Regulatory Council) (2009) *Phytotechnology Technical and Regulatory Guidance and Decision Trees, Revised. PHYTO-3*. Interstate Technology & Regulatory Council, Phytotechnologies Team, Tech Reg Update, Washington, DC. (Available at: www.itrcweb.org, accessed 11 July 2017.)

Janssen, J., Weyens, N., Croes, S., Beckers, B., Meiresonne, L. *et al.* (2015) Phytoremediation of metal contaminated soil using willow: exploiting plant-associated bacteria to improve biomass production and metal uptake. *International Journal of Phytoremediation* 17(11), 1123–1136, DOI: 10.1080/15226514.2015.1045129.

Jeong, S., Moon, H.S. and Nam, K. (2014) Enhanced uptake and translocation of arsenic in Cretan brake fern (*Pteris cretica* L.) through siderophore arsenic complex formation with an aid of rhizospheric bacterial activity. *Journal of Hazardous Materials* 280, 536–543, DOI: 10.1016/j.jhazmat.2014.08.057.

Johnson, D., Anderson, D. and McGrath, S. (2005) Soil microbial response during the phytoremediation of a PAH contaminated soil. *Soil Biology Biochemistry* 37, 2334–2336, DOI: 10.1016/j.soilbio.2005.04.001.

Keswani, C., Bisen, K., Singh, V., Sarma, B.K. and Singh, H.B. (2016) Formulation technology of biocontrol agents: present status and future prospects. In: Arora, N.K., Mehnaz, S. and Balestrini, R. (eds*) Bioformulations: For Sustainable Agriculture*. Springer, New Delhi, India, pp. 35–52.

Kim, J., Kang, S.H., Min, K.A., Cho, K.S. and Lee, I.S. (2006) Rhizosphere microbial activity during phytoremediation of diesel-contaminated soil. *Journal of Environmental Science and Health Part A Toxic/ HazardousSubstances&EnvironmentalEngineering*41(11),2503–2516,DOI:10.1080/10934520600927658.

Kuffner, M., Puschenreiter, M., Wieshammer, G., Gorfer, M. and Sessitsch, A. (2008) Rhizosphere bacteria affect growth and metal uptake of heavy metal accumulating willows. *Plant and Soil* 304, 35–44, DOI: 10.1007/s11104-007-9517-9.

Kuffner, M., De Maria, S., Puschenreiter, M., Fallmann, K., Wieshammer, G. *et al.* (2010) Bacteria associated with Zn and Cd-accumulating *Salix caprea* with differential effects on plant growth and heavy metal availability. *Journal of Applied Microbiology* 108, 1471–1484, DOI: 10.1111/j.1365-2672.2010.04670.x.

Kumar, R., Das, A.J. and Juwarkar, A.A. (2014) Restoration of petrol contaminated soil by PGPR consortium producing rhamnolipids and enhancement of growth and antioxidant activity of *Withania somnifera. Journal of Petroleum & Environmental Biotechnology DOI:10.4172/2157-7463.S5-001*.

Kuppusamy, S., Thavamani, P., Megharaj, M., Venkateswarlu, K., Lee, Y.B. and Naidu, R. (2016) Pyrosequencing analysis of bacterial diversity in soils contaminated long-term with PAHs and heavy metals: implications to bioremediation. *Journal of Hazardous Materials* 317, 169–179, DOI: 10.1016/j.jhazmat.2016.05.066.

Lampis, S., Santi, C., Ciurli, A., Andreolli, M. and Vallini, G. (2015) Promotion of arsenic phytoextraction efficiency in the fern *Pteris vittata* by the inoculation of As-resistant bacteria: a soil bioremediation perspective. *Frontiers in Plant Science* 18(6), 80, DOI: 10.3389/fpls.2015.00080.

Leewis, M-C., Reynolds, C.M. and Leigh, M.B. (2013) Long-term effects of nutrient addition and phytoremediation on diesel and crude oil contaminated soils in subarctic Alaska. *Cold Regions Science and Technology* 96, 129–137, DOI:10.1016/j.coldregions.2013.08.011.

Leewis, M.C., Uhlik, O., Fraraccio, S., McFarlin, K., Kottara, A. *et al.* (2016) Differential impacts of willow and mineral fertilizer on bacterial communities and biodegradation in diesel fuel oil-contaminated soil. *Frontiers in Microbiology* 2(7), 837, DOI: 10.3389/fmicb.2016.00837.

Leigh, M.B., Prouzova, P., Mackova, M., Macek, T., Nagle, D.O. and Fletcher, J.S. (2006) Polychlorinated biphenyl (PCB)-degrading bacteria associated with trees in a PCB-contaminated site. *Applied Environmental Microbiology* 72, 2331–2342, DOI:10.1128/AEM.72.4.2331–2342.2006.

Li, T., Tao, Q., Liang, C., Shohag, M.J., Yang, X. and Sparks, D.L. (2013) Complexation with dissolved organic matter and mobility control of heavy metals in the rhizosphere of hyperaccumulator *Sedum alfredii. Environmental Pollution* 182, 248–255, DOI: 10.1016/j.envpol.2013.07.025.

Li, W., Ye, Z. and Wong, M. (2009) Metal mobilization and production of short-chain organic acids by rhizosphere bacteria associated with a Cd/Zn hyperaccumulating plant, *Sedum alfredii. Plant Soil* 326, 453–467, DOI: 10.1007/s11104-009-0025-y.

Li, Y-M., Chaney, R., Brewer, E., Roseberg, R., Angle, J.S. *et al.* (2003) Development of a technology for commercial phytoextraction of nickel: economic and technical considerations. *Plant and Soil* 249, 107–115, DOI: 10.1023/A:1022527330401.

Liao, C., Xu, W., Lu, G., Liang, X., Guo, C. *et al.* (2015) Accumulation of hydrocarbons by maize (*Zea mays* L.) in remediation of soils contaminated with crude oil. *International Journal of Phytoremediation* 17(7), 693–700. DOI: 10.1080/15226514.2014.964840.

Liu, L., Jiang, C-Y., Liu, X-Y., Wu, J-F., Han, J-G. and Liu, S.-J. (2007) Plant-microbe association for rhizoremediation of chloronitroaromatic pollutants with *Comamonas* sp. strain CNB-1. *Environmental Microbiology* 9, 465–473, DOI: 10.1111/j.1574-6941.2006.00121.x.

Lugtenberg, B. and Kamilova, F. (2009) Plant-growth-promoting rhizobacteria. *Annual Review of Microbiology* 63, 541–556, DOI: 10.1146/annurev.micro.62.081307.162918.

Lumactud, R., Shen, S.Y., Lau, M. and Fulthorpe, R. (2016) Bacterial endophytes isolated from plants in natural oil seep soils with chronic hydrocarbon contamination. *Frontiers in Microbiology* 7, 755, DOI:10.3389/fmicb.2016.00755.

Luo, C., Shen, Z. and Li, X. (2005) Enhanced phytoextraction of Cu, Pb, Zn and Cd with EDTA and EDDS. *Chemosphere* 59, 1–11, DOI: 10.1016/j.chemosphere.2004.09.100.

Ma, Y., Prasad, M.N., Rajkumar, M. and Freitas, H. (2011) Plant growth promoting rhizobacteria and endophytes accelerate phytoremediation of metalliferous soils. *Biotechnology Advances* 29, 248–258, DOI: 10.1016/j.biotechadv.2010.12.001.

Mandal, S., Mandal, M., Das, A., Pati, B. and Ghosh, A. (2009) Stimulation of indoleacetic acid production in a rhizobium isolate of vigna mungo by root nodule phenolic acids. *Archives of Microbiology* 191, 389–393, DOI: 10.1007/s00203-008-0455-6.

Mandal, S.M., Chakraborty, D. and Dey, S. (2010) Phenolic acids act as signaling molecules in plant-microbe symbioses. *Plant Signaling & Behavior* 5, 359–368.

Mangwani, N., Kumari, S. and Das, S. (2015) Involvement of quorum sensing genes in biofilm development and degradation of polycyclic aromatic hydrocarbons by a marine bacterium *Pseudomonas aeruginosa* N6P6. *Applied Microbiology and Biotechnology* 99, 23, DOI: 10.1007/s00253-015-6868-7.

Marques, A.P., Moreira, H., Franco, A.R., Rangel, A.O. and Castro, P.M. (2013) Inoculating *Helianthus annuus* (sunflower) grown in zinc and cadmium contaminated soils with plant growth promoting bacteria – effects on phytoremediation strategies. *Chemosphere* 92(1), 74–83, DOI: 10.1016/j.chemosphere.2013.02.055.

Mathesius, U. and Watt, M. (2010) Rhizosphere signals for plant–microbe interactions: Implications for field-grown plants. In: Ulrich, E., Lüttge, U.E., Beyschlag, W., Burkhard Büdel, B. and Francis, D. (eds) *Progress in Botany* 72. Springer Berlin Heidelberg, Germany, pp. 125–161, DOI: 10.1007/978-3-642-13145-5_5.

McIntosh, P., Kuzovkina, Y.A., Schulthess, C.P. and Guillard, K. (2016) Breakdown of low-level total petroleum hydrocarbons (TPH) in contaminated soil using grasses and willows. *International Journal of Phytoremediation* 2, 18(7), 656–663, DOI: 10.1080/15226514.2015.1109598.

Meliani, A. and Bensoltane, A. (2014) Enhancement of hydrocarbons degradation by use of pseudomonas biosurfactants and biofilms. *Journal of Petroleum & Environmental Biote*chnology 5, 1, DOI:10.4172/2157-7463.1000168.

Menezes Bento, F., de Oliveira Camargo, F.A., Okeke, B.C. and Frankenberger, W.T. Jr. (2005) Diversity of biosurfactant producing microorganisms isolated from soils contaminated with diesel oil. *Microbiological Research* 160, 249–255, DOI: 10.1016/j.micres.2004.08.005.

Meng, L., Qiao, M. and Arp, H.P.H. (2011) Phytoremediation efficiency of a PAH-contaminated industrial soil using ryegrass, white clover and celery as mono- and mixed cultures. *Journal of Soils and Sediments* 11, 482–490, DOI: 10.1007/s11368-010-0319-y.

Michalet, S., Rohr, J., Warshan, D., Bardon, C., Rogy, J.C. *et al.* (2013) Phytochemical analysis of mature tree root exudates in situ and their role in shaping soil microbial communities in relation to tree N-acquisition strategy. *Plant Physiology and Biochemistry* 72, 169–177, DOI: 10.1016/j.plaphy.2013.05.003.

Militon, C., Boucher, D., Vachelard, C., Perchet, G., Barra, V. *et al.* (2010) Bacterial community changes during bioremediation of aliphatic hydrocarbon-contaminated soil. *FEMS Microbiology Ecology* 74, 669–681, DOI: 10.1111/j.1574-6941.2010.00982.x.

Minguzzi, C. and Vergnano, O. (1948) Il contenuto di nichel nelle ceneri di *Alyssum bertolonii. Atti Società Toscana Scienze Naturali* 55, 49–74.

Mishra, S., Singh, A., Keswani, C., Saxena, A., Sarma, B.K. and Singh, H.B. (2015) Harnessing plant-microbe interactions for enhanced protection against phytopathogens. In: Arora, N.K. (ed.) *Plant Microbes Symbiosis: Applied Facets*. Springer, New Delhi, India, pp. 111–125.

Mnif, I., Mnif, S., Sahnoun, R., Maktouf, S., Ayedi, Y. *et al.* (2015) Biodegradation of diesel oil by a novel microbial consortium: comparison between co-inoculation with biosurfactant-producing strain and exogenously added biosurfactants. *Environmental Science and Pollution Research International* 22(19), 14852–14861, DOI: 10.1007/s11356-015-4488-5.

Moreno, F.N., Anderson, C.W.N., Stewart, R.B. and Robinson, B.H. (2004) Phytoremediation of mercury-contaminated mine tailings by induced plant-mercury accumulation. *Environmental Practice* 6, 165–175, DOI: 10.1017/S1466046604000274.

Nagajyoti, P.C., Lee, K.D. and Sreekanth, T.V.M. (2010) Heavy metals, occurrence and toxicity for plants: a review. *Environmental Chemistry Letters* 8, 199–216, DOI 10.1007/s10311-010-0297-8.

Nakamura, T., Motoyama, T., Suzuki, Y. and Yamaguchi, I. (2004) Biotransformation of pentachlorophenol by Chinese chive and a recombinant derivative of its rhizospher-competent microorganism *Pseudomonas gladioli* M-2196. *Soil Biology and Biochemistry* 35, 787–795, DOI:10.1016/j.soilbio.2004.01.006.

Nie, Y., Chi, C-Q., Fang, H., Liang, J-L., Lu, S-L. *et al.* (2014) Diverse alkane hydroxylase genes in microorganisms and environments. *Nature Scientific Reports* 4, 4968, DOI: 10.1038/srep04968.

NRC National Research Council (2002) *Bioavailability of Contaminants in Soils and Sediments: Processes, Tools and Applications*. National Academies, Washington, DC.

Oku, S., Komatsu, A., Tajima, T., Nakashimada, Y. and Kato, J. (2012) Identification of chemotaxis sensory proteins for amino acids in *Pseudomonas fluorescens* Pf0–1 and their involvement in chemotaxis to tomato root exudate and root colonization. *Microbes and Environments* 27, 462–469.

Pagé, A.P., Yergeau, É. and Greer, C.W. (2015) *Salix purpurea* stimulates the expression of specific bacterial xenobiotic degradation genes in a soil contaminated with hydrocarbons. *PLoS One* 10(7), DOI:e0132062. DOI: 10.1371/journal.pone.0132062.

Patten, C.L. and Glick, B.R. (2002) The role of bacterial indoleacetic acid in the development of the host plant root system. *Applied and Environmental Microbiology* 68, 3795–3801, DOI: 10.1128/AEM.68.8.3795-3801.2002.

Pedron, F., Petruzzelli, G., Barbafieri, M. and Tassi, E. (2009) Strategies to use phytoextraction in very acidic soil contaminated by heavy metals. *Chemosphere* 75, 808–814, DOI: 10.1016/j.chemosphere.2009.01.044.

Pedron, F., Petruzzelli, G., Barbafieri, M. and Tassi, E. (2013) Remediation of a mercury-contaminated industrial soil using bioavailable contaminant stripping. *Pedosphere* 23, 104–111, DOI:10.1016/S1002-0160(12)60085-X.

Pedron, F., Rosellini, I., Petruzzelli, G. and Barbafieri, M. (2014) Chelant comparison for assisted phytoextraction of lead in two contaminated soils. *Resources and Environment* 4, 209–214, DOI:10.5923/j.re.20140405.01.

Peijnenburg, W.J.G.M., Zablotskaja, M. and Vijver, M.G. (2007) Monitoring metals in terrestrial environments within a bioavailability framework and a focus on soil extraction. *Ecotoxicology and Environmental Safety* 67, 163–179, DOI: 10.1016/j.ecoenv.2007.02.008.

Petruzzelli, G., Pedron, F., Tassi, E., Franchi, E., Bagatin, R. *et al.* (2014) The effect of thiosulphate on arsenic bioavailability in a multi contaminated soil. A novel contribution to phytoextraction. *Research Journal of Environmental and Earth Sciences* 6, 38–43.

Pezzarossa, B., Gorini, F. and Petruzzelli, G. (2011) Heavy metal and selenium distribution and bioavailability in contaminated sites: a tool for phytoremediation. In: Selim, H.M. (ed.) *Dynamics and Bioavailability of Heavy Metals in the Rootzone*. CRC Press, Boca Raton FL, pp. 93–127.

Phillips, L.A., Greer, C.W., Farrell, R.E. and Germida, J.J. (2012) Plant root exudates impact the hydrocarbon degradation potential of a weathered-hydrocarbon contaminated soil. *Applied Soil Ecology* 52, 56–64, DOI: 10.1016/j.apsoil.2011.10.009.

Prapagdee, B., Chanprasert, M. and Mongkolsuk, S. (2013) Bioaugmentation with cadmium-resistant plant growth-promoting rhizobacteria to assist cadmium phytoextraction by *Helianthus annuus*. *Chemosphere* 92, 659–666, DOI: 10.1016/j.chemosphere.2013.01.082.

Prasad, M.N.V. and Freitas, H.M. (2003) Metal hyperaccumulation in plants. Biodiversity prospecting for phytoremediation technology. *Electronic Journal of Biotechnology* 6, 287–321, DOI: 10.2225/vol6-issue3-fulltext-6.

Prince, R.C., Gramain, A. and McGenity, T.J. (2010) Prokaryotic hydrocarbon degraders. In: Timmis, K.N. (ed.) *Handbook of Hydrocarbon and Lipid Microbiology*. Springer Berlin Heidelberg, Germany, Vol. 19, pp. 1669–1692.

Radwan, S.S., Dashti, N. and El-Nemr, I.M. (2005) Enhancing the growth of *Vicia faba* plants by microbial inoculation to improve their phytoremediation potential for oily desert areas. *International Journal of Phytoremediation* 7, 19–32, DOI:10.1080/16226510590915783.

Radwan, S.S., Dashti, N., El-Nemr, I. and Khanafer, M. (2007) Hydrocarbon utilization by nodule bacteria and plant growth-promoting rhizobacteria. *International Journal of Phytoremediation* 9, 475–486, DOI: 10.1080/15226510701709580.

Rajkumar, M., Prasad, M.N.V., Freitas, H. and Ae, N. (2009) Biotechnological applications of serpentine soil bacteria for phytoremediation of trace elements. *Critical Reviews of Biotechnology* 29, 120–130, DOI: 10.1080/07388550902913772.

Rajkumar, M., Ae, N., Narasimha, M., Prasad, V. and Freitas, H. (2010) Potential of siderophore-producing bacteria for improving heavy metal phytoextraction. *Trends in Biotechnology* 28(3), 142–149, DOI: 10.1016/j.tibtech.2009.12.002.

Rajkumar, M., Sandhya, S., Prasad, M.N.V. and Freitas, H. (2012) Perspectives of plant-associated microbes in heavy metal phytoremediation. *Biotechnology Advances* 30, 1562–1574, DOI:10.1016/j.biotechadv.2012.04.011.

Rascio, N. and Navari-Izzo, F. (2011) Heavy metal hyperaccumulating plants: How and why do they do it? And what makes them so interesting? *Plant Science* 180, 169–181, DOI: 10.1016/j.plantsci.2010.08.016.

Raskin, I. and Ensley, B.D. (1999) *Phytoremediation of Toxic Metals: Using Plants to Clean up the Environment.* Wiley-Interscience, New York.

Rentz, J.A., Alvarez, P.J. and Schnoor, J.L. (2005) Benzo (a) pyrene co-metabolism in the presence of plant root extracts and exudates: Implications for phytoremediation. *Environmental Pollution* 136, 477–484, DOI: 10.1016/j.envpol.2004.12.034.

Rohrbacher, F. and St-Arnaud, M. (2016) Root exudation: the ecological driver of hydrocarbon rhizoremediation. *Agronomy* 6, 19, DOI:10.3390/agronomy6010019.

Rudrappa, T., Czymmek, K.J., Pare, P.W. and Bais, H.P. (2008) Root-secreted malic acid recruits beneficial soil bacteria. *Plant Physiology* 148, 1547–1556, DOI:10.1104/pp.104.900275.

Samardjieva, K.A., Pissarra, J., Castro, P.M. and Tavares, F. (2011) Insights into phytoremediation solutions for environmental recovery. *Recent Patents on Biotechnology* 5(1), 25–39, DOI: 10.2174/187220811795655878.

Santos, F.S., Hernandez-Allica, J., Becerril, J.M., Amaral-Sobrinho, N., Mazur, N. and Garbisu, C. (2006) Chelate induced phytoextraction of metal polluted soils with *Brachiaria decumbens*. *Chemosphere* 65, 43–50, DOI: 10.1016/j.chemosphere.2006.03.012.

Satarug, S., Garrett, S.H., Sens, M.A. and Sens, D.A. (2010) Cadmium, environmental exposure, and health outcomes. *Environmental Health Perspectives* 118(2), 182–190, DOI: 10.1289/ehp.0901234.

Sessitsch, A. and Puschenreiter, M. (2008) Endophytes and rhizosphere bacteria of plants growing in heavy metal contaminated soil. In: Dion, P. and Nautiyal, C.S. (eds) *Microbiology of Extreme Soils*. Springer Berlin Heidelberg, Germany.

Sessitsch, A., Kuffner, M., Kidd, P., Vangronsveld, J., Wenzel, W.W., Fallmann, K. and Puschenreiter, M. (2013) The role of plant-associated bacteria in the mobilization and phytoextraction of trace elements in contaminated soils. *Soil Biology & Biochemistry* 60, 182–194, DOI:10.1016/j.soilbio.2013.01.012.

Seth, C.S., Misra, V., Singh, R.R. and Zolla, L. (2011) EDTA-enhanced lead phytoremediation in sun-flower (*Helianthus annus* L.) hydroponic culture. *Plant and Soil* 347, 231–242, DOI: 10.1007/s11104-011-0841-8.

Shahzad, A., Saddiqui, S. and Bano, A. (2016) The response of maize (*Zea mays* L.) plant assisted with bacterial consortium and fertilizer under oily sludge. *International Journal of Phytoremediation* 18(5), 521–526, DOI: 10.1080/15226514.2015.1115964.

Shelmerdine, P.A., Black, C.R., McGrath, S.P. and Young, S.D. (2009) Modeling phytoremediation by the hyperaccumulating fern, *Pteris vittata*, of soils historically contaminated with arsenic. *Environmental Pollution* 157, 1589–1596, DOI: 10.1016/j.envpol.2008.12.029.

Sheng, X.F., Xia, J.J., Jiang, C.Y., He, L.Y. and Qian, M. (2008) Characterization of heavy metal-resistant endophytic bacteria from rape (*Brassica napus*) roots and their potential in promoting the growth and lead accumulation of rape. *Environmental Pollution* 156, 1164–1170, DOI: 10.1016/j.envpol.2008.04.007.

Silva, M.A., Silva, A.F., Rufino, R.D., Luna, J.M., Santos, V.A. and Sarubbo, L.A. (2016) Production of biosurfactants by pseudomonas species for application in the petroleum industry. *Water Environmental Research* DOI: 10.2175/106143016X14609975746280.

Singh, B.K., Munro, S., Potts, J.M. and Millard, P. (2007) Influence of grass species and soil type on rhizosphere microbial community structure in grassland soils. *Applied Soil Ecology* 36, 147–155, DOI: 10.1016/j.apsoil.2007.01.004.

Slater, H., Gouin, T. and Leigh, M.B. (2011) Assessing the potential for rhizoremediation of PCB contaminated soils in northern regions using native species. *Chemosphere* 84(2), 199–206, DOI:10.1016/j.chemosphere.2011.04.058.

Smith, P.L. (2013) *Indicator Plants: Using Plants to Evaluate the Environment.* Wildtrack Publishing, Sheffield, UK. ISBN 9781904098355.

Somers, E., Vanderleyden, J. and Srinivasan, M. (2004) Rhizosphere bacterial signalling: a love parade beneath our feet. *Critical Reviews in Microbiology* 30, 205–240, DOI: 10.1080/10408410490468786.

Tak, H.I., Ahmad, F. and Babalola, O.O. (2013) Advances in the application of plant growth-promoting rhizobacteria in phytoremediation of heavy metals. In: Whitacre, D.M. (ed.) *Reviews of Environmental Contamination and Toxicology* 223, Springer Science+Business Media, New York. pp. 33–52, DOI 10.1007/978-1-4614-5577-6_2.

Tassi, E., Pedron, F., Barbafieri, M. and Petruzzelli, G. (2004) Phosphate-assisted phytoextraction in As-contaminated soil. *Engineering in Life Science* 4, 341–346, DOI: 10.1002/elsc.200420037.

Tchounwou, P.B., Yedjou, C.G., Patlolla, A.K. and Sutton, D.J. (2012) Heavy metals toxicity and the environment. In: Luch, A. (ed.) *Molecular, Clinical and Environmental Toxicology. Experientia Supplementum.* Springer, Basel, Switzerland. p. 101, DOI:10.1007/978-3-7643-8340-4_6.

Thavamani, P., Malik, S., Beer, M., Megharaj, M. and Naidu, R. (2012) Microbial activity and diversity in long-term mixed contaminated soils with respect to polyaromatic hydrocarbons and heavy metals. *Journal of Environmental Management* 99, 10–17, DOI: 10.1016/j.jenvman.2011.12.030.

Tiwari, S., Sarangi, B.K. and Thul, S.T. (2016) Identification of arsenic resistant endophytic bacteria from *Pteris vittata* roots and characterization for arsenic remediation application. *Journal of Environmental Management* 15(180), 359–365, DOI: 10.1016/j.jenvman.2016.05.029.

Uhlik, O., Jecna, K., Macknova, M., Vlcek, C., Hroudova, M. *et al.* (2009) Biphenyl metabolizing bacteria in the rhizosphere of horseradish and bulk soil contaminated by polychlorinated biphenyls as revealed by stable isotope probing. *Applied Environmental Microbiology* 75, 6471–6477, DOI: 10.1128/AEM.00466-09.

Ullah, A., Heng, S., Munis, M.F.H., Fahad, S. and Yang, X. (2015) Phytoremediation of heavy metals assisted by plant growth promoting (PGP) bacteria: a review. *Environmental and Experimental Botany* 117, 28–40, DOI:10.1016/j.envexpbot.2015.05.001.

US EPA (2001) *Evaluation of Phytoremediation for Management of Chlorinated Solvents in Soil and Groundwater.* EPA 542-R-05-001.

US EPA (2006) *Reference Guide to Non-combustion Technologies for Remediation of Persistent Organic Pollutants in Stockpiles and Soil.* EPA-542-R-05-006.

Vacheron, J., Desbrosses, G., Bouffaud, M-L., Touraine, B., Moënne-Loccoz, Y. *et al.* (2013) Plant growth-promoting rhizobacteria and root system functioning. *Frontiers in Plant Science* 4, 356, DOI: 10.3389/fpls.2013.00356.

Vangronsveld, J., Herzig, R., Weyens, N., Boulet, J., Adriaensen, K. *et al.* (2009) Phytoremediation of contaminated soils and groundwater: lessons from the field. *Environmental Science and Pollution Research* 16, 765–794, DOI: 10.1007/s11356-009-0213-6.

Vázquez, S., Moreno, E. and Carpena, R.O. (2008) Bioavailability of metals and as from acidified multi-contaminated soils: validation of several extraction methods by lupin plants. *Environmental Geochemistry and Health* 30, 193–198, DOI: 10.1007/s10653-008-9143-3.

Vijayakumar, S. and Saravanan, V. (2015) Biosurfactants-Types, sources and applications. *Research Journal of Microbiology* 10(5), 181–192, DOI: 10.3923/jm.2015.181.192.

Wanga, Q., Li, Z., Cheng, S. and Wua, Z. (2010) Effects of humic acids on phytoextraction of Cu and Cd from sediment by *Elodea nuttallii. Chemosphere* 78, 604–608, DOI: 10.1016/j.chemosphere.2009.11.011.

Wei, J., Liu, X., Zhang, X., Chen, X., Liu, S. and Chen, L. (2014) Rhizosphere effect of *Scirpus triqueter* on soil microbial structure during phytoremediation of diesel-contaminated wetland. *Environmental Technology* 35(1–4), 514–520, DOI: 10.1080/09593330.2013.836549.

Weyens, N., van der Lelie, D., Taghavi, S. and Vangronsveld, J. (2009a) Phytoremediation: plant–endophyte partnerships take the challenge. *Current Opinion in Biotechnology* 20, 248–254, DOI: 10.1016/j.copbio.2009.02.012.

Weyens, N., van der Lelie, D., Taghavi, S., Newman, L. and Vangronsveld, J. (2009b) Exploiting plant-microbe partnerships for improving biomass production and remediation. *Trends in Biotechnology* 27, 591–598, DOI: 10.1016/j.tibtech.2009.07.006.

Weyens, N., Schellingen, K., Beckers, B., Janssen, J., Ceulemans, R. *et al.* (2013) Potential of willow and its genetically engineered associated bacteria to remediate mixed Cd and toluene contamination. *Journal of Soils and Sediments* 13(1), 176–188, DOI: 10.1007/s11368-012-0582-1.

Wu, C.H., Wood, T.K., Mulchandani, A. and Chen, W. (2006) Engineering plant-microbe symbiosis for rhizoremediation of heavy metals. *Applied Environmental Microbiology* 72, 1129–1134, DOI: 10.1128/AEM.72.2.1129-1134.2006.

Wuana, R.A. and Okieimen, F.E. (2011) Heavy metals in contaminated soil: a review of sources, chemistry, risks and best available strategies for bioremediation. *ISRN Ecology*, Article ID 402647, DOI:10.5402/2011/402647.

Xu, J-Y., Han, Y-H., Chen, Y., Zhu, L-J. and Ma, L.Q. (2016) Arsenic transformation and plant growth promotion characteristics of As-resistant endophytic bacteria from As-hyperaccumulator *Pteris vittata. Chemosphere* 144, 1233–1240, DOI: 10.1016/j.chemosphere.2015.09.102.

Xu, Y., Sun, G.D., Jin, J.H., Liu, Y., Luo, M. *et al.* (2014) Successful bioremediation of an aged and heavily contaminated soil using a microbial/plant combination strategy. *Journal of Hazardous Materials* 15(264), 430–438, DOI: 10.1016/j.jhazmat.2013.10.071.

Yadav, S.K. (2010) Heavy metals toxicity in plants: an overview on the role of glutathione and phytochelatins in heavy metal stress tolerance of plants. *South African Journal of Botany* 76, 167–179, DOI:10.1016/j.sajb.2009.10.007.

Yergeau, E., Sanschagrin, S., Maynard, C., St-Arnaud, M. and Greer, C.W. (2014) Microbial expression profiles in the rhizosphere of willows depend on soil contamination. *The ISME Journal* 8, 344–358, DOI:10.1038/ismej.2013.163.

Zhu, L.J., Guan, D.X., Luo, J., Rathinasabapathi, B. and Ma, L.Q. (2014) Characterization of arsenic-resistant endophytic bacteria from hyperaccumulators *Pteris vittata* and *Pteris multifida*. *Chemosphere* 113, 9–16, DOI: 10.1016/j.chemosphere.2014.03.081.

18 Role of Plant Growth-Promoting Rhizobacteria (PGPR) in Degradation of Xenobiotic Compounds and Allelochemicals

Deepika Goyal,[1] Janmejay Pandey[1]* and Om Prakash[2#]

[1]*Department of Biotechnology, School of Life Sciences, Central University of Rajasthan, Bandarsindri, NH-8, Kishangarh, Ajmer-305817, Rajasthan, India;* [2]*Microbial Culture Collection, National Centre for Cell Sciences, Pune-411007, Maharashtra, India*

18.1 Xenobiotic Compounds and Allelochemicals – Major Inhibitors of Plant Growth and Productivity

18.1.1 Xenobiotic compounds as priority environmental pollutants

Contamination of Earth's environment with toxic xenobiotic pollutants has been a major cause of concern for several decades. This situation has emerged largely due to non-judicious production, usage and disposal of xenobiotic pollutants during urbanization and activities related to industrialization and agriculture. Xenobiotic compounds are man-made chemicals (such as explosives, pesticides, fungicides, synthesized azo dyes, industrial solvents, alkanes, polycyclic aromatic hydrocarbons, dioxins and furans, polychlorinated biphenyls, chlorinated aromatic compounds and nitro-aromatic compounds, petroleum products, and brominated flame retardants, etc.) that are synthesized for industrial and agricultural application. A majority of the xenobiotic compounds do not have any known natural source and they are characterized by extreme chemical and thermodynamic stability. While this property makes them ideally suited for industrial application and enhances their commercial value, it also makes them extremely persistent in the environment. Furthermore, many of the xenobiotic compounds, e.g. hexachlorocyclohexane (HCH), pentachlorophenol (PCP), polychlorinated biphenols (PCB), etc., also exhibit a strong tendency to bioaccumulation. Therefore, organisms positioned at higher levels in food chains and food webs (including human beings) will tend to have greater accumulation of these toxic compounds compared to those organisms present at the lower levels. Noticeably, these bioaccumulating xenobiotic compounds can be passed from mothers to their children during embryonic development as well as through post-natal breastfeeding. Apart from the tendency to bioaccumulate, a large number of xenobiotic compounds can also impart toxic effects to human beings, ranging from acute toxicity, mutagenicity, carcinogenicity, teratogenicity, etc. In addition, they are harmful due to their ability to poison animals and plants and alter ecosystem functions.

*E-mails: *janmejay@curaj.ac.in and #omprakash@nccs.res.in

© CAB International 2017. *Advances in PGPR Research* (eds H.B. Singh, B.K. Sarma and C. Keswani)

A large number of the xenobiotic pollutants have been identified as priority environmental pollutants by national and international environment monitoring bodies, e.g. United States Environmental Protection Agency (USEPA), European Environmental Agency (EEA), etc. These bodies have formulated stringent regulatory norms for regulating the production, usage and disposal of priority pollutants. A noticeable example of such regulation is that of the United Nations Environment Programme (UNEP), enforced by the Stockholm Convention in 2004. This programme listed twelve xenobiotic organic compounds (including PCBs, nine chlorinated organic pesticides, and dioxins and furans) as Persistent Organic Pollutants (POPS). A list of toxic synthetic organic compounds in environmental soil that are associated with human health risks is given in Table 18.1. The complexity and diversity of chemical structures of these xenobiotic compounds is represented in Fig. 18.1.

Noticeably, the current list of priority pollutants defined by USEPA consists of 126 xenobiotic compounds that are strictly regulated by national discharge regulation (https://www.epa.gov/sites/production/files/2015-09/documents/priority-pollutant-list-epa.pdf). With implementation of such regulation, the worldwide production, usage and disposal of toxic xenobiotics is gradually phasing out. However, due to their past usage and environmental persistence, many of the xenobiotic compounds are still found as residual contaminants in diverse ecological niches including agricultural soil and groundwater.

18.1.2 Effects of xenobiotic compounds on plant growth and productivity

Apart from posing serious human-health-related threats, the contamination of ecological niches by xenobiotic compounds also poses a significant threat of damaging the ecosystem diversity, leading to reduced productivity. Although many of the xenobiotic compounds do not impart significantly adverse effects on plants at lower concentrations, at higher concentrations they are reported to have adverse effects on overall plant growth and productivity. Maliszewska-Kordybach and

Table 18.1. List of synthetic organic xenobiotic compounds identified as priority organic pollutant (POPs) and their representative congeners.

Sl. No.	Synthetic Organic Pollutant/ Xenobiotic Compound	Representative congeners
1	Polychlorinated biphenyl (PCBs) and Synthetic organic pesticides	2,4-dichlorophenoxyacetic acid (2,4-D) Atrazine Tetra-chlorophenol (TCP) Pentachlorophenol (PCP) Tributylin (TBT) Glyphosate
2	Volatile Organic Compounds (VOCs)	Benzene, toluene, ethylbenzene and xylene (BTEX) Methyl tertiary butyl ether (MTBE) Trichloroethylene (TCE)
3	Hydrocarbons	Propane, Benzene, Hexane, Naphthalene, Polypropylene, Polystyrene, Polyaromatic Hydrocarbons Heterocyclic aromatic hydrocarbons Alkyl Polyaromatic Hydrocarbons
4	Explosives	Trinitrotoluene (TNT) Hexahydrotrinitrotriazine (RDX) octahydro-tetranitrotetraocine or High Melting explosive (HMX)
5	Brominated Flame Retardants	Tetrabromobisphenol A, Hexabromocyclododecane, and Polybromodiphenyl ethers

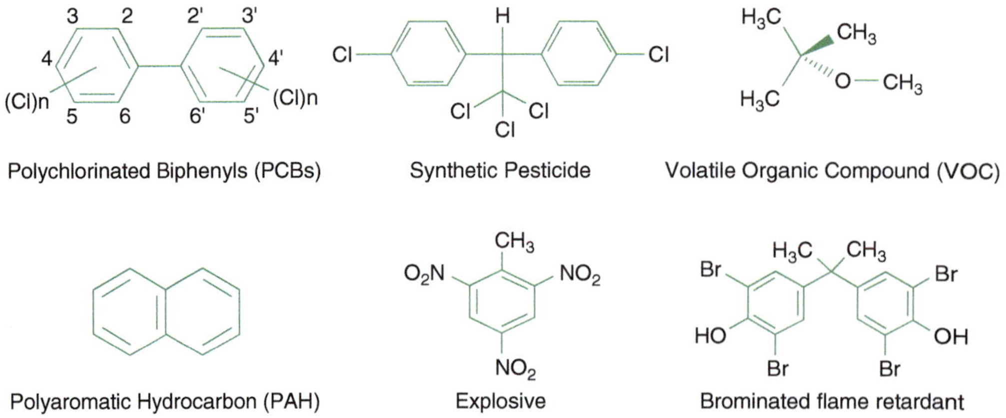

Fig. 18.1. Structural diversity of representative persistent organic pollutants (POPs)/xenobiotic compounds with known human health hazards.

Smreczak (2000) reported that soil contamination with PAHs at levels below 10 mg kg^{-1} stimulated the growth of plants at the early stages of their development, however, with increased concentration (i.e. 20 mg of PAH kg^{-1} of soil) they significantly inhibited plant growth of tomato; the lowest observed level of soil contamination with inhibiting concentrations of PAH exceeded 100 mg kg^{-1} for other plants, viz. wheat, oat, maize, bean and sunflower (Maliszewska-Kordybach and Smreczak, 2000).

Many plants use a three-step detoxification process for mitigating the toxic effects of xenobiotic compounds. The steps include transformation, conjugation and compartmentation. Additionally, many plants also have inherent ability to detoxify some xenobiotic pollutants though co-metabolic transformation, yet they generally lack the complete catabolic pathway for complete degradation/mineralization of these compounds. Additionally, a number of studies have suggested that xenobiotic compounds could induce biochemical and physiological disruption of plant metabolic functions, change the gene expression patterns, disrupt the signalling pathways and render mutations in the plant genomes. To conclude, it could be said that while plants are less susceptible to xenobiotic pollutants compared to animals and human beings, some of the xenobiotic compounds would still induce adverse effects on plant growth and productivity. Therefore, not only is it important to decontaminate the xenobiotic polluted ecological niches for the health and safety of human beings, but also for ensuring the sustained productivity of different ecosystems.

18.1.3 Approaches for decontamination of niches contaminated with xenobiotic compounds

Historically, the approaches used for decontamination of niches contaminated with xenobiotic compounds have fallen under the following broad categories: (i) physico-chemical approaches, and (ii) biological approaches. While the physico-chemical approaches depend upon use of methods such as incineration, excavation, landfilling, vitrification, chemical oxidation, etc., the biological approaches make use of microbial or plant metabolic properties to either transform (to a non-toxic product) or completely metabolize the target pollutant. Both approaches have their well-established advantages and disadvantages, yet the biological approach is regarded as environmentally benign and economical compared to the physico-chemical approach. Some of the major advantages associated with biological treatment approaches include: (i) they could be applied at the site of contamination without the need for removal and transport of the contaminated soil/water to the treatment site; (ii) they have lower greenhouse gas emission.

18.1.4 Biological approaches for degradation of xenobiotic compounds

The biological approaches are segregated on the basis of biocatalyst(s) used for the decontamination process; thus there are two broad categories of biological decontamination approaches, viz. (i) microbial bioremediation (uses microorganisms) and (ii) phytoremediation (uses plants). Furthermore, the microbial bioremediation methods are broadly classified into 2 categories, i.e. (i) biostimulation and (ii) bioaugmentation. Biostimulation involves modification of the environment to stimulate native microbial communities to enable them for decontamination. This can be achieved by addition of various forms of rate limiting nutrients and electron acceptors, such as phosphorus, nitrogen, oxygen, or carbon (e.g. in the form of molasses) – i.e. stimulation of native microbes. On the other hand, bioaugmentation is defined as the technological approach of adding cultured, well characterized microorganism or a microbial consortium into the contaminated soil or groundwater for the purpose of biodegrading specific contaminants. A handful of studies from the recent past have reported successful trials and application of microbial bioremediation approaches for decontamination of xenobiotic pollutants at different scales of operation (e.g. laboratory scale, microcosm scale, pilot scale), yet further technological development of these approaches is required for commercially feasible full-scale application of these approaches.

In comparison to the microbial bioremediation approaches, the plant-based approaches are quite diverse and include phytoextraction, phytotransformation, phytostabilization, phytovolatilization and rhizofiltration. Phytoextraction involves the uptake of pollutants into harvestable biomass for subsequent incineration. Phytotransformation involves enzymatic modification for degradation or immobilization of target pollutant. Phytovolatilization works on the principle of removal of pollutants from soil via evapotranspiration processes. The approach of rhizofiltration makes use of filtering the contaminated water through a mass of plant roots to remove pollutants.

18.1.5 Plant associated microorganisms for degradation of xenobiotic compounds

Although a number of studies have highlighted successful application of plants for carrying out degradation of toxic xenobiotic compounds, the idea of using the phytoremediation-capable plant in combination with the microbial diversity associated with the plant (especially the endophytes and rhizosphere-associated microorganisms) has been suggested to have great remediation potential. Both endophytic microorganisms and rhizosphere-associated microorganisms are naturally occurring, non-pathogenic microorganisms that promote the growth of the plant through an intricate network of multiple mechanisms. As highlighted in Fig. 18.2 the classically defined mechanism involving plant growth-promoting rhizobacteria (PGPR) for plant growth includes mechanisms such as (i) nutrient fixation, (ii) nutrition uptake, (iii) plant growth hormone production, (iv) siderophore production and (v) biological control of pathogens by production of anti-microbial agents (Bisen *et al.*, 2015, 2016; Keswani, 2015; Keswani *et al.*, 2016a, b). While these mechanisms have been reported as being successful in promoting plant growth, recent developments have indicated that plant-associated endophytic microorganisms and, more specifically, PGPR strains can degrade toxic xenobiotic compounds and thus offer a newly suggested mode of action of PGPRs for promotion of microbial growth. Consequently, it could be suggested that PGPRs can support plant growth in ecological niches contaminated with toxic xenobiotic compounds that would otherwise hamper the normal plant growth (Singh *et al.*, 2014; Mishra *et al.*, 2015).

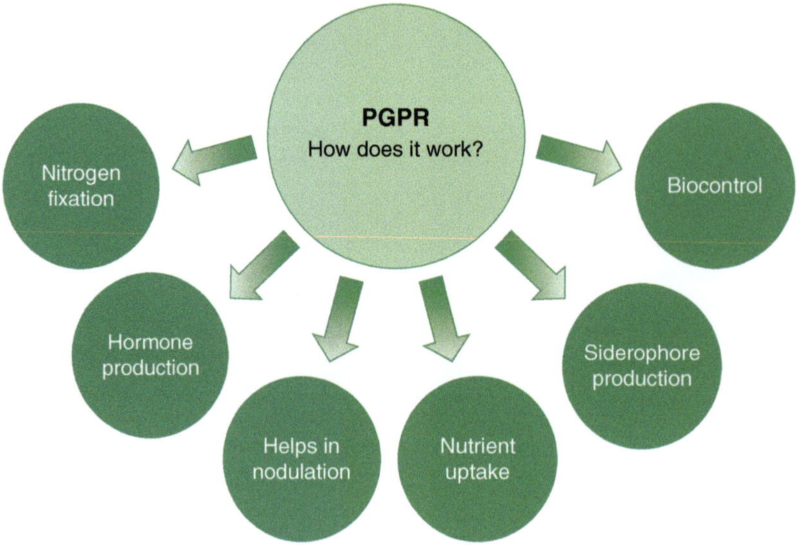

Fig. 18.2. Classically defined mode of actions of Plant Growth-Promoting Rhizobacteria (PGPR) in promotion of plant growth.

18.1.6 Allelochemicals as potential inhibitors of normal plant growth

Allelochemicals are small-molecule secondary metabolites released by plant parts by leaching, root exudation, volatilization, residue decomposition and other processes in both natural and agricultural ecosystems by crop as well as weed plant species; they may exert beneficial or harmful effects on another plant. Traditionally, the allelochemicals were defined as secondary metabolites (e.g. alkaloids, benzoxazinones, cinnamic acid derivatives, cyanogenic compounds, ethylene and other seed germination stimulants, and flavonoids) synthesized but not required for either growth or development of the organism; their primary importance was regarded as important components of plant defence against herbivorous animals (Fraenkel, 1959). However, the notion of allelochemicals was revised after 1974, when E. L. Rice defined allelochemicals as the effect(s) of one plant on other plants through the release of chemical compounds in the environment (Rice, 1979). Interestingly, this definition includes both positive (growth-promoting) and negative (growth-inhibiting) effects of secondary metabolites on plants (Kohli *et al.*, 1997). However, contrary to the widely accepted definition, many plant ecologists consider only the negative effects of these unique metabolites. Other ecologists believe that any direct or indirect effect of one plant on others through the release of small-molecule chemicals may be considered as allelopathy. According to E. L. Rice, there are more than 10,000 different secondary metabolites synthesized by different plant species, yet only a very small fraction of the total small metabolite diversity exerts allelopathy effects.

18.1.7 Structural and functional diversity of allelochemicals and their mode of actions

While our understanding pertaining to allelochemicals and allelopathy continues to grow, it is regarded as playing a pivotal role in defining many aspects of plant ecology, including occurrence, growth, plant succession, the structure of plant communities, dominance, diversity and plant productivity within almost all natural and agricultural plant ecosystems (Kohli *et al.*, 1997). On the basis of the chemical nature, the allelochemicals belong to diverse chemical groups such as: terpenes, benzoquinones, coumarins, flavonoids, terpenoids, strigolactones, phenolic acids, tannins lignin, fatty acids and nonprotein aminoacids. For simplification, the allelochemicals can be classified into ten broad categories (Soltys *et al.*, 2013). The categorization is largely based on the type of the carbon skeleton and number of carbon atoms present in the allelochemicals. As shown in Table 18.2, the number of carbon atoms present within the allelochemicals ranges from 6 to 30. Many of these compounds are synthesized during the shikimate pathway of aromatic amino acid biosynthesis.

With respect to the mode of action, the most commonly cited examples of allelopathy-induced changes in the plants include poor regeneration of plant species, crop damage, yield reductions, etc. The other changes induced by allelopathy effect include reduced seed germination and seedling growth. The allelochemicals impart such adverse effects by targeting critical plant functions including cell division, pollen germination, nutrient uptake, photosynthesis, and specific enzyme functions. Some of the gross morphological effects of allelochemicals as observed on plant growth include: inhibited or retarded germination rate; seed darkening; seed swelling; reduced root and shoot growth; necrosis of root tips; curling of the root axis; discoloration, lack of root hairs; increased number of seminal roots; reduced dry weight accumulation; and lowered reproductive capacity (Ward *et al.*, 2011). The mode of action of allelochemicals is not yet well established at the level of molecular mechanism. One of the studies carried out a controlled allelopathy interaction between the aggressive and allelopathic plant *Sicyos deppei* as the donor plant, and *Lycopersicon esculentum* as the receptor plant; it showed that allelochemicals released by *S. deppei* caused oxidative damage through an increase in reactive oxygen species (ROS) and activation or modification of antioxidant enzymes (Cruz-Ortega *et al.*, 2007).

Table 18.2. The major classes of phenolic allelochemicals in plants (Adapted from Ilori and Ilori, 2012).

Sl. No.	Synthetic Organic Pollutant/ Xenobiotic Compound	Carbon Skeleton	Representative congeners
1	Phenols	C-6	Hydroquinone
	Benzoquinones		Catechol
2	Phenolic Acids	C6-C1	Gallic Acid
			Salicylic Acid
3	Acetophenones	C6-C2	Tyrosol
	Tyrosine derivatives		p-Hydroxyphenylacetic Acid
	Phenyl acetic acid		
4	Hydroxycinnamic Acid	C6-C3	Ferulic Acid,
	Phenylpropenes		Myristicin, Eugenol,
	Coumarins		Aesculectin
	Isocoumarins		Bergenon
	Chromones		Eugenin
5	Naphthoquinones	C6-C4	Juglone, Plumbagin
6	Xanthones	C6-C1-C6	Mangiferin
7	Stilbenes	C6-C2-C6	Resveratrol
	Anthraquninones		Emodin
8	Flavonoids	C6-C3-C6	Cyanidin
	Isoflavonoids		Genistein
9	Lignans	(C6-C3)2	Pinoresinol
	Neolignans		Eusiderin
10	Biflavonoids	(C6-C3-C6)2	Amentoflavone

18.1.8 Role of allelochemicals in ecological success of invasive plants and weeds

With respect to the ecological impact, the allelochemicals are credited to have enabled many weeds and exotic invasive plants to invade agriculturally suitable land. Thus allelopathy may be considered as a feature of their ecological success (Callaway and Aschehoug, 2000; Callaway and Ridenour, 2004; Chengxu *et al.*, 2011). Invasive plant species are often introduced as an exotic, non-indigenous species to that location; subsequently, it grows very aggressively leading to damage to the other plant species and consequently the economic structure of the native place (Mack *et al.*, 2000; Belz, 2007). One recent study showed how *Centaurea diffusa*, a noxious weed in North America, has a stronger negative effect on other grass species due to differences in the effects of its root exudates and how these root exudates affect competition for resources (Callaway and Aschehoug, 2000). The allelochemicals-induced invasion by weeds progresses through plant growth inhibition using a complex mode of allelochemicals action. Sometime it may involve the interaction of different classes of allelochemicals. In general the mixtures of different allelochemicals render a greater effect compared to the individual allelochemical alone. In addition, the effect of allelochemicals on exotic weed invasion also depends upon the physiological and environmental stresses, pests and diseases, solar radiation, herbicide load, nutrient availability, moisture and temperature levels of the ecosystem. Due to the adverse effects of exotic invasive plant-associated allelochemicals, studies focusing on ecological persistence and environmental fate of allelochemicals have become extremely important. Recent studies pertaining to allelopathy have indicated that plant-associated soils and their microbial inhabitants are essential predictors of the overall outcome of allelopathy interactions between plants (Cipollini *et al.*, 2012). Interestingly, a few studies have also suggested that allelopathy chemicals released as plant exudates can also influence the structure and dynamics of soil microbial community.

18.1.9 Invasive plants with allelopathy potential

While allelopathy has been observed with a wide variety of crop and weed species, the best characterized allelopathy potential has been reported from weeds including Congress grass (*Parthenium hysterophorus* L.), Canary grass (*Phalaris minor* Retz.), Russian knapweed (*Acroptilon repens* L.), and Morning glory (*Ipomoea tricolor* Cav.). Nearly every one of these weeds is shown to exert negative effects on agriculture, animal husbandry, ecology and the environment (Maddox *et al.*, 1985; Nath, 1988; Om *et al.*, 2002; Jefferson and Pennacchio, 2003). Detailed studies pertaining to *Parthenium* showed a 30–40% reduction in yield of crop plants when grown on soil containing dried root and leaf material of *Parthenium* (Singh *et al.*, 2003). Similarly, members of the Chenopodiaceae family exhibit allelopathy and inhibit the germination of lettuce seed (Jefferson and Pennacchio, 2003). In another report, Stevens (1986) found that the roots of Russian knapweed inhibited the root growth of many plants including some weed species such as *Lactuca sativa*, *Medicago sativa*, *Echinochloa crusgalli* and *Panicum miliaceum* at 30% concentrations comparable to those found in the soil surrounding *A. repens* plants (Stevens, 1986). In light of the results pertaining to the role of allelochemicals in the invading property of exotic invasive plants, it could be argued that, although the negative results of allelopathic tests are undoubtedly under-reported and understanding of allelopathy in natural ecosystems is rather vague, yet they may have potentially harmful effects towards plants in general and plant growth in particular.

18.1.10 Environmental fate of allelochemicals – natural factors and microbial metabolism

Unlike xenobiotic compounds that are synthetic in nature, the allelochemicals are natural and synthesized as secondary metabolites by a wide variety of plants, yet many allelochemicals are recalcitrant to degradation.

Thus the environmental fate of these chemicals is defined by cumulative outcome of absorption by plants, degradation via photolysis, oxidation and microbial degradation, and to processes of removal or transfer, such as volatilization and adsorption (Blum, 2004). Allelochemicals can be transformed or completely degraded by physical, chemical and microbial processes. The main processes leading to removal of allelochemicals from the soil include volatilization and adsorption (Taylor and Spencer, 1990; Kobayashi, 2004). Additionally, many research studies have also indicated that binding of allelochemicals to soil particles reduces their accessibility in the natural environment (Vidal and Bauman, 1997; Kobayashi, 2004). The relative amounts or proportions of allelochemicals binding/adsorbing to the soil has been shown to be greatly influenced by organic matter (OM) levels and soil types. For example, it has been reported that ferulic acid is adsorbed at 20% on kaolin or gibbsite, 70% on goethite, and 100% on histossol (Vidal and Bauman, 1997). In addition to the physical and chemical processes, the microbiological processes are also critical in determining the environmental fate of allelochemicals. Unlike xenobiotic compounds, allelochemicals are naturally synthesized and lack halogen substitution. Therefore, they are relatively easily degraded by microbial metabolism. A number of studies have indicated degradation of allelochemicals by fungal and bacterial isolates (Schmidt, 1988; Chen, L. *et al.*, 2011; Chen, Y. *et al.*, 2011; Zhu *et al.*, 2011).

18.2 Microbial Degradation of Xenobiotic Compounds and Allelochemicals

18.2.1 Microbial degradation of xenobiotic compounds

A large number of microorganisms are bestowed with the ability of metabolizing recalcitrant synthetic xenobiotic compounds. Furthermore, they use the xenobiotic compounds as a source of carbon and energy, nitrogen and other macro/micronutrients (Bourquin *et al.*, 1981). In some cases, the xenobiotic compounds are also used as the final electron acceptor in the respiratory process (Díaz, 2010). The process of nutrient utilization and energy generation through breakdown of xenobiotic substrate essentially involves complete degradation or mineralization. However, in some cases, the complex xenobiotic substrate is transformed to a less complex organic compound with significantly diminished toxicity (Häggblom and Bossert, 2003). Such processes occur only in the presence of a metabolizable substrate and are referred as "co-metabolic biotransformation" (Häggblom and Bossert, 2003; Rylott *et al.*, 2011). The co-metabolic biotransformation induces modification of the molecular structure of the compound, which results in alteration or complete loss of the characteristics (e.g. solubility, toxicity, etc.) of the original compound. It has often been observed that microbial strains capable of completely mineralizing one type of compound may sometimes fortuitously transform other compounds that are not used as source of nutrient or energy. Furthermore, it is generally accepted that products of co-metabolic biotransformation of xenobiotic compounds leads to detoxification of the original toxic compound (Fetzner, 2002).

Over the past century, a large number of microbial strains (bacteria, archaea and fungi) have been isolated from a wide variety of contaminated environments and characterized for enzymes capable of degrading toxic xenobiotic compounds (Boethling, 1993; Offre *et al.*, 2013; Aranda, 2016). Amongst the isolated microorganisms, bacterial strains belonging to both Gram-positive and Gram-negative domains have been found to have the ability to convert extremely complex xenobiotic compounds (Borja *et al.*, 2005; Rein *et al.*, 2007). While the minute details of microbial degradation of different xenobiotic compounds vary according to the chemical structure of the substrate and metabolic capabilities of the degrading microorganism, the majority of the microbial degradations could be broadly classified into either (i) aerobic or (ii) anaerobic biodegradation. Most of the aerobic degradation processes involve oxidative transformation, whereas the anaerobic degradation process proceeds via reductive transformations (Zhang and Bennett, 2005; Fritsche and Hofrichter, 2008; Díaz *et al.*, 2013).

18.2.2 Degradation of xenobiotic compounds by plant-associated microorganism (endophytes and PGPRs)

Studies carried out in the past for isolation and characterization of xenobiotic-compound-degrading microorganisms grossly overlooked the microbial diversity associated with the plants. Furthermore, for a long time plant-associated microorganisms (endophytes and PGPR) were studied only from the point of view of promoting plant growth (Ryan *et al.*, 2008). However, poor outcomes during phytoremediation studies have indicated that use of plants alone for remediation suffers many limitations. Therefore, in the recent past, the scope of phytoremediation has expanded to include plant-associated endophytes and PGPR strains for degradation of toxic xenobiotic compounds. A number of endophytic bacteria have been identified and characterized for degradation of environmental pollutants. A list of a few representative endophytic bacteria identified for degradation of xenobiotic compounds is presented in Table 18.3.

Amongst the known plant-associated endophytic bacteria identified for degradation of toxic xenobiotic compounds, a majority of the strains belong to the genus *Pseudomonas*. Many studies have indicated that the xenobiotic-degrading plant-associated endophytic bacterial community is dominated by the members of the genus *Pseudomonas*. In one such study, the endophytic microbial community characterized from the plant species *Lolium perenne* was found to be dominated by *Pseudomonas* spp. and exhibited increased PAH degradation (Phillips *et al.*, 2008). With regard to the host plants, the most common xenobiotic compounds degrading endophytic bacteria have been isolated and characterized from poplar trees. Some of the poplar-associated endophytes were found to be methylotrophic and had the ability to mineralize explosives, e.g. TNT, RDX and HMX to CO_2 (Van Aken *et al.*, 2004). Similarly, endophytes isolated from hybrid poplar trees growing on BTEX-contaminated soil were found to be capable of degrading toluene and naphthalene as well as a chlorinated organic herbicide (2,4-D) (Germaine *et al.*, 2006, 2009).

With enhanced understanding of xenobiotic-compound degradation by endophytic microorganisms, it has also been realized that the application of such microorganisms for decontamination of polluted sites may be technologically challenging. The major challenges for development of this approach as a feasible technology include: (i) limited application due to seasonal and regional selectivity of host plant cultivation, (ii) variable bioavailability of target pollutant due to plant mediated adsorption and transportation, (iii) unacceptable transpiration of VOCs

Table 18.3. A list of representative plant-associated endophytic bacteria-mediated degradation of toxic xenobiotic compounds (Adapted from McGuinness and Dowling, 2009).

Xenobiotic Compound	Endophytic Bacterial Strain	Host Plant	Reference
PCBs, TCP	*Herbaspirillum* sp. K1	Wheat	Männistö *et al.* (2001)
Chlorobenzoic acids	*Pseudomonas aeruginosa* R75 *Pseudomonas savastanoi* CB35	Wild rye	Siciliano *et al.* (1998)
Pesticide (2,4-D)	*Pseudomonas putida* VM1450	Pea	Germaine *et al.* (2006)
VOCs	*Burkholderia cepacia G4*	Yellow lupin	Barac *et al.* (2004);
Toluene	*Burkholderia cepacia* Bu61	Poplar	Taghavi *et al.* (2005)
MTBE, BTEX, TCE	(pTOM-Bu61)		
HCs	*Pseudomonas putida* VM1441	Pea	Germaine *et al.* (2009)
Naphthalene	(pNAH7)		
Explosives	*Methylobacterium populi* BJ001	Poplar	Van Aken *et al.* (2004)
TNT, RDX, HMX			

into the atmosphere, etc. On the contrary, there are a few claims that suggest a major advantage of using endophytic bacteria is that the endophytes naturally inhabit the internal tissues of plants and therefore they are significantly less susceptible to problems of competition between bacterial strains.

Despite the limitations allied with the use of endophytic microorganisms for development of bioremediation technology, it would be safe to say that there is a significant role for plant-associated endophytic microorganisms/microbial communities in the decontamination of xenobiotic compounds. The role is all the more significant in mitigation of xenobiotic compounds bioaccumulated in plants via absorption and transpiration. Furthermore, they may also have significant implication in survival and growth of the host plant in ecological niches contaminated by toxic xenobiotic compounds (Afzal *et al.*, 2014).

18.2.3 Degradation of xenobiotic compounds by rhizospheric bacteria and PGPR

Microorganisms residing within the rhizosphere are involved in a variety of host–microbe interactions/processes that impart beneficial effects to the host plants. Many of these processes have been agreed upon for their direct implication in promotion of plant growth. A relatively recent paradigm involving rhizospheric bacteria and PGPR has been their ability to transform/mineralize toxic xenobiotic compounds. The process has been termed as “rhizoremediation” and involves the restoration of contaminated niches via mutual interaction of plant roots and suitable microbial populations occurring in the rhizosphere. Interestingly, it is considered as one of the most evolved processes of natural bioremediation (Wenzel, 2009). The process of rhizoremediation is characterized by a number of natural advantages, e.g. (i) plant root offers a large surface area for microbial colonization and thus leads to ~ 10^2–10^4-fold greater microbial density, (ii) plant exudates act as co-metabolic substrates for rhizospheric bacteria and enable them to survive periods of reduced pollutant availability (Wenzel, 2009; Glick, 2010). On the other hand the rhizospheric bacteria exhibit a diverse range of plant growth-promoting activities that are critical in adaptation of the plant to stressed environments (Dimpka *et al.*, 2009; Ahemad and Khan, 2011; Tak *et al.*, 2013). The rhizospheric microbial communities can also contribute to plant growth by degradation of toxic xenobiotic compounds in the rhizosphere. For example, PGPRs have been reported for degradation of xenobiotic compounds, e.g. explosives, polychlorinated biphenyls (PCBs), synthetic herbicides and hydrocarbons, etc. (Chen *et al.*, 2003; Ramos *et al.*, 2005; Leigh *et al.*, 2006; Kidd *et al.*, 2008). In comparison to the xenobiotic-compound-degrading endophytic microorganisms, the PGPR strains capable of degrading toxic xenobiotic compounds are quite diverse. Consequently, to date a number of toxic organic compounds in soil have been successfully remediated using rhizospheric bacteria/PGPR strains. A list of characterized rhizospheric bacteria/PGPR strains degrading xenobiotic compounds is shown in Table 18.4. Noticeably, some of these strains have been used as natural colonizers of the plant rhizosphere to drive the degradation process, while a few others were engineered to induce the process. It is also worth mentioning that apart from their application in remediation of xenobiotic compounds, the rhizospheric bacteria and PGPR strains associated with different plants are also being characterized and exploited for decontamination of sites polluted by excessive loads of heavy metals (Khan *et al.*, 2009; Marques *et al.*, 2009).

18.2.4 Degradation of xenobiotic compounds via rhizosphere engineering

Over the period of the last 2-3 decades, the metabolic activity of rhizospheric bacteria in general and PGPR in particular for degradation of xenobiotic compounds has been recognized. However, it has also been noted that the natural process of PGPR-driven degradation suffers a number of limitations due to the unique nature of xenobiotic

Table 18.4. List of representative rhizospheric bacteria/PGPR strains used for degradation of xenobiotic compounds. (Adapted from McGuinness *et al.* (2009); Pajuelo *et al.* (2014).

Xenobiotic Compounds	Rhizospheric Bacterial Strain/ PGPR	Interacting Plant	Reference
Polychlorinated biphenyls	*Pseudomonas fluorescence*	*Medicago sativa* *Beta vulgaris* L.	Brazil *et al.* (1995)
	Pseudomonas fluorescence	*Medicago sativa* *Beta vulgaris* L.	Villacieros *et al.* (2005)
	Pseudomonas putida Flav1-1 *Pseudomonas putida* PML2	*Arabidopsis*	Narasimhan *et al.* (2003)
	Rhodococcus, Luteibacter, Williamsia	*Pinus nigra and Salix caprea*	Leigh *et al.* (2006)
Pesticides	*Burkholderia cepacia*	*Hordeum sativam* L.	Jacobsen (1997)
2,4-D	Indigenous degrader Indigenous degrader	*Trifolium pretense* *Lolium perenne*	Shaw and Burns (2004)
Pesticides PCP	*Sphingobium chlorophenolicum*	*Triticum astivum*	Dams *et al.* (2007)
4-Chloro-nitrobenzene	*Comamonas sp. strain CNB-1*	*Medicago sativa*	Liu *et al.* (2007)
VoCs	*Pseudomonas fluorescence*	*Triticumm spp.*	Yee *et al.* (1998)
TCE BTEX & TCE	*Enterobacter, Acinetobacter,*	*Poplar*	Moore *et al.* (2006)
HCs Petroleum product and crude oil	*Azospirillum lipoferum* spp.	*Triticumm spp.*	Shaw and Burns (2004); Muratova *et al.* (2005)
Oil component	*Rhizobium galegae* +*Pseudomonas*	*Galega*	Lindstrom *et al.* (2003)
PAHs	*Pseudomonas putida* PCL1444	*Lolium multiflorum*	Kuiper *et al.* (2001)
Naphthalene Phenanthracene	*Pseudomonas*	*Hordeum sativum* L.	Anokhina *et al.* (2004)

pollutants, type of contaminated soil, interactions of pollutants with soil particles, the content of organic matter, pH, water content, temperature, etc. (McGuinness and Dowling, 2009). To overcome some of these limitations, rhizospheres have been engineered via technological interventions, e.g. organic amendment to achieve the appropriate carbon to nitrogen (C:N) ratio which should range from 12:1 to 20:1 to stimulate the PGPR-mediated degradation of the xenobiotic compounds. It is well established that in the case of synthetic organic compounds, the reduced bioavailability can reduce the rate of bioremediation. Addition of organic amendments and nutrients has been able to enhance the availability of xenobiotics, thereby improving bioremediation rates of hydrocarbons, herbicides, etc. in rhizosphere niches (Lee *et al.*, 2008).

Apart from the amendment of nutrients, the efficient microbial degradation of xenobiotic compounds also needs appropriate concentrations of electron donors and electron acceptors with the micro-environment. This situation has been realized and circumvented by addition of electron acceptors and electron donors, to stimulate naturally occurring microbial populations to degrade the target pollutant or to promote co-metabolism (Miller, 2010; Zawierucha and Malina, 2011). An inappropriate ratio of electron donors and electron acceptors could be one of the major limiting factors in the proliferation and metabolic activity of rhizosphere microorganisms. Soils polluted with organic xenobiotic compounds usually lack electron acceptors (e.g. oxygen). Since, aerobic biodegradation of these pollutants is executed with use of oxygen as the final electron acceptor, in the absence of oxygen, the efficiency of degradation decreases significantly. The degradation efficiency microbial communities within such oxygen-depleted xenobiotic compounds contaminated can be enhanced through biostimulation with supply of dissolved oxygen (Gallizia *et al.*, 2004; Amezcua-Allieri *et al.* 2010). Alternatively, for stimulation of the anaerobic rhizospheric bacteria, nitrate, sulphate, etc. are added as electron acceptors (Amezcua-Allieri *et al.* 2010). Interestingly, a few studies have indicated that plant root exudates may act as the stimulator for the degradative functions of the rhizospheric bacteria (Slater *et al.*, 2011).

In comparison with the biostimulation of the native rhizospheric bacterial community or PGPRs, reports pertaining to use of bioaugmentation (addition of exogenous degradative microbial strains into the contaminated niches) are rather scant. In one of the successful studies, the rhizosphere of *Brassica nigra* was inoculated with rhizosphere bacteria previously isolated from PCB-polluted soil. This intervention resulted in up to 87% PCB removal after 12 weeks of bioaugmentation. Noticeably, only ~ 40% PCB removal was achieved in non-bioaugmented controls (Singer *et al.*, 2003). In another recent example, bioaugmentation of Alfalfa rhizosphere with a *Rhizobium* strain was reported to have a positive influence on degradation of PCB (Xu *et al.*, 2010).

18.2.5 Genetically modified rhizospheric bacteria/PGPR for degradation of xenobiotic compounds

In line with the recombinant DNA technology revolution, the rhizoremediation technology has also adapted and rhizospheric bacteria/PGPR have been genetically engineered for improved bioremediation capacities. Xenobiotic degradation by genetically engineered rhizospheric bacteria/PGPR is widely reported in the case of *Pseudomonas* and *Rhizobium*. Some of the examples of genetically modified rhizospheric bacteria/PGPR that were generated for enhanced degradation of trichloroethylene (Yee *et al.*, 1998) or PCBs (Toure *et al.*, 2003) were in the rhizosphere of plants. In another example study, construction of a metal-resistant and TCE-degrading rhizobacterium was accomplished by expressing the metal-binding peptide EC20 in a TCE-degrading strain (Lee *et al.*, 2006). This genetically engineered agent was found to be able to perform TCE degradation even in metal-polluted soils. More recent examples of genetically engineered rhizospheric bacteria/PGPR being used in degradation of xenobiotic compounds include studies wherein the *bph* operon involved in PCB degradation was integrated into the chromosome of *P. fluorescens* F113 under the regulation of a strongly inducible promoter (NOD box) of *Sinorhizobium meliloti*. Noticeably, the wild type strain F113 was characterized as the excellent root colonizer of many plants including tomato, sugar beet, alfalfa and willow. With chromosomal insertion of the *bph* operon the genetically modified strain F113 could carry out degradation of PCB in a very efficient manner (Brazil *et al.*, 1995).

18.2.6 Microbial degradation of allelochemicals

As described earlier, unlike xenobiotic compounds, the allelochemicals are defined as a set of natural compounds synthesized by plans to carry out synergistic or antagonistic interactions with other plant species. However, in practical terms allelochemicals are made by one plant species for suppression of competing species (Rice, 2013). They are released into the ecosystem as plant root exudates, leaf leachates and products of plant tissue decomposition. Although allelochemicals are not known to directly impart any toxic effect to either the environment or human health, some of the allelochemicals produced by exotic invasive plant species and weeds have still become ecological concerns as they adversely affect the native plant diversity and productivity of ecosystems (Murrell *et al.*, 2011). Many observations have indicated that allelochemicals can impart toxic effects similar to those induced by toxic xenobiotic compounds. Noticeably, just like xenobiotic compounds, the allelochemicals can also persist in the environment over long time spans and therefore they affect not only the neighbouring plants but also the plants cropped or grown/planted at later times.

It has been generally accepted that allelochemicals must accumulate and persist at phototoxic levels in the rhizosphere soil in order to hamper plant growth. However, after their entry into the environment, the persistence, availability and biological activities of allelochemicals are influenced by a range of abiotic factors as well as microbes (Kobayashi, 2004; Granéli and Salomon, 2010; Antunes *et al.*, 2012; Rice, 2013). Plant root exudates can modulate the neighbouring rhizospheric/PGPR microbial community. On the other hand the neighbouring rhizospheric/PGPR microbial activity also acts as a major factor in determining the concentration gradient and flux of the root exudates. As a consequence, the toxicity of allelochemicals secreted in the form of plant root exudates is largely regulated by the metabolic activity of the neighbouring fungal, rhizospheric bacterial and PGPR microbial community (van der Putten *et al.*, 2010; Zhu *et al.*, 2011). A number of fungal isolates (e.g. *Phomopsis liquidambari, Paxillus involutus, Laccaria bicolor* and *Trichoderma harzianum*) have been reported with the ability to degrade natural aromatic compounds including allelochemicals (Zeng and Mallik, 2006; Chen *et al.*, 2011; Xie and Dai, 2015). Similarly, a few bacterial isolates have been reported to be capable of rapidly transforming and/or completely mineralizing allelochemicals (Zhang *et al.*, 2010). In one of the earliest studies on bacterial degradation of allelochemicals, Schmidt reported isolation of a *Pseudomonas spp.* from the soil beneath black walnut trees. This strain could rapidly mineralize the allelochemical juglone (5-hydroxy-1,4-naphthoquinone) as its sole source of carbon and energy (Schmidt, 1988).

In a relatively recent study Zhang *et al.* (2010) reported isolation and characterization of bacterial strains belonging to genera *Pseudomonas* and *Rhodotorula* from rice, pine and bamboo that were capable of degrading a phenolic allelochemical, viz. *p*-coumaric acid (Zhang *et al.*, 2010). Noticeably, in this study it was also shown that the isolated microbes could reverse the inhibitory effect of *p*-coumaric acid on seed germination and seedling growth in culture solutions and soil experiments. This study clearly demonstrates the applicability of microbial degradation of allelochemicals as a potential means for circumventing the harmful effect of allelochemicals contributing to the invasive behaviour of the exotic weeds.

In the context of microbial degradation of allelochemicals, it is worth mentioning that apart from isolation and taxonomic characterization of the degrading microorganism, a number of studies have also shown the characterization of the metabolic pathways involved in the degradation of a few model allelochemicals. For example, in the case of fungal degradation of ferulic acid, the substrate is transformed into either caffeic acid or vanillic acid. Subsequently, these intermediates are transformed to protocatechuic acid, which is broken down to β-carboxy-cis, cis-muconic acid via ring cleavage. Subsequent degradation proceeds via acetic acid and succinic acid to finally yield CO_2, water and energy (Falconnier *et al.*, 1994). In a similar study, Narbad and Gasson (1998) reported degradation of ferulic acid as a sole source of carbon and energy (Narbad and Gasson, 1998). The catabolic pathway was found to be similar to an earlier reported one and proceded via vanillin, vanillic acid and protocatechuic acid as major degradation intermediates. In another report, a strain of *Serratia marcescens* was isolated from soil under coffee cultivation. It was able to degrade caffeine and other methylxanthines (Mazzafera *et al.*, 1996). The catabolic pathway followed is such that caffeine is degraded to theobromine (3,7-dimethylxanthine) and/or paraxanthine (1,7-dimethylxanthine), and subsequently to 7-methylxanthine and xanthine.

18.2.7 Degradation of allelochemicals by rhizospheric bacteria/PGPR

A large number of microorganisms (including bacteria and fungi) have been isolated from diverse ecological sources and identified for their ability to degrade allelochemicals (Dagley, 1971; Falconnier *et al.*, 1994; Mazzafera *et al.*, 1996; Narbad and Gasson, 1998; Blum, 2004; Zhang *et al.*, 2010; Chen *et al.*, 2011). Although there are comparatively fewer discrete studies that have reported isolation and characterization of allelochemicals by rhizospheric bacteria or PGPR strains, in light of the environmental fate and accumulation pattern of the allelochemicals within the plant rhizosphere it could be argued that rhizospheric microbial diversity/PGPR would be relatively rich with respect to degrading potential towards allelochemicals. In a study published by Kaur *et al.* (2009), a comparative investigation was carried out to assess the allelopathic effects of m-tyrosine on three plant species (viz. *Lactuca sativa*, *Phalaris minor* and *Bambusa arundinacea*) in the presence of sterilized soil and non-sterile soil. Noticeably, the allelopathic effects of m-tyrosine were significantly diminished when non-sterile soil was used (Kaur *et al.*, 2009). This result points towards the important role for rhizosphere-specific and bulk soil microbial activity in determining the outcome of allelopathic interactions. Another study pertaining to bacterial degradation of allelochemicals reported degradation of *p*-hydroxybenzoic acid by *Pseudomonas putida* strain CSY-P1 isolated from the rhizosphere of the cucumber (Chen *et al.*, 2015). These reports unmistakably demonstrate that microorganisms belonging to diverse evolutionary domains have the catabolic mechanisms to degrade plant-derived allelochemicals. The primary purpose for such degradation might be defined by the energy and nutrition requirements of the rhizospheric bacteria/PGPR, however, in retrospect this activity may be central to the plant–rhizospheric bacteria or plant–PGPR interaction. It may be the defining mechanism for an ecological phenomenon, e.g. plant community succession, etc.

18.3 Conclusion and Future Perspective

To conclude, it could be said that our understanding regarding the role of rhizospheric bacteria and PGPR in degradation of toxic environmental pollutants and allelochemicals is still not fully mature. However, on the basis of the reports presented by different studies and as summarized in a number of recent reviews it could be proposed that rhizospheric bacteria and PGPR form one of the most important components of the naturally occurring xenobiotic degradation system (McGuinness and Dowling, 2009; Wenzel, 2009; Glick, 2010). They are directly or indirectly involved in promoting plant growth by transformation of toxic xenobiotic compounds to potentially less toxic products or their complete degradation. They might have evolved and established a critical symbiotic association with plants wherein plants have benefited from the xenobiotic compound and allelochemicals degrading ability of the rhizospheric bacteria/PGPR. Thus it could be suggested that rhizospheric bacteria/PGPR not only help plant growth via classically defined mechanisms (e.g. solubilization of insoluble nutrients, production of plant growth hormones, etc.), but also by mitigating the harmful effects of toxic pollutants grossly accumulated in the soil and groundwater. Since the complexity of soil ecosystems is very difficult to control, much more technological advancements are still required to fully harness the metabolic and degrading capabilities of rhizospheric bacteria/PGPR for effective decontamination of polluted ecological niches. The future studies in this direction can focus on isolation of PGPRs from the rhizospheric niches of exotic invasive weed plants and assess them for metabolism of important model xenobiotic compounds and allelochemicals. With development of a remediation technology applying rhizospheric bacteria/PGPR it could be forecast that they would show better survival due to interaction with the plant root system as compared with non-rhizospheric bacteria. This phenomenon would be advantageous in enabling continued long-term degradation of target compounds.

References

Afzal, M., Khan, Q.M. and Sessitsch, A. (2014) Endophytic bacteria: prospects and applications for the phytoremediation of organic pollutants. *Chemosphere* 117, 232–242.

Ahemad, M. and Khan, M.S. (2011) Functional aspects of plant growth promoting rhizobacteria: recent advancements. *Insight Microbiology* 1, 39–54.

Amezcua-Allieri, M., Rodríguez-Dorantes, A. and Melendez-Estrada, J. (2010) The use of biostimulation and bioaugmentation to remove phenanthrene from soil. *International Journal of Oil, Gas and Coal Technology* 3, 39–59.

Anokhina, T., Kochetkov, V., Zelenkova, N., Balakshina, V. and Boronin, A. (2004) Biodegradation of phenanthrene by *Pseudomonas* bacteria bearing rhizospheric plasmids in model plant–microbial associations. *Applied Biochemistry and Microbiology* 40, 568–572.

Antunes, J.T., Leão, P.N. and Vasconcelos, V.M. (2012) Influence of biotic and abiotic factors on the allelopathic activity of the cyanobacterium *Cylindrospermopsis raciborskii* strain LEGE 99043. *Microbial Ecology* 64, 584–592.

Aranda, E. (2016) Promising approaches towards biotransformation of polycyclic aromatic hydrocarbons with *Ascomycota* fungi. *Current Opinion in Biotechnology* 38, 1–8.

Barac, T., Taghavi, S., Borremans, B., Provoost, A., Oeyen, L. *et al.* (2004) Engineered endophytic bacteria improve phytoremediation of water-soluble, volatile, organic pollutants. *Nature Biotechnology* 22, 583–588.

Belz, R.G. (2007) Allelopathy in crop/weed interactions—an update. *Pest Management Science* 63, 308–326.

Bisen, K., Keswani, C., Mishra, S., Saxena, A., Rakshit, A. and Singh, H.B. (2015) Unrealized potential of seed biopriming for versatile agriculture. In: Rakshit, A., Singh, H.B. and Sen, A. (eds) *Nutrient Use Efficiency: From Basics to Advances*. Springer, New Delhi, India, pp. 193–206.

Bisen, K., Keswani, C., Patel, J.S., Sarma, B.K. and Singh, H.B. (2016) *Trichoderma* spp.: efficient inducers of systemic resistance in plants. In: Chaudhary, D.K. and Verma, A. (eds) *Microbial-mediated Induced Systemic Resistance in Plants*. Springer, Singapore, pp. 185–195.

Blum, U. (2004) Fate of phenolic allelochemicals in soils–the role of soil and rhizosphere microorganisms. *Allelopathy: Chemistry and Mode of Action of Allelochemicals* 57–76.

Boethling, R.S. (1993) Biodegradation of xenobiotic chemicals. *Handbook of Hazardous Materials*. Academic Press, San Diego, California, 55–67.

Borja, J., Taleon, D.M., Auresenia, J. and Gallardo, S. (2005) Polychlorinated biphenyls and their biodegradation. *Process Biochemistry* 40, 1999–2013.

Bourquin, A.W., Spain, J. and Pritchard, P. (1981) Microbial degradation of xenobiotic compounds. In: *Proceedings of the 12th Annual Conference on Environmental Toxicology, AFMRL-TR-81–149*, Wright Patterson AFB, Ohio, pp. 354–369.

Brazil, G., Kenefick, L., Callanan, M., Haro, A., De Lorenzo, V. *et al.* (1995) Construction of a rhizosphere pseudomonad with potential to degrade polychlorinated biphenyls and detection of *bph* gene expression in the rhizosphere. *Applied and Environmental Microbiology* 61, 1946–1952.

Callaway, R.M. and Aschehoug, E.T. (2000) Invasive plants versus their new and old neighbors: a mechanism for exotic invasion. *Science* 290, 521–523.

Callaway, R.M. and Ridenour, W.M. (2004) Novel weapons: invasive success and the evolution of increased competitive ability. *Frontiers in Ecology and the Environment* 2, 436–443.

Chen, L., Yang, X., Raza, W., Li, J., Liu, Y. *et al.* (2011) *Trichoderma harzianum* SQR-T037 rapidly degrades allelochemicals in rhizospheres of continuously cropped cucumber. *Applied Microbiology and Biotechnology* 89, 1653–1663.

Chen, S.-Y., Guo, L.-Y., Bai, J.-G., Zhang, Y., Zhang, L. *et al.* (2015) Biodegradation of p-hydroxybenzoic acid in soil by *Pseudomonas putida* CSY-P1 isolated from cucumber rhizosphere soil. *Plant and Soil* 389, 197–210.

Chen, Y.-C., Banks, M.K. and Schwab, A.P. (2003) Pyrene degradation in the rhizosphere of tall fescue (*Festuca arundinacea*) and switchgrass (*Panicum virgatum* L.). *Environmental Science & Technology* 37, 5778–5782.

Chen, Y., Peng, Y., Dai, C.-C. and Ju, Q. (2011) Biodegradation of 4-hydroxybenzoic acid by *Phomopsis Liquidambari*. *Applied Soil Ecology* 51, 102–110.

Chengxu, W., Mingxing, Z. and Xuhui, C.Q. (2011) Review on allelopathy of exotic invasive plants. *Procedia Engineering* 18, 240–246.

Cipollini, D., Rigsby, C.M. and Barto, E.K. (2012) Microbes as targets and mediators of allelopathy in plants. *Journal of Chemical Ecology* 38, 714–727.

Cruz-Ortega, R., Lara-Núñez, A. and Anaya, A.L. (2007) Allelochemical stress can trigger oxidative damage in receptor plants: mode of action of phytotoxicity. *Plant Signaling & Behavior* 2, 269–270.

Dagley, S. (1971) Catabolism of aromatic compounds by micro-organisms. *Advances in Microbial Physiology* 6, 1–46.

Dams, R., Paton, G. and Killham, K. (2007) Rhizoremediation of pentachlorophenol by *Sphingobium chlorophenolicum* ATCC 39723. *Chemosphere* 68, 864–870.

Díaz, E. (2010) Bacterial degradation of aromatic pollutants: a paradigm of metabolic versatility. *International Microbiology* 7, 173–180.

Díaz, E., Jiménez, J.I. and Nogales, J. (2013) Aerobic degradation of aromatic compounds. *Current Opinion in Biotechnology* 24, 431–442.

Dimkpa, C., Weinand, T. and Asch, F. (2009) Plant–rhizobacteria interactions alleviate abiotic stress conditions. *Plant, Cell & Environment* 32, 1682–1694.

Falconnier, B., Lapierre, C., Lesage-Meessen, L., Yonnet, G., Brunerie, P. *et al.* (1994) Vanillin as a product of ferulic acid biotransformation by the white-rot fungus *Pycnoporus cinnabarinus* I-937: identification of metabolic pathways. *Journal of Biotechnology* 37, 123–132.

Fetzner, S. (2002) Biodegradation of xenobiotics. In: Doelle and Da Silva (eds) *Biotechnology*. Encyclopedia of Life Support Systems (EOLSS), Oxford.

Fraenkel, G.S. (1959) The raison d'etre of secondary plant substances. *Science* 129, 1466–1470.

Fritsche, W. and Hofrichter, M. (2008) Aerobic degradation by microorganisms. In: Rehm, H.-J. and Reed, G. (eds) *Biotechnology: Environmental Processes II*, Volume 11b, Wiley-VCH, Weinheim, Germany. Ch6, pp.144–167. doi:10.1002/9783527620951.

Gallizia, I., Vezzulli, L. and Fabiano, M. (2004) Oxygen supply for biostimulation of enzymatic activity in organic-rich marine ecosystems. *Soil Biology and Biochemistry* 36, 1645–1652.

Germaine, K.J., Liu, X., Cabellos, G.G., Hogan, J.P., Ryan, D. and Dowling, D.N. (2006) Bacterial endophyte-enhanced phytoremediation of the organochlorine herbicide 2, 4-dichlorophenoxyacetic acid. *FEMS Microbiology Ecology* 57, 302–310.

Germaine, K.J., Keogh, E., Ryan, D. and Dowling, D.N. (2009) Bacterial endophyte-mediated naphthalene phytoprotection and phytoremediation. *FEMS Microbiology Letters* 296, 226–234.

Glick, B.R. (2010) Using soil bacteria to facilitate phytoremediation. *Biotechnology Advances* 28, 367–374.

Granéli, E. and Salomon, P. S. (2010) Factors influencing allelopathy and toxicity in *Prymnesium parvum*. Journal of the American Water Resources Association, 46, 108–120. doi:10.1111/j.1752-1688.2009.00395.x

Häggblom, M.M. and Bossert, I.D. (eds) (2003) *Microbial Processes and Environmental Applications*. Springer, New York.

Ilori, O. and Ilori, O. (2012) Allelochemicals: types, activities and usage in pest control. *Journal of Science and Science Education Ondo* 3, 106–110.

Jacobsen, C.S. (1997) Plant protection and rhizosphere colonization of barley by seed inoculated herbicide degrading *Burkholderia* (*Pseudomonas*) *cepacia* DBO1 (pRO101) in 2, 4-D contaminated soil. *Plant and Soil* 189, 139–144.

Jefferson, L. and Pennacchio, M. (2003) Allelopathic effects of foliage extracts from four chenopodiaceae species on seed germination. *Journal of Arid Environments* 55, 275–285.

Kaur, H., Kaur, R., Kaur, S. and Baldwin, I.T. (2009) Taking ecological function seriously: soil microbial communities can obviate allelopathic effects of released metabolites. *PloS One* 4, e4700.

Keswani, C. (2015) Proteomics studies of thermotolerant strain of *Trichoderma* spp. Ph.D. Thesis, Banaras Hindu University, Varanasi, India, p. 126.

Keswani, C., Bisen, K., Singh, S.P., Sarma, B.K. and Singh, H.B. (2016a) A proteomic approach to understand the tripartite interactions between plant–*Trichoderma*–pathogen: investigating the potential for efficient biological control. In: Hakeem, K.R. and Akhtar, M.S. (eds) *Plant, Soil and Microbes Vol. 2. Mechanisms and Molecular Interactions*. Springer, Cham, Switzerland, pp. 79–93.

Keswani, C., Bisen, K., Singh, V., Sarma, B.K. and Singh, H.B. (2016b) Formulation technology of biocontrol agents: present status and future prospects. In: Arora, N.K., Mehnaz, S. and Balestrini, R. (eds) *Bioformulations: For Sustainable Agriculture*. Springer, New Delhi, India, pp. 35–52.

Khan, M.S., Zaidi, A., Wani, P.A. and Oves, M. (2009) Role of plant growth promoting rhizobacteria in the remediation of metal contaminated soils. *Environmental Chemistry Letters* 7, 1–19.

Kidd, P., Prieto-Fernández, A., Monterroso, C. and Acea, M. (2008) Rhizosphere microbial community and hexachlorocyclohexane degradative potential in contrasting plant species. *Plant and Soil* 302, 233–247.

Kobayashi, K. (2004) Factors affecting phytotoxic activity of allelochemicals in soil. *Weed Biology and Management* 4, 1–7.

Kohli, R., Batish, D. and Singh, H. (1997) Allelopathy and its implications in agroecosystems. *Journal of Crop Production* 1, 169–202.

Kuiper, I., Bloemberg, G.V. and Lugtenberg, B.J. (2001) Selection of a plant-bacterium pair as a novel tool for rhizostimulation of polycyclic aromatic hydrocarbon-degrading bacteria. *Molecular Plant–Microbe Interactions* 14, 1197–1205.

Lee, S.-H., Oh, B.-I. and Kim, J.G. (2008) Effect of various amendments on heavy mineral oil bioremediation and soil microbial activity. *Bioresource Technology* 99, 2578–2587.

Lee, W., Wood, T.K. and Chen, W. (2006) Engineering TCE degrading rhizobacteria for heavy metal accumulation and enhanced TCE degradation. *Biotechnology and Bioengineering* 95, 399–403.

Leigh, M.B., Prouzová, P., Macková, M., Macek, T., Nagle, D.P. and Fletcher, J.S. (2006) Polychlorinated biphenyl (PCB)-degrading bacteria associated with trees in a PCB-contaminated site. *Applied and Environmental Microbiology* 72, 2331–2342.

Lindstrom, K., Jussila, M.M., Hintsa, H., Kaksonen, A., Mokelke, L. *et al.* (2003) Potential of the Galega-*Rhizobium galegae* system for bioremediation of oil-contaminated soil. *Food Technology and Biotechnology* 41, 11–16.

Liu, L., Jiang, C.Y., Liu, X.Y., Wu, J.F., Han, J.G. and Liu, S.J. (2007) Plant–microbe association for rhizoremediation of chloronitroaromatic pollutants with *Comamonas* sp. strain CNB1. *Environmental Microbiology* 9, 465–473.

Mack, R.N., Simberloff, D., Mark Lonsdale, W., Evans, H., Clout, M. and Bazzaz, F.A. (2000) Biotic invasions: causes, epidemiology, global consequences, and control. *Ecological Applications* 10, 689–710.

Maddox, D.M., Mayfield, A. and Poritz, N.H. (1985) Distribution of yellow star thistle (*Centaurea solstitialis*) and Russian knapweed (*Centaurea repens*). *Weed Science* 315–327.

Maliszewska-Kordybach, B. and Smreczak, B. (2000) Ecotoxicological activity of soils polluted with polycyclic aromatic hydrocarbons (PAHs)-effect on plants. *Environmental Technology* 21, 1099–1110.

Männistö, M.K., Tiirola, M.A. and Puhakka, J.A. (2001) Degradation of 2,3,4,6-tetrachlorophenol at low temperature and low dioxygen concentrations by phylogenetically different groundwater and bioreactor bacteria. *Biodegradation* 12, 291–301.

Marques, A.P., Rangel, A.O. and Castro, P.M. (2009) Remediation of heavy metal contaminated soils: phytoremediation as a potentially promising clean-up technology. *Critical Reviews in Environmental Science and Technology* 39, 622–654.

Mazzafera, P., Olsson, O. and Sandberg, G. (1996) Degradation of caffeine and related methylxanthines by *Serratia marcescens* isolated from soil under coffee cultivation. *Microbial Ecology* 31, 199–207.

McGuinness, M. and Dowling, D. (2009) Plant-associated bacterial degradation of toxic organic compounds in soil. *International Journal of Environmental Research and Public Health* 6, 2226–2247.

Miller, H. (2010) Biostimulation as a form of bioremediation of soil pollutants. *MMG 445 Basic Biotechnology eJournal* 6.

Mishra, S., Singh, A., Keswani, C., Saxena, A., Sarma, B.K. and Singh, H.B. (2015) Harnessing plant-microbe interactions for enhanced protection against phytopathogens. In: Arora, N.K. (ed.) *Plant Microbes Symbiosis: Applied Facets*. Springer, New Delhi, India, pp. 111–125.

Moore, F.P., Barac, T., Borremans, B., Oeyen, L., Vangronsveld, J. *et al.* (2006) Endophytic bacterial diversity in poplar trees growing on a BTEX-contaminated site: the characterisation of isolates with potential to enhance phytoremediation. *Systematic and Applied Microbiology* 29, 539–556.

Muratova, A.Y., Turkovskaya, O., Antonyuk, L., Makarov, O., Pozdnyakova, L. and Ignatov, V. (2005) Oil-oxidizing potential of associative rhizobacteria of the genus *Azospirillum*. *Microbiology* 74, 210–215.

Murrell, C., Gerber, E., Krebs, C., Parepa, M., Schaffner, U. and Bossdorf, O. (2011) Invasive knotweed affects native plants through allelopathy. *American Journal of Botany* 98, 38–43.

Narasimhan, K., Basheer, C., Bajic, V.B. and Swarup, S. (2003) Enhancement of plant-microbe interactions using a rhizosphere metabolomics-driven approach and its application in the removal of polychlorinated biphenyls. *Plant Physiology* 132, 146–153.

Narbad, A. and Gasson, M.J. (1998) Metabolism of ferulic acid via vanillin using a novel CoA-dependent pathway in a newly-isolated strain of *Pseudomonas fluorescens*. *Microbiology* 144, 1397–1405.

Nath, R. (1988) *Parthenium hysterophorus* L. – a review. *Agricultural Reviews* 9, 171–179.

Offre, P., Spang, A. and Schleper, C. (2013) Archaea in biogeochemical cycles. *Microbiology* 67, 437.

Om, H., Dhiman, S., Kumar, S. and Kumar, H. (2002) Allelopathic response of *Phalaris minor* to crop and weed plants in rice–wheat system. *Crop Protection* 21, 699–705.

Pajuelo, E., Rodríguez-Llorente, I., Lafuente, A., Pérez-Palacios, P., Doukkali, B. and Caviedes, M.A. (2014) Engineering the rhizosphere for the purpose of bioremediation: an overview. *CAB Reviews: Perspectives in Agriculture, Veterinary Science, Nutrition and Natural Resources* 9, 001, 1–17.

Phillips, L.A., Germida, J.J., Farrell, R.E. and Greer, C.W. (2008) Hydrocarbon degradation potential and activity of endophytic bacteria associated with prairie plants. *Soil Biology and Biochemistry* 40, 3054–3064.

Ramos, J.L., Gonzalez-Perez, M.M., Caballero, A. and van Dillewijn, P. (2005) Bioremediation of polynitrated aromatic compounds: plants and microbes put up a fight. *Current Opinion in Biotechnology* 16, 275–281.

Rein, A., Fernqvist, M.M., Mayer, P., Trapp, S., Bittens, M. and Karlson, U.G. (2007) Degradation of PCB congeners by bacterial strains. *Applied Microbiology and Biotechnology* 77, 469–481.

Rice, E.L. (1979) Allelopathy—an update. *The Botanical Review* 45, 15–109.

Rice, E.L. (2013) *Allelopathy*, 2nd edn. Academic Press, Cambridge, MA.

Ryan, R.P., Germaine, K., Franks, A., Ryan, D.J. and Dowling, D.N. (2008) Bacterial endophytes: recent developments and applications. *FEMS Microbiology Letters* 278, 1–9.

Rylott, E.L., Lorenz, A. and Bruce, N.C. (2011) Biodegradation and biotransformation of explosives. *Current Opinion in Biotechnology* 22, 434–440.

Schmidt, S. (1988) Degradation of juglone by soil bacteria. *Journal of Chemical Ecology* 14, 1561–1571.

Shaw, L.J. and Burns, R.G. (2004) Enhanced mineralization of [U-14C] 2, 4-dichlorophenoxyacetic acid in soil from the rhizosphere of *Trifolium pratense*. *Applied and Environmental Microbiology* 70, 4766–4774.

Siciliano, S., Goldie, H. and Germida, J. (1998) Enzymatic activity in root exudates of Dahurian wild rye (*Elymus dauricus*) that degrades 2-chlorobenzoic acid. *Journal of Agricultural and Food Chemistry* 46, 5–7.

Singer, A.C., Smith, D., Jury, W.A., Hathuc, K. and Crowley, D.E. (2003) Impact of the plant rhizosphere and augmentation on remediation of polychlorinated biphenyl contaminated soil. *Environmental Toxicology and Chemistry* 22, 1998–2004.

Singh, H., Batish, D., Pandher, J. and Kohli, R. (2003) Assessment of allelopathic properties of *Parthenium hysterophorus* residues. *Agriculture, Ecosystems & Environment* 95, 537–541.

Singh, H.B., Jain, A., Saxena, A., Singh, A., Keswani, C. *et al.* (2014) Deciphering the pathogenic behaviour of phyto-pathogens using molecular tools. In: Neeta, S. (ed.) *Biological Controls for Preventing Food Deterioration: Strategies for Pre- and Postharvest Management*. John Wiley & Sons, Chichester, UK. pp. 377–408.

Slater, H., Gouin, T. and Leigh, M.B. (2011) Assessing the potential for rhizoremediation of PCB contaminated soils in northern regions using native tree species. *Chemosphere* 84, 199–206.

Soltys, D., Krasuska, U., Bogatek, R. and Gniazdowska, A. (2013) Allelochemicals as bioherbicides—present and perspectives. *Herbicides—Current Research and Case Studies in Use, InTech*, 517–542.

Stevens, K.L. (1986) Allelopathic polyacetylenes from *Centaurea repens* (Russian knapweed). *Journal of Chemical Ecology* 12, 1205–1211.

Taghavi, S., Barac, T., Greenberg, B., Borremans, B., Vangronsveld, J. and van der Lelie, D. (2005) Horizontal gene transfer to endogenous endophytic bacteria from poplar improves phytoremediation of toluene. *Applied and Environmental Microbiology* 71, 8500–8505.

Tak, H.I., Ahmad, F. and Babalola, O.O. (2013) Advances in the application of plant growth-promoting rhizobacteria in phytoremediation of heavy metals. *Reviews of Environmental Contamination and Toxicology, Volume 223*. Springer, New York, pp. 33–52.

Taylor, A. and Spencer, W. (1990) Volatilization and vapor transport processes. *Pesticides in the Soil Environment: Processes, Impacts and Modeling*. SSSA Book Series, 2:213-269.doi:10.2136/sssabookser2.c7

Toure, O., Chen, Y. and Dutta, S. (2003) *Sinorhizobium meliloti* electrotransporant containing ortho-dechlorination gene shows enhanced PCB dechlorination. *Fresenius Environmental Bulletin* 12, 320–322.

Van Aken, B., Yoon, J.M. and Schnoor, J.L. (2004) Biodegradation of nitro-substituted explosives 2,4,6-trinitrotoluene, hexahydro-1,3,5-trinitro-1,3,5-triazine, and octahydro-1,3,5,7-tetranitro-1,3,5-tetrazocine by a phytosymbiotic *Methylobacterium* sp. associated with poplar tissues (*Populus deltoids* × *nigra* DN34). *Applied and Environmental Microbiology* 70, 508–517.

van der Putten, W.H. (2010) Impacts of soil microbial communities on exotic plant invasions. *Trends in Ecology & Evolution* 25, 512–519.

Vidal, R.A. and Bauman, T.T. (1997) Fate of allelochemicals in the soil. *Ciência Rural* 27, 351–357.

Villacieros, M., Whelan, C., Mackova, M., Molgaard, J., Sánchez-Contreras, M. *et al.* (2005) Polychlorinated biphenyl rhizoremediation by *Pseudomonas fluorescens* F113 derivatives, using a *Sinorhizobium meliloti* nod system to drive bph gene expression. *Applied and Environmental Microbiology* 71, 2687–2694.

Xie, X.-G. and Dai, C.-C. (2015) Biodegradation of a model allelochemical cinnamic acid by a novel endophytic fungus *Phomopsis liquidambari*. *International Biodeterioration & Biodegradation* 104, 498–507.

Xu, L., Teng, Y., Li, Z.-G., Norton, J.M. and Luo, Y.M.(2010) Enhanced removal of polychlorinated biphenyls from alfalfa rhizosphere soil in a field study: the impact of a rhizobial inoculum. *Science of the Total Environment* 408, 1007–1013.

Yee, D.C., Maynard, J.A. and Wood, T.K. (1998) Rhizoremediation of Trichloroethylene by a Recombinant, Root-Colonizing *Pseudomonas fluorescens* Strain Expressing Toluene ortho-Monooxygenase Constitutively. *Applied and Environmental Microbiology* 64, 112–118.

Zawierucha, I. and Malina, G. (2011) Bioremediation of contaminated soils: effects of bioaugmentation and biostimulation on enhancing biodegradation of oil hydrocarbons. In: Singh, A., Parmar, N. and Kuhad, R.C., (eds) *Bioaugmentation, Biostimulation and Biocontrol*. Springer Berlin Heidelberg, Germany, pp. 187–201.

Zeng, R.S. and Mallik, A.U. (2006) Selected ectomycorrhizal fungi of black spruce (*Picea mariana*) can detoxify phenolic compounds of *Kalmia angustifolia*. *Journal of Chemical Ecology* 32, 1473–1489.

Zhang, C. and Bennett, G.N. (2005) Biodegradation of xenobiotics by anaerobic bacteria. *Applied Microbiology and Biotechnology* 67, 600–618.

Zhang, Z.Y., Pan, L.P. and Li, H.H. (2010) Isolation, identification and characterization of soil microbes which degrade phenolic allelochemicals. *Journal of Applied Microbiology* 108, 1839–1849.

Zhu, X., Zhang, J. and Ma, K. (2011) Soil biota reduce allelopathic effects of the invasive *Eupatorium adenophorum*. *PLoS One* 6, 25393.

19 Harnessing Bio-priming for Integrated Resource Management under Changing Climate

Deepranjan Sarkar,[1] Sumita Pal,[2] H.B. Singh,[2] Ranjeet Singh Yadav[1] and Amitava Rakshit[1]*

[1]Department of Soil Science and Agricultural Chemistry, Institute of Agricultural Sciences, Banaras Hindu University, Varanasi, India; [2]Department of Mycology and Plant Pathology, Institute of Agricultural Sciences, Banaras Hindu University, Varanasi, India

19.1 Introduction

Agriculturists are facing huge pressure with escalating world population, increasing atmospheric CO_2 concentration and growing climate variability, which is difficult to predict by how much and over what time. The drastic impacts will be seen on plant growth by warmer and drier conditions, changes in wind speed, occurrence of new pests and diseases and many more understated variations resulting from altered interactions among the components of crop agro-ecosystems (Smith and Almaraz, 2004). Agrochemical pollution leads to ecological disruptions that cause a loss of ecosystem services, viz. land resources, biodiversity and food sources, which has adverse impacts on human health (Lecours *et al.*, 2012). The global food demand has to be fulfilled with maintenance of sustainability in agricultural production. As soils, plants, atmosphere and climate are very intricately linked, our management practices should purely be based on our scientific knowledge of the environmental system.

Microorganisms participate in many key natural processes like nutrient cycling (Nannipieri *et al.*, 2003), biological control of plant pathogens (Handelsman and Stabb, 1996; Saba *et al.*, 2012) and establishment, development, nutrition and health of plants (Linderman, 1992), which has received increased recognition in agriculture. Exploration of microbial resources in crop production is an urgent need to suppress the toxic effects of chemical inputs in the ecosystem. Integrated resource management has become a necessity to conserve the earth and is the basis of sustainability and a prerequisite for accomplishing sustainable development goals. Different sources of nutrient management changed from time to time as per the requirement and availability of local resources and today bio-priming is an essential component to supplement them. Seed being a costly material, emphasis on varied aspects like treatment, quality and storage are needed so that farmers should be able to harness a good and healthy seedling from each seed sown. Seed enhancement technologies must be

*E-mail: amitavabhu@gmail.com

© CAB International 2017. *Advances in PGPR Research* (eds H.B. Singh, B.K. Sarma and C. Keswani)

adapted for sustainable agriculture and biological seed treatments are expected to be one of the fastest emerging seed treatment sectors in the near future because of their environmental safety, socio-economic aspects and easy registration at different monitoring agencies (Sharma *et al.*, 2015). Rapid, synchronous seedling germination and emergence, greater tolerance to biotic and abiotic stresses and healthy plants are prerequisites for successful crop establishment (Keswani *et al.*, 2013, 2014; Bisen *et al.*, 2015, 2016; Keswani, 2015; Keswani *et al.*, 2016a, b).

All these acts coupled in one process could be termed as bio-priming selected to yield better crops for the conscious world with numerous challenges. Bio-priming is integration of physiological as well as biological aspects of disease control, where the temperature and moisture conditions are optimised during seed imbibition while a microbial protectant becomes established on the seed (Callan *et al.*, 1990). The technique of seed priming improves germination along with rapid seedling emergence and increased performance of seeds (Taylor and Harman, 1990). To harness the benefits of priming, proteomics are the current tools to study the molecular mechanisms behind plant responses to environmental stimuli and the priming phenomenon (Tanou *et al.*, 2012; Mishra *et al.*, 2015). Our understanding of how primed plants can effectively function in the new era could save us from quality deterioration which is a major threat to food security.

19.2 Bio-priming

Priming or bio-priming is an important protective tool to potentiate the plants with better defence responses to combat biotic and abiotic stresses. Bio-priming can also be viewed as a new technique of seed treatment using biological agents to stimulate germination of seed and growth of plant and further protecting the seed from soil- and seedborne diseases (Reddy, 2012). The process involves seed hydration and inoculation of seeds with useful microorganisms (priming agents). A controlled hydration is followed that involves exposing the seeds to low water potentials that hamper germination, but permit pre-germinative physiological and biochemical changes in seeds (Khan, 1992). During imbibition, seeds undergo a number of repair mechanisms like repairing membranes of cells and organelles, as well as protein and enzyme activation to break down the food reserve (McDonald, 1999). Application of microbial inoculants further sensitize the cellular mechanism of plants leading to their improved growth and development under various environmental conditions.

Studies have shown various microbial species as useful primers (Table 19.1). This biological seed treatment has been gaining importance in recent times as an approach for prior sowing or transplanting (Singh, 2016). Seed priming techniques have been popularised for many agricultural and horticultural crops but for some plants (like rice, cabbage, cauliflower, brinjal) seedling bio-priming is recommended and practised. As higher production and productivity of crops is possible only through use of good quality seeds and their proper management practices, bio-priming is an attractive proposition for that.

19.3 Advantages of Bio-priming with Reference to Stress Moderation

Plants are often exposed to various abiotic stresses like extreme temperature (hot or chilling), water stress (drought or flooding), heavy metal contamination, salinity, etc. together with biotic stress, such as attack by harmful pathogens or plant pests. These environmental stresses can reduce plant growth and yield considerably to the extent of 10-15% (Glick *et al.*, 2007b). Bio-priming is a low-cost technique to protect the seed from adverse conditions and offers many advantages for successful crop production (Fig. 19.1).

Ethylene (C_2H_4) is a phytohormone produced under drought and salinity conditions and the level of ethylene concentration is decreased when strains of plant

Table 19.1. Useful primers noted for improved growth and development of plants.

Microorganisms	N-fixer/Organic N transformers	P solubilizers	K solubilizers	Commercial Formulation
Bacteria	*Azospirillum lipoferum*, *A. brasilense*, *A. amazonense*, *Rhizobium* spp., *Azotobacter chroococcum*, *A. vinelandii*, *A. beijerinckii*	*Pseudomonas fluorescens*, *P. putida*, *P. striata*, *Phosphobacteria*, *Bacillus megaterium*, *B. cereus*, *B. pumilus*	*Bacillus mucilaginosus*, *B. edaphicus*, *B. subtilis*, *B. cereus*, *Frateuria aurantia*	Sunrise- *Rhizobium*, Aadhar-*Azospirillum*, Azab-*Azotobacter*, Durga-PSB, Biophos, Ambiphos, Magna ++, Teeka Gold, Shakti-KSB, Mani Dharma Bio Promotor K Mobilizer
Fungus	*Trichoderma harzianum*, *T. viride*, *T. hamatum*, *T. album*, *T. virens*	*Aspergillus niger*, *A. awamori*, *Penicillium purpurogenum*, *Glomus intraradices*, *G. versiforme*, *G. mosseae*, *Acaulospora laevis*, *T. harzanium*	*Glomusmosseae*, *G. intraradices*, *G. constrictum*	CADTricho, Josh-Mycorrhiza, Kalisena- *Aspergillus niger* (AN-27), Trishul (VAM), Mycosignal, ManiDharma VAM, Root Care

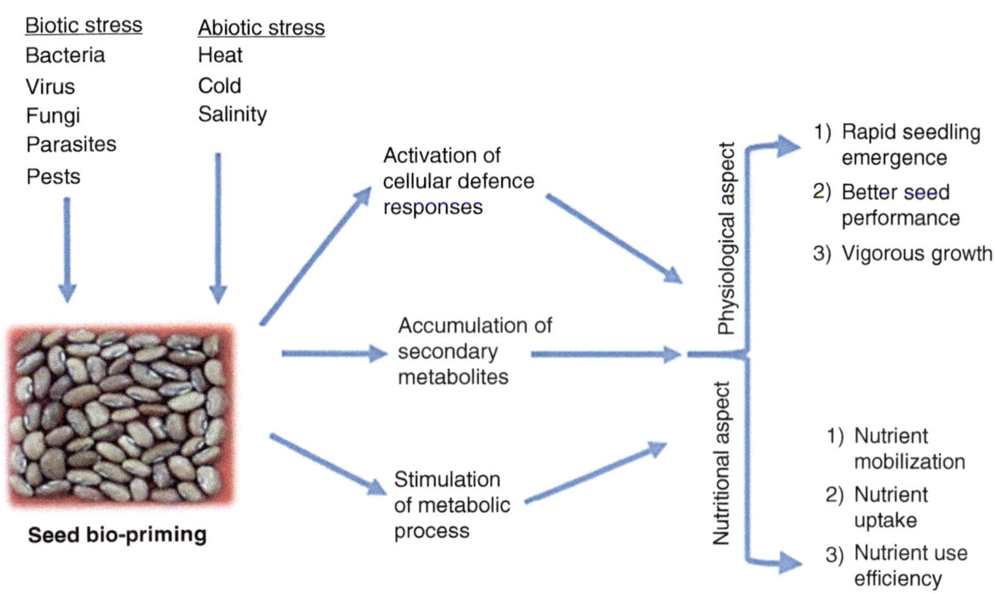

Fig. 19.1. Advantages of bio-priming.

growth-promoting rhizobacteria (PGPR) containing the enzyme 1-aminocyclopropane-1-carboxylate (ACC) deaminase are applied on plants (Zafar-ul-Hye *et al.*, 2014). ACC, which is a precursor of ethylene biosynthesis, is converted into ammonia (NH_3) and α-ketobutyrate by ACC deaminase-containing rhizobacteria (Saleem *et al.*, 2007). Bacterial coatings of seeds support rapid and more uniform seed germination with vigorous plant growth (Moeinzadeh *et al.*, 2010). Priming causes activation of cellular defence responses which helps plants to increase resistance towards biotic and abiotic stress (Conrath *et al.*, 2006). High salt concentrations produce some primary effects (hyper-osmotic stress, ion imbalance) and secondary effects (oxidative stress) due to cellular accumulation of damaging reactive oxygen species (ROS) which damage membrane lipids, proteins and nucleic acids (Mittler, 2002).

Root colonization by *Trichoderma harzianum* results in secretion of plant enzymes, including various peroxidases (PODs), chitinases (CHIs), α-1,3-glucanase (Glc), lipoxygenases (LOXs) and hydroperoxide lyases (HPLs) which further leads to acquisition of compounds like phytoalexins, phenols and other compatible solutes necessary to combat multiple stresses (Mohiddin *et al.*, 2010). Rice seedlings raised from seeds bio-primed with *T. harzianum* had significantly higher proline ($C_5H_9NO_2$), membrane stability index (MSI) and phenol (C_6H_5OH) content than other untreated seeds which alleviated the stress condition and significantly increased length and fresh weight of shoot and root, number of leaves, leaf area, photosynthetic rate, chlorophyll fluorescence, and chlorophyll content (Rawat *et al.*, 2012). Seed hardening and acclimatization of seedlings would reduce the mortality rate before transplanting. Seedling emergence and their proper development can help the better establishment of plant populations under various environmental conditions leading to early flowering and increase in yield.

This bio-initiative not only improves the seed germination rate, vigour and seedling establishment but also induces plant resistance to overcome pathogen infection, minimizes the risk of several plant diseases (Bisen *et al.*, 2015) and enhances nutrient use efficiency of crop species (Rakshit *et al.*, 2015). The enhanced vigour and immunity of crop plants aids in curbing the pathogen at seed and seedling stage itself (Sathya *et al.*, 2016). *Trichoderma* fungi control plant pathogens though parasitism and antibiosis production and stimulate systemic resistance (Harman *et al.*, 2004). Bio-priming has been able to control damping-off of seedlings in many crops, viz. sweet corn, cucumber, pea and soybean (Girolamo and Barbanti, 2012), root rot incidence (caused by *Fusarium solani*, *Macrophomina phaseolina* and *Rhizoctonia solani*) in cowpea (El-Mohamedy *et al.*, 2006), downy mildew in pearl millet (Raj *et al.*, 2004), alternaria blight of sunflower (Rao *et al.*, 2009), and ear rot disease in maize (Chandra Nayaka *et al.*, 2008). The secondary metabolites released from rhizobacteria and plant root system interactions increases the availability of nutrients to the plants with improved ability of plant nitrogen fixation and enhances plant health by biocontrol of plant pathogens (Sturz and Christie, 2003). Thus, the method could be rated as a simple, environmentally safe, long-lasting and effective treatment towards better crop stand and performance.

19.4 Mechanisms Used by Microorganisms for Improved Plant Nutrition

The literature of recent decades reveals that plant growth-promoting rhizobacteria (PGPR) and their interactions with host plants has a wide scope in progress of sustainable agriculture. Their effect in the rhizosphere ecosystem has led researchers to study their ecology, diversity and activity to adapt suitable screening procedures for these beneficial bacteria. The understanding of the PGPR mechanisms which influence plant productivity is very necessary to improve plant growth and maximize the process within the soil system. The direct growth-promoting mechanisms used by PGPR are as follows: (i) nitrogen fixation; (ii) solubilization of phosphorus and potassium; (iii) production of phytohormones such as auxins – indole acetic acid (IAA), cytokinins and gibberellins; (iv) sequestering of iron by production of siderophores; (v) synthesis of hydrolytic enzymes (chitinases, glucanases, proteases, and lipases) able to lyse pathogenic fungal cells; and (vi) lowering of ethylene concentration (Kumar *et al.*, 2011; Ahemad and Kibret, 2014).

Trichoderma being a secondary opportunistic invader, a fast-growing fungus, a strong spore producer, a source of cell-wall-degrading enzymes (cellulases, chitinases, glucanases, etc.) and an important antibiotic producer, use of its different strains can provide numerous benefits: (i) stimulation of root growth and development; (ii) enhanced solubilization of soil nutrients; (iii) control of plant pathogens; (iv) improvement of the plant health; and (v) degradation of hydrocarbons, chlorophenolic compounds, polysaccharides and the xenobiotic pesticides (Harman *et al.*, 2004). Evidence of a naturally occurring diverse group of rhizospheric P-solubilizing microorganisms was available as far back as 1903 (Kucey *et al.*, 1989). Among the whole microbial population in soil, P-solubilizing bacteria (PSB) are reported to constitute 50%, while P-solubilizing fungi (PSF) are only 0.1–0.5% of the total respective population (Chen *et al.*, 2006). Bacteria are considered as more effective than fungi in P solubilization. McGill and Cole (1981) in an extensive study on mechanisms of soil organic P solubilization expressed that the main P-solubilization mechanisms employed by P-solubilizing microorganisms (PSM) include: (a) release of complex or mineral dissolving compounds, e.g., organic anions, siderophores (Fe-chelating molecules), protons, hydroxyl ions, and CO_2, (b) liberation of extracellular enzyme (biochemical P mineralization), and (c) the release of P during substrate degradation (biological P mineralization).

Symbiotic mycorrhizal fungi such as arbuscular mycorrhizal (AM) fungi are known for their ability to take up and transfer P and other growth-limiting nutrients from soils to plants (Elbon and Whalen, 2015). The hyphae of these fungi can spread out several centimetres into the soil and their colonization with roots increases the root surface area for nutrient acquisition (Wu *et al.*, 2005). Enhanced root length and P influx of mycorrhiza-treated plants are important parameters which influence growth and P nutrition of plants (Rakshit and Bhadoria, 2010).

K-solubilizing microorganisms (KSM) are able to release K from K-bearing minerals, such as mica [$KAl_2(Al, Si_3)O_{10}(OH)_2$], illite {$(K,H_3O)(Al,Mg,Fe)_2(Si,Al)_4O_{10}[(OH)_2, (H_2O)]$} and orthoclase ($KAlSi_3O_8$), by excreting organic acids (citric ($C_6H_8O_7$), tartaric ($C_4H_6O_6$) and oxalic ($C_2H_2O_4$) acids) which either directly dissolve rock K or chelate silicon ions to bring the K into solution (Sheng, 2005). Yadav and Tarafdar (2011) found that seed inoculation with the fungi (*Penicillium purpurogenum*) has significantly improved phosphatases (acid (APase) and alkaline (ALPase)), phytase (Phy) and dehydrogenase (DHA) activities compared to non-inoculated fields in a loamy sand soil under arid agro-ecosystems.

19.5 Effect of Bio-priming in Different Crop Species

Use of microbes as priming agents offers several agronomic and environmental benefits for intensive agricultural systems and good responses of bio-priming have been reported for several cereal and vegetable species. Primed seeds, when planted, usually emerge faster with better, uniform, and vigorous crop stand and show persistence even under less than optimum field conditions (Rehman *et al.*, 2011). Various experiments of bio-priming conducted on different crops revealed its beneficial effects on seed germination (rate, percentage and uniformity), seed vigour, root length, shoot length, seedling emergence, stand establishment, seed tolerance to adverse environmental conditions, growth, yield, etc. (Table 19.2).

It has been reported that bio-priming with PGPR increases yield in crops, viz. barley (Çakmakçı *et al.*, 2001), wheat (De Freitas, 2000), maize (Sharifi and Kavazi, 2011), pea (El-Mohamedy and Abd El-Baky, 2008), okra (Pravisya and Jayaram, 2015), broccoli (Tanwar *et al.*, 2013) and safflower (Soleymanifard and Siadat, 2011). Seed priming with *T. viride* and *P. fluorescens* improved seedling emergence of chickpea to 96% and 98% and reduced incidence of dry root rot to 28% and 35%, respectively (Reddy *et al.*, 2011). Increase in plant biomass, relative water content and leaf water potential was found when maize was biotized with *Pseudomonas* spp. (Sandhya *et al.*, 2010).

Seed bio-priming with liquid biofertilizers (*Azospirillum* and Phosphobacteria)

Table 19.2. Effect of bio-priming on growth and productivity across various crop species.

Sl. No.	Author	Crop	Microbes	Mechanism	Improved traits
C_3 plants					
01	Rafi and Dawar (2015)	Chickpea, sunflower, okra and peanut	*Trichoderma harzianum* (Th-6) and *Rhizobium Meliloti* (Rm-5)	Improved resistance of roots towards abiotic stress, improved systemic resistance to diseases, increased uptake of nutrients, increased leaf greenness	Significant enhancement of root length, shoot length, root weight and shoot weight and suppression of root infecting fungi
02	Rawat *et al.* (2012)	Rice (Kalanamak-3131)	*Trichoderma harzianum*	Root colonization results in increased level of plant enzymes like peroxidases (PODs), chitinases (CHIs), α-1,3-glucanase (Glc), lipoxygenases (LOXs), hydroperoxidelyases (HPLs) and such changes in plant metabolism can lead to accumulation of compounds like phytoalexins and phenols to provide durable resistance against any biotic and abiotic stress	Length and fresh weight of shoot and root, number of leaves, leaf area, photosynthetic rate, chlorophyll fluorescence, Chlorophyll content, alleviation of salinity stress
03	Zia-ul-hassan *et al.* (2015)	Wheat cv. Imdad	Rhizobacterial strains, viz. B_1: *Pseudomonas fluorescens* biotype G and B_2: *Pseudomonas fluorescens* biotype F	Phosphate solubilization activity, 1-aminocyclopropane-1-carboxylic acid (ACC) deaminase activity	Growth, yield of wheat under P deficiency stress
04	Reddy *et al.* (2011)	Chickpea JG-11	*Trichoderma viride* and *Pseudomonas fluorescens*	Growth promoting and biocontrol activities	Seedling emergence, reduced incidence of dry root rot, plant biomass
05	Moeinzadeh *et al.* (2010)	Sunflower cv. Azargol	*Pseudomonas fluorescens*	Increased solubilization and uptake of nutrients, production of plant growth regulators, disease resistance, proper colonization	Seed invigoration, seedling growth
06	Sharifi (2012)	Safflower	*Azotobacter chroococcum* strain 5, *Azosprilium* lipoferum strain OF, *Pseudomonas* strain 186	Phytohormone production, enhancing stress resistance, N_2 fixation,increasing the supply or availability of primary nutrients to the host plant	Quantitative, qualitative yield
07	Ananthi *et al.* (2014)	Chilli cv. PKM 1	*Azospirillum* and Phosphobacteria	Production of germinating, accelerating and growth-promoting substances auxins, gibberellins (GAs), cytokinins (CKs), P solubilization, heavy colonization	Germination rate, total germination percentage, seedling growth, vigour

08	Kivi *et al.* (2014)	Spring wheat	*Azotobacter chroococcum* strain 5, *Azosprilium lipoferum* strain OF	Increase in root growth due to growth hormones secreted by the bacteria	N and P use efficiency, root length and weight, yield
09	Akhtar *et al.* (2009)	Wheat	PGPR	Production of various compounds (such as phytohormones, organic acids, siderophores), fix atmospheric nitrogen	Higher N content, N use efficiency, growth, yield
10	El-Mohamedy *et al.* (2015)	Green bean cv. Giza 3	*Trichoderma harzianum*	Induction of physiological changes	Growth, yield, nutritional values and resistance against soilborne pathogens (*Fusarium solani* and *Rhizoctonia solani*)
11	Namvar and Khandan (2014)	Rapeseed	*Azotobacter* sp. and *Azospirillum* spp.	Production of phytohormones	Growth, development, total yield
12	Shaukat *et al.* (2006)	Sunflower	*Azotobacter* spp., *Azospirillum* spp., *Pseudomonas* spp.	Auxin production, peroxidase (POD), acid phosphatise (APase) activity	Growth and yield parameters, soil enzyme activities, protein contents
13	Entesari *et al.* (2013)	Soybean	*Trichoderma* sp. and *Pseudomonas fluorescent*	Increasing antioxidant system, scavenging of ROS	Growth parameters, enzyme activities, nutritional status
14	Rahman *et al.* (2015)	Boro rice	*Trichoderma harzianum*	Increased levels of SOD, increasing ROS scavenging abilities, peroxidise (POD), glutathione reductase (GR), glutathione-s-transferase (GST) and other detoxifying enzymes in leaves	Seedling establishment, yield
15	Namvar *et al.* (2013)	Wheat	*Azotobacter* sp. and *Azospirillum* sp.	Production of phytohormones	Grain yield, yield components, protein content
16	Amara *et al.* (2015)	Wheat	PGPR	Indole-3-acetic acid (IAA), siderophore, catalase (CAT) and oxidase production, phosphate solubilization, *nif*H gene amplification	Root length, shoot length, dry root weight and dry shoot weight
17	Anitha and Jahagivrdar (2015)	Soybean JS 335	*Trichoderma harzianum*, *Pseudomonas fluorescens* and *Bacillus subtilis*	Accumulation of photo-assimilates, antagonism	Increased seed germination, root length, shoot length
18	Pozo *et al.* (1999)	Tomato cv. Earleymech	*Glomus* (*G. mosseae and G. intraradices*)	β-1,3-Glucanase (Glc) activities and enhanced resistance in roots against soilborne pathogens	Protein content of the root extracts, bioprotection against *Phytophthora parasitica*

Continued

Table 19.2. Continued.

Sl. No.	Author	Crop	Microbes	Mechanism	Improved traits
19	Mukhopadhyay and Pan (2012)	Radish	*Trichoderma* (*T. viride* and *T. harzianum*)	Alleviation of stress condition, production of some phenolic compounds, microbial secondary metabolite and enzymes for solubilization of nutrients	Seedling vigour index, length, fresh weight and dry weight of shoot and root, number of leaves, leaf area, photosynthetic rate, chlorophyll fluorescence, chlorophyll content
20	Rai and Basu (2014)	Okra cv. Lalu, Arka Anamika, Ramya, Satsira, Lady Luck, DebpusaJhar, JapaniJhar and BarshaLaxmi	*Trichoderma viride* and *Pseudomonas fluorescens*	Control of minor population of pathogens leading to stronger root growth, secretion of plant growth regulatory factors such as phytohormones, release of soil nutrients and minerals by saprophytic activity of *Trichoderma* in soil	Plant height, number of pods per plant, pod length, pod diameter, seed yield
21	Singh *et al.* (2016a)	Pea cv. NBR Ruchi	*Trichoderma asperellum* BHU T8	Production of phytohormones	Plant growth promotion
22	Singh *et al.* (2016b)	Okra, Tomato, Brinjal, Chilli, Rid e gourd and Guar	*Trichoderma asperellum* BHU T8	Increased PAL, POD, Shikmik acid, Gallic acid, TPC, PPO activity	Plant growth promotion
C_4 plants					
23	Karthika and Vanangamudi (2013)	Maize [COH(M) 5 hybrid]	*Azospirillum* and *Phosphobacteria*	Production of auxins, gibberellins (GAs), cytokinins (CKs), solubilization of insoluble phosphorus	Speed of germination, germination, root length, shoot length, dry matter production, total dry matter production, vigour index (G × SL)
24	Baral and Adhikari (2013)	Maize (variety Rampur composite)	*Azotobacter*	N_2 fixation, phytohormone production, bacterial nitrate reduction	Yield parameters and yield
25	Priya *et al.* (2016)	Maize hybrid (Kargil-900 M) and sorghum hybrid (CSH-16)	*Azospirillum* strain ACD 15, *Fluorescent pseudomonas* strain WGUK 327 (2), *Trichoderma viride* and *Pseudomonas striata*	Production of plant growth-promoting substances (PGPS), extension of extrametrical fungal hyphae of VAM for nutrient absorption of host root	Higher root:shoot ratio, root length, root dry weight

26	Chandra Nayaka *et al.* (2008)	Maize	*Trichoderma harzianum*	Coiling around the hyphae of the *Fusarium verticillioides* and suppressing fumonisin ($C_{34}H_{59}NO_{15}$) synthesis	Seed germination, vigour index, field emergence, yield, thousand seed weight, reduced *F.verticillioides* and fumonisin infection and ear rot disease
27	Sandhya *et al.* (2010)	Maize cv. Kaveri	*Pseudomonas* spp.	IAA and gibberellic acid (GA_3) production, P-solubilization, siderophore, HCN and ammonia (NH_3) producion	Increase in plant biomass, relative water content, leaf water potential, decreased leaf water loss
28	Niranjan Raj *et al.* (2004)	Pearl milletcv. HB3	*Pseudomonas fluorescens*	Production of plant growth regulators such as gibberellins (GAs), cytokinins (CKs) and indole acetic acid (IAA), extensive rooting, incorporation of various phenolic compounds and polymers to the cell wall and secretion of phytoalexins, induction of systemic resistance.	Germination, stand establishment, growth parameters (height, leaf area, tillering capacity) reproductive parameters (number, length and girth of earhead), 1000 seed weight, resistance against downy mildew disease
29	Ghanbari Zarmehri *et al.* (2013)	Maize	*Pseudomonas fluorescens* strain 169 and *Pseudomonas putida* strain 108	Decreasing ethylene (C_2H_4) levels, increased plant IAA concentration results in increased rooting and plant ability for nutrient and water uptake, siderophore production, increased photosynthetic surface via developed leaf area and preventing leaf senescence, and finally more transferred photosynthetic production into ears	Enhancement of forage and grain yield under normal and water deficit stress conditions
30	Zafar-ul-Hye *et al.* (2014)	Maize hybrid DK-6525	*Pseudomonas syringae* and *Pseudomonas fluorescens*	Reduction of ethylene (C_2H_4) stress level by conversion of 1-aminocyclopropane-1-carboxylic acid (ACC) into ammonia (NH_3) and α-ketobutyrate	Yield traits and nutrient uptake (N, P and K) under drought and salinity stress
31	Bangari *et al.* (2012)	Sorghum	*Trichoderma harzianum* and *Pseudomonas fluorescens*	Induction of resistance	Enhanced germination, increased plant height, decreased severity of anthracnose
32	Ghimire *et al.* (2009)	Switchgrass	*Sebacina vermifera*	Ethylene (C_2H_4) oxidation, phytohormone regulation	Seed germination, plant height, root length, biomass production

enhanced the germination rate, total germination percentage, seedling growth and vigour in chilli (Ananthi *et al.*, 2014). Seed inoculation with *Azotobacter* increased 35% grain yield in maize over non-inoculated treatments (Baral and Adhikari, 2013). Inoculation of *Azospirillum brasilense* in rice increased aerial biomass at the tillering and grain-filling stages and the N content accumulation in plants increased by 16 and 50 kg ha^{-1} (García de Salamone *et al.*, 2010). Rapeseed treated with biofertilizer (*Azotobacter* sp. and *Azospirillum* sp.) had greater biomass than plants that were not treated with the inoculum (Namvar and Khandan, 2014). Shaharoona *et al.* (2007) observed increased root elongation, root weight, tillers/pot, seed index and yield of wheat grain and straw in response to rhizobaterial inoculant (*Pseudomonas fluorescens*) under both pot and field conditions.

Yadav and Tarafdar (2011) reported a significant increase in yield and P content of pearl millet and cluster-bean in an arid ecosystem under field conditions due to inoculation with different P-solubilizing fungi (PSF). In pearl millet, an increase in dry matter production by 29–39 % and P concentration in shoots by 14–29%, in roots by 5–7% and seeds by 34–35% were recorded for seed inoculation with PSB isolates. They reported further that a positive response of the PSMs was observed in field soils with high organic matter content and low P availability under the arid ecosystem of Rajasthan, India. Seed bio-priming with *T. harzianum* + 3/4th N and RDF of PK in a pot experiment conducted in alluvial, red and black soils revealed significant increase in effective tillers, chlorophyll content and root length along with enhancement of nitrogen use efficiency (NUE), agronomic use efficiency (AUE) and physiological use efficiency (PUE) of wheat crop (Meena *et al.*, 2016). Singh *et al.* (2014) found that the application of *Trichoderma* in consortium form (BHU51+ BHU105) increased the vigour index, mineral nutrient uptake and reduced the disease incidence of *Rhizoctonia solani* in tomato grown in alluvial soils of Uttar Pradesh.

As a matter of fact, it can be concluded from Table 19.2 that bio-priming leads to better nutrient uptake and stimulation of growth hormones in plants but the secretion of plant enzymes which are responsible for alleviating stress like peroxidases (PODs), chitinases (CHIAs), a-1,3-glucanase (Glc), lipoxygenases (LOXs), hydroperoxidelyases (HPLs), catalases (CATs) are more stimulated in C_3 than C_4. Reduction of ethylene (C_2H_4) stress levels by bio-priming is common in C_4 plants.

19.6 Proteomic Analysis Induced by Bio-Priming

Study of stress-responsive proteins to detect the plant defence mechanism when they are subjected to extreme conditions are essential. The proteomic approach can be used as an important tool for validation of enhanced resistance of plants to stress. Priming facilitates early DNA transcription and RNA and protein synthesis which repair the damaged parts of the seeds and reduce the metabolic exudation (Entesari *et al.*, 2013). To note the changes induced by priming treatment, proteomics are widely used in seed research as a novel tool for protein characterization and function analysis (Rajjou *et al.*, 2006, 2008). Proteomics researches help to track subcellular proteomes and protein complexes (e.g., proteins in the plasma membranes, chloroplasts, mitochondria and nuclei). Transduction of the signal into the cell organelle during the stressed condition represents the primary defence response of plant cells toward stress (Desikan *et al.*, 2003). Protein identification using mass spectrometry has opened a new path for organ and subcellular proteome research. Fundamental information of plant responses to a given stress at the functional level and their related signalling pathways can be obtained through organelle proteome analysis (Hossain *et al.*, 2012).

Knowledge of how cell wall protein (CWP) composition changes along with the differential growth responses to water deficit in different regions of the elongation zone can provide insights into the complexity of mechanisms that regulate root growth during water scarcity (Zhu *et al.*, 2007). The plasma membrane controls many primary cellular functions such as metabolite and ion transport, endocytosis and cell differentiation and proliferation, therefore analysis of the plasma membrane proteome can provide precious information on plant-specific biological processes that can help in formulating strategies to increase the natural defences or tolerance of plants (Komatsu, 2009). Thus, the functions of different plant membrane systems and the subcellular compartments demonstrate cellular adaptation against water deficit conditions (Bhusan *et al.*, 2007). Osmotic adjustment (OA) and cell membrane stability are recognized as effective components of dehydration tolerance in many crops.

Proteomic studies have concluded that the activity of protective enzymes (e.g., peroxidase (POD), catalase (CAT) and superoxide dismutase (SOD)) and the contents of compatible solutes (e.g., malondialdehyde (MDA), proline (Pro) and soluble sugar (SS)) are important metabolic changes during the priming and germination process (Wattanakulpakin *et al.*, 2012). Comparative studies focusing on symbiotic and pathogenic root–microbe interactions are important to note how roots deal with various microorganisms during their exposure in soil (Mathesius, 2009). Transcriptomics in nodules for studying symbiotic root–microbe interactions would help to trace different nutrient transporters (C, N, S, K), metal-binding proteins, aquaporins, ATPases related to nutrient uptake and regulatory proteins in osmoregulation. The investigations of pathogenic relationships between roots and various pathogens would help in finding the microbial signal molecules produced by plants and balancing of defence responses, nutrient exchange and alteration of plant development as intervened by microbes. Such enquiries would reveal plant stress response mechanisms, which are necessary for the creation of genetically engineered stress-tolerant crop plants in a climate-changing world.

19.7 Conclusion

Microbial coating of seeds have enormous and unrealised potential. Primed plants have increased resistance to several biotic or abiotic stresses. Biotization leads to improved plant nutrition. Innovative research in bio-priming will lead to a greater understanding of its multiple roles (plant strengthener, phytostimulator, disease controller, nutrient enhancer, etc.) in progressive agriculture. Biological seed treatment prior to sowing is essential for better performance of plants because it induces biochemical changes in seeds and microbial changes in the rhizosphere, which in turn change and modulate the system of root morphology and nutrient status resulting in persistent growth and development of crop species.

Acknowledgement

The second author thanks the Department of Science and Technology, New Delhi for the award of Woman Scientist Scheme (SR/WOS-A/LS-1199/2015) during the course of study.

References

Ahemad, M. and Kibret, M. (2014) Mechanisms and applications of plant growth promoting rhizobacteria: current perspective. *Journal of King Saudia University – Science* 26, 1–20.

Akhtar, M.J., Asghar, H.N., Shahzad, K. and Arshad, M. (2009) Role of plant growth promoting rhizobacteria applied in combination with compost and mineral fertilizers to improve growth and yield of wheat (*Triticum aestivum* L). *Pakistan Journal of Botany* 41(1), 381–390.

Amara, U., Wang, Y.X., Cui, X.L., Khalid, R., Ali, S. *et al.* (2015) Screening and identification of soil bacteria for growth promotion of wheat (*Triticum aestivum* L.). *Journal of Biodiversity Environment Science* 7(3), 87–99.

Ananthi, M., Selvaraju, P. and Sundaralingam, K. (2014) Study of biopriming using liquid biofertilisers in chilli cv PKM 1 seeds. *Environment Ecology* 32(3A), 1172–1177.

Anitha, U.V.M. and Jahagirdar, S. (2015) Influence of seed priming agents on yield, yield parameters and purple seed stain disease in soybean. *Karnataka Journal of Agriculture Science* 28(1), 20–23.

Bangari, G., Singh, Y. and Singh, V.K. (2012) Assessing biocontrol agents against anthracnose disease and their effect on growth parameter in *Sorghum bicolor* (L.) Moench. *Vegetos* 25(1), 16–20.

Baral, B.R. and Adhikari, P. (2013) Effect of *Azotobacter* on growth and yield of maize. *SAARC Journal of Agriculture* 11(2), 141–147.

Bisen, K., Keswani, C., Mishra, S., Saxena, A., Rakshit, A. and Singh, H.B. (2015) Unrealized potential of seed biopriming for versatile agriculture. In: Rakshit, A., Singh, H.B. and Sen, A. (eds) *Nutrient Use Efficiency: From Basics to Advances*. Springer, New Delhi, India, pp. 193–206.

Bisen, K., Keswani, C., Patel, J.S., Sarma, B.K. and Singh, H.B. (2016) *Trichoderma* spp.: efficient inducers of systemic resistance in plants. In: Chaudhary, D.K. and Verma, A. (eds) *Microbial-mediated Induced Systemic Resistance in Plants*. Springer, Singapore, pp. 185–195.

Çakmakçi, R., Kantar, F. and Sahin, F. (2001) Effect of N_2-fixing bacterial inoculations on yield of sugar beet and barley. *Journal of Plant Nutrition Soil Science* 164(5), 527–531.

Callan, N.W., Mathre, D. and Miller, J.B. (1990) Bio-priming seed treatment for biological control of *Pythium ultimum* preemergence damping-off in sh-2 sweet corn. *Plant Disease* 74, 368–372.

Chandra, N.S., Niranjana, S.R., Uday Shankar, A.C., Niranjan Raj, S. *et al.* (2010) Seed biopriming with novel strain of *Trichoderma harzianum* for the control of toxigenic *Fusarium verticillioides* and fumonisins in maize. *Archives Phytopathology Plant Protection* 43(3), 264–282.

Chen, Y.P., Rekha, P.D., Arunshen, A.B., Lai, W.A. and Young, C.C. (2006) Phosphate solubilizing bacteria from subtropical soil and their tricalcium phosphate solubilizing abilities. *Applied Soil Ecology* 34(1), 33–41.

Conrath, U., Beckers, G.J.M., Flors, V., García-Agustín, P., Jakab, G. *et al.* (2006) Priming: getting ready for battle. *American Phytopathological Society* 19, 1062–1071.

De Freitas, J.R. (2000) Yield and N assimilation of winter wheat (*Triticum aestivum* L., var Norstar) inoculated with rhizobacteria. *Pedobiologia* 44, 97–104.

Desikan, R., Hancock, J. and Neill, S. (2003) Oxidative stress signalling. In: Hirt, H. and Shinozaki, K. (eds) *Plant Responses to Abiotic Stress. Topics in Current Genetics*. Springer, New York, 4, 121–150.

Elbon, A. and Whalen, J.K. (2015) Phosphorus supply to vegetable crops from arbuscular mycorrhizal fungi: a review. *Biology Agriculture Horticulture* 31(2), 73–90.

El-Mohamedy, R.S.R., AbdAlla, M.A. and Badiaa, R.I. (2006) Soil amendment and seed bio-priming treatments as alternative fungicides for controlling root rot diseases on cowpea plants in Nobaria Province. *Research Journal of Agriculture Biological Science* 2, 391–398.

El-Mohamedy, R.S.R. and Abd El-Baky, M.M.H. (2008) Evaluation of different types of seed treatment on control of root rot disease, improvement growth and yield quality of pea plant in Nobaria province. *Research Journal of Agriculture Biological Science* 4(6), 611–622.

El-Mohamedy, R.S.R., Shafeek, M.R. and Fatma, A.R. (2015) Management of root rot diseases and improvement growth and yield of green bean plants using plant resistance inducers and biological seed treatments. *Journal of Agriculture Technology* 11(5), 1219–1234.

García de Salamone, I.E.G., Di Salvo, L.P., Ortega, J.S.E., Sorte, P.M.B., Urquiaga, S. and Teixeira, K.R. (2010) Field response of rice paddy crop to *Azospirillum* inoculation: physiology of rhizosphere bacterial communities and the genetic diversity of endophytic bacteria in different parts of the plants. *Plant Soil* 336(1–2), 351–362.

Ghanbari Zarmehri, S., Moosavi, S.G., Zabihi, H.R. and Seghateslami, M.J. (2013) The effect of plant growth promoting rhizobacteria (PGPR) and zinc fertilizer on forage yield of maize under water deficit stress conditions. *Technology Journal Engineering Applied Science* 3(23), 3281–3290.

Ghimire, S.R., Charlton, N.D. and Craven, K.D. (2009) The mycorrhizal fungus, *Sebacina vermifera*, enhances seed germination and biomass production in switchgrass (*Panicum virgatum* L). *Bioenergy Research* 2, 51–58.

Girolamo, G.D. and Barbanti, L. (2012) Treatment conditions and biochemical processes influencing seed priming effectiveness. *Italian Journal of Agronomy* 7(e25), 178–188.

Glick, B.R., Cheng, Z., Czarny, J. and Duan, J. (2007b) Promotion of plant growth by ACC deaminase-producing soil bacteria. *European Journal of Plant Pathology* 119, 329–339.

Handelsman, J. and Stabb, E.V. (1996) Biocontrol of soilborne plant pathogens. *Plant Cell* 8(10), 1855–1869.

Harman, G.E., Howell, C.R., Viterbo, A., Chet, I. and Lorito, M. (2004) *Trichoderma* species-opportunistic, avirulent plant symbionts. *Nature Review Microbiology* 2, 43–56.

Hossain, Z., Nouri, M.Z. and Komatsu, S. (2012) Plant cell organelle proteomics in response to abiotic stress. *Journal of Proteome Research* 11(1), 37–48.

Karthika, C. and Vanangamudi, K. (2013) Biopriming of maize hybrid COH (M) 5 seed with liquid biofertilizers for enhanced germination and vigour. *African Journal of Agricultural Research* 8(25), 3310–3317.

Keswani, C. (2015). Proteomics studies of thermotolerant strain of *Trichoderma* spp. Ph.D. Thesis, Banaras Hindu University, Varanasi, India, p. 126.

Keswani, C., Singh, S.P. and Singh, H.B. (2013) A superstar in biocontrol enterprise: *Trichoderma* spp. *Biotech Today* 3, 27–30.

Keswani, C., Mishra, S., Sarma, B.K., Singh, S.P. and Singh, H.B. (2014) Unraveling the efficient application of secondary metabolites of various *Trichoderma*. *Applied Microbiology and Biotechnology* 98, 533–544.

Keswani, C., Bisen, K., Singh, S.P., Sarma, B.K. and Singh, H.B. (2016a) A proteomic approach to understand the tripartite interactions between plant-*Trichoderma*-pathogen: investigating the potential for efficient biological control. In: Hakeem, K.R. and Akhtar, M. S. (eds) *Plant, Soil and Microbes Vol. 2. Mechanisms and Molecular Interactions*. Springer, Cham, Switzerland, pp. 79–93.

Keswani, C., Bisen, K., Singh, V., Sarma, B.K. and Singh, H.B. (2016b) Formulation technology of biocontrol agents: present status and future prospects. In: Arora, N.K., Mehnaz, S. and Balestrini, R. (eds) *Bioformulations: For Sustainable Agriculture*. Springer, New Delhi, India, pp. 35–52.

Khan, A.A. (1992) Pre plant physiological seed conditioning. *Horticultural Reviews* 13, 131–181.

Kivi, M.P., Hokmalipour, S. and Darbandi, M.H. (2014) Nitrogen and phosphorus use efficiency of spring wheat (*Triticum aestivum* L.) as affected by seed inoculation with plant growth promoting rhizobacteria (PGPR). *International Journal of Advanced Biological and Biomedical Research* 2(4), 1038–1050.

Kucey, R.M.N., Janzen, H.H. and Leggett, M.E. (1989) Microbially mediated increases in plant-available phosphorus. *Advances in Agronomy* 42, 199–228.

Kumar, A., Prakash, A. and Johri, B.N. (2011) *Bacillus* as PGPR in crop ecosystem. In: Maheshwari, D.K. (ed.) *Bacteria in Agrobiology: Crop Ecosystems*. Springer Berlin Heidelberg, Germany, pp. 37–59.

Lecours, N., Almeida, G.E., Abdallah, J.M. and Novotny, T.E. (2012) Environmental health impacts of tobacco farming: a review of the literature. *Tobacco Control* 21(2), 191–196.

Linderman, R.G. (1992) Vesicular-arbuscular mycorrhizae and soil microbial interactions. In: Bethlenfalvay, G.J. and Linderman, R.G. (eds) *Mycorrhizae in Sustainable Agriculture*. ASA Special Publication, Madison, Wisconsin, USA pp. 45–70.

Mathesius, U. (2009) Comparative proteomic studies of root–microbe interactions. *Journal of Proteomics* 72(3), 353–366.

McDonald, M.B. (1999) Seed deterioration: physiology, repair and assessment. *Seed Science and Technology* 27(1), 177–237.

McGill, W.B. and Cole, C.V. (1981) Comparative aspects of cycling of organic C, N, S and P through soil organic matter. *Geoderma* 26(4), 267–286.

Meena, S.K., Rakshit, A. and Meena, V.S. (2016) Effect of seed bio-priming and N doses under varied soil type on nitrogen use efficiency (NUE) of wheat (*Triticum aestivum* L.) under greenhouse conditions. *Biocatalysis and Agricultural Biotechnology* 6, 68–75.

Mishra, S., Singh, A., Keswani, C., Saxena, A., Sarma, B.K. and Singh, H.B. (2015) Harnessing plant-microbe interactions for enhanced protection against phytopathogens. In: Arora, N.K. (ed.) *Plant Microbes Symbiosis: Applied Facets*. Springer, New Delhi, India, pp. 111–125.

Mittler, R. (2002) Oxidative stress, antioxidants and stress tolerance. *Trends in Plant Science* 7(9), 405–410.

Moeinzadeh, A., Sharif-Zadeh, F., Ahmadzadeh, M. and Tajabadi, F.H. (2010) Biopriming of sunflower (*Helianthus annuus* L.) seed with *Pseudomonas fluorescens* for improvement of seed invigoration and seedling growth. *Australian Journal of Crop Science* 4(7), 564–570.

Mohiddin, F.A., Khan, M.R., Khan, S.M. and Bhat, B.H. (2010) Why *Trichoderma* is considered super hero (super fungus) against the evil parasites? *Plant Pathology Journal* 9, 92–102.

Mukhopadhyay, R. and Pan, S. (2012) Effect of biopriming of radish (*Raphanus sativus*) seed with some antagonistic isolates of *Trichoderma*. *The Journal of Plant Protection Sciences* 4(2), 46–50.

Namvar, A. and Khandan, T. (2014) Biopriming and mineral fertilizers effects on agronomical performance of rapeseed (*Brassica napus* L). *Ekologija* 60(3), 54–63.

Nannipieri, P., Ascher, J., Ceccherini, M., Landi, L., Pietramellara, G. and Renella, G. (2003) Microbial diversity and soil functions. *European Journal of Soil Science* 54(4), 655–670.

Pozo, M.J., Azcón-Aguilar, C., Dumas-Gaudot, E. and Barea, J.M. (1999) β-1,3-glucanase activities in tomato roots inoculated with arbuscular mycorrhizal fungi and/or *Phytophthora parasitica* and their possible involvement in bioprotection. *Plant Science* 141(2), 149–157.

Pravisya, P. and Jayaram, K.M. (2015) Priming of *Abelmoschus esculentus* (L.) Moench (okra) seeds with liquid phosphobacterium: an approach to mitigate drought stress. *Tropical Plant Research* 2(3), 276–281.

Priya, P., Patil, V.C. and Kumar, B.A. (2016) Characterization of grain sorghum [*Sorghum bicolor* (L.) Moench] for root traits associated with drought tolerance. *Research in Environment and Life Sciences* 9(2), 163–165.

Rahman, M., Ali, J. and Masood, M. (2015) Seed Priming and *Trichoderma* Application: a method for improving seedling establishment and yield of dry direct seeded boro (winter) rice in Bangladesh. *Universal Journal of Agricultural Research* 3(2), 59–67.

Rai, A.K. and Basu, A.K. (2014) Presowing seed bio-priming in okara: response for seed production. *The Bioscan* 9(2), 643–647.

Raj, S.N., Shetty, N.P. and Shetty, H.S. (2004) Seed bio-priming with *Pseudomonas fluorescens* isolates enhances growth of pearl millet plants and induces resistance against downy mildew. *International Journal of Pest Management* 50(1), 41–48.

Rajjou, L., Belghazi, M., Huguet, R., Robin, C., Moreau, A. *et al.* (2006) Proteomic investigation of the effect of salicylic acid on *Arabidopsis* seed germination and establishment of early defense mechanisms. *Plant Physiology* 141(3), 910–923.

Rajjou, L., Lovigny, Y., Groot, S.P., Belghazi, M., Job, C. and Job, D. (2008) Proteome-wide characterization of seed aging in arabidopsis: a comparison between artificial and natural aging protocols. *Plant Physiology* 148(1), 620–641.

Rakshit, A. and Bhadoria, P.S. (2010) Role of VAM on growth and phosphorus nutrition of maize with low soluble phosphate fertilization. *Acta Agronómica* 59(1), 111–118.

Rakshit, A., Sunita, K., Pal, S., Singh, A. and Singh, H.B. (2015) Bio-priming mediated nutrient use efficiency of crop species. In: Rakshit, A., Singh, H.B. and Sen, A. (eds) *Nutrient Use Efficiency: From Basics to Advances*. Springer, New Delhi, India, pp. 181–191.

Rao, M.S.L., Kulkarni, S., Lingaraju, S. and Nadaf, H.L. (2009) Bio-priming of seeds: a potential tool in the integrated management of alternaria blight of sunflower. *Helia* 32(50), 107–114.

Rawat, L., Singh, Y., Shukla, N. and Kumar, J. (2012) Seed biopriming with salinity tolerant isolates of *Trichoderma harzianum* alleviates salt stress in rice: growth, physiological and biochemical characteristics. *Journal of Plant Pathology* 94(2), 353–365.

Reddy, A.S.R., Madhavi, G.B., Reddy, K.G., Yellareddygari, S.K. and Reddy, M.S. (2011) Effect of seed bio-priming with *Trichoderma viride* and *Pseudomonas fluorescens* in chickpea (*Cicer arietinum*) in Andhra Pradesh, India. In: *Plant Growth-promoting Rhizobacteria (PGPR) for Sustainable Agriculture, Proceedings of the 2nd Asian PGPR Conference*, Beijing, pp. 324–429.

Reddy, P.P. (2012) Bio-priming of seeds. In: *Recent Advances in Crop Protection*, Springer, New Delhi, India, pp. 83–90.

Rehman, H.U., Basra, S.M.A. and Farooq, M. (2011) Field appraisal of seed priming to improve the growth, yield, and quality of direct seeded rice. *Turkish Journal of Agriculture and Forestry* 35(4), 357–365.

Saba, H., Vibhash, D., Manisha, M., Prashant, K.S., Farhan, H. and Tauseef, A. (2012) *Trichoderma*–a promising plant growth stimulator and biocontrol agent. *Mycosphere* 3(4), 524–531.

Saleem, M., Arshad, M., Hussain, S. and Bhatti, A.S. (2007) Perspective of plant growth promoting rhizobacteria (PGPR) containing ACC deaminase in stress agriculture. *Journal of Industrial Microbiology and Biotechnology* 34(10), 635–648.

Sandhya, V., Ali, S.Z., Grover, M., Reddy, G. and Venkateswarlu, B. (2010) Effect of plant growth promoting *Pseudomonas* spp. on compatible solutes, antioxidant status and plant growth of maize under drought stress. *Plant Growth Regulation* 62(1), 21–30.

Sathya, S., Lakshmi, S. and Nakkeeran, S. (2016) Combined effect of biopriming and polymer coating against chilli damping off. *International Journal of Agricultural Science and Research* 6(3), 45–54.

Shaharoona, B., Jamro, G.M., Zahir, Z.A., Arshad, M. and Memon, K.S. (2007) Effectiveness of various *Pseudomonas* spp. and *Burkholderia caryophylli* containing ACC-Deaminase for improving growth and yield of wheat (*Triticum aestivum* l). *Journal of Microbiology and Biotechnology* 17(8), 1300.

Sharifi, R.S. (2012) Study of nitrogen rates effects and seed biopriming with PGPR on quantitative and qualitative yield of safflower (*Carthamus tinctorius L.*). *Technical Journal of Engineering and Applied Sciences* 2(7), 162–166.

Sharifi, R.S. and Khavazi, K. (2011) Effects of seed priming with plant growth promoting rhizobacteria (PGPR) on yield and yield attribute of maize (*Zea mays* L.) hybrids. *Journal of Food, Agriculture and Environment* 9(3–4), 496–500.

Sharma, K., Singh, U., Sharma, P., Kumar, A. and Sharma, L. (2015) Seed treatments for sustainable agriculture - a review. *Journal of Applied and Natural Science* 7(1), 521–539.

Sheng, X.F. (2005) Growth promotion and increased potassium uptake of cotton and rape by a potassium releasing strain of *Bacillus edaphicus*. *Soil Biology and Biochemistry* 37(10), 1918–1922.

Singh, H.B. (2016) Seed biopriming: a comprehensive approach towards agricultural sustanability. *Indian Phytopathology* 69(3), 203–209.

Singh, S.P., Singh, H.B., Singh, D.K. and Rakshit, A. (2014) *Trichoderma*-mediated enhancement of nutrient uptake and reduction in incidence of *Rhizoctonia solani* in tomato. *Egyptian Journal of Biology* 16(1), 29–38.

Singh, V., Upadhyay, R.S., Sarma, B.K. and Singh, H.B. (2016) *Trichoderma asperellum* spore dose depended modulation of plant growth in vegetable crops. *Microbiological Research* 193, 74–86.

Singh, V., Upadhyay, R.S., Sarma, B.K. and Singh, H.B. (2016) Seed bio-priming with *Trichoderma asperellum* effectively modulate plant growth promotion in pea. *International Journal of Agriculture, Environment and Biotechnology* 9(3), 361–365.

Smith, D.L. and Almaraz, J.J. (2004) Climate change and crop production: contributions, impacts, and adaptations. *Canadian Journal of Plant Pathology* 26(3), 253–266.

Soleymanifard, A. and Siadat, S.A. (2011) Effect of inoculation with biofertilizer in different nitrogen levels on yield and yields components of safflower under dry land conditions. *American-Eurasian Journal of Agriculture and Environmental Sciences* 11(4), 473–477.

Sturz, A.V. and Christie, B.R. (2003) Beneficial microbial allelopathies in the root zone: the management of soil quality and plant disease with rhizobacteria. *Soil and Tillage Research* 72(2), 107–123.

Tanou, G., Fotopoulos, V. and Molassiotis, A. (2012) Priming against environmental challenges and proteomics in plants: update and agricultural perspectives. *Frontiers in Plant Science* 3, 1–5.

Tanwar, A., Aggarwal, A., Kaushish, S. and Chauhan, S. (2013) Interactive effect of AM fungi with Trichoderma viride and Pseudomonas fluorescens on growth and yield of broccoli. *Plant Protection Science* 49(3), 137–145.

Taylor, A.G. and Harman, G.E. (1990) Concepts and technologies of selected seed treatments. *Annual Review of Phytopathology* 28(1), 321–339.

Wattanakulpakin, P., Photchanachai, S., Ratanakhanokchai, K., Kyu, K.L., Ritthichai, P. and Miyagawa, S. (2012) Hydropriming effects on carbohydrate metabolism, antioxidant enzyme activity and seed vigor of maize (*Zea mays* L). *African Journal of Biotechnology* 11(15), 3537–3547.

Wu, S.C., Cao, Z.H., Li, Z.G., Cheung, K.C. and Wong, M.H. (2005) Effects of biofertilizer containing N-fixer, P and K solubilizers and AM fungi on maize growth: a greenhouse trial. *Geoderma* 125(1), 155–166.

Yadav, B.K. and Tarafdar, J.C. (2011) *Penicillium purpurogenum*, unique P mobilizers in arid agro-ecosystems. *Arid Land Research and Management* 25(1), 87–99.

Zafar-ul-Hye, M., Farooq, H.M., Zahir, Z.A., Hussain, M. and Hussain, A. (2014) Application of ACC-deaminase containing rhizobacteria with fertilizer improves maize production under drought and salinity stress. *International Journal of Agriculture and Biology* 16, 591–596.

Zhu, J., Alvarez, S., Marsh, E.L., LeNoble, M.E., Cho, I.J. *et al.* (2007) Cell wall proteome in the maize primary root elongation zone. II. Region-specific changes in water soluble and lightly ionically bound proteins under water deficit. *Plant Physiology* 145(4), 1533–1548.

Zia-ul-hassan, T.S., Shah, A.N., Jamro, G.M. and Rajpar, I. (2015) Biopriming of wheat seeds with rhizobacteria containing ACC-deaminase and phosphorate solubilizing activities increases wheat growth and yield under phosphorus deficiency. *Pakistan Journal of Agriculture Agricultural Engineering and Veterinary Sciences* 31(1), 24–32.

20 Unravelling the Dual Applications of *Trichoderma* spp. as Biopesticide and Biofertilizer

Vivek Singh,[1,2] Shatrupa Ray,[1,2] Kartikay Bisen,[2] Chetan Keswani,[3] R.S. Upadhyay,[1] B.K. Sarma[2] and H.B. Singh[2]*

[1]Department of Botany, Institute of Science, Banaras Hindu University, Varanasi, India; [2]Department of Mycology and Plant Pathology, Institute of Agricultural Sciences, Banaras Hindu University, Varanasi, India; [3]Department of Biochemistry, Institute of Science, Banaras Hindu University, Varanasi, India

20.1 Introduction

The commercial development and market success of biopesticides depend upon formulating biological control agents with a broad spectrum of activity and an easy application technology. Market penetration of biopesticide products for pest control management has increased significantly in recent years (Glare *et al.*, 2012; Singh *et al.*, 2014c), owing largely to increasing awareness in the public of the adverse effects of chemical pesticides on human health and the environment (Gašić and Tanović, 2013). However, major drawbacks that restrict the field application of biopesticides are their relatively slow microbial action and restricted shelf life, along with application techniques that are complicated in comparison to those of chemical pesticides (Frey, 2001). To overcome this problem, research into the development of biopesticides was an extreme priority. However, the term "biopesticide" leads to some confusion with chemical pesticides, as the biocontrol agent used for pest control may not kill the pest, but rather suppress its development. This allows the crop to become sufficiently developed such that the harmful effects of the pest are minimized and do not affect crop productivity (Crump *et al.*, 1999; Hynes and Boyetchko, 2006).

The size of the global biopesticide market is expected to reach US$ 6.6 billion by 2020, while it is expected to attain a compound annual growth rate of 18.8% from 2015 to 2020 (http://www.marketsandmarkets.com/Market-Reports/biopesticides-267.html). In India the demand for biopesticides, in terms of volume and value, is expected to show the growth at compounded annual rates of 18.3% and 19%, respectively, over the 2015–2020 period (http://www.businesswire.com/news/home/20160217005892/en/Indian-Biopesticides-Market-Growth-Trends-Forecast). Increasing demand for safe food is a key driver in enhancing the biopesticide market growth in sustainable agricultural practices.

*E-mail: hbs1@rediffmail.com

© CAB International 2017. *Advances in PGPR Research* (eds H.B. Singh, B.K. Sarma and C. Keswani)

20.2 Commercial Biocontrol Agents

Biopesticides are defined as biocontrol agents, such as bacteria, viruses, fungi, protozoa and nematodes, and their bioactive metabolites which are used to reduce or kill pests, weeds, pathogens and insects (Glare *et al.*, 2012; Gašić and Tanović, 2013; Ray *et al.*, 2016a, b). Some of the more important biopesticides against insects are *Bacillus thuringiensis* (Bt), *Cydia pomonella* granulovirus, Nuclear Polyhedrosis virus, *Beauveria bassiana* and *Metarhizium anisopliae*. The major microbial biocontrol agents used against plant pathogens include *Trichoderma*, *Bacillus*, *Pseudomonas*, *Clonostachys*, *Streptomyces*, yeasts, etc., having a broad spectrum of activity against taxonomically diverse pathogens. On the other hand *Agrobacterium*, *Ampelomyces*, *Coniothyrium*, non-pathogenic *Fusarium*, atoxigenic *Aspergillus*, etc., have narrow spectra of activity against one or a few targeted pathogens (Chandler *et al.*, 2011; Woo *et al.*, 2014; Keswani *et al.*, 2016).

20.3 *Trichoderma* Biodiversity

Trichoderma spp. are free-living fungi which are highly interactive in soil, foliar and root environments of different ecosystems in a wide range of climatic zones (Harman *et al.*, 2004; Kubicek *et al.*, 2008; Singh *et al.*, 2016a, b). The presence of *Trichoderma* species is regulated by several factors such as microclimate and availability of substrates, as well as complex ecological interactions (Hoyos-Carvajal and Bissett, 2011; Singh, 2016). The wider geographical distribution of *Trichoderma* spp. is related to its metabolic diversity and high reproductive capacity along with antagonistic abilities in nature (Lopes *et al.*, 2012; Bisen *et al.*, 2016). *Trichoderma* spp. show high growth rate under *in vitro* conditions and produce varying shades of green conidia (spores), characteristic of this genus, whereas chlamydospores are observed in conditions with submerged mycelium (Figs 20.1, 20.2). At the black side of plate culture, *Trichoderma* colonies are observed as yellow, yellow-green, buff, amber or colourless (Keswani *et al.*, 2013; Saxena *et al.*, 2015).

20.4 *Trichoderma* spp. Identification

Identification of a potent *Trichoderma* isolate for applying as a biocontrol agent is an important step before selecting for field application. It was reported that the *Trichoderma longibrachiatum* behave as an opportunistic pathogen of immunocompromised mammals, including humans (Keswani *et al.*, 2014; Patel *et al.*, 2015), while other reports clearly indicate that *Trichoderma* spp. is also responsible for the epidemics of commercially grown *Agaricus bisporus* (Samuels *et al.*, 2002). In this regard, correct identification of *Trichoderma* fungi has become a top priority. *Trichoderma* spp. has been known since 1865 (Bisby, 1939), and the mycoparasitic and biocontrol potential was first reported by Weindling (1932). The taxonomy and species identification was clear after 1969 (Rifai, 1969). It is quite difficult to

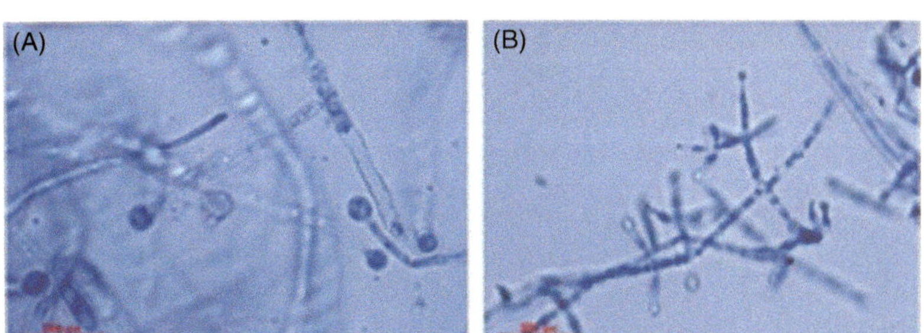

Fig. 20.1. Light microscopic view under 10x magnification. (A) Chlamydospores of *Trichoderma asperellum*, (B) Cylindrical phialides and conidia of *Trichoderma asperellum*.

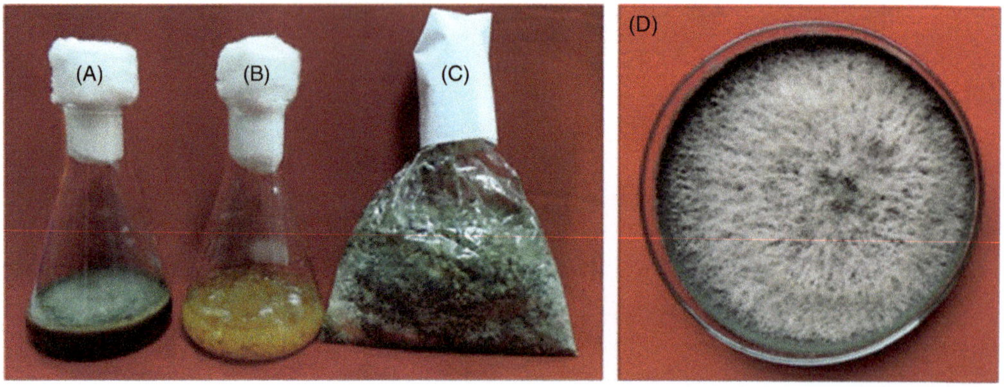

Fig. 20.2. Morphological culture characteristic of *Trichoderma asperellum*. (A) Standing, (B) shaking condition in potato dextrose broth, (C) growth on cereal grains and (D) growth on potato dextrose agar plate.

distinguish *Trichoderma* fungi on the basis of morphological character. However, with the help of gene sequence analysis, phylogenetically distinct species of *Trichoderma* were recognized, reaching up to 100 (Druzhinina *et al.*, 2006) and their number are increasing consistently. In this perspective, modern tools like DNA-barcode systems and genealogical concordance phylogenetic species recognition (GCPSR) play an important role in *Trichoderma* spp. identification based on sequence analysis (Druzhinina *et al.*, 2006).

20.5 *Trichoderma* spp. as Biopesticide

The antagonistic behaviour of *Trichoderma* fungi against numbers of soilborne phytopathogens is well established (Singh *et al.*, 2011; Jain *et al.*, 2015a, b). The main modes of action of *Trichoderma* spp. are mycoparasitism of fungal pathogens with subsequent release of cell-wall-degrading enzymes such as cellulases, glucanases, chitinases, etc. (Kubicek and Harman, 1998; Vinale *et al.*, 2008b), antibiosis by secretion of antimicrobial secondary metabolites and competition for nutrients (Fig. 20.3) (Sarma *et al.*, 2002; Singh *et al.*, 2013a; Keswani *et al.*, 2014; Keswani, 2015).

20.5.1 Mycoparasitism

The mode of the hyphal interaction and parasitism is a complex and sequential process in which *Trichoderma* spp. show directed growth toward other fungi (Chet *et al.*, 1981; Singh, 2014). The growing *Trichoderma* continuously secretes small amounts of an extracellular exochitinase. This diffuses and hydrolyzes the host fungal cell wall, resulting in release of oligomers from target fungi which, in turn sensed by *Trichoderma,* leads to massive production of fungi toxic endochitinases (Brunner *et al.*, 2003). This ultimately enhances the action by diffusing pathogenic fungi cell walls before the attachment of *Trichoderma* to the pathogenic fungi (Brunner *et al.*, 2003). Once the *Trichoderma* attach to the host fungus, it starts secreting different cell-wall-degrading enzymes and other peptaibol antibiotics (Harman *et al.*, 2004). This combined action degrades the pathogenic fungal cell wall, creating holes through which *Trichoderma* hyphae enter the host fungi and get killed due to cytoplamic leakage (Harman *et al.*, 2004).

20.5.2 Antibiosis

Trichoderma produces a number of metabolites with anti-microbial properties against a wide range of phytopathogens. This chemically diverse type of *Trichoderma* metabolite has biocontrol potential. Calistru *et al.* (1997) has demonstrated that hyphal penetration was absent in co-culturing of *Trichoderma* spp. and *Fusarium moniliforme*/*Aspergillus flavus,* pointing out that the inhibitory

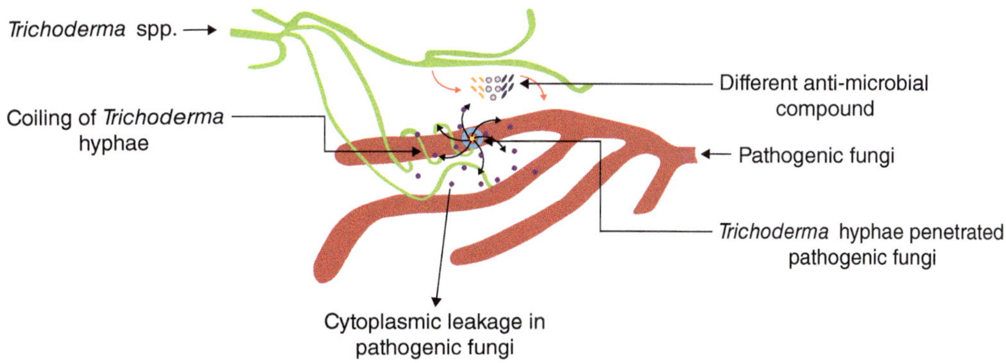

Fig. 20.3. Schematic representation of action taken by *Trichoderma* spp. against phytopathogen.

effect was not solely due to mycoparasitism. Lifshitz *et al.* (1986) showed that the inhibition of *Pythium* species on peas by *T. harzianum* (T-12) and *T. koningii* (T-8) was due to the toxic factor produced in the spermosphere, and not because of mycoparasitism or competition. From this observation it is clear that the metabolites produced by *Trichoderma* spp. play an important role in antibiosis. Major *Trichoderma* spp. producing secondary metabolites against different phytopathogens are described in Table 20.1. Fravel (1988) reported that the purified secondary metabolites of *Trichoderma* spp. act more rapidly in controlling bacterial infections than whole organism application under field conditions.

20.5.3 Competition

Other than mycoparasitism and antibiosis, rhizospheric competition among the biocontrol agent and pathogens is a key mechanism in regulating the existence of either in the rhizosphere. Rhizospheric competition for space and nutrients by the biocontrol agent is important because the utilization of resources indirectly reduces or inhibits the pathogens (Keswani *et al.*, 2014). For example, iron uptake is important for viability of most filamentous fungi and under iron deficiency most fungi secrete siderophores that mobilize environmental iron (Eisendle *et al.*, 2004). *Trichoderma* spp. produced highly proficient siderophores that chelate iron, resulting in inhibition of other fungi (Chet and Inbar, 1994). It was reported that the competition for both rhizophere colonization and nutrients by *T. harzianum* T35 effectively controlled *Fusarium oxysporum*, while the biocontrol activity became more effective as the nutrient concentration decreased (Tjamos *et al.*, 1992; Benítez *et al.*, 2004). On the other hand *Trichoderma* has efficiently utilized and mobilized the nutrient from soil compared to other organisms.

20.6 *Trichoderma* spp. as Biofertilizers

Biofertilizers play a key role in maintaining a soil environment rich in micro-nutrients and macro-nutrients via phosphate and potassium solubilization, nitrogen fixation, excreting plant growth-regulating substances, production of antibiotics and biodegradation of organic matter in the soil (Bhardwaj *et al.*, 2014; Singh *et al.*, 2014b). *Trichoderma* strains easily colonize around plant roots which enhances root growth and development, nutrient uptake and utilization, crop productivity and resistance to abiotic stresses (Benítez *et al.*, 2004; Yadav *et al.*, 2013). It was reported that the application of *Trichoderma* spores on seeds enhanced crop yield (Chet *et al.*, 1997; Jain *et al.*, 2014; Bisen *et al.*, 2015), while the same increase was obtained when seeds and *Trichoderma* were separated by cellophane membrane. This indicates that *Trichoderma* produces some growth factors that result in an enhanced rate of seed germination and

Table 20.1. Major *Trichoderma* secondary metabolite activity against different phytopathogens.

Sl. No.	Major Secondary metabolite	Affective against pathogens	References
1	Pyrone 6-pentyl-2H-pyran-2-one	*Rhizoctonia solani* and *Fusarium oxysporum* f. sp. *lycopersici*	Scarselletti and Faull (1994)
2	6-(1′-pentenyl)-2H-pyran-2-one	*Penicillium* spp., *Aspergillus fumigatus, Candida albicans* and *Cryptococcus neoformans*	Claydon *et al.* (1987); Parker *et al.* (1997)
3	Massoialactone and d-decanolactone	*Botrytis* or *Phytophtora* species, *Aspergillus niger, Candida albicans,* and *Staphylococcus aureus*	Hill *et al.* (1995)
4	Viridepyronone	*Sclerotium rolfsii*	Evidente *et al.* (2003)
5	Koninginins A and B	*Gaeumannomyces graminis* var. *tritici*	Ghisalberti and Rowland (1993)
6	Koninginin D	*Bipolaris sorokiniana, Pythium middleonii, F. oxysporum, Phytophthora cinnamomi* and *R. solani*	Dunlop *et al.* (1989)
7	Viridin	*Stachybotrys atra, Aspergillus niger, Penicillium expansum Colletotrichum lini, Fusarium caeruleum* and *Botrytis allii*	Brian and McGowan (1945)
8	Viridio fungins	*Candida, Aspergillus* and *Cryptococcus* spp.	Harris *et al.* (1993)
9	Harzianopyridone	*R. solani, B. cinerea, Pythium ultimum* and *G. graminis* var. *tritici*	Dickinson *et al.* (1989); Vinale *et al.* (2006)
10	T22-azaphilone, harzianolide and T39 butenolide	*R. solani, P. ultimum* and *G. graminis* var. *tritici, B. cinerea* and *Leptosphaeria maculans*	Almassi *et al.* (1991); Vinale *et al.* (2006); Vinale *et al.* (2008a)
11	Cerinolactone	*P. ultimum, R. solani* and *B. cinerea*	Vinale *et al.* (2011)
12	5-hydoxyvertinolide	*Mycena citricolor*	Andrade *et al.* (1992)
13	Koningic acid	*Bacteroides fragilis*	Itoh *et al.* (1980)
14	Viridepyronone	*S. rolfsii*	Evidente *et al.* (2003)
15	Trichostromaticins A–E	*Moniliophthora perniciosa*	Degenkolb *et al.* (2008)
16	Harzianic acid	*Pythium irregulare, S. sclerotiorum,* and *R. solani*	Vinale *et al.* (2009)

yield (Benítez *et al.*, 1998; Singh *et al.*, 2013b, 2014a). Controlled production of a cytokinin-like compound (zeatyn) and gibberellin GA3 or GA3-related compound by *Trichoderma* spp. would be one of the factors responsible for the biofertilizer activity of *Trichoderma* (Osiewacz, 2002). On the other hand, *Trichoderma* strains produce substances such as gluconic, citric or fumaric acid by metabolism of other carbon sources that increase the acidity in their vicinity (Benítez *et al.*, 2004). In turn, these organic acids solubilize micronutrients, phosphates, and mineral cations including iron, magnesium and manganese (Harman *et al.*, 2004). Therefore, the application of *Trichoderma* in soil results in solubilization of the cations and produces phytohormones that ultimately enhance the yield of crops.

20.7 Commercial Formulations of *Trichoderma* spp.

Due to aforementioned reasons, *Trichoderma* spp. has been widely formulated throughout the world. *Trichoderma*-based production is showing exponential growth in international markets with more than 250 commercially available products. India is the country with greatest distribution of *Trichoderma*-based product, comprising 90% in the Asian market, while South and Central America are emerging rapidly in terms of use of *Trichoderma*-based commercial products, with Brazil one of the main centres for its application. Of the *Trichoderma* spp., *Trichoderma harzianum* is the most frequently used in bioformulation and is marketed throughout the world. On the other hand *T. viride* are largely used as biological control agents in Asia, especially in India with nearly 70% of available products (Woo *et al.*, 2014). *Trichoderma* commercial products are marketed as biofungicides, biopesticides, biostimulants, bio-inoculants, biodecomposers, biofertilizers, bioprotectants, plant growth promoters, etc. About 64.8% of *Trichoderma*-based product is marketed as biofungicides, mainly for controlling root diseases cause by soilborne pathogens such as *Pythium, Sclerotinia, Rhizoctonia, Verticillium, Fusarium, Phytophthora*, etc. There are several types of *Trichoderma*-based formulations commercially available, including wettable powders (55.3%), granular (13.6%), liquid (10.3%) and solids (6.2%), which include coco mat or peat moss, broken corn or cereal grain. Other product formats include emulsion, dry flowable, pellets, powder or talc, and concentrated liquid suspensions (Woo *et al.*, 2014; Saxena *et al.*, 2015).

The shelf life of the biopesticides is a central point for the successful commercialization of product. The short shelf life of *Trichoderma* spp. in formulation creates a major problem for developing commercial formulations; research needs to be done in this area to improve the shelf life and viability of *Trichoderma* spp. in formulation. To date, several types of formulation have been developed and different authors have explained different criteria for the stability of their formulation. According to Cumagun and Ilag (1997) a formulation based on dried conidial pellets of *T. harzianum* is more effective in inhibiting the sclerotia germination of *Rhizoctonia solani* as compared to liquid formulation; this is due to the fact that the formulation of *Trichoderma* spp. based on liquid fermentation is more susceptible to desiccation than the formulation based on solid state fermentation. Sriram *et al.* (2010) reported that addition of chitin in production media and talc formulation of *T. harzianum* increases the shelf-life of the formulation by two month, while additives like osmolyte also increase the shelf life of *T. harzianum* formulation when added to the production media. It was found that the shelf life of formulation was extended to 7 and 12 months by addition of 3% and 6% glycerol to the production media, respectively, compared to formulation without glycerol which gave a shelf life of 4 to 5 months (Sriram *et al.*, 2011).

Furthermore, immobilization of microorganisms was an effective method to improve shelf life and field efficacy. Microencapsulation is an important immobilization technology that increases shelf life of microorganisms as compared to other types of formulations, and controlled release of microbes from this type of formulation also leads to increased field application (John *et al.*, 2011). Spray-dried *Trichoderma* conidia microencapsulated with sugar such as sucrose, molasses or glycerol, extensively increases the survival percentages of conidia after drying (Jin and Custis, 2011). On the other hand microencapsulation of *T. harzianum* conidia in 1:1 blend of maltodextrin–gum Arabic polymeric matrix gave eleven-fold higher conidia survival compared to non-encapsulated conidia after spray-drying (Muñoz-Celaya *et al.*, 2012). Al-Taweil *et al.* (2010) reported an increase in shelf life of *Trichoderma* by using alginate and paraffin oil formulation.

Survivability of *Trichoderma* spp. conidia has increased in an alginate pellets formulation supplemented with 10% cellulose (Shaban and El-Komy, 2001). Kolombet *et al.* (2008) have studied the effect of different amendments in formulating the *Trichoderma asperellum*. These amendments include: addition of starch as a food base, lowering the pH and addition of small amounts of copper to reduce metabolic activity and give a shelf life of six months for the developed formulation. *T. harzianum* SQRT037 conidia formulated with organic fertilizers showed an increase in controlling *Fusarium* wilt of cucumber compared to the treatment comprised of a formulation containing only conidial suspension (Yang *et al.*, 2011). Formulation with bentonite-vermiculite maintained the colony-forming unit of *T. harzianum* for 8 weeks and enhanced the shoot weight of melon plants while it also provided resistance to *Fusarium* wilt disease (Martínez-Medina *et al.*, 2009). *T. harzianum* isolate Th-10 was effective in controlling *Fusarium* wilt of banana. It was found that the dried banana leaf was the best carrier material for growth with high density of propagules of *T. harzianum*, while addition of jaggery to dried banana leaves increased its multiplication which led to a shelf life of more than six months on the stored substrate (Thangavelu *et al.*, 2004).

20.8 Conclusion and Future Prospects

Development of new microbial formulations is a challenge for biopesticide industries and shortcuts usually result in the failure of the effectiveness of formulated microbes in field conditions. Understanding the formulated biocontrol agent and improvements in formulations are key for the success of biopesticide industries. As biopesticides mainly contain live organisms, so it is highly important to develop formulations that maintain the microbial population and efficacy from production to application. Application of less expensive inert materials for developing formulations directly reduces the manufacturing cost. Selecting new adjuvants for developing formulations to retain stability, and give greater shelf life and performance of microbes in field conditions are major areas of research.

Acknowledgments

The authors are grateful to the Department of Biotechnology, New Delhi, India for awarding project grant (BT/PR5990/AGR/5/587/2012).

References

Almassi, F., Ghisalberti, E.L., Narbey, M.J. and Sivasithamparam, K. (1991) New antibiotics from strains of *Trichoderma harzianum*. *Journal of Natural Products* 54, 396–402.

Al-Taweil, H.I., Osman, M.B., Abdulhamid, A., Mohammad, N. and Yussof, W.M.W. (2010) Comparison of different delivery system of *Trichoderma* and *Bacillus* as biofertilizer. *Advances in Environmental Biology* 31–34.

Andrade, R., Ayer, W.A. and Mebe, P.P. (1992) The metabolites of *Trichoderma longibrachiatum*. Part 1. Isolation of the metabolites and the structure of trichodimerol. *Canadian Journal of Chemistry* 70, 2526–2535.

Benítez, T., Delgado-Jarana, J., Rincón, A.M., Rey, M. and Limón, M.C. (1998) Biofungicides: *Trichoderma* as a biocontrol agent against phytopathogenic fungi. *Recent Research Developments in Microbiology* 2, 129–150.

Benítez, T., Rincón, A.M., Limón, M.C. and Codón, A.C. (2004) Biocontrol mechanisms of *Trichoderma* strains. *International Microbiology* 7, 249–260.

Bhardwaj, D., Ansari, M.W., Sahoo, R.K. and Tuteja, N. (2014) Biofertilizers function as key player in sustainable agriculture by improving soil fertility, plant tolerance and crop productivity. *Microbial Cell Factories* 13, 1.

Bisby, G.R. (1939) *Trichoderma viride* Pers. ex Fries, and notes on *Hypocrea*. *Transactions of the British Mycological Society* 23, 149–168.

Bisen, K., Keswani, C., Mishra, S., Saxena, A., Rakshit, A. and Singh, H.B. (2015) Unrealized potential of seed biopriming for versatile agriculture. In: Rakshit, A., Singh, H.B. and Sen, A. (eds) *Nutrient Use Efficiency: From Basics to Advances*. Springer, New Delhi, India, pp. 193–206.

Bisen, K., Keswani, C., Patel, J.S., Sarma, B.K. and Singh, H.B. (2016) *Trichoderma* spp.: efficient inducers of systemic resistance in plants. In: Chaudhary, D.K. and Verma, A. (eds) *Microbial-mediated Induced Systemic Resistance in Plants*. Springer, Singapore, pp. 185–195.

Brian, P.W. and McGowan, J.C. (1945) Viridin: a highly fungistatic substance produced by *Trichoderma viride*. *Nature* 156, 144–145.

Brunner, K., Peterbauer, C.K., Mach, R.L., Lorito, M., Zeilinger, S. and Kubicek, C.P. (2003) The Nag1 N-acetylglucosaminidase of *Trichoderma atroviride* is essential for chitinase induction by chitin and of major relevance to biocontrol. *Current Genetics* 43, 89–295.

Calistru, C., McLean, M. and Berjak, P. (1997) *In vitro* studies on the potential for biological control of *Aspergillus flavus* and *Fusarium moniliforme* by *Trichoderma* species. *Mycopathologia* 137, 115–124.
Chandler, D., Bailey, A.S., Tatchell, G.M., Davidson, G., Greaves, J. and Grant, W.P. (2011) The development, regulation and use of biopesticides for integrated pest management. *Philosophical Transactions of the Royal Society B: Biological Sciences* 366, 1987–1998.
Chet, I. and Inbar, J. (1994) Biological control of fungal pathogens. *Applied Biochemistry and Biotechnology* 48, 37–43.
Chet, I., Harman, G.E. and Baker, R. (1981) *Trichoderma hamatum*: its hyphal interactions with *Rhizoctonia solani* and *Pythium* spp. *Microbial Ecology* 7, 29–38.
Chet, I., Inbar, J. and Hadar, Y. (1997) Fungal antagonists and mycoparasites. In: Wicklow, D.T. and Söderström, B. (eds) *The Mycota, Vol. IV – Environmental and Microbial Relationships.* Springer Berlin Heidelberg, Germany, pp. 165–184.
Claydon, N., Allan, M., Hanson, J.R. and Avent, A.G. (1987) Antifungal alkyl pyrones of *Trichoderma harzianum. Transactions of the British Mycological Society* 88, 503–513.
Crump, N.S., Cother, E.J. and Ash, G.J. (1999) Clarifying the nomenclature in microbial weed control. *Biocontrol Science and Technology* 9, 89–97.
Cumagun, C.J.R. and Ilag, L.L. (1997) Parasitism of sclerotial bodies of *Rhizoctonia solani* Kuehn by *Trichoderma harzianum* Rifai and *Penicillium oxalicum* Currie and Thom. *Phil. Phytopathology* 33, 17–26.
Degenkolb, T., Von Doehren, H., Fog Nielsen, K., Samuels, G.J. and Brueckner, H. (2008) Recent advances and future prospects in peptaibiotics, hydrophobin, and mycotoxin research, and their importance for chemotaxonomy of *Trichoderma* and *Hypocrea. Chemistry & Biodiversity* 5, 671–680.
Dickinson, J.M., Hanson, J.R., Hitchcock, P.B. and Claydon, N. (1989) Structure and biosynthesis of harzianopyridone, an antifungal metabolite of *Trichoderma harzianum. Journal of the Chemical Society, Perkin Transactions* 1, 1885–1887.
Druzhinina, I.S., Kopchinskiy, A.G. and Kubicek, C.P. (2006) The first 100 *Trichoderma* species characterized by molecular data. *Mycoscience* 47, 55–64.
Dunlop, R.W., Simon, A., Sivasithamparam, K. and Ghisalberti, E.L. (1989) An antibiotic from *Trichoderma koningii* active against soilborne plant pathogens. *Journal of Natural Products* 52, 67–74.
Eisendle, M., Oberegger, H., Buttinger, R., Illmer, P. and Haas, H. (2004) Biosynthesis and uptake of siderophores is controlled by the PacC-mediated ambient-pH regulatory system in *Aspergillus nidulans. Eukaryotic Cell* 3, 561–563.
Evidente, A., Cabras, A., Maddau, L., Serra, S., Andolfi, A. and Motta, A. (2003) Viridepyronone, a new antifungal 6-substituted 2 h-pyran-2-one produced by *Trichoderma viride. Journal of Agricultural and Food Chemistry* 51, 6957–6960.
Fravel, D.R. (1988) Role of antibiosis in the biocontrol of plant diseases. *Annual Review of Phytopathology* 26, 75–91.
Frey, P.M. (2001) Biocontrol agents in the age of molecular biology. *Trends in Biotechnology* 19, 432–433.
Gašić, S. and Tanović, B. (2013) Biopesticide formulations, possibility of application and future trends. *Pesticides and Phytomedicine* 28, 97–102.
Ghisalberti, E.L. and Rowland, C.Y. (1993) Antifungal metabolites from *Trichoderma harzianum. Journal of Natural Products* 56, 1799–1804.
Glare, T., Caradus, J., Gelernter, W., Jackson, T., Keyhani, N. *et al.* (2012) Have biopesticides come of age? *Trends in Biotechnology* 30, 250–258.
Harman, G.E., Howell, C.R., Viterbo, A., Chet, I. and Lorito, M. (2004) *Trichoderma* species—opportunistic, avirulent plant symbionts. *Nature Reviews Microbiology* 2, 43–56.
Harris, G.H., Jones, E.T.T., Meinz, M.S., Nallin-Omstead, M., Helms, G.L. *et al.* (1993) Isolation and structure elucidation of viridiofungins A, B and C. *Tetrahedron Letters* 34, 5235–5238.
Hill, R.A., Cutler, H.G. and Parker, S.R. (1995) *Trichoderma* and metabolites as control agents for microbial plant diseases. *PCT International Applications* 9520879.
Hoyos-Carvajal, L. and Bissett, J. (2011) Biodiversity of *Trichoderma* in neotropics. In: Grillo, O. and Venora, G. (eds) *The Dynamical Process of Biodiversity - Case Studies of Evolution and Spatial Distribution.* InTech, pp 303–320.
Hynes, R.K. and Boyetchko, S.M. (2006) Research initiatives in the art and science of biopesticide formulations. *Soil Biology and Biochemistry* 38, 845–849.
Itoh, Y., Takahashi, S., Haneishi, T. and Arai, M. (1980) Structure of heptelidic acid, a new sesquiterpene antibiotic from fungi. *The Journal of Antibiotics* 33, 525–526.
Jain, A., Singh, A., Chaudhary, A., Singh, S., Singh, H.B. (2014) Modulation of nutritional and antioxidant potential of seeds and pericarp of pea pods treated with microbial consortium. *Food Research International* 64, 275–282.

Jain, A., Singh, A., Singh, S., Sarma, B.K., Singh, H.B. (2015a) Biocontrol agents-mediated suppression of oxalic acid induced cell death during *Sclerotinia sclerotiorum*–pea interaction. *Journal of Basic Microbiology* 55, 601–606.

Jain, A., Singh, A., Singh, S. and Singh, H.B. (2015b) Biological management of *Sclerotinia sclerotiorum* in pea using plant growth promoting microbial consortium. *Journal of Basic Microbiology* 55, 961–972.

Jin, X. and Custis, D. (2011) Microencapsulating aerial conidia of *Trichoderma harzianum* through spray drying at elevated temperatures. *Biological Control* 56, 202–208.

John, R.P., Tyagi, R.D., Brar, S.K., Surampalli, R.Y. and Prévost, D. (2011) Bio-encapsulation of microbial cells for targeted agricultural delivery. *Critical Reviews in Biotechnology* 31, 211–226.

Keswani, C. (2015) Proteomics studies of thermotolerant strain of *Trichoderma* spp. Ph.D. Thesis, Banaras Hindu University, Varanasi, India, p. 126.

Keswani, C., Singh, S.P. and Singh, H.B. (2013) A superstar in biocontrol enterprise: *Trichoderma* spp. *Biotech Today* 3, 27–30.

Keswani, C., Mishra, S., Sarma, B.K., Singh, S.P. and Singh, H.B. (2014) Unraveling the efficient applications of secondary metabolites of various *Trichoderma* spp. *Applied Microbiology and Biotechnology* 98, 533–544.

Keswani, C., Bisen, K., Singh, V., Sarma, B.K., Singh, H.B. (2016) Formulation technology of biocontrol agents: present status and future prospects. In: Arora, N.K., Mehnaz, S. and Balestrini, R. (eds) *Bioformulations: For Sustainable Agriculture*. Springer, New Delhi, India, pp. 35–52.

Kolombet, L.V., Zhigletsova, S.K., Kosareva, N.I., Bystrova, E.V., Derbyshev, V.V. *et al.* (2008) Development of an extended shelf-life, liquid formulation of the biofungicide *Trichoderma asperellum*. *World Journal of Microbiology and Biotechnology* 24, 123–131.

Kubicek, C.P. and Harman, G.E. (eds) (1998) *Trichoderma and Gliocladium. Basic Biology, Taxonomy and Genetics Vol.* 1. Taylor & Francis, London.

Kubicek, C.P., Komon-Zelazowska, M. and Druzhinina, I.S. (2008) Fungal genus *Hypocrea/Trichoderma*: from barcodes to biodiversity. *Journal of Zhejiang University Science B* 9, 753–763.

Lifshitz, R., Windham, M.T. and Baker, R. (1986) Mechanism of biological control of pre-emergence damping-off of pea by seed treatment with *Trichoderma* spp. *Phytopathology* 76, 720–725.

Lopes, F.A.C., Steindorff, A.S., Geraldine, A.M., Brandão, R.S., Monteiro, V.N. *et al.* (2012) Biochemical and metabolic profiles of *Trichoderma* strains isolated from common bean crops in the Brazilian Cerrado, and potential antagonism against *Sclerotinia sclerotiorum*. *Fungal Biology* 116, 815–824.

Martínez-Medina, A., Roldán, A. and Pascual, J.A. (2009) Performance of a *Trichoderma harzianum* bentonite–vermiculite formulation against *Fusarium* wilt in seedling nursery melon plants. *HortScience* 44, 2025–2027.

Muñoz-Celaya, A.L., Ortiz-García, M., Vernon-Carter, E.J., Jauregui-Rincón, J., Galindo, E. and Serrano-Carreón, L. (2012) Spray-drying microencapsulation of *Trichoderma harzianum* conidias in carbohydrate polymers matrices. *Carbohydrate Polymers* 88, 1141–1148.

Osiewacz, H.D. (2002) *Molecular and Biology Fungal Development*. Marcel Dekker, New York.

Parker, S.R., Cutler, H.G., Jacyno, J.M. and Hill, R.A. (1997) Biological activity of 6-pentyl-2 H-pyran-2-one and its analogs. *Journal of Agricultural and Food Chemistry* 45, 2774–2776.

Patel, J.S., Sarma, B.K., Singh, H.B., Upadhyay, R.S., Kharwar, R.N. and Ahmed, M. (2015) *Pseudomonas fluorescens* and *Trichoderma asperellum* enhance expression of Gα subunits of the pea heterotrimeric G-protein during *Erysiphe pisi* infection. *Frontiers in Plant Science* 6.

Ray, S., Singh, S., Sarma, B.K. and Singh, H.B. (2016a) Endophytic *Alcaligenes* isolated from horticultural and medicinal crops promotes growth in okra (*Abelmoschus esculentus*). *Journal of Plant Growth Regulation* 35, 401–412.

Ray, S., Singh, V., Singh, S., Sarma, B.K. and Singh, H.B. (2016b) Biochemical and histochemical analyses revealing endophytic *Alcaligenes faecalis* mediated suppression of oxidative stress in *Abelmoschus esculentus* challenged with *Sclerotium rolfsii*. *Plant Physiology and Biochemistry* 109, 430–441.

Rifai, M.A. (1969) A revision of the genus *Trichoderma*. *Mycological Paper* 116, 1–56.

Samuels, G.J., Dodd, S.L., Gams, W., Castlebury, L.A. and Petrini, O. (2002) *Trichoderma* species associated with the green mold epidemic of commercially grown *Agaricus bisporus*. *Mycologia* 94, 146–170.

Sarma, B.K., Singh, D.P., Mehta, S., Singh, H.B. and Singh, U.P. (2002) Plant growth-promoting rhizobacteria-elicited alterations in phenolic profile of chickpea (*Cicer arietinum*) infected by *Sclerotium rolfsii*. *Journal of Phytopathology* 150, 277–282.

Saxena, A., Raghuwanshi, R. and Singh, H.B. (2015) *Trichoderma* species mediated differential tolerance against biotic stress of phytopathogens in *Cicer arietinum* L. *Journal of Basic Microbiology* 55, 195–206.

Scarselletti, R. and Faull, J.L. (1994) In vitro activity of 6-pentyl-α-pyrone, a metabolite of *Trichoderma harzianum*, in the inhibition of *Rhizoctonia solani* and *Fusarium oxysporum* f. sp. *lycopersici*. *Mycological Research* 98, 1207–1209.

Shaban, G.M. and El-Komy, H.M. (2001) Survival and proliferation of alginate encapsulated *Trichoderma* spp. in Egyptian soil in comparison with allyl alcohol soil fumigation. *Mycopathologia* 151, 139–146.

Singh, A., Jain, A., Sarma, B.K., Upadhyay, R.S. and Singh, H.B. (2013a) Rhizosphere microbes facilitate redox homeostasis in *Cicer arietinum* against biotic stress. *Annals of Applied Biology* 163, 33–46.

Singh, A., Sarma, B.K., Upadhyay, R.S. and Singh, H.B. (2013b) Compatible rhizosphere microbes mediated alleviation of biotic stress in chickpea through enhanced antioxidant and phenylpropanoid activities. *Microbiological Research* 168, 33–40.

Singh, A., Jain, A., Sarma, B.K., Upadhyay, R.S. and Singh, H.B. (2014a) Rhizosphere competent microbial consortium mediates rapid changes in phenolic profiles in chickpea during *Sclerotium rolfsii* infection. *Microbiological Research* 169, 353–360.

Singh, A., Sarma, B.K., Singh, H.B. and Upadhyay, R.S. (2014b) *Trichoderma*: A silent worker of plant rhizosphere. In: Gupta, V.K., Schmoll, M., Herrera-Estrella, A., Upadhyay, R.S., Druzhinina, I. and Tuohy, M.G. (eds) *Biotechnology and Biology of Trichoderma*. Elsevier, Amsterdam, pp. 533–542.

Singh, B.N., Singh, A., Singh, S.P. and Singh, H.B. (2011) *Trichoderma harzianum*-mediated reprogramming of oxidative stress response in root apoplast of sunflower enhances defence against *Rhizoctonia solani*. *European Journal of Plant Pathology* 131, 121–134.

Singh, H.B. (2014) Management of plant pathogens with microorganisms. *Proceedings of the Indian National Science Academy* 80, 443–454.

Singh, H.B. (2016) Seed biopriming: a comprehensive approach towards agricultural sustainability. *Indian Phytopathology* 69, 203–209.

Singh, H.B., Singh, A. and Sarma, B.K. (2014c) *Trichoderma viride* 2% WP (Strain No. BHU-2953) formulation suppresses tomato wilt caused by *Fusarium oxysporum* f. sp. *lycopersici* and chilli damping-off caused by *Pythium aphanidermatum* effectively under different agroclimatic conditions. *International Journal of Agriculture, Environment and Biotechnology* 7, 313–320.

Singh, V., Upadhyay, R.S., Sarma, B.K. and Singh, H.B. (2016a) *Trichoderma asperellum* spore dose depended modulation of plant growth in vegetable crops. *Microbiological Research* 193, 74–86.

Singh, V., Upadhyay, R.S., Sarma, B.K. and Singh, H.B. (2016b) Seed bio-priming with *Trichoderma asperellum* effectively modulate plant growth promotion in pea. *International Journal of Agriculture, Environment and Biotechnology* 9, 361–365.

Sriram, S., Palanna, K.B. and Ramanujam, B. (2010) Effect of chitin on the shelf-life of *Trichoderma harzianum* in talc formulation. *Indian Journal of Agricultural Sciences* 80, 930.

Sriram, S., Roopa, K.P. and Savitha, M.J. (2011) Extended shelf-life of liquid fermentation derived talc formulations of *Trichoderma harzianum* with the addition of glycerol in the production medium. *Crop Protection* 30, 1334–1339.

Thangavelu, R., Palaniswami, A. and Velazhahan, R. (2004) Mass production of *Trichoderma harzianum* for managing *fusarium* wilt of banana. *Agriculture, Ecosystems & Environment* 103, 259–263.

Tjamos, E.C., Papavizas, G.C. and Cook, R.J. (1992) *Biological Control of Plant Diseases Progress and Challenges for the Future*. Plenum, New York (in cooperation with NATO Scientific Affairs Division).

Vinale, F., Marra, R., Scala, F., Ghisalberti, E.L., Lorito, M. and Sivasithamparam, K. (2006) Major secondary metabolites produced by two commercial *Trichoderma* strains active against different phytopathogens. *Letters in Applied Microbiology* 43, 143–148.

Vinale, F., Sivasithamparam, K., Ghisalberti, E.L., Marra, R., Barbetti, M.J. *et al.* (2008a) A novel role for *Trichoderma* secondary metabolites in the interactions with plants. *Physiological and Molecular Plant Pathology* 72, 80–86.

Vinale, F., Sivasithamparam, K., Ghisalberti, E.L., Marra, R., Woo, S.L. and Lorito, M. (2008b) *Trichoderma*–plant–pathogen interactions. *Soil Biology and Biochemistry* 40, 1–10.

Vinale, F., Flematti, G., Sivasithamparam, K., Lorito, M., Marra, R. *et al.* (2009) Harzianic acid, an antifungal and plant growth promoting metabolite from *Trichoderma harzianum*. *Journal of Natural Products* 72, 2032–2035.

Vinale, F., Girona, I.A., Nigro, M., Mazzei, P., Piccolo, A. *et al.* (2011) Cerinolactone, a hydroxy-lactone derivative from *Trichoderma cerinum*. *Journal of Natural Products* 75, 103–106.

Weindling, R. (1932) *Trichoderma lignorum* as a parasite of other soil fungi. *Phytopathology* 22, 837–845.

Woo, S.L., Ruocco, M., Vinale, F., Nigro, M., Marra, R. *et al.* (2014) *Trichoderma*-based products and their widespread use in agriculture. *The Open Mycology Journal* 8.

Yadav, S.K., Dave, A., Sarkar, A., Singh, H.B. and Sarma, B.K. (2013) Co-inoculated biopriming with *Trichoderma, Pseudomonas* and *Rhizobium* improves crop growth in *Cicer arietinum* and *Phaseolus vulgaris*. *International Journal of Agriculture, Environment and Biotechnology* 6, 255–259.

Yang, X., Chen, L., Yong, X. and Shen, Q. (2011) Formulations can affect rhizosphere colonization and biocontrol efficiency of *Trichoderma harzianum* SQR-T037 against *Fusarium* wilt of cucumbers. *Biology and Fertility of Soils* 47, 239–248.

21 Genome Insights into Plant Growth-Promoting Rhizobacteria, an Important Component of Rhizosphere Microbiome

Vasvi Chaudhry,[1*] Niladri Chaudhry[2] and Shrikant S. Mantri[3]
[1]*Bacterial Genomics & Evolution Laboratory, CSIR-Institute of Microbial Technology, Chandigarh-160036, India;* [2]*Department of Pathology, Institute of Medical Sciences, Banaras Hindu University, Varanasi, Uttar Pradesh 221005, India;* [3]*Computational Biology Laboratory, National Agri-Food Biotechnology Institute (NABI), Mohali, Punjab-160071, India*

21.1 Introduction

Plants interact with the environment and their associated microbial communities in both above- and belowground ecosystems. This assemblage of plant with environment and associated microorganisms together comprises the "plant microbiome" similarly to the way a human being possesses its microbiome (Turner *et al.*, 2013; Berg *et al.*, 2014). The plant microbiome has been considered as one of the key determinants of plant health and productivity (Hartmann *et al.*, 2009). The diverse niches that make up the plant microbiome are the phyllosphere, rhizosphere and internal tissues (Turner *et al.*, 2013; Andreote *et al.*, 2014). The rhizosphere is the zone located around the plant below ground, a compartment comprising the microbes and soil in the vicinity of the roots. The term "rhizosphere" was first used by Hiltner (1904) to describe the zone of soil under the influence of roots, which is considered one of the most complex, diverse and active environments on earth (Hinsinger and Marschner, 2006; Pierret *et al.*, 2007; Jones and Hinsinger, 2008; Hinsinger *et al.*, 2009; Raaijmakers *et al.*, 2009; Mendes *et al.*, 2013; Keswani *et al.*, 2016). Diverse kinds of organisms are present in the rhizospheric region, including bacteria, fungi, oomycetes, nematodes, protozoa, algae, viruses, archaea and arthropods (Lynch, 1990; Metting, 1992; Bonkowski *et al.*, 2009; Buee *et al.*, 2009; Raaijmakers *et al.*, 2009; Bisen *et al.*, 2016). Owing to the nutrients provided by plant root exudates, the rhizosphere harbours microbial diversity and represents a zone where maximum microbial activity takes place (Vacheron *et al.*, 2013). Just like the human gut, the rhizosphere is an environment where highly diverse microbial communities perform important functions such as disease protection and nutrient uptake for the betterment of the host (Ramírez-Puebla *et al.*, 2013; Adam *et al.*, 2016).

The total microbial population present in the rhizosphere is also known as the "rhizospheric microbiome" or "rhizobiome" that is formed by plant root exudates as

*E-mail: dr.vasvi.india@gmail.com

© CAB International 2017. *Advances in PGPR Research* (eds H.B. Singh, B.K. Sarma and C. Keswani)

nutrients from the complex microbial community in the soil. The result of microbial interactions in the rhizosphere ranges from plant beneficial ("friend"), plant pathogenic ("foe") to human pathogenic ("alien") microorganisms in the rhizosphere which comprises nitrogen-fixing bacteria, mycorrhizal fungi, root endophytes, plant growth-promoting rhizobacteria (PGPR), mycoparasitic fungi and protozoa; all of these have positive impact on plant growth and nutrient acquisition (van der Heijden *et al.*, 2008). On the other hand, microorganisms such as the pathogenic fungi, oomycetes, bacteria and nematodes can invade the microbial community and create a negative impact on plant growth and fitness once they overcome plant defence responses. The other group of microorganisms present in the rhizosphere is human pathogens (van Baarlen *et al.*, 2007; Tyler and Triplett, 2008; Holden *et al.*, 2009). Therefore, it is a major challenge to differentiate among "the friend", "the foe" and "the alien" microorganisms and how they interact with plant roots and influence plant growth. Moreover, a comprehensive understanding of the mechanisms that govern selection and activity of beneficial bacteria by plant roots will provide new opportunities to increase crop production.

In recent years studies on the rhizosphere microbiome have gained interest due to technological advances, including next-generation sequencing, bioinformatics and metagenomics that enabled sequence-based analyses of the plant microbiome functions during their association with the host (Lundberg *et al.*, 2012; Bulgarelli *et al.*, 2013). To understand the association of PGPR with roots, certain questions need to be addressed, such as 'Which bacteria colonize the root?' and 'What are they doing in the root environment?', 'How do they interact with other bacteria present there?' and 'How do they affect plant growth?' Therefore, unearthing the answers to these questions will give us better understanding of the PGPR associated with plant roots, as well as strong support for their application as biofertilizers for agricultural sustainability.

21.2 Bacterial Rhizobiome

The rhizosphere microbiome or rhizobiome comprises the total microbial community present in the rhizosphere. However, the bacterial rhizobiome comprises the total bacterial community colonizing the roots: they may be plant beneficial, plant pathogenic, or human pathogenic and the total bacterial population lies in the range of $\sim 10^8$–10^9 colony-forming units (CFU) gm^{-1} of rhizospheric soil (Fig. 21.1). The term *rhizobacteria* is used to describe a subset of rhizosphere bacteria able to colonize the root environment. Within the total bacterial community a specialized class of rhizobacteria that colonize the root environment are known as the plant growth-promoting rhizobacteria (PGPR) (Kloepper *et al.*, 1991; Kloepper, 1994). These PGPR are naturally-occurring soil bacteria that reside in the rhizosphere under the influence of plant root exudates and make an important contribution to the overall plant growth and development. PGPR were first defined by Kloepper and Schroth (1978) to describe soil bacteria that colonize the roots of plants enhancing plant growth. Beneficial rhizobacteria should have at least three characteristics to be classified as PGPR: (i) they must be able to colonize the root, (ii) they must survive, multiply and compete in the rhizosphere and (iii) they must promote plant growth (Lugtenberg and Kamilova, 2009).

According to their interactions with plants, PGPR can be divided into two groups, the extracellular plant growth-promoting rhizobacteria (ePGPR) and the intracellular plant growth-promoting rhizobacteria (iPGPR) (Dimkpa *et al.*, 2009; Viveros *et al.*, 2010). The ePGPR are present in the rhizosphere and rhizoplane whereas the iPGPR generally colonize the specialized nodular structures of root cells. The bacterial genera belonging to *Agrobacterium*, *Arthrobacter*, *Azotobacter*, *Azospirillum*, *Bacillus*, *Burkholderia*, *Caulobacter*, *Chromobacterium*, *Erwinia*, *Flavobacterium*, *Micrococcus*, *Pseudomonas*, *Pantoea*, *Paenibacillus* and *Serratia* are classified as ePGPR (Ahemad and Kibret, 2014; Pérez-Montaño *et al.*, 2014). The iPGPR belongs to the family of Rhizobiaceae that

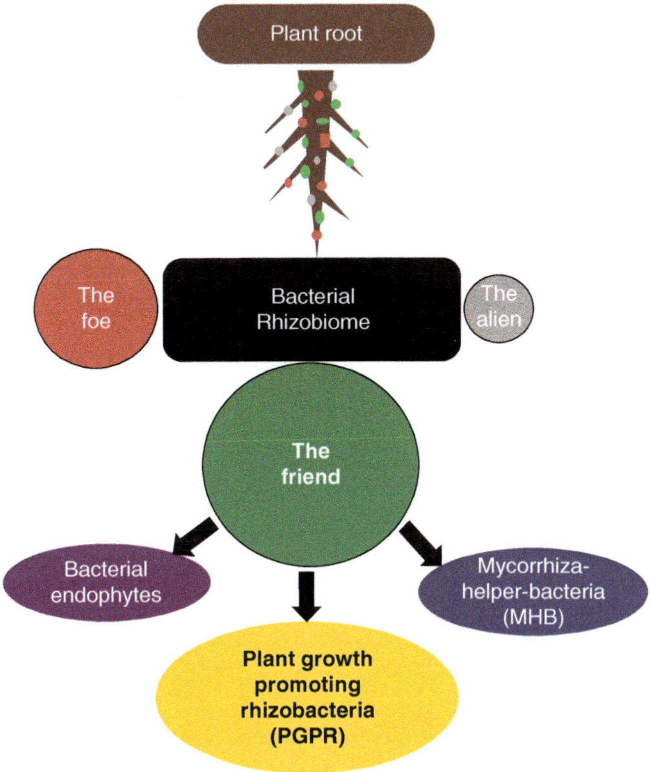

Fig. 21.1. Schematic drawing representing the bacterial rhizobiome associated with rhizosphere, plant-beneficial ("the friend"), plant-pathogenic ("the foe"), and human-pathogenic bacteria ("the alien") associated to the host plant. The terms "the friend", "the foe", and "the alien" are used to describe the complex bacterial interactions in the rhizosphere environment. Further, the beneficial bacteria may be grouped into endophytes, MHB and PGPR.

includes *Allorhizobium*, *Bradyrhizobium*, *Mesorhizobium*, *Rhizobium* and bacterial endophytes (Bhattacharyya and Jha, 2012). The PGPR function as a consortium, together protecting plants from various seed and soilborne diseases (Kloepper *et al.*, 2004), by providing plants with essential nutrients and stimulating plant growth by producing various plant growth-promoting factors (Viveros *et al.*, 2010).

21.3 Mechanisms of PGPR

The mechanisms by which PGPR can influence plant growth differ between bacterial species and strains, so typically there is not a single mechanism for promoting plant growth. Several mechanisms are documented in these PGPR that can help in facilitating plant growth. They are broadly classified as direct and indirect mechanisms. Direct effects on plants may involve enhanced availability of nutrients, stimulation of root system development via production of phytohormones such as indole acetic acid, cytokinins, etc., and inhibition of the plant's ethylene synthesis (Somers *et al.*, 2004; Glick, 2005; Blaha *et al.*, 2006; Lugtenberg and Kamilova, 2009; Drogue *et al.*, 2012). Indirect beneficial effects of PGPR on plants include competition or antagonism towards phytopathogens and induced systemic resistance (Haas and Défago, 2005; Zamioudis and Pieterse, 2012). Figure 21.2 shows important direct and indirect mechanisms of plant growth promotion by PGPR.

The mechanisms by which PGPR stimulate plant growth involve the availability of nutrients originating from genetic processes, such as biological nitrogen fixation, phosphate solubilization, stress alleviation through the modulation of 1-aminocyclopropane-1-carboxylate (ACC) deaminase expression production of phytohormones and siderophores.

PGPR model strains have been extensively studied over the last decade, revealing

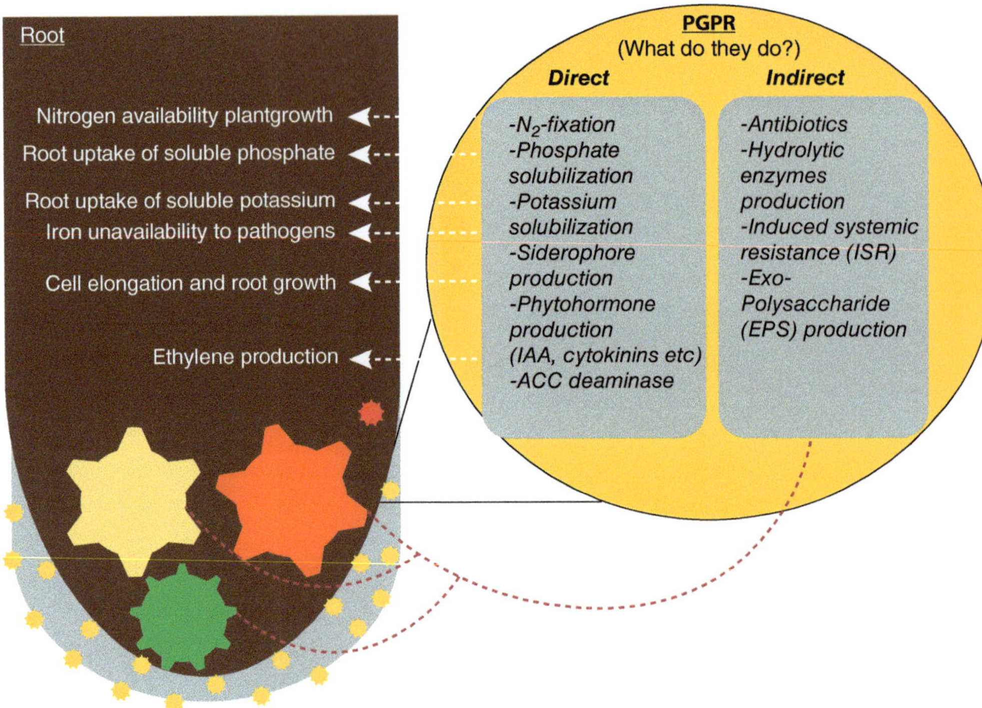

Fig. 21.2. Schematic representation showing PGPR–plant root interactions and important mechanisms in PGPR involved in plant growth promotion.

the molecular mechanism of their plant-beneficial traits. Studies have shown that many PGPR strains possess more than one plant-beneficial property (Haas and Défago, 2005; Almario *et al.*, 2014). So far, only a general description of the occurrence of plant beneficial genes has been documented. However, availability of whole genomes of PGPR has brought fundamental insights into the potential associations of plant-beneficial traits in PGPR, and in-depth knowledge can be achieved based on comparative genome analysis and phylogenetic analysis (Kim and Price, 2011; Martiny *et al.*, 2013).

21.4 NGS Technologies and Genome Assembly

Bacterial genome sequencing is now 20 years old, since the first bacterial genome of *Haemophilus influenzae* Rd was sequenced by Fleischmann *et al.* (1995). Before the invention of next-generation sequencing (NGS) in 2005, the use of genome sequencing in bacteria was a high-cost, labour-intensive and time-consuming process by the conventional Sanger sequencing method. With the advancement of NGS systems a massive amount of sequencing data (from gigabases to terabases) can be generated with less cost and time. The NGS platforms are classified as second- and third-generation sequencing technologies (Schadt *et al.*, 2010; Niedringhaus *et al.*, 2011; Liu *et al.*, 2012). The second-generation sequencing technology includes the Roche-454, Illumina platforms, the Life Technologies system, sequencing by oligonucleotide ligation and detection (SOLiD) and ion torrent platforms. The third-generation sequencing platform currently available is the PacBio RS by Pacific Biosciences. The first NGS platform that was widely used in microbiology was the Roche-454, which adopted the principle of pyrosequencing (Margulies *et al.*, 2005). Illumina DNA sequencing technology was based on bridge-amplification and reversible terminators (Bentley, 2006), and is provided as instruments including HiSeq and its bench-top version called MiSeq. Using these NGS platforms, the sequencing process has seen remarkable change as the sequencing projects that used to take years can now be completed in a few days in a cost-effective way.

With sequencing no longer a bottleneck, the result was a massive amount of sequence data. However, in genome-sequencing using NGS, different depths of sequencing coverage are used to obtain the final assembled sequence (called contigs or scaffolds) from raw reads. Every NGS technology has its advantages and disadvantages based on read length, accuracy, sequencing errors, its ability to produce single-end or paired-end reads as well as cost-effectiveness. Therefore, sometimes it is important to use two or more NGS systems simultaneously to gain high-quality genome data. The approach of assembling reads generated from different platforms is called 'hybrid genome assembly'. Use of these NGS technologies provides the whole genome sequence of bacteria which led to the comprehensive understanding of the molecular genetics of many bacterial species (Schuster, 2008; MacLean *et al.*, 2009).

21.5 Genome-based Taxonomy and Phylogenomics

Bacterial taxonomy is critically important in different fields, particularly PGPR studies; rhizobacteria strain characterization is crucial before a strain can find application as biofertilizer. However, 16S rRNA gene sequence data have served as the "gold standard" for bacterial classification for more than 30 years and therefore huge numbers of 16S rRNA sequences are available in public repositories (Fox *et al.*, 1977; DeSantis *et al.*, 2006; Pruesse *et al.*, 2007). However, owing to the conserved nature of the 16S rRNA gene, its resolution is too low to resolve two different species and sometimes even two different genera (Tindall *et al.*, 2010; Kampfer, 2012). Whole-genome sequencing (WGS) allows an intensive assessment of relationships and unequivocal designation of bacterial groups into taxonomic schemes based on genome sequence information, such as the Karlin genomic signatures, average amino acid identity (AAI), average nucleotide identity (ANI), and *in silico* genome-to-genome distance hybridization (GGDH) (Konstantinidis and Stackebrandt, 2013). Owing to the decreasing cost of sequencing technologies, the genomic signatures present in genome data have become routine for bacterial identification and in-depth characterization. Therefore, molecular taxonomy is more accepted, rather than relying on time-consuming and laborious classical polyphasic taxonomy (Thompson *et al.*, 2011; Thompson *et al.*, 2013).

In databases, the complete and draft genome sequence of hundreds of PGPR strains associated with different plants/crops have been reported and made publicly available. Several PGPR genomic studies were mainly focused on crop species such as wheat, chickpea, *Miscanthus*, and pepper (Ma *et al.*, 2011; Mathimaran *et al.*, 2012; Song *et al.*, 2012; Chaudhry *et al.*, 2013). However, PGPR from soil have also been sequenced (Matilla *et al.*, 2011). Some studies on genome sequence of PGPR that belong to *Enterobacter cloacae* and *Pseudomonas putida* isolated from plantation crops such as coconut, cocoa and areca nut were also documented in the literature (Gupta *et al.*, 2014). Table 21.1 summarizes the list of PGPR strains whose genomes have been sequenced.

21.6 Genome Mining of Plant Beneficial Genes in PGPR

Systematic analysis of whole genome data and further identification and characterization of genes that contribute to the beneficial properties of PGPR is critically important for the effective understanding and manipulation of the association between the plant host and rhizobacteria. PGPR genome analysis has provided a new way to closely view the adaptation of PGPR in plant roots and to reveal the colonization features to harbour

Table 21.1. List of important PGPR whose genomes have been sequenced and their accession numbers.

S. No.	Strain name	Accession no.
1	*Agrobacterium radiobacter* K84	NC_011987, NC_011985, NC_011994, NC_011990, NC_011983
2	*Azoarcus* sp. BH72	NC_008702
3	*Azospirillum brasilense* CBG497	AzospirilluScope project in Mage database
4	*Azospirillum brasilense* Sp245	NC_016617, NC_016594, NC_016618, NC_016595, NC_016596, NC_016619, NC_016597
5	*Azospirillum lipoferum* 4B	NC_016622, NC_016585, NC_016586, NC_016623, NC_016587, NC_016624, NC_016588,
6	*Azospirillum* sp. B510	NC_013854, NC_013855, NC_013856, NC_013857, NC_013858, NC_013859, NC_013860
7	*Azospirillum amazonense Y2*	AFBX00000000
8	*Bacillus amyloliquefaciens* BS006	LJAU00000000
9	*Bacillus amyloliquefaciens* Co1-6	CVPA00000000
10	*Bacillus subtilis* ALBA01	LVYH00000000
11	*Bacillus subtilis* UD1022	CP011534
12	*Bacillus cereus* 905	LSTW00000000
13	*Bacillus methylotrophicus* FKM10	LNTG00000000
14	*Bacillus pumilus* WP8	CP010075
15	*Bacillus subtilis* XF-1	CP004019
16	*Bacillus sp.* JS	CP003492
17	*Brevibacillus brevis* DZQ7	LDZV00000000
18	*Burkholderia ambifaria* AMMD	NC_008392, NC_008391, NC_008390, NC_008385
19	*Burkholderia ambifaria* RZ2MS16	LKPJ00000000
20	*Burkholderia cepacia* 383	NC_007510, NC_007511, NC_007509
21	*Burkholderia phytofirmans* PsJN	NC_010681, NC_010679, NC_010676
22	*Delftia tsuruhatensis* MTQ3	LCZH00000000
23	*Enterobacter cloacae* GS1	NZ_AJXP00000000
24	*Enterobacter* sp. 638	NC_009436, NC_009425
25	*Gluconacetobacter diazotrophicus* PAI5	NC_010125, NC_010124, NC_010123
26	*Herbaspirillum seropedicae* SmR1	NC_014323
27	*Klebsiella* sp. D5A	LOAR00000000
28	*Pantoea vagans* C9 1	NC_014562, NC_014561, NC_014563, NC_014258
29	*Paenibacillus mucilaginosus* 3016	CP003235
30	*Paenibacillus polymyxa* M-1	HE577054, HE577055
31	*Paenibacillus polymyxa* CF05	CP009909
32	*Paenibacillus polymyxa* SC2	CP002213, CP002214.
33	*Pseudomonas aurantiaca* JD37	CP009290
34	*Pseudomonas brassicacearum* NFM421	NC_015379
35	*Pseudomonas fluorescens* A506	NC_021361, NC_017911
36	*Pseudomonas fluorescens* WCS374	CP007638
37	*Pseudomonas fluorescens* F113	NC_016830
38	*Pseudomonas fluorescens* SBW25	NC_012660
39	*Pseudomonas fluorescens* PS006	LRMR00000000
40	*Pseudomonas fluorescens* ET76	LNAB00000000
41	*Pseudomonas protegens* Pf-5	NC_004129
42	*Pseudomonas putida* KT2440	NC_002947
43	*Pseudomonas putida* W619	NC_010501
44	*Pseudomonas putida* WCS358	NZ_JMIT00000000
45	*Pseudomonas putida* MTCC5279	AMZE00000000
46	*Pseudomonas putida* BIRD-1	CP002290
47	*Pseudomonas stutzeri* A1501	NC_009434

Continued

Table 21.1. Continued.

S. No.	Strain name	Accession no.
48	*Pseudomonas simiae* WCS417	CP007637
49	*Pseudomonas trivialis* IHBB745	CP011507
50	*Pseudomonas sp.* FeS53a	JYFT00000000
51	*Rhizobium* sp. NT26	FO082820, FO082821, FO082822
52	*Serratia proteamaculans* 568	NC_009832, NC_009829
53	*Serratia fonticola* AU-P3(3)	ASZB00000000

plant roots as its habitat (Chaudhry *et al.*, 2013). Genomic studies of PGPR revealed important genes for plant-beneficial traits such as genes for the production of indole-3-acetic acid (IAA), ACC deaminase, chitinase, mineral phosphate solubilization, siderophers synthesis, acetoin and 2,3-butanediol, suppression of pathogenic fungi, resistance to oxidative stress and ability to break down toxic compounds (Liu *et al.*, 2016). These traits explain the role of PGPR in nutrient cycling as well as their ability to colonize plant roots. Therefore, genome mining and identification of genes that have potential benefits to hosts, such as growth promotion, nutrition and biocontrol have also been used for modulating plant health by selecting beneficial bacterial populations based on genome sequence information. Further investigations examined whether host genetic factors also control the establishment of symbiont including root-nodules and endophyte; therefore host plant genome studies are also important in studies of the PGPR associated with the hosts (Zgadzaj *et al.*, 2015).

21.7 Comparative Genomic Analyses

The NGS revolution made a large number of PGPR genome sequences data available. Genome analysis has transitioned from single to multiple genomes and the field of comparative genomics can now contribute to advancing knowledge of PGPR, their adaptation to plant roots, effect on plant growth and development. The genetic repertoire in the PGPR genome reflects the root colonization lifestyle of these rhizobacteria. Likewise, colonization of a single bacterial species in diverse habitats from soil and water to animal or plant hosts, correlates with the ability to utilize different nutrient sources and a high potential for adaptation to changing environmental conditions. In order to gain insights into the genetic determinants specifically present in PGPR genomes, comparative genome analysis of multiple niche bacteria of same species will give clues to the genes important for root colonization. Such studies require pan-genome analysis. A pan-genome includes the full complement of genes in a species, which consists of the "core genome" containing genes present in all strains, a "dispensable or accessory genome" containing genes present in two or more strains, and "unique genes" which are specific to a single strain (Medini *et al.*, 2005). Therefore, it provides a framework for estimating the genomic diversity of bacterial strains.

Despite extensive literature on PGPR modes of action, the molecular features that define a PGPR remain difficult to understand. PGPR can occupy different microhabitats ranging from saprophytic soil bacteria that colonize the rhizosphere to bacteria that can also colonize internal root tissues as endophyte. Moreover, several bacteria inhabit plant roots as alternate ecological niches and at that time function as PGPR. Therefore, it is essential to first distinguish between true PGPR and non-PGPR strains from diverse ecological niches, using comparative genomics studies with pan-genome analysis. So as of now, sequencing of the PGPR genome is a fundamental step for developing a potential PGPR strain to serve as an efficient biological control agent and plant growth promoter.

21.8 Conclusion and Future Prospects

The availability of complete and draft genomic sequences of PGPR have allowed research on these rhizobacteria to move at a faster pace. Undoubtedly, the NGS technologies have deepened our understanding of PGPR communities through provision of high-throughput sequencing at low cost; however, the functioning of the plant microbiome still remains unclear. Thus, obtaining information on microbial communities by using different multi-omics and meta-omics technologies should be the goal of future research. Such multiple approaches will further help in gaining in-depth knowledge on differentially expressed genes of PGPR-mediated plant processes under different environmental conditions. This advance knowledge would be helpful in modulating plant microbiomes to reduce plant diseases and enhance crop productivity.

References

Adam, E., Groenenboom, A.E., Kurm, V., Rajewska, M., Schmidt, R. *et al.* (2016) Controlling the microbiome: microhabitat adjustments for successful biocontrol strategies in soil and human gut. *Frontiers in Microbiology* 7, 1079.

Ahemad, M. and Kibret, M. (2014) Mechanisms and applications of plant growth promoting rhizobacteria: current perspective. *Journal of King Saud University - Science* 26, 1–20.

Almario, J., Gobbin, D., Défago, G., Moënne-Loccoz, Y. and Rezzonico, F. (2014) Prevalence of type III secretion system in effective biocontrol pseudomonads. *Research in Microbiology* 165, 300–304.

Andreote, F.D., Gumiere, T. and Durrer, A. (2014) Exploring interactions of plant microbiomes. *Scientia Agricola* 71, 528–539.

Bentley, D.R. (2006) Whole genome re-sequencing. *Current Opinion in Genetics & Development* 16, 545–552.

Berg, G., Grube, M., Schloter, M. and Smalla, K. (2014) The plant microbiome and its importance for plant and human health. *Frontiers in Microbiology* 5, 491.

Bhattacharyya, P.N. and Jha, D.K. (2012) Plant growth-promoting rhizobacteria (PGPR): emergence in agriculture. *World Journal of Microbiology and Biotechnology* 28, 1327–1350.

Bisen, K., Keswani, C., Patel, J.S., Sarma, B.K. and Singh, H.B. (2016) *Trichoderma* spp.: efficient inducers of systemic resistance in plants. In: Chaudhary, D.K. and Verma, A. (eds) *Microbial-mediated Induced Systemic Resistance in Plants*. Springer, Singapore, pp. 185–195.

Blaha, D., Prigent-Combaret, C., Mirza, M.S. and Moënne-Loccoz, Y. (2006) Phylogeny of the 1-aminocyclopropane-1-carboxylic acid deaminase-encoding gene acdS in phytobeneficial and pathogenic Proteobacteria and relation with strain biogeography. *FEMS Microbiology Ecology* 56, 455–470.

Bonkowski, M., Villenave, C. and Griffiths, B. (2009) Rhizosphere fauna: the functional and structural diversity of intimate interactions of soil fauna with plant roots. *Plant and Soil* 321, 213–233.

Buee, M., De Boer, W., Martin, F., van Overbeek, L. and Jurkevitch, E. (2009) The rhizosphere zoo: an overview of plant-associated communities of microorganisms, including phages, bacteria, archaea, and fungi, and of some of their structuring factors. *Plant and Soil* 321, 189–212.

Bulgarelli, D., Schlaeppi, K., Spaepen, S., Ver Loren van Themaat, E. and Schulze-Lefert, P. (2013) Structure and functions of the bacterial microbiota of plants. *Annual Review of Plant Biology* 64, 807–838.

Chaudhry, V., Asif, M.H., Bag, S.K., Goel, R., Mantri, S.S. *et al.* (2013) Draft genome sequence of *Pseudomonas putida* strain MTCC5279. *Genome Announcement* 2013, 1.

DeSantis, T.Z., Hugenholtz, P., Larsen, N., Rojas, M., Brodie, E.L. *et al.* (2006) Greengenes, a chimera-checked 16S rRNA gene database and workbench compatible with ARB. *Applied and Environmental Microbiology* 72, 5069–5072.

Dimkpa, C., Weinand, T. and Asch, F. (2009) Plant–rhizobacteria interactions alleviate abiotic stress conditions. *Plant, Cell & Environment* 32(12), 1682–1694.

Drogue, B., Doré, H., Borland, S., Wisniewski-Dyé, F. and Prigent-Combaret, C. (2012) Which specificity in cooperation between phytostimulating rhizobacteria and plants? *Research in Microbiology* 163, 500–510.

Fleischmann, R.D., Adams, M.D., White, O., Clayton, R.A., Kirkness, E.F. *et al.* (1995) Whole-genome random sequencing and assembly of *Haemophilus influenzae* Rd. *Science* 269, 496–512.

Fox, G.E., Peckman, K.J. and Woese, C.E. (1977) Comparative cataloging of 16S ribosomal ribonucleic acid: molecular approach to prokaryotic systematics. *International Journal of Systematic and Evolutionary Microbiology* 27, 44–57.

Glick, B.R. (2005) Modulation of plant ethylene levels by the bacterial enzyme ACC deaminase. *FEMS Microbiology Letters* 251, 1–7.

Gupta, A., Gopal, M., Thomas, G.V., Manikandan, V., Gajewski, J. *et al.* (2014) Whole genome sequencing and analysis of plant growth promoting bacteria isolated from the rhizosphere of plantation crops coconut, cocoa and arecanut. *PLoS One* 9(8), e104259.

Haas, D. and Défago, G. (2005) Biological control of soil-borne pathogens by fluorescent pseudomonads. *Nature Reviews Microbiology* 3, 307–319.

Hartmann, A., Schmid, M., Van Tuinen, D. and Berg, G. (2009) Plant-driven selection of microbes. *Plant and Soil* 321, 235–257.

Hiltner, L. (1904) Uber neuere Erfahrungen und Probleme auf dem Gebiete der Bodenbakteriologie unter besonderden berucksichtigung und Brache. *Arbeiten der Deutschen Landwirtschafts-Gesellschaft* 98, 59–78.

Hinsinger, P. and Marschner, P. (2006) Rhizosphere – perspectives and challenges – a tribute to Lorenz Hiltner. *Plant and Soil* 283, vii–viii.

Hinsinger, P., Bengough, A.G., Vetterlein, D. and Young, I.M. (2009) Rhizosphere: biophysics, biogeochemistry and ecological relevance. *Plant and Soil* 321, 117–152.

Holden, N., Pritchard, L. and Toth, I. (2009) Colonization outwith the colon: plants as an alternative environmental reservoir for human pathogenic enterobacteria. *FEMS Microbiology Reviews* 33, 689–703.

Jones, D. and Hinsinger, P. (2008) The rhizosphere: complex by design. *Plant and Soil* 312, 1–6.

Kampfer, P. (2012) Systematics of prokaryotes: the state of the art. *Antonie Van Leeuwenhoek* 101, 3–11.

Keswani, C., Bisen, K., Singh, S.P., Sarma, B.K. and Singh, H.B. (2016) A proteomic approach to understand the tripartite interactions between plant–*Trichoderma*–pathogen: investigating the potential for efficient biological control. In: Hakeem, K.R. and Akhtar, M.S. (eds) *Plant, Soil and Microbes Vol. 2. Mechanisms and Molecular Interactions*. Springer, Cham, Switzerland, pp. 79–93.

Kim, P.J. and Price, N.D. (2011) Genetic co-occurrence network across sequenced microbes. *PLOS Computational Biology* 7, e1002340.

Kloepper, J.W. (1994) Plant growth-promoting rhizobacteria (other systems). In: Okon, Y. (eds) *Azospirillum/ Plant Associations*. CRC Press, Boca Raton, FL, USA, pp. 111–118.

Kloepper, J.W. and Schroth, M.N. (1978) Plant growth promoting rhizobacteria on radishes. In: *Proceeding of the 4th International Conference on Plant Pathogenic Bacteria*. Vol. 2, Station de Pathologie Vegetale et Phytobacteriologie, INRA, Angers, France, pp. 879–882.

Kloepper, J.W., Zablotowick, R.M., Tipping, E.M. and Lifshitz, R. (1991) Plant growth promotion mediated by bacterial rhizosphere colonizers. In: Keister, D.L. and Cregan, P.B. (eds) *The Rhizosphere and Plant Growth*. Kluwer, Dordrecht, The Netherlands, pp. 315–326.

Kloepper, J.W., Ryu, C.M. and Zhang, S. (2004) Induced systemic resistance and growth promotion of plant growth by *Bacillus* spp. *Phytopathology* 94, 1259–1266.

Konstantinidis, K.T. and Stackebrandt, E. (2013) Defining taxonomic ranks. In: Rosenberg, E., DeLong, E.F., Lory, S., Stackebrandt, E. and Thompson, F.L. (eds) *The Prokaryotes – Prokaryotic Biology and Symbiotic Associations*, 4th edn. Springer, New York, p. 229.

Liu, L., Li, Y., Li, S., Hu, N., He, Y. *et al.* (2012) Comparison of next-generation sequencing systems. *Journal of Biomedicine and Biotechnology* 2012, 251–364.

Liu, W., Wang, Q., Hou, J., Tu, C., Luo, Y. and Christie, P. (2016) Whole genome analysis of halotolerant and alkalotolerant plant growth-promoting rhizobacterium *Klebsiella* sp. D5A. *Scientific Reports* 6, 26710.

Lugtenberg, B. and Kamilova, F. (2009) Plant-growth-promoting rhizobacteria. *Annual Review of Microbiology* 63, 541–556.

Lundberg, D.S., Lebeis, S.L., Paredes, S.H., Yourstone, S., Gehring, J. *et al.* (2012) Defining the core *Arabidopsis thaliana* root microbiome. *Nature* 488, 86–90.

Lynch, J.M. (ed.) (1990) *The Rhizosphere*. John Wiley, Chichester, UK.

Ma, M., Wang, C., Ding, Y., Li, L., Shen, D. *et al.* (2011) Complete genome sequence of *Paenibacillus polymyxa* SC2, a strain of plant growth-promoting Rhizobacterium with broad-spectrum antimicrobial activity. *Journal of Bacteriology* 193, 311–312.

MacLean, D., Jones, J.D. and Studholme, D.J. (2009) Application of 'next-generation' sequencing technologies to microbial genetics. *Nature Reviews Microbiology* 7, 287–296.

Margulies, M., Egholm, M., Altman, W.E., Attiya, S., Bader, J.S. *et al.* (2005) Genome sequencing in microfabricated high-density picolitre reactors. *Nature* 437, 376–380.
Martiny, A.C., Treseder, K. and Pusch, G. (2013) Phylogenetic conservatism of functional traits in microorganisms. *The ISME Journal* 7, 830–838.
Mathimaran, N., Srivastava, R., Wiemken, A., Sharma, A.K. and Boller, T. (2012) Genome sequences of two plant growth-promoting fluorescent *Pseudomonas* strains, R62 and R81. *Journal of Bacteriology* 194, 3272–3273.
Matilla, M.A., Pizarro-Tobias, P., Roca, A., Fernandez, M., Duque, E. *et al.* (2011) Complete genome of the plant growth-promoting rhizobacterium *Pseudomonas putida* BIRD-1. *Journal of Bacteriology* 193, 1290.
Medini, D., Donati, C., Tettelin, H., Masignani, V. and Rappuoli, R. (2005) The microbial pan-genome. *Current Opinion in Genetics & Development* 15, 589–594.
Mendes, R., Garbeva, P. and Raaijmakers, J.M. (2013) The rhizosphere microbiome: significance of plant beneficial, plant pathogenic, and human pathogenic microorganisms. *FEMS Microbiology Reviews* 37(5), 634–663.
Metting, F.B. Jr (1992) *Soil Microbial Ecology: Applications in Agricultural and Environmental Management.* Marcel Dekker, Inc, USA.
Niedringhaus, T.P., Milanova, D., Kerby, M.B., Snyder, M.P. and Barron, A.E. (2011). Landscape of next-generation sequencing technologies. *Analytical Chemistry* 83, 4327–4341.
Pérez-Montaño, F., Alías-Villegas, C., Bellogín, R.A., Del Cerro, P., Espuny, M.R. *et al.* (2014) Plant growth promotion in cereal and leguminous agricultural important plants: from microorganism capacities to crop production. *Microbiological Research* 169(5), 325–336.
Pierret, A., Doussan, C., Capowiez, Y., Bastardie, F. and Pages, L. (2007) Root functional architecture: a framework for modeling the interplay between roots and soil. *Vadose Zone Journal* 6, 269–281.
Pruesse, E., Quast, C., Knittel, K., Fuchs, B.M., Ludwig, W. *et al.* (2007) SILVA: a comprehensive online resource for quality checked and aligned ribosomal RNA sequence data compatible with ARB. *Nucleic Acids Research* 35, 7188–7196.
Raaijmakers, J.M., Paulitz, T.C., Steinberg, C., Alabouvette, C. and Moenne-Loccoz, Y. (2009) The rhizosphere: a playground and battlefield for soilborne pathogens and beneficial microorganisms. *Plant and Soil* 321, 341–361.
Ramírez-Puebla, S.T., Servín-Garcidueñas, L.E., Jiménez-Marín, B., Bolaños, L.M., Rosenblueth, M. *et al.* (2013) Gut and root microbiota commonalities. *Applied and Environmental Microbiology* 79, 2–9.
Schadt, E.E., Turner, S. and Kasarskis, A. (2010) A window into third-generation sequencing. *Human Molecular Genetics* 19, R227–R240.
Schuster, S.C. (2008) Next-generation sequencing transforms today's biology. *Nature Methods* 5, 16–18.
Somers, E., Vanderleyden, J. and Srinivasan, M. (2004) Rhizosphere bacterial signalling: a love parade beneath our feet. *Critical Reviews in Microbiology* 30, 205–240.
Song, J.Y., Kim, H.A., Kim, J.S., Kim, S.Y., Jeong, H. *et al.* (2012) Genome sequence of the plant growth-promoting Rhizobacterium *Bacillus* sp. strain JS. *Journal of Bacteriology* 194, 3760–3761.
Thompson, C.C., Vieira, N.M., Vicente, A. and Thompson, F. (2011) Towards a genome based taxonomy of Mycoplasmas. *Infection, Genetics and Evolution* 11, 1798–1804.
Thompson, C.C., Emmel, V.E., Fonseca, E.L., Marin, M.A., Vicente, A.C.P. (2013) Streptococcal taxonomy based on genome sequence analyses. *F1000Research* 67, 1–9.
Tindall, B.J., Rossello-Mora, R., Busse, H.J., Ludwig, W. and Kampfer, P. (2010) Notes on the characterization of prokaryote strains for taxonomic purposes. *International Journal of Systematic and Evolutionary Microbiology* 60, 249–266.
Turner, T.R., James, E.K. and Poole, P.S. (2013) The plant microbiome. *Genome Biology* 14(6), 1.
Tyler, H.L. and Triplett, E.W. (2008) Plants as a habitat for beneficial and/or human pathogenic bacteria. *Annual Review of Phytopathology* 46, 53–73.
Vacheron, J., Desbrosses, G., Bouffaud, M.L., Touraine, B., Moënne-Loccoz, Y. *et al.* (2013) Plant growth-promoting rhizobacteria and root system functioning. *Frontiers in Plant Science* 4, 356.
van Baarlen, P., van Belkum, A., Summerbell, R.C., Crous, P.W. and Thomma, B. (2007) Molecular mechanisms of pathogenicity: how do pathogenic microorganisms develop cross-kingdom host jumps? *FEMS Microbiology Reviews* 31, 239–277.
van der Heijden, M.G.A., Bardgett, R.D., van Straalen, N.M. (2008) The unseen majority: soil microbes as drivers of plant diversity and productivity in terrestrial ecosystems. *Ecology Letters* 11(3), 296–310.

Viveros, O.M., Jorquera, M.A., Crowley, D.E., Gajardo, G. and Mora, M.L. (2010) Mechanisms and practical considerations involved in plant growth promotion by rhizobacteria. *Journal of Soil Science and Plant Nutrition* 10, 293–319.

Zamioudis, C. and Pieterse, C.M.J. (2012) Modulation of host immunity by beneficial microbes. *Molecular Plant-Microbe Interactions Journal* 25, 139–150.

Zgadzaj, R., James, E.K., Kelly, S., Kawaharada, Y., de Jonge, N. *et al.* (2015) A legume genetic framework controls infection of nodules by symbiotic and endophytic bacteria. *PLOS Genetics* 11(6), e1005280.

22 Plant Growth-Promoting Rhizobacteria (PGPR): Mechanism, Role in Crop Improvement and Sustainable Agriculture

Pallavi Mittal,[1] Madhu Kamle,[2] Shubhangini Sharma,[1] Pooja Choudhary,[1] Devendra Pratap Rao[3] and Pradeep Kumar[2]*

[1]*Department of Biotechnology, ITS Paramedical College, Ghaziabad, India;* [2]*Department of Forestry, North Eastern Regional Institute of Science and Technology (Deemed University), Nirjuli (Itanagar)-791109, Arunachal Pradesh, India;* [3]*Department of Chemistry, D. A.V. College, Uttar Pradesh, Kanpur 208001, India*

22.1 Introduction

In today's world sustainable agriculture is critically important in accomplishing the demand for food for a fast-growing population. Traditional and conventional agricultural techniques are not sufficient to meet future agricultural needs. However, modern approaches also contain most of the chemical pollutants which, via extensive use of synthetic chemicals, fertilisers and pesticides, can cause damage to environments and human health. Agriculture is still facing the destructive activities of several pests and pathogens from early times, leading to losses of crops and their aesthetic values (Bhattacharjee and Dey, 2014). Extensive economic losses are reported every year due to the occurrence of diseases in plants. At the same time there are investigations into microbial diversity and the cultivation of useful micro-organisms from almost all possible habitats on earth to control diseases, environmental functions and various biotechnological applications (Singh, 2015).

Bacteria associated with plants can be characterized into beneficial, toxic and neutral groups on the basis of their effects on plant growth (Dobbelaere *et al.*, 2003). The valuable free-living soil bacteria are usually known as plant growth-promoting rhizobacteria (Kloepper *et al.*, 1989); the term "rhizobacteria" was originally described by Kloepper and Schroth (1978). These soil bacterial populations competitively colonise plant roots and enhance plant growth by reducing the movements of soilborne pathogen populations. There are self-governing mechanisms of vegetal growth promotion, PGPRs colonizing the rhizosphere, the rhizoplane (root surface) or the root itself (within radicular tissues) (Gray and Smith, 2005).

Many substances produced by rhizobacteria play dynamic roles as biocontrol agents and indirectly raise plant growth. In the current scenario for sustainable agriculture biological approaches for the improvement of crop production are gaining strong interest among agriculturists and ecologists following the integrated plant

*E-mail: pkbiotech@gmail.com

© CAB International 2017. *Advances in PGPR Research* (eds H.B. Singh, B.K. Sarma and C. Keswani)

nutrient managements system. Consequently, there is ongoing specific research worldwide with ever greater extent to explore a vast range of rhizobacteria having unique characteristics such as heavy-metal detoxification potential, pesticide tolerance/degradation, biological control of pathogens and insects, along with plant growth-promoting properties. These plant growth promoters include substances such as phytohormones, siderophores, 1-aminocyclopropane-1-carboxylate, hydrogen cyanate (HCN), and ammonia production, nitrogenase activity, phosphate solubilization, etc. In plants, induced systemic resistance (ISR) resembles pathogen-induced systemic acquired resistance (SAR) under conditions where the induced bacteria and the exciting pathogen remain spatially separated (Bisen *et al.*, 2015, 2016; Keswani *et al.*, 2016a, b); about 1–2% bacteria support plant growth in the rhizosphere (Antoun and Kloepper, 2001). Different diverse genera of bacteria have been recognised as PGPRs, some symbiotic (*Rhizobium, Bradyrhizobium, Mesorhizobium*), others non-symbiotic (*Pseudomonas, Bacillus, Klebsiella, Azotobacter, Azospirillum, Azomonas*). These rhizobacteria are now being used worldwide as bio-inoculants to promote plant growth and development under various biotic and abiotic stresses.

22.2 History

Grossbard (1948–1952), Wright (1952–1957), and other well known researchers proved that antibiotics were produced in soil by *Pencillium, Aspergillus, Trichoderma* and *Streptomyces* ssp. Norman Borlaug's green revolution in the 1960s described the concepts of the rhizosphere in plant research and its future. The importance of siderophores formed by *Erwinia carotovora* was determined by Kloepper *et al.* (1980). PGPR such as *Azospirillum* have been shown to generate various types of plant growth-promoting substances and nitrogen fixation which increases plant productivity (Dobbelaere *et al.*, 2003).

22.3 Plant Growth-Promoting Rhizobacteria (PGPR)

Plant diseases negatively affect the capacity to maintain the quality and abundance of food, feed and fibre produced by growers around the world. Diverse mitigation approaches and effective management strategies may be employed to control plant diseases. Beyond good agronomic and horticultural practices, growers frequently depend on application of chemical fertilizers and pesticides which contributed considerably to the impressive improvements in crop productivity and quality over the past 100 years (Junaid *et al.*, 2013). Many abiotic and biotic factors influence plant growth in agriculture. Plant growth-promoting rhizobacteria (PGPR) are a diverse group of free-living bacteria that reside in soil and can initiate in the rhizosphere, at root surfaces or in association with roots, and which can develop the coverage or quality of plant growth directly or indirectly.

Rhizobacteria that can improve plant growth and crop production by specific mechanisms like siderophores and phytohormone production, inorganic phosphate solubilization and anti-fungal activities, are often referred to as plant growth-promoting rhizobacteria (Singh, 2015). They constitute a group of valuable plant bacteria, which are potentially useful for improved plant growth and increased crop yields (Saharan and Nehra, 2011).

PGPR are characterized by the following unique properties:

1. PGPR should be proficient to colonize the root surface;
2. they should survive, multiply and compete with other microbiota, to express their improved plant growth and biocontrol activity; and
3. they should possess efficient plant growth promoting ability.

PGPR provide an effective alternative to the needless use of chemical fertilizers, pesticides and other supplements regularly used in agriculture. The variety and characterization of PGPRs in the rhizosphere might be a dependable tool for sustainable agriculture and livelihoods (Singh, 2015). PGPR directly support plant growth by producing phytohormones, solubilizing phosphate or fixing nitrogen, and also improve plant growth during pathogen attack (Rayazsayyed *et al.*, 2014). The rhizosphere as a soil partition is influenced by plant growth (Hiltner, 1904) and the outcome from the release from the plant of organic materials, a phenomenon known as rhizodeposition, consists mostly of plant metabolites (the exudates) and plant debris (dead cells), loss of mucilage, etc. (Dessaux *et al.*, 2016). Numerous PGPR have been recognized since the role of the rhizosphere as an ecosystem has expanded to a significant role in the functioning of the biosphere. A variety of species of bacteria like *Pseudomonas, Azospirillum, Azotobacter, Klebsiella, Enterobacter, Alcaligenes, Arthrobacter, Burkholderia, Bacillus* and *Serratia* have been reported to enhance plant growth (Saharan and Nehra, 2011).

22.4 Role of PGPR in Soil

Soil is the natural habitat of several microorganisms of which some are beneficial and some are harmful for the community of plants. The activity of the microbes plays a vital role for quality of soil and thus affects plants' efficiency. The PGPR are beneficial as they promote plant growth by proliferation of root hairs, deformation in root hairs and branching, increase in seedling rise, early nodule formation and functioning, enhanced leaf surface area, vigour, biomass, phytohormone, nutrient, water and air uptake; they also promote accumulation of carbohydrates, and yield in various plant species (Podile and Kishore, 2006). The use of PGPR to enhance plant growth is observed and also the mixing of one or more soils and intercropping with legumes in the cultivated crops proved beneficial towards strength and fertility of soil (Tisdale and Nelson, 1975; Chew, 2002). In soil amendment procedures where the use of biochar (a fine-grained, highly porous charcoal) permitted the organization of fertile soils in tropical areas despite atmospheric nitrogen being available in inert form that cannot be utilized by the plants. Soil bacteria, especially the leguminous crops, have the capability to convert this inert nitrogen to a usable form for plants (Hellriegel and Wilfarth, 1888).

The main impact of the root microbes on plant health is shown most clearly in disease-repressive soils. Most soilborne pathogens are saprophytic in nature and grow in the rhizosphere to reach their host before they can contaminate the host tissues and effectively escape the rhizosphere battle zone (Berendsen *et al.*, 2012). Root exudates include different substances, mostly categorized into amino acids, organic acids, fatty acids, sterols, proteins, soluble sugars, and also perform diverse functions in the rhizosphere and are classified according to their beneficial effects.

Plants do not secrete one substance, but combinations of molecules. This combination of exudates depends on external factors, such as plant size, photosynthetic activity and soil conditions, but also species or even genotype-specific type (Mommer *et al.*, 2016). Microbes are a budding source for the biotechnology industry which offers countless new genes and biochemical pathways to probe for enzymes, antibiotics, and other useful molecules. The application of PGPR holds great potential for development and establishment of sustainable agriculture.

22.5 PGPR and their Interaction with Plants

Diverse ranges of resources are provided to the soil organisms by roots and their exudates, and the rhizosphere is an extremely miscellaneous habitat. The rhizosphere contains root herbivores such as insect larvae and nematodes, their natural enemies and a broad range of microbes including symbiotic, pathogenic, and saprophytic fungi and protozoa. The richness of these soil bacteria depends on abiotic conditions such as soil pH, temperature and moisture content (Bardgett, 2005). Thus, it is extremely important to isolate, identify and confirm the capacity of plant growth promotion (PGP). The activity of PGP can be performed by employing serial dilution and plating techniques on various artificial nutrient media, and biochemical tests for screening of plant growth-promoting traits can be divided into direct and indirect tests. Direct biochemical tests are used for estimation of nitrogen fixation, phosphate solubilization, siderophore production and phytohormones (IAA, ethylene, gibberellin and cytokinins). Indirect biochemical testing can be used to analyze the potential and capability of the PGPR to suppress soilborne pathogens and in antibiosis in the production of secondary metabolites, induced resistance, parasitism, predation and competition for nutrients. The present assessment is a preliminary effort to address diversity and the latest biochemical and molecular methods which are used for the isolation and characterization of PGPRs from rhizosphere soil. The most recent developments in PGPRs reported from various agro-climatic zones will be strengthened by employing biochemical and molecular tools for the characterization.

Rhizobacteria are the dominant soil microbes, critically important for maintenance of soil fertility and recycling of soil nutrients (Glick, 2012). On the basis of the plant–microbe interaction type the PGPRs are separated into symbiotic bacteria (living inside the plants and exchanging metabolites with plants directly) and rhizobacteria (free-living bacteria existing outside the plants).

22.6 Mechanism of PGPR

PGPR influence plant growth and yield in several ways. The improvement in reproductive and vegetative growth is recognised in various crops. Treatments with PGPR increased germination percentage, seedling vigour, development, plant stand, root and shoot growth, total biomass of the plants, seed weight, early flowering, grains filling, fodder and fruit yields, etc. (van Loon *et al.*, 1998; Ramamoorthy *et al.*, 2001).

Various mechanisms employed by the PGPR as biocontrol agents in resisting plant diseases are broadly classified into:

1. direct antagonisms.
2. indirect antagonisms.

Plant growth promotion can be achieved by both direct and indirect interaction of host and microbes: direct interaction between the beneficial microbes and their host plant, and indirectly due to the microbes' antagonistic activity against plant pathogens. However, the precise mechanisms of plant growth promotion of PGPRs are not all clearly revealed. These include increased nitrogen fixation, production of auxins, gibberellins, cytokinins, ethylene, solubilization of phosphorus, oxidation of sulfur, increased nitrate availability, extracellular production of antibiotics, lytic enzymes, hydrocyanic acid, increase in root permeability, strict competition for the available nutrients and root sites, suppression of deleterious rhizobacteria and enhancement in the uptake of essential plant nutrients, etc. (Pal *et al.*, 1999; Enebak and Carey, 2000; Mishra *et al.*, 2015).

Plant diseases are based on connections between the components of the disease triangle, i.e. host, pathogen and environment. Biocontrol agents are organisms that cooperate with the mechanisms of the disease triangle to manage the disease. Induced Systemic Resistance (ISR) of the plants to pathogens is a widespread phenomenon relating to the fundamental signalling pathways and their possible use in plant protection. Plants respond in salicylic-acid-dependent signalling flip-flop which exploits the broad range of host disease resistance effective for numerous pathogens. Salicylic acid (SA) plays a significant role in the signalling pathway leading to induced systemic response: SA is synthesized in response to infection caused both locally and systemically; *de novo* production of SA in non-infected plant parts contributes to systemic expression of ISR (Saharan and Nehra, 2011). Mechanisms of some biocontrol agents are now available in the literature (Zhang *et al.*, 2002) and knowing the mechanism of action of a biocontrol agent may reinforce reliability either by improving the mechanism or by using the biocontrol agents under the right conditions.

22.6.1 Direct antagonism

The major contribution involved in supporting plant growth is by the direct mechanisms which influence growth in a direct and straightforward manner. These mechanisms influence the plant growth activity directly, but the ways in which influence is exerted will vary from species to species, as well as strain to strain. Direct antagonism results from physical contact and/or a high degree of selectivity for the pathogens by biological control agents.

Hyperparasitism

Direct parasitism or lysis of a fatal pathogen by other microorganisms is described as hyperparasitism (Bhattachrjee and Dey, 2014). It is the most effective and direct form of antagonism (Pal *et al.*, 2006). It mainly involves the tropic growth of biocontrol agents towards the target organism, coiling, final attack and dissolution of the target pathogen cell wall or membrane by the activity of enzymes (Tewari, 1996). It is one of the main mechanisms involved in *Trichoderma* (Sharma, 1996), as with *Trichoderma harzianum* which exhibits mycoparasitic activity against *Rhizoctonia solani* hyphae (Altomare *et al.*, 1999), mycoparasitism being under the control of enzymes. Harman (2000) described the contribution of chitinase and β-1,3 glucanase in the *Trichoderma* mediated by biological control. Since enzymes are the products of genes, insignificant changes in the structure of a gene can lead to the production of different enzyme (Keswani *et al.*, 2013, 2014). The lesser ability of a *Trichoderma* strain to produce endochitinase led to a reduced ability to control *Botryyis cineria*, however it was better able to rapidly control *Rhizoctonia solani* (Gupta *et al.*, 1995). Single pathogens can be attacked by multiple hypoparasites, e.g. *Acremonium altenatum, Acrodontium crateriforme, Ampelomyces quisqualis* and *Gliocladium virens* are a few of the fungi that have the capability to parasitize powdery mildew pathogens (Kiss, 2003).

Nitrogen fixation

Nitrogen (N) is an essential element for all forms of life and the main imperative nutrient for plant growth, development and productivity: it is a constituent of nucleotides, membrane lipids and amino acids (enzymatic and structural proteins). While nitrogen is available as 78% of the atmosphere, its residue is unavailable to the plants. It is thus the main limiting nutrient for plant growth, as no plant species is proficient at fixing atmospheric nitrogen into ammonia and applying it directly for its growth. The microorganisms that play a main role in nitrogen fixation are:

- symbiotic nitrogen-fixing bacteria, i.e. those that can form symbiosis with leguminous plants (Ahemad and Khan, 2012); and
- non-symbiotic fixers, which are free-living and endophytic microbes that provide only a small amount of fixed nitrogen needed by the plants (Bhattacharyya and Jha, 2012; Glick, 2012).

The genes that are dependable for nitrogen fixation are recognized as *nif* genes which are found in both symbiotic as well as in non-symbiotic organisms (Kim and Rees, 1994).

Nitrogen-fixing bacteria include both free rhizospheric prokaryotes (e.g. *Achromobacter, Acetobacter, Alcaligenes, Arthrobacter, Azospirillum, Azotobacter, Azomonas, Bacillus, Beijerinckia, Clostridium, Corynebacterium, Derxia, Enterobacter, Herbaspirillum, Klebsiella, Pseudomonas, Rhodospirillum, Rhodopseudomonas* and *Xanthobacter* (Tilak *et al.*, 2005) and symbiotic rhizospheric prokaryotes that fix nitrogen only in association with certain plants. This latter group comprises rhizobia (*Rhizobium, Bradyrhizobium, Sinorhizobium, Azorhizobium, Mesorhizobium* and *Allorhizobium*) associated with leguminous plants and *Frankia* strains, filamentous sporulating bacteria associated with actinorhizal plants (Gray and Smith, 2005).

In most ecosystems and through this process, the fixing microorganisms participate in the accumulation of nitrogen compounds over time (Vitousek *et al.*, 2002). This process is then sufficient to maintain the stock of nitrogen compounds in the ecosystem and to restore the losses.

Phosphate solubilization

Phosphorus (P) is the most significant source of plant nutrition after nitrogen. It plays an important role in approximately all major metabolic mechanisms in plants, including photosynthesis, energy transfer, macromolecular biosynthesis, signal transduction and respiration (Reed *et al.*, 2011). It is widely available in soils in both organic and inorganic forms, but plants are incapable of using most of it because 95–99% of phosphate is present in the insoluble, immobilized and precipitated form (Khan *et al.*, 2010). Plants absorb phosphate in just two soluble forms: the monobasic (H_2PO^4) and the dibasic (HPO_4^{2-}) ions (Ahemad and Kibret, 2014). Plant growth-promoting rhizobacteria present in the soil utilize diverse strategies to make use of available forms of phosphorus and make it obtainable for plants to absorb. Phosphate-solubilizing bacteria are beneficial bacteria capable of hydrolyzing insoluble inorganic phosphorus into soluble organic phosphorus which can be absorbed as a nutrient by the plants. Phosphate-solubilizing PGPR integrated in the genera *Arthrobacter, Bacillus, Beijerinckia, Burkholderia, Erwinia, Enterobacter, Flavobacterium, Microbacterium, Pseudomonas, Rhodococcus, Rhizobium* and *Serratia* have attracted the attention of agriculturists as soil inoculums to recover plant growth and yield (Ahemad and Kibret, 2014).

Potassium solubilization

Potassium (K) is an essential macronutrient required for plant growth and development. The concentrations of soluble potassium in the soil are typically very low and further, 90% of potassium in the soil exists in the form of insoluble silicate minerals and rocks (Zaidi *et al.*, 2009). Additionally, due to excessive fertilizer application, potassium shortage is becoming one of the most major constraints to crop production. Lacking satisfactory potassium, the plants will have unsuccessfully developed roots, grow slowly, produce small seeds and have minor yields.

Phytohormone production

This is plant growth stimulation by PGPR also referred to as phytostimulators or plant growth regulators. The phytohormones are found in minute quantities but significantly influence the biochemical, physiological and morphological processes in plants, and their synthesis is smoothly regulated. Phytohormones which are synthesized exogenously by natural and synthetic means are known as plant growth regulators. Some of the microbes have the ability to produce or alter the concentration of growth regulators such as IAA, GA, cytokinins and ethylene.

A broad range of microorganisms growing in the rhizosphere are proficient to produce substances that adjust plant growth and development. Plant growth-promoting rhizobacteria generate phytohormones such as auxins, cytokinins, gibberellins and ethylene which can influence cell propagation in the root construction by overproduction of lateral roots and root hairs, with a consequent increase in nutrients and water uptake (Ma, 2005).

22.6.2 Indirect mechanisms

Phytopathogenic microorganisms are the foremost constraint on sustainable agriculture and ecosystems. These microbes' presence subverts the soil ecology, degrades soil fertility, disrupts the environment, and hence shows some harmful effects on human health, along with pollution to groundwater. Plant growth-promoting rhizobacteria are capable of a sustainable and ecofriendly approach to achieve productiveness of the soil and, indirectly, plant growth. Thus this approach became motivation for the extensive range of utilization of plant growth-promoting rhizobacteria. This would further lead to falling need for agrochemicals (fertilizers and pesticides) for developing soil fertility by a variety of mechanisms, via production of antibiotics, siderophores, HCN, hydrolytic enzymes, etc. (Iqbal *et al.*, 2012; Tariq *et al.*, 2014).

Antibiosis

Antibiosis is a type of biological interaction between two or more organisms involving metabolic substances used by one against the other or mediated by specific or non-specific metabolism of a microorganism by lytic enzymes (enhanced growth) or by other toxic components (Battacharjee and Dey, 2014). The production of one or more antibiotics is the mechanism most commonly associated with the ability of plant growth-promoting bacteria to act as antagonistic agents against phytopathogens (Glick *et al.*, 2007). The process of producing antibiotics is one of the most powerful biocontrol mechanisms of plant growth-promoting rhizobacteria against phytopathogens and has become increasingly better understood over the past two decades (Shilev, 2013).

A diverse range of antibiotics is recognized which includes compounds like amphisin, 2,4-diacetylphloroglucinol (DAPG), oomycin A, phenazine, pyrrolnitrin, pyoluteorin, tropolone, tensin, and cyclic lipopeptides formed by pseudomonads, and kanosamine, zwittermicin A, and xanthobaccin produced by *Bacillus*, *Stenotrophomonas* sp., and *Streptomyces*. In aiming to control the, mainly fungal, plant pathogens (Loper and Gross, 2007), the basis of antibiosis and the biocontrol activity that kills or reduces the growth of the unwanted pathogen have become well understood over the past two decades (Dowling and O'Gara, 1994; Whipps, 2001; Lugtenberg and Kamilova, 2009).

Siderophore production

Siderophores are high-affinity iron-chelating compounds secreted by fungi, bacteria and grasses (Saharan and Nehra, 2011). Iron plays a lead role in the energy metabolism of aerobic and semi-anaerobic microorganisms. Siderophores are secreted to solubilize iron from their surrounding environments, forming a complex ferric-siderophore that can move by diffusion and be returned to the cell surface (Andrews *et al.*, 2003). The active transport system through the membrane begins with the recognition of the ferric-siderophore by specific membrane receptors of Gram-negative and Gram-positive bacteria (Boukhalfa and Crumbliss, 2002).

Its accessibility in soil for microorganisms and plants drops significantly with increases in pH above 6. Microorganisms (some actinomycetes, bacteria and fungi) then contend for iron by releasing siderophores which, having of small molecular weight (M_r = 500-1000) possessing a high affinity for ferric iron ($k_f > 1030$) and biosynthesis of siderophores is induced by low levels of iron (Singh, 2015).

Siderophore production activity in soil played a vital role in determination of different microorganisms for enhanced plant development. Iron uptake in plants is enhanced by microbial siderophores by recognizing the bacterial-siderophore complex.

Burkholderia phytofirmans

Burkholderia phytofirmans is a novel plant-linked bacterium. It belongs to the beta-proteobacteria isolated from *Glomus vesciculiferum*-infected onion roots (Nowak *et al.*, 1997). Other strains which belong to this species have been isolated from soils and rhizosphere. The morphology of these bacteria is Gram-negative, they are non-sporulating rods that show growth at various types of sugars. This bacteria strain does not produce nitrite or nitrate. PsJN (*Burkholderia* spp. Strain) produces ACC deaminase activity, perhaps contributing to the plant growth-promoting abilities of the strain (Ait *et al.*, 2000).

Burkholderia phytofirmans acts as a role model for understanding the mechanisms of plant interaction with bacteria under salt stress conditions. It enhances the plant growth and development. There are many different PGPR inoculants which have recently been commercialized that appear to promote growth through at least one mechanism; suppression of plant disease mainly known as bioprotectants and enhanced nutrient acquisition such as biofertilizers, or phytohormone production (biostimulants). Inoculant improvement has been most successful in carrying biological control agents of plant disease, i.e. organisms that are capable of killing other organisms that are pathogenic or disease-causing to crops. This bacterium has mainly been studied and increasingly marketed, as the biological control agents include the genera *Burkholderia*. They suppress plant disease through at least one mechanism; induction of systemic resistance (ISR), and production of siderophores or antibiotics.

Role of Burkholderia phytofirmans as PGPR

This bacterial species belonging to the genus *Burkholderia* is associated with the plant rhizosphere and able to exert a valued effect on plant growth. It induces positive effects in plants by helping with increased growth, development and reduced stress susceptibility. The various plant development points of PsJN strains can be found in the rhizosphere and also in colonizing their internal tissue. The PsJN strain increases several growth parameters and accelerates the growth rate of plants (Maria *et al.*, 2013). Plants played a significant role by selecting and enriching the bacterial type by the release of the constituents of root exudates. The bacterial community found in the rhizosphere utilizes the organic constituents of root exudates as source of energy for the rhizosphere development. Various types of bacteria are present in the soil, rhizoplane and plant tissues and have proficient systems for uptake and catabolism of the root exudates. PGPR are generally considered as inoculum sources for plant growth stimulation and furthermore offer attractive and ecofriendly agricultural practices. Use of biofertilizers and bioenhancers including bioagents is a cost-effective approach for increased productivity and yield of agriculture crops. This approach helps in overcoming the unwanted use of chemical pesticides and fertilizers and thus reduces pollution through eco-friendly agriculture. Hence, the application of rhizosphere bacteria for plant growth promotion and development is advantageous for sustainable agriculture.

PGPR or combinations of PGPR could advance the nutrient use efficiency of fertilizers and help in yield and production status of crops like chickpea and rice cultivation. Similarly, a plant growth-promoting consortium comprising two species, *Burkholderia* sp. MSSP and *Sinorhizobium meliloti* PP3 with abilities to produce IAA when employed for *Cajanus cajan*, shows stupendous increase in seedling growth (Battacharjee and Dey, 2014).

Microbe–microbe signalling in the rhizosphere

Burkholderia phytofirmans produces the quorum signalling compound 3-hydroxy-C8-homoserine lactone. Strain PsJN colonizes a variety of plants endophytically and in the rhizosphere plant growth promotion is reported in plants such as potato, grapevine and tomato (Frommel *et al.*, 1991).

The ability of microorganisms to coordinate their gene expression in a population-density-dependent manner is termed as Quorum Sensing (QS). Significantly, various breaking studies on QS were explored by using models of plant–microbe associations such as the formation of biofilms, conjugation, and production of virulence factors, motility and synthesis of secondary metabolites. The signals produced by microbes belongs to an extensive range of chemical classes, and several QS systems using diverse types of signals frequently occur within a single organism. This type of signalling among microbes is likely to play a vital role in significant stabilization of the rhizosphere microbial community as well as distressing plant development (Venturi and Keel, 2016).

22.7 Future Prospective

In present scenarios with rapid expansion of the human population, high-yielding and enhanced production of crops is essential to satisfy the rising demand for food. To accomplish the goal it is necessary to treasure eco-friendly and cost-effective approaches for sustainable crop production that employ PGPR. Interactions between members of different microorganisms often result in the enhancement of significant processes benefiting plant growth and health. PGPR shows positive influence on crop productivity and ecosystem management in terms of biofertilization, biocontrol and bioremediation. Hence the value of exploring rhizoengineering based on favourable partitioning of exotic biomolecules, which create a unique setting for the interaction between plant and microbes. This approach deserves prime importance, utilizing biotechnological interventions to increase information on rhizosphere biology to enable integrated management of soil microbial populations and its implementation in agriculture.

Further understanding of the broad mechanism of PGPR can help in gaining and exploring new PGPR strains that can provide novel genetic constituents and bioactive chemicals for varied application in agriculture and environment sustainability. Advances in plant growth-promoting consortia (PGPC) possibly will be a viable strategy for augmented action of plant growth-promoting rhizobacteria.

22.8 Conclusion

A chronic and major threat for ecosystem constancy and sustainable agriculture worldwide is pathogenic microorganisms. The chemicals previously used to manage this damage and encourage yield were found to be effective against detrimental microorganisms, but simultaneously causing damage to the ecosystem. Thus PGPR is regarded as a promising tool for providing extensive benefits to agriculture. PGPR play an important role for increasing soil fertility, plant growth promotion, and suppression of phytopathogens for development of eco-friendly and sustainable agriculture. PGPR are an admirable approach for exploring novel genetic components and bioactive chemicals that prove both advantageous for plants and are environment-friendly. Application of appropriate PGPR for specific plants and favourable environments might gain outstanding results in the future of sustainable environment and agriculture.

Acknowledgements

The authors (PK and MK) are highly grateful to the Director and Head, Department of Forestry, NERIST (Deemed University), Arunachal Pradesh, India.

References

Ahemad, M. and Khan, M.S. (2012) Effects of pesticides on plant growth promoting traits of *Mesorhizobium* strain MRC4. *Journal of the Saudi Society of Agricultural Sciences* 11, 63–71.

Ahemad, M. and Kibret, M. (2014) Mechanisms and applications of plant growth promoting rhizobacteria: current perspective. *Journal of King Saud University – Science* 26, 1–20.

Andrews, S.C., Robinson, A.K. and Rodríguez-Quiñones, F. (2003) Bacterial iron homeostasis. *FEMS Microbiology Reviews* 27, 215–237.

Antoun, H. and Kloepper, J.W. (2001) Plant growth promoting rhizobacteria (PGPR). In: Brenner, S. and Miller, J.H. (eds) *Encyclopedia of Genetics*. Academic Press, New York, pp. 1477–1480.

Ashrafuzzaman, M., Hossen, F.A., Ismail, M.R., Hoque, A., Islam, M.Z. *et al.* (2009) Efficiency of plant growth-promoting rhizobacteria (PGPR) for the enhancement of rice growth. *African Journal of Biotechnology* 8, 1247–1252.

Badri, D.V., Weir, T.L., Van der Lelie, D. and Vivanco, J.M. (2009) Rhizosphere chemical dialogues: plant–microbe interactions. *Current Opinion in Biotechnology* 20, 642–650.

Bardgett, R.D. (2005) *The Biology of Soil: A Community and Ecosystem Approach*. Oxford University Press, Oxford.

Berendsen, R. L., Pieterse, C.M.J., Bakker, P.A.H.M. (2012) The rhizosphere microbiome and plant health. *Trends in Plant Science*, 17, 478–486

Bhattacharjee, R. and Dey, U. (2014) An overview of fungal and bacterial biopesticides to control plant pathogens/diseases. *African Journal of Microbiology Research* 8, 1749–1762.

Bhattacharyya, P.N. and Jha, D.K. (2012) Plant growth-promoting rhizobacteria (PGPR): emergence in agriculture. *World Journal of Microbiology and Biotechnology* 28, 1327–1350.
Bisen, K., Keswani, C., Mishra, S., Saxena, A., Rakshit, A. and Singh, H.B. (2015) Unrealized potential of seed biopriming for versatile agriculture. In: Rakshit, A., Singh, H.B. and Sen, A. (eds) *Nutrient Use Efficiency: from Basics to Advances*. Springer, New Delhi, India, pp. 193–206.
Bisen, K., Keswani, C., Patel, J.S., Sarma, B.K. and Singh, H.B. (2016) *Trichoderma* spp.: efficient inducers of systemic resistance in plants. In: Chaudhary, D.K. and Verma, A. (eds) *Microbial-mediated Induced Systemic Resistance in Plants*. Springer, Singapore, pp. 185–195.
Boukhalfa, H. and Crumbliss, A.L. (2002) Chemical aspects of siderophore mediated iron transport. *Biometals* 15, 325–339.
Broeckling, C.D., Broz, A.K., Bergelson, J., Manter, D.K. and Vivanco, J.M. (2008) Root exudates regulate soil fungal community composition and diversity. *Applied and Environmental Microbiology* 74, 738–744.
Chew, K. (2002) Advances in plant health management in the twentieth century. *Annual Review of Phytopathology* 38, 95–116.
Dessaux, Y., Grandclément, C. and Faure, D. (2016) Engineering the rhizosphere. *Trends in Plant Science* 21, 266–278.
Dobbelaere, S.V., Erleyden, J. and Okon, Y. (2003) Plant growth-promoting effects of diazotrophs in the rhizosphere. *Critical Reviews in Plant Sciences* 22, 107–149.
Doornbos, R.F., van Loon, L.C. and Bakker, P.A. (2012) Impact of root exudates and plant defense signaling on bacterial communities in the rhizosphere. A review. *Agronomy for Sustainable Development* 32, 227–243.
Dowling, D.N. and O'Gara, F. (1994) Metabolites of *Pseudomonas* involved in the biocontrol of plant disease. *Trends in Biotechnology* 12, 133–141.
Glick, B.R. (2012) Plant growth-promoting bacteria: mechanisms and applications. *Scientifica* 2012.
Glick, B.R., Cheng, Z., Czarny, J., and Duan, J. (2007) Promotion of plant growth by ACC deaminase-producing soil bacteria. *European Journal of Plant Pathology* 119, 329–339.
Gray, E.J. and Smith, D.L. (2005) Intracellular and extracellular PGPR: commonalities and distinctions in the plant–bacterium signaling processes. *Soil Biology and Biochemistry* 37, 395–412.
Gupta, R., and Mukerji, K.G. (2002) Root exudate-Biology. In: *Techniques in Mycorrhizal Studies*. Springer, Netherlands, pp. 103–131.
Harman, G.E. (2000) Myths and dogmas of biocontrol: changes in perceptions derived from research on *Trichoderma harzianum* T- 22. *Plant Diseases* 84, 377–393.
Hellriegel, H. and Wilfarth, H. (1888) Untersuchungen uber die Stickstoffnahrung der Gramineen und Leguminosen. *Biblio Life* 238.
Hiltner, L. (1904) Uber neuere Erfahrungen und Probleme auf dem Gebiete der Bodenbakteriologie unter besonderer Bercksichtigung der Grndngung und Brache. *Arbeiten der Deutschen L and wirtschaftlichen Gesellschaft* 98, 5978.
Howell, C.R., Beier, R.C. and Stipanovic, R.D. (1988) Production of ammonia by *Enterobacter cloacae* and its possible role in the biological control of *Pythium* preemergence damping-off by the bacterium. *Phytopathology* 78, 1075–1078.
Innerebner, G., Knief, C. and Vorholt, J.A. (2011) Protection of *Arabidopsis thaliana* against leaf-pathogenic *Pseudomonas syringae* by *Sphingomonas* strains in a controlled model system. *Applied and Environmental Microbiology* 77, 3202–3210.
Iqbal, M.A., Khalid, M., Shahzad, S.M., Ahmad, M., Soleman, N. and Akhtar, N. (2012) Integrated use of *Rhizobium leguminosarum*, plant growth promoting rhizobacteria and enriched compost for improving growth, nodulation and yield of lentil. *Chilean Journal of Agricultural Research* 72, 104.
Junaid, J.M., Dar, N.A., Bhat, T.A., Bhat, A.H. and Bhat, M.A. (2013) Commercial biocontrol agents and their mechanism of action in the management of plant pathogens. *International Journal of Modern Plant and Animal Sciences* 1, 39–57.
Kang, S.M., Joo, G.J., Hamayun, M., Na, C.I., Shin, D.H. *et al.* (2009) Gibberellin production and phosphate solubilization by newly isolated strain of *Acinetobacter calcoaceticus* and its effect on plant growth. *Biotechnology Letters* 31, 277–281.
Kapulnik, Y., Okon, Y., Kigel, J., Nur, I. and Henis, Y. (1981) Effects of temperature, nitrogen fertilization and plant age on nitrogen fixation by *Setaria italica* inoculated with *Azospirillum brasilense* (strain Cd). *Plant Physiology* 68, 340–343.
Keswani, C., Singh, S.P. and Singh, H.B. (2013) A superstar in biocontrol enterprise: *Trichoderma* spp. *Biotech Today* 3, 27–30.

Keswani, C., Mishra, S., Sarma, B.K., Singh, S.P. and Singh, H.B. (2014) Unraveling the efficient application of secondary metabolites of various *Trichoderma*. *Applied Microbiology and Biotechnology* 98, 533–544.

Keswani, C., Bisen, K., Singh, S.P., Sarma, B.K. and Singh, H.B. (2016a) A proteomic approach to understand the tripartite interactions between plant-*Trichoderma*-pathogen: investigating the potential for efficient biological control. In: Hakeem, K.R. and Akhtar, M. S. (eds) *Plant, Soil and Microbes Vol. 2. Mechanisms and Molecular Interactions*. Springer, Cham, Switzerland, pp. 79–93.

Keswani, C., Bisen, K., Singh, V., Sarma, B.K. and Singh, H.B. (2016b) Formulation Technology of Biocontrol Agents: Present Status and Future Prospects. In: Arora, N.K., Mehnaz, S. and Balestrini, R. (eds) *Bioformulations: For Sustainable Agriculture*. Springer, New Delhi, India, pp. 35–52.

Khan, M.S., Zaidi, A., Ahemad, M., Oves, M. and Wani, P.A. (2010) Plant growth promotion by phosphate solubilizing fungi–current perspective. *Archives of Agronomy and Soil Science* 56, 73–98.

Kim, J. and Rees, D.C. (1994) Nitrogenase and biological nitrogen fixation. *Biochemistry* 33, 389–397.

Kiss, L. (2003) A review of fungal antagonists of powdery mildews and their potential as biocontrol agents. *Pest Management Science* 59, 475–483.

Kloepper, J.W. (1994) Plant growth-promoting rhizobacteria (other systems). *Azospirillum/Plant Associations* 137–166.

Kloepper, J.W. and Schroth, M.N. (1978) Plant growth-promoting rhizobacteria on radishes. In: *Proceedings of the 4th International Conference on Plant Pathogenic Bacteria, Angers, France*. INRA Angers, France pp. 879–882.

Kloepper, J.W., Leong, J., Teintze, M. and Schroth, M.N. (1980) Enhanced plant growth by siderophores produced by plant growth-promoting rhizobacteria. *Nature* 286, 885–886.

Kloepper, J.W., Lifshitz, R. and Zablotowicz, R.M. (1989) Free-living bacterial inocula for enhancing crop productivity. *Trends in Biotechnology* 7, 39–44.

Loper, J.E. and Gross, H. (2007) Genomic analysis of antifungal metabolite production by *Pseudomonas fluorescens* Pf-5. *European Journal of Plant Pathology* 119, 265–278.

Lugtenberg, B. and Kamilova, F. (2009) Plant-growth-promoting rhizobacteria. *Annual Review of Microbiology* 63, 541–556.

Ma, J.F. (2005) Plant root responses to three abundant soil minerals: silicon, aluminum and iron. *Critical Reviews in Plant Sciences* 24, 267–281.

Mishra, S., Singh, A., Keswani, C., Saxena, A., Sarma, B.K. and Singh, H.B. (2015) Harnessing plant-microbe interactions for enhanced protection against phytopathogens. In: Arora, N.K. (ed.) *Plant Microbes Symbiosis: Applied Facets*. Springer, New Delhi, India, pp. 111–125.

Mommer, L., Kirkegaard, J. and Van Ruijven, J. (2016) Root–Root interactions: towards a rhizosphere framework. *Trends in Plant Science* 21, 209–217.

Nowak, J., Asiedu, S.K., Bensalim, S., Richards, J., Stewart, A. *et al.* (1997) From laboratory to applications: challenges and progress with in vitro dual cultures of potato and beneficial bacteria. In: *Pathogen and Microbial Contamination Management in Micropropagation*. Springer, Netherlands, 321–329.

Ongena, M., Jacques, P., Touré, Y., Destain, J., Jabrane, A. and Thonart, P. (2005) Involvement of fengycin-type lipopeptides in the multifaceted biocontrol potential of *Bacillus subtilis*. *Applied Microbiology and Biotechnology* 69, 29–38.

Pal, K.K., Dey, R., Bhatt, D.M. and Chauhan, S. (1999) Enhancement of groundnut growth and yield by plant growth-promoting rhizobacteria. *Arachis Newsletter* 51–53.

Pinton, R., Varanini, Z. and Nannipieri, P. (2007) *The Rhizosphere: Biochemistry And Organic Substances At The Soil–Plant Interface*. CRC Press, Boca Raton, FL.

Podile, A.R. and Kishore, G.K. (2006) Plant growth-promoting rhizobacteria. In: *Plant-Associated Bacteria*. Springer, Netherlands, pp. 195–230.

Ramamoorthy, V., Viswanathan, R., Raguchander, T., Prakasam, V. and Samiyappan, R. (2001) Induction of systemic resistance by plant growth promoting rhizobacteria in crop plants against pests and diseases. *Crop protection* 20, 1–11.

Reed, S.C., Cleveland, C.C. and Townsend, A.R. (2011) Functional ecology of free-living nitrogen fixation: a contemporary perspective. *Annual Review of Ecology, Evolution, and Systematics* 42, 489–512.

Saharan, B.S. and Nehra, V. (2011) Plant growth promoting rhizobacteria: a critical review. *Life Sciences and Medicine Research* 1–30.

Sharma. G. (1996) Studies on the integrated management of banded leaf and sheath blight of maize caused by *Rhizoctonia solani*. MSc (Ag.) thesis, G.B. Pant University of Agriculture and Technology, Pantnagar, India, p.65.

Shilev, S. (2013) Soil rhizobacteria regulating the uptake of nutrients and undesirable elements by plants. In: *Plant Microbe Symbiosis: Fundamentals and Advances*. Springer, New Delhi, India, pp. 147–167.

Singh, B.P. (2015) Screening and characterization of plant growth promoting rhizobacteria (PGPR): an overview. *Bulletin of Environmental and Scientific Research* 4, 1–14.

Singh, H.B., Jain, A., Saxena, A., Singh, A., Keswani, C. *et al.* (2014) Deciphering the pathogenic behaviour of phyto-pathogens using molecular tools. In: Neeta, S. (ed.) *Biological Controls for Preventing Food Deterioration: Strategies for Pre-and Postharvest Management*. John Wiley & Sons, Chichester, UK. pp. 377–408.

Singh, R.P., Dhania, G., Sharma, A. and Jaiwal, P.K. (2007) Biotechnological approaches to improve phytoremediation efficiency for environment contaminants. In: *Environmental Bioremediation Technologies*. Springer Berlin Heidelberg, Germany, pp. 223–258.

Tariq, M., Hameed, S., Yasmeen, T., Zahid, M. and Zafar, M. (2014) Molecular characterization and identification of plant growth promoting endophytic bacteria isolated from the root nodules of pea (*Pisum sativum* L.). *World Journal of Microbiology and Biotechnology* 30, 719–725.

Tilak, K.V.B.R., Ranganayaki, N., Pal, K.K., De, R., Saxena, A.K. *et al.* (2005) Diversity of plant growth and soil health supporting bacteria. *Current Science* 89, 136–150.

Tisdale, S.L. and Nelson, W.L. (1975) Soil fertility and fertilizers. *Collier Macmillan Publication*, London, New York.

Tiwari, A.K. (1996) Biological control of chickpea wilt complex using different formulations of *Gliocladium virens* through seed treatment. PhD thesis, G.B. Pant University of Agriculture and Technology, Pantnagar, India, p. 167.

Vacheron, J., Desbrosses, G., Bouffaud, M.L., Touraine, B., and Prigent-Combaret, C. (2014) Plant growth-promoting rhizobacteria and root system functioning. *Ecophysiology of Root Systems-environment Interaction* 166.

van Loon, L.C., Bakker, P.A.H.M. and Pieterse, C.M.J. (1998) Systemic resistance induced by rhizosphere bacteria. *Annual Review of Phytopathology* 36, 453–483.

Venturi, V. and Keel, C. (2016) Signaling in the rhizosphere. *Trends in Plant Science* 21, 187–198.

Vitousek, P.M., Cassman, K., Clevel, C., Crews, T., Field, C.B. *et al.* (2002) Towards an ecological understanding of biological nitrogen fixation. *Biogeochemistry* 57, 1–45.

Whipps, J.M. (2001) Microbial interactions and biocontrol in the rhizosphere. *Journal of Experimental Botany* 52, 487–511.

Yazdani, M., Bahmanyar, M.A., Pirdashti, H. and Esmaili, M.A. (2009) Effect of phosphate solubilization microorganisms (PSM) and plant growth promoting rhizobacteria (PGPR) on yield and yield components of corn (*Zea mays* L.). *Proceedings of World Academy of Science, Engineering and Technology* 37, 90–92.

Zaidi, A., Khan, M.S., Ahemad, M., Oves, M. and Wani, P.A. (2009) Recent advances in plant growth promotion by phosphate-solubilizing microbes. In: *Microbial Strategies for Crop Improvement*, Springer Berlin Heidelberg, Germany, pp. 23–50.

Zhang, S., Moyne, A.L., Reddy, M.S. and Kloepper, J.W. (2002) The role of salicylic acid in induced systemic resistance elicited by plant growth-promoting rhizobacteria against blue mold of tobacco. *Biological Control* 25, 288–296.

23 PGPR: A Good Step to Control Several of Plant Pathogens

Laith K. Tawfeeq Al-Ani[1,2]

[1]*School of Biological Sciences, Universiti Sains Malaysia, 11800 Pinang, Malaysia;*
[2]*Department of Plant Protection, College of Agriculture-Baghdad University, Baghdad 10071, Iraq*

23.1 Introduction

Plant growth-promoting rhizobacteria (PGPR) are able to play a very important role in protecting plants from infection, as well as promoting plant growth through colonizing the roots. PGPRs are a beneficial group of soil microorganisms that very efficiently colonize the rhizoplane and rhizosphere. One third of the crops produced globally get damaged due to infection from diseases, irrespective of the use of several protective measures. The prime factor is the use of synthetic chemicals that protects plants from numerous diseases, but in contrast severely affect the environment, including humans, animals, plants, beneficial microorganisms, rivers, lakes, etc. The environment is already exposed to residues of chemicals that are sprayed to control plant pathogens. The synthetic chemicals that are used to control plant pathogens comprise viricides for plant viruses, bactericides for plant bacteria, fungicides for plant fungus and nematicides for plant nematodes. All these pesticides can have severe side effects as they can leave residues that, whether either higher or lower in amount, gradually cause damage to the environment.

A second factor is the decreased susceptibility of plant pathogens to pesticides. For example, downy mildew caused by the Oomycota exclusively belongs to Peronosporaceae that are obligate parasites of plants, having the potential to cause major diseases globally. Downy mildew of cucurbits caused by *Pseudoperonospora cubensis* is extremely epidemic. Moreover, the systemic fungicides used against this fungus bear a very high risk of resistance development (Lebeda and Cohen, 2011). Downy mildew was one of the ten plant pathogens accepted by FRAC as bearing a high risk of resistance development to fungicides (Pathogen risk list, 2005, http:// FRAC.info).

Thirdly, the pesticides are quite expensive. For instance, the fungicides applied against downy mildews in 1996 amounted to 1.2 billion SFr in value (Urban and Lebeda, 2006). These are simply the main reasons (and there are more that need not be mentioned here). We therefore need to look for other methods to control plant pathogens that replace the chemical factor. PGPR as biocontrol agents are a good way to control plant pathogens.

PGPR can protect several crops such as cereals, vegetables, legumes and others from

*E-mail: cmv_virus2002@yahoo.com; laith.kt77@gmail.com

© CAB International 2017. *Advances in PGPR Research* (eds H.B. Singh, B.K. Sarma and C. Keswani)

infection by different diseases caused by viruses, fungi, bacteria, and nematodes as well as ones caused by nutrient deficiencies. Though PGPR can use multiple modes of action in controlling plant pathogens, the entire process can be summarized in two basic mechanisms, i.e. direct and indirect (Glick, 1995; Gupta *et al.*, 2015; Bisen *et al.*, 2016). The mechanisms of direct effect on plant pathogens include production of antibiotics like pyocyanine, pyrrolnitrin, 2,4-diacetylphloroglucinol (Pierson and Thomashow, 1992), production of siderophores (O'Sullivan and O'Gara, 1992), the synthesis of hydrogen cyanide (HCN) (Glick, 1995), production enzymes that can hydrolyze the cell walls of plant pathogens (Mauch *et al.*, 1988), competition for colonization sites and for nutrients (O'Sullivan and O'Gara, 1992; Prasad *et al.*, 2015), as well as degradation of the pathogenicity factors of substances such as toxins and enzymes (Podile and Kishore, 2006; Prasad *et al.*, 2015). On the other hand, the indirect effect includes induction of resistance and promotion of growth in plants against pathogens (Glick, 1995).

Thus, it is evident that though different mechanisms control several plant pathogens comprising viruses, bacteria, fungi and nematodes, the mode of action depends on the type of plant pathogens. The production of antibiotics by biocontrol agents like PGPR is the first step to control the plant pathogens. Antibiotics are produced from known PGPR strains such as *Bacillus*, *Enterobacter*, *Paenibacillus*, *Burkholderia*, *Pseudomonas*, *Azotobacter*, *Klebsiella*, *Azospirillum*, *Streptomyces*, *Serratia* and *Rhizobium* species (Kloepper *et al.*, 1989; Glick, 1995; Joseph *et al.*, 2007; Saharan and Nehra, 2011). *Pseudomonas* and *Bacillus* are characteristically unique from other genera of PGPR in biocontrol of plant pathogens. *Pseudomonas* strains are particularly able to induce systemic resistance in *Arabidopsis*, radish, cucumber, tobacco and carnation (van Loon and Bakker, 2005). *Bacillus* strains are among the most commonly reported PGPR strains (Vessey, 2003; Compant *et al.*, 2005).

The production of antibiotics by PGPR involve different chemicals such as phenazine-1-carboxyclic acid, 2,4-diacetyl phloroglucinol, oomycin, pyoluteorin, pyrrolnitrin, kanosamine, zwittermycin-A and pantocin (Fernando *et al.*, 2005). Antibiotics play a very important role in management of plant diseases and are also important characteristics of PGPR strains through which we particularly screen certain isolates from others. Many bacteria belonging to PGPR can be isolated from the rhizosphere and then tested against plant pathogens like fungi in a dual culture test. Also the study of culture filtrates for the presence of antibiotics is the step to separate the microorganisms during isolation of PGPR. Volatile and non-volatile compounds are major types of antibiotics.

Then, the production of siderophores is also a very important characteristic for the PGPR strains. Siderophores provide some advantages that include conferring a competitive edge to PGPR strains colonizing the soil, rhizosphere and roots. The ability of PGPR to produce siderophores can be assessed in solid or liquid media. Also, siderophores can supply iron and promote growth of plants.

The ability to synthesize hydrogen cyanide (HCN) inhibits growth of several plant pathogens. Hydrogen cyanide is a volatile antibiotic compound produced by some rhizobacteria (Fernando *et al.*, 2005). Some isolates of rhizobacteria were reported to produce HCN, such as isolates of *Bacillus subtilis*. By contrast, some other isolates of *Bacillus subtilis* produced HCN but no deleterious effects were observed (Saha *et al.*, 2012; Reetha *et al.*, 2014). Therefore, HCN is depending on the species of rhizobacteria or the strain.

The production of hydrolytic enzymes to degrade cell walls include mechanisms antagonistic to plant pathogens. There are several hydrolytic enzymes, such as proteases, β-1,3-glucanases, chitinases, etc. (Kim *et al.*, 2008; Zhang *et al.*, 2015; Keswani *et al.*, 2016). Also, the production of enzymes is different among isolates and species of rhizobacteria. *Pseudomonas sp.* Psp.8D-45 *and Bacillus subtilis* Bs 8B-1 did not produce chitinases, but *Pseudomonas fluorescens* Pf 9A-14 can produce chitinases (Khabbaz *et al.*, 2015).

The rhizobacteria and pathogenic microbes are in constant competition for colonization of the rhizosphere/rhizoplane region (Dutta and Podile, 2010). *Pseudomonas fluorescens,* a rhizosphere-competent bacteria has the site-specific recombinase gene that it transfers into other *Pseudomonas* strains (rhizosphere-incompetent) thereby enhancing their ability to colonize root tips (Dekkers *et al.*, 2000). Further, the competition for nutrients basically depends upon the composition of root exudates, comprising chiefly free oxygen, ions, enzymes, mucilage, water, as well as secondary metabolites and primary carbon (Uren, 2000; Bertin *et al.*, 2003).

Then, the degradation of pathogenicity factors is a very important mechanism in suppressing plant pathogens and decreasing plant diseases. For example, the rhizobacteria of *Burkholderia cepacia* and *Ralstonia solanacearum* are able to detoxify fusaric acid produced by *Fusarium oxysporum* (Toyoda *et al.*, 1988). Some PGPR are capable of producing several enzymes inactivating plant pathogens. *Bacillus megaterium* B153-2-2 secretes several enzymes that inactivate four enzymes of *Rhizoctonia solani.* The mechanism of detoxification and inactivation of plant pathogen enzymes are unique for some PGPR.

On the other hand, indirect effects like induction of plant defence play a very important role in control of plant pathogens. Induction of plant defence comprises induced systemic resistance (ISR), and in certain instances can induce the systemic acquired resistance (SAR). PGPR enhances the natural resistance mechanism of the host (Nehl *et al.*, 1997) by activation of certain pathogenesis-related (PR) genes in plants similar to pathogen induced SAR (Wang *et al.*, 2005). The activation of defence responses in plants by PGPR against several plant pathogens is limited for some strains of PGPR (Podile and Kishore, 2006). The mechanisms of ISR include: (1) increase of physiological tolerance through reduction of symptoms; (2) enhancement of the growth of the host; and (3) systemic-resistance by induction of phytoalexins, "priming" of defence responses and pathogenesis related proteins, as well as induction of reinforcement of the cell wall (Prasad *et al.*, 2015). *Pseudomonas putida* strain 89B-27 and *Serratia marcescens* strain 90-166 are able to induce plant defence against fusarium wilt of cucumber (Liu *et al.*, 1995). *Bacillus subtilis* GB0 and *B. pumilus* INR7 were reported to reduce angular leaf spot in cucumber by PGPR-mediated resistance. (Raupach *et al.*, 2000). *B. subtilis* AF1 protected peanut seedlings against *Aspergillus niger* that caused crown rot disease by altering the phytoalexin metabolism (Sailaja and Podile, 1998).

Mostly, PGPR enhances the growth of plants through providing essential minerals like nitrogen, phosphorus, and others. Essential minerals improving plant health through significantly increasing total chlorophyll content, soil enzyme activities, nutrient uptake, plant dry weight, shoot length, shoot weight, coefficient of velocity of germination, seed yield and seed protein, etc. (Ahemad and Khan, 2010; Sharma *et al.*, 2011; Jahanian *et al.*, 2012). Therefore, the improvement of plant growth usually protects the plants from infection, and enhances the tolerance of infection of plants to disease. For example, tomato was treated with a PGPR strain, *Bacillus subtilis* 21-1 (BS21-1), which contributed to growth promotion of tomato and induces resistance against *Botrytis cinerea* (Lee *et al.*, 2014). PGPR can affect plant growth by providing nutrients like N, P, K, iron and others, that facilitate nutrient uptake from the root environment as well as aid in production and liberation of secondary metabolites, thereby causing the injurious effects to phytopathogenic organisms in the rhizosphere (Dardanelli *et al.*, 2010). Also, PGPR have the ability to enhance plant growth by producing auxins, gibberellins, cytokinins and ethylene (Frankenberger and Arshad, 1995).The importance of soil-plant-microbe interactions in development of efficient inoculants with PGPR can lead to improvement of plant growth and protection of plants from disease (Souza *et al.*, 2015).

23.2 Biocontrol of Plant Virus

Viruses are plant pathogens and obligate parasites with a size less than 200 nm and generally having no viricidal or therapeutic agents. Therefore, plant viruses are very difficult to control and require either induction of natural resistances in plant or other indirect methods. The induced resistances in plant were mentioned by Kuc (1995) and van Loon *et al.* (1998) as two main factors capable of motivating or activating the natural host resistance. The natural resistances in plants are induced through several factors such as non-pathogenic fungi, non-pathogenic bacteria (PGPR), phytohormones, proteins and chemical activators (Zhou and Niu, 2009; Al-Ani, 2010; Lian *et al.*, 2011; Al-Ani *et al.*, 2013a, b; Zhao *et al.*, 2013; Alazem and Lin, 2015; Al-Ani, 2016).

PGPR are able to induce systemic resistance (ISR) and systemic acquired resistance (SAR) in plants. Biocontrol of plant viruses by plant growth-promoting agents is a mechanism to improve plant growth and reduce damage caused by virus infections that includes dwarfness by reduction of plant growth hormones like auxin and gibberellin concentrations, yellowing of leaf by decrease in photosynthesis, disturbance in host cell metabolism, higher phenol oxidase activity, increased respiration rate, and abnormal accumulation of metabolic products (Waller, 2002). This suggests that the plant viruses differ from some other pathogens such as bacteria, fungi, and plant parasites in that they do not induce disease through production of a translocatable toxin (Waller, 2002), or parasite by appressorium.

PGPR strains are reported to induce resistance in cucumber against cucumber mosaic virus (CMV) (Kokalis-Burelle, 2002). *Bacillus pumilus* strain SE34, *Kluyvera cryocrescens* strain IN114, *Bacillus amyloliquefaciens* strain IN937a and *Bacillus subtilis* strain IN937b were reported to express induced resistance against CMV infection when applied to tomato plants by soil drench mechanism (Zehnder *et al.*, 2000). The leaf, soil, and seeds of cucumber were treated with *P. fluorescens* f5 strain against CMV, and it was found that disease severity was drastically reduced in field conditions (Al-Ani, 2006).

The seed treatment of cucumber and tomato with two strains of PGPR such as *P. fluorescens* strain 89B-27 and *S. marcescens* delayed the development of symptoms of cucumber mosaic virus (CMV) (Raupach *et al.*, 1996). Also, *Bacillus amyloliquefaciens* 937a, *B. subtilis* 937b and *B. pumilus* SE34 used to treat seeds of tomato expressed reduced disease severity and incidence of Tomato mottle virus (ToMoV) under field conditions (Murphy *et al.*, 2000). The root treatment of tobacco with *Pseudomonas fluorescens* CHA0 enhanced capacity to induce PR proteins (PR-1 group proteins, β-glucanases (GUS) and endochitinase) and resistance to leaf necrosis caused by tobacco necrosis virus (TNV) (Maurhofer *et al.*, 1994; Maurhofer *et al.*, 1998). The protection of *Arabidopsis* against CMV by *Serratia marcescens* strain 90-166 included resistance through induction of *PDF1.2* gene as a JA signalling indicator, but independent of *SA* and *NPR1* (Ryu *et al.*, 2004). PGPR induce systemic resistance by increase in activity of two enzymes such as peroxidase and β-1,3-glucanase (El-Borollosy and Oraby, 2012). A mix of two PGPR strains, *Pseudomonas aeruginosa* and *Stenotrophomonas rhizophilia*, enhance systemic defences against CMV in tomato in the greenhouse (Dashti *et al.*, 2012).

A new antiviral ribonuclease (RNase) purified from *Bacillus cereus* ZH14 inhibited 90% of tobacco mosaic virus (TMV) particles (Zhou and Niu, 2009). *Bacillus pumilus* strain EN16 and *Bacillus subtilis* strain SW1 were able to induce systemic resistance in tobacco against TMV by increasing the amounts of defence enzymes and PR proteins (Lian *et al.*, 2011). *B. amyloliquefaciens* Ba33 is reported to reduce the local necrotic lesion number and disease index caused by tobacco mosaic virus (TMV), on leaves of *Nicotiana tabacum* cv. *Samsum* NN and *N. tabacum* NC89 plants, *in vitro* suppression of TMV by Ba33 was confirmed due to inactivation of TMV particles (Shen *et al.*, 2013). Treatment of tomato plants by *Enterobacter asburiae* BQ9 showed an increase in fresh mass of plant, reduced disease severity and high efficacy in biocontrol. BQ9 can induce plant defence against TYLCV through increasing the expression of PR genes such as *PR1a* and *PR1b*, and the induced resistance was mechanistically connected to expression of H_2O_2 and antioxidant enzymes such as lyase, peroxidase, catalase and superoxide dismutase, and increase in activities of phenylalanine ammonia (Hongwei *et al.*, 2016).

23.3 Biocontrol of Plant Bacteria

In the past year, there have been few studies related to PGPR application in controlling plant bacterial diseases. There are several methods that can control the plant bacterial diseases, such as resistant varieties, cultural practices, chemical synthesis, and biocontrol agents like PGPR (Saddler, 2002). PGPR can control plant bacterial diseases using methods that are direct or indirect, but both the mechanisms may cumulatively contribute to fight against plant bacterial diseases. Direct control may operate through production of antibiotics, siderophores and enzymes, as well as degradation of pathogenicity factors, competition for specific sites and nutrients, while indirect methods include induction of resistance and growth promotion.

PGPR reduce natural root populations of *Erwinia carotovora,* the causal organism of potato blackleg and soft rot diseases, by production of antibiotics (Kloepper, 1983). Five PGPR of cotton were reported to be antagonistic to 32 races of *Xanthomonas campestris* pv. *malvacearum* owing to production of siderophores and HCN (Mondal *et al.*, 2000). *Pseudomonas fluorescens* CRb-26 was reported to produce four major phenolic metabolites (one was identified as 2,4-diacetylphloroglucinol), inhibiting the growth of bacterial blight affecting growth of cotton plants (*Gossypium hirsutum* and *G. barbadense*). Two isolates of *Bacillus* spp. induced resistance to the bacterial blight pathogen in leaves of a susceptible cotton (Mondal and Verma, 2002). *Bacillus* sp. reduced the pathogenicity of *Erwinia carotovora* (*Pectobacterium carotovorum*) by inactivating N-acylhomoserine lactones through production of enzymes (Dong *et al.*, 2000).

Streptomyces diastatochromogenes sk-6, was reported to reduce disease severity of soft rot caused by *Erwinia carotovora* (Doolotkeldieva *et al.*, 2016). Two PGPR isolates of *Pseudomonas* sp. RBL 101 and RSI 125 reduced disease severity of bacterial wilt caused by *Ralstonia solanacearum* through production of siderophores (Jagadeesh *et al.*, 2001). Cronin *et al.* (1997) reported biocontrol of the potato soft rot pathogen, *Erwinia carotovora* sub sp. *atroseptica* (van Hall) Dye, by *Pseudomonas fluorescens* (Trevisan) Migula F113 due to production of the antibiotic 2,4-diacetylphloroglucinol (DAPG). Similarly, *Pantoea agglomerans* strain E325, *Pseudomonas fluorescens* strain A506 and *P. agglomerans* strain C9-1 were effective in suppressing *E. amylovora*, causal organism of fire blight disease, due to competition for specific sites or nutrients (Pusey, 2002).

23.4 Biocontrol of Plant Fungi

PGPR act on plants via numerous mechanisms, and can be used to control plant fungi. The effectivity of PGPR depends on the specific genus, strain, species and the pathosystem type of plant fungal pathogen. The *Pseudomonas* and *Bacillus* genera of PGPR are the most effective as compared to other genera. Normally the direct biological mechanism of PGPR that suppresses the pathogen is antibiosis inhibiting growth of spore or mycelia, degradation or deformation of hyphal wall through enzymes by production of lytic enzymes such as chitinase, β-1,3-glucanase, protease and lipase, production of volatile compounds, hydrocyanic acid (HCN), prevention of spore growth or penetration on the root by attaching and covering the root with bacteria, and competition on nutrients such as carbon and iron.

PGPR can also act indirectly by induction of plant defence systems and plant growth promotion that impact on changes in cell wall structure, production of pathogenesis-related (PR) proteins, phytoalexin, induction of molecules eliciting defence, such as jasmonic acid (JA), salicylic acid

(SA), and ethylene (ET), enhancing plant growth from nutrient acquisition by hormonal production of indole-3-acetic (IAA), acid (auxin), cytokinins, gibberellins, nitrogen fixation, phosphate solubilization and other minerals.

Pseudomonas corrugata 13 or *Pseudomonas aureofaciens* 63±28 were reported to induce plant defence against *Pythium aphanidermatum* in cucumber root by increasing activates of peroxidase enzyme (PO) (Chen *et al.*, 2000). Basha and Ulaganathan (2002) reported *Bacillus* strain BC121 as a producer of chitinase protein that causes hyphal lysis and degradation of the cell wall of *Curvularia lunata* thereby affecting mycolytic activity of the fungus. *Bacillus amyloliquefaciens* 7079 was reported to suppress disease rates of Fusarium wilt of tomato and Phytophthora blight of peppers by inducing systemic resistance and enhancing plant growth (Soohee and Kim, 2005). Similarly, *Bacillus megaterium* DE BARY TRS-4 was shown to reduce disease intensity through a combination of several mechanisms such as an increase in polyphenolics, phenylalanine ammonia lyase, chitinase, β-1,3-glucanase and peroxidase, as well as enhancement in plant growth-promoting factors, such as solubilizing phosphate, producing IAA, antifungal metabolites and siderophores (Chakraborty *et al.*, 2006). *Pseudomonas* sp. FQA PB-3 protect chilli and tomato seedlings from infestation by *Pythium* and *Phytophthora* species, associated with damping-off diseases through increased activity of peroxidase and phenylalanine ammonia-lyase (Sharma *et al.*, 2007). *Paenibacillus* sp. is able to counter growth of *Rhizoctonia bataticola*, by production of peptide antifungal metabolites causing abnormal contraction of fungal cytoplasm, granulation and fragmentation of hyphal mycelia and cell lysis (Senthilkumar *et al.*, 2007).

Pseudomonas fluorescens WM35 and *P. putida* WM06 were able to protect bean plants from rust infection by induced systemic resistance (Abeysinghe, 2009). *P. aeruginosa* TO3 was found to exhibit biocontrol activity against *Macrophomina phaseolina*, the causal organism of charcoal rot disease, by direct and indirect mechanisms that included production of siderophores and HCN, as well as enhancement of plant growth by production of indole-3-acetic acid (IAA) (Khare and Arora, 2010). *Pseudomonas* spp., UOM ISR 17 through seed treatment, were found to protect pearl millet against downy mildew (Jogaiah *et al.*, 2010).

Pseudomonas fluorescens strain pa4 was capable of controlling *Fusarium oxysporum* f. sp. *cubense* by producing siderophores, HCN, lytic enzymes, and antifungal compounds (Mohammed *et al.*, 2011). Saha *et al.* (2012) reported two strains of *Bacillus subtilis* AI01 and AI03 as being antagonist to *Fusarium solani*. The biological mechanism of AI01 and AI03 included secretion of siderophores, indole acetic acid (IAA) as well as many hydrolytic enzymes such as chitinase, protease, lipase and amylase. Also, scanning electron microscopic studies of interaction between the isolated *Bacillus* strains and *F. solani* exhibited deformation of fungal mycelium, and attachment of bacterial colonies to the hyphae at the zone of interaction (Saha *et al.*, 2012). According to Singh *et al.* (2012) *in vitro* cultures of *P. fluorescens* 4 and *P. fluorescens* 4 (new) inhibited *Sclerotium rolfsii* due to production of eight phenolic acids such as caffeic, vanillic, salicylic, tannic, ferulic, gallic, o-coumeric and cinnamic acid.

PGPR inoculants suppress plant disease through one or more mechanisms, due to PGPR that inhabit plant roots exerting a positive effect ranging from direct effectiveness mechanisms to an indirect effect (Ramjegathesh *et al.*, 2013). *Bacillus subtilis* strain S25 inhibited *Phytophthora capsici* that caused root and crown rot of tomato due to production of siderophores, lytic and antifungal enzymes (Sharma *et al.*, 2015). *Bacillus subtilis* GB03 can biocontrol against *Botrytis cinerea* infection of *Arabidopsis* seedlings by an indirect mechanism, which elicited induction of systemic resistance in the plant through production of volatile organic compounds (VOCs) (Sharifi and Ryu, 2016). *Bacillus vallismortis* strain EXTN-1 reduced disease severity caused by *Phytophthora capsici* due to production of antifungal iturin A analogues (Park *et al.*, 2016).

Pantoea eucalypti (isolate NT6), *Bacillus methylotrophicus* (isolate MT3), *Pseudomonas veronii* (isolates BT4 and NT2), and *P. rhodesiae* (isolate BT2) *in vitro* produced antimicrobial compounds and reduced leaf damage caused by *B. cinerea*, and also enhanced growth of tomato plants, produced the auxin indole-3-acetic, siderophores and solubilized inorganic phosphate (Romero *et al.*, 2016). Three strains of *Bacillus subtilis*, 5B6, 8D4, and 8G12, were reported to reduce disease severity caused by *Colletotrichum graminicola* and *Botrytis cinerea* (Chung and Ryu, 2016). Thus, these are some of the confirmatory reports that suggest that PGPR have several mechanisms which can suppress more than one pathogen at the same time.

23.5 Biocontrol of Plant Nematode

Plant nematodes are eukaryotic organisms, inhabiting the soil and attacking roots and leaves of plants. Management of plant nematode diseases uses several methods such as chemical, physical and biological control. Biological controls include biocontrol agents comprising fungi, bacteria or PGPR, and viruses. PGPR are already reported as biocontrol agents for nematodes and other plant pathogens. Some bacteria are obligate parasites on nematode while others may be saprophytic but also have interactions with nematode. PGPR are able to control the plant nematode pathogens through direct or indirect biological mechanisms that were mentioned previously.

Usually, the mode of action of nematodes for biocontrol mechanisms include some of the following actions.

1. Production of lytic enzymes like chitinase and other enzymes that degrade cell walls: for example, reducing numbers of *Heterodera glycines* caused soybean cyst nematode by five isolates of bacteria that have chitinolytic activity (Tian *et al.*, 2000).
2. Production of HCN that causes paralysis of nematode: for instance, *Pseudomonas aeruginosa* PAO1 causing paralysis and death of the nematode *Caenorhabditis elegans* through production of HCN that caused inhibition of mitochondrial cytochrome oxidase in the nematode (Gallagher and Manoil, 2001).
3. Production of toxin; for instance *P. aeruginosa* strain PA14 kills *Caenorhabditis elegans* through production of a toxin thereby killing the nematode (Mahajan-Miklos *et al.*, 1999).
4. Reduction of the number of eggs, masses, galls, cysts and juvenile population: for example, five isolates of PGPR like *Serratia marcescens*, *Bacillus amyloliquefaciens*, *P. putida*, *P. fluorescens* strain Pf1, *P. fluorescens* and *B. cereus* were reported to reduce the masses, eggs and galls of root knot nematode (Smith, 1994; Santhi and Sivakumar, 1995; Almaghrabi *et al.*, 2013) and *P. aeruginosa* strain LPT5 reduced cysts and juvenile populations of the nematode *Heterodera cajani* in *Sesame indicum*; also, the strain LPT5 was also found to produce IAA, HCN, chitinase, glucanase and siderophore, and also solubilized inorganic phosphate under *in vitro* conditions (Kumar *et al.*, 2009).
5. Improving plant growth with addition of fertilizer is playing a role in management of plant nematode diseases: for instance, the management of nematode *Meloidogyne incognita* on tomato was done by reducing the galls effect on the multiplication and growth of nematode, and at the same time improvement in plant growth was achieved by treatment with three combinations comprising compost, organic goat dung and *P. fluorescens*; poultry manure and *P. fluorescens*; and poultry manure with *Azotobacter chroococcum* (Siddiqui, 2004).
6. Inducing systemic resistance: for example, *P. fluorescens* strain CHA0 induced juvenile deaths of nematode *Meloidogyne javanica* by promoting systemic resistance through SA accumulation in roots (Siddiqui and Shaukat, 2004).

23.6 Biocontrol of Plant Parasite

Parasitic weeds constitute a small but very important group of plant pathogens. Biocontrol of plant parasitic weeds by PGPR involves different mechanisms, such as inhibition of seed germination and radical growth. PGPR can produce several phytohormones like indole-3-acetic acid (IAA). Striga seeds formed short swollen germination tubes under high hormonal concentrations (Logan and Stewart, 1991). Two strains of *Azospirillum brasilense* were found to inhibit germination of *Striga hermonthica* (Del.) (plant parasitic), owing to production of lipophilic molecules during bacterial growth (Miché *et al.*, 2000). Moreover, *Azospirillum brasilense* was found to inhibit germination and radicle growth of conditioned seeds of *Orobanche aegyptiaca* owing to production of synthetic peptides competing for the site of binding of germination stimulant (Dadon *et al.*, 2004). *P. fluorescens* Bf7-9 was found to suppress pre-emergence activity of *Orobanche foetida* and *O. crenata* without affecting the growth of the faba bean (*Vicia faba* L.) (Zermane *et al.*, 2007). Four PGPR of *Bacillus licheniformis, B. pumilus, B. amyloliquefaciens* and *B. megaterium*, were also reported to inhibit seed germination of field dodder (*Cuscuta campestris* Yunck.) (Marija and Dragana, 2009).

23.7 Biocontrol of Phytoplasma

Phytoplasmas are wall-less plant pathogens inhabiting the plant phloem and hemolymph of insect vectors (Bertaccini *et al.*, 2012), and causing several important plant diseases globally. Generally, the disease management of phytoplasmas includes several methods, such as controlling the vectors, genetic resistance, plant growth promotion and induction of plant defences. The ability of PGPR to control phytoplasmas is not managed directly. Theoretically, biocontrol of phytoplasma by PGPR may be by biological mechanisms, such as direct or indirect interaction. *Pseudomonas putida* S1Pf1Rif was reported to inhibit chrysanthemum yellows (CY) phytoplasma infection of chrysanthemum (Jarausch and Ester Torres, 2014).

23.8 Conclusion

PGPR have several biological mechanisms that are used to control several kinds of plant pathogens like virus, bacteria, fungi, nematodes, parasites and phytoplasmas. PGPR uses direct or indirect mechanisms according to which kind of plant pathogen is involved and the ability of that pathogen to evolve and adapt to a changing environment. Therefore, PGPR determines the enemy and the action suitable to kill or stop the pathogens, but how? There is competition for an ecological niche among the organisms. The competitions for nutrients like carbon and iron are at colonization sites. But this does not mean that all PGPR can adapt or compete in accordance with plant pathogens in an ecological niche, as described by the "gene-for gene" theory. This difference includes genus, species, strains and ecological niche. For example, two genera of PGPR like *Pseudomonas* and *Bacillus* are very prominently used to control plant pathogens. Finally, the success of PGPR to control plant pathogen disease severity must include strains that could compete with different plant pathogens at the same time using several biological mechanisms to kill or inhibit the growth of plant pathogens so as to enhance the efficacy of sustainable agriculture.

References

Abeysinghe, S. (2009) Induced systemic resistance (ISR) in bean (*Phaseolus vulgaris* L.) mediated by rhizobacteria against bean rust caused by *Uromyces appendiculatus* under greenhouse and field conditions. *Archives of Phytopathology and Plant Protection* 42(11), 1079–1087.

Ahemad, M. and Khan, M.S. (2010) Phosphate-solubilizing and plant growth-promoting *Pseudomonas aeruginosa* PS1 improves greengram performance in quizalafop-p-ethyl and clodinafop amended soil. *Archives of Environmental Contamination and Toxicology* 58, 361–372.

Al-Ani, L.K.T. (2006) Induce Resistance against Cucumber Mosaic Virus by *Pseudomonas fluorescens* Migula. M.Sc., Department of Plant Protection, College of Agriculture, University of Baghdad, Iraq, p. 90.

Al-Ani, L.K.T. (2010) Biological control of fusarium wilt of banana by non pathogenic *Fusarium oxysporum. PPSKH colloquium*, Pust Pengajian Sains Kajihayat/School of Biological Sciences, Universiti Sains Malaysia, Malaysia June, 2010, p. 10.

Al-Ani, L.K.T. (2016) Potential of utilizing biological and chemical agents in the control of Fusarium Wilt of banana. PhD, School of Biology Science, Universiti Sains Malaysia, Pulau Pinang, Malaysia, p. 259.

Al-Ani, L.K.T., Salleh, B., Mohammed, A.M., Ghazali, A.H.A., Al-Shahwany, A.W. and Azuddin, N.F. (2013) Biocontrol of Fusarium Wilt of banana by non-pathogenic *Fusarium* spp. *International symposium on tropical fungi*, ISTF, IPB International Convention Center, Bogor, Indonesia, 09/2013, pp. 50–51.

Alazem, M. and Lin, N.-S. (2015) Roles of plant hormones in the regulation of host–virus interactions. *Molecular Plant Pathology* 16(5), 529–540.

Almaghrabi, O.A., Massoud, S.I. and Abdelmoneim, T.S. (2013) Influence of inoculation with plant growth promoting rhizobacteria (PGPR) on tomato plant growth and nematode reproduction under greenhouse conditions. *Saudi Journal of Biological Sciences* 20, 57–61.

Basha, S. and Ulaganathan, K. (2002) Antagonism of *Bacillus* species (strain BC121) towards *Curvularia lunata. Current Science* 82(12), 1457–1436.

Bertaccini, A., Paltrinieri, S., Contaldo, N., Fránová, J., Mogens, N. and Duduk, B. (2012) Phytoplasma detection and identification: from 16S ribosomal gene to multiple identification. *QBOL-EPPO Conference on DNA Barcoding and Diagnostic Methods for Plant Pests.* Programme, participant list and summaries of presentations and posters. May 21-25, 2012, Haarlem (NL), p. 12.

Bertin, C., Yang, X.H. and Weston, L.A. (2003) The role of root exudates and allelochemicals in the rhizosphere. *Plant and Soil* 256, 67–83.

Bisen, K., Keswani, C., Patel, J.S., Sarma, B.K. and Singh, H.B. (2016) *Trichoderma* spp.: Efficient Inducers of Systemic Resistance in Plants. In: Chaudhary, D.K. and Verma, A. (eds) *Microbial-mediated Induced Systemic Resistance in Plants.* Springer, Singapore, pp. 185–195.

Chakraborty, U., Chakraborty, B. and Basnet, M. (2006) Plant growth promotion and induction of resistance in *Camellia sinensis* by *Bacillus megaterium. Journal of Basic Microbiology* 46(3), 186–195.

Chen, C., Belanger, R.R., Benhamou, N. and Paulitz, T.C. (2000) Defense enzymes induced in cucumber roots by treatment with plant growth-promoting rhizobacteria (PGPR) and *Pythium aphanidermatum. Physiological and Molecular Plant Pathology* 56, 13–23.

Chung, J.-H. and Ryu, C.-M. (2016) Disease management in road trees and pepper plants by foliar application of *Bacillus* spp. *Research in Plant Disease* 22(2), 81–93.

Compant, S., Duffy, B., Nowak, J., Clement, C. and Barka, E.A. (2005) Use of plant growth-promoting bacteria for biocontrol of plant diseases: principles, mechanisms of action, and future prospects. *Applied and Environmental Microbiology* 71, 4951–4959.

Cronin, D., Moenne-Loccoz, Y., Fenton, A., Dunne, C., Dowling, D.N. and O'Gara, F. (1997) Ecological interaction of a biocontrol *Pseudomonas fluorescens* strain producing 2,4-diacetyl-phloroglucinol with the soft rot potato pathogen *Erwinia carotovora* subsp. *atroseptica. FEMS Microbiology Ecology* 23, 95–106.

Dadon, T., Nun, N.B. and Mayer, A.M. (2004) A factor from *Azospirillum brasilense* inhibits germination and radicle growth of Orobanche aegyptiaca. *Israel Journal of Plant Sciences* 52, 83–86.

Dardanelli, M.S., Carletti, S.M., Paulucci, N.S., Medeot, D.B., Cáceres, E.A.R. *et al.* (2010) Benefits of plant growth-promoting rhizobacteria and rhizobia in agriculture. In: Maheshwari, D.K. (ed.) *Plant Growth and Health Promoting Bacteria.* Springer Berlin Heidelberg, Germany, pp. 1–20.

Dashti, N.H., Ali, N.Y., Cherian, V.M. and Montasser, M.S. (2012) Application of plant growth-promoting rhizobacteria (PGPR) in combination with a mild strain of Cucumber mosaic virus (CMV) associated with viral satellite RNAs to enhance growth and protection against a virulent strain of CMV in tomato. *Canadian Journal of Plant Pathology* 34(2), 177–186.

Dekkers, L.C., Mulders, I.H., Phoelich, C.C., Chin-A-Woeng, T.F.C., Wijfjes, A.H. and Lugtenberg, B.J.J. (2000) The colonization gene of the tomato- *Fusarium oxysporum* f. sp. *radicis-lycopersici* biocontrol strain *Pseudomonas fluorescens* WCS365 can improve root colonization of other wild-type *Pseudomonas* spp. bacteria. *Molecular Plant-Microbe Interactions* 13, 1177–1183.

Dong, Y.H., Xu, J.L., Li, X.Z. and Zhang, L.H. (2000) AiiA, an enzyme that inactivates the acylhomoserine lactone quorum-sensing signal and attenuates the virulence of *Erwinia carotovora. Proceedings of the National Academy of Sciences of the United States of America* 97, 3526–3531.

Doolotkeldieva, T., Bobusheva, S. and Suleymankisi, A. (2016) Biological control of *erwinia carotovora* ssp. *carotovora* by *streptomyces* species. *Advances in Microbiology* 6, 104–114.

Dutta, S. and Podile, A.R. (2010) Plant growth promoting rhizobacteria (PGPR): the bugs to debug the root zone. *Critical Reviews in Microbiology* 36(3), 232–244.

El-Borollosy, A.M. and Oraby, M.M. (2012) Induced systemic resistance against Cucumber mosaic cucumovirus and promotion of cucumber growth by some plant growth-promoting rhizobacteria. *Annals of Agricultural Science* 57(2), 91–97.

Fernando, W.G.D., Nakkeeran, S. and Zhang, Y. (2005) Biosynthesis of antibiotics by PGPR and its relation in biocontrol of plant diseases. In: Siddiqui, Z.A. (ed.) *PGPR: Biocontrol and Biofertilization*. Springer, Dordrecht, The Netherlands, pp. 67–109.

Frankenberger, W.T. and Arshad, M. (1995) Phytohormones in Soils; Microbial Production and Function. Marcel Dekker, New York.

Gallagher, L.A. and Manoil, C. (2001) *Pseudomonas aeruginosa* PAO1 kills *Caenorhabditis elegans* by cyanide poisoning. *Journal of Bacteriology* 183(21), 6207–6214.

Glick, B.R. (1995) The enhancement of plant growth by free-living bacteria. *Canadian Journal of Microbiology* 41, 109–117.

Gupta, G., Parihar, S.S., Ahirwar, N.K., Snehi, S.K. and Singh, V. (2015) Plant growth promoting rhizobacteria (PGPR): current and future prospects for development of sustainable agriculture. *Journal of Microbial and Biochemical Technology* 7(2), 096–102.

Hongwei, L.I., Xueling, D.I.N.G., Chao, W.A.N.G., Hongjiao, K.E., Zhou, W.U. *et al.* (2016) Control of Tomato yellow leaf curl virus disease by *Enterobacter asburiae* BQ9 as a result of priming plant resistance in tomatoes. *Turkish Journal of Biology* 40, 150–159.

Jagadeesh, K.S., Kulkarni, J.H. and Krishnaraj, P.U. (2001) Evaluation of the role of fluorescent siderophore in the biological control of bacterial wilt in tomato using Tn5 mutants of fluorescent *Pseudomonas* sp. *Current Science* 81(8), 882–883.

Jahanian, A., Chaichi, M.R., Rezaei, K., Rezayazdi, K. and Khavazi, K. (2012) The effect of plant growth promoting rhizobacteria (pgpr) on germination and primary growth of artichoke (Cynara scolymus). *International Journal of Agriculture and Crop Sciences* 4, 923–929.

Jarausch, W. and Ester Torres, E. (2014) Management of phytoplasma-associated diseases. In: Bertaccini, A. (ed.) *Phytoplasmas and Phytoplasma Disease Management: How to Reduce Their Economic Impact*. IPWG - International Phytoplasmologist Working Group, Bologna, pp. 199–208.

Jogaiah, S., Shivanna, R.K., Gnanaprakash, P.H. and Hunthrike, S.S. (2010) Evaluation of plant growth-promoting Rhizobacteria for their efficiency to promote growth and induce systemic resistance in pearl millet against downy mildew disease. *Archives of Phytopathology and Plant Protection* 43(4), 368–378.

Joseph, B., Patra, R.R. and Lawrence, R. (2007) Characterization of plant growth promoting Rhizobacteria associated with chickpea (*Cicer arietinum* L). *International Journal of Plant Production* 1 (Suppl 2), 141–152.

Keswani, C., Bisen, K., Singh, S.P., Sarma, B.K. and Singh, H.B. (2016) A proteomic approach to understand the tripartite interactions between plant-*Trichoderma*-pathogen: investigating the potential for efficient biological control. In: Hakeem, K.R. and Akhtar, M.S. (eds) *Plant, Soil and Microbes Vol. 2. Mechanisms and Molecular Interactions*. Springer, Switzerland, pp. 79–93.

Khabbaz, S.E., Zhang, L., Cáceres, L.A., Sumarah, M., Wang, A. and Abbasi, P.A. (2015) Characterisation of antagonistic *Bacillus* and *Pseudomonas* strains for biocontrol potential and suppression of damping-off and root rot diseases. *Annals of Applied Biology* 166, 456–471.

Khare, E. and Arora, N.K. (2010) Effect of Indole-3-Acetic Acid (IAA) Produced by *Pseudomonas aeruginosa* in Suppression of Charcoal Rot Disease of Chickpea. *Current Microbiology* 61, 64–68.

Kim, Y.C., Jung, H. and Kim, K.Y. (2008) An effective biocontrol bioformulation against phytophthora blight of pepper using growth mixtures of combined chitinolytic bacteria under different field conditions. *European Journal of Plant Pathology* 120, 373–382.

Kloepper, J.W. (1983) Effect of seed piece inoculation with plant growth-promoting rhizobacteria on populations of *Erwinia carotovora* on potato roots and daughter tubers. *Phytopathology* 73, 217–219.

Kloepper, J.W., Lifshitz, R. and Zablotowicz, R.M. (1989) Free-living bacterial inocula for enhancing crop productivity. *Trends in Biotechnology* 7, 39–43.

Kokalis-Burelle, N. (2002) Biological control of tomato diseases. In: Gnanamanickam, S.S. (ed.) *Biological Control of Crop Diseases*. Marcel Dekker, New York, pp. 231–268.

Kuc, J. (1995) Phytoalexins, stress metabolism, and disease resistance in plants. *Annual Review of Phytopathology* 33, 275–297.

Kumar, T., Wahla, V., Pandey, P., Dubey, R.C. and Maheshwari, D.K. (2009) Rhizosphere competent *Pseudomonas aeruginosa* in the management of *Heterodera cajani* on sesame. *World Journal of Microbiology and Biotechnology* 25, 277–285.

Lebeda, A. and Cohen, Y. (2011) Cucurbit downy mildew (*Pseudoperonospora cubensis*)—biology, ecology, epidemiology, host-pathogen interaction and control. *European Journal of Plant Pathology* 129, 157–192.

Lee, S.-W., Lee, S.-H., Balaraju, K., Park, K.-S., Nam, K.-W. *et al.* (2014) Growth promotion and induced disease suppression of four vegetable crops by a selected plant growth-promoting rhizobacteria (PGPR) strain Bacillus subtilis 21-1 under two different soil conditions. *Acta Physiologiae Plantarum* 36, 1353–1362.

Lian, L., Xie, L., Zheng, L. and Lina, Q. (2011) Induction of systemic resistance in tobacco against Tobacco mosaic virus by *Bacillus* spp. *Biocontrol Science and Technology* 21(3), 281–292.

Liu, L., Kloepper, J.W. and Tuzun, S. (1995) Induction of systemic resistance in cucumber against Fusarium wilt by plant growth-promoting rhizobacteria. *Phytopathology* 85, 695–698.

Logan, D.C. and Stewart, G.R. (1991) Role of ethylene in the germination of the hemiparasite *Striga hermonthica*. *Plant physiology* 97, 1435–1438.

Mahajan-Miklos, S., Tan, M.-W., Rahme, L.G. and Ausubel, F.M. (1999) Molecular mechanisms of bacterial virulence elucidated using a *Pseudomonas aeruginosa–Caenorhabditis elegans* pathogenesis model. *Cell* 96, 47–56.

Marija, S. and Dragana, B. (2009) Effect of plant growth-promoting Rhizobacteria on the germination of field dodder (*Cuscuta campestris* Yunck.) and alfalfa. *Plant Protection (Zaštita bilja)* 60(4), 227–236.

Mauch, F., Mauch-Mani, B. and Boller, T. (1988) Antifungal hydrolase in pea tissue. II. Inhibition of fungal growth by combinations of chitinase and β-1,3-glucanases. *Plant Physiology* 88, 936–942.

Maurhofer, M., Hase, C., Meuwly, P., Métraux, J.P. and Défago, G. (1994) Induction of systemic resistance of tobacco to tobacco necrosis virus by the root-colonizing *Pseudomonas fluorescens* strain CHA0: influence of the gacA gene and of pyoverdine production. *Phytopathology* 84, 139–146.

Maurhofer, M., Reimmann, C., Sacherer, S.P., Heeb, S., Haas, D. and Defago, G. (1998) Salicylic acid biosynthetic genes expressed in *Pseudomonas fluorescens* strain P3 improve the induction of systemic resistance in tobacco against tobacco necrosis virus. *Phytopathology* 88, 678–684.

Miché, L., Bouillant, M.-L., Rohr, R., Sallė, G. and Bally, R. (2000) Physiological and cytological studies on the inhibition of Striga seed germination by the plant growth-promoting bacterium *Azospirillum brasilense*. *European Journal of Plant Pathology* 106, 347–351.

Mohammed, A.M., AL-Ani, L.K.T., Bekbayeva, L. and Salleh, B. (2011) Biological Control of *Fusarium oxysporum* f. sp. *cubense* by *Pseudomonas fluorescens* and BABA *in vitro*. *World Applied Sciences Journal* 15(2), 189–191.

Mohammed, A.M., Al-Ani, L.K.T. and Salleh, B. (2013) Potential management of Fusarium oxysporum f. sp. cubense, the banana wilt pathogen by using pseudomonas and beta-amino-butyric acid (BABA). *International Symposium on Tropical Fungi*, ISTF, IPB International Convention Center, Bogor, Indonesia; 09/2013, p. 37.

Mondal, K.K. and Verma, J.P. (2002) Biological control of cotton diseases. In: Gnanamanickam, S.S. (ed.) *Biological Control of Crop Diseases*, Marcel Dekker, New York, pp. 97–119.

Mondal, K.K., Singh, R.P., Dureja, P., Verma, J.P. (2000) Secondary metabolites of cotton rhizobacteria in the suppression of bacterial blight of cotton. *Indian Phytopathology* 53, 22–27.

Murphy, J.F., Zehnder, G.W., Schuster, D.J., Sikora, E.J., Polston, J.E., and Kloepper, J.W. (2000) Plant growth-promoting rhizobacterial mediated protection in tomato against Tomato mottle virus. *Plant Disease* 84, 779–784.

Nehl, D.B., Allen, S.J. and Brown, J.F. (1997) Deleterious rhizosphere bacteria: an integrating perspective. *Applied Soil Ecology* 5, 1–20.

O'Sullivan, D.J. and O'Gara, F. (1992) Traits of fluorescent *Pseudomonas* spp. involved in suppression of plant root pathogens. *Microbiological Reviews* 56, 662–676.

Park, K., Park, Y.-S., Ahamed, J., Ryutta, S., Ryu, H. *et al.* (2016) Elicitation of induced systemic resistance of chili pepper by iturin A analogs derived from *Bacillus vallismortis* EXTN-1. *Canadian Journal of Plant Science* 96, 564–570.

Pierson, L.S. and Thomashow, L.S. (1992) Cloning and heterologous expression of the phenazine biosynthetic locus from *Pseudomonas aureofaciens*. *Molecular Plant-Microbe Interactions* 5, 330–339.

Podile, A.R. and Kishore, G.K. (2006) Plant growth-promoting rhizobacteria. In: Gnanamanickam, S.S. (ed.) *Plant-associated Bacteria*. Springer, Netherlands, pp. 205–230.

Prasad, R., Kumar, M. and Varma, A. (2015) Role of PGPR in soil fertility and plant health. In: Egamberdieva, D., Shrivastava, S. and Varma, A. (eds) *Plant-growth-Promoting Rhizobacteria (PGPR) and Medicinal Plants*. Springer, Cham, Switzerland, pp. 247–260.

Pusey, P.L. (2002) Biological control agents for fire blight of apple compared under conditions limiting natural dispersal. *Plant Disease* 86, 639–644.

Ramjegathesh, R., Samiyappan, R., Raguchander, T., Prabakar, K. and Saravanakumar, D. (2013) Plant–PGPR interactions for pest and disease resistance in sustainable agriculture. In: Maheshwari, D.K. (ed.) *Bacteria in Agrobiology: Disease Management*. Springer Berlin Heidelberg, Germany, pp. 293–320.

Raupach, G.S., Liu, L., Murphy, J.F., Tuzun, S. and Kloepper, J.W. (1996) Induced systemic resistance of cucumber and tomato against cucumber mosaic virus using plant growth-promoting rhizobacteria (PGPR). *Plant Disease* 80, 891–894.

Raupach, G.S., Taensa, I. and Kloepper, J.W. (2000) Biocontrol of cucumber diseases in the field by plant growth-promoting rhizobacteria with and without methyl bromide fumigation. *Plant Disease* 84, 1073–1075.

Reetha, A.K., Pavani, S.L. and Mohan, S. (2014) Hydrogen cyanide production ability by bacterial antagonist and their antibiotics inhibition potential on *Macrophomina phaseolina* (tassi.) goid. *International Journal of Current Microbiology and Applied Sciences* 3(5), 172–178.

Romero, F.M., Marina, M. and Pieckenstain, F.L. (2016) Novel components of leaf bacterial communities of field-grown tomato plants and their potential for plant growth promotion and biocontrol of tomato diseases. *Research in Microbiology* 167(3), 222–233, doi: 10.1016/j.resmic.2015.11.001.

Ryu, C.-M., Murphy, J.F., Mysore, K.S. and Kloepper, J.W. (2004) Plant growth-promoting rhizobacteria systemically protect *Arabidopsis thaliana* against Cucumber mosaic virus by a salicylic acid and NPR1-independent and jasmonic acid-dependent signaling pathway. *The Plant Journal* 39, 381–392.

Saddler, G. (2002). Bacteria and plant disease. In: Waller, J.M., Lenné, J.M. and Waller, S.J. (eds) *Plant Pathologist's Pocketbook*. CAB International, Wallingford, UK, pp. 94–107.

Saha, D., Purkayastha, G.D., Ghosh, A., Isha, M. and Saha, A. (2012) Isolation and characterization of two new *Bacillus subtilis* strains from the rhizosphere of eggplant as potential biocontrol agents. *Journal of Plant Pathology* 94 (1), 109–118.

Saharan, B.S. and Nehra, V. (2011) Plant growth promoting rhizobacteria: a critical review. *Life Sciences and Medicine Research* LSMR-21, 1–30.

Sailaja, P.R. and Podile, A.R. (1998) A phytoalexin is modified to less fungistatic substances by crown rot pathogen in groundnut (Arachis hypogaea L.). *Indian Journal of Experimental Biology* 36, 631–634.

Santhi, A. and Sivakumar, V. (1995) Biocontrol potential of *Pseudomonas fluorescens* (Migula) against root-knot nematode, *Meloidogyne incognita* (Kofoid and White) Chitwood, 1949 on tomato. *Journal of Biological Control* 9, 113–115.

Senthilkumar, M., Govindasamy, V. and Annapurna, K. (2007) Role of antibiosis in suppression of charcoal rot disease by soybean endophyte *Paenibacillus* sp. HKA-15. *Current Microbiology* 55, 25–29.

Sharifi, R. and Ryu, C.-M. (2016) Are bacterial volatile compounds poisonous odors to a fungal pathogen botrytis cinerea, alarm signals to Arabidopsis seedlings for eliciting induced resistance, or both? *Frontiers in Microbiology* 7, 196. Doi: 10.3389/fmicb.2016.00196.

Sharma, A., Pathak, A., Sahgal, M., Meyer, J.-M., Wray, V. and Johri, B.N. (2007) Molecular characterization of plant growth promoting rhizobacteria that enhance peroxidase and phenylalanine ammonia-lyase activities in chili (*Capsicum annuum* L.) and tomato (*Lycopersicon esculentum* Mill.). *Archives of Microbiology* 188, 483–494.

Sharma, R., Chauhan, A. and Shirkot, C.K. (2015) Characterization of plant growth promoting *Bacillus* strains and their potential as crop protectants against *Phytophthora capsici* in tomato. *Biological Agriculture and Horticulture* 31(4), 240–244.

Sharma, S.K., Johri, B.N., Ramesh, A., Joshi, O.P. and Prasad, S.V.S. (2011) Selection of plant growth-promoting Pseudomonas spp. that enhanced productivity of soybean-wheat cropping system in central India. *Journal of Microbiology and Biotechnology* 21, 1127–1142.

Shen, L., Wang, F., Liu, Y., Qian, Y., Yang, J. and Sun, H. (2013) Suppression of tobacco mosaic virus by *Bacillus amyloliquefaciens* strain Ba33. *Journal of Phytopathology* 161, 293–294.

Siddiqui, I.A. and Shaukat, S.S. (2004) Systemic resistance in tomato induced by biocontrol bacteria against the root- knot nematode, *Meloidogyne javanica* is independent of salicylic acid production. *Journal of Phytopathology* 152, 48–54.

Siddiqui, Z.A. (2004) Effects of plant growth promoting bacteria and composed organic fertilizers on the reproduction of *Meloidogyne incognita* and tomato growth. *Bioresource Technology* 95, 223–227.

Singh, A., Maurya, S., Singh, R. and Singh, U.P. (2012) Antibiotic potential of plant growth promoting rhizobacteria (PGPR) against *Sclerotium rolfsii*. *Archives of Phytopathology and Plant Protection* 45(14), 1655–1662.

Smith, L.J. (1994) The potential of rhizobacteria as biological control agents of root-knot nematode. Thesis, University of Queensland, Australia.

Soohee, C. and Kim, S.-D. (2005) Biological control of phytopathogenic fungi by bacillus amyloliquefaciens 7079; suppression rates are better than popular chemical fungicides. *Journal of Microbiology and Biotechnology* 15(5), 1011–1021.

Souza, R.D., Ambrosini, A. and Passaglia, L.M.P. (2015) Plant growth-promoting bacteria as inoculants in agricultural soils. *Genetics and Molecular Biology* 38(4), 01–419.

Tian, H., Riggs, R.D. and Crippen, D.L. (2000) Control of soybean cyst nematode by chitinolytic bacteria with chitin substrate. *Journal of Nematology* 32(4), 370–376.

Toyoda, H., Hashimoto, H., Utsumi, R., Kobayashi, H. and Ouchi, S. (1988) Detoxification of fusaric acid by a fusaric acid-resistant mutant of *Pseudomonas solanacearum* and its application to biological control of Fusarium wilt of tomato. *Phytopathology* 78, 1307–1311.

Urban, J. and Lebeda, A. (2006) Fungicide resistance in cucurbit downy mildew—methodological, biological and population aspects. *Annals of Applied Biology* 149, 63–75.

Uren, N.C. (2000) Types, amounts and possible functions of compounds released into the rhizosphere by soil grown plants. In: Pinton, R., Varanini, Z. and Nannipieri, P. (eds) *The Rhizosphere, Biochemistry and Organic Substances at the Soil Interface*. Marcel Dekker, New York, 19–40.

van Loon, L.C. and Bakker, P.A.H.M. (2005) Induced systemic resistance as a mechanism of disease suppression by rhizobacteria. In: Siddiqui, Z.A. (ed.) *PGPR: Biocontrol and Biofertilization*. Springer, Netherlands, pp. 39–66.

van Loon, L.C., Bakker, P.A.H. and Pieterse, C.M.J. (1998) Systemic resistance induced by rhizosphere bacteria. *Annual Review of Phytopathology* 36, 453–483.

Vessey, J.K. (2003) Plant growth promoting rhizobacteria as biofertilizers. *Plant Soil* 255, 571–586.

Waller, J.M. (2002) Virus diseases. In: Waller, J.M., Lenné, J.M. and Waller, S.J. (eds) *Plant Pathologist's Pocketbook*. CAB International, Wallingford, UK, pp. 108–125.

Wang, Y., Ohara, Y., Nakayashiki, H., Tosa, Y. and Mayama, S. (2005) Microarray analysis of the gene expression profile induced by the endophytic plant growth-promoting rhizobacteria, *Pseudomonas fluorescens* FPT9601-T5 in Arabidopsis. *Molecular Plant-Microbe Interactions* 18, 385–396.

Zehnder, G.W., Yao, C., Murphy, J.F., Sikora, E.R. and Kloepper, J.W. (2000) Induction of resistance in tomato against cucumber mosaic cucumovirus by plant growth-promoting rhizobacteria. *BioControl* 45, 127–137.

Zermane, N., Souissi, T., Kroschel, J. and Sikora, R. (2007) Biocontrol of broomrape (*Orobanche crenata* Forsk. and *Orobanche foetida* Poir.) by *Pseudomonas fluorescens* isolate Bf7-9 from the faba bean rhizosphere. *Biocontrol Science and Technology* 17(5), 483–497.

Zhang, L., Khabbaz, S.E., Wang, A., Li, H. and Abbasi, P.A. (2015) Detection and characterization of *broad-spectrum antipathogen* activity of novel rhizobacterial isolates and suppression of Fusarium crown and root rot disease of tomato. *Journal of Applied Microbiology* 118(3), 685–703.

Zhao, P.P., Shang, J., Guo, Z.C., Xie, H.F., Xi, D.H. *et al*. (2013) Temperature-related effects of treatments with jasmonic and salicylic acids on Arabidopsis infected with cucumber mosaic virus. *Russian Journal of Plant Physiology* 60(5), 672–680.

Zhou, W.-W. and Niu, T.N. (2009) Purification and some properties of an extracellular ribonuclease with antiviral activity against tobacco mosaic virus from *Bacillus cereus*. *Biotechnology Letters* 31, 101–105.

24 Role of *Trichoderma* Secondary Metabolites in Plant Growth Promotion and Biological Control

Jyoti Singh,[1] Rahul Singh Rajput,[2] Kartikay Bisen,[2] Surendra Singh[1] and H.B. Singh[2]*
[1]*Department of Botany, Institute of Science, Banaras Hindu University, Varanasi-221005, India;* [2]*Department of Mycology and Plant Pathology, Institute of Agricultural Sciences, Banaras Hindu University, Varanasi-221005, India*

24.1 Introduction

After World War II ended, the world's population started rising due to instant increase in population of less developed countries. The resultant effects of this tremendous growth will be observed on living standards, resource use and the environment for a long span of time (UNPD, 2008). Cultivation of plants is closely linked to the development of human civilization, which has been ongoing for more than 10,000 years, and therefore plant diseases have been a major concern to mankind for a long time. In agriculture, the maximum loss is due to plant diseases, which is a major challenge faced by cultivators of each crop at any time. Insect pests and soilborne fungal diseases are important biotic factors causing detrimental effects to agricultural and livestock products. Approximately 10–30% of produced crop loss is due to pests, diseases and weeds (Kumar and Gupta, 2012). The losses can be direct or indirect, including reduction in quality and quantity of crop, decrease in crop production and loss of natural resources. In today's competitive environment reacting to the above-mentioned problems, there is an urgent need for increased production of quality crops with no blemishes, free of disease and pests, which has motivated excessive use of chemical fertilizers, leading to serious environmental problems.

For such crops, management of diseases is required which should be focused on preventing the establishment of disease and minimizing the development and spread of established disease. This control is crucial for reliable production of agricultural outputs such as food and fodder; it should also cause significant reduction in the required usage of land, water, fuel and other inputs. Each and every plant, either in natural or in cultivated populations, possesses an inherent disease resistance, but there are also various devastating plant diseases (like Irish Potato Famine, Chestnut blight) and severe diseases (like Rice blast, Soybean cyst nematode, Citrus canker) which are quite recurrent. There is need for continuous advancement in this field of plant pathology to improve the management of diseases and also to cope with the pressure caused by spreading of plant pathogens, changes in agricultural practices,

*E-mail: hbs1@rediffmail.com

© CAB International 2017. *Advances in PGPR Research* (eds H.B. Singh, B.K. Sarma and C. Keswani)

and progressing evolution. There are diverse methods for management of these diseases. Once we perceive the category of disease, then the handling and management of disease will be easier. There are basically five methods of plant disease management, viz. regulatory, cultural, biological, physical and chemical/synthetic (Agrios, 2005). The chemical method is routinely applied as the requisite strategy to constrain plant disease and is a relatively effortless, uncomplicated and affordable technique (Tweedy, 1983; Lenteren, 2000; Agrios, 2005). It is applied with proliferative vigour because of its potency in disease management (Delp, 1983). Inorganic compounds (sulfur, copper, bicarbonates), synthetic compounds (petroleum derived), botanical (neem and cinnamaldehyde), and beneficial microorganisms (mycofungicides) constitute numerous divergent types of fungicides that are available for the control of fungal diseases (Dreistadt *et al.*, 2004).

Regardless of their efficiency in restricting disease, chemical fungicides are causing a menace to human health and polluting the environment (Herr, 1995). Latterly, plenty of synthetic fungicides have been banned in the western world because of their undesirable attributes such as high and acute toxicity, long deterioration phases, accumulation in food chains and extension of their power to devastate useful organisms (Strange, 1993). These products also induce innumerable issues unanticipated during their application in the field: subtle and persistent poisoning of the applier, field labour, and even consumers; destruction of aquatic life, birds and other wildlife (Marco *et al.*, 1987; Forget *et al.*, 1993; Perry *et al.*, 1998; NRC, 2000). New restrictions have been enforced on management practices as environmental pollution began to threaten to be a pandemic hazard. To reduce these complications, government agencies are setting limits on application of pesticides. Subsequently, chemicals for disease control inexorably may be subjected to further constraints (Baker, 1992). Taking into consideration these undesirable attributes of synthetic fungicides, it is now a top priority to encourage auxiliary treatments that are less hazardous to humans, plants and animals, and have less impact on the environment. A solution to these problems regarding synthetic fungicides is the application of biological compounds that control microorganisms.

One of the most productive approaches of biological control is employing microbes in agriculture. There is an influential group of fungi and bacteria that have antagonistic effects on other microorganisms and this property can be harnessed as a form of biological control of plant pathogens (Orietta and Larrea, 2001). The after-effects of utilizing beneficial microbes are strain dependent and their benefits for the treated plant include: (i) inception of a hostile microbial community in the rhizosphere; (ii) removal of pathogens; (iii) comprehensive betterment of plant health; (iv) growth promotion; (v) elevated nutrient access and uptake; and (vi) upgraded host resistance to both biotic and abiotic stresses (Harman, 2000; Harman *et al.*, 2004; Vinale *et al.*, 2008; Keswani *et al.*, 2013, 2014; Bisen *et al.*, 2015; Keswani, 2015).

Biocontrol agents (BCAs) offer one of the most well-grounded strategies to attain by themselves the objective of disease control, or combined with minimal doses of chemicals in the management of plant pathogens to leave the smallest possible impact of toxic chemical residues on the ecosystem (Chet and Inbar, 1994; Harman and Kubicek, 1998; Bisen *et al.*, 2016; Keswani *et al.*, 2016a, b). There are two varieties of biocontrol agents: general (capable of constraining a massive number of taxonomically different pathogens, e.g. *Trichoderma, Bacillus, Pseudomonas*, yeast, etc.) and specialist (capable of controlling only targeted species, e.g. *Agrobacterium, Aspergillus*, etc.). So far, plenty of BCAs have been recorded and are available as commercial products, including strains belonging to bacterial genera such as *Pseudomonas, Agrobacterium, Streptomyces* and *Bacillus*, and fungal genera such as *Trichoderma, Gliocladium, Ampelomyces, Candida* and *Coniothyrium*.

24.2 *Trichoderma*: An Overview

Fungi are an extremely diverse group of organisms, including about 230,000 species encompassing an enormous diversity with discrete ecological niches, life-cycle approaches, and morphologies. Among the 1.5 million species estimated, only about 5% were formally classified and of them only a limited number are considered as effective biocontrol agents. *Trichoderma* spp. are ubiquitous soil fungi, members of the filamentous ascomycete genus *Trichoderma* (teleomorph *Hypocrea*, Ascomycota, Dikarya), opportunistic, avirulent plant symbionts; they are among the most continually isolated biotrophic and saprotrophic fungi as they decompose organic matter, function as parasites and antagonists of many phytopathogenic fungi and so protect plants from disease. These are often encountered on other fungi, on dead wood and bark, in soil and rhizosphere as a component of the plant root ecosystem. However, these species also have the potential to colonize plant roots and rhizosphere. The ability of the fungi of parasitizing plant pathogenic fungi makes them applicable as biofungicides (Hjeljord and Tronsmo, 1998; Mukhopadhyay *et al.*, 1992; Mukhopadhyay and Mukherjee, 1996; Chet *et al.*, 1998; Harman and Bjorkmann, 1998; Singh *et al.*, 2016a, b). The prospective of *Trichoderma* species to aid increased growth response was demonstrated both in greenhouse experiments and in the hydroponic system. Observations were made of 30% increase in seedling emergence, and those plants also exhibited a 95% increase in root area. Similarly in *Trichoderma*-inoculated plants, a striking increment in phosphorus and iron concentration was observed. *Trichoderma* can promote priming for upgraded defense in plants. The promotion of plant growth up to 300% has been reported in the case of *Trichoderma*. Also the production of organic acids such as gluconic, fumaric and citric acids have been announced in *Trichoderma* spp. that can lower soil pH and allow solubilization of phosphates, and additionally micro- and macronutrients such as iron, manganese and magnesium that play a crucial role in plant metabolism. Benefits including biological control of plant diseases, induced systemic resistance, enhanced nutrient availability and uptake, promoting plant growth, upgraded crop yields and degrading xenobiotic pesticides have been observed (Harman, 2000). The major mechanisms applied by fungal antagonists to restrict the growth of plant pathogens are: antibiosis, competition and parasitism, and they also induce systemic resistance (van Loon *et al.*, 1998; Bisen *et al.*, 2015).

Among all the above-mentioned, the biocontrol mechanisms that *Trichoderma* mostly utilizes in direct confrontation with fungal pathogens are: (a) mycoparasitism, which is a complex process comprising recognition of the host, attack and subsequent penetration by involvement of chitinase and ß-1,3-glucanase and killing due to degradation of the cell wall leading to the lysis of hyphae; and (b) antibiosis, which is the process of secretion of anti-microbial compounds to suppress or kill pathogenic fungi in the vicinity of its growth area by disruption of cell membranes, inhibition of metabolic activity and induction of plant defense systems. *Trichoderma* is documented as the emitter of varied types and quantities of secondary metabolites with different biological activities (Ghisalberti and Sivasithamparam, 1991; Sivasithamparam and Ghisalberti, 1998; Keswani *et al.*, 2013). Antibiosis is adjudged the most conspicuous, in which assemblages of secondary metabolites such as antibiotics and toxins are obtained, which impart the antagonistic activity of fungal biocontrol agents against plant pathogens (Figure 24.1). It has been reported that several fungal biocontrol agents exhibit these antimicrobial secondary metabolites (Gottlieb and Shaw, 1970; Fries, 1973; Hutchinson, 1973; Sivasithamparam and Ghisalberti, 1998; Vyas and Mathur, 2002; Keswani *et al.*, 2014). Globally, *Trichoderma* is present at the preeminent region in the moiety of fungal biocontrol agents (Whipps and Lumsden, 2001). *Trichoderma* is the most studied biocontrol agent for management of plant diseases and commercially marketed as biopesticides, biofertilizers and soil amendments in many places. Some strains of *Trichoderma* play a crucial role in the bioremediation of contaminated soils and can also be applied in integrated pest management and phytoremediation. Several integrated pest management (IPM) applications, combining biological and chemical methods, have been suggested for the biocontrol of

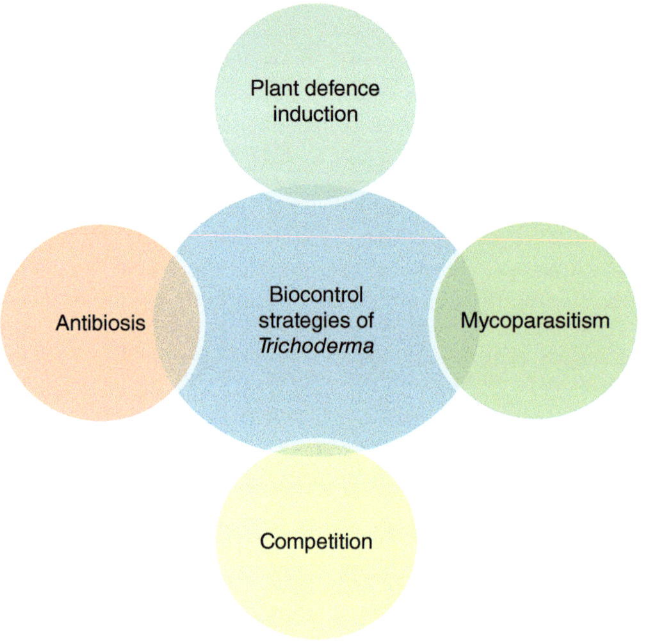

Fig. 24.1. Biocontrol mechanisms of *Trichoderma*.

fungal plant diseases. In this context our cardinal concern will be the ability of secondary metabolites obtained from different *Trichoderma* spp.

24.3 Secondary Metabolites

In every cell of living organisms a process takes place that assembles and deploys key organic compounds, i.e. metabolites (Mumpuni et al., 1998). The process which is responsible for disintegration of food and other chemicals into energy and materials needed for physical wellness, development and reproduction, is termed metabolism, and the metabolites can be the originator materials, midway materials, or final products of these chemical reactions. A convoluted amalgamation of various metabolites and reactions are linked in the production of all that is essential for an organism to sustain life. The metabolites that are directly involved in normal growth, development and reproduction are considered as primary metabolites. The most essential among them are carbohydrates, proteins, nucleic acids, and lipids. Secondary metabolites are organic compounds that are obliquely involved in the normal growth, development, or reproduction of an organism; their omission does not result in immediate death, as in the case of primary metabolites. But under adverse conditions, they transform into quite essential elements required for the survival of the organism. More than a century ago Kossel (1891) characterised secondary metabolites negatively – as those compounds that do not belong to a class of primary metabolites, a definition that has attracted criticism ever since.

Secondary metabolites comprise natural compounds which are chemically varied and have relatively low molecular weight (in most cases <3 kDa), and that are mainly produced by microorganisms and plants with a high degree of specificity, as their production is confined to a group of species or genera. Secondary metabolites are the products of specialized pathways, biosynthesized from primary metabolites (i.e. polyketides or mevalonate pathways from Acetyl Coenzyme A, or amino acids). Genes constrained for the production of these metabolites are clustered together and their expression is induced by the existence of some regulators (Herbert, 1989). More often the genes involved in biosynthesis of secondary metabolites are found in the genome; frequently as heterochromatin, therefore the putative products remain unknown. The basic reason is that under standard laboratory cultivation conditions, many of these gene clusters are not expressed, as they do not face any stress (Brakhage and Schroeckh, 2011). Production of secondary metabolites in fungi is a complex procedure associated with morphological development. During the normal life cycle of some fungi, the expression of secondary metabolites may occur at a predictable point where it is necessary to survive in unfavourable conditions, also produced by various microbes during processes of development and sporulation (Keller *et al.*, 2005). These products are of tremendous importance in biotechnological applications, including antibiotics, and are natural artefacts that can accomplish the inhibition of microbial growth (Mapleston *et al.*, 1992; Stone and Williams, 1992; Sekiguchi and Gaucher, 1977). Although not essential for their primary metabolic processes, microbes, and particularly fungi, produce various secondary metabolites, including compounds of industrial and economic relevance (Herbert, 1989). A wide range of secondary metabolites, which are structurally highly diverse, each of them effecting various biological effects such as competition, symbiosis, metal transport, differentiation, etc. are produced by various fungi; the survival of the organisms is aided by these compounds, often at very low concentrations, and they can be regarded as carriers of chemical communication in plant microbe and soil interactions, with prominent roles in signal transduction, development and cohabitation with other organisms (Demain and Fang, 2000; Keller *et al.*, 2005; Hoffmeister and Keller, 2007; Karlovsky, 2008; Osbourn, 2010).

The specificity of secondary metabolism incited the botanists and mycologists to adopt secondary metabolite production as a taxonomical attribute in plants (Smith, 1976) and fungi (Frisvad *et al.*, 1998). These compounds are extensively exploited for agricultural, medical or pharmaceutical purposes owing to their chemical and biological properties (Calvo *et al.*, 2002). Secondary metabolites are studied as the object of natural product chemistry and their enormous structural variability attracted and enhanced the curiosity of chemists. The pharmaceutical industries have been inspired by the biological activities of these natural compounds, to explore the microbial cultures and plant extracts for lead structures. This strategy proved highly successful as impressive numbers of compounds have been purified and their structures elucidated in the past four decades.

24.4 Secondary Metabolites of *Trichoderma*

Numerous strains of *Trichoderma* spp. are well reported as producing diverse secondary metabolites having low molecular weight in abundance, which also include antibacterial and antifungal antibiotics (Ghisalberti and Rowland, 1993; Sivasithamparam and Ghisalberti, 1998; Reino *et al.*, 2008; Vinale *et al.*, 2008, 2009) (Fig. 24.2). The quantitative production of these metabolites shows variation depending on the strain or species confronted, any particular compound or the endurance of the other microorganism (Vinale *et al.*, 2009). Some species/strains are reported to play an important part in plant growth promotion and induce the systemic resistance in plants (Harman *et al.*, 2004; Vinale *et al.*, 2008; Singh *et al.*, 2016a, b). The key factors which mainly contribute to the antagonistic character of these species are faster metabolic rates, anti-microbial metabolites, and physiological conformation. Mycoparasitism, spatial and nutrient competition, antibiosis by enzymes and secondary metabolites and activation of plant defence systems, are typical biocontrol mechanisms of these fungi. The secondary metabolites produced by *Trichoderma* spp. are strain dependent and include antifungal substances belonging to a variety of classes of chemical compounds. These metabolites

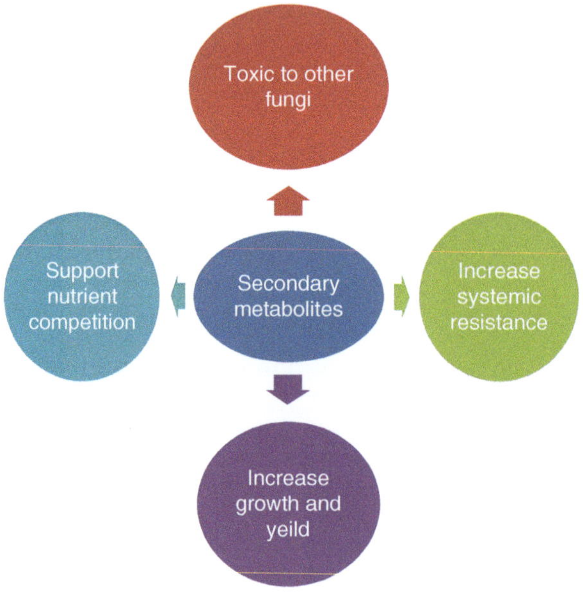

Fig. 24.2. Different functions of secondary metabolites obtained from *Trichoderma*.

have also been documented to affect the target-specific processes such as hyphal elongation and sporulation (Keller *et al.*, 2005). Some of the secondary metabolites obtained from *Trichoderma* strains at low concentration act as microbe associated molecular patterns (MAMPs) but show antimicrobial activity at high concentrations (Vinale *et al.*, 2008). Ghisalberti and Sivasithamparam (1991) classified these metabolites into three categories: (i) volatile antibiotics, i.e. 6-pentyl-a-pyrone (6PP) and most of the isocyanide derivates; (ii) water-soluble compounds, i.e. heptelidic acid or koningic acid; and (iii) peptaibols, which are linear oligopeptides of 12–22 amino acids rich in a-aminoisobutyric acid, N-acetylated at the N-terminus and containing an amino alcohol (Pheol or Trpol) at the C-terminus (Le Doan *et al.*, 1986; Rebuffat *et al.*, 1989). Thousands of secondary metabolite structures have been studied and published to date. Several other metabolites, viz. trichocaranes (Macias *et al.*, 2000), demethylsorbicillin, oxosorbicillinol (Abe *et al.*, 2000), trichodenones, harzialactone A and B, (*R*)-mevalonolactone (Amagata *et al.*, 1998), 6-*n*-pentyl pyrone, isonitrile acid (Brewer and Taylor, 1981; Graeme-Cook and Faull, 1991), trichoviridin, 3-(3-isocyano-6-oxabicyclo[3,1,0] hex-2-en-5-yl) acrylic acid and 3-(3-isocyanocyclopent-2- enylidene) propionic acid (Brewer *et al.*, 1982) have already been reported to be produced by *Trichoderma* spp.

24.5 Adequacy of Secondary Metabolites Inferred from *Trichoderma*

The production of a wide range of non-volatile and volatile antibiotics by *Trichoderma* spp. has been well documented (Weindling and Emerson, 1936; Sivasithamparam and Ghisalberti, 1998; Vyas and Mathur, 2002). Some of the secondary metabolites are phytotoxins that attack plants, mycotoxins that act on fungal pathogens, pigments with antioxidant activity and antibiotics inhibiting or killing microbial competitors. Pyrones, volatile metabolites responsible for coconut aroma, exhibit antifungal activities in both *in vivo* and *in vitro* conditions towards numerous plant pathogenic fungi. There is an interconnection between the production of pyrones and biocontrol activity of the microorganism

(Scarselletti and Faull, 1994; Worasatit *et al.*, 1994). These compounds also possess plant growth-promoting factors and enhance the activities of polyphenoloxidase and ß-1,3-glucanase in both root and shoot tissue leading to induction of defence responses in plants (El-Hasan and Buchenauer, 2009). Cytosporone S, another pyrone isolated from a *Trichoderma* spp. was documented recently to have *in vitro* antibiotic activity. Antifungal pyrones isolated from the culture filtrates of *T. harzianum* and *Trichoderma koningii* check the growth of many fungal pathogens, including *Bipolaris sorokiniana, Fusarium oxysporum, Gaeumannomyces graminis* var. *tritici, Phytophthora cinnamomi, Pythium middletonii* and *R. solani* (Claydon *et al.*, 1987; Simon *et al.*, 1988).

Koninginins are complex pyranes that possess *in vitro* antibiotic activity against *Gaeumannomyces graminis* var. *tritici*, and are also able to hinder the growth of soilborne pathogens including *Rhizactonia solani, Phytopthora cinnamomi, Pythium middletonnii, Fusarium oxysporum* and *Bipolaris sorokiniana* (Dunlop *et al.*, 1989).

Viridin, a steroidal metabolite isolated from diverse *Trichoderma* spp., possesses potential to inhibit spore germination in *Botrytis allii, Colletotrichum lini, Fusarium caeruleum, Penicillium expansum, Aspergillus niger* and *Stachybotrys atra* (Reino *et al.*, 2008). Viridiol produced by *T. viride* and *T. hamatum* in substrate with high C/N ratios, also shows similar antifungal and phytotoxic properties *in vivo* (Moffatt *et al.*, 1969; Howell and Stipanovic, 1994) and also aids as plant growth inhibitor.

Harzianopyridone, a nitrogen heterocyclic compound,and harzianic acid obtained from *T. harzianum* show antibiotic activity against *Rhizactonia solani, Pythium ultimatum* and *Sclerotinia sclerotiorum*. Harzianic acid also has quite good affinity for essential metals such as Fe^{3+} (Vinale *et al.*, 2013).

T22 azaphilones isolated from *T. harzianum* showed a remarkable inhibition in the growth of *R. solani, P. ultimatum* and *G. Graminis* var. *tritici*. *T. harzianum* also provides hazianolide, T39 butenolide and dehydroharzianolide which possess *in vitro* antifungal activities against several phytopathogenic agents (Almassi *et al.*, 1991; Vinale *et al.*, 2006).

Cerinolactone, a novel hydroxyl-lactone derivative isolated from *T. cerinum*, confers *in vitro* antifungal activity against *R. solani, P. ultimatum* and *B.cinerea* (Vinale *et al.*, 2012). Isocyano metabolites, i.e. dermadin and isonitrile trichoviridin, isolated from *T. viride* are difficult to obtain due to their instability, but possess antibiotic activity. A fungistatic metabolite, gliotoxin, and an inhibitor of oomycete pathogens, gliovrin, isolated from *T. virens* belong to the diketopiperazine class and are recorded to have potential as a biocontrol mechanism against soilborne fungal pathogens such as *R. solani* and *P. ultimatum* by antibiotic production (Howell and Stipanovic, 1983), and are also capable of inhibiting the spore germination of *B. cinerea* (Di Pietro *et al.*, 1993).

Alamethicin, a peptaibol obtained from *T. viride* inhibits ß-glucan synthase activity and prevents reconstruction of the cell wall in pathogenic fungi (Lorito *et al.*, 1996). It also induces a defence response in *Phaseolus lunatus* (Engelberth *et al.*, 2000) and *Arabidopsis thaliana* (Chen *et al.*, 2003). Peptaibols are produced by many species of *Trichoderma*, viz. *T. asperellum, T. harzianum, T. koningii, T. virens* and *T. viride* (Iida *et al.*, 1995; Wada *et al.*, 1995; Huang *et al.*, 1996; Landreau *et al.*, 2002; Chutrakul and Peberdy, 2005; Szekeres *et al.*, 2005; Wei *et al.*, 2005; Xiao-Yan *et al.*, 2006). Recently the biosynthesis and biological properties of peptaibols were reviewed (Szekeres *et al.*, 2005). *T. virens* produce the peptaibols antibiotic and also confer their role in biocontrol activity (Wiest *et al.*, 2002).

One of the approaches for disease management is detoxification of fungal toxins. *Trichoderma harzianum* hydrolases and *Trichoderma viride* are capable of degrading aflatoxin B1 (AFB1) and ochratoxin A (OTA) in *in vitro* conditions (Mann and Rehm, 1976). *B.cinerea* is inhibited by *Trichoderma* protease as it degrades hydrolytic enzymes involved in infection by pathogens (Elad and Kapat, 1999).

Among different *Trichoderma* strains, *T. harzianum* produces the highest number of siderophores, which is beneficial to plants from two perspectives: (a) it solublizes unavailable iron for uptake by plants, and (b) it suppresses growth of pathogens by depriving them of iron sources (Leong, 1986; Lehner *et al.*, 2013).

Trichodermin and viridin, produced by *Trichoderma* sp. even at very low concentrations, inhibited pathogenic fungal growth (Weindling and Emerson, 1936; Weindling, 1941). It was reported by Dennis and Webster (1971) that *T. polysporum* and *T. viride* also produced trichodermin and *T. hamatum* produced peptide antibiotics, and further they demonstrated the fungicidal effects of *Trichoderma* metabolites on phytopathogenic *Pythium*. Trichodermin-4, an antibiotic produced by *T. lignorum* was used to control plant diseases (Fedorinchik *et al.*, 1975). The germination of uredospore of the rust pathogen of groundnut, *Puccinia arachidis*, was inhibited by a phenol-like compound isolated from *T. harzianum* (Govindasamy and Balasubramanian, 1989). Seven *Trichoderma* spp. were evaluated for antagonistic activity against *Fusarium oxysporum*, *F. equiseti*, *F. solani*, *Sclerotinia sclerotiorum*, *S. minor*, *Rhizoctonia* sp. and *S. rolfsii* and in almost 80% of the interactions, the presence of diffusible metabolites in the medium was observed (Monaco *et al.*, 1994). Five different metabolites were produced by a strain of *T. harzianum* isolated from wheat roots and from the obtained metabolites, three new octaketide-derived compounds exhibited antifungal activity against *G. graminis* var. *tritici*, the causal agent of take-all disease of wheat (Ghisalberti and Rowland, 1993). The growth of *Fusarium moniliforme* and *Aspergillus flavus* was inhibited by isolates of *T. viride* and *T. harzianum* by producing inhibitory volatile compounds (Calistru *et al.*, 1997).

The volatile secondary metabolites produced from *Trichoderma pseudokoningii*, *T. viride* and *Trichoderma aureoviride* inhibited the process of protein synthesis and mycelial growth in two isolates of *Serpula lacrymans* in varying degrees (Humphris *et al.*, 2002). Four different *T. harzianum* isolates, along with the high secretion of chitinases, are involved in the biological control of the tomato root pathogen *Pyrenochaeta lycopersici* by the mechanism of production of nonvolatile metabolites (Perez *et al.*, 2002).

The variation in secondary metabolite production by *Trichoderma* sp. due to the effect of temperature was studied (Mukherjee and Raghu, 1997). They observed that high concentrations of fungi-toxic metabolites from *Trichoderma* were produced at high temperatures. However, *Trichoderma* sp. was not effective in suppressing *S. rolfsii* at temperatures above 30°C.

Secondary metabolites from two commercialized strains of *T. harzianum*, T22 and T39, were isolated for the first time by Vinale *et al.* (2006). Under *in vitro* conditions strain T22 produces three major bioactive compounds, from which one is a new azaphilone that showed remarkable antifungal activity against *R. solani*, *P. ultimum* and *G. graminis* var. *tritici* (Table 24.1).

24.6 Conclusion

The success rate of biocontrol agents relies upon the interactions of beneficial microbes established with pathogens and plants by the assistance of their metabolites. *Trichoderma* is taken into account as a promising candidate for inhibiting the phytopathogens because these species perform well in both biocontrol and plant growth promotion. But they have certain impediments, such as being area-specific because all strains cannot equally and fully flourish in all environmental conditions and larger spore count with higher viability is required in the formulations used in the field. Better understanding of processes is required not only for the application of safer and less expensive methods to protect plants and increase crop yield. The difficulties associated with the use of living microbes can be overcome by the introduction of new biopesticides and biofertilizers, i.e. based on the metabolites or bioactive compounds. Target specificity regardless of geographical location, longer shelf life and requirement of fewer amounts are the convenience associated with secondary metabolites. They can also be produced in large quantities on an industrial scale, easily separated from the fungal biomass, dried and

Table 24.1. Representative list of *Trichoderma* secondary metabolites.

Functions	Involved secondary metabolites	Structures
Biological Control	Harzianic acid	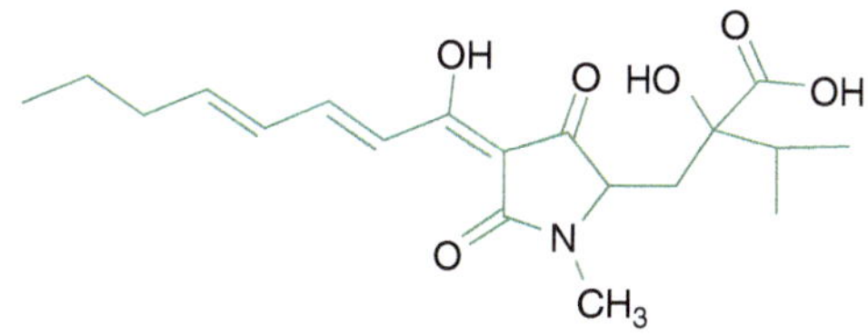
	Harzianopyridone	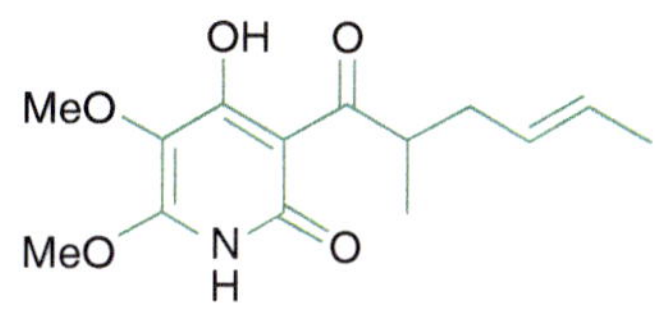
	Harzianolide	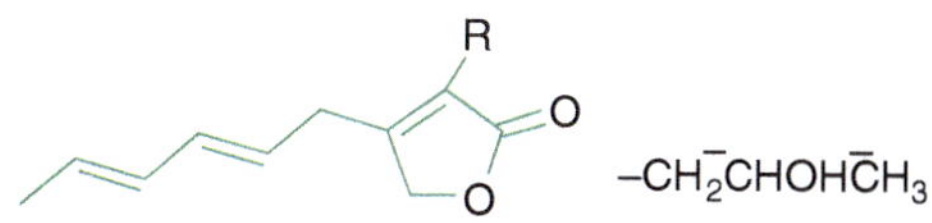
	Viridin	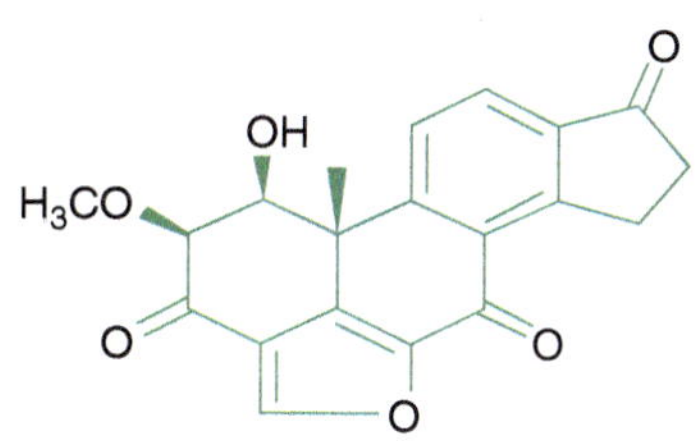
	6-pentyl-2H-pyran-2-one	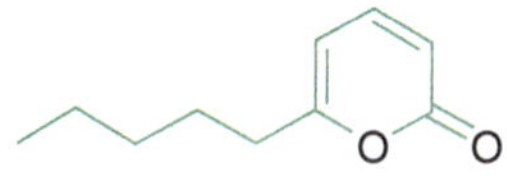
	Koninginins	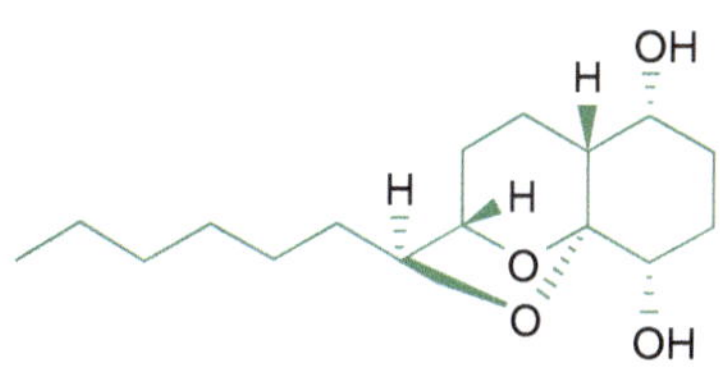
Plant Growth Promotion	Hazianolide	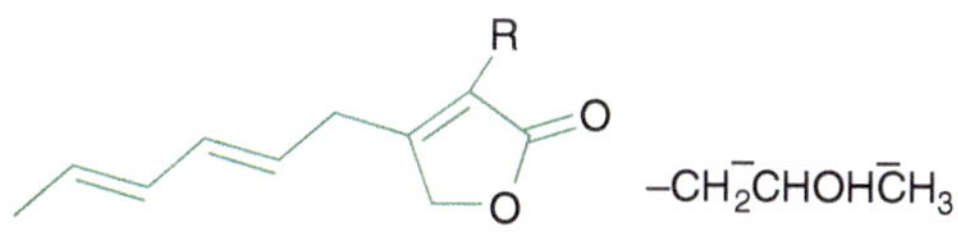
	Viridiol	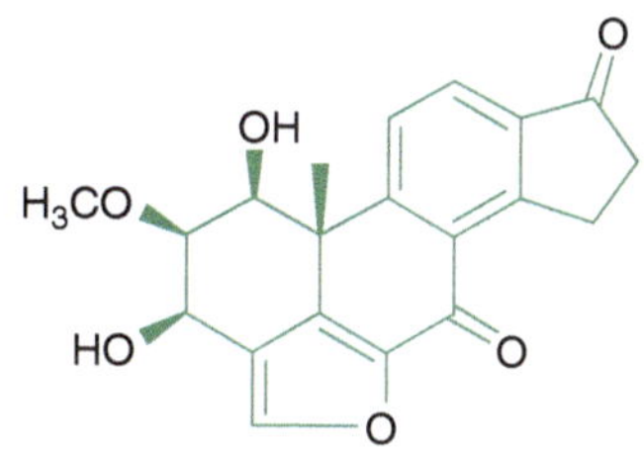

Continued

Table 24.1. Continued.

Functions	Involved secondary metabolites	Structures
	6-pentyl-2H-pyran-2-one	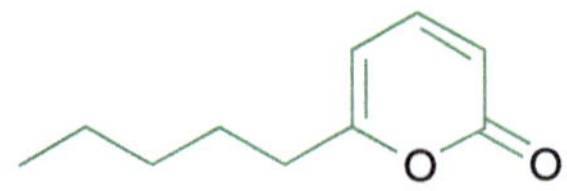
	Koninginis	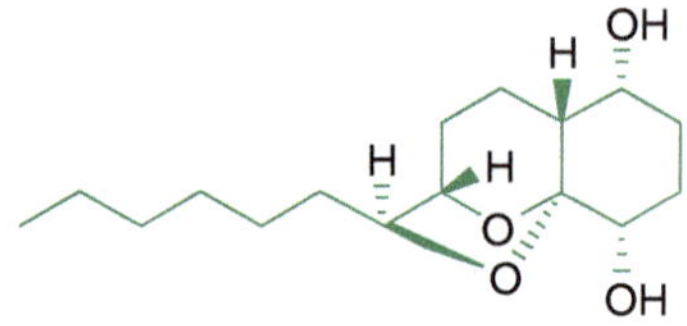
	Trichosetin	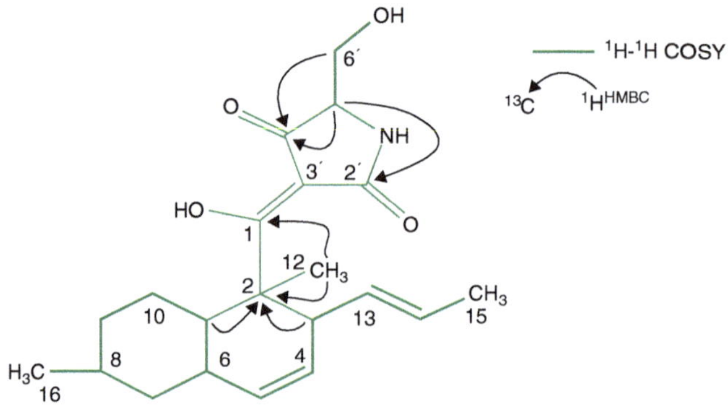
Herbicidal	Viridiol	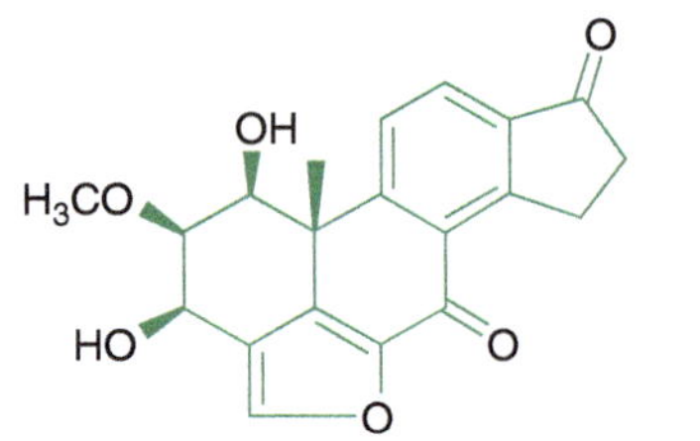
	(3H)-benzoxazolinone	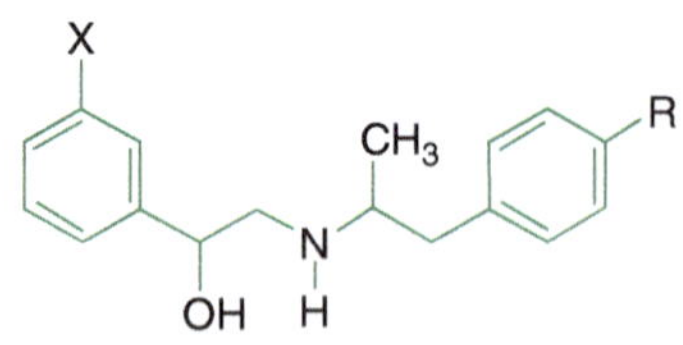
Flavour and Aroma	6-pentyl-2H-pyran-2-one	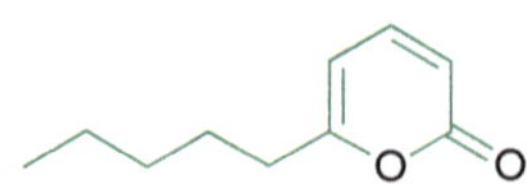
	3-octanone	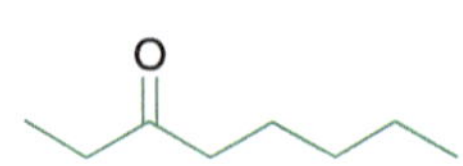

Continued

Table 24.1. Continued.

Functions	Involved secondary metabolites	Structures
	1-octen-3-ol	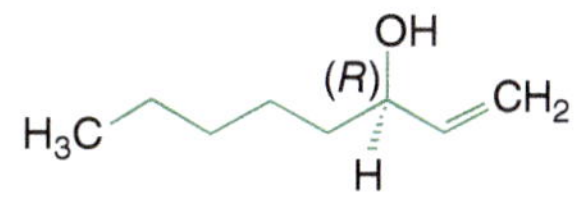
Anti-ageing	R-mevalonolactone,	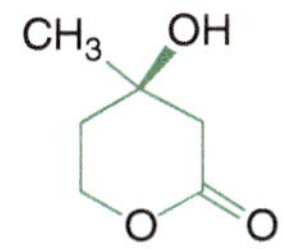
	Mevastatin	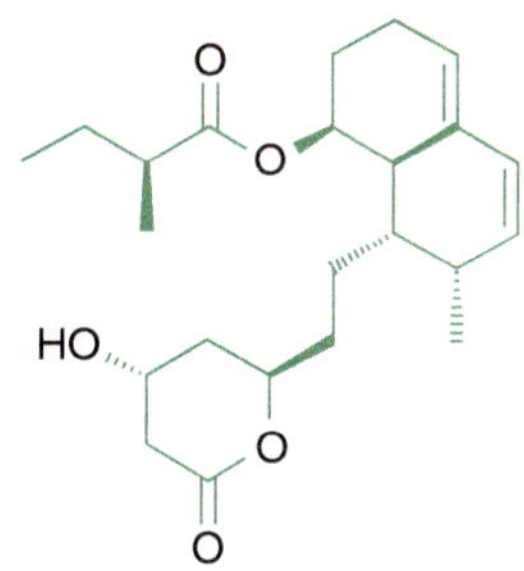
Anticancer	Viridin	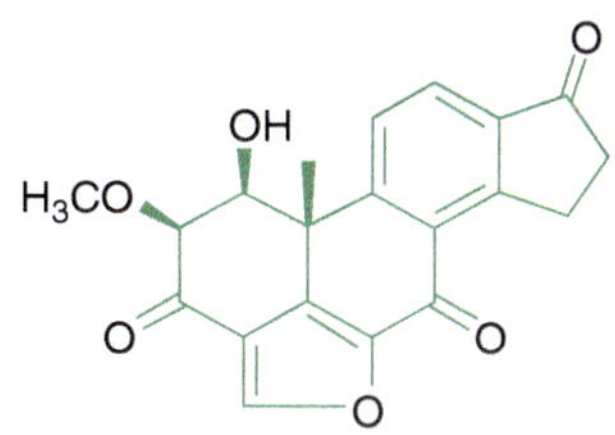
	Trichodenones	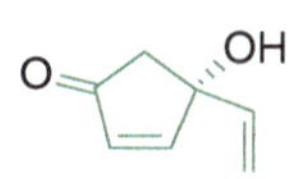
	Harzianum A	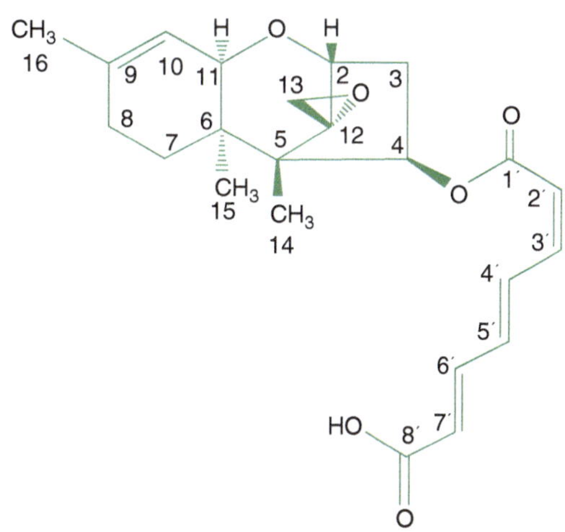

Continued

Table 24.1. Continued.

Functions	Involved secondary metabolites	Structures
	Harziphilone	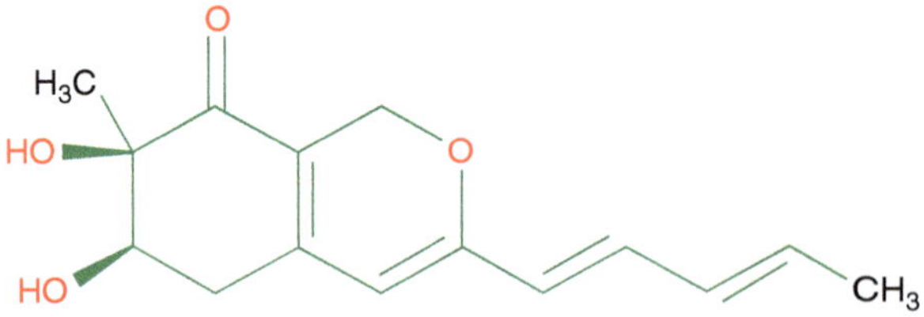

formulated for spray or drench application, hence their popularity is growing by the day and they can become an integral part of the crop management practices. However, most of the information which has been published is limited to the laboratory or greenhouse experiments. In the current scenario, these metabolites should be tested under field conditions in order to promote them as fungicides, as they have promising advantages over whole-organism formulations. It has been documented by Fravel (1988) that secondary metabolites purified from *Trichoderma* spp. can potentially control bacterial infections more rapidly and can be more effective than whole-organism application under field conditions. If we consider the worst extremes, *Trichoderma* as a whole organism can pose a hazard just like chemical fungicides owing to unidentified toxic metabolites, but probably its secondary metabolites can be a promising and specific solution. So as a deduction, conventional formulations of biopesticides can be replaced by developing next-generation secondary metabolites for management of phytopathogens.

Acknowledgment

JS and RSR are grateful to CSIR and UGC respectively, for providing financial assistance.

References

Abe, N., Yamamoto, K. and Hirota, A. (2000) Novel fungal metabolites, demethylsorbicillin and oxosorbicillinol, isolated from *Trichoderma* sp. USF-2690. *Bioscience, Biotechnology and Biochemistry* 64, 620–622.

Agrios, G.N. (2005) *Plant Pathology,* Fifth Edition. Academic Press, New York, p.948.

Almassi, F., Ghisalberti, E.L., Narbey, M.J. and Sivasithamparam, K. (1991) New antibiotics from strains of *Trichoderma harzianum. Journal of Natural Products* 54, 396–402.

Amagata, T., Minoura, K. and Numata, A. (1998) Cytotoxic metabolites produced by a fungal strain from a *Sargassum* algae. *Journal of Antibiotics* 51, 432–434.

Baker, R. (1992) Biological control of diseases of crops grown in covered and environmentally controlled structures. In: Tjamos, E.C., Papavizas, G.C. and Cook, R.J. (eds) *Biological Control of Plant Diseases: Progress and Challenges for the Future.* Plenum Press, New York, pp. 231–241.

Bisen, K., Keswani, C., Mishra, S., Saxena, A., Rakshit, A. and Singh, H.B. (2015) Unrealized potential of seed biopriming for versatile agriculture. In: Rakshit, A., Singh, H.B. and Sen, A. (eds) *Nutrient Use Efficiency: From Basics to Advances*. Springer, New Delhi, India, pp. 193–206.

Bisen, K., Keswani, C., Patel, J.S., Sarma, B.K. and Singh, H.B. (2016) *Trichoderma* spp.: efficient inducers of systemic resistance in plants. In: Chaudhary, D.K. and Verma, A. (eds) *Microbial-mediated Induced Systemic Resistance in Plants*. Springer, Singapore, pp. 185–195.

Brakhage, A.A. and Schroeckh, V. (2011) Fungal secondary metabolites – strategies to activate silent gene clusters. *Fungal Genetics Biology* 48, 15–22.

Brewer, D. and Taylor, A. (1981) *Trichoderma hamatum* isolated from pasture soil. *Mycopathologia* 76, 167–173.

Brewer, D., Feicht, A., Taylor, A., Keeping, J.W., Taha, A.A. and Thaller, V. (1982) Production of experimental quantities of isocyanide metabolites of *Trichoderma hamatum*. *Canadian Journal of Microbiology* 28, 1252–1260.

Calistru, C., McLean, M. and Berjak, P. (1997) In vitro studies on the potential for biological control of *Aspergillus flavus* and *Fusarium moniliforme* by *Trichoderma* species. *Mycopathologia* 137, 115–124.

Calvo, A.M., Wilson, R.A., Bok, J.W. and Keller, P.N. (2002) Relationship between secondary metabolism and fungal development. *Microbiology and Molecular Biology Reviews* 66, 447–459.

Chen, F., D'Auria, J.C. and Tholl, D. (2003) An *Arabidopsis thaliana* gene for methylsalicylate biosynthesis, identified by a biochemical genomics approach, has a role in defence. *The Plant Journal* 36, 577–588.

Chet, I. and Inbar, J. (1994). Biological control of fungal pathogens. *Applied Biochemistry and Biotechnology* 48, 37–43.

Chet, I., Benhamou, N. and Haran, S. (1998) Mycoparasitism and lytic enzymes. In: Harman, G.E. and Kubicek, C.P. (eds) *Trichoderma and Gliocladium, Vol 2. Enzymes, Biological Control and Commercial Applications.* Taylor & Francis, London, pp. 153–171.

Chutrakul, C. and Peberdy, J.F. (2005) Isolation and characterisation of a partial peptide synthetas gene from *Trichoderma asperellum*. *FEMS Microbiology Letters* 252, 257–265.

Claydon, N., Allan, M., Hanson, J.R. and Avent, A.G. (1987) Antifungal alkyl pyrones of *Trichoderma harzianum*. *Transactions of the British Mycological Society* 88, 505–513.

Delp, C.J. (1983) Changing emphasis in disease management. In: Kommedahl, T. and Williams, P.H. (eds) *Challenging Problems in Plant Health. The American Phytopathology Society*, St. Paul, MN, pp. 416–421

Demain, A.L. and Fang, A. (2000) The natural functions of secondary metabolites. In: Scheper, T. (ed.) *Advances in Biochemical Engineering/Biotechnology, Vol. 69*. Springer Berlin Heidelberg, Germany, pp. 1–39.

Dennis, L. and Webster, J. (1971) Antagonistic properties of species group of *Trichoderma* III hyphal interactions. *Transactions of the British Mycological Society* 57, 363–369.

Di Pietro, A., Lorito, M., Hayes, C.K., Broadway, R.M. and Harman, G.E. (1993) Endochitinase from *Gliocladium virens*: Isolation, characterization and synergistic antifungal activity in combination with gliotoxin. *Phytopathology* 83, 308–313.

Dreistadt, S., Clark, J. and Flint, M. (2004). *Pests of Landscape Trees and Shrubs: an Integrated Pest Management Guide*, Second Edition, 151pp. University of California, Division of Agriculture and Natural Resources, Oakland, California.

Dunlop, R.W., Simon, A., Sivasithamparam, K. and Ghisalberti, E.L. (1989) An antibiotic from *Trichoderma koningii* active against soilborne plant pathogens. *Journal of Natural Products* 52, 67–74.

Elad, Y. and Kapat, A. (1999) The role of *Trichoderma harzianum* protease in the biocontrol of *Botrytis cinerea*. *European Journal of Plant Pathology* 105, 177–189.

El-Hasan, A. and Buchenauer, H. (2009) Actions of 6-pentyl-alpha-pyrone in controlling seedling blight incited by *Fusarium moniliforme* and inducing defence responses in maize. *Journal of Phytopathology* 157, 697–707.

Engelberth, J., Koch, T., Schuler, G., Bachmann, N., Rechtenbach, J. and Boland, W. (2000) Ion channel-forming alamethicin is a potent elicitor of volatile biosynthesis and tendril coiling. Cross talk between jasmonate and salicylate signaling in lima bean. *Journal of Plant Physiology* 125, 369–377.

Fedorinchik, N.S., Tarunina, T.A., Tyutyunnikov, M.G. and Kudryatseva, K.I. (1975) Trichodermin-4, a new biological preparation for plant disease control. *Plant Protection* 3, 67–72.

Forget, G., Goodman, T. and de Villiers, A. (eds) (1993) *Impact of Pesticide Use on Health in Developing Countries*. International Development Research Centre, Ottawa, Canada, p.335.

Fravel, D.R. (1988) Role of antibiosis in the biocontrol of plant diseases. *Annual Review of Phytopathology* 26, 75–91.

Fries, N. (1973) Effects of volatile organic compounds on the growth and development of fungi. *Transactions of the British Mycological Society* 60, 1–21.

Frisvad, J.C., Thrane, U. and Filtenborg, O. (1998). In: Frisvad, J. (ed.) *Chemical Fungal Taxonomy*. CRC, London, pp. 289–321.

Ghisalberti, E.L. and Rowland, C.Y. (1993) Antifungal metabolites from *Trichoderma harzianum*. *Journal of Natural Products* 56, 1799–1804.

Ghisalberti, E.L. and Sivasithamparam, K. (1991) Antifungal antibiotics produced by *Trichoderma* spp. *Soil Biology & Biochemistry* 23, 1011–1020.

Gottlieb, D. and Shaw, P.D. (1970) Mechanism of action of antifungal antibiotics. *Annual Review of Phytopathology* 8, 371–402.

Govindasamy, V. and Balasubramanian, R. (1989) Biological control of groundnut rust, *Puccinia arachidism* by *Trichoderma harzianum. Z Pflanzenkr Pflanzenschutz* 96, 337–345.

Graeme-Cook, K.A. and Faull, J.L. (1991) Effect of ultraviolet-induced mutants of *Trichoderma harzianum* with altered antibiotic production on selected pathogens in vitro. *Canadian Journal of Microbiology* 37, 659–664.

Harman, G.E. (2000) Myths and dogmas of biocontrol: changes in perceptions derived from research on *Trichoderma harzianum* T- 22. *Plant Disease* 84, 377–393.

Harman, G.E. and Bjorkmann, T. (1998) Potential and existing uses of *Trichoderma* and research on *Trichoderma harzianum* T-22. *Plant Disease* 84, 377–393.

Harman, G.E. and Kubicek, C.P. (1998) *Trichoderma* and *Gliocladium*. Taylor & Francis, London, p. 278.

Harman, G.E., Howell, C.R., Viterbo, A., Chet, I. and Lorito, M. (2004) *Trichoderma* species - opportunistic, avirulent plant symbionts. *Natural Review of Microbiology* 2, 43–56.

Herbert, R.B. (1989) *The Biosynthesis of Secondary Metabolites*, 2nd edn, Chapman & Hall, London, p. 232.

Herr, L.J. (1995) Biological control of *Rhizoctonia solani* by binucleate *Rhizoctonia* spp. and hypovirulent *R. solani* agents. *Crop Protection* 14(3), 179–186.

Hjeljord, L. and Tronsmo, A. (1998) *Trichoderma* and *Gliocladium* in biological control: an overview. In: Harman, G.E. and Kubicek, C.P. (eds) *Trichoderma and Gliocladium, Vol 2. Enzymes, Biological Control and Commercial Applications*. Taylor & Francis, London, pp. 129–155.

Hoffmeister, D. and Keller, N.P. (2007) Natural products of filamentous fungi: enzymes, genes, and their regulation. *Natural Products Reports* 24, 393–416.

Howell, C.R. and Stipanovic, R.D. (1983) Gliovirin, a new antibiotic from *Gliocladium* virens and its role in the biological control of *Pythium ultimum*. *Canadian Journal of Microbiology* 29, 321–324.

Howell, C.R. and Stipanovic, R.D. (1994) Effect of sterol biosynthesis inhibitors on phytotoxin (viridiol) production by *Gliocladium virens* in culture. *Phytopathology* 84, 969–972.

Huang, Q., Tezuka, Y., Hatanaka, Y., Kikuchi, T., Nishi, A. and Tubaki, K. (1996) Studies on metabolites of mycoparasitic fungi. V. Ion-spray ionization mass spectrometric analysis of trichokonin-II, a peptaibol mixture obtained from the culture broth of *Trichoderma koningii. Chemical and Pharmaceutical Bulletin* (Tokyo) 44, 590–593.

Humphris, S.N., Bruce, A., Buultjens, E. and Wheatley, R.E. (2002) The effects of volatile microbial secondary metabolites on protein synthesis in *Serpula lacrymans. FEMS Microbiology Letter* 210, 215–219.

Hutchinson, S.A. (1973) Biological activities of volatile fungal metabolites. *Annual Review of Phytopathology* 11, 223–246.

Iida, A., Sanekata, M., Wada, S., Fujita, T., Tanaka, H. *et al.* (1995) Fungal metabolites. XVIII. New membrane-modifying peptides, trichorozins I–IV, from the fungus *Trichoderma Harzianum. Chemical and Pharmaceutical Bulletin* (Tokyo) 43, 392–397.

Karlovsky, P. (ed.) (2008) *Secondary Metabolites in Soil Ecology. Soil Biology 14.* 1 Springer Berlin Heidelberg, Germany.

Keller, N.P., Turner, G. and Bennett, J.W. (2005) Fungal secondary metabolism – from biochemistry to genomics. *Nature Review of Microbiology* 3, 937–947.

Keswani, C. (2015). Proteomics studies of thermotolerant strain of *Trichoderma* spp. Ph.D. Thesis, Banaras Hindu University, Varanasi, India, p. 126.

Keswani, C., Singh, S.P. and Singh, H.B. (2013) A Superstar in Biocontrol Enterprise: *Trichoderma* spp. *Biotech Today* 3, 27–30.

Keswani, C., Mishra, S., Sarma, B.K., Singh, S.P. and Singh, H.B. (2014) Unraveling the efficient application of secondary metabolites of various *Trichoderma*. *Applied Microbiology and Biotechnology* 98, 533–544.

Keswani, C., Bisen, K., Singh, V., Sarma, B.K. and Singh, H.B. (2016a) Formulation technology of biocontrol agents: present status and future prospects. In: Arora, N.K., Mehnaz, S. and Balestrini, R. (eds) *Bioformulations: For Sustainable Agriculture*. Springer, New Delhi, India, pp. 35–52.

Keswani, C., Bisen, K., Singh, S.P., Sarma, B.K. and Singh, H.B. (2016b) A proteomic approach to understand the tripartite interactions between plant-*Trichoderma*-pathogen: investigating the potential for efficient biological control. In: Hakeem, K.R. and Akhtar, M.S. (eds) *Plant, Soil and Microbes Vol. 2. Mechanisms and Molecular Interactions*. Springer, Cham, Switzerland, pp. 79–93.

Kossel, A. (1891) Ueber die Chemische Zusammensetzung der Zelle. *Archives of Physiology*, pp. 181–186.

Kumar, S. and Gupta, O. (2012) Expanding dimension of plant pathology. *JNKVV Research Journal* 46(3), 286–293.

Landreau, A., Pouchus, Y.F., Sallenave-Namont, C., Biard, J.F., Boumard, M.C. *et al.* (2002) Combined use of LC/MS and a biological test for rapid identification of marine mycotoxins produced by *Trichoderma koningii. Journal of Microbiology and Methods* 48, 181–194.

Le Doan, T., El-Hajii, M., Rebuffat, S., Rajeswari, M.R. and Bodo, B. (1986) Fluorescein studies on the interaction of trichorzianine A IIIc with model membranes. *Biochimica et Biophysica Acta* 858, 1–5.

Lehner, S.M., Atanasova, L. and Neumann, N.K. (2013) Isotope-assisted screening for iron-containing metabolites reveals high diversity among known and unknown siderophores produced by *Trichoderma* spp. *Applied and Environmental Microbiology* 79, 18–31.

Lenteren, J.C.V. (2000) A greenhouse without pesticides: fact or fantasy? *Crop Protection* 19, 375–384.

Leong, J. (1986) Siderophores: their biochemistry and possible role in the biocontrol of plant pathogens. *Annual Review of Phytopathology* 24, 187–209.

Lorito, M., Farkas, V., Rebuffat, S., Bodo, B. and Kubicek, C.P. (1996) Cell wall synthesis is a major target of mycoparasitic antagonism by *Trichoderma harzianum*. *Journal of Bacteriology* 178, 6382–6385.

Macias, F.A., Varela, R.M., Simonet, A.M., Cutler, H.G., Cutler, S.J. *et al.* (2000) Bioactive carotanes from *Trichoderma virens*. *Journal of Natural Products* 63, 1197–2000.

Mann, R. and Rehm, H.J. (1976) Degradation products from aflatoxin B1 by *Corynebacterium rubrum, Aspergillus niger, Trichoderma virid,* and *Mucor ambiguus*. *European Journal of Applied Microbiology* 2, 297–306.

Mapleston, R.A., Stone, M.J. and Williams, D.H. (1992) The evolutionary role of secondary metabolites. *Gene* 115, 151–157.

Marco, G.J., Hollingworth, R.M. and Durham, W. (eds) (1987) *Silent Spring Revisited*. American Chemical Society, Washington, DC, p. 214.

Moffatt, J.S., Bu'Lock, J.D. and Yuen, T.H. (1969). Viridiol, a steroid-like product from *Trichoderma viride*. *Journal of the Chemical Society, Chemical Communication* 14, 839.

Monaco, C., Perello, A. and Rollan, M.C. (1994) In vitro tests of the antagonistic behavior of *Trichoderma* spp. against pathogenic species of the horticultural region of La Plata, Argentina. *Microbiologia* 10, 423–428.

Mukherjee, P.K. and Raghu, K. (1997) Effect of temperature on antagonistic and biocontrol potential of *Trichoderma* sp. on *Sclerotium rolfsii*. *Mycopathologia* 139, 151–155.

Mukhopadhyay, A.N. and Mukherjee, P.K. (1996) Fungi as fungicides. *International Journal of Tropical Plant Diseases* 14, 1–17.

Mukhopadhyay, A.N., Shrestha, S.M. and Mukherjee, P.K. (1992) Biological seed treatment for control of soilborne plant pathogens. *FAO Plant Protection Bulletin* 40, 21–30.

Mumpuni, A., Sharma, H.S.S. and Averil, E.B. (1998) Effect of metabolites produced by *Trichoderma harzianum* biotypes and *Agaricus bisporus* on their respective growth radii in culture. *Applied Environmental Microbiology* 64, 5053–5056.

National Research Council (2000) *The Future Role of Pesticides in US Agriculture*. National Academies Press, Washington, DC, p. 301.

Orietta, F. and Larrea, V. (2001) Antagonistic microorganisms for phytosanitary control. *Pest Manejo Integrado Spanish* 62, 96–100.

Osbourn, A. (2010) Secondary metabolic gene clusters: evolutionary toolkits for chemical innovation. *Trends Genetics* 26, 449–457.

Perez, L.M., Besoain, X., Reyes, M., Pardo, G. and Montealegre, J. (2002) The expression of extracellular fungal cell wall hydrolytic enzymes in different *Trichoderma harzianum* isolates correlates with their ability to control *Pyrenochaeta lycopersici*. *Biological Research* 35, 401–410.

Perry, A.S., Yamamoto, I., Ishaaya, I. and Perry, R.Y. (1998) *Insecticides in Agriculture and Environment: Retrospects and Prospects*. Springer Berlin Heidelberg, Germany, p. 261.

Rebuffat, S., El Hajji, M., Hennig, P., Davoust, D. and Bodo, B. (1989) Isolation, sequence and conformation of seven trichorzianines B from *Trichoderma harzianum*. *International Journal of Peptide and Protein Research* 34, 200–210.

Reino, J.L., Guerriero, R.F., Hernandez-Gala, R. and Collado, I.G. (2008) Secondary metabolites from species of the biocontrol agent *Trichoderma*. *Phytochemistry Review* 7, 89–123.

Scarselletti, R. and Faull, J.L. (1994) *In Vitro* activity of 6-pentyl-a-pyrone, a metabolite of *Trichoderma harzianum*, in the inhibition of *Rhizoctonia solani* and *Fusarium oxysporum* f. sp. *lycopersici*. *Mycological Research* 98, 1207–1209.

Sekiguchi, J. and Gaucher, G.M. (1977) Conidiogenesis and secondary metabolism in *Penicillium urticae*. *Applied Environmental Microbiology* 33, 147–158.

Simon, A., Dunlop, R.W., Ghissalberti, E.L. and Sivasithamparam, K. (1988) *Trichoderma koningii* produces a pyrone compound with antibiotic properties. *Soil Biology and Biochemistry* 20, 263–264.

Singh, V., Upadhyay, R.S., Sarma, B.K. and Singh, H.B. (2016a) *Trichoderma asperellum* spore dose depended modulation of plant growth in vegetable crops. *Microbiological Research* 193, 74–86.

Singh, V., Upadhyay, R.S., Sarma, B.K. and Singh, H.B. (2016b) Seed bio-priming with *Trichoderma asperellum* effectively modulate plant growth promotion in pea. *International Journal of Agriculture, Environment and Biotechnology* 9, 361.

Sivasithamparam, K. and Ghisalberti, E.L. (1998). Secondary metabolism in *Trichoderma* and *Gliocladium*. In: Harman, G.E. and Kubicek, C.P. (eds) *Trichoderma and Gliocladium, Vol. 1. Basic Biology, Taxonomy and Genetics*. Taylor & Francis, London, pp. 139–191.

Smith, P.M. (1976) *The Chemotaxonomy of Plants*. Arnold, London.

Stone, M.J. and Williams, D.H. (1992) On the evolution of functional secondary metabolites (natural products). *Molecular Microbiology* 6, 29–34.

Strange, R.N. (1993) *Plant Disease Control: Towards Environmentally Acceptable Methods*. Chapman & Hall, London.

Szekeres, A., Leitgeb, B., Kredics, L., Antal, Z., Hatvani, L. *et al.* (2005) Peptaibols and related peptaibiotics of *Trichoderma*—a review. *Acta Microbiologica et Immunologica Hungarica* 52, 137–168.

Tweedy, B.J. (1983) The future of chemicals for controlling plant diseases. In: Kommedahl, T. and Williams, P.H. (eds) Challenging problems in plant health. *The American Phytopathology Society*, St. Paul, MN, pp. 405–415.

United Nations Population Division (2008) *World Population Prospects: The 2008 Revision*. UN, New York, p.107. Available at http://www.un.org/esa/population/publications/wpp2008/wpp2008_highlights.pdf.

van Loon, L.C., Bakker, P.A.H.M. and Pieterse, C.M.J. (1998) Systemic resistance induced by rhizospherebacteria. *Annual Review of Phytopathology* 36, 453–483.

Vinale, F., Marra, R., Scala, F., Ghisalberti, E.L., Lorito, M. and Sivasithamparam, K. (2006) Major secondary metabolites produced by two commercial *Trichoderma* strains active against different phytopathogens. *Letter of Applied Microbiology* 43, 143–148.

Vinale, F., Sivasithamparam, K., Ghisalberti, E.L., Marra, R., Woo, S.L. and Lorito, M. (2008) *Trichoderma*-plant-pathogen interactions. *Soil Biology and Biochemistry* 40, 1–10.

Vinale, F., Ghisalberti, E.L., Sivasithamparam, K., Marra, R., Ritieni, A. *et al.* (2009) Factors affecting the production of *Trichoderma harzianum* secondary metabolites during the interaction with different plant pathogens. *Letters in Applied Microbiology* 48, 705–711.

Vinale, F., Arjona, G.I. and Nigro, M. (2012) Cerinolactone, a hydroxylactone derivative from *Trichoderma cerinum*. *Journal of Natural Products* 75, 103–106.

Vinale, F., Nigro, M., Sivasithamparam, K., Flematti, G., Ghisalberti, E.L. *et al.* (2013) Harzianic acid: a novel siderophore from Trichoderma harzianum. *FEMS Microbiology Letters* 347, 123–129.

Viterbo, A., Ramot, O., Chemin, L. and Chet, I. (2002) Significance of lytic enzymes from *Trichoderma* spp in the biocontrol of fungal plant pathogens. *Antonie Van Leeuwenhoek* 81, 549–556.

Vyas, R.K. and Mathur, K. (2002) *Trichoderma* spp. in cumin rhizosphere and their potential in suppression of wilt. *Indian Phytopathology* 55, 455–457.

Wada, S., Iida, A., Akimoto, N., Kanai, M., Toyama, N. and Fujita, T. (1995) Fungal metabolites. XIX. Structural elucidation of channel-forming peptides, trichorovins-I-XIV, from the fungus *Trichoderma viride*. *Chemical and Pharmaceuticals Bulletin* (Tokyo) 43, 910–915.

Wei, X., Yang, F. and Straney, D.C. (2005) Multiple non-ribosomal peptide synthetase genes determine peptaibol synthesis in *Trichoderma virens*. *Canadian Journal of Microbiology* 51, 423–429.

Weindling, R. (1941) Experimental consideration of the mould toxins of *Gliocladium* and *Trichoderma*. *Phytopathology* 31, 991–1003.

Weindling, R. and Emerson, O.H. (1936) The isolation of a toxic substance from the culture of a *Trichoderma*. *Phytopathology* 26, 1068–1070.

Whipps, J.M. and Lumsden, R.D. (2001) Commercial use of fungi as plant disease biological control agents: status and prospects. In: Butt, T., Jackson, C. and Magan, N, (eds) *Fungal Biocontrol Agents: Progress, Problems and Potential*. CAB International, Wallingford, UK, pp. 9–22.

Wiest, A., Crzegorski, D., Xu, B.W., Goulard, C., Rebuffat, S. *et al.* (2002) Identification of peptaibols from *Trichoderma virens* and cloning of a peptaibol synthetase. *Journal of Biological Chemistry* 277, 20862–20868.

Worasatit, N., Sivasithamparam, K., Ghisalberti, E.L. and Rowland, C. (1994) Variation in pyrone production, pectic enzymes and control of rhizoctonia root rot of wheat among single-spore isolates of *Trichoderma koningii*. *Mycological Research* 98, 1357–1363.

Xiao-Yan, S., Qing-Tao, S., Shu-Tao, X., Xiu-Lan, C., Cai-Yun, S. and Yu-Zhong, Z. (2006) Broad-spectrum antimicrobial activity and high stability of trichokonins from *Trichoderma koningii* SMF2 against plant pathogens. *FEMS Microbiology Letters* 260, 119–125.

25 PGPR-Mediated Defence Responses in Plants under Biotic and Abiotic Stresses

Gagan Kumar,[1] Jai Singh Patel,[2] Anupam Maharshi,[1] Arpan Mukherjee,[2] Chetan Keswani,[3] S.P. Singh,[3] H.B. Singh[1] and B.K. Sarma[1]*

[1]Department of Mycology and Plant Pathology, Institute of Agricultural Sciences, Banaras Hindu University, Varanasi-221005, India; [2]Department of Botany, Institute of Science, Banaras Hindu University, Varanasi-221005, India; [3]Department of Biochemistry, Institute of Science, Banaras Hindu University, Varanasi-221005, India

25.1 Introduction

The rhizospheric region of different plants can be colonized by plant growth-promoting rhizobacteria (PGPRs) and provide beneficial effects such as plant growth promotion, resistance against diseases caused by phytopathogenic bacteria, fungi and nematodes (Kloepper *et al.*, 2004). A report by van Loon *et al.* (1998) suggested elicitation of physical or chemical changes towards plant defence and the process is known as induced systemic resistance (ISR). PGPRs are used as bio-inoculants for the purpose of phytostimulation, biofertilization and biocontrol. Generally PGPRs are known for their growth promotion activities. Several mechanisms are reported for PGPRs, such as modulation of root architecture, root and shoot growth by production of phytohormones such as cytokinins and auxins, etc. (Fig. 25.1). Many other indirect mechanisms include the action of secondary metabolites produced by PGPRs, such as hydrogen cyanide and antibiotics, which can inhibit the effect of deleterious phytopathogens and promote plant growth and yield. PGPRs can also trigger defence mechanisms via ISR and reduce inoculum densities of phytopathogens (Mantelin and Touraine, 2004; López-Bucio *et al.*, 2007; Bisen *et al.*, 2015, 2016). PGPR-induced ISR can enhance plant defence activities both in field and greenhouse conditions (Kloepper *et al.*, 2004; van Loon *et al.*, 1998).

Very few reports are published regarding activities of PGPR in eliciting abiotic tolerance such as against salinity, drought, and nutrient deficiency or excess. Earlier reviews (Glick, 2003; Zhuang *et al.*, 2007) suggested elicitation of heavy metal tolerance by PGPRs. PGPRs have the ability to modulate different types of plant growth promotion activities by regulation of cell division, differentiation and expansion. Promotion of such events requires complex networking processes of signalling molecules between root and shoot under the effect of both abiotic and biotic challenges. Novel agricultural applications may be developed after knowing the signalling mechanisms involved in interaction of plants with microbes in the rhizospheric regions. Plant root exudation includes certain biomolecules such as sugars, organic acids and vitamins, which are involved in

*E-mail: birinchi_ks@yahoo.com

© CAB International 2017. *Advances in PGPR Research* (eds H.B. Singh, B.K. Sarma and C. Keswani)

Fig. 25.1. Plant growth promotion activities by PGPR in pea. (A) PGPR inoculated and (B) PGPR non-inoculated.

the recruitment process of appropriate microorganisms in the rhizosphere by acting as attractants. However, PGPRs can release phytohormones, which include volatile or non-volatile molecules, that act either directly or indirectly to modulate the host plant immunity and regulate plant growth components (Ortíz-Castro *et al.*, 2009).

25.2 PGPR in Abiotic Stress Management

Presently the world population is increasing continuously and it is very tough to balance crop production according to the needs of the growing population. Such food demand may be fulfilled by checking the significant losses caused due to abiotically challenged soils. One of the major abiotic factors is salinity that affects plant productivity worldwide. When plants are exposed to high salinity, imbalance in ionic concentrations is created inside the plant and such imbalances cause a spatial type of water deficit. Various mechanisms are adopted by plants for excluding the effect of salinity such as production of osmolytes, synthesis of polyamines, reduction in reactive oxygen species by production of antioxidant enzymes, production of ions and ion compartmentalization, etc. A report by Huang *et al.* (2012) showed that PGPRs elevate stress specific adoptive responses against various environmental stresses in plants.

Soil salinity is an important limiting factor of plant growth in arid regions. A report by Mayak *et al.* (2004a) suggested that PGPR can play a role in eliciting salt tolerance by plants when they are applied. They showed that application of *Achromobacter piechaudii,* which has ACC deaminase activity, can increase salt tolerance in plants. *A. piechaudii*, the producer of ACC deaminase, is able to increase tomato growth by up to 66% at higher salt concentration. IST (induced systemic tolerance) to salt has been reported (Zhang *et al.*, 2008) in *Arabidopsis* by using a strain of *Bacillus subtilis* GB03, which is released commercially as a biological control agent. Ryu *et al.* (2004) reported the production of certain organic volatile compounds by *B. subtilis* GB03 for induced systemic tolerance. Transcriptome analysis of 600 *Arabidopsis* genes showed that the expression of *HIGH-AFFINITY K^+ TRANSPORTER 1* (*HKT1*), which is involved in Na^+ import in roots was decreased. It is known that HKT1 has the ability to adjust Na^+ and K^+ levels, and the activity varies from tissue to tissue. However, exposure of *athtk1* mutant plants to bacterial volatile compounds resulted in representation of typical salt stress phenotypes like stunting, and also resulted in inhibition of seedling growth. Such experimentation proved that bacterial volatile compounds have the ability to down-regulate *HKT1* expression in *Arabidopsis* roots, but up-regulate in shoot tissues, and so be involved in regulation of lower Na^+ in the whole plant. The Na^+ export mutant *salt overly sensitive3* (*sos3*) shows no difference in IST towards salt stress, which suggested that *HKT1* functions in shoots to retrieve Na^+ from the xylem, and by similar mechanism facilitate recirculation of Na^+ from shoot-root. Furthermore the results suggest that tissue-specific regulation of *HKT1* by bacterial volatile compounds can control Na^+ homeostasis under salt stress in Arabidopsis.

A strain of *Pseudomonas* sp. (AKM-P6) has been identified as thermotolerant and also to have PGPR activity; it was isolated from the rhizospheric region of pigeon pea (Ali *et al.*, 2009). The same strain was also reported to produce high molecular weight proteins in sorghum in its leaves. Further, some specific biomolecules, such as the amino acid proline, are produced in high amounts in plants treated with PGPR as a marker of abiotic stress (Ali *et al.*, 2009). Some exopolysaccharides have also been predicted to play a role in thermotolerance when treated with *Pseudomonas* sp. (AKM-P6). Kumar, *et al.* (2014) reported that under high temperature *Pseudomonas* also reduces the incidence of stem gall disease in coriander. PGPRs are also reported to increase cold tolerance in host plants. Similar activity has also been reported in *Vitis vinifera* when inoculated with *Burkholderia phytofermans* during cold stress (Barka *et al.*, 2006). Drought stress can also be minimized by application of PGPR (Yang *et al.*, 2009). The bacteria *Achromobacter piechaudii* (ARV8) have been reported for increasing biomass of plants under water stress (Mayak *et al.*, 2004b). Another report (Cho *et al.*, 2008) suggested the role of *Pseudomonas chlororaphis* (06) in inducing drought stress in host crops.

25.3 PGPR in Biotic Stress Management

Beneficial microbes can reduce disease incidence by triggering resistance in host plants. The phenomenon is known as Induced Systemic Resistance (ISR). During this elevated state of resistance, several defence pathways become active in host plants and this elevated defence level is effective in reducing development of various pathogens (van Loon *et al.*, 1998). In carnations, when ISR was induced by *Pseudomonas fluorescens* strain WCS417r the plants were protected systemically against the soilborne pathogen *Fusarium oxysporum* f. sp. *dianthi* (Van Peer *et al.*, 1991). Similarly, PGPRs shielded the leaves of cucumber plants from damage by *Colletotrichum orbiculare*, the causal agent of anthracnose (Wei *et al.*, 1991). PGPR-induced ISR is very similar to pathogen-induced systemic acquired resistance (SAR) where in both cases the plants are able to defend themselves against an invading pathogen effectively. Interestingly, in both ISR and SAR the uninfected plant parts also show enhanced resistance to the invading pathogens (Van Wees *et al.*, 1997; van Loon *et al.*, 1998; Keswani *et al.*, 2016a, b). Fluorescent pseudomonads are reported to have ISR activities that are effective against a large number of pathogens. Application of *P. fluorescens* inhibited mycelial growth of the rice sheath blight pathogen *Rhizoctonia solani* in rice plants due to triggering of ISR (Radjacommare *et al.*, 2004). The mechanisms behind *P. fluorescens*-mediated suppression of the sheath blight fungus is demonstrated to be due to enhanced activity of chitinase genes in rice (Nandakumar *et al.*, 2001).

Another potential PGPR strain *Serratia marcescens* strain B2 with biocontrol ability reduced disease development by some soil-borne pathogens in greenhouse conditions, but the strain was not inhibitory to mycelial growth of the same pathogens in plates during the *in vitro* antagonistic test. The results point out clearly that the PGPR strains stimulated and activated systemic resistance as the pathogens were not antagonized (Someya *et al.*, 2002). Similar results were also obtained in bean plants when treated with the PGPR strain *P. aeruginosa* which induced ISR against the infection of *Collehotricum lindemuthianum* (Bigirimana and Hofte, 2002). In contrast, during the experiment with *Pisum sativum* it was observed that the PGPR strain *P. fluorescens* strain 63-28 when inoculated into pea roots secreted more chitinase at the penetration site of *F. oxysporum* f. sp. *pisi* and antagonism was thought to be the main reason behind suppression of the pathogen (Benhamou *et al.*, 1996).

Plants are always in contact with microbes that are mostly beneficial. It is important for plants to discriminate between beneficial microbes and harmful ones and respond accordingly, either to maintain a relationship with the beneficial ones or to keep away the harmful ones. Plants need to deploy a prompt and effective defence response to protect them from damage by checking the growth of pathogens. Therefore physiological events that lead to recognition of and discrimination between beneficial and pathogenic microbes are very important for subsequent response (de Leon and Montesano, 2013). Due to the diversity of stresses plants encounter, plants have evolved various mechanisms to live with distinct abiotic and biotic stresses during the process of evolution. Interestingly, plants activate their defence responses initially in a similar way for both pathogenic and beneficial microbes during the interaction in either rhizosphere or in phyllosphere (Shoresh *et al.*, 2010). However, sustenance of those responses will depend on the type of microbes that the plant is interacting with.

The application technique for PGPR is very important to get the desired effect on plants. Among the various delivery techniques, seed biopriming of plants with PGPRs is gaining popularity as it enhances the time duration of contact with the host plants and also increases the level of stimulation of defence responses compared to the non-bioprimed plants. The consortium of PGPRs with other compatible microbial strains also increased the defence responses of bioprimed pea plants during interaction with the pathogen *Sclerotinia sclerotiorum* to that of plants bioprimed with the individual microbes. The compatible microbial consortium of PGPRs induced the antioxidant enzyme activities and phenylpropanoid pathway simultaneously, leading to enhanced accumulation of total phenolic content, proline content and pathogenesis-related (PR) proteins during the pathogen attack. The phenolic accumulation was enhanced up to 1.4-4.6-fold in plants bioprimed with a consortium of PGPRs compared with that of non-bioprimed control plants (Jain *et al.*, 2012). However, the population of these microbes in the rhizosphere decreases up to 50% with plant maturity. In a study conducted to manage blast disease in rice using an integrated approach, a consortium of two compatible PGPR strains, *P. fluorescens* Aur 6 and *Chryseobacterium balustinum* Aur 9, was used (Lucas *et al.*, 2009). Results from this study showed that application of PGPRs in consortium reduced disease intensity up to 50% in comparison to their individual application under field conditions. Similarly, the number of galls per root was also reduced more than 20% when *Rhizobium* strain was applied in consortium to *Paecilomyces lilacinus* KIA to that of its individual application.

A single strain of PGPR is generally able to induce resistance against various pathogens in a single host (Somers *et al.*, 2004). The most studied rhizobacteria are *Pseudomonas* and *Bacillus* species for their ISR effect on host plants against invading pathogens (Kloepper *et al.*, 2004; Van Wees *et al.*, 2008). Induced resistance via ISR and SAR occurs through two different signalling pathways. The SAR is mediated via a salicylic acid (SA) signalling pathway, whereas ISR is mediated through ethylene (ET) and jasmonic acid (JA) signalling pathways (van Loon *et al.*, 1998). The signalling molecules accumulating during the cascades when applied exogenously were also able to induce sufficient resistance against the target pathogens (Ryals *et al.*, 1996). The development of necrotic lesions is commonly observed and found to be an essential feature in SAR-mediated resistance (Vleesschauwer and Höfte, 2009), but in some cases it was also induced without development of any necrotic lesions and the phenomenon was well studied in *Arabidopsis thaliana* (Mishina and Zeier, 2007). SA induces the expression of certain sets of defence-responsive genes known as pathogenesis-related genes (PRs) (van Loon, 2007) while ISR is not related to the induction of PR genes. The tobacco roots treated with *P. fluorescens* CHA0 induce the accumulation of PR proteins in the leaves induced by SA (Maurhofer *et al.*, 1994). These PRs act as signature molecules of SAR in various plant species and also contribute to the level of resistance achieved (Vleesschauwer and Höfte, 2009). Some examples of these PRs are 1-3-glucanases and chitinases having the ability to hydrolyze the cell wall of fungal pathogens, others are yet to be explored. The PRs related to SAR express an important share of the enhanced defence responsive ability of induced tissues (van Loon *et al.*, 1998). The expression of *PR1* gene or protein seems to be induced by SA and is also used as molecular marker that demonstrates the induction of SAR (van Loon and Bakker, 2006).

Lignifications are enhanced during pathogen attack and also represent an adapted procedure to block the entry of pathogen because of their non-degradable and antimicrobial properties (Rogers and Campbell, 2004). PGPRs enhance a high and homogeneous deposition of lignin polymers in the cambial cells in chickpea during attack by *Sclerotium rolfsii* (Singh *et al.*, 2013a). The phloem cells also showed enhanced and broader deposition of lignin in sclerenchyma cap in the PGPR treated plants. PGPRs enhanced the expression of the enzyme phenylalanine ammonia lyase (PAL), the first enzyme of the phenylpropanoid pathway, which leads to better accumulation of phenolic compounds, an initial step of lignification. Similarly, PGPR application also enhanced the activity of antioxidants related to ISR response and saved plants from a variety of pathogens (Jetiyanon, 2007; Singh *et al.*, 2011). However, the efficacy of such defence responses is dependent on the PGPR strains that are applied. ISR-mediated host defence response has gained more attention in recent years because the ISR inducer molecules are required in very small quantities to trigger ISR responses and are able to save distant parts of plants from pathogens as well.

ISR is normally regulated by a series of interconnected signal transduction processes, where polyphenols play an important role, like an alarm that helps to block pathogen development. Plant phenolics are synthesized only when a plant recognizes a pathogen through pattern recognition receptors (PRRs) and distinguishes pathogens from non-pathogens by pathogen-associated molecular patterns (PAMPs) that help in activation of the phenylpropanoid pathway. Concentration of the phenolics and its activation pathway was increased when the plants were inoculated with PGPRs (Sarma *et al.*, 2002; Singh *et al.*, 2011). Phenolic compounds had many different roles like repelling and attracting different organisms in the plant surface by playing multiple roles such as protection from pathogens and protective compounds and signal molecules. Among all the phenolics shikimic acid accumulation is most effective during PGPR application with chickpea (Keswani *et al.*, 2013, 2014; Singh *et al.*, 2014; Keswani, 2015). Synthesis of shikimic acid in PGPR-treated plants was significantly higher compared to untreated plants. Other different phenolics such as *t*-chlorogenic acid, myricetin, ferulic acid, syringic acid and quercetin were also accumulated in higher amounts in the leaves of PGPR-treated chickpea plants after pathogen infection compared to the untreated plants.

Plants sense the ISR elicitors present in PGPR that ultimately react through exaggerated immune response. PGPR-mediated ISR response also helps in enzyme mobilization as well as reprogramming it for host defence mechanisms through synthesis of PR proteins, activation of the enzymes PAL, peroxidases (PO), superoxide dismutase (SOD) and polyphenol oxidases (PPO) (Jetiyanon, 2007). It also helps the plant to synthesize proline and phenols. Increased activities of ISR show the potentiality of microorganisms either individually or in consortium to modify the plant's gene expression process that finally helps in reduction of pathogen effects (Jain *et al.*, 2012). The phenylpropanoid pathway is highly activated only when a consortium of compatible microbes comprising PGPR strains was inoculated. Such consortia of compatible microbes with various properties had appropriate scientific reasons for them to be applied together to maximize the benefits by minimizing the pathogenic attack and increasing plant growth.

Direct interaction between *npr1* and a specialized TGA transcription factor is essential for making complex elements that bind within promoter region of *PR* gene (Després *et al.*, 2000; Fan and Dong, 2002). Higher expression of the *npr1* gene leads to an increase in the pathogen resistance mechanism in the plant (Friedrich *et al.*, 2001). Mutation of *npr1* in *Arabidopsis* does not produce *PR* gene and is unable to exhibit SAR. PGPR-induced ISR is independent of SA and is not associated with synthesis of PR proteins. Therefore, it can be anticipated that a SA mutant strain would be able to produce ISR. Interestingly, *Arabidopsis* mutant of *npr1* is unable to display *P. fluorescens* WCS417r-mediated ISR activities. This shows that *npr1* regulates the defence network by modulating various steps of SAR and ISR signalling pathways (van Loon *et al.*, 1998). Pieterse *et al.* (1996, 1998, 2000) suggested that rhizobacterial strain *P. fluorescens* WCS417r in *Arabidopsis* produced the ISR response with the help of *npr1*, JA and ethylene but without participation of the SA responsive pathway.

For plant defence the oxidative burst is the most popular, universal and earliest resistant mechanism against invading pathogens. Reactive Oxygen Species (ROS) are regulated by plants in a way where the ROS, such as H_2O_2, is sensitive to pathogenic organisms but not to the plant cell. H_2O_2 also provides enhancement to the plant cell wall by helping the formation of lignifications, adding cross linkages in cell wall components such as lignin polymers. It also increases the concentration of defence-related components like proline and hydroxyproline-rich protein products at the time of pathogen attack, which help to prevent the growth and development of invading pathogens. Some pathogens like *Sclerotinia* are able to break down the host oxidative burst barrier by releasing oxalate that results in the compromising of host defence systems (Cessna *et al.*, 2000). But in the case of PGPR-treated pea plants, a dramatic increase in H_2O_2 concentration was shown that was able to regulate the plant defence responses against *Sclerotinia* (Jain *et al.*, 2013, 2014). Production of H_2O_2 significantly increased in pea plants after 24 h of pathogen attack, reaching more than 250% higher in PGPR-inoculated plants than the untreated control plants. Modifications in antioxidant enzymes lead to increases in stress tolerance in plants. It was believed that plants develop antioxidant protective mechanisms to suppress and neutralize the oxidative damage that is caused by high concentrations of ROS in plant cells. Many scientific studies suggested that activities of antioxidant enzymes were increased when plants were treated with PGPR (van Loon *et al.*, 2006). The antioxidant activities were up-regulated severalfold during PGPR treatment of chickpea plants under *S. rolfsii* infection compared to the untreated control plants (Singh *et al.*, 2013a). ROS is able to trigger different signalling pathways like ET, JA and SA, and activation of these pathways also leads to expression of several defence-related genes against pathogen attack (Conklin and Barth, 2004). It was reported that the activity of PAL and PO was higher in the PGPR-treated plants compared to the control plants (Karthikeyan *et al.*, 2006). In the case of PGPR-treated plants the concentration of total phenol increased up to 30% compared to control untreated plants.

Researchers also proposed that production of chitinase and generation of ROS work together with compounds able to move systemically which enhances production of ROS in the lower part of plants where infection takes place. ROS also helps in induction of some defence-related genes at the time of pathogen attack resulting in lowering of pathogen development. However, until now the perfect relation of ROS generation and chitinase production is not clear. The synergistic consortium of microorganisms helps to activate chitinase production and induce the accumulation of phenol in chickpea in comparison to both single and double microorganism application (Singh *et al.*, 2013b). SAR is the ideal and desirable property of plant defence that increases the resistance level of plant cells for a few months during the later period of pathogen infection. The mechanisms help to activate defence responses from the site of pathogen infection to the entire plant through the activation of systemic responses that also help to defend uninfected plant tissues from further pathogen attack (Pieterse *et al.*, 2009).

SAR also activates resistance mechanisms in the entire foliage and increases production of the defence-related signal molecule SA for better protection of the plants (Kachroo and Robin, 2013). Both the defence-related compounds JA and SA were synthesized at the time of pathogen attacks. The JA-mediated pathway is stimulated during a necrotropic pathogen attack and herbivores attack, but the SA-mediated pathway is activated during a biotropic pathogen attack (Thaler *et al.*, 2012). The *npr1* regulatory protein (negative expresser of PR1 protein) is essential for SA transduction and works as co-activator of *PR* gene expression. Studies showed that both chitinase and β-1, 3 glucanases (all are pathogenesis-related proteins) work against fungal pathogen infections synergistically in different plants (van Loon, 1997).

It is also reported that pathogenesis-related proteins are synthesized during specific pathogen attack in plants by application of different strains of PGPRs. Mixed microbial cultures of *Trichoderma harzianum*, *Bacillus subtilis* and *P. aeruginosa* enhanced production of chitinases and β-1,3 glucanases in pea plants during infection of *S. sclerotiorum* (Jain *et al.*, 2012). The chitinase activity increased up to 1.4 to 1.8-fold and the activity of β-1,3 glucanases increased up to 1.4 to 4.6-fold in PGPR treatments in comparison to the non-PGPR-treated control pea plants. It was reported that chitinases and β-1,3 glucanases activities also increased up to double in palm trees during treatment with PGPR mixed with chitin as compared with the control plants at the time of pathogen attack (Karthikeyan *et al.*, 2006). PGPR-induced ISR is well established (Kloepper *et al.*, 2004; van Loon *et al.*, 2006; Bakker *et al.*, 2007) and the ISR response in plants helps to repress soilborne as well as foliar pathogens (Fig. 25.2).

25.4 Conclusion

The plant rhizosphere microbiome consists of both beneficial and harmful microorganisms. An imbalance between beneficial and pathogenic microorganisms can affect the survival of plants in a natural ecosystem. The rhizosphere is the region around the plant roots, where resident microbes interact with each other as well as with the host plant via exchange of signalling molecules. The microorganisms compete with their nearest neighbours for nutrients, space and water, and owing to such interactions an informal association is developed with the host plant. The outcome of plant–microbe interactions can be both beneficial as well as harmful, and it is dependent on the host genotype, microbiome structure and surrounding environmental conditions. Protecting plants from biotic and abiotic stresses, increasing yield and nutritional security are some of the priorities for current researchers. Chemical pesticides can protect plants from pathogens and chemical fertilizers can enhance crop production, but usually it is achieved by compromising human health. Therefore, the use of agriculturally important rhizosphere microbes has been greatly appreciated.

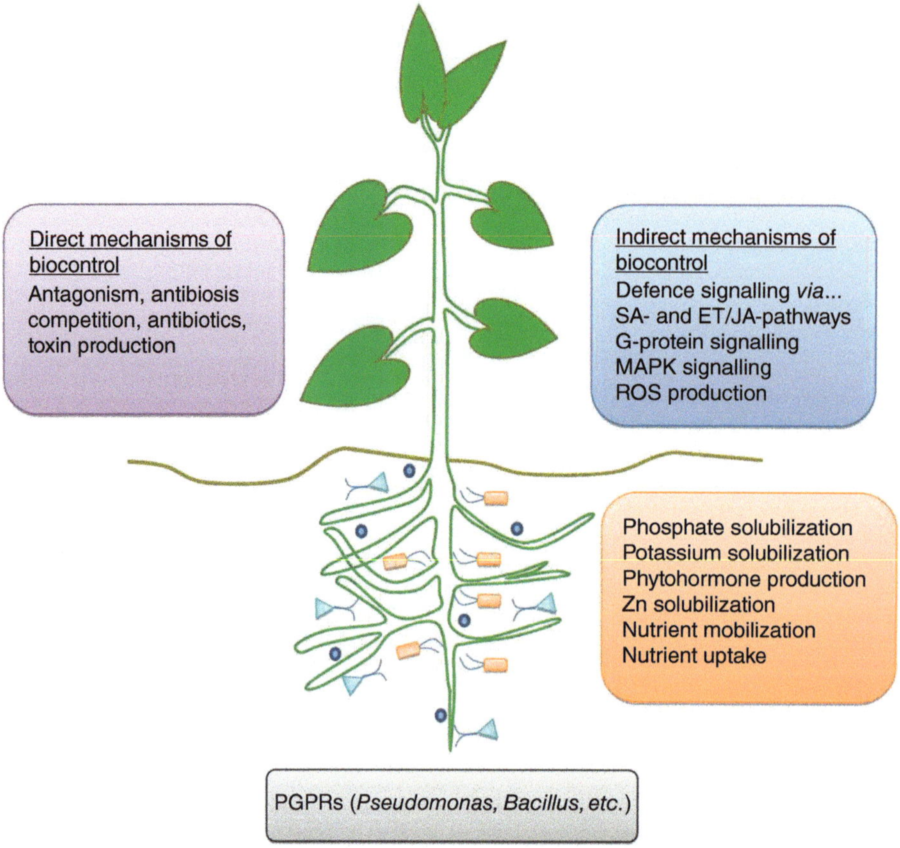

Fig. 25.2. Mechanisms of biocontrol by PGPR strains.

References

Ali, S.Z., Sandhya, V., Grover, M., Kishore, N., Venkateswar, R.L. and Venkateswarlu, B. (2009) *Pseudomonas* sp. strain AKM-P6 enhances tolerance of sorghum seedlings to elevated temperatures. *Biology and Fertility of Soils* 46, 45–55.

Bakker, P.A.H.M., Pieterse, C.M.J. and van Loon, L.C. (2007) Induced systemic resistance by fluorescent *Pseudomonas* spp. *Phytopathology* 97, 239–243.

Barka, E.A., Nowak, J. and Clement, C. (2006) Enhancement of chilling resistance of inoculated grapevine plantlets with a plant growth-promoting rhizobacterium, *Burkholderia phytofirmans* strain PsJN. *Applied Environmental Microbiology* 72, 7246–7252.

Benhamou, N.R., Belanger, R. and Paulitz, T.C. (1996) Induction of differential host responses by *Pseudomonas fluorescens* in Ri T-DNA transformed pea roots after challenge with *Fusarium oxysporum f. sp. pisi* and *Pythuim ultimum*. *Phytopathology* 86, 114–178.

Bigirimana, J. and Hofte, M. (2002) Induction of systemic resistance to *Colletotrichum lindemuthianum* in bean by a benzothiadiazole derivative and rhizobacteria. *Phytoparasitica* 30, 159–168.

Bisen, K., Keswani, C., Mishra, S., Saxena, A., Rakshit, A. and Singh, H.B. (2015) Unrealized potential of seed biopriming for versatile agriculture. In: Rakshit, A., Singh, H.B. and Sen, A. (eds) *Nutrient Use Efficiency: From Basics to Advances*. Springer, New Delhi, India, pp. 193–206.

Bisen, K., Keswani, C., Patel, J.S., Sarma, B.K. and Singh, H.B. (2016) *Trichoderma* spp.: efficient inducers of systemic resistance in plants. In: Chaudhary, D.K. and Verma, A. (eds) *Microbial-mediated Induced Systemic Resistance in Plants*. Springer, Singapore, pp. 185–195.

Cessna, S.G., Sears, V.E., Dickman, M.B. and Low, P.S. (2000) Oxalic acid, a pathogenicity factor for *Sclerotinia sclerotiorum* suppresses the oxidative burst of the host plant. *The Plant Cell* 12, 2191–2200.

Cho, S.M., Kang, B.R., Han, S.H., Anderson, A.J., Park, J.Y. *et al.* (2008) 2R,3R-butanediol, a bacterial volatile produced by *Pseudomonas chlororaphis* O6, is involved in induction of systemic tolerance to drought in *Arabidopsis thaliana*. *Molecular Plant–Microbe Interaction* 21(8), 1067–1075.

Conklin, P.L. and Barth, C. (2004) Ascorbic acid, a familiar small molecule intertwined in the response of plants to ozone, pathogens, and the onset of senescence. *Plant, Cell and Environment* 27, 959–970.

de Leon, I.P. and Montesano, M. (2013) Activation of defense mechanisms against pathogens in mosses and flowering plants. *International Journal of Molecular Sciences* 14, 3178–3200.

Després, C., DeLong, C., Glaze, S., Liu, E. and Fobert, P.R. (2000) The *Arabidopsis* NPR1/NIM1 protein enhances the DNA binding activity of a subgroup of the TGA family of bZIP transcription factors. *Plant Cell* 12, 279–290.

Fan, W. and Dong, X. (2002) *In vivo* interaction between NPR1 and transcription factor TGA2 leads to salicylic acid-mediated gene activation in *Arabidopsis*. *Plant Cell* 14, 1377–1389.

Friedrich, L., Lawton, K., Dietrich, R., Willits, M., Cade, R. and Ryals, J. (2001) NIM1 overexpression in *Arabidopsis* potentiates plant disease resistance and results in enhanced effectiveness of fungicides. *Molecular Plant–Microbe Interaction* 14, 1114–1124.

Glick, B.R. (2003) Phytoremediation: synergistic use of plants and bacteria to clean up the environment. *Biotechnology Advances* 21, 383–393.

Huang, G.T., Ma, S.L., Bai, L.P., Zhang, L., Ma, H. *et al.* (2012) Signal transduction during cold, salt, and drought stresses in plants. *Molecular Biology Reports* 39(2), 969–987.

Jain, A., Singh, S., Sarma, B.K. and Singh, H.B. (2012) Microbial consortium-mediated reprogramming of defence network in pea to enhance tolerance against *Sclerotinia sclerotiorum*. *Journal of Applied Microbiology* 112, 537–550.

Jain, A., Singh, A., Singh, S. and Singh, H.B. (2013) Microbial consortium-induced changes in oxidative stress markers in pea plants challenged with *Sclerotinia sclerotiorum*. *Journal of Plant Growth Regulation* 32, 388–398.

Jain, A., Singh, A., Singh, S., Sarma, B.K. and Singh, H.B. (2014) Biocontrol agents mediated suppression of oxalic acid induced cell death during *Sclerotinia sclerotiorum* pea interaction. *Journal of Basic Microbiology* 55, 601–606.

Jetiyanon, K. (2007) Defensive-related enzyme response in plants treated with a mixture of *Bacillus* strains (IN937a and IN937b) against different pathogens. *Biological Control* 42, 178–185.

Kachroo, A. and Robin, G.P. (2013) Systemic signaling during plant defense. *Current Opinion in Plant Biology* 16, 527–533.

Karthikeyan, M., Radhika, K., Mathiyazhagan, S., Bhaskaran, R., Samiyappan, R. and Velazhahan, R. (2006) Induction of phenolics and defense-related enzymes in coconut (*Cocos nucifera* L.) roots treated with biocontrol agents. *Brazilian Journal of Plant Physiology* 18, 367–377.

Keswani, C. (2015) Proteomics studies of thermotolerant strain of *Trichoderma* spp. Ph.D. Thesis, Banaras Hindu University, Varanasi, India, p. 126.

Keswani, C., Singh, S.P. and Singh, H.B. (2013) A superstar in biocontrol enterprise: *Trichoderma* spp. *Biotech Today* 3, 27–30.

Keswani, C., Mishra, S., Sarma, B.K., Singh, S.P. and Singh, H.B. (2014) Unraveling the efficient application of secondary metabolites of various *Trichoderma*. *Applied Microbiology and Biotechnology* 98, 533–544.

Keswani, C., Bisen, K., Singh, S.P., Sarma, B.K. and Singh, H.B. (2016a) A proteomic approach to understand the tripartite interactions between plant–*Trichoderma*–pathogen: investigating the potential for efficient biological control. In: Hakeem, K.R. and Akhtar, M.S. (eds) *Plant, Soil and Microbes Vol. 2. Mechanisms and Molecular Interactions*. Springer, Switzerland, pp. 79–93.

Keswani, C., Bisen, K., Singh, V., Sarma, B.K. and Singh, H.B. (2016b) Formulation Technology of Biocontrol Agents: Present Status and Future Prospects. In: Arora, N.K., Mehnaz, S. and Balestrini, R. (eds) *Bioformulations: For Sustainable Agriculture*. Springer, New Delhi, India, pp. 35–52.

Kloepper, J.W., Ryu, C.-M. and Zhang, S. (2004) Induced systemic resistance and promotion of plant growth by *Bacillus* spp. *Phytopathology* 94, 1259–1266.

Kumar, G., Saxena, R., Yadav, S.K., Patel, J.S. and Sarkar, A. (2014) Effect of Temperature on the Efficacy of Fungicide and Biocontrol Agent with Incidence of Stem Gall Disease of Coriander. *Trends in Bioscience* 7(19), 3038–3040.

López-Bucio, J., Campos-Cuevas, J.C., Hernandez-Calderon, E., Velasquez-Becerra, C., Farias-Rodriguez, R. *et al.* (2007) *Bacillus megaterium* rhizobacteria promote growth and alter root-system architecture through an auxin- and ethylene-independent signaling mechanism in *Arabidopsis thaliana. Molecular Plant-Microbe Interaction* 20, 207–217.

Lucas, J.A., Solano, B.R., Montes, F., Ojeda, J., Megias, M. and Gutierrez Ma~nero, F.J. (2009) Use of two PGPR strains in the integrated management of blast disease in rice (Oryza sativa) in Southern Spain. *Field Crop Research* 114, 404–410.

Mantelin, S. and Touraine, B. (2004) Plant growth-promoting bacteria and nitrate availability: impacts on root development and nitrate uptake. *Journal of Experimental Botany* 55, 27–34.

Maurhofer, M., Hase, C., Meuwly, P., Metraux, J.P. and Defago, G. (1994) Induction of systemic resistance of tobacco to tobacco necrosis virus by the root-colonizing *Pseudomonas fluorescens* strain CHA0: Influence of the *gacA* gene and of pyoverdine production. *Phytopathology* 84, 139–146.

Mayak, S., Tirosh, T. and Glick, B.R. (2004a) Plant growth-promoting bacteria confer resistance in tomato plants to salt stress. *Plant Physiology and Biochemistry* 42(6), 565–572.

Mayak, S., Tirosh, T. and Glick, B.R. (2004b) Plant growth-promoting bacteria that confer resistance to water stress in tomatoes and peppers. *Plant Science* 166, 525–530.

Mishina, T.E. and Zeier, J. (2007) Pathogen-associated molecular pattern recognition rather than development of tissue necrosis contributes to bacterial induction of systemic acquired resistance in *Arabidopsis. The Plant Journal* 50, 500–513.

Nandakumar, R., Babu, S., Viswanathan, R., Raguchander, T. and Samiyappan, R. (2001) Induction of systemic resistance in rice against sheath blight disease by *Pseudomonas fluorescens. Soil Biology and Biochemistry* 33, 603–612.

Ortíz-Castro, R., Contreras-Cornejo, H.A., Macias-Rodriguez, L. and Lopez-Bucio, J. (2009) The role of microbial signals in plant growth and development. *Plant Signaling and Behavior* 4(8), 701–712.

Pieterse, C.M.J., Van Wees, S.C.M., Hoffland, E., Van Pelt, J.A. and van Loon, L.C. (1996) Systemic resistance in *Arabidopsis* induced by biocontrol bacteria is independent of salicylic acid accumulation and pathogenesis-related gene expression. *Plant Cell* 8, 1225–1237.

Pieterse, C.M.J., Van Wees, S.C.M., Van Pelt, J.A., Knoester, M., Laan, R. *et al.* (1998) A novel signaling pathway controlling induced systemic resistance in *Arabidopsis. Plant Cell* 10, 1571–1580.

Pieterse, C.M.J., Van Pelt, J.A., Ton, J., Parchmann, S., Mueller, M.J. *et al.* (2000) Rhizobacteria-mediated induced systemic resistance (ISR) in *Arabidopsis* requires sensitivity to jasmonate and ethylene but is not accompanied by an increase in their production. *Physiological and Molecular Plant Patholology* 57, 123–134.

Pieterse, C.M.J., Leon-Reyes, A., Van der Ent, S. and Wees, S.C.M.V. (2009) Networking by small-molecule hormones in plant immunity. *Nature Chemical Biology* 5, 308–316.

Radjacommare, R., Kandan, A., Nandakumar, R. and Samiyapan, R. (2004) Association of the hydrolytic enzyme chitinase against *Rhizoctonia solani* in rhizobacteria treated rice plants. *Journal of Phytopathology* 152, 365–370.

Rogers, L.A. and Campbell, M.M. (2004) The genetic control of lignin deposition during plant growth and development. *New Phytologist* 164, 17–30.

Ryals, J.A., Neuenschwander, U.H., Willits, M.G., Molina, A., Steiner, H.Y. and Hunt, M.D. (1996) Systemic acquired resistance. *Plant Cell* 8, 1808–1819.

Ryu, C.M., Farag, M.A., Hu, C.H., Reddy, M.S., Kloepper, J.W. and Pare, P.W. (2004) Bacterial volatiles induce systemic resistance in Arabidopsis. *Plant Physiology* 134, 1017–1026.

Sarma, B.K., Singh, D.P., Mehta, S. and Singh, H.B. (2002) Plant growth promoting rhizobacteria-elicited alterations in phenolic profile of chickpea (*Cicer arietinum*) infected by *Sclerotium rolfsii. Journal of Phytopathology* 150, 277–282.

Shoresh, M., Harman, G.E. and Mastouri, F. (2010) Induced systemic resistance and plant responses to fungal biocontrol agents. *Annual Review of Phytopathology* 48, 1–23.

Singh, A., Sarma, B.K., Upadhyay, R.S. and Singh, H.B. (2013a). Compatible rhizosphere microbes mediated alleviation of biotic stress in chickpea through enhanced antioxidant and phenylpropanoid activities. *Microbiological Research* 168, 33–40.

Singh, A., Jain, A., Sarma, B.K., Upadhyay, R.S. and Singh, H.B. (2013b) Rhizosphere microbes facilitate redox homeostasis in *Cicer arietinum* against biotic stress. *Annals of Applied Biology* 163, 33–46.

Singh, A., Jain, A., Sarma, B.K., Upadhyay, R.S. and Singh, H.B. (2014) Rhizosphere competent microbial consortium mediates rapid changes in phenolic profiles in chickpea during *Sclerotium rolfsii* infection. *Microbiological Research* 169, 353–360.

Singh, B.N., Singh, A., Singh, S.P. and Singh, H.B. (2011) *Trichoderma harzianum* mediated reprogramming of oxidative stress response in root apoplast of sunflower enhances defence against *Rhizoctonia solani*. *European Journal of Plant Pathology* 131, 121–134.

Somers, E., Vanderleyden, J. and Srinivasan, M. (2004) Rhizosphere bacterial signalling, a love parade beneath our feet. *Critical Reviews in Microbiology* 30, 205–235.

Someya, N., Nakajima, M., Hibi, T., Yamaguchi, I. and Akutsu, K. (2002) Induced resistance to rice blast by antagonistic bacterium *Serratia marcescens* strain B2. *Journal of General Plant Patholology* 68, 177–182.

Stockwell, V.O., Johnson, K.B., Sugar, D. and Loper, J.E. (2011) Mechanistically compatible mixtures of bacterial antagonists improve biological control of fire blight of pear. *Phytopathology* 101, 113–123.

Thaler, J.S., Humphrey, P.T. and Whiteman, N.K. (2012) Evolution of jasmonate and salicylate signal crosstalk. *Trends in Plant Science* 17, 260–270.

van Loon, L.C. (2007) Plant responses to plant growth-promoting rhizobacteria. *European Journal of Plant Patholology* 119, 243–254.

van Loon, L.C. and Bakker, P.A.H.M. (2006) Root-associated bacteria inducing systemic resistance. In: Gnanamanickam, S.S. (ed.) *Plant-Associated Bacteria*. Springer, Dordrecht, Netherlands, pp. 269–316.

van Loon, L.C., Bakker, P.A.H.M. and Pieterse C.M.J. (1998) Systemic resistance induced by rhizosphere bacteria. *Annual Review of Phytopathology* 36, 453–483.

van Loon, L.C., Rep, M. and Pieterse, C.M.J. (2006) Significance of inducible defense related proteins in infected plants. *Annual Review of Phytopathology* 44, 135–162.

van Peer, R., Niemann, G.J. and Schippers, B. (1991) Induced resistance and phytoalexin accumulation in biological control of fusarium wilt of carnation by *Pseudomonas* sp. Strain WCS417r. *Phytopathology* 91, 728–734.

Van Wees, S.C.M., Pieterse, C.M.J., Trijssenaar, A., Van Westend, Y.A.M., Hartog, F. and Van Loon, L.C. (1997) Differential induction of systemic resistance in *Arabidopsis* by biocontrol bacteria. *Molecular Plant Microbe Interaction* 10, 716–724.

Van Wees, S.C.M., Van der Ent, S. and Pieterse, C.M.J. (2008) Plant immune responses triggered by beneficial microbes. *Current Opinion in Plant Biology* 11, 443–448.

Vleesschauwer, D. and Höfte, M. (2009) Rhizobacteria-induced systemic resistance. *Advances in Botanical Research* 51, 223–281.

Wei, G., Kloepper, J.W. and Tuzun, S. (1991) Induction of systemic resistance of cucumber to *Colletotrichum orbiculare* by select strains of plant growth-promoting rhizobacteria. *Phytopathology* 81, 1508–1512.

Yang, J., Kloepper, J.W. and Ryu, C.M. (2009) Rhizosphere bacteria help plants tolerate abiotic stress. *Trends in Plant Science* 14(1), 1–4.

Zhang, H., Kim, M.S., Sun, Y., Dowd, S.E., Shi, H. and Pare, P.W. (2008) Soil bacteria confer plant salt tolerance by tissue-specific regulation of the sodium transporter HKT1. *Molecular Plant-Microbe Interactions* 21(6), 737–744.

Zhuang, X., Chen, J., Shim, H. and Bai, Z. (2007) New advances in plant growth-promoting rhizobacteria for bioremediation. *Environment International* 33, 406–413.

Index